Lecture Notes in Computer Science   7062

Commenced Publication in 1973
Founding and Former Series Editors:
Gerhard Goos, Juris Hartmanis, and Jan van Leeuwen

Bao-Liang Lu   Liqing Zhang
James Kwok (Eds.)

Neural
Information Processing

18th International Conference, ICONIP 2011
Shanghai, China, November 13-17, 2011
Proceedings, Part I

 Springer

Volume Editors

Bao-Liang Lu
Shanghai Jiao Tong University
Department of Computer Science and Engineering
800, Dongchuan Road, Shanghai 200240, China
E-mail: bllu@sjtu.edu.cn

Liqing Zhang
Shanghai Jiao Tong University
Department of Computer Science and Engineering
800, Dongchuan Road, Shanghai 200240, China
E-mail: zhang-lq@cs.sjtu.edu.cn

James Kwok
The Hong Kong University of Science and Technology
Department of Computer Science and Engineering
Clear Water Bay, Kowloon, Hong Kong, China
E-mail: jamesk@cse.ust.hk

ISSN 0302-9743   e-ISSN 1611-3349
ISBN 978-3-642-24954-9   ISBN 978-3-642-24955-6 (eBook)
DOI 10.1007/978-3-642-24955-6
Springer Heidelberg Dordrecht London New York

Library of Congress Control Number: 2011939737

CR Subject Classification (1998): F.1, I.2, I.4-5, H.3-4, G.3, J.3, C.1.3, C.3

LNCS Sublibrary: SL 1 – Theoretical Computer Science and General Issues

Typesetting: Camera-ready by author, data conversion by Scientific Publishing Services, Chennai, India

Printed on acid-free paper

Springer is part of Springer Science+Business Media (www.springer.com)

Preface

This book and its sister volumes constitute the proceedings of the 18th International Conference on Neural Information Processing (ICONIP 2011) held in Shanghai, China, during November 13–17, 2011. ICONIP is the annual conference of the Asia Pacific Neural Network Assembly (APNNA). ICONIP aims to provide a high-level international forum for scientists, engineers, educators, and students to address new challenges, share solutions, and discuss future research directions in neural information processing and real-world applications.

The scientific program of ICONIP 2011 presented an outstanding spectrum of over 260 research papers from 42 countries and regions, emerging from multidisciplinary areas such as computational neuroscience, cognitive science, computer science, neural engineering, computer vision, machine learning, pattern recognition, natural language processing, and many more to focus on the challenges of developing future technologies for neural information processing. In addition to the contributed papers, we were particularly pleased to have 10 plenary speeches by world-renowned scholars: Shun-ichi Amari, Kunihiko Fukushima, Aike Guo, Lei Xu, Jun Wang, DeLiang Wang, Derong Liu, Xin Yao, Soo-Young Lee, and Nikola Kasabov. The program also includes six excellent tutorials by David Cai, Irwin King, Pei-Ji Liang, Hiroshi Mamitsuka, Ming Zhou, Hang Li, and Shanfeng Zhu. The conference was followed by three post-conference workshops held in Hangzhou, on November 18, 2011: "ICONIP2011Workshop on Brain – Computer Interface and Applications," organized by Bao-Liang Lu, Liqing Zhang, and Chin-Teng Lin; "The 4th International Workshop on Data Mining and Cybersecurity," organized by Paul S. Pang, Tao Ban, Youki Kadobayashi, and Jungsuk Song; and "ICONIP 2011 Workshop on Recent Advances in Nature-Inspired Computation and Its Applications," organized by Xin Yao and Shan He.

The ICONIP 2011 organizers would like to thank all special session organizers for their effort and time high enriched the topics and program of the conference. The program included the following 13 special sessions: "Advances in Computational Intelligence Methods-Based Pattern Recognition," organized by Kai-Zhu Huang and Jun Sun; "Biologically Inspired Vision and Recognition," organized by Jun Miao, Libo Ma, Liming Zhang, Juyang Weng and Xilin Chen; "Biomedical Data Analysis," organized by Jie Yang and Guo-Zheng Li; "Brain Signal Processing," organized by Jian-Ting Cao, Tomasz M. Rutkowski, Toshihisa Tanaka, and Liqing Zhang; "Brain-Realistic Models for Learning, Memory and Embodied Cognition," organized by Huajin Tang and Jun Tani; "Clifford Algebraic Neural Networks," organized by Tohru Nitta and Yasuaki Kuroe; "Combining Multiple Learners," organized by Younès Bennani, Nistor Grozavu, Mohamed Nadif, and Nicoleta Rogovschi; "Computational Advances in Bioinformatics," organized by Jonathan H. Chan; "Computational-Intelligent Human–Computer Interaction," organized by Chin-Teng Lin, Jyh-Yeong Chang,

John Kar-Kin Zao, Yong-Sheng Chen, and Li-Wei Ko; "Evolutionary Design and Optimization," organized by Ruhul Sarker and Mao-Lin Tang; "Human-Originated Data Analysis and Implementation," organized by Hyeyoung Park and Sang-Woo Ban; "Natural Language Processing and Intelligent Web Information Processing," organized by Xiao-Long Wang, Rui-Feng Xu, and Hai Zhao; and "Integrating Multiple Nature-Inspired Approaches," organized by Shan He and Xin Yao.

The ICONIP 2011 conference and post-conference workshops would not have achieved their success without the generous contributions of many organizations and volunteers. The organizers would also like to express sincere thanks to APNNA for the sponsorship, to the China Neural Networks Council, International Neural Network Society, and Japanese Neural Network Society for their technical co-sponsorship, to Shanghai Jiao Tong University for its financial and logistic supports, and to the National Natural Science Foundation of China, Shanghai Hyron Software Co., Ltd., Microsoft Research Asia, Hitachi (China) Research & Development Corporation, and Fujitsu Research and Development Center, Co., Ltd. for their financial support.

We are very pleased to acknowledge the support of the conference Advisory Committee, the APNNA Governing Board and Past Presidents for their guidance, and the members of the International Program Committee and additional reviewers for reviewing the papers. Particularly, the organizers would like to thank the proceedings publisher, Springer, for publishing the proceedings in the *Lecture Notes in Computer Science Series*. We want to give special thanks to the Web managers, Haoyu Cai and Dong Li, and the publication team comprising Li-Chen Shi, Yong Peng, Cong Hui, Bing Li, Dan Nie, Ren-Jie Liu, Tian-Xiang Wu, Xue-Zhe Ma, Shao-Hua Yang, Yuan-Jian Zhou and Cong Xie for checking the accepted papers in a short period of time. Last but not least, the organizers would like to thank all the authors, speakers, audience, and volunteers.

November 2011

Bao-Liang Lu
Liqing Zhang
James Kwok

ICONIP 2011 Organization

Organizer

Shanghai Jiao Tong University

Sponsor

Asia Pacific Neural Network Assembly

Financial Co-sponsors

Shanghai Jiao Tong University
National Natural Science Foundation of China
Shanghai Hyron Software Co., Ltd.
Microsoft Research Asia
Hitachi (China) Research & Development Corporation
Fujitsu Research and Development Center, Co., Ltd.

Technical Co-sponsors

China Neural Networks Council
International Neural Network Society
Japanese Neural Network Society

Honorary Chair

Shun-ichi Amari   Brain Science Institute, RIKEN, Japan

Advisory Committee Chairs

Shoujue Wang   Institute of Semiconductors,
                             Chinese Academy of Sciences, China
Aike Guo   Institute of Neuroscience, Chinese Academy of
                             Sciences, China
Liming Zhang   Fudan University, China

Advisory Committee Members

Sabri Arik	Istanbul University, Turkey
Jonathan H. Chan	King Mongkut's University of Technology, Thailand
Wlodzislaw Duch	Nicolaus Copernicus University, Poland
Tom Gedeon	Australian National University, Australia
Yuzo Hirai	University of Tsukuba, Japan
Ting-Wen Huang	Texas A&M University, Qatar
Akira Hirose	University of Tokyo, Japan
Nik Kasabov	Auckland University of Technology, New Zealand
Irwin King	The Chinese University of Hong Kong, Hong Kong
Weng-Kin Lai	MIMOS, Malaysia
Min-Ho Lee	Kyungpoor National University, Korea
Soo-Young Lee	Korea Advanced Institute of Science and Technology, Korea
Andrew Chi-Sing Leung	City University of Hong Kong, Hong Kong
Chin-Teng Lin	National Chiao Tung University, Taiwan
Derong Liu	University of Illinois at Chicago, USA
Noboru Ohnishi	Nagoya University, Japan
Nikhil R. Pal	Indian Statistical Institute, India
John Sum	National Chung Hsing University, Taiwan
DeLiang Wang	Ohio State University, USA
Jun Wang	The Chinese University of Hong Kong, Hong Kong
Kevin Wong	Murdoch University, Australia
Lipo Wang	Nanyang Technological University, Singapore
Xin Yao	University of Birmingham, UK
Liqing Zhang	Shanghai Jiao Tong University, China

General Chair

Bao-Liang Lu	Shanghai Jiao Tong University, China

Program Chairs

Liqing Zhang	Shanghai Jiao Tong University, China
James T.Y. Kwok	Hong Kong University of Science and Technology, Hong Kong

Organizing Chair

Hongtao Lu	Shanghai Jiao Tong University, China

Workshop Chairs

Guangbin Huang   Nanyang Technological University, Singapore
Jie Yang   Shanghai Jiao Tong University, China
Xiaorong Gao   Tsinghua University, China

Special Sessions Chairs

Changshui Zhang   Tsinghua University, China
Akira Hirose   University of Tokyo, Japan
Minho Lee   Kyungpoor National University, Korea

Tutorials Chair

Si Wu   Institute of Neuroscience, Chinese Academy of Sciences, China

Publications Chairs

Yuan Luo   Shanghai Jiao Tong University, China
Tianfang Yao   Shanghai Jiao Tong University, China
Yun Li   Nanjing University of Posts and Telecommunications, China

Publicity Chairs

Kazushi Ikeda   Nara Institute of Science and Technology, Japan
Shaoning Pang   Unitec Institute of Technology, New Zealand
Chi-Sing Leung   City University of Hong Kong, China

Registration Chair

Hai Zhao   Shanghai Jiao Tong University, China

Financial Chair

Yang Yang   Shanghai Maritime University, China

Local Arrangements Chairs

Guang Li   Zhejiang University, China
Fang Li   Shanghai Jiao Tong University, China

Secretary

Xun Liu   Shanghai Jiao Tong University, China

Program Committee

Shigeo Abe	Takio Kurita
Bruno Apolloni	Minho Lee
Sabri Arik	Chi Sing Leung
Sang-Woo Ban	Chunshien Li
Jianting Cao	Guo-Zheng Li
Jonathan Chan	Junhua Li
Songcan Chen	Wujun Li
Xilin Chen	Yuanqing Li
Yen-Wei Chen	Yun Li
Yiqiang Chen	Huicheng Lian
Siu-Yeung David Cho	Peiji Liang
Sung-Bae Cho	Chin-Teng Lin
Seungjin Choi	Hsuan-Tien Lin
Andrzej Cichocki	Hongtao Lu
Jose Alfredo Ferreira Costa	Libo Ma
Sergio Cruces	Malik Magdon-Ismail
Ke-Lin Du	Robert(Bob) McKay
Simone Fiori	Duoqian Miao
John Qiang Gan	Jun Miao
Junbin Gao	Vinh Nguyen
Xiaorong Gao	Tohru Nitta
Nistor Grozavu	Toshiaki Omori
Ping Guo	Hassab Elgawi Osman
Qing-Long Han	Seiichi Ozawa
Shan He	Paul Pang
Akira Hirose	Hyeyoung Park
Jinglu Hu	Alain Rakotomamonjy
Guang-Bin Huang	Sarker Ruhul
Kaizhu Huang	Naoyuki Sato
Amir Hussain	Lichen Shi
Danchi Jiang	Jochen J. Steil
Tianzi Jiang	John Sum
Tani Jun	Jun Sun
Joarder Kamruzzaman	Toshihisa Tanaka
Shunshoku Kanae	Huajin Tang
Okyay Kaynak	Maolin Tang
John Keane	Dacheng Tao
Sungshin Kim	Qing Tao
Li-Wei Ko	Peter Tino

Ivor Tsang
Michel Verleysen
Bin Wang
Rubin Wang
Xiao-Long Wang
Yimin Wen
Young-Gul Won
Yao Xin
Rui-Feng Xu
Haixuan Yang
Jie Yang

Yang Yang
Yingjie Yang
Zhirong Yang
Dit-Yan Yeung
Jian Yu
Zhigang Zeng
Jie Zhang
Kun Zhang
Hai Zhao
Zhihua Zhou

Reviewers

Pablo Aguilera
Lifeng Ai
Elliot Anshelevich
Bruno Apolloni
Sansanee
    Auephanwiriyakul
Hongliang Bai
Rakesh Kr Bajaj
Tao Ban
Gang Bao
Simone Bassis
Anna Belardinelli
Yoshua Bengio
Sergei Bezobrazov
Yinzhou Bi
Alberto Borghese
Tony Brabazon
Guenael Cabanes
Faicel Chamroukhi
Feng-Tse Chan
Hong Chang
Liang Chang
Aaron Chen
Caikou Chen
Huangqiong Chen
Huanhuan Chen
Kejia Chen
Lei Chen
Qingcai Chen
Yin-Ju Chen

Yuepeng Chen
Jian Cheng
Wei-Chen Cheng
Yu Cheng
Seong-Pyo Cheon
Minkook Cho
Heeyoul Choi
Yong-Sun Choi
Shihchieh Chou
Angelo Ciaramella
Sanmay Das
Satchidananda Dehuri
Ivan Duran Diaz
Tom Diethe
Ke Ding
Lijuan Duan
Chunjiang Duanmu
Sergio Escalera
Aiming Feng
Remi Flamary
Gustavo Fontoura
Zhenyong Fu
Zhouyu Fu
Xiaohua Ge
Alexander Gepperth
M. Mohamad Ghassany
Adilson Gonzaga
Alexandre Gravier
Jianfeng Gu
Lei Gu

Zhong-Lei Gu
Naiyang Guan
Pedro Antonio Gutiérrez
Jing-Yu Han
Xianhua Han
Ross Hayward
Hanlin He
Akinori Hidaka
Hiroshi Higashi
Arie Hiroaki
Eckhard Hitzer
Gray Ho
Kevin Ho
Xia Hua
Mao Lin Huang
Qinghua Huang
Sheng-Jun Huang
Tan Ah Hwee
Kim Min Hyeok
Teijiro Isokawa
Wei Ji
Zheng Ji
Caiyan Jia
Nanlin Jin
Liping Jing
Yoonseop Kang
Chul Su Kim
Kyung-Joong Kim
Saehoon Kim
Yong-Deok Kim

Jianxin Wu
Qiang Wu
Si Wu
Wei Wu
Wen Wu
Bin Xia
Chen Xie
Zhihua Xiong
Bingxin Xu
Weizhi Xu
Yang Xu
Xiaobing Xue
Dong Yang
Wei Yang
Wenjie Yang
Zi-Jiang Yang
Tianfang Yao
Nguwi Yok Yen
Florian Yger
Chen Yiming
Jie Yin
Lijun Yin
Xucheng Yin
Xuesong Yin

Jiho Yoo
Washizawa Yoshikazu
Motohide Yoshimura
Hongbin Yu
Qiao Yu
Weiwei Yu
Ying Yu
Jeong-Min Yun
Zeratul Mohd Yusoh
Yiteng Zhai
Biaobiao Zhang
Danke Zhang
Dawei Zhang
Junping Zhang
Kai Zhang
Lei Zhang
Liming Zhang
Liqing Zhang
Lumin Zhang
Puming Zhang
Qing Zhang
Rui Zhang
Tao Zhang
Tengfei Zhang

Wenhao Zhang
Xianming Zhang
Yu Zhang
Zehua Zhang
Zhifei Zhang
Jiayuan Zhao
Liang Zhao
Qi Zhao
Qibin Zhao
Xu Zhao
Haitao Zheng
Guoqiang Zhong
Wenliang Zhong
Dong-Zhuo Zhou
Guoxu Zhou
Hongming Zhou
Rong Zhou
Tianyi Zhou
Xiuling Zhou
Wenjun Zhu
Zhanxing Zhu
Fernando José Von Zube

Table of Contents – Part I

Perception, Emotion and Development

Bioinformatics

Biologically Inspired Vision and Recognition

Bio-medical Data Analysis

Brain Signal Processing

Brain-Computer Interfaces

Brain-Like Systems

Brain-Realistic Models for Learning, Memory and Embodied Cognition

Clifford Algebraic Neural Networks

Combining Multiple Learners

Computational Advances in Bioinformatics

Computational-Intelligent Human Computer Interaction

Table of Contents – Part II

Evolutionary Design and Optimisation

Graphical Models

Human-Originated Data Analysis and Implementation

Information Retrieval

Integrating Multiple Nature-Inspired Approaches

Kernel Methods and Support Vector Machines

Learning and Memory

Table of Contents – Part III

Multi-agent Systems

Natural Language Processing and Intelligent Web Information Processing

Neural Encoding and Decoding

Neural Network Models

Neuromorphic Hardware and Implementations

Object Recognition

Visual Perception Modelling

Advances in Computational Intelligence Methods Based Pattern Recognition

Stable Fast Rewiring Depends on the Activation of Skeleton Voxels

Sanming Song and Hongxun Yao*

School of Computer Science and Technology, Harbin Institute of Technology,
150001, Harbin, China
{ssoong,h.yao}@hit.edu.cn

Abstract. Compared with the relatively stable structural networks, the functional networks, defined by the temporal correlation between remote neurophysiological events, are highly complex and variable. However, the transitions should never be random. So it was proposed that some stable fast rewiring mechanisms probably exist in the brain. In order to probe the underlying mechanisms, we analyze the fMRI signal in temporal dimension and obtain several heuristic conclusions. 1) There is a stable time delay, 7~14 seconds, between the stimulus onset and the activation of corresponding functional regions. 2) In analyzing the biophysical factors that support stable fast rewiring, it is, to our best knowledge, the first to observe that skeleton voxels may be essential for the fast rewiring process. 3) Our analysis on the structure of functional network supports the scale-free hypothesis.

Keywords: fMRI, rewiring, functional network.

1 Introduction

Though $\sim 3 \times 10^7$ synapses are lost in our brain each year, it can be neglected when taking the total number of synapses, $\sim 10^{14}$, into consideration. So the structural network is relatively stable. But the functional networks, which relates to the cognitive information processing, are highly variable. So a problem arising: how does a relatively stable network generate so many complex functional networks? It has puzzled intellects for years [1]. We don't want to linger on this enduring pitfall, but only emphasize the functional network, especially the transition between functional networks. For, though incredible advances have been obtained in disclosing the structural network with the help of neuroanatomy, we know little about the functional networks [2].

Gomez Portillo et al. [3-4] recently proposed that there may be a fast rewiring process in the brain, and they speculated that the scale-free characteristics might be determined by a local-and-global rewiring mechanism by modeling the brain as an adaptive oscillation network using Kuramoto oscillator. But their assumptions are problematic. Since human cognitions are continuous process, the random assumption about the initial states is infeasible. What's worse, the rewiring

* This work is supported by National Natural Science Foundation of China (61071180).

B.-L. Lu, L. Zhang, and J. Kwok (Eds.): ICONIP 2011, Part I, LNCS 7062, pp. 1–8, 2011.

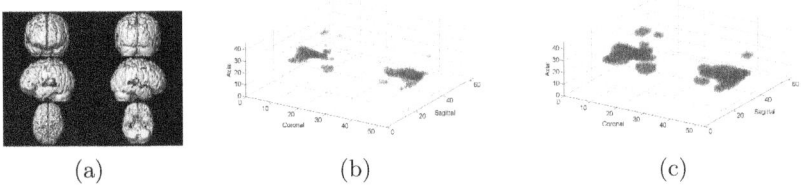

(a)   (b)   (c)

Fig. 1. Morphological dilation to ROI: (a) the activated region given by SPM; (b) the extracted ROI; (c) the mask obtained by ROI dilation

is too slow to satisfy the swift and precise information process requirements of the brain. (Also, in personal communication, Gomez Portillo explained that the underlying biophysical basis is not clear at present.) Obviously, our brain has been experiencing "oriented" rewiring instead of "random" rewiring, so the rewiring is stable, and which should be oriented to response precisely to the stimulus.

To disclose the biophysical mechanisms underlying fast rewiring, we conduct explorative analysis to the fMRI signal in temporal dimension and many important and heuristic conclusions are drawn. The experiment material is introduced in Section 2. In Section 3 and 4, we analyze the temporal stability and skeleton voxels. Finally, we conclude this paper in Section 5.

2   Material

Analysis were conducted based on an auditory dataset available at the SPM site [5], which comprises whole brain BOLD/EPI images and is acquired by a modified 2 Tesla Siemens MAGNETOM Vision system. Each acquisition consisted of 64 contiguous slices (64×64×64 3mm×3mm×3mm voxels). Data acquisition took 6.05s per volume, with the scan to scan repeat time set to 7s. 96 acquisitions were made (TR=7s) from a single subject, in blocks of 6, giving 16 42s blocks. The condition for successive blocks alternated between rest and auditory stimulation, starting with rest. Auditory stimulation was with bi-syllabic words presented binaurally at a rate of 60 per minute [6]. The first 12 scans were discarded for T1 effects, leaving with 84 scans for analysis. SPM8 was the main tool for image pre-processing. All volumes were realigned to the first volume and a mean image was created using the realigned volumes. A structural image, acquired using a standard three-dimensional weighted sequence ($1×1×3mm^3$ voxel size) was co-registered to this mean (T2) image. Finally all the images were spatially normalized [7] to a standard Tailarach template [8] and smoothed using a 6mm full width at half maximum (FWHM) isotropic Gaussian kernel. And we use the default (FWE) to detect the activated voxels.

3   Temporal Stability

Though SPM can be used to detect the activated regions, as shown in Fig.1 (a), we know little on the temporal relationship between the stimulus and the

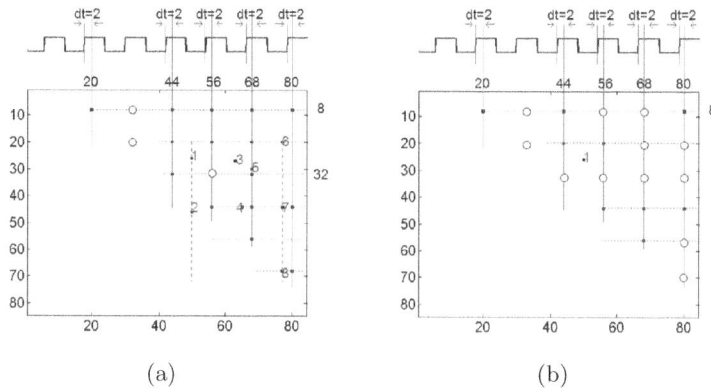

(a)   (b)

Fig. 2. The temporal stability of fast rewiring. For convience, only the upper triangle part of the binary similarity matrix is show, and the threshod coefficients are (a)$\eta = 0.4$,(b)$\eta = 0.5$. The rectangle wave above each matrix denotes the stimulus onset in the block design. The local maximum (denoted by black dots) show that there is a huge jump between the first and the second scan after each stimulus onset. And the periodicality of local maximums, which is consistent with the periodicality of stimulus onset, shows that the delay between the stimulus onset and the corresponding cortical activation is very stable. And the delay is about 7 14 seconds (dt=2 scans, TR=7 seconds). Not that there are some "noise", like the missing of local maximums (red circles) or priming phenonmenon (red numbers), please see the text for explainations.

response. So, in order to facilitate the analysis, we need to extract those activated voxels for post-processing. And the extracted activated regions, also known as region of interest (ROI), is shown in Fig.1 (b).

The ROI provided by SPM are the most activated voxels responding to the stimulus. In order to reduce the noise and also facilitate the latter analysis, we expand the ROI at firstly, because the essential voxels that play an important role in rewiring may be not included in the ROI, as it shown in latter sections. We adopt the morphological dilation [9]and the 3×3×3 cubic structural element is used. The dilated ROI, denoted as M, is shown in Fig.1 (c).

Due to the adaptation effects or the descending of blood oxygenated level of the functional regions, the energy intensity of fMRI signal varies with large amplitude. To facilitate the comparison between scans, we have to normalize the energy of eachscan. Let f_i be the ith scan, and g_i be the normalized image, then the normalization can be depicted as

$$g_i = f_i \frac{\max_i \{\langle f_i \rangle\}}{\langle f_i \rangle} \tag{1}$$

where $\langle \cdot \rangle$ stands for the mean energy. Then we can extracted the voxels of interests using the mask M,

$$G_i = [g_i]_M \tag{2}$$

where G_i is the masked image.

Even when the brain is at the resting state, the intensities of voxels are very large. So the baseline activities of voxels are very high, and the stimulus only makes the voxels more active or less active. Compared with the mean activities, the energy changes are very small, so it is better not to use the original voxels intensities. We differentiate the images as follow,

$$dG_i = G_i - G_{i-1} \qquad (3)$$

By the way, the baseline activity of each voxel is described by a constant factor in the design matrix. Threshold the differential images with some θ and transform the images into binary vectors, then we can obtain the similarity matrix by calculating the inner product of any two binary vectors. That is,

$$dist\,(G_i, G_j) = \langle \mathrm{sgn}\,(dG_i - \theta)\,, \mathrm{sgn}\,(dG_i - \theta)\rangle \qquad (4)$$

where $\theta = \eta \cdot \max\,\{dist\,(G_i, G_j)\}_{ij}$ and η is a coefficient.

If the onset of stimulus can induce remarkable change in fMRI signal, there would be a steep peak in the corresponding differential vector. And if the change is also stable, then there would be periodic local maximums. In Fig.2, we show the local maximums using different thresholds. For convenience, we only show the upper triangle part of the binary similarity matrix. See the caption for details.

As it can be seen from Fig.2, there are periodic local maximums in the similarity matrix. So the periodic stimulus can induce the periodic enhancement and recession of fMRI signal. So the dynamic rewiring of functional network is temporal stable.

What's more, we found that the local maximums don't appear immediately when the stimulus set, and there is always 2-period delay. Because the scanning period, 7 seconds, is too large when compared with the cognitive processing speed of the brain, and the volumes were pre-processed by slice-timing, the time delay is estimated to be about 7~14s, which coincides with the theoretical value [6]. The stable time delay phenomenon reminds us that we can study the mechanisms and rules underlying the fast rewiring by scrutinizing these first 2-scans after each stimulus onset, which would be described in the next section.

It is also interesting to note that there are two kinds of noise in the similarity matrix. The first one is the pseudo local maximums, like 1-4 and 6-8 in Fig.2 (a) and 1 in Fig.2 (b). The pre-activation behaviors may be relevant to the priming phenomenon [10]. By the way, in SPM, these activities are captured by the derivative of HRF function [11].

The second is the absence of some local maximums. It is worth nothing that the absences can't be understood as the deactivation of these regions. There are two ways to look at the issue: the first is that the results largely depend on the threshold, with lower ones produces more local maximums; and the second is that the basic assumption, the brain is at resting state when stimulus disappears, may not hold some time, due to the activity changes induced by the adaptation or any physiological noise.

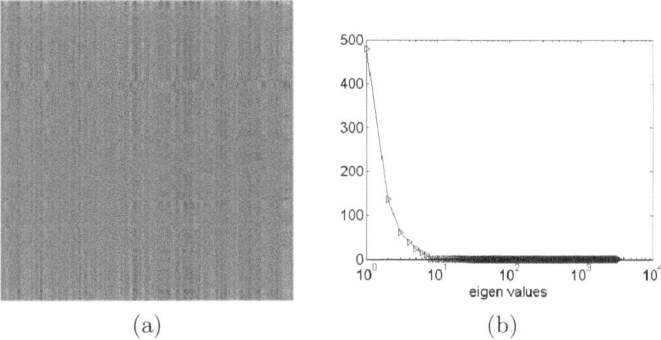

(a)   (b)

Fig. 3. The transition matrix(a) and its eign values(b). Only one eigen value is much large than other eigen values. So approximately, there exists an linear relationships between the column of the transformation matrix.

4   Skeleton Voxels

Let G_i^j be the jth scanned fMRI image after the onset of the ith stimulus. All the first scans are denoted as $G^1 = [G_1^1, \cdots, G_S^1]$, the second scans as $G^2 = [G_1^2, \cdots, G_S^2]$, where S is the number of stimulus. The transition matrix is A, then

$$G^2 = AG^1, \tag{5}$$

Using the simple pseudo-inverse rule, we have

$$A = G^2 G^{1T} \left(G^1 G^{1T} \right)^{-1}. \tag{6}$$

The transition matrix is shown in Fig.3(a). To analyze the characteristics of the transition matrix, we decompose the matrix by SVD (singular matrix decomposition) and the eigen values are plotted in Fig.3(b). The first eigen value is much larger than any other eigen values. So the matrix nearly has only one principle direction. It demonstrates that there is an approximately linear relationship between the columns. It accords with what we see from Fig.3, the rows of the transition matrix are very similar, but the columns are very different.

Turn formula (8a) to its add-and-sum form,

$$G_{in}^2 = \sum_{m=1}^{K} a_{nm} G_{im}^1, \tag{7}$$

where a_{nm} can be seen as the weight of the m th voxel.

As described in Section 2, there is a rest block before each stimulus onset, so in statistics, we consider the activity of each voxel to be random. Then formula (8b) can be simply reduced to

$$G_{in}^2 \approx \overline{G_{i\cdot}^1} \sum_{m=1}^{K} a_{nm}, \tag{8}$$

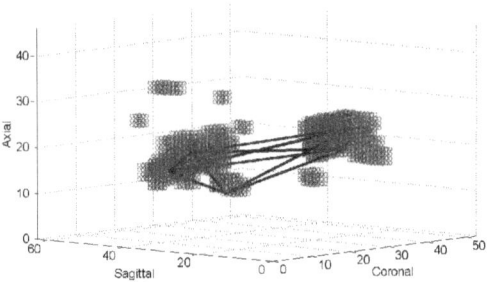

Fig. 4. Skeleton voxels and the connections between them. Note that the connections are obtained by threshod the connection weights a_{nm}.

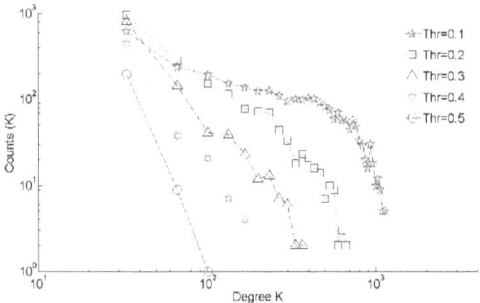

Fig. 5. Voxel Degree follows truncated exponential distribution. Thr stands for different threshold coefficients.

where $\overline{G_i^1}$ stands for the mean energy of the first scan after the onset of the ith stimulus. Then a_{nm} can also be seen as the contributions of the mth voxel. As discussed above, the columns of the trasition matrix differs much, so it is probably that only a subset of voxels, which have larger contributions, play essential roles. Threshold the transform matrix shows that only several voxels are very important in function network rewiring, as shown in Fig.4. We named them "skeleton voxels".

Similarly, a_{nm} can be seen as the connection weight between voxel m and n. If we threshold the weight and only take the presence or absence of these connections into consideration, the degree distribution of voxels can also be obtained. As shown in Fig.5, the degrees follow a truncated exponential distribution, which supports the conclusion of Achard et al. [12]. It says that only a small subset of voxels have large degree, while the others have small degree, and which is the typical characters of scale-free network. We also found that most of the voxels with large degree belong to skeleton voxels, so skeleton voxels might support stable fast rewiring by widespread connections.

We also analyzed the neighbor distribution of single voxels. In most cases, their activations are mainly determined by few local neighboring voxels and their counterparts in the opposite hemisphere, which supports the sparse coding hypothesis and the connection role played by the callose. A sample is shown in Fig.6. It should be noted that skeleton voxels have more neighbors, and we think it may have something to do with the stability requirements.

Of course, the "closest" neighbors correlate with the specific tasks brain undertakes. But our analysis shows that some voxels may be essential for the stable fast rewiring.

5 Discussions and Conclusions

Our brain experiences complex changes every day. When stimulus set, peripheral sense organs send information into the central neural system. Then we formulate the cognition by functional complex network dynamics [13]. Cognition to different objects or concepts relies on different functional works. So analyzing the switching mechanisms between networks is not only essential for disclosing the formulation of function networks, but also for revealing the relationship between structural network and function networks. They are all heuristic for artificial networks.

The emergence of fMRI technology provides us the opportunity to analyze the network mechanisms in a micro-macro scale. The paper mainly discusses the stability of fast rewiring in functional network by analyzing the fMRI signal, and many interesting and heuristic conclusions are drawn. Firstly, we verified that there is a 2-scans time delay between the stimulus onset and the activation of corresponding functional region. The most important is that we found that the delay is very stable. Secondly, we proposed for the first time that there might be skeleton voxels that induces the stable fast rewiring. What's more, our analysis on the degree distribution of voxels supports the scale-free hypothesis.

Also recently, we note that Tomasi et al [14][15] proposed that some "hubs" areas are essential for the brain network architecture by calculating the density of functional connectivity. Our analysis supports their conclusions. But our method provides a new way to analyze the activation pattern of each sub-network. Our future work will focus on the statistical test of skeleton voxels, exploration of skeleton voxels in other modals and reasoning the origination skeleton voxels.

Acknowledgments. We thank *Welcome* team for approval use of the MoAE dataset. Special thanks go to *He-Ping Song* from *Sun Yat-Sen University* and another two annoymous viewers for their helpful comments for the manuscript.

References

1. Deligianni, F., Robinson, E.C., Beckmann, C.F., Sharp, D., Edwards, A.D., Rueckert, D.: Inference of functional connectivity from structural brain connectivity. In: 10th IEEE International Symposium on Biomedical Imaging, pp. 460–463 (2010)
2. Chai, B., Walther, D.B., Beck, D.M., Li, F.-F.: Exploring Functional Connectivity of the Human Brain using Multivariate Information Analysis. In: Neural Information Processing Systems (2009)

3. Gomez Portillo, I.J., Gleiser, P.M.: An Adaptive Complex Network Model for Brain Functional Networks. PLoS ONE 4(9), e6863 (2009), doi:10.1371/journal.pone.0006863

4. Gleiser, P.M., Spoormaker, V.I.: Modeling hierarchical structure in functional brain networks. Philos. Transact. A Math. Phys. Eng. Sci. 368(1933), 5633–5644 (2010)

5. Friston, K.J., Rees, G.: Single Subject epoch auditory fMRI activation data, http://www.fil.ion.ucl.ac.uk/spm/

6. Friston, K.J., Holmes, A.P., Ashburner, J., Poline, J.B.: SPM8, http://www.fil.ion.ucl.ac.uk/spm/

7. Friston, K.J., Ashburner, J., Frith, C.D., Poline, J.B., Heather, J.D., Frackowiak, R.S.J.: Spatial registration and normalization of images. Human Brain Mapping 2, 1–25 (1995)

8. Talairach, P., Tournoux, J.: A stereotactic coplanaratlas of the human brain. Thieme Verlag (1988)

9. Frank, Y.S.: Image Processing and Mathematical Morphology: Fundaments and Applications. CRC Press (2009)

10. Whitney, C., Jefferies, E., Kircher, T.: Heterogeneity of the left temporal lobe in semantic representation and control: priming multiple vs. single meanings of ambiguous words. Cereb. Cortex (2010), doi: 10.1093/cercor/bhq148 (advance access published August 23)

11. Friston, K.J.: Statistical parametrical mapping: the analysis of functional brain images. Academic Press (2007)

12. Achard, S., Salvador, R., Whitcher, B., Suckling, J., Bullmore, E.: A resilient, low-frequency, small-world human brain functional network with highly connected association cortical hubs. The Journal of Neuroscience 26(1), 63–72 (2006)

13. Kitano, K., Yamada, K.: Functional Networks Based on Pairwise Spike Synchrony Can Capture Topologies of Synaptic Connectivity in a Local Cortical Network Model. In: Köppen, M., Kasabov, N., Coghill, G. (eds.) ICONIP 2008, Part I. LNCS, vol. 5506, pp. 978–985. Springer, Heidelberg (2009)

14. Tomasi, D., Volkow, N.D.: Functional Connectivity Density Mapping. Proc. Natl. Acad. Sci. USA 107, 9885–9890 (2010)

15. Tomasi, D., Volkow, N.D.: Association between Functional Connectivity Hubs and Brain Networks Cereb. Cortex (in press, 2011)

A Computational Agent Model
for Hebbian Learning of Social Interaction

Jan Treur

VU University Amsterdam, Agent Systems Research Group
De Boelelaan 1081, 1081 HV, Amsterdam, The Netherlands
treur@cs.vu.nl
http://www.cs.vu.nl/~treur

Abstract. In social interaction between two persons usually a person displays understanding of the other person. This may involve both nonverbal and verbal elements, such as bodily expressing a similar emotion and verbally expressing beliefs about the other person. Such social interaction relates to an underlying neural mechanism based on a mirror neuron system, as known within Social Neuroscience. This mechanism may show different variations over time. This paper addresses this adaptation over time. It presents a computational model capable of learning social responses, based on insights from Social Neuroscience. The presented model may provide a basis for virtual agents in the context of simulation-based training of psychotherapists, gaming, or virtual stories.

Keywords: Hebbian learning, ASD, computational model, social interaction.

1 Introduction

Showing mutual empathic understanding is often considered a form of glue between persons within a social context. Recent developments within Social Neuroscience have revealed that a mechanism based on *mirror neurons* plays an important role in generating and displaying such understanding, both in nonverbal form (e.g., smiling in response to an observed smile) and in verbal form (e.g., attributing an emotion to the other person); cf. [11, 19]. Such empathic responses vary much over persons. For example, when for a person these responses are low or nonexistent, often the person is considered as 'having some autistic traits'. Within one person such differences in responding may occur as well over time, in the sense of learning or unlearning to respond. This is the focus of this paper.

It is often claimed that the mirroring mechanism is not (fully) present at birth, but has to be shaped by experiences during lifetime; for example, [3, 11, 14]. For persons (in particular children) with low or no social responses, it is worth while to offer them training sessions in imitation so that the mirror neuron system and the displayed social responses may improve. This indeed turns out to work, at least for the short term, as has been reported in, for example [7, 13]. Thus evidence is obtained that the mirror neuron system has a certain extent of plasticity due to some learning mechanism. In [14] it is argued that Hebbian learning (cf. [8, 10]) is a good candidate for such a learning mechanism.

B.-L. Lu, L. Zhang, and J. Kwok (Eds.): ICONIP 2011, Part I, LNCS 7062, pp. 9–19, 2011.

In this paper a Hebbian learning mechanism is adopted to obtain an adaptive agent model showing plasticity of the agent's mirror neuron system. The model realises learning (and unlearning) of social behaviour (in particular, empathic social responses), depending on a combination of innate personal characteristics and the person's experiences over time obtained in social context. A person's experiences during lifetime may concern self-generated experiences (the person's responses to other persons encountered) or other-generated experiences (other persons' responses to the person). By varying the combination of innate characteristics and the social context offering experiences, different patterns of learning and unlearning of socially responding to other persons are displayed.

In Section 2 the adaptive agent model for Hebbian learning of social behaviour is presented. In Section 3 some simulation results are discussed, for different characteristics and social contexts. In Section 4 a mathematical analysis of the learning behaviour is made. Section 5 concludes the paper.

2   The Adaptive Agent Model Based on Hebbian Learning

The basic (non-adaptive) agent model (adopted from [20]) makes use of a number of internal states for the agent self, as indicated by the nodes in Fig. 1. A first group of states consists of the sensory representations of relevant external aspects: a sensory representation of a body state (labeled by) b, of a stimulus s, and of another agent B, denoted by srb, sr_s, sr_B, respectively. Related sensor states are ss_b, ss_s, ss_B, which in turn depend on external world states ws_b, ws_s, ws_B. Moreover, pb_b and $pc_{B,b}$ denote preparation states for bodily expression of b and communication of b to agent B. Following [5], the preparation for bodily expression b is considered to occur as an *emotional response* on a sensed stimulus s. Feeling this emotion is based on the sensory representation sr_b of b. These b's will be used as labels for specific emotions. Communication of b to B means communication that the agent *self* believes that B feels b; for example: 'You feel b', where b is replaced by a word commonly used for the type of emotion labeled in the model by b.

The states indicated by $ps_{c,s,b}$ are considered *control* or *super mirror* states (cf. [11], pp. 200-203, [12], [16]) for context c, stimulus s and body state b; they provide control for the agent's execution of (prepared) actions, such as expressing body states or communications, or regulation of the gaze. Here the context c can be an agent B, which can be another agent (self-other distinction), or the agent *self*, or c can be *sens* which denotes enhanced sensory processing sensitivity: a trait which occurs in part of the population, and may affect social behaviour (e.g., [1, 4]). One reason why some children do not obtain a sufficient amount of experiences to shape their mirror neuron system, is that they tend not to look at other persons due to enhanced sensory processing sensitivity for face expressions, in particular in the region of the eyes; e.g., [4, 15]. When observing the face or eyes of another person generates arousal which is experienced as too strong, as a form of emotion regulation the person's own gaze often is taken away from the face or eyes observed; cf. [9]. Such an avoiding behavioural pattern based on emotion regulation may stand in the way of the development of the mirror neuron system. In summary, three types of super mirroring states may (nonexclusively) occur to exert control as follows:

- if a super mirror state for agent *B* occurs (*self-other distinction*), a prepared communication will be performed and directed to *B*
- if it occurs for *self*, the agent will *execute* the related prepared actions
- if it occurs for *sens*, the agent will regulate some aspects of functioning to *compensate for enhanced sensitivity*: to suppress preparation and expression of related bodily responses, and to adapt the gaze to avoid the stimulus *s*.

Expressing body state *b* is indicated by effector state eb_b, communication of *b* to *B* by $ec_{b,B}$, and regulated gaze to avoid stimulus *s* by eg_s. These effector states result in a modified body state indicated by ws_b and an adapted gaze avoiding *s* indicated by wg_s.

In case the stimulus *s* is another agent *B*'s body expression for *b* (denoted by $s_{b,B}$, for example, a smiling face), then *mirroring* of this body state means that the agent prepares for the same body expression *b*; e.g., [11, 16, 19]. If this prepared body state is actually expressed, so that agent *B* can notice it, then this contributes an *empathic nonverbal response*, whereas communication of *b* to *B* is considered an *empathic verbal response*. The bodily expression of an observed feeling *b* together with a communication of *b* to *B* occurring at the same time is considered a *full empathic response* of *self* to *B*. These two elements for empathic response are in line with the criteria for empathy explicated in [6], p. 435 (assuming true, faithful bodily and verbal expression): (1) presence of an affective state in a person, (2) isomorphism of the person's own and the other person's affective state, (3) elicitation of the person's affective state upon observation or imagination of the other person's affective state, (4) knowledge of the person that the other person's affective state is the source of the person's own affective state.

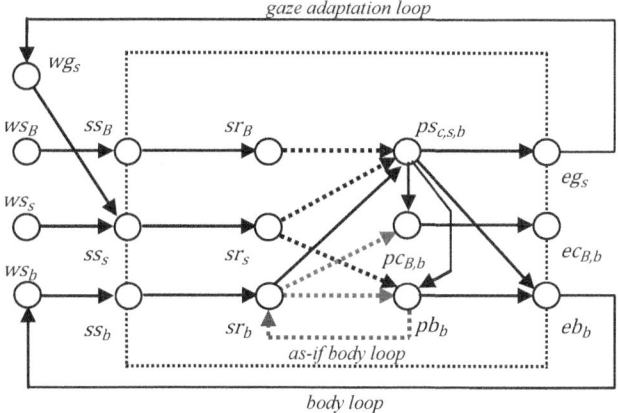

Fig. 1. Overview of the adaptive agent model

The arrows connecting the nodes in Fig. 1 indicate the dynamical relationships between the states. Most of these connections have been given strength *1*, but six of them (indicated by dotted arrows) have a dynamical strength, adapted over time according to Hebbian learning. Note that the graph of the model shown in Fig. 1 shows three loops: the body loop to adapt the body, the as-if body loop to adapt the internal body representation and integrate felt emotions in preparations for responses, and the gaze adaptation loop. The effect of these loops is that for any new external situation encountered, a (numerical) approximation process takes place until the internal states reach an equilibrium (assuming that the external situation does not change too fast). However, as will be seen in Section 3, it is also possible that a (static) external situation leads to periodic oscillations (limit cycle behaviour).

The connection strengths are indicated by ω_{ij} with the node labels i and j (the names of the nodes as indicated in Fig. 1) as subscripts. A distinction is made between expression states and the actual states for body and gaze. The first type of states are the agent's effector states (e.g., the muscle states), whereas the body and gaze states result from these. The sensory representation of a body state b is not only affected by a corresponding sensor state (via the body loop), but also by the preparation for this body state (via the as-if body loop). Preparation for a verbal empathic communication depends on feeling a similar emotion, and adequate self-other distinction.

Super mirroring for an agent A (*self* or B) generates a state indicating on which agent (*self-other distinction*) the focus is, and whether or not to act. Super mirroring for *enhanced sensory processing sensitivity*, generates a state indicating in how far the stimulus induces a sensory body representation level experienced as inadequately high. To cover regulation to compensate for enhanced sensory processing sensitivity (e.g., [1]), the super mirroring state for this is the basis for three possible regulations: of the prepared and expressed body state, and of the gaze.

A first way in which regulation takes place, is by a suppressing effect on preparation of the body state (note that the connection strength $\omega_{pssens,s,bpbb}$ from node $ps_{sens,s,b}$ to node pb_b is taken negative). Such an effect can achieve, for example, that even when the agent feels the same as the other agent, an expressionless face is prepared. In this way a mechanism for *response-focused regulation* (suppression of the agent's own response) to compensate for an undesired level of emotion is modelled; cf. [9]. Expressing a prepared body state depends on a super mirroring state for self and a super mirroring state for enhanced sensitivity with a suppressing effect (note that $\omega_{pssens,s,bebb}$ is taken negative). This is a second way in which a mechanism for response-focused regulation is modelled to compensate for an undesired level of arousal. A third type of regulation to compensate for enhanced sensory processing sensitivity, a form of *antecedent-focused regulation* (*attentional deployment*) as described in [9], is modelled by directing the own gaze away from the stimulus. Note that node eg_s for avoiding gaze for stimulus s has activation level 1 for total avoidance of the stimulus s, and 0 for no avoidance (it indicates the extent of avoidance of s). To generate a sensor state for stimulus s, the gaze avoidance state for s is taken into account: it has a suppressing effect on sensing s (note that ω_{wgsss} is taken negative).

The model has been specified in dynamical system format (e.g., [18]) as follows. Here for a node label k, by $a_k(t)$ the activation level (between 0 and 1) of the node labeled by k at time t is denoted, by *input(k)* the set of node labels is denoted that provides input (i.e., have an incoming arrow to node k in Fig. 1), and *th(W)* is a threshold function.

$$\frac{d\,a_k(t)}{dt} = \gamma[\,th(\Sigma_{j \in input(k)}\,\omega_{jk}\,a_j(t)) - a_k(t)\,] \tag{1}$$

The parameter γ is an update speed factor, which might differ per connection, but has been given a uniform value *0.8* in Section 3. The following logistic threshold function *th(W)* with $\sigma > 0$ a steepness and $\tau \geq 0$ a threshold value has been used in the simulations (except for the sensor states):

$$th(W) = \left(\frac{1}{1+e^{-\sigma(W-\tau)}} - \frac{1}{1+e^{\sigma\tau}}\right) / \left(1 - \frac{1}{1+e^{\sigma\tau}}\right) \quad \text{or} \quad th(W) = \frac{1}{1+e^{-\sigma(W-\tau)}} \quad (2)$$

The former threshold function can be approximated by the simpler latter expression for higher values of $\sigma\tau$ (e.g., σ higher than $20/\tau$). For the sensor states for *b* and *B* the identity function has been used for *th(W)*, and for the sensor state of *s* the update equation has been taken more specifically to incorporate the effect of gaze on the sensor state (note that the connection strength $\omega_{wg_s ss_s}$ from the world gaze state to the sensor state is taken negative):

$$\frac{d\,a_{ss_s}(t)}{dt} = \gamma\,[\omega_{ws_s ss_s}\,a_{ws_s}(t)(1 + \omega_{wg_s ss_s}\,a_{wg_s}(t)) - a_{ss_s}(t)\,] \quad (3)$$

Hebbian Learning
The model as described above was adopted from [20]; as such it has no adaptive mechanisms built in. However, as put forward, for example, in [3, 11, 14] learning plays an important role in shaping the mirror neuron system. From a Hebbian perspective [10], strengthening of a connection over time may take place when both nodes are often active simultaneously ('neurons that fire together wire together'). The principle goes back to Hebb [10], but has recently gained enhanced interest by more extensive empirical support (e.g., [2]), and more advanced mathematical formulations (e.g., [8]). In the adaptive agent model the connections that play a role in the mirror neuron system (i.e., the dotted arrows in Fig. 1) are adapted based on a Hebbian learning mechanism. More specifically, such a connection strength ω is adapted using the following *Hebbian learning rule*, taking into account a maximal connection strength *1*, a *learning rate* η, and an *extinction rate* ζ (usually small):

$$\frac{d\omega_{ij}(t)}{dt} = \gamma[\eta\,a_i(t)a_j(t)(1 - \omega_{ij}(t)) - \zeta\omega_{ij}(t)] = \gamma[\eta\,a_i(t)a_j(t) - (\eta\,a_i(t)a_j(t) + \zeta)\,\omega_{ij}(t)] \quad (4)$$

A similar Hebbian learning rule can be found in [8], p. 406. By the factor $1 - \omega_{ij}(t)$ the learning rule keeps the level of $\omega_{ij}(t)$ bounded by *1* (which could be replaced by any other positive number); Hebbian learning without such a bound usually provides instability. When the extinction rate is relatively low, the upward changes during learning are proportional to both $a_1(t)$ and $a_2(t)$ and maximal learning takes place when both are *1*. Whenever one of them is *0* (or close to *0*) extinction takes over, and ω slowly decreases (unlearning). This learning principle has been applied (simultaneously) to all six connections indicated by dotted arrows in Fig. 1. In principle, the adaptation speed factor γ, the learning rate η and extinction rate ζ, could be taken differently for the different dynamical relationships. In the example simulations discussed in Section 3 uniform values have been used: $\gamma = 0.8$, $\eta = 0.2$ and $\zeta = 0.004$.

3   Example Simulations of Learning Processes

A number of simulation experiments have been conducted for different types of scenarios, using numerical software. For the examples discussed here the values for the threshold and steepness parameters are as shown in Table 1. Note that first the value *3* for sensitivity super mirroring threshold was chosen so high that no enhanced sensitivity occurs. The speed factor γ was set to *0.8*, the learning rate $\eta = 0.2$ and extinction rate $\zeta = 0.004$. The step size Δt was set to *1*. All nonadapted connection strengths have been given value *1*, except those for suppressing connections

$$\omega_{ps_{sens,s,bpb_b}}, \quad \omega_{ps_{sens,s,beb_b}} \text{ and } \omega_{wg_{sss}}$$

which have been given the value *-1*. The scenario was chosen in such a way that after every 100 time units another agent is encountered for a time duration of 25 units with a body expression that serves as stimulus. Initial values for activation levels of the internal states were taken *0*. A first pattern, displayed in Fig. 2, is that in normal circumstances, assuming initial strengths of the learned connections of *0.3*, the model is indeed able to learn the empathic responses as expected. Here (and also in Fig. 3) time is on the horizontal axis and activation levels at the vertical axis.

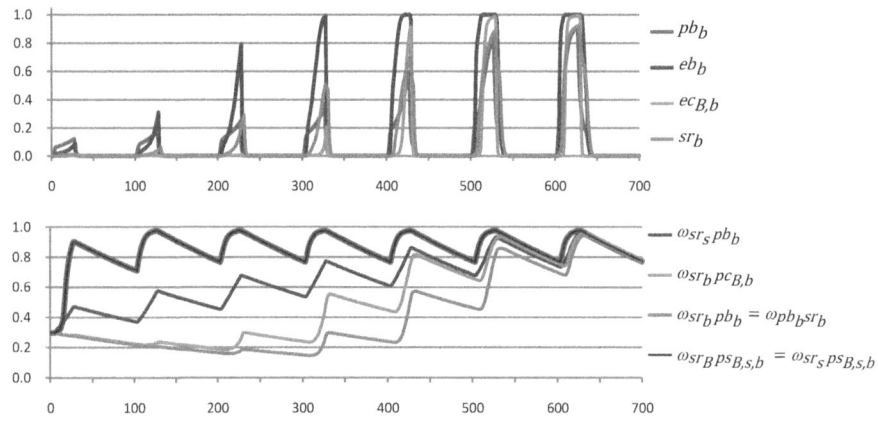

Fig. 2. Example scenario of the Hebbian learning process

The upper graph shows levels for body representation, body preparation, expressed body states and communication. The lower graph shows the learning patterns for the connections (the dotted arrows in Fig. 1). Note that the two connections

$$\omega_{sr_b pb_b} \text{ (for emotion integration)} \quad \text{and} \quad \omega_{pb_b sr_b} \text{ (as-if body loop)}$$

have the same values, as they connect the same nodes sr_b and pb_b, and have been given the same initial values. Moreover, also the connections

$$\omega_{sr_B ps_{B,s,b}} \text{ and } \omega_{sr_s ps_{B,s,b}}$$

have the same values, as in the considered scenario the input nodes for sr_B and sr_s have been given the same values, and also the initial values for the connections. This can easily be varied. In Fig. 2 it is shown that when regular social encounters take place, the connections involved in responding empathically are strengthened to values that approximate *1*. Notice that due to the relatively low initial values of the connections chosen, for some of them first extinction dominates, but later on this downward trend is changing into an upward trend. Accordingly the empathic responses become much stronger, which is in line with the literature; e.g., [7], [13].

Table 1. Settings for threshold and steepness parameters

		τ	σ
representing body state	sr_b	1	3
super mirroring B	$ps_{B,s,b}$	0.7	30
super mirroring sensitivity	$ps_{sens,s,b}$	3	30
mirroring/preparing body state	pb_b	1	3
preparing communication	$pc_{b,B}$	0.8	3
expressing body state	eb_b	1.2	30
expressing communication	$ec_{b,B}$	0.8	30
expressing gaze avoidance state	eg_s	0.6	30

How long the learned patterns will last will depend on the social context. When after learning the agent is isolated from any social contact, the learned social behaviours may vanish due to extinction. However, if a certain extent of social contact is offered from time to time, the learned behaviour is maintained well. This illustrates the importance of the social context. When zero or very low initial levels for the connections are given, this natural learning process does not work. However, as other simulations show, in such a case (simulated) imitation training sessions (starting with the therapist imitating the person) still have a positive effect, which is also lasting when an appropriate social context is available. This is confirmed by reports that imitation training sessions are successful; e.g., [7], [13].

In addition to variations in social environment, circumstances may differ in other respects as well. From many persons with some form of autistic spectrum disorder it is known that they show enhanced sensory processing sensitivity; e.g., [1], [4]; this was also incorporated in the model. Due to this, their regulation mechanisms to avoid a too high level of arousal may interfere with the social behaviour and the learning processes. Indeed, in simulation scenarios for this case it is shown that the adaptive agent model shows an unlearning process: connection levels become lower instead of higher. This pattern is shown in Fig. 3. Here the same settings are used as in Table 1, except the sensitivity super mirroring threshold which was taken *1* in this case, and the initial values for the connection weights, which were taken *0.7*. It is shown that the connections

$\omega_{sr_s pb_b}$ (for mirroring) and $\omega_{sr_b pb_b}$ and $\omega_{pb_b sr_b}$ (for emotion integration)

are decreasing, so that the responses become lower over time.

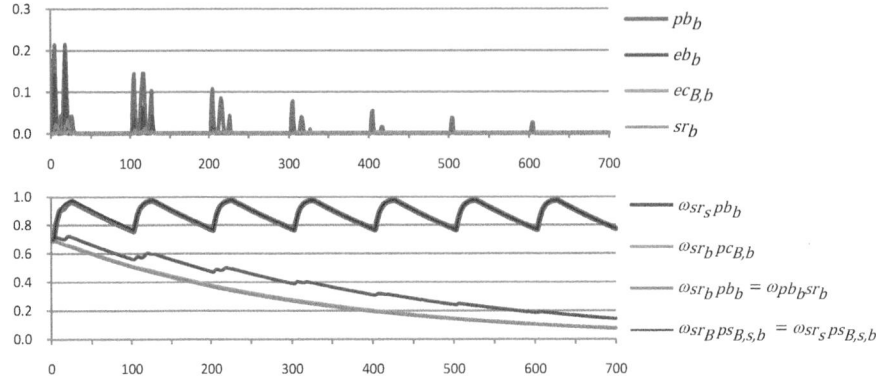

Fig. 3. Learning under enhanced sensory processing sensitivity

This is due to the downregulation which, for example, leads to a gaze that after a short time is taken away from the stimulus, and returns after the arousal has decreased, after which the same pattern is repeated; this is shown in the upper graph (the two or three peaks per encounter). Note that the values related to super mirroring of and communication to another agent stay high: the downregulation as modelled does not have a direct effect on these processes. When downregulation is also applied to communication, also these connections will extinguish. When for such a case imitation training sessions are offered in a simulation, still the connection levels may be strengthened. However, these effects may not last in the natural context: as soon as these sessions finish, the natural processes may start to undo the learned effects. To maintain the learned effects for this case such training sessions may have to be repeated regularly.

4 Formal Analysis

The behaviour of the agent's adaptation process can also be investigated by formal analysis, based on the specification for the connection strength $\omega = \omega_{ij}$ from node i to node j.

$$\frac{d\omega(t)}{dt} + \gamma\,(\eta a_i(t)a_j(t) + \zeta)\,\omega(t) = \gamma\eta a_i(t)a_j(t) \tag{5}$$

This is a first-order linear differential equation with time-dependent coefficients: a_i and a_j are functions of t which are considered unknown external input in the equation for ω. An analysis can be made for when equilibria occur:

$$\frac{d\omega(t)}{dt} = 0 \iff (\eta a_i a_j + \zeta)\,\omega = \eta a_i a_j \iff \omega = \frac{\eta a_i a_j}{\eta a_i a_j + \zeta} \tag{6}$$

One case here is that $\omega = 0$ and one of a_i and a_j is 0. When a_i and a_j are nonzero, (6) can be rewritten as (since $a_i a_j \leq 1$):

$$\omega = 1/(1 + \zeta/\eta a_i a_j) \leq 1/(1 + \zeta/\eta) \tag{7}$$

This shows that when no extinction takes place ($\zeta = 0$), an equilibrium for ω of 1 is possible, but if extinction is nonzero, only an equilibrium < 1 is possible. For example, when $\eta = 0.2$ and $\zeta = 0.004$ as in Section 3, then an equilibrium value will be ≤ 0.98, as also shown in the example simulations.

Further analysis can be made by obtaining an explicit analytic solution of the differential equation in terms of the functions a_i and a_j. This can be done as follows. Take

$$W(t) = \int_{t_0}^{t} a_i(u)a_j(u)\, du \tag{8}$$

the accumulation of $a_i(t)a_j(t)$ over time from t_0 to t; then

$$\frac{dW(t)}{dt} = a_i(t)a_j(t) \tag{9}$$

Given this, the differential equation (5) for ω can be solved by using an integrating factor as follows:

$$\frac{d e^{\chi(\eta W(t)+\zeta(t-t_0))}\omega(t)}{dt} = \gamma a_i(t)a_j(t)\ e^{\chi(\eta W(t)+\zeta(t-t_0))} \tag{10}$$

from which it can be obtained:

$$\omega(t) = \omega(t_0)\ e^{-\chi(\eta W(t)+\zeta(t-t_0))} + \eta \int_{t_0}^{t} a_i(u)a_j(u)\ e^{-\chi(\eta(W(t)-W(u))+\zeta(t-u))}du \tag{11}$$

For the special case of constant $a_i a_j = c$, from (11) explicit expressions can be obtained, using $W(t) = c(t-t_0)$ and $W(t)-W(u) = c(t-u)$:

$$\int_{t_0}^{t} a_i(u)a_j(u)\ e^{-\chi(\eta(W(t)-W(u))+\zeta(t-u))}du = \int_{t_0}^{t} ce^{-\chi(\eta c+\zeta)(t-u)}du$$

$$= \frac{1}{\chi(\eta c+\zeta)}[1 - e^{-\chi(\eta c+\zeta)(t-t_0)}] \tag{12}$$

Although in a simulation usually $a_i a_j$ will not be constant, these expressions are still useful in a comparative manner. When $a_i a_j \geq c$ on some time interval, then by monotonicity the above expressions (11) for ω with $a_i a_j = c$ provide a lower bound for ω. From these expressions it can be found that

$$\eta c/(\eta c+\zeta) - \omega(t) = [\eta c/(\eta c+\zeta) - \omega(0)]\ e^{-\chi(\eta c+\zeta)t} \tag{13}$$

which shows the convergence rate to an equilibrium for constant $a_i a_j = c$, provides an upper bound for the deviation from the equilibrium. This has half-value time

$$\ln(2)/\chi(\eta c+\zeta) = 0.7/\chi(\eta c+\zeta) \tag{14}$$

When $a_i a_j \geq c$ on some time interval, then by the monotonicity mentioned earlier, the upward trend will be at least as fast as described by this expression. For example, for the settings in Section 3 with $c = 0.2$ this provides half-value time *20*. This bound indeed is shown in simulations (e.g., in Figs 2 and 3) in time periods with $a_i a_j$ around or above *0.2*.

For scenarios in which encounters with other agents alternate with periods when nobody is there, as in Figs 2 and 3, a fluctuating learning curve is displayed. A question is how the balance between the different types of episodes should be in order to keep the learned effects at a certain level. Given the indications (14) above a rough estimation can be made of how long a time duration td_1 of increase should last to compensate a time duration td_2 of decrease:

$$ e^{-\gamma(\eta c + \zeta)td_1} = e^{-\gamma\zeta td_2} \qquad\qquad td_2/td_1 = (\eta c + \zeta)/\zeta = 1 + \eta c/\zeta \qquad (15) $$

For example, when $\eta = 0.2$ and $\zeta = 0.004$, as in Section 3, for $c = 0.2$ this provides: $td_2/td_1 = 11$. This means that for this case under normal circumstances around 9% of the time an encounter with another agent should take place leading to $a_i a_j \geq 0.2$ to maintain the empathic responses. This indeed corresponds to what was found by simulation experiments varying the intensity of encounters.

5 Discussion

To function well in social interaction it is needed that a person displays a form of empathic understanding, both by nonverbal and verbal expression. Within Social Neuroscience it has been found how such empathic social responses relate to an underlying neural mechanism based on a mirror neuron system. It is often suggested that innate factors may play a role, but also that a mirror neuron system can only function after a learning process has taken place (e.g., [3], [11], [14]): the strength of a mirror neuron system may change over time within one person. In this paper an adaptive agent model was presented addressing this aspect of adaptation over time, based on knowledge from Social Neuroscience.

The notion of empathic understanding taken as a point of departure is in line with what is formulated in [6]. The learning mechanism used is based on Hebbian learning, as also suggested by [14]. It is shown how under normal conditions by learning the empathic responses become better over time, provided that a certain amount of social encounters occur. The model also shows how imitation training (e.g., [7], [13]) can strengthen the empathic responses. Moreover, it shows that when enhanced sensory processing sensitivity [1] occurs (e.g., as an innate factor), the natural learning process is obstructed by avoidance behaviour to downregulate the dysproportional arousal [9].

In [17] a computational model for a mirror neuron system for grasp actions is presented; learning is also incorporated, but in a biologically implausible manner, as also remarked in [14]. In contrast, the presented model is based on a biologically plausible Hebbian learning model, as also suggested by [14]. The presented agent model provides a basis for the implementation of virtual agents, for example, for simulation-based training of psychotherapists, or of human-like virtual characters.

References

1. Baker, A.E.Z., Lane, A.E., Angley, M.T., Young, R.L.: The Relationship Between Sensory Processing Patterns and Behavioural Responsiveness in Autistic Disorder: A Pilot Study. J. Autism Dev. Disord. 38, 867–875 (2008)
2. Bi, G., Poo, M.: Synaptic modification by correlated activity: Hebb's postulate revisited. Annu. Rev. Neurosci. 24, 139–166 (2001)
3. Catmur, C., Walsh, V., Heyes, C.: Sensorimotor learning configures the human mirror system. Curr. Biol. 17, 1527–1531 (2007)
4. Corden, B., Chilvers, R., Skuse, D.: Avoidance of emotionally arousing stimuli predicts social-perceptual impairment in Asperger syndrome. Neuropsychologia 46, 137–147 (2008)
5. Damasio, A.: The Feeling of What Happens. Body and Emotion in the Making of Consciousness. Harcourt Brace, New York (1999)
6. DeVignemont, F., Singer, T.: The empathic brain: how, when and why? Trends in Cognitive Science 10, 437–443 (2006)
7. Field, T., Sanders, C., Nadel, J.: Children with autism display more social behaviors after repeated imitation sessions. Autism 5, 317–323 (2001)
8. Gerstner, W., Kistler, W.M.: Mathematical formulations of Hebbian learning. Biol. Cybern. 87, 404–415 (2002)
9. Gross, J.J.: Antecedent- and response-focused emotion regulation: divergent consequences for experience, expression, and physiology. J. of Pers. and Social Psychology 74, 224–237 (1998)
10. Hebb, D.: The Organisation of Behavior. Wiley (1949)
11. Iacoboni, M.: Mirroring People: the New Science of How We Connect with Others. Farrar, Straus & Giroux (2008)
12. Iacoboni, M.: Mesial frontal cortex and super mirror neurons. Beh. Brain Sci. 31, 30–30 (2008)
13. Ingersoll, B., Lewis, E., Kroman, E.: Teaching the Imitation and Spontaneous Use of Descriptive Gestures in Young Children with Autism Using a Naturalistic Behavioral Intervention. J. Autism Dev. Disord. 37, 1446–1456 (2007)
14. Keysers, C., Perrett, D.I.: Demystifying social cognition: a Hebbian perspective. Trends Cogn. Sci. 8, 501–507 (2004)
15. Kirchner, J.C., Hatri, A., Heekeren, H.R., Dziobek, I.: Autistic Symptomatology, Face Processing Abilities, and Eye Fixation Patterns. Journal of Autism and Developmental Disorders (2010), doi:10.1007/s10803-010-1032-9
16. Mukamel, R., Ekstrom, A.D., Kaplan, J., Iacoboni, M., Fried, I.: Single-Neuron Responses in Humans during Execution and Observation of Actions. Current Biology 20, 750–756 (2010)
17. Oztop, E., Arbib, M.A.: Schema design and implementation of the grasp-related mirror neuron system. Biol. Cybern. 87, 116–140 (2002)
18. Port, R., van Gelder, T.J.: Mind as Motion: Explorations in the Dynamics of Cognition. MIT Press (1995)
19. Rizzolatti, G., Sinigaglia, C.: Mirrors in the Brain: How Our Minds Share Actions and Emotions. Oxford University Press (2008)
20. Treur, J.: A Cognitive Agent Model Displaying and Regulating Different Social Response Patterns. In: Walsh, T. (ed.) Proceedings of the Twenty-Second International Joint Conference on Artificial Intelligence, IJCAI 2011, pp. 1735–1742 (2011)

An Information Theoretic Approach to Joint Approximate Diagonalization

Yoshitatsu Matsuda[1] and Kazunori Yamaguchi[2]

[1] Department of Integrated Information Technology,
Aoyama Gakuin University,
5-10-1 Fuchinobe, Chuo-ku, Sagamihara-shi, Kanagawa, 252-5258, Japan
matsuda@it.aoyama.ac.jp
[2] Department of General Systems Studies,
Graduate School of Arts and Sciences, The University of Tokyo,
3-8-1, Komaba, Meguro-ku, Tokyo, 153-8902, Japan
yamaguch@graco.c.u-tokyo.ac.jp

Abstract. Joint approximate diagonalization (JAD) is a solution for
blind source separation, which can extract non-Gaussian sources without any other prior knowledge. However, because JAD is based on an
algebraic approach, it is not robust when the sample size is small. Here,
JAD is improved by an information theoretic approach. First, the "true"
probabilistic distribution of diagonalized cumulants in JAD is estimated
under some simple conditions. Next, a new objective function is defined
as the Kullback-Leibler divergence between the true distribution and the
estimated one of current cumulants. Though it is similar to the usual JAD
objective function, it has a positive lower bound. Then, an improvement
of JAD with the lower bound is proposed. Numerical experiments verify
the validity of this approach for a small number of samples.

1 Introduction

Independent component analysis (ICA) is a widely-used method in signal processing [5,4]. It solves blind source separation problems under the assumption
that source signals are statistically independent of each other. In the linear model
(given as $X = AS$), it estimates the $N \times N$ mixing matrix $A = (a_{ij})$ and the
$N \times M$ source signals $S = (s_{im})$ from only the observed signals $X = (x_{im})$.
N and M correspond to the number of signals and the sample size, respectively. Joint approximate diagonalization (denoted by JAD) [3,2] is one of the
efficient methods for estimating A. C_{pq} is defined as an $N \times N$ matrix whose
(i,j)-th element is κ^x_{ijpq}, which is the 4-th order cumulant of the observed signals X. It can be easily proved that $\Delta_{pq} = VC_{pq}V'$ is a diagonal matrix
for any p and q if $V = (v_{ij})$ is equal to the separating matrix A^{-1} except
for any permutation and scaling. In other words, A can be estimated as the
inverse of V which diagonalizes $\bar{C}_{pq}$ for every pair of p and q, where $\bar{C}_{pq}$
is an estimation of the true C_{pq} on the observed signals X. Thus, an error
function for any p and q is defined as $\sum_{i,j>i} (\nu_{ijpq})^2$ where each ν_{ijpq} is the

B.-L. Lu, L. Zhang, and J. Kwok (Eds.): ICONIP 2011, Part I, LNCS 7062, pp. 20–27, 2011.
© Springer-Verlag Berlin Heidelberg 2011

(i, j)-th element of $\boldsymbol{V}\bar{\boldsymbol{C}}_{pq}\boldsymbol{V}'$ $(\nu_{ijpq} = \sum_{k,l} v_{ik}v_{jl}\kappa^x_{klpq})$. Then, the average of the squares of ν_{ijpq} over i, $j > i$, p, and $q > p$ is defined as the objective function $\Psi = \left(\sum_{i,j>i}\sum_{p,q>p}(\nu_{ijpq})^2\right) / \left(\frac{N(N-1)}{2}\right)^2$. Note that a pair of the same index $p = q$ is excluded. This exclusion is needed in Section 2.1. Lastly, the estimated separating matrix $\widehat{\boldsymbol{V}}$ is given as $\widehat{\boldsymbol{V}} = \mathrm{argmin}_{\boldsymbol{V}}\Psi$. One of the significant advantages of JAD is that it does not depend on the specific statistical properties of source signals except for non-Gaussianity. Because JAD utilizes only the linear algebraic properties on the cumulants, JAD is guaranteed to generate the unique and accurate estimation of $\boldsymbol{V}$ if each cumulant κ^x_{ijpq} is estimated accurately [2]. However, from the viewpoint of the robustness, JAD lacks the theoretical foundation. Because many other ICA methods are based on some probabilistic models, the estimated results are guaranteed to be "optimal" in the models. On the other hand, JAD is theoretically valid only if the objective function Ψ is equal to 0. In other words, it is *not* guaranteed in JAD that $\boldsymbol{V}$ with smaller (but non-zero) Ψ is more "desirable." This theoretical problem often causes a deficiency of robustness in practical applications.

In this paper, an information theoretic approach to JAD is proposed. In order to realize this approach, the "true" probabilistic distribution of ν_{ijpq} in Ψ is theoretically estimated under the conditions that $\boldsymbol{V}$ is accurately estimated (namely, $\boldsymbol{V} = \boldsymbol{A}^{-1}$). Then, a new objective function is defined as the Kullback-Leibler (KL) divergence between this true distribution and the estimated one of current ν_{ijpq}. Lastly, by this approach, we can derive an appropriate termination condition of minimization of Ψ which increases the efficiency. In our previous work [7], the termination condition on JAD has been improved by a similar approach. However, their estimation of the distribution of cumulants was relatively rough and did not involve the effect of the sample size explicitly. Besides, it used model selection instead of the direct KL divergence. Here, we propose more rigorous estimation of the KL divergence involving the sample size. This paper is organized as follows. In Section 2.1, the true probabilistic distribution of ν_{ijpq} is estimated theoretically. In Section 2.2, the KL divergence between the true distribution and the estimated one of the current cumulants is derived as a new objective function. Besides, a positive lower bound is derived. In Section 2.3, an improvement of JAD utilizing the lower bound is proposed. Section 3 shows numerical results on artificial datasets. This paper is concluded in Section 4.

2   Method

2.1   Estimation of True Distribution

Here, the true probabilistic distribution of ν_{ijpq} is estimated theoretically under the four simple conditions: (1) a linear ICA model; (2) a large number of samples; (3) a random mixture; (4) a large number of signals. In this paper, "true" means that $\boldsymbol{V}$ is completely equivalent to $\boldsymbol{A}^{-1}$. First, the following condition is assumed.

Condition 1 (Linear ICA Model). *The linear ICA model* $\boldsymbol{X} = \boldsymbol{AS}$ *holds, where the mean and the variance of each source* s_{im} *are 0 and 1, respectively.*

This is the fundamental condition in JAD and many other ICA methods. Then, $\nu_{ijpq}(i < j)$ is transformed by the multilinearlity of cumulants and $V = A^{-1}$ as follows:

$$
\begin{aligned}
\nu_{ijpq} &= \sum_{k,l} v_{ik}\bar{\kappa}^x_{klpq}v_{jl} = \sum_{k,l} v_{ik}v_{jl} \sum_{r,s,t,u} a_{kr}a_{ls}a_{pt}a_{qu}\bar{\kappa}^s_{rstu} \\
&= \sum_{r,s,t,u} \delta_{ir}\delta_{js}a_{pt}a_{qu}\bar{\kappa}^s_{rstu} = \sum_{t,u} a_{pt}a_{qu}\bar{\kappa}^s_{ijtu}
\end{aligned}
\tag{1}
$$

where δ_{ij} is the Kronecker delta and $\bar{\kappa}^s_{ijkl}$ is an estimator of the 4-th order cumulant on S. Because the first and second order statistics κ^s_i and κ^s_{ij} are accurately determined by $\kappa^s_i = 0$ and $\kappa^s_{ij} = \delta_{ij}$ under Condition 1, $\bar{\kappa}^s_{ijkl}(i < j)$ is given as

$$
\bar{\kappa}^s_{ijkl} = \frac{\sum_m s_{im}s_{jm}s_{km}s_{lm}}{M} - \delta_{ik}\delta_{jl} - \delta_{il}\delta_{jk}.
\tag{2}
$$

If the estimation is accurate ($\bar{\kappa}^s_{ijkl} = \kappa^s_{ijkl}$), $\nu_{ijpq}(i < j)$ is 0 under Condition 1. That is the principle of JAD. However, the estimator (or the estimation error) $\bar{\kappa}^s_{ijkl}$ is different from the ideal $\kappa^s_{ijkl} = 0$ in practice. Now, the following condition is introduced.

Condition 2 (Large Number of Samples). *The sample size M is sufficiently large so that the central limit theorem (CLT) holds.*

Then, each estimation error $\bar{\kappa}^s_{ijkl}$ is given according to a normal distribution with the mean of 0 by CLT. Because ν_{ijpq} is a linear combination of the estimation errors (Condition 1), each ν_{ijpq} is also given according to a normal distribution with the mean of 0. Then, the following lemma on the distribution of each ν_{ijpq} holds:

Lemma 1. *Each ν_{ijpq} follows the normal distribution with the mean of 0 under Conditions 1 and 2.*

Thus, the second order statistics (covariances) on ν_{ijpq} ($i < j$ and $p < q$) completely determine the distribution of ν_{ijpq}'s. However, it is difficult to estimate those statistics when A is unknown. So, the following condition is added.

Condition 3 (Random Mixture). *Each element a_{ij} in A is given randomly and independently, whose mean and variance are 0 and $1/N$, respectively.*

This condition is easily satisfied by pre-whitening X and multiplying it by a random matrix whose elements are given randomly according to the normal distribution with the mean of 0 and the variance of $1/N$. Note that this condition is consistent with the orthogonality ($\sum_k a_{ik}a_{jk} = \delta_{ij}$) when N is sufficiently large. Then, the covariances on ν_{ijpq} can be estimated by $E_A (E_S (\nu_{ijpq}\nu_{klrs}))$ where $E_A ()$ and $E_S ()$ are the expectation operators on A and S, respectively. By Eq. (1), the expectation is rewritten as

$$
E_A (E_S (\nu_{ijpq}\nu_{klrs})) = \sum_{t,u,v,w} E_A (a_{pt}a_{qu}a_{rv}a_{sw}) E_S (\bar{\kappa}^s_{ijtu}\bar{\kappa}^s_{klvw}).
\tag{3}
$$

Under Condition 3, $E_A \left(a_{pt} a_{qu} a_{rv} a_{sw} \right)$ is given by $\left(\delta_{pr} \delta_{qs} \delta_{tv} \delta_{uw} \right) / N^2$ because of $p < q$ and $r < s$. On the other hand, $E \left(\bar{\kappa}_{ijtu}^{s} \bar{\kappa}_{klvw}^{s} \right)$ is given as

$$
E_S \left(\bar{\kappa}_{ijtu}^{s} \bar{\kappa}_{klvw}^{s} \right) = \frac{E_S \left(s_{im} s_{jm} s_{tm} s_{um} s_{km} s_{lm} s_{vm} s_{wm} \right)}{M}
$$
$$
- \frac{\left(\delta_{it} \delta_{ju} + \delta_{iu} \delta_{jt} \right) \left(\delta_{kv} \delta_{lw} + \delta_{kw} \delta_{lv} \right)}{M} \tag{4}
$$

where Eq. (2) and the following two properties were utilized: (1) s_{im} is independent of s_{jn} for $i \neq j$ or $n \neq m$; (2) $E_S \left(\sum_m s_{im} s_{jm} s_{tm} s_{um} / M \right) = \delta_{it} \delta_{ju} + \delta_{iu} \delta_{jt}$ for $i \neq j$. Then, Eq. (3) ($i < j$ and $k < l$) is transformed further into

$$
E_A \left(E_S \left(\nu_{ijpq} \nu_{klrs} \right) \right) = \frac{\delta_{pr} \delta_{qs} \sum_{t,u} E_S \left(s_{im} s_{jm} s_{km} s_{lm} s_{tm} s_{tm} s_{um} s_{um} \right)}{M N^2}
$$
$$
- \frac{2 \delta_{pr} \delta_{qs} \delta_{ik} \delta_{jl}}{M N^2}. \tag{5}
$$

By a basic relation between cumulants and moments, the expectation term in Eq. (5) is given as a sum of products of cumulants over all the partitions of the set of the subscripts [8]. Then, the following equation holds:

$$
\sum_{t,u} E_S \left(s_{im} s_{jm} s_{km} s_{lm} s_{tm} s_{tm} s_{um} s_{um} \right) = \sum_{t,u} \sum_{P \in \Pi_{ijklttuu}} \prod_{B \in P} \kappa^s [B] \tag{6}
$$

where $\Pi_{ijklttuu}$ is the set of all the partitions of the subscripts $\{i, j, k, l, t, t, u, u\}$, P is a partition, B is a subset of subscripts in P, and $\kappa^s[B]$ is the cumulant of S on the subscripts in B. Under Condition 1, every term on a partition including a singleton vanishes because the mean of every source is 0. In addition, the subscripts must be identical in the same subset in a partition because the sources are independent of each other. Now, the multiplicity of the summation is focused on. The original form of each term on a partition P in Eq. (6) is a double summation on t and u. However, if t (or u) belongs to a subset including the other subscripts in P, the term on P is a single summation or a single value because t (or u) is bound to the other subscripts. Therefore, only the partitions of $\{\cdots, \{t, t\}, \{u, u\}\}$ correspond to double summations. Because every cumulant does not depend on N, the value of a term without a double summation is of the order of N. Thus, Eq. (6) is rewritten as

$$
\sum_{P \in \Pi_{ijklttuu}} \prod_{B \in P} \kappa^s [B] = \left(\sum_{t,u} \delta_{uu} \delta_{tt} \sum_{P \in \Pi_{ijkl}} \prod_{B \in P} \kappa^s [B] \right) + O(N)
$$
$$
= N^2 \delta_{ik} \delta_{jl} + O(N) \tag{7}
$$

where $i < j$ and $k < l$ are utilized. Now, the following condition is introduced.

Condition 4 (Large Number of Signals). *The number of signals N is sufficiently large.*

Then, Eq. (5) is rewritten as

$$E_A\left(E_S\left(\nu_{ijpq}\nu_{klrs}\right)\right) = \frac{\delta_{ik}\delta_{jl}\delta_{pr}\delta_{qs} + O\left(\frac{1}{N}\right)}{M} \simeq \frac{\delta_{ik}\delta_{jl}\delta_{pr}\delta_{qs}}{M}, \tag{8}$$

which means that every ν_{ijpq} is independent of each other. In addition, the variance of every ν_{ijpq} is $1/M$. Consequently, the following theorem is derived:

Theorem 1. *Under the above four conditions, each ν_{ijpq} ($i < j$ and $p < q$) is expected to be an independent and identically distributed random variable according to the normal distribution with the variance of $1/M$.*

2.2   Objective Function

By utilizing Theorem 1, a new objective function for JAD is defined as the KL divergence between an estimated distribution of the current ν_{ijpq} (denoted by $P\left(\nu\right)$) and the true distribution $G_{\mathsf{t}}\left(\nu\right) = \exp\left(-\nu^2 M/2\right)/\sqrt{2\pi/M}$. The KL divergence D between $P\left(\nu\right)$ and $G_{\mathsf{t}}\left(\nu\right)$ is given as

$$D = \int P\left(\nu\right)\log\frac{P\left(\nu\right)}{G_{\mathsf{t}}\left(\nu\right)}\mathrm{d}\nu = \int P\left(\nu\right)\left(\log P\left(\nu\right) - \log G_{\mathsf{t}}\left(\nu\right)\right)\mathrm{d}\nu$$

$$\simeq -H\left(P\right) + \frac{\log\left(2\pi/M\right)}{2} - \frac{4}{N^2\left(N-1\right)^2}\sum_{i,j>i,p,q>p}\frac{-M\nu_{ijpq}^2}{2}$$

$$= -H\left(P\right) + \frac{\log\left(2\pi/M\right)}{2} + \frac{M\Psi}{2} \tag{9}$$

where $H\left(P\right) = -\int P\left(\nu\right)\log P\left(\nu\right)\mathrm{d}\nu$ is the entropy of $P\left(\nu\right)$. Though it is generally difficult to estimate $H\left(P\right)$ from given samples ν_{ijpq}'s, a rough Gaussian approximation with the mean of 0 is employed in this paper. In other words, $P\left(\nu\right)$ is approximated by a normal distribution with the mean of 0. Then, the variance of the distribution is given as the average of the squares of ν. Therefore, the variance is estimated as the usual JAD objective function Ψ. Consequently, $P\left(\nu\right)$ and $H\left(P\right)$ are approximated as $\exp\left(-\nu^2/2\Psi\right)/\sqrt{2\pi\Psi}$ and $\log\left(2\pi e\Psi\right)/2$, respectively. Now, D depends on only Ψ and M, which is given as

$$D\left(\Psi\right) = \frac{-1 - \log M}{2} - \frac{\log\Psi}{2} + \frac{M\Psi}{2}. \tag{10}$$

Because the optimal value $\widehat{\Psi}$ satisfies $D'\left(\widehat{\Psi}\right) = -\frac{1}{2\Psi} + \frac{M}{2} = 0$, $\widehat{\Psi}$ is given by $\widehat{\Psi} = 1/M$. It is easily shown that $D\left(\Psi\right)$ is minimized if and only if $\Psi = \widehat{\Psi}$. Therefore, $\widehat{\Psi} = 1/M$ can be regarded as a lower bound of Ψ.

2.3   Improvement of JADE with Lower Bound

Here, one widely-used algorithm of the usual JAD named JADE [3] is improved by incorporating the lower bound $\widehat{\Psi}$. In JADE, $\boldsymbol{X}$ is pre-whitened and $\boldsymbol{V}$ is estimated by the Jacobi method under the orthogonal constraints. Note again that the orthogonality is consistent with Condition 3 when N is sufficiently large. The Jacobi method minimizes Ψ by sweeping the optimization over every pair of signals, where each pair optimization corresponds to a rotation which can be calculated analytically. Only the termination condition on each rotation is modified in the proposed method. In the usual JADE, each rotation is actually carried out if the rotation is greater than a fixed small threshold. Therefore, the usual JADE tends to converge slowly. In the proposed method, each rotation is actually carried out only if the current Ψ on i and j $\left(= \left(\sum_{p,q>p} (\nu_{ijpq})^2\right) / \left(\frac{N(N-1)}{2}\right)\right)$ is greater than the lower bound $\widehat{\Psi} = 1/M$. Therefore, the proposed method can inhibit unnecessary optimizations. In summary, the improved JADE with the lower bound is given as follows:

1. *Initialization.* Multiply X by a random matrix, whiten X, and calculate ν_{ijpq}.
2. *Sweep.* For every pair (i, j), rotate actually ν_{ijpq} for optimizing Ψ only if $\left(\sum_{p,q>p} (\nu_{ijpq})^2\right) / \left(\frac{N(N-1)}{2}\right) > 1/M$.
3. *Convergence decision.* If no pair has been actually rotated in the current sweep, end. Otherwise, go to the next sweep.

3   Results

Here, the two types of numerical experiments were carried out. In the first experiment, the distribution of ν_{ijpq} with $V = A^{-1}$ is compared with $G_t(\nu)$ in order to verify the validity of Theorem 1. In the second experiment, the proposed improvement of JADE in Section 2.3 is compared with the usual JADE for various sample size. Regarding datasets, artificial sources were used, a half of which were generated by the Laplace distribution (super-Gaussian) and the other half by the uniform distribution (sub-Gaussian). JAD is known to be effective especially for such cases where sub- and super-Gaussian sources are mixed. A was given as an orthogonalized random matrix. Fig. 1 shows the comparative results of the histograms of ν_{ijpq} and the true distribution $G_t(\nu)$ for $N = 10$, 20, and 30 and M = 100 and 10000. The KL divergence (averaged over 10 runs) is also shown. It shows that $G_t(\nu)$ can approximate the actual distribution of ν_{ijpq} highly accurately in all the cases. The results verify the validity of Theorem 1. Though the accuracy of the approximation slightly deteriorated when M was large (10000), it seems negligible. Because $G_t(\nu)$ includes no free parameters (even the variance is fixed), the accuracy is quite surprising. Fig. 2 shows the transitions of the objective function Ψ at the convergence (a), the computation time (b), and the final error (c) along the sample size M by the proposed method with the lower bound (solid curves) and the usual JADE (dashed). N was set to 20. Fig. 2-(a) shows that the optimized value of the total objective function Ψ in the proposed method was slightly larger than the usual JADE. It is consistent with the adoption of the lower bound. On the other hand, Fig. 2-(b) shows that the proposed method converged much more rapidly than JADE for a small number of samples. Moreover, Fig. 2-(c) shows that the proposed method slightly

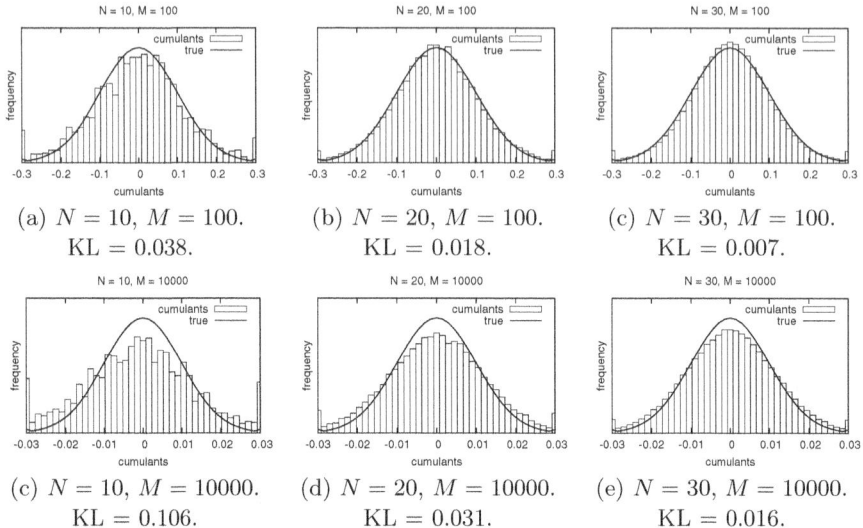

Fig. 1. Comparison of distributions of ν_{ijpq} with the true distribution $G_t(\nu)$: Each histogram shows an experimental distribution of ν_{ijpq} with $V = A^{-1}$. Each curve displays $G_t(\nu)$. The KL divergence is shown below each figure.

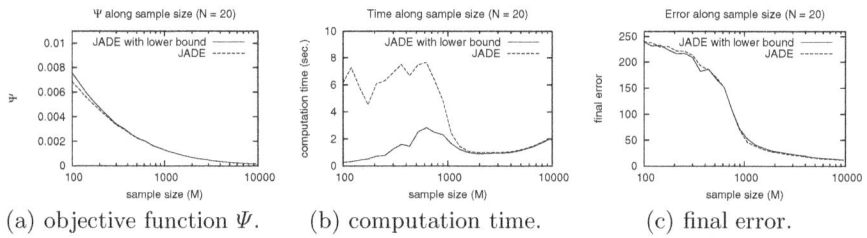

Fig. 2. Objective function, computation time, and final error along the sample size: The curves show the transitions of the objective function Ψ at the convergence (a), the computation time (b), and the final error (c) along the sample size (on a log scale). Two curves in each figure were displayed by the improved JADE with the lower bound (solid curve) and the usual JADE with a fixed small threshold $\epsilon = 10^{-6}$ (dashed). Artificial datasets were given similarly in Fig. 1 ($N = 20$). The final error is measured by Amari's separating error [1] at the convergence, which is defined as the sum of normalized non-diagonal elements of the product of the estimated separating matrix and the given (accurate) mixing one. All the results were averaged over 10 runs.

outperformed the usual JADE on the final separating error when the sample size was small. These results verify the efficiency of the proposed method at least for a small number of samples.

4 Conclusion

In this paper, we proposed an information theoretic approach to JAD. It was shown that the true distribution of cumulants with the accurate separating matrix is the normal distribution whose variance is the inverse of the sample size. Then, the KL divergence between the true distribution and the estimated one of current cumulants is defined as an objective function. The function is similar to Ψ in JAD, except that it has a positive lower bound. In addition, a simple improvement of JADE with the lower bound was proposed in order to avoid unnecessary optimizations. Numerical experiments on artificial datasets verified the efficiency of this approach for a small number of samples. We are now planning to develop this approach further in order to achieve the drastic improvement of not only efficiency but accuracy in the estimation of the separating matrix. Though the entropy $H(P)$ is estimated by a Gaussian approximation in this paper, this approximation is quite rough and wasteful of an important clue to estimate V. By approximating $H(P)$ more accurately, the construction of a more robust JAD algorithm is expected. We are also planning to compare this method with other ICA methods such as the extended infomax algorithm [6]. In addition, we are planning to apply this method to various practical applications as well as many artificial datasets.

References

1. Amari, S., Cichocki, A.: A new learning algorithm for blind signal separation. In: Touretzky, D., Mozer, M., Hasselmo, M. (eds.) Advances in Neural Information Processing Systems, vol. 8, pp. 757–763. MIT Press, Cambridge (1996)
2. Cardoso, J.F.: High-order contrasts for independent component analysis. Neural Computation 11(1), 157–192 (1999)
3. Cardoso, J.F., Souloumiac, A.: Blind beamforming for non Gaussian signals. IEE Proceedings-F 140(6), 362–370 (1993)
4. Cichocki, A., Amari, S.: Adaptive Blind Signal and Image Processing: Learning Algorithms and Applications. Wiley (2002)
5. Hyvärinen, A., Karhunen, J., Oja, E.: Independent Component Analysis. Wiley (2001)
6. Lee, T.W., Girolami, M., Sejnowski, T.J.: Independent component analysis using an extended infomax algorithm for mixed subgaussian and supergaussian sources. Neural Computation 11(2), 417–441 (1999)
7. Matsuda, Y., Yamaguchi, K.: An adaptive threshold in joint approximate diagonalization by assuming exponentially distributed errors. Neurocomputing 74, 1994–2001 (2011)
8. McCullagh, P., Kolassa, J.: Cumulants. Scholarpedia 4(3), 4699 (2009), http://www.scholarpedia.org/article/Cumulants

Support Constraint Machines

Marco Gori and Stefano Melacci

Department of Information Engineering
University of Siena, 53100 Siena, Italy
{marco,mela}@dii.unisi.it

Abstract. The significant evolution of kernel machines in the last few years has opened the doors to a truly new wave in machine learning on both the theoretical and the applicative side. However, in spite of their strong results in low level learning tasks, there is still a gap with models rooted in logic and probability, whenever one needs to express relations and express constraints amongst different entities. This paper describes how kernel-like models, inspired by the parsimony principle, can cope with highly structured and rich environments that are described by the unified notion of constraint. We formulate the learning as a constrained variational problem and prove that an approximate solution can be given by a kernel-based machine, referred to as a *support constraint machine* (SCM), that makes it possible to deal with learning tasks (functions) and constraints. The learning process resembles somehow the unification of Prolog, since the learned functions yield the verification of the given constraints. Experimental evidence is given of the capability of SCMs to check new constraints in the case of first-order logic.

Keywords: Kernel machines, Learning from constraints, Support vector machines.

1 Introduction

This paper evolves a general framework of learning aimed at bridging logic and kernel machines [1]. We think of an intelligent agent acting in the perceptual space $\mathcal{X} \subset I\!R^d$ as a vectorial function $f = [f_1, \ldots, f_n]'$, where $\forall j \in I\!N_n : \ f_j \in W^{k,p}$ belongs to a Sobolev space, that is to the subset of L^p whose functions f_j admit weak derivatives up to some order k and have a finite L^p norm. The functions $f_j : \ j = 1, \ldots, n$, are referred to as the "tasks" of the agent. We can introduce a norm on f by the pair (P, γ), where P is a pseudo-differential operator and $\gamma \in I\!R^n$ is a vector of non-negative coordinates

$$R(f) = \| f \|_{P_\gamma}^2 = \sum_{j=1}^{n} \gamma_j < Pf_j, Pf_j >, \tag{1}$$

which is used to determine smooth solutions according to the parsimony principle. This is a generalization to multi-task learning of what has been proposed

B.-L. Lu, L. Zhang, and J. Kwok (Eds.): ICONIP 2011, Part I, LNCS 7062, pp. 28–37, 2011.

in [2] for regularization networks. The more general perspective suggests considering objects as entities picked up in $\mathcal{X}^{p,\star} = \bigcup_{i\leq p} \bigcup_{|\alpha_i|\leq p^{\underline{i}}} \mathcal{X}_{\alpha_{1,i}} \times \mathcal{X}_{\alpha_{2,i}}, \ldots, \mathcal{X}_{\alpha_{i,i}}$ where $\alpha_i = \{\alpha_{1,i}, \ldots, \alpha_{i,i}\} \in \mathcal{P}(p, i)$ is any of the $p^{\underline{i}} = p(p-1)\ldots(p-i+1)$ (falling factorial power of p) i-length sequences without repetition of p elements. In this paper, however, we restrict the analysis to the case in which the objects are simply points of a vector space. We propose to build an interaction amongst different tasks by introducing constraints of the following types[1]

$$\forall x \in \mathcal{X} : \quad \phi_i(x, y(x), f(x)) = 0, \quad i \in \mathbb{N}_m$$

where $\mathbb{N}_m$ is the set of the first m integers, and $y(x) \in \mathbb{R}$ is a *target function*, which is typically defined only on samples of the probability distribution. This makes it possible to include the classic supervised learning, since pairs of labelled examples turns out to be constraints given on a finite set of points. Notice that one can always reduce a collection of constraints to a single equivalent constraint. For this reason, in the reminder of the paper, most of the analysis will focus on single constraints. In some cases the constraints can be profitably relaxed and the index to be minimized becomes

$$R(f) = \| f \|_{P_\gamma}^2 + C \cdot 1' \int_{\mathcal{X}} \Xi(x, y(x), f(x)). \tag{2}$$

where $C > 0$ and the function Ξ penalizes how we depart from the perfect fulfillment of the vector of constraints ϕ, and 1 is a vector of ones. If $\phi(x, y(x), f(x)) \geq 0$ then we can simply set $\Xi(x, y(x), f(x)) := \phi(x, y(x), f(x))$, but in general we need to set the penalty properly. For example, the check of a bilateral constraint can be carried out by posing $\Xi(x, y(x), f(x)) := \phi^2(x, y(x), f(x))$.

Of course, different constraints can represent the same admissible functional space $\mathcal{F}_\phi$. For example, constraints $\check{\phi}_1(f, y) = \epsilon - |y - f| \geq 0$ and $\check{\phi}_2(f, y) = \epsilon^2 - (y - f)^2 \geq 0$ where f is a real function, define the same $\mathcal{F}_\phi$. This motivates the following definition.

Definition 1. *Let $\mathcal{F}_{\phi_1}, \mathcal{F}_{\phi_2}$ be the admissible spaces of ϕ_1 and ϕ_2, respectively. Then we define the relation $\phi_1 \sim \phi_2$ if and only if $\mathcal{F}_{\phi_1} = \mathcal{F}_{\phi_2}$.*

This notion can be extended directly to pairs of collection of constraints, that is $\mathcal{C}_1 \sim \mathcal{C}_2$ whenever there exists a bijection $\mathcal{C}_1 \overset{\nu}{\to} \mathcal{C}_2$ such that $\forall \phi_1 \in \mathcal{C}_1 \ \nu(\phi_1) \sim \phi_1$. Of course, $\sim$ is an equivalent relation. We can immediately see that $\phi_1 \sim \phi_2 \Leftrightarrow \forall f \in \mathcal{F} : \ \exists P(f) : \quad \phi_1(f) = P(f) \cdot \phi_2(f)$, where P is any positive real function. Notice that if we denote by $[\phi]$ a generic representative of $\sim$, than the quotient set $\mathcal{F}_\phi / \sim$ can be constructed by

$$\mathcal{F}_\phi / \sim = \{\phi \in \mathcal{F}_\phi : \ \phi = P(f) \cdot [\phi](f)\}.$$

Of course we can generate infinite constraints equivalent to $[\phi]$. For example, if $[\phi(f, y)] = \epsilon - |y - f|]$, the choice $P(f) = 1 + f^2$ gives rise to the equivalent

---

[1] We restrict the analysis to universally-quantified constraints, but a related analysis can be carried out when involving existential quantifiers.

constraint $\phi(f, y) = (1+f^2) \cdot (\epsilon - |y - f|)$. The quotient set of any single constraint ϕ_i suggests the presence of a *logic structure*, which makes it possible to devise reasoning mechanisms with the representative of the relation $\sim$. Moreover, the following notion of entailment naturally arises:

Definition 2. *Let $\mathcal{F}_{\overline{\phi}} = \{ f \in \mathcal{F} : \ \overline{\phi}(f) \geq 0 \}$. A constraint $\overline{\phi}$ is entailed by $\mathcal{C} = \{ \phi_i, i \in I\!N_m \}$, that is $\mathcal{C} \models \overline{\phi}$, if $\mathcal{F}_{\mathcal{C}} \subset \mathcal{F}_{\overline{\phi}}$.*

Of course, for any constraint ϕ that can be formally deduced from the collection $\mathcal{C}$ (premises), we have $\mathcal{C} \models \overline{\phi}$. It is easy to see that the entailment operator states invariant conditions in the class of equivalent constraints, that is if $\mathcal{C} \sim \mathcal{C}'$, $\mathcal{C} \models \phi$, and $\phi \sim \phi'$ then $\mathcal{C}' \models \phi'$. The entailment operator also meets the classic chain rule, that is if $\mathcal{C}_1 \models \mathcal{C}_2$ and $\mathcal{C}_2 \models \mathcal{C}_3$ then $\mathcal{C}_1 \models \mathcal{C}_3$.

2   SCM for Constraint Checking

A dramatic simplification of the problem of learning from constraints derives from sampling the input space $\mathcal{X}$, so as to restrict their verification on the set $[\mathcal{X}]_\ell := \{ \boldsymbol{x}_\kappa \in \mathcal{X}, \ \kappa \in I\!N_\ell \}$. This typically cannot guarantee that the algorithm will be able to satisfy the constraint over the whole input space. However, in this work we consider that there is a marginal distribution $\mathcal{P}_\mathcal{X}$ that underlies the data in $\mathcal{X}$, as it is popularly assumed by the most popular machine learning algorithms, so that the constraint satisfaction will holds with high probability.

Theorem 1. *Given a constraint ϕ, let us consider the problem of learning from*

$$\forall \kappa \in I\!N_\ell : \ \phi(x_\kappa, y(x_\kappa), f(x_\kappa)) = 0. \tag{3}$$

There exist a set of real constants λ_κ, $\kappa \in I\!N_\ell$ such that any weak extreme of functional (1) that satisfies (3) is also a weak extreme of $E_\phi(f) = \| f \|_{P_\gamma}^2 + \sum_{\kappa \in I\!N_\ell} \lambda_\kappa \cdot \phi(x_\kappa, f(x_\kappa))$. The extreme $f^\star$ becomes a minima if the constraints are convex, and necessarily satisfy the Euler-Lagrange equations $Lf^\star(x) + \sum_{\kappa=1}^\ell \lambda_\kappa \cdot \nabla_f \phi(x_\kappa, y(x_\kappa), f^\star(x_\kappa)) \delta(x - x_\kappa) = 0$ being $L := P'P$, where P' is the adjoint of P. Moreover, let us assume that $\forall x \in \mathcal{X} : \ g(x, \cdot)$ be the Green function of L. The solution $f^\star$ admits the representation

$$f^\star(x) = \sum_{\kappa \in I\!N_\ell} a_\kappa \cdot g(x, x_\kappa) + f_P(x), \tag{4}$$

where $a_\kappa = -\lambda_\kappa \nabla_f \phi(x_\kappa, y(x_\kappa), f(x_\kappa))$. The uniqueness of the solution arises, that is $f_P = 0$, whenever $Ker P = \{0\}$. If we soften the constraint (3) then all the above results still hold when posing $\forall \kappa \in I\!N_\ell : \ \lambda_\kappa = C$.

PROOF: (Sketch)
Let $\mathcal{X} = I\!\!R^d$ be and let $\{ \zeta_h(x) \}_{h=1}^\infty$ be a sequence of mollifiers and choose $\epsilon := 1/h$, where $h \in I\!N$. Then $\{ \zeta_h(x) \}_{h=1}^\infty \xrightarrow{weakly} \delta(x)$ converges in the classic weak limit sense to the delta distribution. Given the single constraint ϕ let us

consider the sampling $\mathcal{F}_\phi \xrightarrow{S} \mathcal{F}_\phi : \phi \to [\phi]_h$ carried out by the mollifiers ζ_h on $[\mathcal{X}]_\ell$

$$[\phi]_h(x, y(x), f(x)) := \sum_{\kappa \in I\!N_\ell} \phi(x, y(x), f(x)) \zeta_h(x - x_\kappa) = 0.$$

Of course, $[\phi]_h$ is still a constraint and, it turns out that as $h \to \infty$ it is equivalent to $\forall \kappa \in I\!N_\ell : \phi(x_\kappa, y(x_\kappa), f(x_\kappa)) = 0$. Now the proof follows by expressing the overall error index on the finite data sample for $[\phi]_h(x, y(x), f(x))$.

We can apply the classic Euler-Lagrange equations of variational calculus with subsidiary conditions for the case of holonomic constraints ([3] pag. 97-110), so as any weak extreme of (1) that satisfies (3) is a solution of the Euler-Lagrange equation

$$Lf^\star(x) + \sum_{\kappa \in I\!N_\ell} \lambda_i(x) \cdot \nabla_f \phi_i(x, y(x), f^\star(x)) \zeta_h(x - x_\kappa) = 0, \qquad (5)$$

where $L := [\gamma_1 L, \ldots, \gamma_n L]'$ and ∇_f is the gradient w.r.t. f. The convexity of the constraints guarantees that the extreme is a minima and $KerP = \{0\}$ ensures strict convexity and, therefore, uniqueness. Finally, (4) follows since $\forall x \in \mathcal{X}$: $g(x, \cdot)$ is the Green function of L. The case of soft-constraints can be treated by similar arguments. ∎

From (4), which give the representation of the optimal solution we can collapse the dimensionality of $\mathcal{F}$ and search for solutions in a finite space. This is stated in the following theorem.

Theorem 2. *Let us consider the learning under the sampled constraints (3). In the case $kerP = \{0\}$ we have $f_P = 0$ and the optimization is reduced to the finite-dimensional problem*

$$\min_a \left\{ \sum_{\kappa \in I\!N_n} \gamma_j a_j' G a_j + \sum_{\kappa \in I\!N_\ell} \lambda_k \phi(x_\kappa, y(x_\kappa), f^\star(x_\kappa)) \right\} \qquad (6)$$

that must hold jointly with (3). If $\phi \geq 0$ holds for an equality soft-constraint ϕ then the above condition still holds and, moreover, $\forall \kappa = 1, \ldots, \ell : \lambda_\kappa = C$.

PROOF: The proof comes out straightforwardly when plugging the expression of $f^\star$ given by (4) into $R(f)$. ∎

For a generic bilateral soft-constraint we need to construct a proper penalty. For example we can find $\arg\min_a \left\{ \sum_{j=1}^n \gamma_j a_j' G a_j + C \sum_{\kappa=1}^\ell \phi^2(x_\kappa, f(x_\kappa)) \right\}$. Then, the optimal coefficients a can be found by gradient descent, or using any other efficient algorithm for unconstrained optimization. In particular, we used an adaptive gradient descent to run the experiments. Note that when ϕ is not convex, we may end up in local minima.

Theorem 2 can be directly applied to classic formulation of learning from examples in which $n = 1$ and $\phi = \Xi$ is a classic penalty and yields the classic optimization of $\arg\min_a \{a'Ga + C\sum_{\kappa=1}^{S_\ell} \Xi(x_\kappa, y(x_\kappa), f(x_\kappa))\}$. Our formulation of learning leads to discovering functions f compatible with a given collection of constraints that are as smooth as possible. Interestingly, the shift of focus on constraints opens the doors to the following *constraint checking problem*

Definition 3. *Let us consider the collection of constraints* $\mathcal{C} = \{\phi_i, i \in \mathbb{N}_m\} = \mathcal{C}_p \cup \mathcal{C}_c$ *where* $\mathcal{C}_p \cap \mathcal{C}_c = \emptyset$. *The* constraint checking problem *is the one of establishing whether or not* $\forall \phi_i \in \mathcal{C}_c$ $\mathcal{C}_p \models \phi_i$ *holds true. Whenever we can find* $f \in \mathcal{F}$ *such that this happens, we say that* $\mathcal{C}_c$ *is entailed by* $\mathcal{C}_p$, *and use the notation* $\mathcal{C}_p \models \mathcal{C}_c$.

Of course, the entailment can be related to the quotient set $\mathcal{F}_\phi / \sim$ and its analysis in the space $\mathcal{F}_\phi$ can be restricted to the representative of the defined equivalent class. Constraint checking is somehow related to *model checking* in logic, since we are interested in checking the constraints $\mathcal{C}_s$, more than in exhibiting the steps which leads to the proof.

Now, let $\mathcal{C}_p \models \phi$ and $f^* = \arg\min_{f\in\mathcal{F}_p} \| f \|_{P_\gamma}$. Then it is easy to see that $\phi(f^*) = 0$. Of course, the vice versa does not hold true. That if $f^* = \arg\min_{f\in\mathcal{F}_p} \| f \|_{P_\gamma}$ and $\phi(f^*) = 0 : \mathcal{C}_p \not\models \phi$. For example, consider the case in which the premises are the following collection of supervised examples $\mathcal{S}_1 := \{(x_\kappa, y_k)\}_{\kappa=1}^{\ell}$ and $\mathcal{S}_2 := \{(x_\kappa, -y_k)\}_{\kappa=1}^{\ell}$ given on the two functions f_1, f_2. It is easy to see that we can think of $\mathcal{S}_1$ and $\mathcal{S}_2$ in terms of two correspondent constraints ϕ_1 and ϕ_2, so as we can set $\mathcal{C}_p := \{\phi_1, \phi_2\}$. Now, let us assume that $\phi(f^*) = f_1^* - f_2^* = 0$. This holds true whenever $a_{\kappa,1} = -a_{\kappa,2}$ Of course, the deduction $\mathcal{C} \models \phi$ is false, since f can take any value in outside the condition forced on supervised examples[2]. This is quite instructive, since it indicates that even though the deduction is formally false, the generalization mechanism behind the discovering of f^* yields a sort of approximate deduction.

Definition 4. *Let* $f^* = \arg\min_{f\in\mathcal{F}_p} \| f \|_{P_\gamma}$ *be and assume that* $\phi(f^*) = 0$ *holds true. We say that* ϕ *is* formally checked *from* $\mathcal{C}_p$ *and use the notation* $\mathcal{C}_p \vdash \phi$.

Interestingly, the difference between $\mathcal{C}_p \models \phi$ and $\mathcal{C}_p \vdash \phi$ is rooted in the gap between deductive and inductive schemes. While $\models$ does require a sort of unification by checking the property $\phi(f) = 0$ for all $f \in \mathcal{F}_p$, the operator $\vdash$ comes from the computation of f^* that can be traced back to the parsimony principle. Whenever we discover that $\mathcal{C}_p \vdash \phi$, it means that either $\mathcal{C}_p \models \phi$ or $f^* \in \mathcal{F}_p \cap \mathcal{F}_\phi \subset \mathcal{F}_p$, where $\subset$ holds in strict sense. Notice that if we use soft-optimization then the notion of *simplification* strongly emerges which leads to a decision process in which more complex constraints are sacrificed because of the preference of simple constraints. We can go beyond $\vdash$ by relaxing the need to check $\phi(f^*) = 0$ thanks to the introduction of the following notion of *induction from constraints*.

---

[2] Notice that the analysis is based on the assumption of hard-constraints and that in case of soft-constraints, which is typical in supervised learning, the claim that the deduction is false is even reinforced.

Definition 5. *Let $\epsilon > 0$ be and $[\mathcal{X}]_u \subset \mathcal{X}$ be a sample of u unsupervised examples of $\mathcal{X}$. Given a set of premises $\mathcal{C}_p$ on $[\mathcal{X}]_u$ and let $\mathcal{F}_p^u$ be the correspondent set of admissible functions. Furthermore, let $f^\star = \arg\min_{f \in \mathcal{F}^u} \| f \|_{P_\gamma}$ be and denote by $[f^\star]_u$ its restriction to $[\mathcal{X}]_u$. Now assume that $\| \phi([f^\star]_u) \| < \epsilon$ holds true. Under these conditions we say that ϕ is induced from $\mathcal{C}_p$ via $[\mathcal{X}]_u$, and use the notation $(\mathcal{C}_p, [\mathcal{X}]_u) \vdash^\star \phi$.*

Notice that the adoption of special loss functions, like the classic hinge function, gives rise to support vectors, but also to *support constraints*. Given a collection of constraints (premises) $\mathcal{C}_p$, then ϕ is a support constraint for $\mathcal{C}_p$ whenever $\mathcal{C}_p \nvdash \phi$. When the opposite condition holds, we can either be in presence of a formal deduction $\mathcal{C}_p \models \phi$ or of the more general checking $\mathcal{C}_p \vdash \phi$ in the environment condition.

3   Checking First-Order Logic Constraints

We consider the semi-supervised learning problem (6) composed of a set of constraints that include information on labeled data and prior knowledge on the learning environment in the form of First-Order Logic (FOL) clauses. Firstly, we will show how to convert FOL clauses in real-valued functions. Secondly, using an artificial benchmark, we will include them in our learning framework to improve the quality of the classifier. Finally, we will investigate the constraint induction mechanism, showing that it allows us to formally check other constraints that were not involved in the training stage (Definitions 4 and 5).

First-Order Logic (FOL) formula can be associated with real-valued functions by classic *t-norms* (triangular norms [4]). A t-norm is function $T : [0,1] \times [0,1] \to I\!\!R$, that is commutative, associative, monotonic and that features the neutral element 1. For example, given two unary predicates $a_1(\boldsymbol{x})$ and $a_2(\boldsymbol{x})$, encoded by $f_1(\boldsymbol{x})$ and $f_2(\boldsymbol{x})$, the product norm, which meets the above conditions on T-norms, operates as follows: $a_1(\boldsymbol{x}) \wedge a_2(\boldsymbol{x}) \longmapsto f_1(\boldsymbol{x}) \cdot f_2(\boldsymbol{x})$, $a_1(\boldsymbol{x}) \vee a_2(\boldsymbol{x}) \longmapsto 1 - (1 - f_1(\boldsymbol{x})) \cdot (1 - f_2(\boldsymbol{x}))$, $\neg a_1(\boldsymbol{x}) \longmapsto 1 - f_1(\boldsymbol{x})$, and $a_1(\boldsymbol{x}) \Rightarrow a_2(\boldsymbol{x}) \longmapsto 1 - f_1(\boldsymbol{x}) \cdot (1 - f_2(\boldsymbol{x}))$. Any formula can be expressed by the CNF (Conjunctive Normal Form) so as to transform it to a real-valued constraint step by step. In the experiment we focus on universally quantified ($\forall$) logic clauses, but the extension to cases in which the existential quantifier is involved is possible.

We consider a benchmark based on 1000 bi-dimensional points belonging to 4 (partially) overlapping classes. In particular, 250 points for each class were randomly generated with uniform distribution. The classes a_1, a_2, a_3, a_4 can be thought of the characteristic functions of the domains $\mathcal{D}_1, \mathcal{D}_2, \mathcal{D}_3, \mathcal{D}_4$ defined as $\mathcal{D}_1 = \{(x_1, x_2) \in I\!\!R^2 : x_1 \in (0,2) \wedge x_2 \in (0,1)\}$, $\mathcal{D}_2 = \{(x_1, x_2) \in I\!\!R^2 : x_1 \in (1,3) \wedge x_2 \in (0,1)\}$, $\mathcal{D}_3 = \{(x_1, x_2) \in I\!\!R^2 : x_1 \in (1,2) \wedge x_2 \in (0,2)\}$, and $\mathcal{D}_4 = \{(x_1, x_2) \in I\!\!R^2 : (x_1 \in (1,2) \wedge x_2 \in (0,1)) \vee (x_1 \in (0,2) \wedge x_2 \in (1,2))\}$. Then the appropriate multi-class label was assigned to the collection of 1000 points by considering their coordinates (see Fig. 1). A multi-class label is a binary vector of p components where 1 marks the membership to the i-th category (for example, $[0,1,1,0]$ for a point of classes a_2 and a_3). Four binary classifiers were trained using the associated functions f_1, f_2, f_3, f_4. The decision of each classifier on an input $\boldsymbol{x}$ is $o_j(\boldsymbol{x}) = \mathbf{1}(f_j(\boldsymbol{x}) - b_j)$ where b_j is the bias term of the j-th classifier and $\mathbf{1}(\cdot)$ is the Heaviside function.

We simulate a scenario in which we have access to the whole data collection, where ℓ points ($\ell/4$ for each class) are labeled, and to domain knowledge expressed by the following FOL clauses,

$$\forall \boldsymbol{x} \quad a_1(\boldsymbol{x}) \wedge a_2(\boldsymbol{x}) \Rightarrow a_3(\boldsymbol{x}) \tag{7}$$

$$\forall \boldsymbol{x} \quad a_3(\boldsymbol{x}) \Rightarrow a_4(\boldsymbol{x}) \tag{8}$$

$$\forall \boldsymbol{x} \quad a_1(\boldsymbol{x}) \vee a_2(\boldsymbol{x}) \vee a_3(\boldsymbol{x}) \vee a_4(\boldsymbol{x}). \tag{9}$$

While the first two clauses express relationships among the classes, the last clause specifies that a sample must belong to at least one class. For each of the ℓ labeled training points, we assume to have access to a *partial labeling*, such as, for example, $[0, ?, 1, ?]$, that means that we do not have any information on classes 2 and 4. This setup emphasizes the role of the FOL clauses in the learning process. We performed a 10-fold cross-validation and measured the average classification accuracy on the out-of-sample test sets. A small set of partially labeled data is excluded from the training splits, and it is only used to validate the classifier parameters (that were moved over $[0.1, 0.5, \ldots, 12]$ for the width of a Gaussian kernel, and $[10^{-4}, 10^{-3}, \ldots, 10^2]$ for the regularization parameter λ of soft-constrained SCM). We compared SCMs that include constraints on labeled points only (SCL_L) with SCMs that also embed constraints from the FOL clauses (that we indicate with SCM_{FOL}). In Fig. 1 we report a visual comparison of the two algorithms, where the outputs of f_1, f_2, f_3, f_4 are plotted ($\ell = 16$).

The introduction of the FOL clauses establish a relationship among the different classification functions and positively enhance the inductive transfer among the four tasks. As a matter of fact, the output of SCM_{FOL} is significantly closer to the real class boundaries (green dashed lines) than in SCM_L. The missing label information is compensated by the FOL rules, and injected on the whole data distribution by the proposed learning from constraint scheme. Note that the missing labels cannot be compensated by simply applying the FOL clauses on the partially labeled vectors. Interestingly the classifier has learned the "shape" of the lower portion of class 3 by the rule of (7), whereas the same region on class 4 has been learned thanks to the inductive transfer from (8).

We iteratively increased the number of labeled training points ℓ from 8 to 320, and we measured the classification macro accuracies. The output vector on a test point $\boldsymbol{x}$ (i.e. $[o_1(\boldsymbol{x}), o_2(\boldsymbol{x}), o_3(\boldsymbol{x}), o_4(\boldsymbol{x})]$) is considered correctly predicted if it matches the full label vector that comes with the ground truth. We also computed the accuracy of the SCM_L classifier whose output is post-processed by applying the FOL rules, in order to fix every incoherent prediction. The results are reported in Fig. 2. The impact of the FOL clauses in the classification accuracy is appreciable when the number of labeled points is small, whereas it is less evident when the information on the FOL clauses can be learned from the available training labels, as expected. Clearly the results becomes less significant when the information on the FOL clauses can be learned from the available

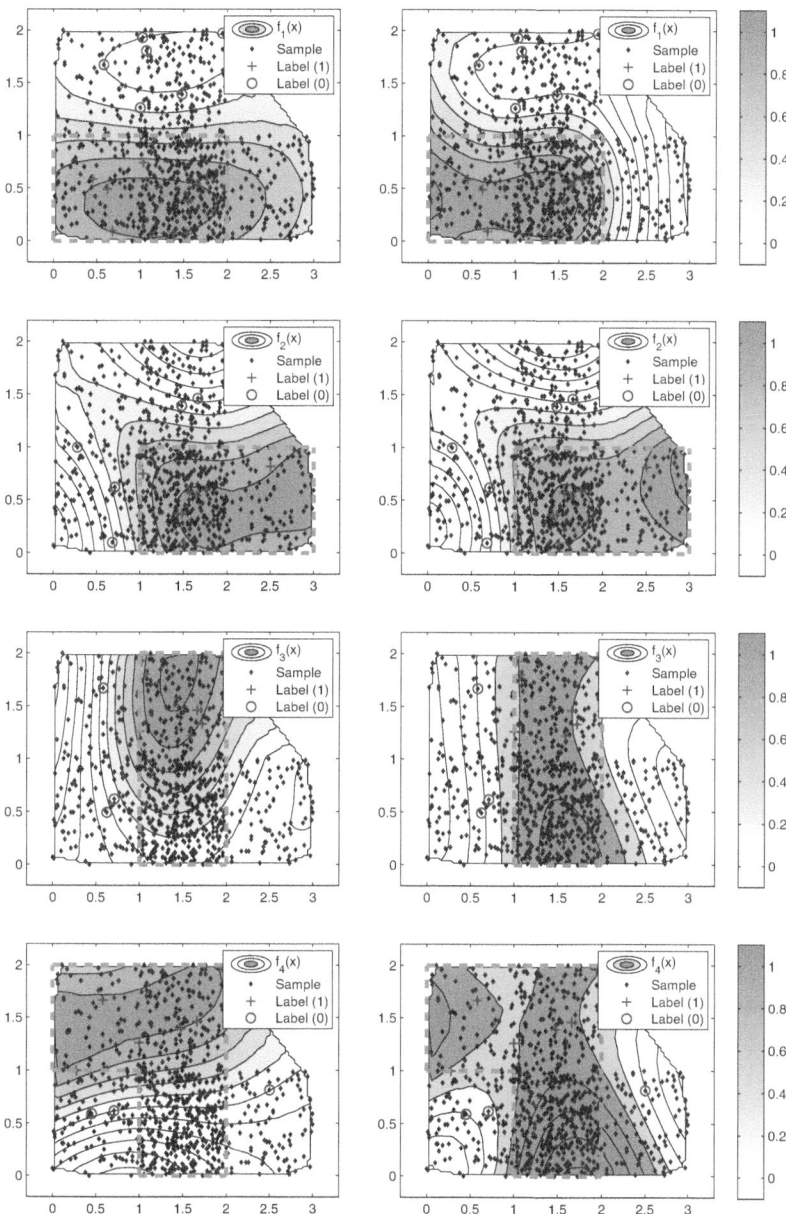

Fig. 1. The functions f_1, f_2, f_3, f_4 in the data collection where a set of FOL constraints applies. The j-th row, $j = 1, \ldots, 4$, shows the outcome of the function f_j in SCMs that use labeled examples only (left) and in SCMs with FOL clauses, SCM$_{FOL}$ (right). The green dashed lines shows the real boundaries of the j-th class.

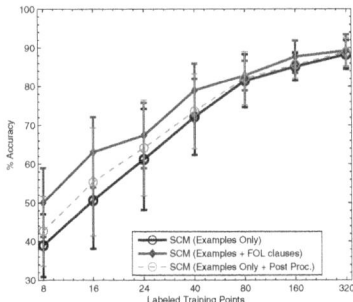

Fig. 2. The average accuracy (and standard deviation) of the SCM classifier: using labeled examples only (SCL$_L$), using examples and FOL clauses (SCM$_{FOL}$), using examples only and post processing the classifier output with the FOL rules

training labels, so that in the rightmost region of the graph the gap between the curves becomes smaller. In general, the output of the classifier is also more stable when exploiting the FOL rules, showing a averagely smaller standard deviation.

Given the original set of rules that constitutes our Knowledge Base (KB) and that are fed to the classifier, we distinguish between two categories of logic rules that can be deducted from the trained SCM$_{FOL}$. The first category includes the clauses that are related to the geometry of the data distribution, and that, in other words, are strictly connected to the topology of the environment in which the agent operates, as the ones of (7-9). The second category contains the rules that can be logically deducted by analyzing the FOL clauses that are available at hand. The classifier should be able to learn both the categories of rules even if not explicitly added to the knowledge base.

The mixed interaction of the labeled points and the FOL clauses of the KB leads to an SCM agent that can check whether a *new* clause holds true in our environment. Note that the checking process is not implemented with a strict decision on the truth value of a logic sentence (*holds true or false*), since there are some rules that are verified only on some (possibly large) regions of the input space, so that we have to evaluate the *truth degree* of a FOL clause. If it is over a reasonably high threshold, the FOL sentence can be assumed to hold true. In Table 1 we report the degree of satisfaction of different FOL clauses and the Mean Absolute Error (MAE) on the corresponding t-norm-based constraints. We used the SCM$_{FOL}$ trained with $\ell = 40$. Even if it is simple to devise them when looking at the data distribution, it is not possible to do this as the input space dimension increases, so that we can only "ask" the trained SCM if a FOL clause holds true. This allow us to rebuild the hierarchical structure of the data, if any, and to extract compact information from the problem at hand. The rules belonging to the KB are accurately learned by the SCM$_{FOL}$, as expected. The SCM$_{FOL}$ is also able to deduct all the other rules that are supported in the entire data collection. The ones that do not hold for all the data points have the same truth degree as the percentage of points for which they should hold true,

Table 1. Mean Absolute Error (MAE) of the t-norm based constraints and the percentage of points for which a clause is marked *true* by the SCM (Average Truth Value), and their standard deviations (in brackets). Logic rules belong to different categories (Knowledge Base - KB, Environment - ENV, Logic Deduction - LD). The percentage of Support indicates the fraction of the data on which the clause holds true.

FOL clause	Category	Support	MAE	Truth Value
$a_1(\boldsymbol{x}) \wedge a_2(\boldsymbol{x}) \Rightarrow a_3(\boldsymbol{x})$	KB	100%	0.0011 (0.00005)	98.26% (1.778)
$a_3(\boldsymbol{x}) \Rightarrow a_4(\boldsymbol{x})$	KB	100%	0.0046 (0.0014)	98.11% (2.11)
$a_1(\boldsymbol{x}) \vee a_2(\boldsymbol{x}) \vee a_3(\boldsymbol{x})$	KB	100%	0.0049 (0.002)	96.2% (3.34)
$a_1(\boldsymbol{x}) \wedge a_2(\boldsymbol{x}) \Rightarrow a_4(\boldsymbol{x})$	LD	100%	0.0025 (0.0015)	96.48% (3.76)
$a_1(\boldsymbol{x}) \wedge a_3(\boldsymbol{x}) \Rightarrow a_2(\boldsymbol{x})$	ENV	100%	0.017 (0.0036)	91.32% (5.67)
$a_3(\boldsymbol{x}) \wedge a_2(\boldsymbol{x}) \Rightarrow a_1(\boldsymbol{x})$	ENV	100%	0.024 (0.014)	91.7% (4.57)
$a_1(\boldsymbol{x}) \wedge a_3(\boldsymbol{x}) \Rightarrow a_4(\boldsymbol{x})$	ENV	100%	0.0027 (0.0014)	96.13% (3.51)
$a_2(\boldsymbol{x}) \wedge a_3(\boldsymbol{x}) \Rightarrow a_4(\boldsymbol{x})$	ENV	100%	0.0025 (0.0011)	96.58% (4.13)
$a_1(\boldsymbol{x}) \wedge a_4(\boldsymbol{x})$	ENV	46%	0.41 (0.042)	45.26% (5.2)
$a_2(\boldsymbol{x}) \vee a_3(\boldsymbol{x})$	ENV	80%	3.39 (0.088)	78.26% (6.13)
$a_1(\boldsymbol{x}) \vee a_2(\boldsymbol{x}) \Rightarrow a_3(\boldsymbol{x})$	ENV	65%	0.441 (0.0373)	68.28% (5.86)
$a_1(\boldsymbol{x}) \wedge a_2(\boldsymbol{x}) \Rightarrow \neg a_4(\boldsymbol{x})$	LD	0%	0.26 (0.06)	3.51% (3.76)
$a_1(\boldsymbol{x}) \wedge \neg a_2(\boldsymbol{x}) \Rightarrow a_3(\boldsymbol{x})$	ENV	0%	0.063 (0.026)	27.74% (18.96)
$a_2(\boldsymbol{x}) \wedge \neg a_3(\boldsymbol{x}) \Rightarrow a_1(\boldsymbol{x})$	ENV	0%	0.073 (0.014)	5.71% (5.76)

whereas rules that do not apply to the given problem are correctly marked with a significantly low truth value.

4   Conclusions

This paper gives insights on how to fill the gap between kernel machines and models rooted in logic and probability, whenever one needs to express relations and express constraints amongst different entities. The *support constraint machines* (SCMs), are introduced that makes it possible to deal with learning functions in a multi-task environment and to check constraints. In addition to the impact in multi-task problems, the experimental results provide evidence of novel inference mechanisms that nicely bridge formal logic reasoning with supervised data. It is shown that logic deductions that do not hold formally can be fired by samples of labeled data. Basically SCMs provide a natural mechanism under which logic and data complement each other.

References

1. Diligenti, M., Gori, M., Maggini, M., Rigutini, L.: Bridging logic and kernel machines. Machine Learning (to appear, on–line May 2011)
2. Poggio, T., Girosi, F.: A theory of networks for approximation and learning. In: Technical report. MIT (1989)
3. Giaquinta, M., Hildebrand, S.: Calculus of Variations I, vol. 1. Springer, Heidelberg (1996)
4. Klement, E., Mesiar, R., Pap, E.: Triangular Norms. Kluwer Academic Publisher (2000)

Human Activity Inference Using Hierarchical Bayesian Network in Mobile Contexts

Young-Seol Lee and Sung-Bae Cho

Dept. of Computer Science, Yonsei University
134 Shinchon-dong, Seodaemoon-gu, Seoul 120-749, Korea
tiras@sclab.yonsei.ac.kr,
sbcho@cs.yonsei.ac.kr

Abstract. Since smart phones with diverse functionalities become the general trend, many context-aware services have been studied and launched. The services exploit a variety of contextual information in the mobile environment. Even though it has attempted to infer activities using a mobile device, it is difficult to infer human activities from uncertain, incomplete and insufficient mobile contextual information. We present a method to infer a person's activities from mobile contexts using hierarchically structured Bayesian networks. Mobile contextual information collected for one month is used to evaluate the method. The results show the usefulness of the proposed method.

Keywords: Bayesian network, mobile context, activity inference.

1 Introduction

Smartphones, such as Apple iPhone and Google Android OS based phones, with various sensors are becoming a general trend. Such phones can collect contextual information like acceleration, GPS coordinates, Cell ID, Wi-Fi, etc. Many context-aware services are introduced and tried to provide a user with convenience using them. For example, Foursquare includes a location-based social networking service. It ranks the users by the frequency of visiting a specific location and encourages them to check in the place. Loopt service recommends some visited locations for the users, and Whoshere service shows friends' locations. Davis et al. tried to use temporal, spatial, and social contextual information to help manage consumer multimedia content with a camera phone [1]. Until now, most of such services use only raw data like GPS coordinates.

Many researchers have attempted to infer high-level semantic information from raw data collected in a mobile device. Belloti *et al.* tried to infer a user's activities to recommend suitable locations or contents [2]. Chen proposed intelligent location-based mobile news service as a kind of location based service [3]. Santos *et al.* studied user context inference using decision trees for social networking [4]. Most of the research used various statistical analysis and machine learning techniques like

B.-L. Lu, L. Zhang, and J. Kwok (Eds.): ICONIP 2011, Part I, LNCS 7062, pp. 38–45, 2011.
© Springer-Verlag Berlin Heidelberg 2011

probabilistic model, fuzzy logic, and case based reasoning. However, it is practically difficult to infer high level context because mobile environment includes uncertainty and incompleteness. This paper presents a method to infer human activities from mobile contexts using hierarchical Bayesian networks.

2 Related Works

Some researchers have collected contextual information in the mobile environment. VTT research center has developed technologies to manage contextual information and infer higher-level context abstractions from raw measurement data [5]. An adaptive user interface has been also developed by the VTT research center [6]. Helsinki University developed a ContextPhone framework which collected contexts (GPS, GSM cell ID, call history, SMS history, and application in use) on the Nokia 60 series [7].

Some researchers studied object based activity recognition using RFID tags. Patterson *et al.* examined reasoning with globally unique object instances detected by a Radio Frequency Identification (RFID) glove [8]. They constructed a model with hidden Markov model (HMM) and applied the model to identify what they are cooking. Wyatt *et al.* studied object based activity recognition using RFID tags, a RFID reader, and models mined from the Web [9]. It is used to solve a fundamental problem in recognizing human activities which need labeled data to learn models for the activities.

On the other hand, there are location based activity recognition algorithms. In the approaches, it is assumed that a person's environment affects his activity. Therefore, they often focused on the accurate place detection algorithms rather than activity recognition for practical applications. Liao *et al.* used relational Markov network and conditional random field (CRF) [10] to extract activities from location information. Anderson and Muller attempted to recognize activities using HMM from GSM Cell-ID data [11].

In this paper, we propose a hierarchical Bayesian network based method to automate activity inference. It raises scalability and precision in recognizing activities by modeling hierarchical Bayesian network with intermediate nodes.

3 Proposed Method

In this section, the whole process of activity recognition will be described. The whole system is composed of four components, which are context collection, statistical analysis, activity recognition with hierarchical Bayesian networks, and a user interface as shown in Fig. 1. The hierarchical Bayesian networks are designed based on the context hierarchy [12] and Hwang and Cho's work [13].

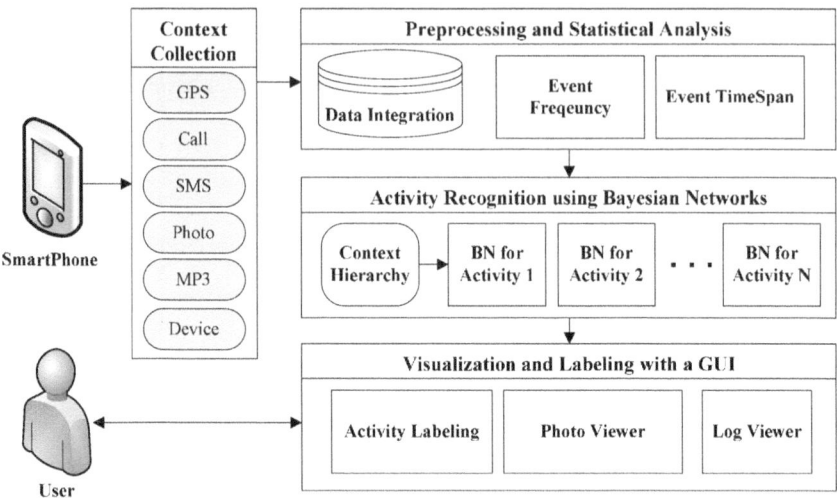

Fig. 1. System Overview

3.1   Context Collection and Preprocessing

Context collection is the first step to infer activity. Table 1 shows the information from a mobile phone. In the process, raw data are preprocessed by simple analysis (frequency, elapsed time, etc).

Table 1. Contextual information from smartphone

Data	Attributes
Location	GPS coordinates, Speed, Altitude, Place name
Activity	User's activity (reported by manual labeling)
Music	MP3 title, start time, end time, singer
Photograph	Time, captured object
Call history	Start time, end time, person (friend/lover/family/others)
SMS history	Time, person (friend/lover/family/others)
Battery	Time, whether charging or not (Yes/No), charge level (%)

There are many practical difficulties to recognize user's activities using mobile contextual information in a smartphone. The fundamental problem is that mobile context in a smartphone does not reflect user's activities perfectly. User's activities generate various contextual information and only parts of them are observable in a smartphone. Moreover, important context in a smartphone is often damaged and uncertain. For instance, GPS signals may be invalid or be lost in some locations, especially indoor environment. Schedule stored in a smartphone may be different with the facts because it seems difficult to expect exact coincidence between schedule and user's activities in real life. Sometimes, unexpected events and accidents may occur. On the other hand, call history, SMS history, and photographs have important personal information, but are difficult to understand the semantics automatically. We applied several data mining techniques (decision tree, association rules, etc) to analyze and interpret the context. As a result, we distinguish the useful context.

Mobile data can be classified into three types such as useful, partially useful and useless contexts to infer a user's activity. In this paper, we ignore useless context, and use other contexts to infer activity.

3.2   Bayesian Network Modeling for Activity Using Context Hierarchy

According to Kaenampornpan *et al.*, many researchers defined context and elements of context for context awareness. The goal of a context classification is to build a conceptual model of a user's activity. Activity Theory is a valuable framework used to analyze and model human activities by providing a comprehensive types and relationships of context [14]. It consists of six components which are subject, object, community, rules, division of labor, and artifact. Kofod-Petersen *et al.* built context taxonomy from six elements found in Activity Theory and applied it to their semantic networks as five context categories: environmental, personal, social, task, spatio-temporal context [12]. Fig. 2 presents our context model structure which is based on context hierarchy to recognize the activities.

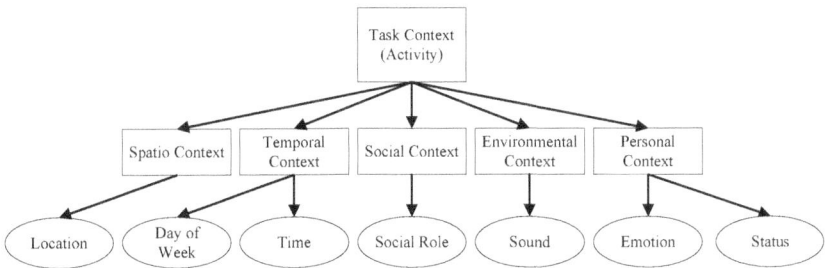

Fig. 2. Context model for activity recognition

The hierarchical context model allows us to implicitly decompose a user's activities into simpler contexts. Intuitively, it might be easier for the model to design probabilistic network modules with hierarchical tree structure rather than the complex one. Further, it is advantageous to break a network for an activity into smaller sub-trees and then build models for the sub-trees related to specific context. Fig. 3 shows basic structure of our Bayesian network model for each activity.

Bayesian network is a directed acyclic graphical model that is developed to represent probabilistic dependencies among random variables [13]. It relies on Bayes rule like (1) and conditional independence to estimate the distribution over variables.

$$P(X \mid Y) = \frac{P(Y \mid X)P(X)}{P(Y)} \tag{1}$$

The nodes in the Bayesian network represent a set of observations (*e.g.*, locations, day of week, etc), denoted as $O = \{o_1, o_2, o_3, ..., o_n\}$, and corresponding activities (*e.g.,* walking, studying, etc), denoted as $A = \{a_1, a_2, a_3, ..., a_m\}$. These nodes, along with the connectivity structure imposed by directed edges between them, define the conditional probability distribution $P(A|O)$ over the target activity A. Equation (2) shows that a_i, which is a specific activity to recognize, can be inferred from observations.

$$P(A \mid O) \rightarrow P(a_i \mid o_1, o_2, o_3, ..., o_n) \qquad (2)$$

The proposed model has intermediate nodes, denoted as $C = \{c_1, c_2, ..., c_m\}$, to represent hidden variables for activity inference.

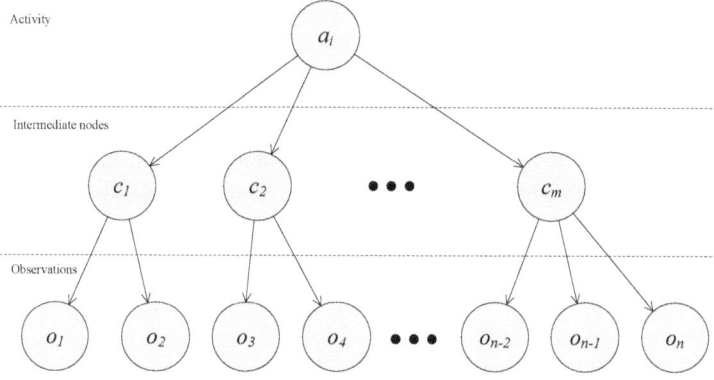

Fig. 3. Bayesian network structure for activity recognition

3.3   Hierarchical Bayesian Networks for Activity Class

In the previous section, Bayesian networks for each activity make it easy to extract activities with only related observations. However, a person can do independent activities at the same time. For instance, he can watch a television and eat something simultaneously. The simultaneous activities cause confusion and mistake to recognize activities because of mixed contextual information. Each Bayesian network module for an activity cannot deal with the situation effectively. We define a global domain for similar activities as activity class. The recognition for global domain of activities is a summary of all similar activities. For instance, Fig. 4 shows the conceptual structure. Let E be the set of all evidences in a given Bayesian network. Then, a Bayesian network factorizes the conditional distribution like $P(a|E)$, where a is an activity and every $E = \{e_1, e_2, e_3, ..., e_n\}$. It is assumed that e_{i+1}, e_{i+2}, and e_{i+4} are the observed nodes in a given environment. Under the evidences, a_i and a_{i+2} are the most probable activities, and we may have the possibility of confusion between a_i and a_{i+2}. Intuitively, activity class as a global activity captures the "compatibility" among the variables. Using the hierarchical structure, the conditional distribution over the activity class A is written as $P(A|a_i, a_{i+1}, ...)$.

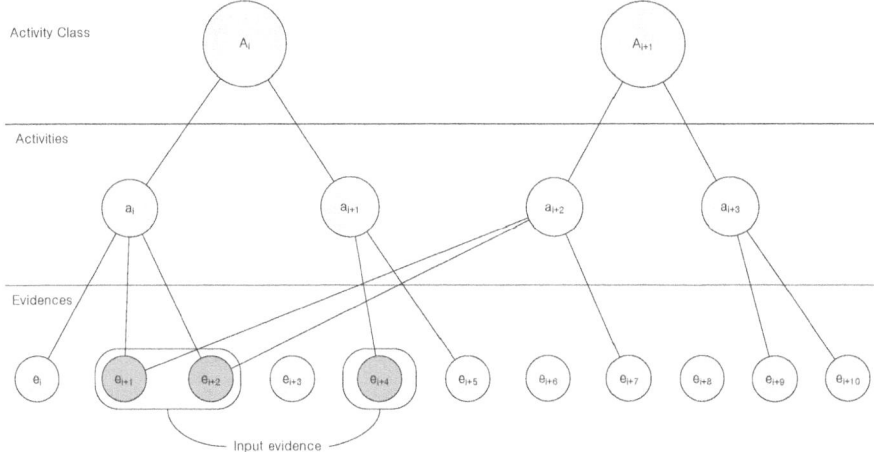

Fig. 4. Structure for Activity class from activities

4   Experimental Result

We define 23 activities according to GSS (General Social Survey on Time Use, 1998) which is a statistical survey for daily activities on time usage in Canada (Statistics Canada, 2005). The activities are suitable to be selected from contextual information regarding user's environment because GSS provides activity types related to location, time and activity purposes. We compare reported activities and inferred activities using our Bayesian networks to calculate the hit rate of the recognition.

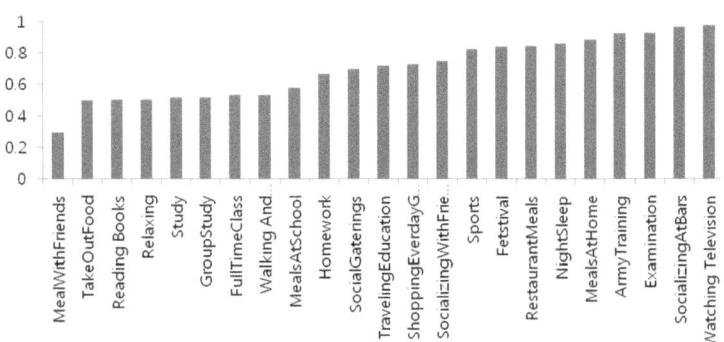

Fig. 5. Accuracy of activity recognition

According to our analysis, time and location are the most important factors to estimate the probability of each activity. That is, location and time dependent activities tend to be detected well as shown in Fig. 5. For example, 'Night Sleep' occurs at specific location (mainly home) and time (mostly night). 'Examination' has fixed dates in a semester. Korean regular army training is one of duties in South Korea, of which schedule is determined by the Ministry of National Defense. It is also easy to estimate the occurrence. On the while, location and time independent activities such as 'Relaxing' and 'Reading books' are difficult to detect automatically from contextual information. Fig. 6 illustrates the result of activity class recognition.

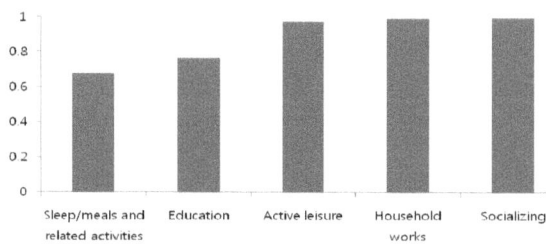

Fig. 6. Accuracy of activity class recognition

Finally, we introduce an interface to annotate a user's activity for some images. It helps a user to check recognized activity related to a photograph. It reduces fatigue of manual annotation. In this point of view, we have developed a prototype annotation tool as shown in Fig. 7. The probability of activity class tended to be more accurate than each activity. A user can easily use their activity to annotate their photographs taken by a mobile phone.

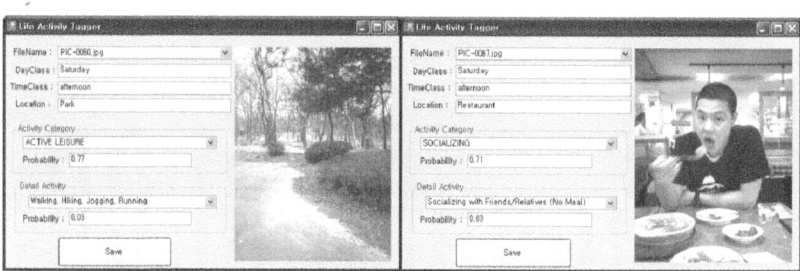

Fig. 7. User interface screenshots for activity annotation

5   Summary and Discussion

In this paper, we have proposed a method to recognize activities using hierarchical probabilistic models. The system is composed of 4 components, which are context collection, preprocessing and feature extraction, activity recognition, and an interface for visualization and labeling. Bayesian network models for activity refer to context hierarchy and activity hierarchy. It is evaluated with the data collected in real mobile environment. The proposed interface makes labeling easy with effective visualization in order to support recognized activities more accurately.

Our future research must include more diverse personal information and improve the performance of activity recognition. In addition, various menu interfaces have to be developed for user's convenience.

Acknowledgments. This research was supported by the Original Technology Research Program for Brain Science through the National Research Foundation of Korea (NRF) funded by the Ministry of Education, Science and Technology (2010-0018948).

References

1. Davis, M., Canny, J., House, V.N., Good, N., King, S., Nair, R., Burgener, C., Strickland, R., Campbell, G., Fisher, S., Reid, N.: MMM2: Mobile Media Metadata for Media Sharing. In: ACM MM 2005, Singapore, pp. 267–268 (2005)
2. Bellotti, V., Begole, B., Chi, H.E., Ducheneaut, N., Fang, J., Isaacs, E., King, T., Newman, W.M., Partridge, K., Price, B., Rasmussen, P., Roberts, M., Schiano, J.D., Walendowski, A.: Activity-Based Serendipitous Recommendations with the Magitti Mobile Leisure Guide. In: Proc. of the 26th Annual SIGCHI Conf. on Human Factors in Computing Systems (CHI 2008), pp. 1157–1166 (2008)
3. Chen, C.-M.: Intelligent Location-Based Mobile News Service System with Automatic News Summarization. Expert Systems with Applications 37(9), 6651–6662 (2010)
4. Santos, C.A., Cardoso, M.P.J., Ferreira, R.D., Diniz, C.P., Chaínho, P.: Providing User Context for Mobile and Social Networking Applications. Pervasive and Mobile Computing 6(3), 324–341 (2010)
5. Korpip, P., Mantyjarvi, J., Kela, J., Keranen, H., Malm, E.-J.: Managing Context Information in Mobile Devices. IEEE Pervasive Computing 2(3), 42–51 (2003)
6. Korpip, P., Koskinen, M., Peltola, J., Makela, S.-M., Seppanen, T.: Bayesian Approach to Sensor-Based Context-Awareness. Personal and Ubiquitous Computing 7(2), 113–124 (2003)
7. Raento, M., Oulasvirta, A., Petit, R., Toivonen, H.: ContextPhone - A Prototyping Platform for Context-Aware Mobile Applications. IEEE Pervasive Computing 4(2), 51–59 (2005)
8. Patterson, J.D., Fox, D., Kautz, H., Philipose, M.: Fine-Grained Activity Recognition by Aggregating Abstract Object Usage. In: 9th IEEE International Symposium on Wearable Computers, pp. 44–51 (2005)
9. Wyatt, D., Philipose, M., Choudhury, T.: Unsupervised Activity Recognition using Automatically Mined Common Sense. In: Proc. of the 20th National Conference on Artificial Intelligence, pp. 21–27 (2005)
10. Liao, L., Fox, D., Kautz, H.: Extracting Places and Activities from GPS Traces using Hierarchical Conditional Random Fields. International Journal of Robotics Research 26(1), 119–134 (2007)
11. Anderson, I., Muller, H.: Department of Computer Science: Practical activity recognition using GSM data. Technical report CSTR-06-016, University of Bristol (2006)
12. Kofod-Petersen, A., Cassens, J.: Using Activity Theory to Model Context Awareness. In: Roth-Berghofer, T.R., Schulz, S., Leake, D.B. (eds.) MRC 2005. LNCS (LNAI), vol. 3946, pp. 1–17. Springer, Heidelberg (2006)
13. Hwang, K.-S., Cho, S.-B.: Landmark Detection from Mobile Life Log using a Modular Bayesian Network Model. Expert Systems with Applications 36(3), 12065–12076 (2009)
14. Kaenampornpan, M., O'Neill, E.: Modelling Context: An Activity Theory Approach. In: Markopoulos, P., Eggen, B., Aarts, E., Crowley, J.L. (eds.) EUSAI 2004. LNCS, vol. 3295, pp. 367–374. Springer, Heidelberg (2004)

Estimation System for Human-Interest Degree while Watching TV Commercials Using EEG

Yuna Negishi[1,2], Zhang Dou[1], and Yasue Mitsukura[2]

[1] Graduate School of Bio-Applications Systems Engineering,
Tokyo University of Agriculture and Technology
2-24-16 Naka, Koganei, Tokyo Japan
{50010401226,50011401225}@st.tuat.ac.jp
[2] Graduate School of Science and Technology, Keio University
3-14-1 Hiyoshi, Kohoku, Yokohama, Kanagawa Japan
mitsukura@sd.keio.ac.jp

Abstract. In this paper, we propose an estimation system for the
human-interest degree while watching TV commercials using the elec-
troencephalogram(EEG). When we use this system, we can estimate the
human-interest degree easily, sequentially, and simply. In particular, we
measure the EEG using a simple electroencephalograph and survey the
human-interest degree using questionnaires in a scale. For construction
this estimation system, we investigate the relationship between the EEG
and the result of questionnaires. In order to evaluate our estimation system, we show results of experiments using real TV commercials.

Keywords: human-interest, electroencephalogram(EEG), TV commercial.

1 Introduction

In companies, they advertise their products or services to consumers using various advertisements, (i.e. TV commercial, radio advertising, newspaper flyer, ban-
ner advertising on Internet). Companies use appropriate advertisements which
match target or scene. In Japan, consumers often contact TV commercials, because a penetration level of TV is higher than other advertising media. Therefore,
when consumers buy goods or choose services, they are influenced by what they
see on TV commercials. In addition, using TV commercials, it is possible that
companies feed visual and audio information to passive audience. For these reasons, TV commercials are very important advertising strategy. As for companies,
they need to produce effective TV commercials which consumers are interested in
TV commercials and products/services. For production of effective TV commercials, it is important to get in touch with what audience has an interest. When
we investigate a consumer's interest, we often use questionnaires or interviews.
It is easy to understand consumer's interest because we can ask them "Are you
interested in this commercial?", or "How do you rate your interest on a scale?".
On the other hand, it is difficult to evaluate sequential the human-interest which

B.-L. Lu, L. Zhang, and J. Kwok (Eds.): ICONIP 2011, Part I, LNCS 7062, pp. 46–53, 2011.
© Springer-Verlag Berlin Heidelberg 2011

is associated with changing scene of TV commercials. The human-interest for TV commercials are needed to evaluate sequentially because investigators want to know which scene of TV commercials audiences are interested in. Recently, people study the investigation of latent human mind using biological signals. In previous study[1], it became obvious that we can obtain a result with reliability using them. Moreover, it is possible that we get a sequential evaluation result, because we can measure these signals sequentially. In particular, in the evaluation of TV commercials, there are some previous studies using biological signals[2],[3]. In these studies, they use some biological signals(EEG, galvanic skin response, and electromyogram/heart rate) and questionnaires/interviews. Accordingly, for both subjects and investigators, measuring more than one biological signal is great burden. Furthermore, measuring some signals interfere with watching TV commercials. We regard that an estimation system for the human-interest degree should be clearly/easily, sequentially, and simply, especially for TV commercials. Thus, we focus on the EEG for estimating the human-interest degree. The EEG reflects brain activity. It seems that brain activity is suitable to estimate the human-interest degree because it regulates the conscious and the feelings. We can measure a part of brain activity easily and cheaply by using the EEG. In addition, the EEG reflects the changes of psychological states in real-time because it responds fast to external stimuli. Therefore, the EEG is suitable as index to estimate the human-interest degree for TV commercials.

The contribution of this study in this paper is to estimate "the human-interest degree" while watching TV commercials to only measure the EEG. In particular, we focus on the sequentially investigation by using the EEG. We investigate the relationship between the EEG while watching TV commercials and the result of questionnaires. For measuring the EEG, we use a simple electroencephalograph. Accordingly, our system includes not only the clear/easy result of questionnaire and the sequential acquisition of data but also the simple EEG measurement.

2   Estimation System for Human-Interest Degree

In this section, we explain the estimation system for human-interest degree in detail. First, we show the system summary. Then, we describe the method of constructing our system. In particular, we explain the method of EEG measurement and feature extraction, an EEG clustering, and a multiple regression analysis.

2.1   Overview of Our System

Fig. 1. shows the overview of estimation system for human-interest degree. Our system estimates the human-interest degree while watching TV commercials sequentially. First of all, we acquire the EEG data while watching TV commercials using the simple electroencephalograph. Moreover, we apply the EEG data to our system. After that, we get the sequential estimation result for human-interest degree.

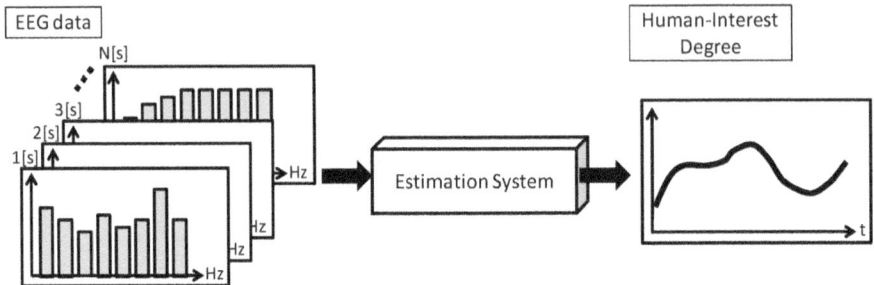

Fig. 1. Overview of Estimation System for Human-Interest Degree

2.2   EEG Measurement and Feature Extraction

EEG is electrical signal caused by the activity of cerebral cortex. We can measure human brain activity noninvasively, high temporal resolution, and flexibility by using EEG. It is well known that EEG changes corresponding to mental condition, cerebration, and emotion. EEG measurement point is defined by the international 10-20 system. Generally, we measure the EEG based on these points using multiple-electrode electroencephalograph having many electrodes. However, there are some problems. It takes more than 30 minutes to wear. Also, subjects need to use gel for electrolyte and that would stress them. For these reasons, multiple-electrode electroencephalograph is impractical. Therefore, in this experiment, we use a simple electroencephalograph which is formed of band type. This electroencephalograph is less demanding method for subjects because it is easy to wear and less stressing by nothing of tightness. Moreover, this device measures the EEG activity at Fp1(left frontal lobe) in the international 10-20 system. There is very low noise caused by hair at this measurement point. Furthermore, the EEG changes occur in prefrontal area largely[4]. Thus, we consider that this simple electroencephalograph is an appropriate method of measuring the EEG to estimate the human-interest. In addition, measurement methodology is referential recording: reference electrode is arranged at the left ear lobe and exploring electrode is at Fp1. Using this device, sampling frequency is 128Hz. The EEG data is analyzed by using fast fourier transform per one sec., and we can obtain the amplitude spectra at 1–64Hz. We ought to consider the effective EEG band frequency and characteristic of the simple electroencephalograph. Although we apply the 4–22Hz at 1Hz interval the effective frequency bands by using the band pass filter, taking account of EEG characteristics.

2.3   Construction of Estimation System for Human-Interest Degree

First of all, we show about the EEG features and the questionnaire. Then, we classify the subjects by the EEG feature using a cluster analysis because we consider the individual characteristic of the EEG. We explain the cluster analysis.

Finally, in order to estimate the human-interest degree, we apply the EEG features and the result of questionnaire to a multiple regression analysis. We show the multiple regression analysis.

Questionnaire. The questionnaire shows human's emotion or cerebration by linguistic expression. Moreover, this is called subjective assessment. In this experiment, the subjects evaluate the human-interest degree for each TV commercial on a scale of zero to ten. We use this degree when we analysis the human-interest.

EEG Features. The EEG features are the EEG data of amplitude spectra of 4–22Hz. In this experiment, we suppose that the EEG features show the change of human's emotion or cerebration while watching TV commercials. In addition, when we construct the system, we use the time average of the spectrum while watching TV commercials.

Cluster Analysis. The cluster analysis is used for clustering the subjects using the individual characteristic of EEG. In particular, we apply the hierarchical algorithm. We begin with each data and proceed to divide it into successively smaller clusters. Moreover, the distance function is important factor for classification. There are some distance functions however we adopt the Euclidean distance because this distance is most common method. In this paper, we classify into the subjects depending on the EEG features by the cluster analysis. Furthermore, we decide the effective cluster number by investigation the relationship between the number of data and the number of variable(EEG features)[5].

Multiple Regression Analysis. We use the multiple regression analysis to quantitatively evaluate the degree of human-interest. The multiple regression analysis is one of the multivariate analysis to have been constructed the estimation model from a set of training data. The multiple regression equation as estimation model is constructed by using some explanatory variables and an objective variable. In this paper, we apply the EEG features as the explanatory variables and the result of questionnaire as the objective variable. That is, we estimate the result of questionnaire using the EEG features.

3   Experiment of Watching TV Commercials

In this section, we show about the experiment of watching TV commercials. We measured the EEG while watching each TV commercial and investigated the degree of human-interest for each TV commercial using questionnaires. In this experiment, the subjects are set up as 54 people, including 28 males and 26 females. Each subject watch 10 different kinds of TV commercials that consist of 15 seconds. We measure the EEG while watching TV commercials using the simple electroencephalograph. After watching each TV commercial, we investigate the degree of human-interest for each TV commercial using questionnaires.

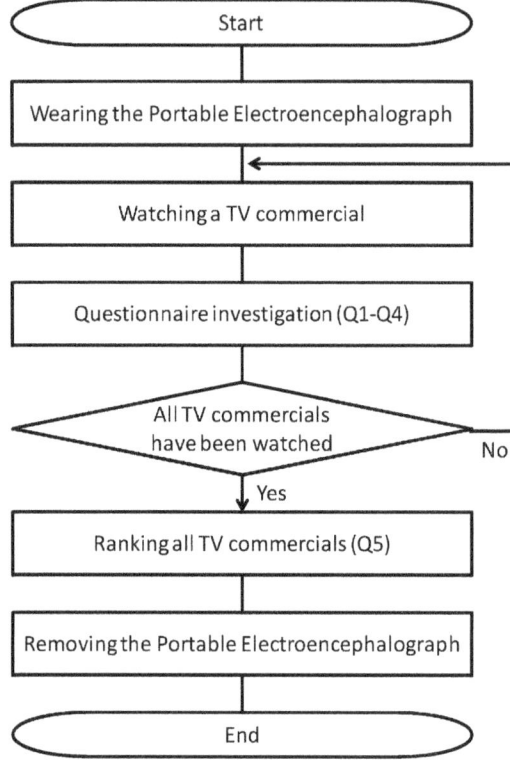

Fig. 2. The Procedure of an Experiment

Then, Fig. 2. shows the experimental procedure. Moreover, TABLE 1 represents the experimental condition and data. Furthermore, TABLE 2 is used for the questionnaires in this experiment.

4   Simulations

We explain the results of simulations for evaluation of our system. We conducted two kinds of simulations in order to discuss the effectiveness of our system. The first is the estimation of human-interest degree for each TV commercial. The second is the sequential estimation of human-interest degree for TV commercials. We describe the each result.

4.1   Estimation of Human-Interest Degree for Each TV Commercial

We verify the effectiveness and generalization of the estimation system, particularly about the subject clustering with the EEG features. In order to evaluate our system, we define *accuracy rate*[%] as follows:

Table 1. Experimental Condition and Data

		Condition and Data
Number of subjects	54	male: 28, female: 26
		20's: 23, 30's: 23, 40's: 5, 50's: 3
Object of watch		10 kinds of TV commercials(Japanese)
Watching time		15 seconds each TV commercial
Order of watch		in random order
Experimental period		25–27th, August, 2010

Table 2. The Questionnaire Items

	Questions
Q1	How do you rate your interest degree for the TV commercial on a scale of 0 to 10?
Q2	Which scenes are you interested in? (Choose some scenes from the pictures)
Q3	How many times have you seen the TV commercial? (Choose i. 0, ii. 1–3, iii. 4–)
Q4	How many times have you use/touch the products/services? (Choose i. 0, ii. 1–3, iii. 4–)
Q5	Rank TV commercials in descending order of human-interest degree.

$$accuracy\ rate[\%] = \frac{CORRECT}{TOTAL} \times 100. \tag{1}$$

We use the most high-scoring and the most low-scoring TV commercials by the questionnaire(Q2) to calculate *accuracy rate*[%]. Then, we define the human-interest degree which is estimated using the EEG by A, and the score of questionnaires by B. We determine the situation which the higher TV commercial of A matches higher one of B by *CORRECT*. In addition, *TOTAL* indicates the number of all data.

Incidentally, TABLE 3 shows the relationship between the number of subject cluster and the accuracies of ranking of 2 TV commercials by proposed system. According to TABLE 3, we find that 2 clusters is the highest accuracy rate. In contrast, we considered that more number of cluster was higher accuracy

Table 3. The Number of Cluster and Recognition Accuracies

Number of Cluster	Accuracy Rate[%]
1	74.1
2	79.6
3	74.1
4	75.9

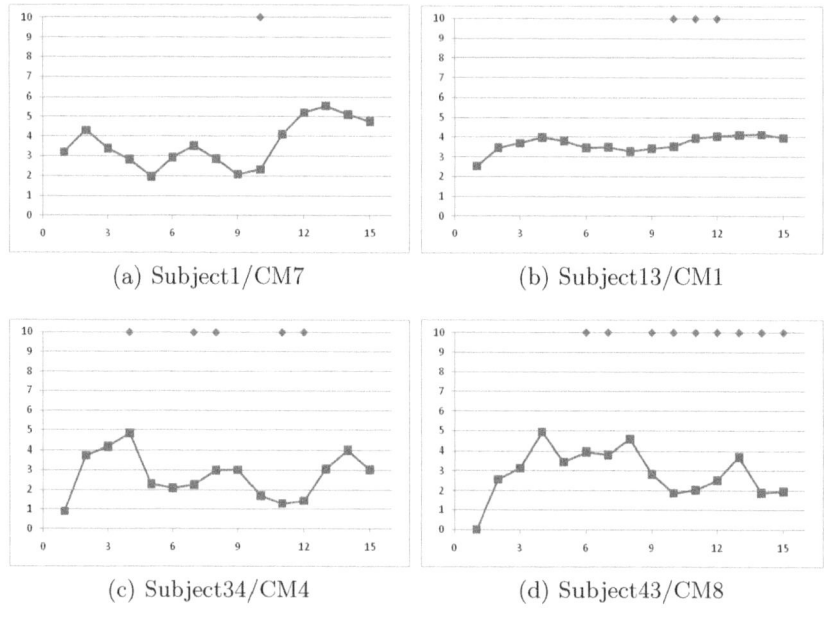

(a) Subject1/CM7   (b) Subject13/CM1

(c) Subject34/CM4   (d) Subject43/CM8

◆: Human-Interest from Questionnaire   ■■■ : Human-Interest Degree from Estimating System

The vertical axis: Human-interest degree obtained by the proposed system.
The horizontal axis: The time[second]

Fig. 3. Results by the Proposed System

rate because the number of data and variance of data were smaller. However, in this simulation, we regard that we classify the subjects to 2 clusters because accuracy rate is not proportional to the number of cluster. Furthermore, we show the effectiveness of estimation of human-interest degree for each TV commercial by the result of accuracy rate 79.6%.

4.2 Sequential Estimation of Human-Interest Degree for TV Commercials

We attempt to estimate the human-interest degree for TV commercials sequentially. In previous section, we show the effectiveness of our estimation system. Accordingly, in order to estimate human-interest degree sequentially, we use that system. When we measure the EEG using the simple electroencephalograph, the EEG signals of 1 second are transformed into frequency components. For this reason, we estimate the human-interest degree for TV commercials every 1 seconds. However, we apply moving average to the EEG features because we need consideration perception time, remainder time, and measuring error. Fig. 3. shows the example of sequential estimation. Based on the graph of Fig. 3. (a) or (c), we confirm an upward tendency around the interest scene from the questionnaire. Moreover, there are some subjects which the estimation results have a few ups and downs, (see Fig. 3. (b)). Furthermore, according to the graph of Fig. 3. (d), there are a lot of scenes which the subject are interested in by the questionnaire. On the other hand, human-interest degree by the estimating system has a lot of ups and downs. According to these results, we consider that we show the fluctuation of human-interest degree which we cannot understand human-interest by the questionnaire. Furthermore, we found the personal difference of the questionnaire.

5   Conclusions

In this study, we propose the estimation system for the human-interest degree while watching TV commercials using the EEG. In particular, we consider that (1)easy-to-understand result, (2)sequential estimation, and (3)simple system. Firstly, we use the questionnaires for easy-to-understand result. Secondly, we apply the EEG for sequential estimation. Finally, we adopt the simple electroencephalograph to construct simple system. However it became evident that we should investigate the personal difference of the questionnaire. We will consider that not only the individual characteristic of the EEG but also the personal difference of the questionnaire.

Acknowledgments. We thank all subjects in Dai Nippon Printing Co., Ltd for participating in our experiment.

References

1. Dijksterhuis, A., Bos, M.W., Nordgren, L.F., van Baaren, R.B.: On Making the Right Choice: The Deliberation-Without-Attention Effect. Science 311(5763), 1005–1007 (2006)
2. Ohme, R., Matuskin, M., Szczurko, T.: Neurophysiology uncovers secrets of TV commercials. Der. Markt. 49(3-4), 133–142 (2010)
3. Vecchiato, G., Astolfi, L., Fallani, F.D.V., Cincotti, F., Mattia, D., Salinari, S., Soranzo, R., Babiloni, F.: Changes in Brain Activity During the Observation of TV Commercials by Using EEG, GSR and HR Measurements. Brain Topography 23(2), 165–179 (2010)
4. Sutton, S.K., Davidson, R.J.: Prefrontal brain electrical asymmetry predicts the evaluation of affective stimuli. Neuropsychologia 38(13), 1723–1733 (2000)
5. Lotte, F., Congedo, M., Lecuyer, A., Lamarche, F., Arnaldi, B.: A review of classification algorithms for EEG-based brain-computer interfaces. Journal of Neural Engineering 4(2), R1–R13 (2007)

Effects of Second-Order Statistics
on Independent Component Filters

André Cavalcante[1], Allan Kardec Barros[2], Yoshinori Takeuchi[1],
and Noboru Ohnishi[1]

[1] School of Information Science, Nagoya University, Japan
[2] Laboratory for Biological Information Processing,
Universidade Federal do Maranhao, Brazil

Abstract. It is known that independent component analysis (ICA) generates filters that are similar to the receptive fields of primary visual cortex (V1) cells. However, ICA fails to yield the frequency tuning exhibited by V1 receptive fields. This work analysis how the shape of IC filters depend on second-order statistics of the input data. Specifically, we show theoretically and through experimentation how the structure of IC filters change with second-order statistics and different types of data preprocessing. Here, we preprocess natural scenes according to four conditions: *whitening*, *pseudo-whitening*, *local-whitening* and *high-pass filtering*. As results, we show that the filter structure is strongly modulated by the inverse of the covariance of the input signal. However, the distribution of size in frequency domain are similarly biased for all preprocessing conditions.

Keywords: Independent component analysis, second-order statistics, receptive fields.

1 Introduction

An important result in computational neuroscience is that the independent component analysis (ICA) of natural scenes yields filters similar to the Gabor-like receptive fields of simple cells in the primary visual cortex (V1)[1]. However, it is known ICA fails to generate filters that match the spatial frequency tuning and orientation distribution observed for V1 receptive fields. Specifically, the ICA filters normally have higher central frequencies and are not as broadly distributed in orientation as V1 receptive fields. In this way, it is important to determine precisely how the shape of IC filters is adapted.

Baddeley [2] suggested that since natural scenes have local variance, the process of sparseness maximization (which is closely related to independence maximization in case of natural signals) would generate zero DC filters. On the other hand, Thomson [3] suggested that both power and phase structure would require high-frequency filters for sparseness maximization. Specifically, the Fourier transform of the fourth-order correlation function of whitened images showed that oblique-high-frequency components have higher contributions to kurtosis

B.-L. Lu, L. Zhang, and J. Kwok (Eds.): ICONIP 2011, Part I, LNCS 7062, pp. 54–61, 2011.

than vertical and horizontal-low-frequencies components. Furthermore, it was shown for natural sounds that the optimal coding bandwidth is similar for kurtosis calculated from either the second spectrum or from the phase-only second spectrum [4].

On the other hand, Lewicki [5] showed that distribution of filter bandwidths is strikingly similar to the inverse of the power spectrum of the input signal when the efficient coding algorithm has second-order constraints. In this work, we analyze the influence of second-order statistics on independent component filters by modifying the power spectra of natural scenes. To our best knowledge, no work has determined precisely what is the role of the amplitude information on ICA when there are second-order constraints involved. The importance of this work is that we are the first to show mathematically why ICA fails to fails to generate filters with the same characteristics of V1 receptive fields.

This work is divided as follows: in the methods section we provide a analysis on how the structure of ICA filters is formed; section "Results" describes the experiments and outcomes; the section "Discussion" provides information about the relation of these results to physiological studies; a conclusion section provides information about remaining challenges.

2 Methods

2.1 The Generative Model of IC Filters

Independent component filters are optimized so that their responses to the input signal are maximally statistically independent. In this way, ICA filters represents a "privileged" transform that (maximally) reduce both second and higher-order correlations for the specific statistical distribution of the input [6]. Although second-order uncorrelatedness alone is computationally inexpensive, second-order correlations impose a heavy computational cost on the ICA adaptive process. Whitening the input signal before learning and maintaining the filters orthogonalized ensures both low-computational cost and second-order uncorrelatedness during the adaptation.

Taking these constraints into account, it is easy to show that each ICA filter, represented here by $\mathbf{w}_i$, can be found as the linear combination

$$\mathbf{w}_i = b_{i1} \cdot \mathbf{v}_1 + b_{i2} \cdot \mathbf{v}_2 + \cdots + b_{in} \cdot \mathbf{v}_n, \tag{1}$$

where the set of vectors $\mathbf{V} = \{\mathbf{v}_1, \mathbf{v}_2, \ldots, \mathbf{v}_n\}$ (not necessarily an orthogonal set) represents the decorrelation transform applied on the input signal, and the filter coefficients $b_{i1}, b_{i2}, \ldots, b_{in}$ are normally optimized based on a non-linear transformation of the input. The goal of this non-linear transformation is to make $\mathbf{w}_i$ able to reduce higher-order correlations as well. It is important to notice that this generative model may be only valid for ICA strategies that perform second-order decorrelation by means of a pre-whitening transform.

In this way, filters $\mathbf{w}_i$ only reduce as much higher-order correlations as possible for a combination of second-order decorrelation vectors. The immediate question how precisely the vectors $\mathbf{v}_j$ compromise the ICA higher-order decorrelation.

2.2 FastICA

In case of the FastICA algorithm [7], $\mathbf{V}$ is the principal component analysis (PCA) solution given by the eigen-decomposition of the covariance matrix of the input, i.e.,

$$\mathbf{V} = \mathbf{D}^{-\frac{1}{2}}\mathbf{E}^{\mathrm{T}}, \tag{2}$$

where $\mathbf{D}$ is the diagonal matrix of eigenvalues and $\mathbf{E}$ is the matrix of eigenvectors. In this case, the set $\mathbf{V}$ will form a base of the white space of the input signal for which complete second-order decorrelation is guaranteed. Furthermore, one can establish a semantical order to the base vectors $\mathbf{v}_j$. Specially in case of natural signals with "$1/f$" Fourier spectra, these vectors will normally reflect the frequency components organization.

The generative model can therefore be rewritten as

$$\mathbf{w}_i = \sum_{ij} b_{ij} \cdot d_j^{-\frac{1}{2}} \cdot \mathbf{e}_j, \tag{3}$$

where d_j is the eigenvalue corresponding to the eigenvector $\mathbf{e}_j$. Therefore, the magnitude of the product $b_{ij} \cdot d_j^{-\frac{1}{2}}$ defines the contribution of $\mathbf{e}_j$ to the structure of $\mathbf{w}_i$.

Not surprisingly, in case of "$1/f$" data, high-frequency eigenvectors will be normally accompanied by much larger $d_j^{-\frac{1}{2}}$ than those of low-frequency eigenvectors so that in the final $\mathbf{w}_i$ is a high-frequencies are emphasized.

In case of the FastICA algorithm, it is also possible to estimate another set of "filters" called basis functions. Each basis function is given by

$$\mathbf{a}_i = \sum_{ij} b_{ij} \cdot d_j^{\frac{1}{2}} \cdot \mathbf{e}_j. \tag{4}$$

Now, high-frequency eigenvectors will be accompanied by lower values $d_j^{\frac{1}{2}}$ than those of low-frequency eigenvectors for "$1/f$" data. Therefore, $\mathbf{a}_i$ is a "lower frequency" filter than $\mathbf{w}_i$.

3   Results

The natural scenes dataset was obtained from the McGill Calibrated Colour Image Database [8]. This database consists of TIFF formated non-compressed images. The resolution of image files is 576 x 768 pixels. In order to build a dataset as general as possible, 150 scenes were selected from natural image categories such as forests, landscapes, and natural objects such as animals. Images containing man-made objects were not used.

3.1   Preprocessing Conditions

Natural scenes had their amplitude spectrum modified according to four preprocessing criteria: *whitening, pseudo − whitening, local − whitening* and *high −*

pass filtering. Notice that for all conditions, the PCA whitening step in Fas-tICA is performed assuring stability. In the *whitening* condition, the amplitudes of all frequency components were set the unit. In the *pseudo − whitening* condition, all images were zero-phase filtered by the frequency response $R(f) = f \exp[-(f/f_o)^4]$, where $f_o = 0.4$ cycles/pixel.

In the *local − whitening*, whitening was applied to non-overlapping image patches used for learning of independent component filters. In the *high − pass filtering*, all images were high-pass filtered according the filter $R(f) = - \exp[-(f/f_o)^4] + 1$, where f_o is the same as for the pseudo-whitening filter.

3.2   Learned Filters and Basis Functions

In order to learn filters for each preprocessing condition, 100,000 non-overlapping image patches of 16×16 pixels were extracted from the natural scenes dataset. Here, we use the FastICA algorithm where the working non-linearity was the hyperbolic tangent and the number of iterations was set to 250. Examples of learned filters and respective basis functions (columns of the inverse of filter matrix) are shown in Figure 1.

In Figure 1, the panel in the first row $(1/f)$ exhibits the filters and basis functions for natural scenes with raw $1/f$ amplitude spectra. The second row (whitening) shows the filters learned from natural scenes with whitened amplitude spectra. The third-row $(pseudo − whitening)$ shows the filters and bases learned after the amplitude spectra of the scenes have been filtered by using Eq. 4. The last panel shows the results for high-pass filtered data.

As expected, $1/f$ data yields visually high-frequency (sharp) filters and corresponding low-frequency (smooth) basis functions. Whitened data generates filters which are similar to the basis functions. Pseudo-whitening generates very-high-frequency filters and cannot be visualized in square pixel sample (pixelation issue), whereas the basis functions are well-defined. In case of the *local − whitening* condition, filters are super localized in the spatial window. Filters and basis functions are also very similar. In contrary to $1/f$ data, high-pass filtered data generates low-frequency filters and high-frequency basis functions.

In order to provide a quantitative analysis of the learned filters, they have been fitted by real-valued Gabor functions. Figure 2 shows polar plots representing the parameters of the best fits for each preprocessing condition. In the polar plots, each filter is represented by a colored circle whose orientation and distance to the plot origin represents the preferred orientation (degrees) and spatial frequency of the filter (cycles/pixel), respectively. The color of the circle represents the area occupied by the filter in frequency domain. Here, this area is given in terms of the product of the horizontal and vertical frequency lengths which are given by $\sigma_u = 1/(2\pi \cdot \sigma_x)$ and $\sigma_v = 1/(2\pi \cdot \sigma_y)$, respectively.

The polar plot for $1/f$ images shows that the great majority of filters are localized in the highest frequency part of Fourier domain. Furthermore, the concentration of filters in the polar plots is visibly higher in oblique orientations. The color of the circles in the polar plots also demonstrate that filters have small products between lengths in frequency domain. For *Whitening* data, filters are

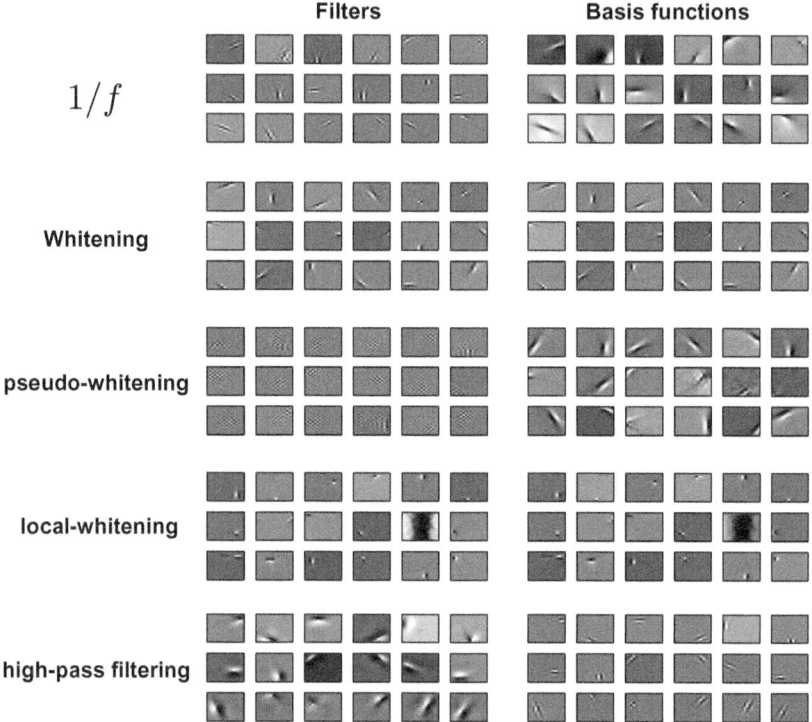

Fig. 1. Examples of learned ICA filters and basis functions. The panels in each row show examples of filters and basis functions learned for each preprocessing condition. For comparison, the first row $(1/f)$ shows filters and basis functions for natural scenes with original amplitude spectrum.

fairly distributed over the Fourier domain in terms of both spatial frequency and orientation. The color of circles also demonstrate that these filters can have larger areas in frequency domain than filters learned from $1/f$ data.

As shown in Figure 1, the structure of the filters learned for the *pseudo – whitening* condition contains much energy in very high frequencies. In this way, a fitting analysis of these filters is not reliable. Further information on the results for this preprocessing condition data is however provided in the Discussion section. Similar to the *Whitening* preprocessing, the *Local – whitening* condition generates filters fairly scattered over the frequency domain. The difference is that these filters have larger areas in frequency domain as demonstrated by the product of frequency lengths than those for *Whitening* condition. The filters estimated for *High – pass* filtered data are concentrated at low frequencies. However, similar to $1/f$ filters they have in majority very small areas in frequency domain. In order to provide a better analysis of the filter areas in frequency domain, Figure 3 shows the histograms of the products between horizontal and vertical frequency lengths for each preprocessing condition. For all preprocessing

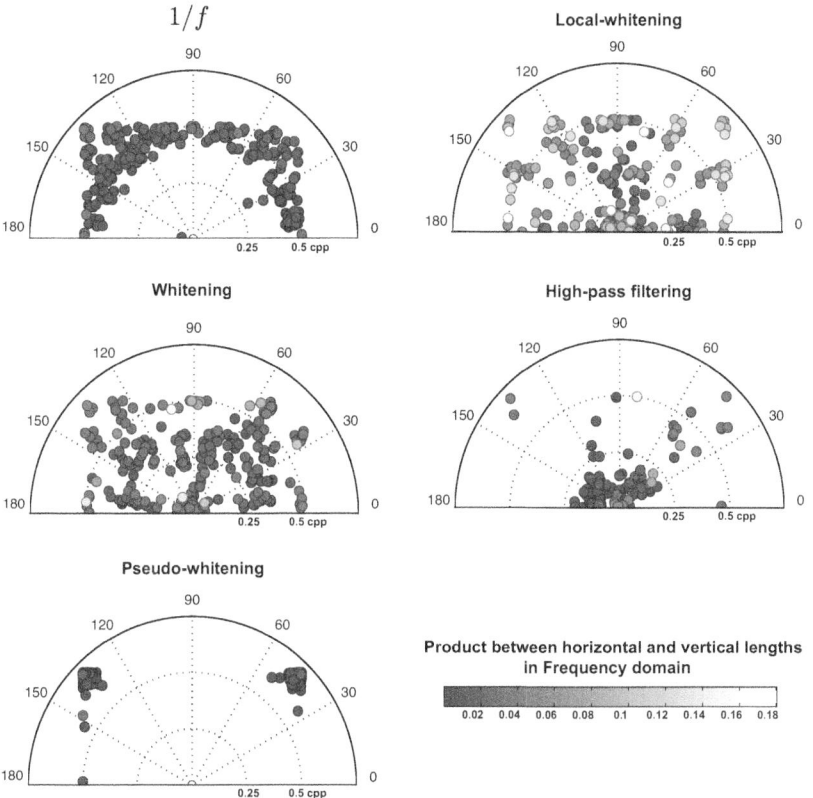

Fig. 2. Quantitative analysis of the IC filters. The parameters of best Gabor fits are represented by polar plots for each preprocessing condition. Each gray colored circle in the polar plots represents an IC filter. The orientation of the circle represents the orientation of the filter (degrees). The distance of the circle to the plot origin represents the spatial frequency given in cycles/pixel (cpp). The circle color represents the product between the filter's horizontal and vertical lengths in frequency domain. The associated graymap is show on the right bottom of the figure.

conditions, histograms have been calculated using 32 bins. Interestingly, these histograms shows that filters are biased towards small values of frequency area independent of the preprocessing condition. However, it is also possible to observe that *whitening* and *local — whitening* can generates higher values more commonly than other conditions. This is because the "flat" covariance of white data forces the filters to cover the frequency domain more uniformly.

4 Discussion

Firstly, let us discuss the effects of second-order statistics on independent component analysis by analyzing Eqs. 3 and 4. Both equations are basically formed

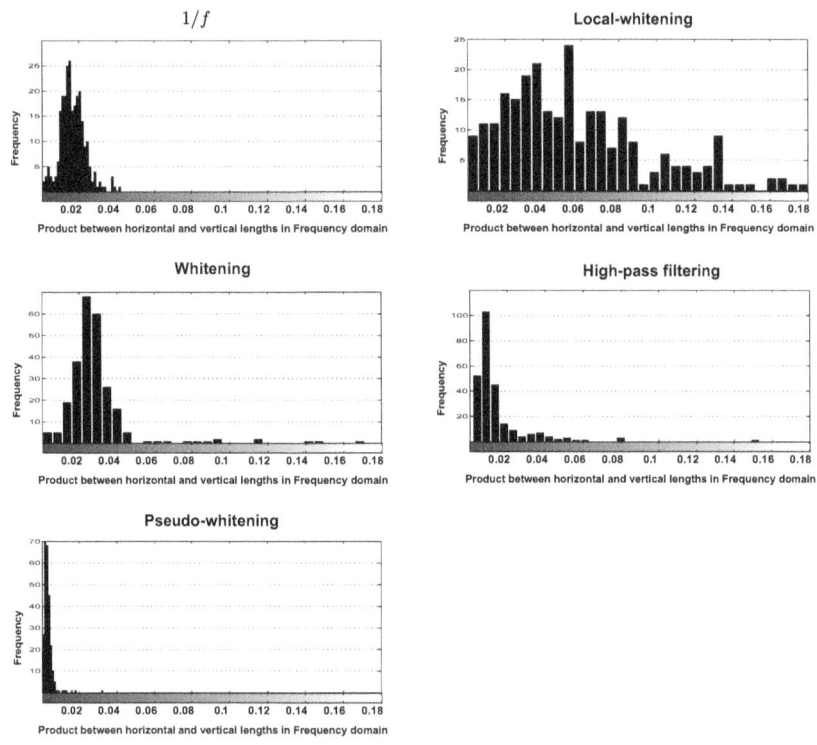

Fig. 3. Histogram of frequency area. Each histogram correspond to the products between filter's horizontal and vertical lengths in frequency domain for each polar plot in Figure 2. All histograms have been calculated using 32 bins.

of two main terms, a higher-order term b_{ij} and a second-order term formed by eigenvalue d_j and eigenvector $\mathbf{e}_j$. It is important to notice that regarding higher-order information from the input, there are no differences between an IC filter $\mathbf{w}_i$ and its respective basis function $\mathbf{a}_i$ since both are formed by the same values b_{ij}. The only difference between them is the exponent of the eigenvalue d_j. However this does have an impact on the structure of $\mathbf{w}_i$ and $\mathbf{a}_i$. Specifically, the energy composition of filters will be modulated by the inverse of the covariance of the input data whilst the basis functions will be directly modulated by the input covariance.

This effect can be observed by the filters and basis functions shown in Figure 1. For instance, $1/f$ data yields high-frequency filters and low-frequency basis functions but with similar orientation, phase and position in the spatial window. However, both *whitening* and *local − whitening* preprocessing conditions, which flat the amplitude spectra of the images, generate filters and basis functions that are visually very similar. This also suggests that the *pseudo − whitening* condition yields very high-frequency filters because after preprocessing by Eq. 5, the very high-frequencies are strongly attenuated in the resulting images.

The quantitative analysis of the filters also demonstrates that filters will be localized in the frequency areas where the input energy has low energy. For instance, for $1/f$ data, the concentration of filters is higher at oblique high-frequency areas which for raw natural scenes are the regions with lowest energy. This behavior is also observed for the preprocessing conditions used in this work. It is important to notice that the dependence of ICA on second-order statistics can not be removed by simply flatting the amplitude spectrum of the input data.

These results explain why ICA fails to generate filters that match the frequency tuning observed for V1 receptive fields. It also demonstrates that the ICA may generate this frequency tuning in case the input data or the preprocessing technique is well chosen. In this regard, Lewicki [9] showed a methodology to generate filters which are very similar to the cat's auditory receptive fields. However, he emphasizes that the specific choice of input signal is essential to obtain auditory-like receptive fields.

5   Conclusion

In this work, we have shown the extent of the effects of second-order statistics on the shape of ICA filters. Specifically, the center frequency of IC filters is determined by the inverse the covariance of the input data. On the other hand, characteristics such as filter size in frequency domain are less dependent on amplitude information. We propose to extend this work to also analyze how the response behavior of the IC filters depend on the second-order statistics.

References

1. Hateren, J.H., van der Schaaf, A.: Independent component filters of natural images compared with simple cells in primary visual cortex. Proc. R. Soc. Lond. B 265, 359–366 (1998)
2. Baddeley, R.: Searching for Filters With "Interesting" Output Distributions: An Uninteresting Direction to Explore? Network 7, 409–421 (1996)
3. Thomson, M.G.A.: Higher-order structure in natural scenes. JOSA A 16(7), 1549–1533 (1999)
4. Thomson, M.G.A.: Sensory coding and the second spectra of natural signals. Phys. Rev. Lett. 86, 2901–2904 (2001)
5. Lewicki, M.S.: Efficient coding of natural sounds. Nature Neuroscience 5(4), 356–363 (2002)
6. Bell, A.J., Sejnowski, T.J.: The "independent components" of natural scenes are edge filters. Vision Research 37(23), 3327–3338 (1997)
7. Hyvrinen, A.: Fast and Robust Fixed-Point Algorithms for Independent Component Analysis. IEEE Transactions on Neural Networks 10(3), 626–634 (1999)
8. McGill Color Image Database, `http://tabby.vision.mcgill.ca/`
9. Smith, E., Lewicki, M.S.: Efficient Auditory Coding. Nature 439, 70–79 (2006)

Neural Model of Auditory Cortex
for Binding Sound Intensity
and Frequency Information in Bat's Echolocation

Yoshitaka Mutoh[1] and Yoshiki Kashimori[1,2]

[1] Dept. of Engineering Science, Univ. of Electro-Communications, Chofu,
Tokyo 182-8585 Japan
[2] Graduate School of Information Systems, Univ. of Electro-Communications,
Chofu, Tokyo 182-8585 Japan
kashi@pc.uec.ac.jp

Abstract. Most species of bats making echolocation use the sound pressure level (SPL) and Doppler-shifted frequency of ultrasonic echo pulse to measure the size and velocity of target. The neural circuits for detecting these target features are specialized for amplitude and frequency analysis of the second harmonic constant frequency (CF2) component of Doppler-shifted echoes. The neuronal circuits involved in detecting these features have been well established. However, it is not yet clear the neural mechanism by which these neuronal circuits detect the amplitude and frequency of echo signals. We present here neural models for detecting SPL amplitude and Doppler-shifted frequency of echo sound reflecting a target. Using the model, we show that the tuning property of frequency is changed depending on the feedback connections between cortical and subcortical neurons. We also show SPL amplitude is detected by integrating input signals emanating from ipsi and contralatreal subcortical neurons.

Keywords: auditory system, echolocation, frequency tuning, SPL amplitude, neural model.

1   Introduction

Mustached bats emit ultrasonic pulses and listen to returning echoes for orientation and hunting flying insects. The bats analyze the correlation between the emitted pulses and their echoes and extract the detailed information about flying insects based on the analysis. This behavior is called echolocation. The neuronal circuits underlying echolocation detect the velocity of target with accuracy of 1 cm/sec and the distance of target with accuracy of 1 mm. To extract the various information about flying insects, mustached bats emit complex biosonar that consists of a long-constant frequency (CF) component followed by a short frequency-modulated (FM) component [1]. Each pulse contains four harmonics and so eight components represented by (CF1, CF2, CF3, CF4, and FM1, FM2, FM3, FM4). The information of target distance and velocity are processed separately along the different pathways in the brain by using four FM components and four CF components, respectively [2].

B.-L. Lu, L. Zhang, and J. Kwok (Eds.): ICONIP 2011, Part I, LNCS 7062, pp. 62–69, 2011.
© Springer-Verlag Berlin Heidelberg 2011

In natural situation, large natural objects in environment, like bushes or trees, produce complex stochastic echoes, which can be characterized by the echo roughness. The echo signal reflecting from a target insect is embedded in the complex signal. Even in such an environment, bats can detect accurately the detailed information of flying insect. To extract the information about insects, it is needed to encode efficiently the amplitude and frequency information of echo sounds and to combine them.

To investigate the neural mechanism underlying the encoding of amplitude and frequency information, we study the neural pathway for detecting target size and velocity, which consists of cochlea, inferior colliculus (IC), and Doppler-shifted constant frequency (DSCF) area. The cochlea is remarkably specialized for fine-frequency analysis of the second harmonic CF component (CF2) of Doppler-shifted echoes. The information about echo CF2 is transmitted to IC, and the size and relative velocity of target insect are detected in DSCF area by analyzing the sound pressure level (SPL) amplitude and Doppler-shifted frequency of echo signals. There are several experimental results on IC and DSCF neurons. It was reported that DSCF neurons responds to a specific range of SPL amplitude, as well as best frequency (BF) [3]. Xia and Suga [4] demonstrated that electric stimulation of DSCF neurons evokes the BF shifts of IC neurons away from the BF of the stimulated DSCF neuron (centrifugal BF shift) and bicuculine (an antagonist of inhibitory GABA receptors) applied to the stimulation site changes the centrifugal BF shifts into the BF shifts towards the BF of stimulated DSCF neurons (centripetal BF shift). This indicates that the BF modulation elicited by top-down signal may play a crucial role in frequency coding of IC and DSCF neurons. However, it is not yet clear how the bat's auditory system detects the SPL amplitude and Doppler-shifted frequency of echo signals.

In the present study, we propose neural network models for detecting SPL amplitude and Doppler-shifted frequency of sound echoes. Using the model, we show how these sound features are represented in bat's auditory cortex.

2   Model

The auditory system of bat's brain contains choclea (Ch), inferior colliculus (IC), and Doppler-shifted constant frequency (DSCF) processing area in each hemisphere, as shown in Fig. 1. The Ch neurons project to contralateral IC neurons and DSCF neurons receive the output of contralateral IC neurons. The Ch neurons have frequency map by which sound frequency is encoded. The amplitude of SPL is encoded in to firing rate of Ch neurons. The IC neurons have frequency map and also encode SPL in to firing rate. DSCF area has a specific map detecting sound frequency and SPL, in which frequency is represented along the radial axis [3,5,6] and the amplitude of SPL is represented along a circular axis [3,7]. The frequency axis is overrepresented with the frequencies of 60.6 kHz to 62.3 kHz, corresponding to echo frequencies of second harmonics. The neurons in the DSCF area are excited by the output of contralateral IC neurons and inhibited by that of ipsilateral IC neurons [8]. To investigate the neural mechanism underlying auditory information processing of bat's brain, we developed neural models for detecting Doppler-shifted frequency and SPL amplitude of echo signals.

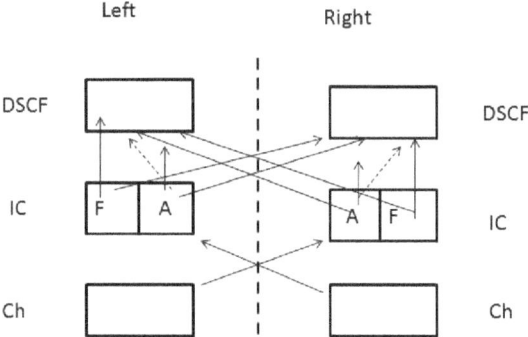

Fig. 1. Auditory pathway involved in SPL amplitude and Doppler-shifted frequency information processing. The left and right hemispheres contain Ch, IC, and DSCF area, respectively. The regions for encoding SPL amplitude and Doppler-shifted frequency are denoted by 'A' and 'F' in ICs, respectively. The solid and dashed lines indicate the excitatory and inhibitory connections, respectively. The vertical dashed line indicates the corpus callosum (CC). Ch: Choclea, IC: inferior colliculus, and DSCF: Doppler-shifted constant frequency processing area.

2.1   Model for Detecting Doppler-Shifted Frequency of Echo Signals

Figure 2 illustrates a model for detecting Doppler-shifted frequency of echo sounds. The model consists of three layers, right (R) -Ch, left (L) -IC, and R-DSCF layers. For simplicity, we do not consider here the model for other layers, L-Ch, R-IC, and L-DSCF, because they have the similar connections to the model shown in Fig. 2 and differ only from reverse relationship in right and left side. The R-Ch layer has one-dimensional array of neurons, each of which is tuned to a frequency of sounds. The L-IC and R-DSCF layer have also tonotopical maps corresponding to tuning property of R-Ch layer. The neurons in each layer were modeled with the leaky integrate-and fire (LIF) neuron model [9]. The details of the network model were described in Ref. 10.

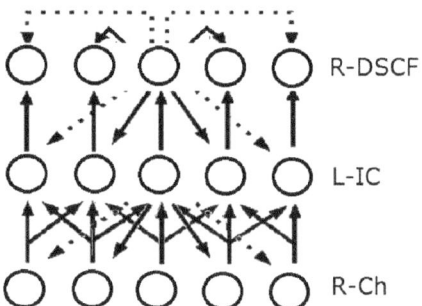

Fig. 2. Neural network model for detecting Doppler-shifted frequency of echo sound. The model consists of three layers, right (R)-choclear, left (L)-inferior colliculus, and R-DSCF layers.

2.2   Model for Detecting SPL Amplitude of Echo Signals

Figure 3 illustrates the model for detecting SPL amplitude. The model consists of five layers, right-(R-) and left-(L-) Chs, R- and L-ICs, and R-DSCF layer. To investigate the essential mechanism underlying the detection of SPL amplitude, we do not consider here L-DSCF layer that has the similar network to that of R-DSCF layer.

The R-DSCF neurons can detect SPL amplitude as a peak location of activated DSCF neurons. The excitatory input from L-IC neurons elicits the firing of DSCF neurons, but the inhibitory input from R-IC neurons suppresses the firing of DSCF neurons. This allows the DSCF neurons to exhibit a firing region in R-DSCF network, in which neurons, located in the left side of the neuron that has membrane potential just under the firing threshold, are activated, because the inhibitory synaptic weights are gradually increased as the position of neurons moves to the right side of the DSCF network, as shown in Fig. 3. Then the lateral inhibition across DSCF neurons enables the DSCF network to activate only the neurons nearby the edge between activated and silent neurons in spatial activity of DSCF neurons, leading to detection of SPL amplitude.

The mathematical descriptions of the model are as follows.

2.2.1   Choclea

The models of R- and L-Ch have one-dimensional arrays of neurons, each neuron of which is tuned to a specific frequency of sound. The Ch neuron was modeled with the LIF neuron model. The membrane potentials of ith X-Ch neurons (X= R, L) are determined by

$$\tau_{X-Ch} \frac{dV_i^{X-Ch}}{dt} = -V_i^{X-Ch} + I_i, \quad (X = R, L), \tag{1}$$

$$I_i = I_0 e^{-(i-i_0)^2/\sigma_{Ch}^2}, \tag{2}$$

where τ_{X-Ch} is the time constant of V_i^{X-Ch} and I_i is the input to ith X-Ch neuron, which is described by the Gaussian function that has a maximum response of i_0th Ch neuron.

2.2.2   R- and L-IC

IC neurons integrate the outputs of contralateral Ch neurons and encode the information of SPL amplitude in to firing rate of IC neurons. The R- and L-IC layers have one-dimensional array of neurons, respectively The model of IC neuron was based on the LIF model. The membrane potentials of ith X-IC neurons (X= R, L) are determined by

$$\tau_{X-IC} \frac{dV_i^{X-IC}}{dt} = -V_i^{X-IC} + \sum_j w(ij; X-IC; Y-Ch) X_j^{Y-Ch}(t),$$

$$(X, Y) = (R, L) \text{ or } (L, R), \tag{3}$$

where τ_{X-IC} is the time constant and $w(ij; X - IC; Y - Ch)$ is the synaptic weight of the connection from jth Y-Ch neuron to ith X-Ch one. $X_j^{Y-Ch}(t)$ is the output of jth Y-Ch neuron, described with α-function [9].

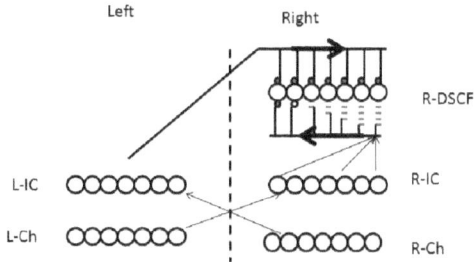

Fig. 3. Neural network model for detecting SPL amplitude of echo sound. The black circles and short horizontal bars in R-DSCF indicate excitatory and inhibitory synapses, respectively. The inhibitory synaptic weights are monotonically increased along the array of neurons from the left to right sides.

2.2.3   R-DSCF

R-DSCF neurons receive excitatory inputs from L-IC neuron in contralateral side and excitatory or inhibitory inputs from R-IC neurons in the ipsilateral side, as shown in Fig. 1. The synaptic weights of the contralateral connections from L-IC neurons to R-DSCF ones have a constant value, and those of the ipsilateral connections from R-IC neurons to R-DSCF ones change gradually from a positive value to a negative value. The DSCF neuron was modeled with the LIF model. The membrane potential of R-DSCF neuron is determined by

$$\tau_{R-DSCF}\frac{dV_i^{R-DSCF}}{dt} = -V_i^{R-DSCF} + w(i; R - DSCF; L - IC)S_{L-IC}(t)$$
$$+ w(i; R - DSCF; R - IC)S_{R-IC}(t)$$
$$+ \sum_j w(ij; R - DSCF; R - DSCF)X_j^{R-DSCF}(t) + w_e S_{L-IC}^f(t),$$

$$\tag{4}$$

$$S_{X-IC}(t) = \sum_j X_j^{X-IC}, \quad (X = R, L), \tag{5}$$

$$w(i; R - DSCF; L - IC) = w_0(1 - \alpha(i - 1)), \tag{6}$$

where, τ_{R-DSCF} is the time constant of V_i^{R-DSCF}. $w(i; R - DSCF; Y - IC)$ (Y=L, R) are the synaptic weights of ith R-DSCF neuron receiving the output of Y-IC neurons $S_{Y-IC}(t)$, $w(ij; R - DSCF; R - DSCF)$ is the synaptic weight of the inhibitory connection between ith and jth R-DSCF neurons, and w_e is the weight of the excitatory synapses mediating the output of L-IC neurons, $S_{L-IC}^{f}(t)$, in frequency map described in section 2.1. w_0 is constant, and α the decreasing rate of inhibitory synaptic weight.

3   Results

3.1   BF Shifts of IC Neurons Caused by Feedback Signals

Figure 4a shows the change in the tuning property of IC neurons in the case where tone stimulus was delivered and electric stimulus (ES) was applied to DSCF neuron tuned to 60.6 kHz. The result was calculated by using the network model shown in Fig. 2. The ES resulted in the BF shift of the IC neurons away from the BF of the stimulated DSCF neuron, that is, centrifugal BF shift. Before the ES, the IC neurons maximally responded to 60.6 kHz (vertical dashed line). When DSCF neuron tuned to 61.1 kHz was electrically stimulated, the BF of IC neuron was shifted from 60.6 kHz to 60.5 kHz. That is, the IC neurons showed a centrifugal shift.

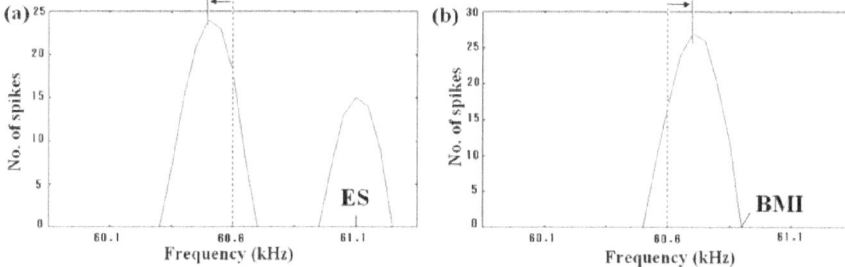

Fig. 4. Tuning properties of IC neurons. Centrifugal (a) and centripetal (b) BF shift. The dashed lines indicate the center of firing pattern in control. The arrows indicate BF shifts caused by electric stimulation (ES) and application of bicuculine, respectively. BMI: bicuculine.

Figure 4b shows the response properties of IC neurons when the antagonist of GABA, bicuculine, was applied to the DSCF neurons tuned to 61.1 kHz. The inhibition of GABA led to the BF shift of the IC neuron towards the BF of the bicuculine-injected DSCF neuron. The BF of IC neurons shifted from 60.6 kHz to 60.8 kHz. That is, the IC neurons showed a centripetal BF shift.

Our model reproduces well the experimental results measured by Xia and Suga [4].

3.2   Neural Mechanism for Determining Directions of BF Shifts

Figure 5a illustrate the changes in the synaptic potentials of the top-down connections from the DSCF neuron tuned to 61.6kHz to IC neurons, that is, the receptive field of the DSCF neuron, in the case of electric stimulation of a DSCF neuron. The ES made the peak position of the synaptic potential shift away from the position of ES, resulting in the BF away from the BF of the stimulated DSCF neuron, that is, centrifugal BF shift. In contrast to the centrifugal BF shift, the application of bicuculine, an antagonist of GABA receptors, modulated the shape of the receptive field, as shown in Fig. 5b. The injection of bicuculine made the peak position shift towards the injection site of bicuculine, leading to the BF shift towards the BF of the bicuculine-injected DSCF neuron.

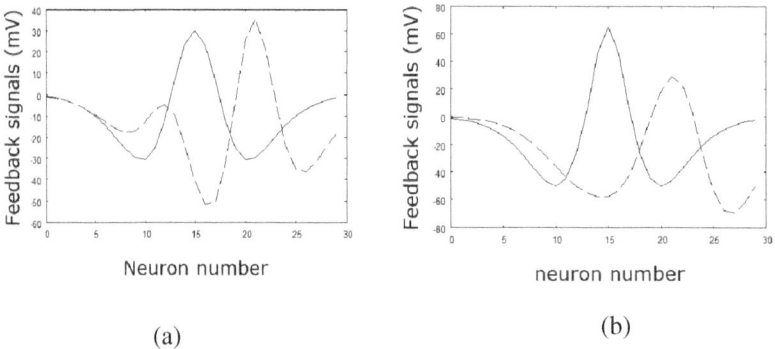

(a)   (b)

Fig. 5. The changes in synaptic potentials of top-down connections from a DSCF neuron to IC neurons. The solid lines indicate the synaptic potentials in control, and the dashed lines indicate those under ES (a) and application of bicuculine (b), respectively.

3.3   Response Properties of DSCF Neurons for SPL Amplitude

Figure 6a shows the firing pattern of R-DSCF neurons in the network model shown in Fig.3 for two different SPL amplitudes of echo signals. The SPL amplitude is detected by a peak location of activated neurons. As shown in Fig. 6a, the peak location of firing pattern moved to right side of DSCF network as SPL amplitude is increased.

The R-DSCF neuron has a window of response, with a lower and upper threshold for SPL amplitude, as shown in Fig. 6b. The response property of DSCF neurons to a specific region of SPL amplitude enables the DSCF neurons to encode amplitude information of echo sound.

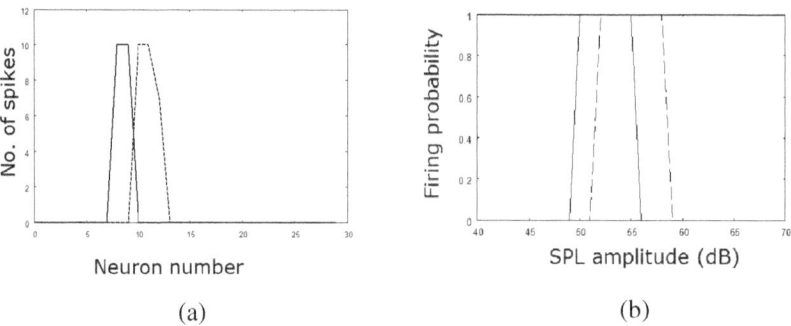

(a)   (b)

Fig. 6. Response properties of R-DSCF neurons for SPL amplitude. (a) Neuronal activities evoked by to echo signals with different SPL amplitudes. (b) Response windows. The solid and dashed lines, respectively, represent the response properties for two different SPL amplitudes.

4   Conclusion

We have presented here the neural mechanisms for detecting SPL amplitude and Doppler-shifted frequency of echo sounds. We show that Doppler-shifted frequency is encoded by the tonotopical map, in which tuning property of subcortical neurons is adequately modulated by corticofugal signals. We also show that the amplitude of echo sound is encoded by integrating the outputs of ipsi and contralateral subcortical neurons.

References

1. Suga, N.: Biosonar and neural computation in bats. Sci. Amer. 262, 60–68 (1990)
2. Suga, N.: Echolocation: Choleotopic and computational maps. In: Arbib, M.A. (ed.) The Handbook of Brain Theory and Neural Networks, 2nd edn., pp. 381–387. MIT Press (2003)
3. Suga, N., Manabe, T.: Neural bases of amplitude-spectrum representation in auditory cortex of the mustached bat. J. Neurophysiol. 47, 225–255 (1982)
4. Xiao, Z., Suga, N.: Reorganization of chocleotopic map in the bat's auditory system by inhibition. Proc. Natl. Acad. Sci. USA 99, 15743–15748 (2002)
5. Suga, N., Jen, P.: Peripheral control of acoustic signals in the auditory system of echolocating bats. J. Exp. Biol. 62, 277–311 (1975)
6. Suga, N., Niwa, H., Taniguchi, I., Margoliash, D.: The personalized auditory cortex of the mustached bat: adaptation for echolocation. J. Neurophysiol. 58, 643–654 (1987)
7. Suga, N.: Amplitude spectrum representation in the Doppler-shifted-CF processing area of the auditory cortex of the mustache bat. Science 196, 64–67 (1977)
8. Liu, W., Suga, N.: Binaural and commissural organization of the primary auditory cortex of the mustached bat. J. Comp. Physiol. A 81, 599–605 (1997)
9. Koch, C.: Biophysics of Computation. Oxford Univ. Press, New York (1999)
10. Nagase, Y., Kashimori, Y.: Modulation of Corticofugal Signals by Synaptic Changes in Bat's Auditory System. In: Wong, K.W., Mendis, B.S.U., Bouzerdoum, A. (eds.) ICONIP 2010, Part I. LNCS, vol. 6443, pp. 124–131. Springer, Heidelberg (2010)

Naive Bayesian Multistep Speaker Recognition Using Competitive Associative Nets

Shuichi Kurogi, Shota Mineishi, Tomohiro Tsukazaki, and Takeshi Nishida

Kyusyu Institute of technology, Tobata, Kitakyushu,
Fukuoka 804-8550, Japan
{kuro@,mineishi@kurolab.}cntl.kyutech.ac.jp
http://kurolab.cntl.kyutech.ac.jp/

Abstract. This paper presents a method of multistep speaker recognition using naive Bayesian inference and competitive associative nets (CAN2s). We have been examining a method of speaker recognition using feature vectors of pole distribution extracted by the bagging CAN2, where the CAN2 is a neural net for learning piecewise linear approximation of nonlinear function, and bagging CAN2 is the bagging (bootstrap aggregating) version. In order to reduce the recognition error, we formulate a multistep recognition using naive Bayesian inference. After introducing several modifications for reasonable recognition, we show the effectiveness of the present method by means of sereral experiments using real speech signals.

Keywords: Multistep speaker recognition, Bayesian inference, Competitive associative net.

1 Introduction

This paper describes a method of multistep speaker recognition using Bayesian inference and competitive associative nets (CAN2s). Here, the CAN2 is an artificial neural net for learning efficient piecewise linear approximation of nonlinear function by means of using competitive and associative schemes [1–6]. The effectiveness has been shown in several applications involving learning and analyzing speech signals. We have shown that the speech time-series is reproduced and recognized with high precision by the bagging (bootstrap aggregating) version of the CAN2 [4]. Recently, we have shown that the poles of piecewise linear predictive coefficients obtained by the bagging CAN2 are effective for speaker recognition [6, 7]. Here, note that among the previous research studies of speaker recognition, the most common way to characterize the speech signal is short-time spectral analysis, such as Linear Prediction Coding (LPC) and Mel-Frequency Cepstrum Coefficients (MFCC) [9–12]. Namely, these methods extract multi-dimensional features from each of consecutive intervals of speech, where a speech interval spans 10-30ms of the speech signal which is called a frame of speech. Thus, a single feature vector of the LPC or the MFCC corresponds to the average of multiple piecewise linear predictive coefficients of the bagging CAN2. Namely, the

B.-L. Lu, L. Zhang, and J. Kwok (Eds.): ICONIP 2011, Part I, LNCS 7062, pp. 70–78, 2011.

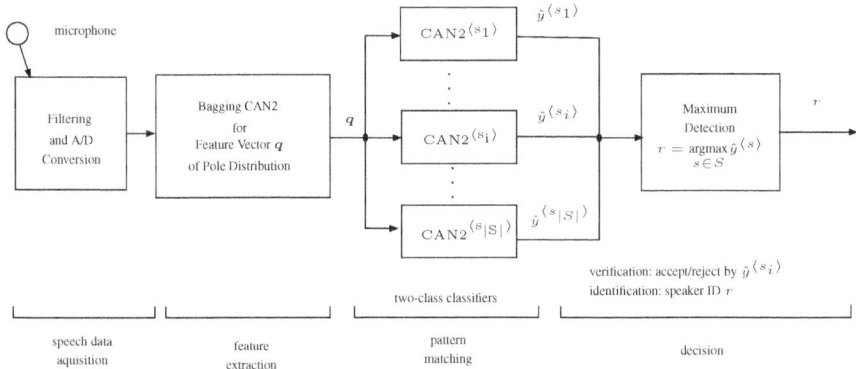

Fig. 1. Speaker recognition system using the CAN2s

bagging CAN2 has stored more precise information on the speech signal so that it can reproduce speech signal with high precision.

In our most recent research [7], we have shown a method of speaker recognition using the feature vector of pole distribution of piecewise linear coefficients extracted by the bagging CAN2, and presented a simple example of multistep recognition using Bayes' rule to reduce the recognition error. This paper focuses on the latter topic. Namely, we formulate the Bayesian multistep recognition in **2**, and show the performance using real speech data in **3**.

2   Naive Bayesian Multistep Speaker Recognition Using CAN2

2.1   Overview of Speaker Recognition

Fig. 1 shows the present speaker recognition system using the CAN2s. The speaker recognition system, in general, consists of four steps: speech data acquisition, feature extraction, pattern matching, and making a decision. Furthermore, the speaker recognition is classified into verification and identification, where the former is the process of accepting or rejecting the identity claim of a speaker, which is regarded as two-class classification. The latter, on the other hand, is the process of determining which registered speaker provides a given utterance, which is regarded as multi-class classification. In addition, speaker recognition has two schemes: text-dependent and text-independent schemes. The former require the speaker to say key words or sentences with the same text for both training and recognition phases, whereas the latter do not rely on a specific text being spoken.

2.2   Singlestep Speaker Recognition

In this study, we use a feature vector of pole distribution obtained from a speech signal (see [7] for details). Let $Q^{\langle s \rangle}$ be a set of feature vectors $q = (q_1, q_2, \cdots, q_k)^T$ of a speech signal from a speaker $s \in S = \{s_i | i \in I_S\}$, where $I_S = \{1, 2, \cdots, |S|\}$. We use a learning machine called CAN2, which learns to approximate the following target function:

$$f^{\langle s \rangle}(q) = \begin{cases} 1, & \text{if } q \in Q^{\langle s \rangle}, \\ -1, & \text{otherwise.} \end{cases} \tag{1}$$

Let $\text{CAN2}^{\langle s \rangle}$ be the learning machine for the speaker s. Then, with a number of training data $(q, f^{\langle s \rangle}(q))$ for $q \in Q^{\langle s \rangle}$, we train $\text{CAN2}^{\langle s \rangle}$ to approximate the above function by a continuous function as $\hat{y}^{\langle s \rangle} = \hat{f}^{\langle s \rangle}(q)$. So, we can execute a singlestep verification with the binarization of the output given as

$$v^{\langle s \rangle} = \begin{cases} 1, & \text{if } \hat{y}^{\langle s \rangle} = \hat{f}^{\langle s \rangle}(q) \geq y_\theta, \\ -1, & \text{otherwise.} \end{cases} \tag{2}$$

Namely, we accept the speaker s_i if $v^{\langle s \rangle} = 1$, and reject otherwise. Here, a threshold y_θ is introduced for adjusting the recognition performance shown below. We execute a singlestep identification by the maximum detection given by

$$r = \underset{i \in I_S}{\text{argmax}} \; \{\hat{y}^{\langle s_i \rangle} = \hat{f}^{\langle s_i \rangle}(q)\}. \tag{3}$$

Namely, we identify the speech signal is of the rth speaker with the above r.

2.3   Naive Bayesian Multistep Speaker Recognition

For speaker verification, let $p_V(v^{\langle s_i \rangle} | s)$ be the probability of the output $v^{\langle s_i \rangle}$ of $\text{CAN2}^{\langle s_i \rangle}$ to classify the feature vector q of a speaker s. Here, note that $p_V(v^{\langle s_i \rangle} = 1 | s) + p_V(v^{\langle s_i \rangle} = -1 | s) = 1$ for every s_i and s in S. Let $v_{1:t}^{\langle s_i \rangle} = v_1^{\langle s_i \rangle}, \cdots, v_t^{\langle s_i \rangle}$ be an output sequence of $\text{CAN2}^{\langle s_i \rangle}$ for a sequence of feature vectors obtained from a speaker s. Then we estimate whether the speaker s is s_i or $\bar{s}_i$ by means of naive Bayesian inference given by

$$p_V(s | v_{1:t}^{\langle s_i \rangle}) = \frac{p_V(v_t^{\langle s_i \rangle} | s) p_V(s | v_{1:t-1}^{\langle s_i \rangle})}{p_V(v_t^{\langle s_i \rangle} | s) p_V(s | v_{1:t-1}^{\langle s_i \rangle}) + p_V(v_t^{\langle s_i \rangle} | \bar{s}) p_V(\bar{s} | v_{1:t-1}^{\langle s_i \rangle})} \tag{4}$$

where $p_V(s | v_{1:t}^{\langle s_i \rangle}) + p_V(\bar{s} | v_{1:t}^{\langle s_i \rangle}) = 1$ and $s = s_i$ or $\bar{s}_i$. Here, we employ conditional independence assumption given by $p_V(v_t^{\langle s_i \rangle} | s, v_{1:t-1}^{\langle s_i \rangle}) = p_V(v_t^{\langle s_i \rangle} | s)$, which is shown effective in many real world applications of naive Bayes classifier [8]. We execute the Bayesian speaker verification as follows; we accept the speaker s_i when $p_V(s = s_i | v_{1:t}^{\langle s_i \rangle}) \geq p_{V\theta}$ is satisfied at $t (= t_V)$ for the increase of t until $t = T_V$, and reject otherwise. Here, $p_{V\theta}$ and T_V are constants.

   For speaker identification, let $p_I(v^{\langle S \rangle} | s) = p_I(v^{\langle s_1 \rangle}, \cdots, v^{\langle s_{|S|} \rangle} | s)$ be the joint probability of $v^{\langle S \rangle} = (v^{\langle s_1 \rangle}, \cdots, v^{\langle s_{|S|} \rangle})$ responding to a feature vector q obtained from a speaker $s \in S$. We assume $p_{\mathrm{I}}(\boldsymbol{v}^{\langle S \rangle}|s) = \prod_{s_i \in S} p_{\mathrm{I}}(v^{\langle s_i \rangle}|s)$ because any two probabilities $p_{\mathrm{I}}(v^{\langle s_i \rangle}|s)$ and $p_{\mathrm{I}}(v^{\langle s_j \rangle}|s)$ for $i \neq j$ are supposed to be independent. Let $\boldsymbol{v}_{1:t}^{\langle S \rangle} = \boldsymbol{v}_1^{\langle S \rangle}, \cdots, \boldsymbol{v}_t^{\langle S \rangle}$ be a sequence of $\boldsymbol{v}^{\langle S \rangle}$ obtained from a speaker s, then we have the Bayesian inference given by

$$p_{\mathrm{I}}(s|\boldsymbol{v}_{1:t}^{\langle S \rangle}) = \frac{p_{\mathrm{I}}(\boldsymbol{v}_t^{\langle S \rangle}|s)p_{\mathrm{I}}(s|\boldsymbol{v}_{1:t-1}^{\langle S \rangle})}{\sum_{s_i \in S} p_{\mathrm{I}}(\boldsymbol{v}_t^{\langle S \rangle}|s_i)p_{\mathrm{I}}(s_i|\boldsymbol{v}_{1:t-1}^{\langle S \rangle})}. \tag{5}$$

where we use conditional independence assumption $p_{\mathrm{I}}(\boldsymbol{v}_t^{\langle S \rangle}|s, \boldsymbol{v}_{1:t-1}^{\langle S \rangle}) = p_{\mathrm{I}}(\boldsymbol{v}_t^{\langle S \rangle}|s)$. We examine two cases to execute the multistep speaker identification as follows;

Case I_1: We identify the rth speaker holding $r = \operatorname*{argmax}_{i \in I_S} p_{\mathrm{I}}(s_i|\boldsymbol{v}_{1:t}^{\langle S \rangle})$ at $t = T_1$, where T_1 is a constant (we set $T_1 = 7$ in the experiments shown below).

Case I_2: We identify the rth speaker at $t (= t_1)$ when $p_{\mathrm{I}}(s = s_r|\boldsymbol{v}_{1:t}^{\langle S \rangle}) \geq p_{\mathrm{I}\theta}$ is satisfied for the increase of t until T_1, where $p_{\mathrm{I}\theta}$ is a threshold. In this case, there is a possibility that no speaker is identified.

2.4   Problems and Modifications for Reasonable Recognition

First of all, let us introduce the following error metrics to evaluate the performance of the classifiers and the recognition.

$$E_{\mathrm{FN}}(s_i) \triangleq p_{\mathrm{V}}\left(v^{\langle s_i \rangle} = -1 \mid s_i\right), \tag{6}$$

$$E_{\mathrm{FP}}(s_i) \triangleq p_{\mathrm{V}}\left(v^{\langle s_i \rangle} = 1 \mid \bar{s}_i\right), \tag{7}$$

$$E_{\mathrm{V}} \triangleq \frac{1}{|S|} \sum_{s_i \in S} \left(p_{\mathrm{V}}\left(v^{\langle s_i \rangle} = -1 \mid s_i\right) + \sum_{s_j \in S \setminus \{s_i\}} p_{\mathrm{V}}\left(v^{\langle s_j \rangle} = 1 \mid s_i\right) \right), \tag{8}$$

$$E_{\mathrm{I}} \triangleq \frac{1}{|S|} \sum_{s_i \in S} \sum_{s_i \in S} p_{\mathrm{I}}(\boldsymbol{v}^{\langle S \rangle} \neq \boldsymbol{v}_i^*|s_i) \tag{9}$$

where FN and FP represent false negative and false positive, respectively, often used in binary classification problems, and $\boldsymbol{v}_i^*$ is the desired vector whose ith element is 1 and the other elements are -1,

Modification of the original probability distribution: We have to estimate the original probability distribution as $\widehat{p}(v^{\langle s_i \rangle}|s)$ from a limited number of speech data. Then, some estimated probabilities may happen to become 0 or 1, which leads to a stuck of the multistep inference by Eq.(4) and Eq.(5). So, we employ the modification given by $\widehat{p}(v^{\langle s_i \rangle}|s) := (1-2p_0)\widehat{p}(v^{\langle s_i \rangle}|s)+p_0$ in order for the new $\widehat{p}(v^{\langle s_i \rangle}|s)$ to be in $[p_0, 1-p_0]$, where p_0 is a small constant (we use $p_0 = 0.05$ in the experiments shown below).

Modification of multistep probability: Similarly as above, we can see that $p_V(s|v_{1:t}^{\langle s_i \rangle})$ and $p_l(s|v_{1:t}^{\langle S \rangle})$ once become 0 or 1, they will not change any more. So we truncate the probability to be in $[p_1, 1 - p_1]$, where p_1 is a constant (we set $p_1 = 0.01$ in the experiments shown blow).

Tuning the threshold of binary classifiers: By means of the threshold y_θ in Eq.(2), we can tune the ratio of FN and FP, because the original learning machine, i.e. CAN2, is an approximator of continuous function. We set y_θ to be the center of the mean output of the CAN2s for true positives (TP) and that of true negatives (TN). Note that we could not have examined this strategy in detail so far, this tuning may largely affect the performance of the recognition as shown below.

Introducing "void" speaker: We introduce "void" speaker for reasonable identification. Namely, there is a case where no classifier, i.e. $\mathrm{CAN2}^{\langle s_i \rangle}$, provides positive output. In such a case, we might have to do an exceptional processing because the above inference method does not consider such cases. Here, we define the probability for the void speaker by the mean verification error as follows:

$$p_l(v^{\langle S \rangle}|s) = \begin{cases} E_V & \text{if } v^{\langle S \rangle} \neq v_{\text{void}}^* \wedge s = \text{void}, \\ E_V & \text{if } v^{\langle S \rangle} = v_{\text{void}}^* \wedge s \neq \text{void}. \\ 1 - E_V & \text{if } v^{\langle S \rangle} = v_{\text{void}}^* \wedge s = \text{void}, \\ 1 - E_V & \text{if } v^{\langle S \rangle} \neq v_{\text{void}}^* \wedge s \neq \text{void}, \end{cases} \tag{10}$$

3   Experimental Results and Remarks

3.1   Experimental Setting

We have used speech signals sampled with 8kHz of sampling rate and 16 bits of resolution in a silent room of our laboratory. They are from five male speakers: $S =\{$SM, SS, TN, WK, YM$\}$. We have examined five texts of Japanese words: $W =\{$/kyukodai/, /daigaku/, /kikai/, /fukuokaken/, /gakusei/$\}$ where each utterance duration of the words is about 1s. For each speaker and each text, we have ten samples of speech data, $L = \{1, 2, \cdots, 10\}$. Namely, we have speech data $x = x_{s,w,l}$ for $s \in S$, $w \in W$ and $l \in L$.

In order to evaluate the performance of the present method, we use the leave-one-set-out cross-validation (LOOCV). Precisely, for text-dependent tasks, we evaluate the performance with test dataset $X(S, w, l) = \{x_{s,w,l} \mid s \in S\}$ and training dataset $X(S, w, L_{\bar{l}}) = \{x_{s,w,i} \mid s \in S, i \in L\backslash\{l\}\}$ for each $w \in W$ and $l \in L$. On the other hand, for text-independent tasks, we use test dataset $X(S, w, l)$ and training dataset $X(S, W_{\bar{w}}, L) = \{x_{s,u,i} \mid s \in S, u \in W\backslash\{w\}, i \in L\}$ for each $w \in W$ and $l \in L$.

3.2   Experimental Results

We have conducted experiments for each text and we have the error rate shown in Table 1. From (a) for text-dependent recognition, we can see that the original singlestep

Table 1. Error rates, E_V and E_I, and the mean of decision step numbers, $\langle t_V \rangle$ and $\langle t_I \rangle$, of speaker recognition for each text. The superscripts "ss" and "ms" of the error indicate the singlestep (original) and multistep (Bayesian) recognition, respectively. We terminate the Bayesian steps at $t = T_V = T_I = 7$. The result is obtained with $p_{V\theta} = p_{I\theta} = 0.95$.

(a) text-dependent speaker recognition

	/kyukodai/	/daigaku/	/kikai/	/fukuokaken/	/gakusei/
E_V^{ss}	0.028 (7/250)	0.020 (5/250)	0.032 (8/250)	0.044 (11/250)	0.016 (4/250)
E_I^{ss}	0.080 (4/50)	0.060 (3/50)	0.060 (3/50)	0.100 (5/50)	0.040 (2/50)
E_V^{ms}	0.000 (0/25)	0.000 (0/25)	0.000 (0/25)	0.000 (0/25)	0.000 (0/25)
$\langle t_V \rangle$	2.2	2.0	2.4	2.6	2.0
$E_{I_1}^{ms}$	0.000 (0/5)	0.000 (0/5)	0.000 (0/5)	0.000 (0/5)	0.000 (0/5)
$E_{I_2}^{ms}$	0.000 (0/25)	0.000 (0/25)	0.000 (0/25)	0.000 (0/25)	0.000 (0/25)
$\langle t_I \rangle$	1.8	1.6	1.8	2.2	1.4

(b) text-independent speaker recognition

	/kyukodai/	/daigaku/	/kikai/	/fukuokaken/	/gakusei/
E_V^{ss}	0.092 (23/250)	0.056 (14/250)	0.096 (24/250)	0.140 (35/250)	0.132 (33/250)
E_I^{ss}	0.200 (10/50)	0.200 (10/50)	0.280 (14/50)	0.400 (20/50)	0.420 (21/50)
E_V^{ms}	0.000 (0/25)	0.040 (1/25)	0.000 (0/25)	0.160 (4/25)	0.040 (1/25)
$\langle t_V \rangle$	3.2	2.2	2.6	3.6	4.0
$E_{I_1}^{ms}$	0.000 (0/5)	0.000 (0/5)	0.000 (0/5)	0.000 (0/5)	0.200 (1/5)
$E_{I_2}^{ms}$	0.000 (0/25)	0.040 (1/25)	0.000 (0/25)	0.000 (0/25)	0.080 (1/25)
$\langle t_I \rangle$	3.0	2.2	2.6	3.0	3.0

method achieves not zero but small error rate, while the Bayesian multistep method has achieved correct verification and identification for all texts. On the other hand, from (b) for text-independent speaker recognition, we can see that the error of the singlestep method is bigger than the error of the above text-dependent case, which causes the error of the multistep method. However, the multistep method has smaller error than the singlespep method in almost all cases, and achieved correct verification and identification for /kyukodai/ and /kikai/.

The mean of the decision step numbers, $\langle t_V \rangle$ for verification and $\langle t_I \rangle$ for "**Case I_2**" identification, are less than or equal to 4 steps, which may be reasonable in real applications of the present method. Now, let us examine the errors in detail for the text-independent recognition of the text /gakusei/. We show the state variable values through the recognition steps in Table 2 and Fig. 2 to see the results numerically and visually. From the table, we can see that the identification error is occured for the speech signal of WK, wher ethe Bayesian probability p_V and p_I do not reach the threshold $p_{V\theta} = p_{I\theta} = 0.95$. From the table and figure, it is supposed that the reduction of FN (false negative) of CAN2$^{\langle WK \rangle}$ as well as FP (false positive) of CAN2$^{\langle TN \rangle}$ is necessary for eliminating this error. Actually, we can see this fact from the table of (E_{FN}, E_{FP}) shown in Fig. 3. Although it is not so easy to solve this problem, it is important and valuable to know this causality.

Table 2. State variable values in text-independent speaker recognition for /gakusei/. The second row indicates $\mathrm{CAN2}^{\langle s \rangle}$ for s=SM, SS, TN, WK, YM (and void for p_I), and each row below the second shows the responses to the tth speech signal of speakers shown on the left. Percentage values are shown for p_V and p_I.

| | t | y_i | | | | | v_i | | | | | p_V | | | | | p_I | | | | | |
		SM	SS	TN	WK	YM	SM	SS	TN	WK	YM	SM	SS	TN	WK	YM	SM	SS	TN	WK	YM	void
SM	0	–	–	–	–	–	–	–	–	–	–	50	50	50	50	50	17	17	17	17	17	17
	1	+0.30	-0.90	-0.93	-0.89	-0.36	+1	-1	-1	-1	-1	83	14	23	41	5	82	1	6	5	1	6
	2	-0.37	-0.77	-0.93	-0.73	+0.30	-1	-1	-1	+1	+1	59	3	8	66	41	96	1	1	1	2	1
	3	-0.26	-0.92	-0.96	-0.75	+0.08	-1	-1	-1	+1	-1	30	1	3	85	4	96	1	1	3	1	1
	4	+0.19	-0.92	-0.95	-0.84	-0.17	+1	-1	-1	-1	-1	67	1	1	80	1	99	1	1	1	1	1
	5	-0.00	-0.84	-0.96	-0.72	-0.19	+1	-1	-1	+1	-1	90	1	1	92	1	99	1	1	1	1	1
	6	+0.28	-0.91	-0.93	-0.82	-0.42	+1	-1	-1	-1	-1	98	1	1	89	1	99	1	1	1	1	1
	7	+0.11	-0.98	-0.97	-0.85	-0.10	+1	-1	-1	-1	-1	99	1	1	84	1	99	1	1	1	1	1
SS	0	–	–	–	–	–	–	–	–	–	–	50	50	50	50	50	17	17	17	17	17	17
	1	+0.05	-0.22	-0.69	-0.98	-0.96	+1	+1	-1	-1	-1	83	95	23	41	5	10	88	1	1	1	1
	2	-0.56	+0.21	-0.53	-0.97	-0.98	-1	+1	-1	-1	-1	59	99	8	33	1	1	99	1	1	1	1
	3	-0.68	-0.23	-0.06	-0.95	-0.93	-1	+1	+1	-1	-1	30	99	19	25	1	1	99	1	1	1	1
	4	-0.52	-0.15	-0.81	-0.89	-0.78	-1	+1	-1	-1	-1	11	99	6	19	1	1	99	1	1	1	1
	5	-0.58	-0.65	+0.18	-0.97	-1.00	-1	-1	+1	-1	-1	4	94	15	14	1	1	84	11	4	1	1
	6	-0.56	-0.29	-0.07	-0.96	-0.95	-1	+1	-1	-1	-1	1	99	32	10	1	1	99	1	1	1	1
	7	-0.46	-0.43	-0.08	-0.94	-0.95	-1	+1	+1	-1	-1	1	99	54	7	1	1	99	1	1	1	1
TN	0	–	–	–	–	–	–	–	–	–	–	50	50	50	50	50	17	17	17	17	17	17
	1	-0.24	-0.75	+0.10	-0.95	-0.92	-1	-1	+1	-1	-1	23	14	72	41	5	1	5	65	21	1	8
	2	-0.55	-0.93	+0.54	-0.86	-0.86	-1	-1	+1	-1	-1	8	3	87	33	1	1	1	89	9	1	1
	3	-0.04	-0.94	-0.43	-0.70	-0.85	+1	-1	-1	+1	-1	30	1	67	58	1	20	1	51	28	1	1
	4	-0.45	-0.67	-0.11	-0.93	-0.89	-1	-1	+1	-1	-1	11	1	84	49	1	1	1	85	15	1	1
	5	-0.78	-0.81	-0.08	-0.75	-0.36	-1	-1	+1	+1	-1	4	1	93	73	1	1	1	74	25	1	1
	6	-0.68	-0.73	+0.06	-0.87	-0.29	-1	-1	+1	-1	-1	1	1	97	65	1	1	1	90	10	1	1
	7	-0.57	-0.61	-0.22	-0.79	-0.84	-1	-1	+1	-1	-1	1	1	99	57	1	1	1	96	3	1	1
WK	0	–	–	–	–	–	–	–	–	–	–	50	50	50	50	50	17	17	17	17	17	17
	1	-0.07	-0.73	-0.60	-0.77	-0.59	+1	-1	-1	-1	-1	83	14	23	41	5	82	1	6	5	1	6
	2	-0.60	-0.94	-0.08	-0.54	-0.61	-1	-1	+1	+1	-1	59	3	44	66	1	3	1	35	58	1	5
	3	-0.63	-0.94	+0.04	-0.87	-0.42	-1	-1	+1	-1	-1	30	1	67	58	1	1	1	64	35	1	1
	4	-0.49	-0.85	-0.23	-0.53	-0.82	-1	-1	+1	+1	-1	11	1	84	80	1	1	1	49	51	1	1
	5	-0.97	-0.95	+0.39	-0.78	-0.70	-1	-1	+1	-1	-1	4	1	93	73	1	1	1	74	25	1	1
	6	-0.09	-0.99	-0.84	-0.90	-0.02	+1	-1	-1	-1	-1	15	1	80	65	1	12	1	67	20	1	1
	7	-0.22	-0.93	-0.21	-0.35	-0.91	-1	-1	+1	+1	-1	5	1	91	84	1	1	1	63	36	1	1
YM	0	–	–	–	–	–	–	–	–	–	–	50	50	50	50	50	17	17	17	17	17	17
	1	-0.95	-0.96	-0.98	-0.87	+0.81	-1	-1	-1	-1	+1	23	14	23	41	93	1	1	1	1	98	1
	2	-0.94	-0.99	-0.99	-0.96	+0.95	-1	-1	-1	-1	+1	8	3	8	33	99	1	1	1	1	99	1
	3	-0.92	-0.99	-0.97	-0.91	+0.95	-1	-1	-1	-1	+1	3	1	3	25	99	1	1	1	1	99	1
	4	-0.93	-0.99	-0.99	-0.94	+0.83	-1	-1	-1	-1	+1	1	1	1	19	99	1	1	1	1	99	1
	5	-0.84	-0.90	-0.94	-0.96	+0.79	-1	-1	-1	-1	+1	1	1	1	14	99	1	1	1	1	99	1
	6	-0.88	-0.75	-0.98	-0.92	+0.78	-1	-1	-1	-1	+1	1	1	1	10	99	1	1	1	1	99	1
	7	-0.79	-0.95	-0.99	-0.97	+0.83	-1	-1	-1	-1	+1	1	1	1	7	99	1	1	1	1	99	1

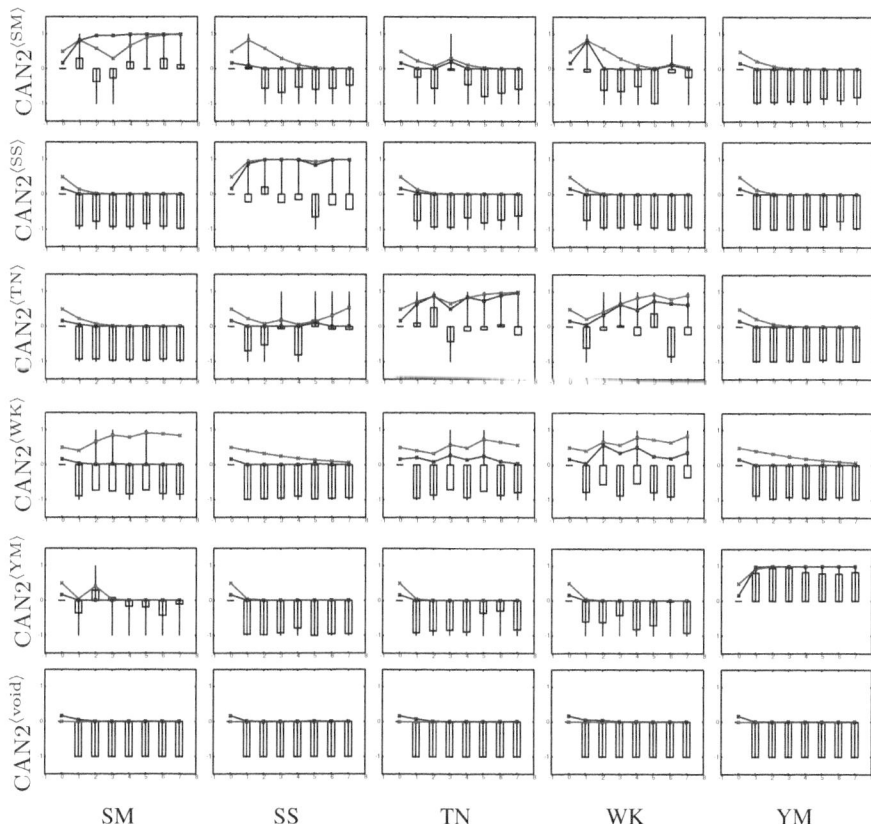

Fig. 2. Experimental result of text-independent speaker recognition for /gakusei/. The black boxes indicate y_i by its height, the black unit impulses v_i, the red broken line p_V and the blue broken lines p_I.

Table 3. FN and FP error rate, (E_{FN}, E_{FP}) [%], of the classifiers

	/kyukodai/	/daigaku/	/kikai/	/fukuokaken/	/gakusei/
CAN2$^{\langle SM \rangle}$	(30.0, 2.5)	(30.0, 5.0)	(0.0, 2.5)	(0.0, 20.0)	(20.0, 12.5)
CAN2$^{\langle SS \rangle}$	(20.0, 2.5)	(0.0, 5.0)	(40.0, 15.0)	(30.0, 22.5)	(10.0, 0.0)
CAN2$^{\langle TN \rangle}$	(10.0, 7.5)	(10.0, 5.0)	(20.0, 10.0)	(20.0, 7.5)	(20.0, 27.5)
CAN2$^{\langle WK \rangle}$	(20.0, 15.0)	(10.0, 5.0)	(10.0, 5.0)	(20.0, 10.0)	(40.0, 17.5)
CAN2$^{\langle YM \rangle}$	(10.0, 7.5)	(0.0, 2.5)	(10.0, 7.5)	(10.0, 7.5)	(0.0, 2.5)

4   Conclusion

We have formulated a method of multistep speaker recognition using naive Bayesian inference and CAN2s. After introducing several modifications for reasonable recognition, we have shown the effectiveness of the present method using real speech signals. Namely, the error rate of the single step method is reduced by the present multistep method. In order to reduce the multistep recognition error much more, we have to reduce both FN and FP errors of the original classifiers, i.e. CAN2s, for all texts.

This work was partially supported by the Grant-in Aid for Scientific Research (C) 21500217 of the Japanese Ministry of Education, Science, Sports and Culture.

References

1. Ahalt, A.C., Krishnamurthy, A.K., Chen, P., Melton, D.E.: Competitive learning algorithms for vector quantization. Neural Networks 3, 277–290 (1990)
2. Kohonen, T.: Associative Memory. Springer, Heidelberg (1977)
3. Kurogi, S., Ueno, T., Sawa, M.: A batch learning method for competitive associative net and its application to function approximation. In: Proc. SCI 2004, vol. V, pp. 24–28 (2004)
4. Kurogi, S., Nedachi, N., Funatsu, Y.: Reproduction and recognition of vowel signals using single and bagging competitive associative nets. In: Ishikawa, M., Doya, K., Miyamoto, H., Yamakawa, T. (eds.) ICONIP 2007, Part II. LNCS, vol. 4985, pp. 40–49. Springer, Heidelberg (2008)
5. Kurogi, S.: Improving generalization performance via out-of-bag estimate using variable size of bags. J. Japanese Neural Network Society 16(2), 81–92 (2009)
6. Kurogi, S., Sato, S., Ichimaru, K.: Speaker recognition using pole distribution of speech signals obtained by bagging CAN2. In: Leung, C.S., Lee, M., Chan, J.H. (eds.) ICONIP 2009. LNCS, vol. 5863, pp. 622–629. Springer, Heidelberg (2009)
7. Kurogi, S., Mineishi, S., Sato, S.: An analysis of speaker recognition using bagging CAN2 and pole distribution of speech signals. In: Wong, K.W., Mendis, B.S.U., Bouzerdoum, A. (eds.) ICONIP 2010, Part I. LNCS, vol. 6443, pp. 363–370. Springer, Heidelberg (2010)
8. Zhang, H.: The optimality of naive Bayes. In: Proc. FLAIRS 2004 Conference (2004)
9. Campbell, J.P.: Speaker Recognition: A Tutorial. Proc. the IEEE 85(9), 1437–1462 (1997)
10. Furui, S.: Speaker Recognition. In: Cole, R., Mariani, J., et al. (eds.) Survey of the state of the art in human language technology, pp. 36–42. Cambridge University Press (1998)
11. Hasan, M.R., Jamil, M., Rabbani, M.G., Rahman, M.S.: Speaker identification using Mel frequency cepstral coefficients. In: Proc. ICECE 2004, pp. 565–568 (2004)
12. Bocklet, T., Shriberg, E.: Speaker recognition using syllable-based constraints for cepstral frame selection. In: Proc. ICASSP (2009)

Medial Axis for 3D Shape Representation

Wei Qiu and Ko Sakai

Graduate School of Systems and Information Engineering, University of Tsukuba,
1-1-1 Tennodai, Tsukuba, Ibaraki, 305-8573 Japan
kyu@cvs.cs.tsukuba.ac.jp, sakai@cs.tsukuba.ac.jp
http://www.cvs.cs.tsukuba.ac.jp/

Abstract. Cortical representation of shape is a crucial problem in vision science. Recent physiological studies on monkeys have reported that neurons in the primary visual cortex (V1) represent 2D shape by Medial Axis (MA). Physiology has also shown that a set of smooth surfaces represents 3D shape in a higher stage (IT). Based on the physiological evidence, we propose that monocular retinal images yield 2D-MAs that represent 2D-surfaces in V1, and the 2D-MAs are fused to yield 3D-MA that represents 3D-surfaces in IT. To investigate this hypothesis, we developed a computational model based on the physiological constraints, and evaluated its power on the shape encoding. The model represented a variety of 3D-shapes including natural shapes, with reconstruction errors of around 0.2 regardless of the shape complexity. The results support the visual system encodes monocular 2D-MAs in V1 and fuses them into 3D-MA in IT so that 3D-shape is represented by smooth surfaces.

Keywords: Medial axis, MA, Shape representation, 3D, Binocular disparity.

1   Introduction

Symmetry is an important queue for shape perception. According to physiological, psychological and computational studies [e.g., 1,2,3], Medial Axis (MA) that encodes shape using local symmetry axis plays an important role for object representation. A recent physiological study has shown that neurons in higher-order visual areas are tuned to 3D spatial configuration that appears to be the basis for the representation of 3D shape [4]. However, it has not been clarified how the visual system fuses 2D information to translate them into 3D shape.

We see the world with left and right eyes that are horizontally separated by about 60-70mm. Because of the separation, the retinal images of 3D objects are slightly different. Our brain uses this difference that is called binocular disparity to perceive depth of objects. A numerous physiological studies have reported that a majority of neurons in the primary visual cortex (V1) show selectivity for binocular disparity within their small receptive-fields. This local disparity is integrated to represent global depth in higher-level visual areas. This process appears to be crucial in the representation of 3D shape [5]. Neurons in higher-level visual areas such as inferior temporal cortex (IT) have been reported to represent 3D surfaces even if only binocular disparity along contour is provided [4, 6], though disparity along contour may not be capable of representing surfaces because contour-based disparity cannot

B.-L. Lu, L. Zhang, and J. Kwok (Eds.): ICONIP 2011, Part I, LNCS 7062, pp. 79–87, 2011.

encode the surface inside the contours – surface texture is also needed. It has not been clarified how binocular disparities in V1 are converted into surfaces, and further to a 3D shape.

In the present study, we investigated cortical representation of 3D shape with specific interests on the transformation of 2D representation of shape into 3D representation. We propose that 2D-MAs are computed independently for the left and right retinal images in the early stage of visual pathway, and then 3D-MA is obtained from the fusion of these 2D-MAs. It means that 2D surfaces are formed for each eye and 3D shape is computed from the disparity of the 2D surfaces, not from the disparity of contours. If so, a smooth 3D shape could be formed from binocular disparity along contours. In this paper, in order to investigate the plausibility of this hypothesis, we developed a computational model based on the visual neural system, and observed its behavior. The model consists of two major stages: (1) that computes 2D-MAs from the left and right images, and (2) that computes the depth (3D-MA) from the disparity of the left and right 2D-MAs. We assumed that the detection of 2D-MA is based on the onset synchronization of border-ownership-selective neurons in V2, and that the size of the receptive field (including surrounding regions) is preserved simultaneously with the MA [3,7]. 3D-shape was reconstructed by overlapping spheres with their centers located along the 3D-MA and radii equal to the receptive-field size. As the first step toward examining this hypothesis, we carried out the simulations of the model with a variety of stimuli with distinct shapes, including natural objects, to investigate how plausible is the model in the representation of 3D objects from realistic 2D retinal images. We compared the reconstructed 3D-shape and the original. The results showed fairly federate reconstruction of 3D-shape from the fusion of 2D-MAs.

2   The Model and the Methods

The model consists of three stages. A schematic diagram of the model is shown in Fig.1. We gave a pair of stereo images as input stimuli. The representation of shape went through the following steps: (a) computing 2D-MAs from left and right images, (b) calculating depth from the disparity between the detected 2D-MAs to compute 3D-MA, and (c) measuring the distances between every point on the 3D-MA and the nearest point on a contour. For the reconstruction of shape, we overlapped a number of spheres along the 3D-MA with their radius equals to the measured distance.

2.1   Input Stimulus

We generated the left and right images of artificial objects and natural objects. The simple artificial shapes were generated by computer graphics (CG), and natural objects were taken by a 3D digital camera. CG images were convergent stereo images that were taken from two imaginary cameras located at equi-distance from the object center (Shade 11, E frontier). Natural images were parallel stereo images taken by FinePix REAL 3DW1 (Fujifilm). The images were 200x200 pixel with the range of disparity set between -200~200 pixel. We binarized the images with the inside of an object (figure) set to 1(white) and the outside 0 (black).

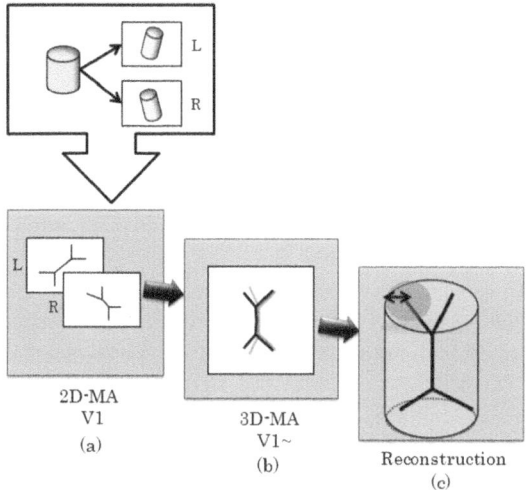

2D-MA
V1
(a)

3D-MA
V1~
(b)

Reconstruction
(c)

Fig. 1. An illustration of the model consisting of three stages

2.2 MA Detection

2D-MA is computed in Stage (a) as follows: generating a number of circles that contact with a contour at more than two points, and binding these centres (MATLAB, 2-D Medial Axis Computation Package). 3D-MA is computed in Stage (b) with three steps.

(1) The first step computes binocular disparity from corresponding feature-points in a conventional way:

$$disparity = \frac{x_r - x_l}{f}, \tag{1}$$

where (x_l, y_l) on the left image matches to (x_r, y_r) on the right, and f is the focal length. The ends and intersections of 2D-MAs were chosen as the feature points for the sake of simplicity. Note that the feature points were discarded if no correspondence was found on the other eye. Depth of each point (d) is determined by:

$$d^2 - (\frac{delta}{disparity} + 2 fix)d + fix^2 = 0, \tag{2}$$

where fix represents the distance from the fixation point, and $delta$ represents the distance between two eyes.

(2) The second step computes the coordinate of each feature-point in (x, y, z), which is defined as:

$$x = \frac{x_r + x_l}{2}, y = y_r (= y_l), z = \frac{d}{s}, \tag{3}$$

where s represents a scaling coefficient that approximates the ratio of the object size in an image with respect to the real size.

(3) The third step connects the feature-points by short lines to establish 3D-MA.

2.3   Reconstruction

We reconstructed a 3D shape from the computed 3D-MA to examine the capability of the model in 3D-shape representation. In Stage (c), we computed the distance from every point on the 3D-MA to a nearby 2D contour (on z=0), and overlapped spheres along the MA with their radius set to the measured distance. This approximation in distance appears to be valid since z=0 is set to the object center.

2.4   Reconstruction Error

In order to evaluate our algorithm quantitatively, we calculated the error in reconstruction. We defined the error as the difference between the original object ($I(x,y,z)$) and the reconstructed 3D-shape ($R(x,y,z)$):

$$Error = \frac{\sum_{x,y,z}[I(x,y,z) - R(x,y,z)]^2}{\sum_{x,y,z}[I(x,y,z) + R(x,y,z)]^2} \quad . \tag{4}$$

We divided $(I-R)^2$ by $(I+R)^2$ to normalize the error with respect to object volume.

3   Simulation Results

We carried out the simulations of the model with various stimuli to test whether the model is capable of representing 3D shape with 3D-MA. In this section, we present a few but typical examples of the simulation results.

3.1   Stimuli and Method

The stimuli included CG images of three typical shapes (Capsule, Cone and Torus) and two Natural images (Eggplant and Bear). The model computed 3D-MA of these stimuli. We reconstructed 3D shape from the computed MA, and determined the reconstruction error with respect to the original shape.

3.2   Results

First, we carried out the simulations of a capsule-shaped object that appears to be easily encoded by MA. Fig.2 shows the results for the frontal view of the object. 2D-MAs are slightly different between the left and right ones as expected. The computed MA is shown in the x-y plane (b-1) and in x-y-z (b-2) with the arrow indicating the direction of view. The computed depth (-Z) becomes farther from the viewpoint as the value of Y grows. The reconstructed shape is shown in the three sub-panels in (c) as viewed from three distinct directions (for a presentation purpose). The upside of the capsule in the original (input) image is tilted toward far off. This feature is similar to that of the reconstructed shape. The reconstruction error that was computed by equation (4) was 0.18, indicating fairly accurate representation of the shape.

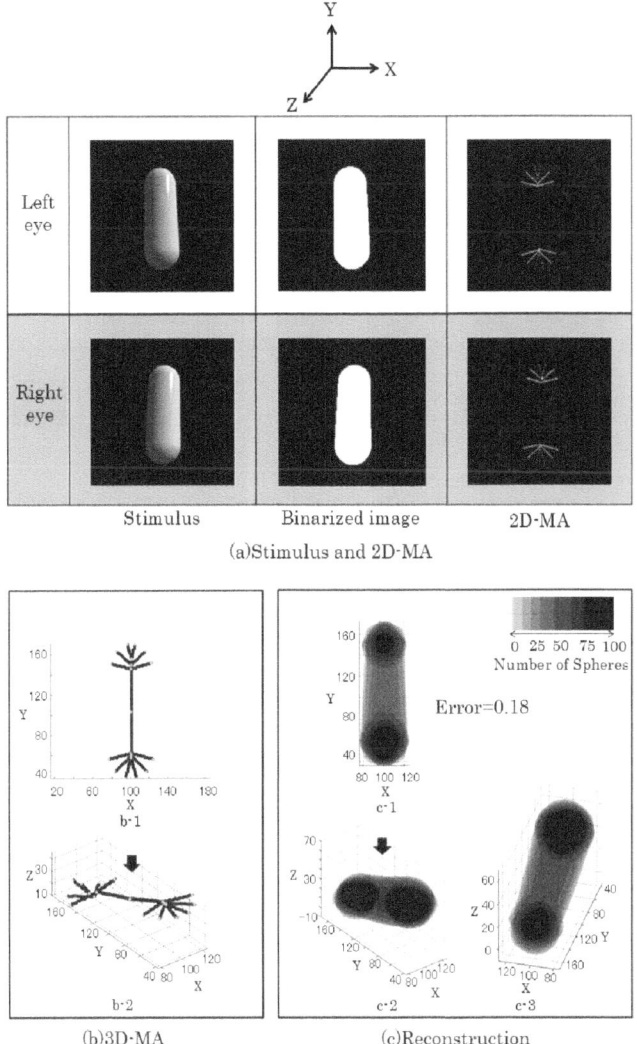

Fig. 2. The computation of 3D-MA and the shape reconstruction from it for Capsule (front view). Panel (a) shows the input images (left), the binary images (middle), and the computed 2D-MAs(right). Panel (b) shows the computed 3D-MA viewed from distinct directions. The input images were shot from Z direction (indicated by an arrow). Panel (c) shows the reconstructed shape represented by a number of overlapping spheres, viewed from distinct directions.

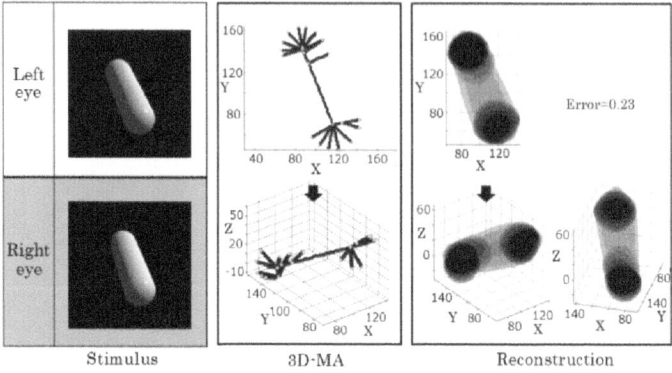

Fig. 3. The results for the Capsule (Side view). The conventions are the same as Fig. 2.

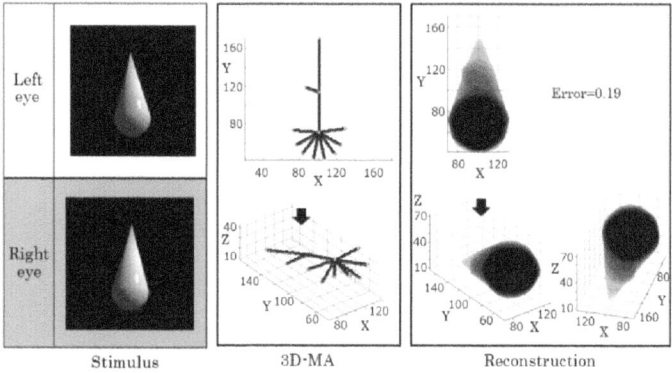

Fig. 4. The results for Cone. The conventions are the same as Fig. 2.

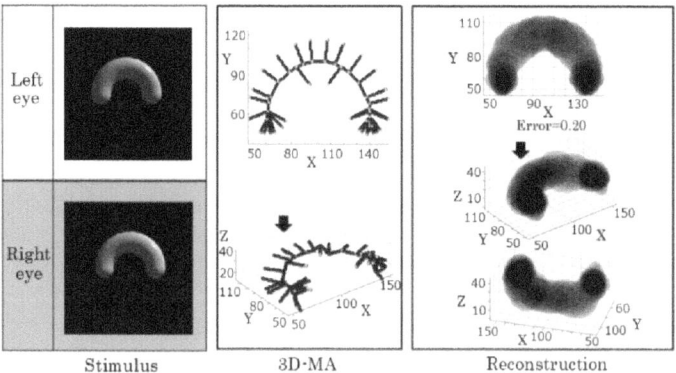

Fig. 5. The results for Torus. The conventions are the same as Fig. 2.

Fig.3 shows the results for a side-view of the same capsule-shaped stimulus, together with the computed 3D-MA and the reconstructed shape. We obtained shape and depth as similar to the frontal view, despite the distinct viewing direction. The reconstructed shape appears to correspond to the original one. The reconstruction error was 0.23. These results indicate that, using the disparity of 2D-MAs, we can reconstruct 3D shape with smooth surfaces, which is difficult for contour-based stereopsis (the depth of contours can be computed but not smooth surfaces inside the contours).

Next, we tested Cone-shape stimulus with a sharp corner and varying thickness. Fig.4 shows the results in which the sharp corner and the varying thickness are reproduced successfully with the reconstruction error of 0.19 that is as low as the Capsule. These results indicate that the model is capable of representing various 3D shapes with sharp corners and varying thickness such as those included in a cone and a polyhedron in general.

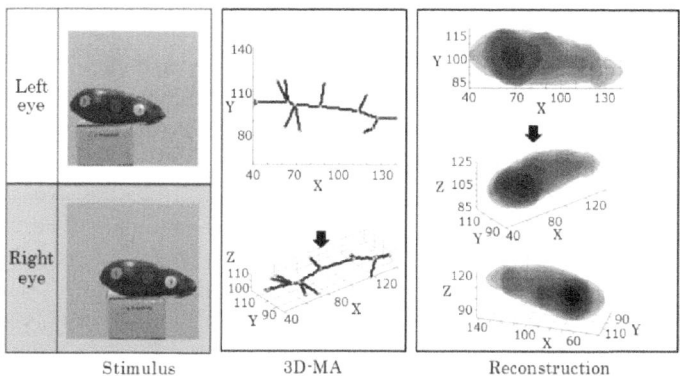

Fig. 6. The reconstruction result of a natural image (Eggplant). The conventions are the same as Fig. 2.

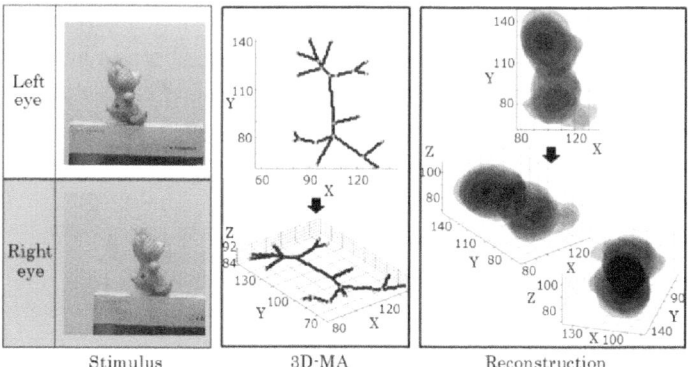

Fig. 7. The reconstruction result of a natural image (Bear). The conventions are the same as Fig. 2.

We also tested the model with Half-Torus-shape stimulus that has a large curvature in the direction perpendicular to a small curvature. Fig.5 shows the results. The curved surface is reproduced smoothly, and the reconstructed shape resembles the original with the reconstruction error of 0.20. These results indicate that the model is capable of encoding objects with complex curves.

Next, we carried out simulations with natural images. Fig.6 shows the result for Eggplant stimulus, as an example. Unlike the CG images, it has an irregular contour and an overall curvature along the major (X) axis. The model reproduced these features, suggesting that our model is capable of representing natural shapes with relatively simple contours such as an eggplant.

Finally, we tested the stimulus of Bear, another natural image that is more complex than eggplant, including all features we have tested so far. Fig.7 shows the results.

The model reconstructed successfully the features, such as a sharp nose, curved ears and a smoothly rounded body. These results indicate that the model can extract fairly accurate 3D shape by fusing two 2D-MAs even for natural objects in natural scenes.

We listed the reconstructed shapes and their errors in table 1 for an overall evaluation. In all cases, including basic shapes of Capsule, Cone and Torus, and natural shapes, the model represented successfully 3D shape with the error between 0.18 and 0.23, indicating fairly accurate representation of 3D shape by 3D-MA that is constructed from 2D-MA.

Table 1. Reconstruction error

Stimulus	Original shape+Reconstruction (blue) (black)	Error
Capsule Front view		0.18
Capsule Side view		0.23
Cone		0.19
Torus		0.20

4   Conclusions and Discussions

We proposed that 3D shape is represented in the cortex by 3D-MA that can be computed from the fusion of two 2D-MAs that are obtained independently from the left and right retinal images. As the first step to examine the plausibility of this hypothesis, we developed a computational model, and carried out the simulations of the model with various stimuli. In this paper, we showed, as examples, the results for three basic shapes of Capsule, Cone and Torus, and two natural shapes (eggplant and bear). In order to test the accuracy of the 3D representation by the model, we reconstructed the shape from the encoded information retrieved from the model. The model represented successfully 3D shape through MA representation with reconstruction error of around 0.2. The results support the hypothesis that 3D objects can be represented by 3D-MA that is computed from the disparity between 2D-MAs.

These results suggest that 2D-MAs representing surfaces are computed independently from our binocular retinal images in an early visual stage, and then, 3D-MA is computed based on the disparity between the 2D-MAs. In particular, our results reproduced the physiological fact that shape representation based on curved surfaces in the cortex can be made even if only binocular disparities of contours are provided. 3D-MA representation from the fusion of 2D-MAs is expected to advance the further understanding of the shape representation in the cortex including how binocular disparities in V1 neurons are transformed to surfaces and to a 3D shape.

References

1. Lee, T.S., Mumford, D., Romero, R., Lamme, V.A.F.: The role of the primary visual cortex in higher level vision. Vision Res. 38, 2429–2454 (1998)
2. Kovacs, I., Julesz, B.: Perceptual sensitivity maps within globally defined visual shapes. Nature 370, 644–646 (1994)
3. Hatori, Y., Sakai, K.: Robust Detection of Medial-Axis by Onset Synchronization of Border-Ownership Selective Cells and Shape Reconstruction from Its Medial-Axis. In: Köppen, M., Kasabov, N., Coghill, G. (eds.) ICONIP 2008. LNCS, vol. 5506, pp. 301–309. Springer, Heidelberg (2009)
4. Yamane, Y., Carlson, E.T., Bowman, K.C., Wang, Z., Connor, C.E.: A neural code for three dimensional shape in macaque inferotemporal cortex. Nat. Neurosci. 11, 1352–1360 (2008)
5. Parker, A.J.: Binocular depth perception and the cerebral cortex. Nat. Rev. Neurosci. 8, 379–391 (2007)
6. Heydt, R., Zhou, H., Friedman, H.S.: Representation of stereoscopic edges in monkey visual cortex. Vision Res. 40, 1955–1967 (2000)
7. Sakai, K., Nishimura, H.: Surrounding suppression and facilitation in the determination of border-ownership. J. Cognitive Neurosci. 18, 562–579 (2006)

A Biologically Inspired Model for Occluded Patterns

Mohammad Saifullah

Department of Computer and Information Science
Linkoping University, Sweden
Mohammad.saifullah@liu.se

Abstract. In this paper a biologically-inspired model for partly occluded patterns is proposed. The model is based on the hypothesis that in human visual system occluding patterns play a key role in recognition as well as in reconstructing internal representation for a pattern's occluding parts. The proposed model is realized with a bidirectional hierarchical neural network. In this network top-down cues, generated by direct connections from the lower to higher levels of hierarchy, interact with the bottom-up information, generated from the un-occluded parts, to recognize occluded patterns. Moreover, positional cues of the occluded as well as occluding patterns, that are computed separately but in the same network, modulate the top-down and bottom-up processing to reconstruct the occluded patterns. Simulation results support the presented hypothesis as well as effectiveness of the model in providing a solution to recognition of occluded patterns. The behavior of the model is in accordance to the known human behavior on the occluded patterns.

Keywords: Vision, Neural network model, Occluded patterns, Biologically-inspired.

1   Introduction

Occlusion is one of the major sources of trouble for pattern recognition systems. When a pattern is occluded by another pattern a recognition system has to face two problems. First, the system receives features from the occluded as well as occluding pattern and needed to distinguish the two. Second, system is deprived of some features belonging to the occluded pattern that might be very crucial for recognition. Performance of a system for recognizing occluded patterns is largely depended on solution to these two problems.

An interesting case of the occlusion arises when occluded part of a pattern has discriminatory role between two or more pattern classes. For example, patterns 'E' and 'F' are indistinguishable when lower edge '_' of 'E' is occluded. In this paper this occlusion is called 'critical occlusion'. This situation is unsolvable for simple pattern recognition system. When we humans confronts with such a situation we need some additional information (e.g., contextual cues) to predict the true identity of such patterns. Without additional information this problem is unsolvable. Humans utilize different contextual cues to recognize critically as well as simple occluded or ambiguous patterns [1], [2]. At the core of this recognition ability lies the architecture of the endowed visual information processing system.

B.-L. Lu, L. Zhang, and J. Kwok (Eds.): ICONIP 2011, Part I, LNCS 7062, pp. 88–96, 2011.

The visual information processing in humans can be easily understood in terms of two stream hypothesis [3]. These streams are the ventral stream or 'What' pathway and the dorsal stream or 'Where' pathway. The ventral stream is responsible for recognizing and identifying objects while the dorsal deals with the spatial information associated with objects. These two pathways interact with each other to solve different visual tasks.

Now, consider the fig. 1 to recapitulate the most common human behavior on the task of occluding patterns. The first two patterns (from left) are unconcluded and can be easily recognized. The third pattern is an occluded one but we can recognize this pattern and also perceive the occluded parts of the pattern. The fourth pattern has some missing parts. This pattern can be recognized but it is comparatively difficult to perceive the missing parts. The fifth one is a critically-occluded pattern; In this case the exact identity of the pattern cannot be determined. It can be any of the two alphabets 'O' or 'Q'. This type of occlusion requires some additional information (e.g., contextual or semantic cues) to predict the exact identity of the pattern. The last pattern is also critically-occluded but there is a difference comparing to the previous case. The occluding pattern is similar to the potentially occluded feature, i.e., the feature that make alphabet 'O' to 'Q'. Though in this case it is still not possible to decide about the identity of the exact pattern without any extra information, but one feels biased to consider it as a character 'Q'. On the basis of above discussion we hypothesize that: i) Occluding patterns play an important role in recognition of occluded patterns and perception of occluding parts, 2) Occluding patterns helps reconstruct the representation of the occluded pattern in the visual system, which is not the case with pattern having missing parts. On the basis of the above mentioned hypothesis and known architecture of the human visual system a model for the occluded pattern is presented in this paper.

The work in this paper is related to the earlier work [4], [5]. Especially, the work in this paper is close to that of Fukushima [6] [7], where he used the biologically-inspired hierarchical structure but with only bottom-up processing of information. In that work, the occluding patterns were removed to avoid the unwanted effects of irrelevant features. In this paper, occluding patterns are utilized to recognize and reconstruct the occluding parts in a bidirectional hierarchical neural network.

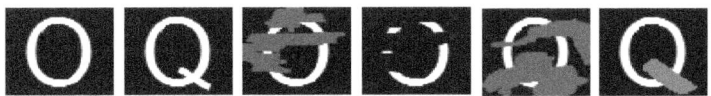

Fig. 1. A few example patterns

2   Approach

Our approach in this work is inspired by the modular and parallel processing architecture of the human brain, where different individual processes go on in parallel having different local goals, but at the same time these processes interact with each other for achieving a common, global goal. More specifically, we model the visual information processing along the two pathways of the human visual system as a solution for the recognition of occluding objects. According to this approach the ventral pathway store the representation of the objects along its hierarchical structure. The visible parts of the occluded patterns are processed along the ventral pathway for recognition. This recognition process is facilitated by the top-down cues provided by the direct connections from the lower to higher parts of the hierarchy of the ventral stream. The dorsal pathway encodes the positional information and modulates the ventral pathway to reconstruct the incomplete representation of the occluding parts. This approach requires information to flow not only in bottom-up or top-down direction but all possible direction. In this work this kind of omnidirectional processing is realized by developing a fully recurrent neural network.

3 The Model

The model presented in this paper is a modified and extended version of the biologically-inspired models for object recognition [8], [9], based on the ventral stream of the human visual system. It has a hierarchical structure of layers of units that interact with each other through feed-forward as well as recurrent connections. The model encodes size, position and shape invariant representation along its hierarchy through learning.

The model proposed in this paper is shown in Fig. 2. The first layer, layer at the lowest level of the hierarchy, is Input layer. The next layer in the model hierarchy is V1. This layer is organized into group of units or hypercolumns such that each group of units looks only at a part of the input or receptive field. This layer sense and encodes the oriented edges in the input images but only from unoccluded parts of the pattern. After V1 the model is divided into two information processing pathways. These two pathways are named after the visual processing pathways of the human brain as the ventral and the dorsal pathways. The ventral pathway process the shape information while the dorsal pathway takes care of the positional information of the input pattern. The ventral pathway is further divided into two information processing channels, i.e., the main channel and the direct channel. The main channel contains the layers V2, V4 and Pat_ID while the direct channel is composed of two layers, namely, Direct_Ch and Pat_Category. The V2 layer is also organized into group of units and these groups extract input from the contiguous groups of unit from V1 and thus have somewhat topographic representation of the input pattern. The V4 layer is a single group layer and has receptive field that encompass the whole input image. The last layer in the main channel is the Pat_ID layer that interacts with the V4 layer to compute and display the result of processing in this channel. The Direct_Ch get its input from the V1 and project to Pat_Category layer. This channel provides a less refined, in terms of discriminatory ability, but fast computing channel to the model for category recognition. The Pat_Category and Pat_ID layers are mutually connected to each other. The direct channel computes the object category information that is used as top-down cues to modulate processing along the main channel. This top-down modulation facilitates the processing in the main channel by limiting the output choices for the information processing along the main channel.

The dorsal pathway is composed of only one layer, Pat_Saliency. This layer gets its input from the Input layer and projects to the V2 layer in a topographic manner.

Ideally, this pathway should get its input from V1 layer but to contain the complexity of the model it is connected to input directly. The Pat_Saliency layer also has a self-connection to help generate a positional-template of the input pattern. This layer modulates the units of the V2 layer according to the positional-templates of the input images which contains the critically occluded part of the pattern as well. This modulation allows the units in the V2 layer to reconstruct the representation at the occluded locations of the input patterns and hence facilitate the correct recognition of the object identity. Interaction of the Pat_Saliency layer with the V2 layer is based on our hypothesis that how occluding patterns help perceive the occluded pattern in a biologically inspired way.

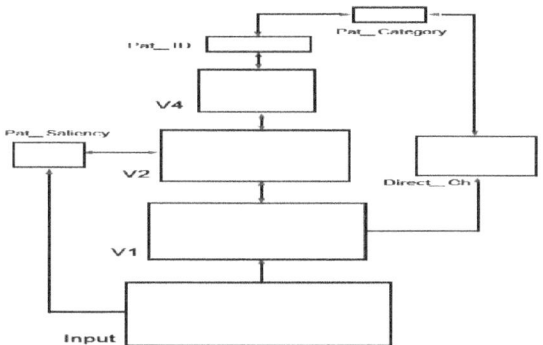

Fig. 2. The proposed biologically-inspired model

3.1 Network Algorithm

An interactive (bidirectional, recurrent) neural network is developed to realize the model. The network was developed in Emergent [10], using the biological plausible algorithm Leabra [8]. Each unit of the network had a sigmoid-like activation function. Learning in the network was based on a combination of Conditional Principal Component Analysis (CPCA), which is a Hebbian learning algorithm and Contrastive Hebbian learning (CHL), which is a biologically-based alternative to back propagation of error, applicable to bidirectional networks [11].

4 Data for Training and Testing

To simplify the training and analysis of the results, gray level images of only six alphabetical character classes (fig. 3) are used for training and testing. These characters are further divided into three categories, such that each category contains two different classes of characters. These categories are created manually on the basis of shape resemblance of classes with each other, so that simulations for critically-occluded objects can be easily demonstrated according to the hypothesis presented in this paper.

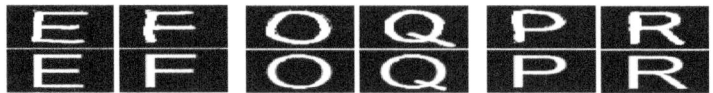

Fig. 3. Patterns used for training. Six different classes of patterns are grouped together into three categories. Close pattern classes in the figure falls in the same category.

Tabel 1. Simulation results, when direct channel recognize the category of the object correctly

No.	Pattern Specification	Avg. Correct (%)	Remarks
1	No occlusion	100	---
2	Non-critical occlusion	97	A few wrong recognitions when occlusion is close to critical occlusion
3	Missing Parts	73	---
4	Critical occlusion	---	Arbitrary recognition of a class, mostly in favor of pattern with occluded feature

5   Simulation and Results

Training for the two network channels of the ventral pathway, i.e., the main channel and the direct channel, is performed separately. First, the training of the direct channel was performed for learning the three input categories. Second, the main channel was trained on six different classes and three categories. In the second part of the training, the main channel learns representation of each of the six classes as well as a mapping between the Pat_ID and Pat_Category. Each unit of Pat_ID and Pat_Category represent one pattern class and one category respectively in the order they appear in the fig. 3. The dorsal pathway does not require any learning.

After completion of the training, the network is simulated on different occluded patterns, Table 1. In the following section a few simulation results are presented with analysis:

5.1   Pattern without Any Occlusion

Figure 4 shows the behavior of the model when an un-occluded pattern is presented to the network. Since the pattern is complete therefore the main channel can easily recognize it without any additional help from the dorsal pathway or the direct channel. When the pattern is presented to the network, the direct channel activates the correct category of the object (cycle: 6) and modulate the processing in the main ventral pathway, in a top-down manner, by interacting with Pat_ID layer. This action of the direct channel bias limits the choices of the main channel and biases its processing towards class-representations belonging to a specific category (cycle: 13-16). Meanwhile, the Pat_Saliency layer of the dorsal pathway encodes the template like positional cues of the input pattern and modulates the V2 layer. This result in completing the representation of the pattern first at V2 layer and then in rest of the network (cycle: 5-39). The interaction between the top-down and bottom-up cues as well as with thetemplate like positional cues from the dorsal pathway result in the activation of correct pattern class at the Pat_ID layer (cycle: 17).

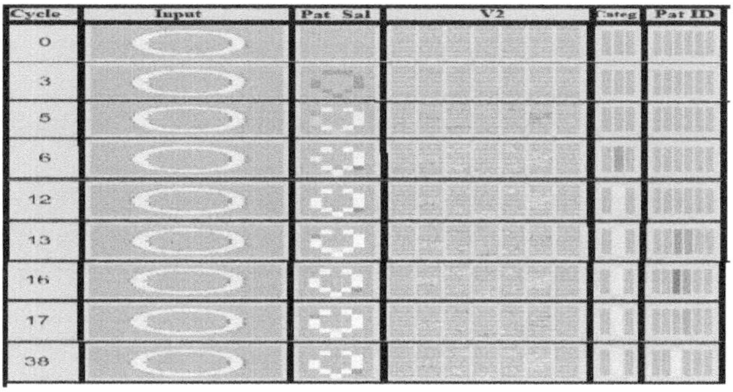

Fig. 4. Snapshots of activations in the various network layers (each layer is made up of a matrix of units, and the activation values of these matrices are shown here). The recorded changes in activation for different processing cycles illustrates how the top-down and bottom-up interactions within ventral pathway and interaction between the ventral and the dorsal pathway lead to desired specific behavior. For each graph, in the order from left to right, the columns represent: Number of processing cycle (how far computation of activation has gone), activations in Input layer, Pat_Saliency layer, V2 layer, Pat_Category layer and Pat_ID layer of the network. Yellow (light) colors denote high activation values, red (dark) colors low activation. Gray (neutral) color means no activation.

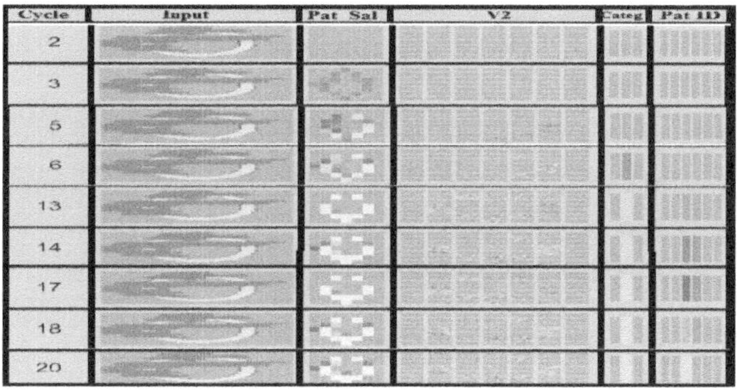

Fig. 5. Cycle-wise activation of various network layers while processing patterns with non-critical occlusion

5.2   Pattern with Non-critical Occlusion

In this simulation an image of occluded pattern is presented to the network and the corresponding behavior of the network is shown in the fig. 5. The category of the pattern is correctly recognized at Pat_Category layer (cycle: 13). The role of the dorsal channel is not critical in this case as pattern is not critically-occluded and there is no confusion about the identity of the object. But, Pat_Saliency layer help to reconstruct the representations of the occluding parts of the pattern (cycle: 13-20). Moreover, the result of interaction among different information processing channels in the network lead to correct identification of the pattern class at Pat_ID layer (cycle:14-20).

5.3 Pattern with Missing Parts

Figure 6 shows the behavior of the network when a pattern with missing parts is presented to the network. It is the same pattern that is used in the previous simulation but after removing the occluding patterns. Patterns of activations of Pat_ID and Pat_Category layers are similar to the previous case. The main difference of this case from the previous one is the activation patterns at Pat_Saliency and V2 layers. Due to absence of occluding patterns the Pat_Saliency layer does not encode the positional template of the missing parts of the patterns (cycle: 3-51). Consequently, interaction between the Pat_Saliency and the V2 layer does not reconstruct the representation of the missing parts of the occluding pattern (cycle: 5-51). It supports our hypothesis about the role of occluding patterns in reconstructing the occluding parts of the patterns through the dorsal pathway.

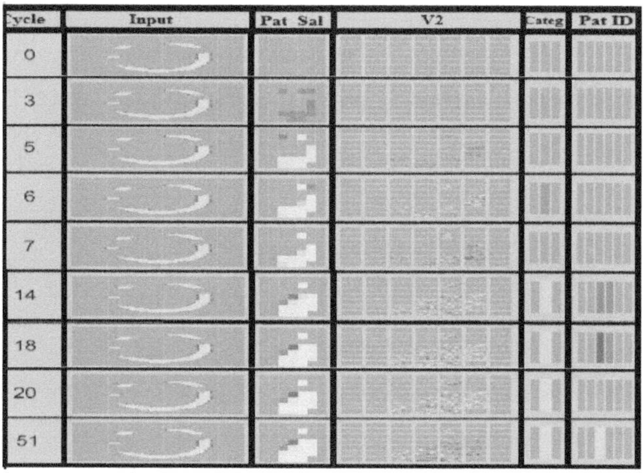

Fig. 6. Cycle-wise activation of various network layers while processing pattern with missing parts

5.4 Pattern with Critical Occlusion

Figure 7 shows processing of the network for a critically-occluded pattern. Soon after the input is presented to the network the Pat_Category layer categorizes the pattern (cycle: 7) and strengthens the units of the Pat_ID layer that represent the objects of this particular category (cycle: 13-18). The Pat_ID layer in turns modulates the main channel by biasing it for two specific classes of patterns. In this case, the bottom-up cues along the main network belong to the whole category and not to a specific object, as the critical part of the pattern are occluded. In the same way, the top-down cues also belongs to the whole category. In this situation bottom-up and top-down cues will interact along the hierarchy of the main network and activates the representations of the features that are common in both classes. Consequently, the most probable object at the output layer should be the pattern with minimum occluded features, in this case that would be pattern 'O'. However, the positional cues in the form of pattern template from the dorsal channel make a difference here. Since this template contains the occluded parts of the pattern as well, and modulates the V2 layer units that represent the occluded part of pattern 'Q'. The modulation of critically occluded pattern especially interacts with the top-down category cues and activates the units representing the critically-occluded patterns. The resulting representation in the main network is tilted towards the pattern 'Q' that is accordingly displayed at the layer Pat_ID. But, as representation of the both classes is present in the network therefore it is just a matter of chance that which pattern will win and accordingly be declared as a final output. It depends on the shape of occluding patterns as well as on the accuracy with which Pat_Saliency layer encodes the positional cues. Sometime, this interaction results in a state where the network cannot decide about the exact class of the pattern and oscillate between the two choices. In such cases some kind of contextual cues are required to bias the result towards a specific class.

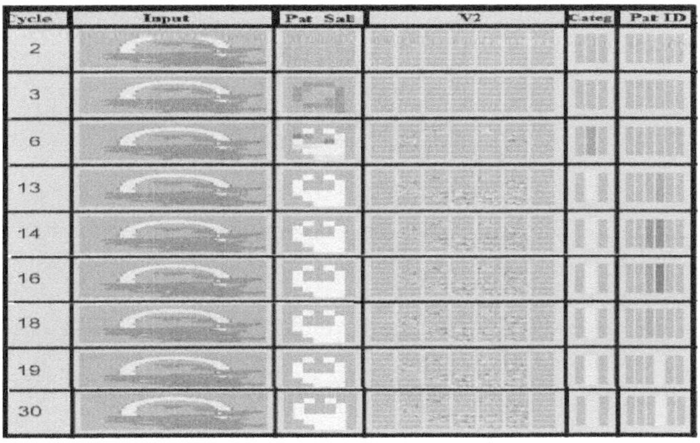

Fig. 7. Cycle-wise activation of various network layers while processing critically occluded pattern

6 Conclusion

In this paper a biologically-inspired model for recognition of occluded pattern is presented. The information processing strategy is based on the hypothesis that occluding patterns provide the important cues for recognition as well as reconstructing the representation of occluded parts. The architecture of the model is based on the two stream hypothesis of human visual system. The computers simulations with the model demonstrate that it provides a satisfactory solution to the occluded patterns recognition as well as produce a behavior that is in accordance to the known human behavior on recognizing partly occluded objects.

References

1. Palmer, E.: The effects of contextual scenes on the identification of objects. Memory & Cognition 3, 519–526 (1975)
2. Bar, M.: Visual objects in context. Nature Reviews Neuroscience 5, 617–629 (2004)

3. Ungerleider, L.G., Haxby, J.V.: 'What' and 'where' in the human brain. Current Opinion in Neurobiology 4, 157–165 (1994)
4. Grossberg, S., Mingolla, E.: Neural dynamics of perceptual grouping: Textures, boundaries, and emergent segmentations. Attention, Perception, & Psychophysics 38, 141–171 (1985)
5. Lee, J.S., Chen, C.H., Sun, Y.N., Tseng, G.S.: Occluded objects recognition using multiscale features and hopfield neural network. Pattern Recognition 30, 113–122 (1997)
6. Fukushima, K.: Recognition of partly occluded patterns: a neural network model. Biological Cybernetics 84, 251–259 (2001)
7. Fukushima, K.: Neural Network Model Restoring Partly Occluded Patterns. In: Palade, V., Howlett, R.J., Jain, L. (eds.) KES 2003. LNCS, vol. 2774, Springer, Heidelberg (2003)
8. O'Reilly, R.C., Munakata, Y.: Computational explorations in cognitive neuroscience: Understanding the mind by simulating the brain. The MIT Press, Massachusetts (2000)
9. Saifullah, M.: Exploring Biologically-Inspired Interactive Networks for Object Recognition. Licentiate theses. Linkoping University electronic press (2011)
10. Aisa, B., Mingus, B., O'Reilly, R.: The emergent neural modeling system. Neural Networks 21, 1146–1152 (2008)
11. O'Reilly, R.C., Munakata, Y.: Computational explorations in cognitive neuroscience: Understanding the mind by simulating the brain. The MIT Press, Massachusetts (2000)

Dynamic Bayesian Network Modeling of Cyanobacterial Biological Processes via Gene Clustering

Nguyen Xuan Vinh[1], Madhu Chetty[1], Ross Coppel[2], and Pramod P. Wangikar[3]

[1] Gippsland School of Information Technology, Monash University, Australia
{vinh.nguyen,madhu.chetty}@monash.edu
[2] Department of Microbiology, Monash University, Australia
Ross.Coppel@monash.edu
[3] Chemical Engineering Department, Indian Institute of Technology, Mumbai, India
wangikar@iitb.ac.in

Abstract. Cyanobacteria are photosynthetic organisms that are credited with both the creation and replenishment of the oxygen-rich atmosphere, and are also responsible for more than half of the primary production on earth. Despite their crucial evolutionary and environmental roles, the study of these organisms has lagged behind other model organisms. This paper presents preliminary results on our ongoing research to unravel the biological interactions occurring within cyanobacteria. We develop an analysis framework that leverages recently developed bioinformatics and machine learning tools, such as genome-wide sequence matching based annotation, gene ontology analysis, cluster analysis and dynamic Bayesian network. Together, these tools allow us to overcome the lack of knowledge of less well-studied organisms, and reveal interesting relationships among their biological processes. Experiments on the *Cyanothece* bacterium demonstrate the practicability and usefulness of our approach.

Keywords: cyanobacteria, *Cyanothece*, dynamic Bayesian network, clustering, gene ontology, gene regulatory network.

1 Introduction

Cyanobacteria are the only prokaryotes that are capable of photosynthesis, and are credited with transforming the anaerobic atmosphere to the oxygen-rich atmosphere. They are also responsible for more than half of the total primary production on earth and found the base of the ocean food web. In recent years, cyanobacteria have received increasing interest, due to their efficiency in carbon sequestration and potential for biofuel production. Although their mechanism of photosynthesis is similar to that of higher plants, cyanobacteria are much more efficient as solar energy converters and CO_2 absorbers, essentially due to their simple cellular structure. It is estimated that cyanobacteria are capable of producing 30 times the amount oil per unit area of land, compared to terrestrial oilseed crops such as corn or palm[14]. These organisms therefore may hold the

B.-L. Lu, L. Zhang, and J. Kwok (Eds.): ICONIP 2011, Part I, LNCS 7062, pp. 97–106, 2011.

key to solve two of the most fundamental problems of our time, namely climate change and the dwindling fossil fuel reserves.

Despite their evolutionary and environmental importance, the study of cyanobacteria using modern high throughput tools and computational techniques has somewhat lagged behind other model organisms, such as yeast or *E. coli* [18]. This is reflected partly by the fact that none of the cyanobacteria has an official, effective gene annotation in the Gene Ontology Consortium repository as of May 2011 [20]. Nearly half of *Synechocystis* sp. PCC 6803's genes, the best studied cyanobacterium, remain unannotated. Of the annotated genes, the lack of an official, systematic annotating mechanism, such as that currently practiced by the Gene Ontology Consortium, make it hard to verify the credibility of the annotation as well as to perform certain type of analysis, e.g., excluding a certain annotation evidence code.

In this paper, to alleviate the difficulties faced when studying novel, less well-studied organisms such as cyanobacteria, we develop an analysis framework for building network of biological processes from gene expression data, that leverages several recently developed bioinformatics and machine learning tools. The approach is divided into three stages:

- Filtering and clustering of genes into clusters which have coherent expression pattern profiles. For this, we propose using an automated scheme for determining a suitable number of clusters for the next stages of analysis.
- Assessment of clustering results using functional enrichment analysis based on gene ontology. Herein, we propose using annotation data obtained from two different sources: one from the Cyanobase cyanobacteria database [11], and another obtained by means of computational analysis, specifically by amino sequence matching, as provided by the Blast2GO software suite [5].
- Building a network of interacting clusters. This is done using the recently developed formalism of dynamic Bayesian network (DBN). We apply our recently proposed GlobalMIT algorithm for learning the globally optimal DBN structure from time series microarray data, using an information theoretic based scoring metric.

It is expected that the network of interacting clusters will reveal the interactions between biological processes represented by these clusters. However, when doing analysis on the cluster (or biological process) level, we lose information on individual genes. Obtaining such information is possible if we apply network reverse engineering algorithms directly to the original set of genes without clustering, resulting in the underlying gene regulatory network (GRN). Nevertheless, with a large number of genes and a limited number of experiments as often seen in microarray data, GRN-learning algorithms face severe difficulties in correctly recovering the underlying network. Also, a large number of genes (including lots of unannotated genes) makes the interpretation of the results a difficult task. Analysis at the cluster level serves two purposes: (i) to reduce the number of variables, thus making the network learning task more accurate, (ii) to facilitate interpretation. Similar strategies to this approach have also been employed in [7, 16, 18].

In the rest of this paper, we present our detailed approach for filtering and clustering of genes, assessing clustering results, and finally building network of interacting clusters. For an experimental cyanobacterial organism, we chose the diazotrophic unicellular *Cyanothece* sp. strain ATCC 51142, hereafter *Cyanothece*. This cyanobacterium represents a relatively less well-studied organism, but with a very interesting capability of performing both nitrogen fixation and photosynthesis within a single cell, the two processes that are at odds with each other [19].

2   Filtering and Clustering of Genes

Cyanobacteria microarray data often contain measurements for 3000 to 6000 genes. Many of these genes, such as house keeping genes, are not expressed, or expressed at a constant level throughout the experiments. For analysis, it is desirable to filter out these genes, and retain only genes which are differentially expressed. There are various methods for filtering genes such as the threshold filter, Student's t-test or analysis of variance (ANOVA) [4]. In this work, we implement a simple but widely employed threshold filter to remove genes that are not differentially expressed above a certain threshold throughout the experimental process, e.g., 1.5-fold or 2-fold change.

Next, we cluster the selected genes into groups of similar pattern profiles. In the recent years, there has been dozens of clustering algorithms specifically developed for the purpose of clustering microarray data. Some of the most popular methods include K-means, hierarchical clustering, self organizing map, graph theoretic based approaches (spectral clustering, CLICK, CAST), model based clustering (mixture models), density based approaches (DENCLUE) and affinity propagation based approaches [9]. In this work, we implement the widely used K-means with log-transformed microarray data.

A crucial parameter for K-means type algorithms is the number of clusters K. For our purpose in this paper, K will control the level of granularity of the next stages of analysis. We use our recently developed Consensus Index for automatically determining the relevant number of clusters from the data [24]. The Consensus Index (CI) is a realization of a class of methods for model selection by stability assessment [17], whose main idea can be summarized as follows: for each value of K, we generate a set of clustering solutions, either by using different initial starting points for K-means, or by a certain perturbation scheme such as sub-sampling or projection. In regard to the set of clusterings obtained, when the specified number of clusters coincides with the "true" number of clusters, this set has a tendency to be less diverse—an indication of the robustness of the obtained cluster structure. The Consensus Index was developed to quantify this diversity. Specifically, given a value of K, suppose we have generated a set of B clustering solutions $\mathcal{U}_K = \{U_1, U_2, ..., U_B\}$, each with K clusters. The consensus index of $\mathcal{U}_K$ is defined as:

$$\mathrm{CI}(\mathcal{U}_K) = \frac{\sum_{i<j} \mathrm{AM}(\mathbf{U}_i, \mathbf{U}_j)}{B(B-1)/2} \tag{1}$$

where the agreement measure AM is a clustering similarity measure. In this work, we use the Adjusted Rand Index (ARI) and the Adjusted Mutual Information (AMI—which is the adjusted-for-chance version of the widely used Normalized Mutual Information) as clustering similarity measures [23]. The optimal number of clusters K^* is chosen as the one that maximizes CI, i.e., $K^* = \arg\max_{K=2...K_{max}} \mathrm{CI}(\mathcal{U}_K)$ where K_{max} is the maximum number of clusters to be considered.

3   Assessment of Clustering Results

Having obtained a reasonable clustering solution, we next investigate the major biological functions of each cluster. In this work, this is done by means of functional enrichment analysis using gene ontology (GO), where every GO terms appearing in each cluster is assessed to find out whether a certain functional category is significantly over-represented in a certain cluster, more than what would be expected by chance. To do this, first of all, we need a genome-wide annotation of genes in the organism of interest. As stated previously, one of the difficulties working with less well-studied organisms is that there is not an official annotation database. To address this challenge, we propose gathering annotation data from two different sources: one from the Cyanobase database [11], and another from genome-wide amino sequence matching using the Blast2GO software suit [5]. We describe each source below.

The Cyanobase maintains, for each cyanobacterium in its database, an annotation file which was obtained by IPR2GO, a manually-curated mapping of InterPro terms to GO terms that is maintained by the InterPro consortium [21]. Although being the result of a manual curation process, surprisingly, it has been reported that the accuracy of this mapping can be considerably lower than some automated algorithms, such as that reported in [10]. Moreover, the number of annotated genes normally accounts for just less than half of the genome, eg. in the case of *Cyanothece*, there are currently only annotations for 2566 genes out of 5359 genes (as of May 2011).

Thus, in order to supplement the Cyanobase IPR2GO annotation, we employ Blast2GO, a software suit for automated gene annotation based on sequence matching [5]. Blast2GO uses BLAST search to find similar sequences to the sequence of interest. It then extracts the GO terms associated to each of the obtained hits and return the GO annotation for the query. For *Cyanothece*, Blast2GO was able to supplement the annotation for almost another one thousand genes. In this work, we aggregate Cyanobase IPR2GO and Blast2GO annotation into a single pool, then use BiNGO [12] for GO functional category enrichment analysis. For BiNGO, we use the filtered gene set as the reference set, the hypergeometric test as the test for functional over-representation, and False Discovery Rate (FDR) as the multiple hypothesis testing correction scheme.

4   Building a Network of Interacting Clusters

Our next stage is to build a network of clusters, in order to understand the interactions occurring between the biological processes represented by these clusters. We perform this modeling task using the recently developed dynamic Bayesian network (DBN) formalism [8, 13, 26, 27]. The simplest model of this type is the first-order Markov stationary DBN, in which both the structure of the network and the parameters characterizing it are assumed to remain unchanged over time, such as the one exemplified in Figure 1a. In this model, the value of a variable at time $(t + 1)$ is assumed to depend only on the value of its parents at time (t). DBN addresses two weaknesses of the traditional static Bayesian network (BN) model: (i) it accounts for the temporal aspect of time-series data, in that an edge must always direct forward in time (i.e., cause must precede consequence); and (ii) it allows feedback loops (Fig. 1b).

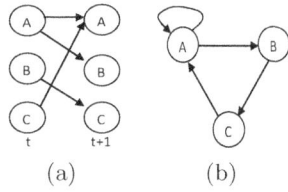

(a)   (b)

Fig. 1. Dynamic Bayesian Network: (a) a 1st order Markov stationary DBN; (b) its equivalent folded network

Recent work in machine learning has progressed to allow more flexible DBN models, such as one with, either parameters [6], or both structure and parameters [3, 15] changing over time. It is worth noting that more flexible models generally require more data to be learned accurately. In situations where training data are scarce, such as in microarray experiments where the data size can be as small as a couple of dozen samples, a simpler model such as the first-order Markov stationary DBN might be a more suitable choice. Moreover, it has been recently shown that the globally optimal structure of a DBN can be efficiently learned in polynomial time [2, 22]. Henceforth, in this work we choose the first order Markov DBN as our modeling tool.

For a DBN structure scoring metric, we propose using a recently introduced information theoretic criterion named MIT (Mutual Information Test) [1]. MIT has been previously shown to be effective for learning static Bayesian network, yielding results competitive to other popular scoring metrics, such as BIC/MDL, K2 and BD, and the well-known constraint-based approach PC algorithm. Under the assumption that every variable has the same cardinality—which is generally valid for dicretized microarray data—our algorithm recently developed in [22][1] can be employed for finding the globally optimal DBN structure, in polynomial time.

---

[1] see our report and software at `http://code.google.com/p/globalmit/`

5   Experiments on *Cyanothece* sp. strain ATCC 51142

In this section, we present our experimental results on *Cyanothece*. We collected two publicly available genome-wide microarray data sets of *Cyanothece*, performed in alternating light-dark (LD) cycles with samples collected every 4h over a 48h period: the first one starting with 1h into dark period followed by two DL cycles (DLDL), and the second one starting with two hours into light period, followed by one LD and one continuous LL cycle (LDLL) [25]. In total, there were 24 experiments.

Filtering and clustering of genes: Using a threshold filter with a 2-fold change cutoff, we selected 730 genes for further analysis. We first used the Consensus Index to determine the number of clusters in this set. Fig. 2(a) show the CI with $K \in [2, 50]$. It can be seen that the CI with both the ARI and AMI strongly suggests $K = 5$ (corresponding to the global peak). Also, a local peak is present at $K = 9$. As discussed in [24], the local peak may correspond be the result of the hierarchical clustering structure in the data. We performed K-means clustering with both $K = 5$ and $K = 9$, each for 1 million times with random initialization, and picked the best clustering results, presented in Fig. 2(b,c).

Assessment of clustering results: From the visual representation in Fig. 2(b), it can be seen that the clusters have distinct pattern profiles. GO analysis of the clustering results are presented in Tables 1 and 2. From Table 1, of our particular interest is cluster C5, which is relatively small but contains genes exclusively involved in the nitrogen fixation process. It is known that *Cyanothece* sp. strain ATCC 51142 is among the few organisms that are capable of performing both oxygenic photosynthesis and nitrogen fixation in the same cell. Since the nitrogenase enzyme involved in N_2 fixation is inactivated when exposed to oxygen, *Cyanothece* separates these processes temporally, so that oxygenic photosynthesis occurs during the day, and nitrogen fixation during the night. Cluster C4 is also of our interest, since its contains a large number of genes involved in photosynthesis. As the experimental condition involves alternative light-dark condition, it could be expected that the genes involved in nitrogen fixation and photosynthesis will strongly regulate each other, in the sense that the up-regulation of N_2 fixation genes will lead to the down-regulation of photosynthesis genes, and vice-versa.

Building a network of interacting clusters: We apply the GlobalMIT algorithm [22] to learn the globally optimal DBN structure, first to the 5-cluster clustering result. We take the mean expression value of each cluster as its representative. There are thus 5 variables over 24 time-points fed into GlobalMIT. The globally optimal DBN network as found by GlobalMIT is presented on Fig. 3(a). It is readily verifiable the fact that nitrogen fixation genes and photosynthesis genes strongly regulate each other, since there is a link between cluster C4 (photosystem) and C5 (nitrogen fixation).

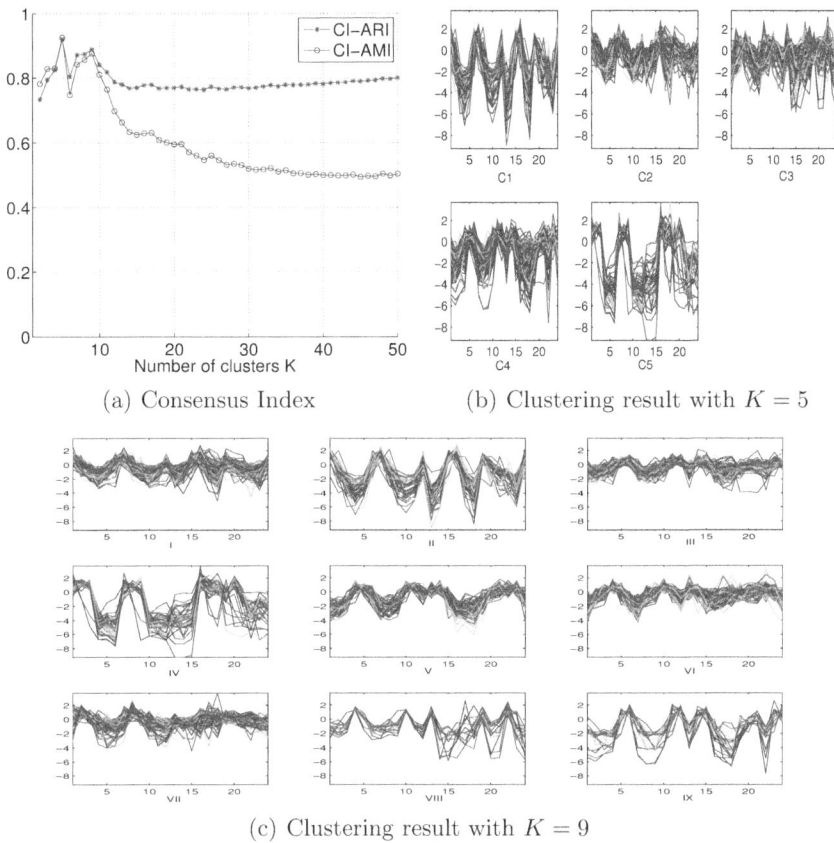

(a) Consensus Index

(b) Clustering result with $K = 5$

(c) Clustering result with $K = 9$

Fig. 2. Cluster analysis of *Cyanothece* microarray data

We perform a similar analysis on the 9-cluster clustering result. The DBN for this 9-cluster set is presented in Fig. 3(a). Note that clusters I and VII are disconnected. We are interested in verifying whether the link between the photosynthesis cluster and the nitrogen fixation cluster remains at this level of granularity. Visually, it it easily recognizable from Fig. 2(b-c) that cluster C5 in the 5-cluster set corresponds to cluster IV in the 9-cluster set. GO analysis on cluster IV confirms this observation (Table 2). We therefore pay special attention to clusters VI, since there are a link between cluster VI and IV. Not surprisingly, GO analysis reveal that cluster VI contains a large number of genes involved in photosynthesis. The structural similarity between the two graphs is also evident from Fig. 3. At a higher level of granularity, the clusters become more specialized. The links between cluster VIII and {IX, III} are also of interest, since cluster VIII is a tightly co-regulated group which contains several genes with regulation activities, which might regulate genes involving transport and photosynthesis (clusters III and IX).

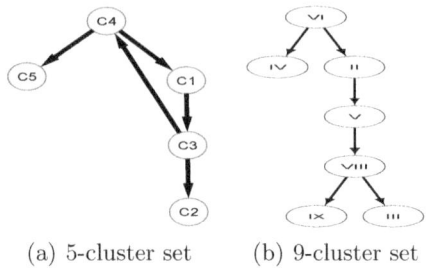

(a) 5-cluster set   (b) 9-cluster set

Fig. 3. DBN analysis of *Cyanothece* clustered data

Table 1. GO analysis of the 5-cluster clustering results

Cluster	Size	GO ID	Description	#Genes	Corrected P-value
C1	54	8746	NAD(P)+transhydrogenase activity	3	0.98%
		70469	respiratory chain	3	2.9%
C2	206	8652	cellular amino acid biosynthesis	18	2.1%
		4518	nuclease activity	8	4.1%
C3		32991	macromolecule complex	61	2E-10
		30529	ribonucleoprotein complex	24	2.9E-7
	236	6412	translation	29	1.5E-6
		44267	cellular protein metabolic process	46	5.4E-5
		19538	protein metabolic process	50	0.14%
C4	196	71944	cell periphery	35	6.8E-5
		9512	photosystem	20	6.5E-3
		6022	aminoglycan metabolic process	6	2%
C5	38	9399	nitrogen fixation	19	5.7E-22
		51536	iron-sulfur cluster binding	12	9.6E-6
		16163	nitrogenase activity	5	1.5E-5

Table 2. GO analysis of the 9-cluster clustering results

Cluster	Size	GO ID	Description	#Genes	Corrected P-value
I	157	8652	cellular amino acid biosynthesis	17	4.5E-3
		46394	carboxylic acid biosynthesis	17	1.6%
II	48	55114	oxidation reduction process	14	17%
		15980	energy derivation by oxidation	6	17%
III	127	15979	photosynthesis	17	45%
		6810	transport	27	45%
IV	36	9399	nitrogen fixation	19	1.5E-24
V	68	6022	aminoglycan metabolic process	5	2.1%
		7049	cell cycle	4	3.8%
VI	158	15979	photosynthesis	36	3.6E-10
VII	101	6412	translation	27	6.7E-13
VIII	16	65007	biological regulation	5	7.7%
		51171	regulation of nitrogen compound	3	7.7%
IX	19	15706	nitrate transport	3	2.2E-3
		6810	transport	9	2.2%

6   Discussion and Conclusion

In this paper, we have presented an analysis framework for unraveling the inter-actions between biological processes of novel, less well-studied organisms such as cyanobacteria. The framework harnesses several recently developed bioinformatics and data mining tools to overcome the lack of information of these organisms. Via Blast2GO and IPR2GO, we could collect annotation information for a large number of genes. Cluster analysis helps to bring down the number of variables for the subsequent network analysis phase, and also facilitates interpretation. We have demonstrated the applicability of our framework on *cyanothece*. Our future work involves further analysis of other cyanobacteria that are potential for carbon sequestration and biofuel production.

Acknowledgments. This project is supported by an Australia-India strategic research fund (AISRF).

Availability: Our implementation of the GlobalMIT algorithm used in this paper in Matlab and C++ is available at `http://code.google.com/p/globalmit`.

References

1. de Campos, L.M.: A scoring function for learning bayesian networks based on mutual information and conditional independence tests. Mach. Learn. Res. 7, 2149–2187 (2006)
2. Dojer, N.: Learning Bayesian Networks Does Not Have to Be NP-Hard. In: Královič, R., Urzyczyn, P. (eds.) MFCS 2006. LNCS, vol. 4162, pp. 305–314. Springer, Heidelberg (2006)
3. Dondelinger, F., Lebre, S., Husmeier, D.: Heterogeneous continuous dynamic bayesian networks with flexible structure and inter-time segment information sharing. In: ICML, pp. 303–310 (2010)
4. Elvitigala, T., Polpitiya, A., Wang, W., Stockel, J., Khandelwal, A., Quatrano, R., Pakrasi, H., Ghosh, B.: High-throughput biological data analysis. IEEE Control Systems 30(6), 81–100 (2010)
5. Gotz, S., Garcia-Gomez, J.M., Terol, J., Williams, T.D., Nagaraj, S.H., Nueda, M.J., Robles, M., Talon, M., Dopazo, J., Conesa, A.: High-throughput functional annotation and data mining with the Blast2GO suite. Nucleic Acids Research 36(10), 3420–3435 (2008)
6. Grzegorczyk, M., Husmeier, D.: Non-stationary continuous dynamic Bayesian networks. In: NIPS 2009, (2009)
7. de Hoon, M., Imoto, S., Kobayashi, K., Ogasawara, N., Miyano, S.: Inferring gene regulatory networks from time-ordered gene expression data of bacillus subtilis using differential equations. In: Pac. Symp. Biocomput., pp. 17–28 (2003)
8. Husmeier, D.: Sensitivity and specificity of inferring genetic regulatory interactions from microarray experiments with dynamic Bayesian networks. Bioinformatics 19(17), 2271–2282 (2003)
9. Jiang, D., Tang, C., Zhang, A.: Cluster analysis for gene expression data: a survey. IEEE Transactions on Knowledge and Data Engineering 16(11), 1370–1386 (2004)
10. Jung, J., Thon, M.: Automatic Annotation of Protein Functional Class from Sparse and Imbalanced Data Sets. In: Dalkilic, M.M., Kim, S., Yang, J. (eds.) VDMB 2006. LNCS (LNBI), vol. 4316, pp. 65–77. Springer, Heidelberg (2006)
11. Kazusa DNA Research Institute: The cyanobacteria database (2011), `http://genome.kazusa.or.jp/cyanobase`

12. Maere, S., Heymans, K., Kuiper, M.: BiNGO: a cytoscape plugin to assess over-representation of gene ontology categories in biological networks. Bioinformatics 21(16), 3448–3449
13. Murphy, K., Mian, S.: Modelling gene expression data using dynamic bayesian networks. Tech. rep., Computer Science Division. University of California, Berkeley, CA (1999)
14. Oilgea Inc.: Comprehensive oilgae report (2011), http://www.oilgae.com
15. Robinson, J., Hartemink, A.: Learning Non-Stationary Dynamic Bayesian Networks. The Journal of Machine Learning Research 11, 3647–3680 (2010)
16. Segal, E., Shapira, M., Regev, A., Pe'er, D., Botstein, D., Koller, D., Friedman, N.: Module networks: identifying regulatory modules and their condition-specific regulators from gene expression data. Nature Genetics 34, 166–176 (2003)
17. Shamir, O., Tishby, N.: Model selection and stability in k-means clustering. In: COLT 2008, Springer, Heidelberg (2008)
18. Singh, A., Elvitigala, T., Cameron, J., Ghosh, B., Bhattacharyya-Pakrasi, M., Pakrasi, H.: Integrative analysis of large scale expression profiles reveals core transcriptional response and coordination between multiple cellular processes in a cyanobacterium. BMC Systems Biology 4(1), 105 (2010)
19. Stockel, J., Welsh, E.A., Liberton, M., Kunnvakkam, R., Aurora, R., Pakrasi, H.B.: Global transcriptomic analysis of cyanothece 51142 reveals robust diurnal oscillation of central metabolic processes. Proceedings of the National Academy of Sciences 105(16), 6156–6161 (2008)
20. The Gene Ontology Consortium: Current annotations (2011), http://www.geneontology.org
21. The InterPro Consortium: Interpro: An integrated documentation resource for protein families, domains and functional sites. Briefings in Bioinformatics 3(3), 225–235 (2002)
22. Vinh, N.X., Chetty, M., Coppel, R., Wangikar, P.P.: Polynomial Time Algorithm for Learning Globally Optimal Dynamic Bayesian Network. In: Zhang, L., Kwok, J. (eds.) ICONIP 2011, Part I, vol. 7062. Springer, Heidelberg (2011)
23. Vinh, N.X., Epps, J., Bailey, J.: Information theoretic measures for clusterings comparison: is a correction for chance necessary? In: Proceedings of the 26th Annual International Conference on Machine Learning, ICML 2009, pp. 1073–1080. ACM, New York (2009)
24. Vinh, N.X., Epps, J., Bailey, J.: Information theoretic measures for clusterings comparison: Variants, properties, normalization and correction for chance. Journal of Machine Learning Research 11, 2837–2854 (2010)
25. Wang, W., Ghosh, B., Pakrasi, H.: Identification and modeling of genes with diurnal oscillations from microarray time series data. IEEE/ACM Transactions on Computational Biology and Bioinformatics 8(1), 108–121 (2011)
26. Yu, J., Smith, V.A., Wang, P.P., Hartemink, A.J., Jarvis, E.D.: Advances to Bayesian network inference for generating causal networks from observational biological data. Bioinformatics 20(18), 3594–3603 (2004)
27. Zou, M., Conzen, S.D.: A new dynamic Bayesian network (DBN) approach for identifying gene regulatory networks from time course microarray data. Bioinformatics 21(1), 71–79 (2005)

Discrimination of Protein Thermostability Based on a New Integrated Neural Network

Jingru Xu and Yuehui Chen*

Computational Intelligence Lab,
School of Information Science and Engineering
University of Jinan, 106 Jiwei Road, 250022 Jinan, P.R. China
yhchen@ujn.edu.cn

Abstract. The research of protein thermostability has been vigorously studied in the field of biophysical and biological technology. What is more, protein thermostability in the level of amino acid sequence is still a challenge in the research of the protein pattern recognition. In this paper, we try to use new integrated feedforward artificial neural network which was optimized by particle swarm optimization (PSO-NN) to recognize the mesophilic and thermophilic proteins. Here, we adopted Genetic Algorithm based Selected Ensemble (GASEN) as our integration methods. A better accuracy was got by GASEN. So, the integrated methods were proved to be effectual.

Keywords: protein thermostability, integrated, Genetic Algorithm based Selected Ensemble, particle swarm optimization, artificial neural network.

1   Introduction

Thermophilic and mesophilic proteins are polymers that are made up of the same 20 kinds of amino acids [1], but in the condition of high-temperature, there is a remarkable contrast between thermostable enzymes and normal temperature enzymes. This phenomenon affords perplexity to us. If we know the protein thermostability very well, it would be helpful to know better the folding mechanism and the function of protein. What is more, it would be an important assistant to reconstruct the stability of protein [2]. If we want to reconstruct the stability of protein, we must make clear the protein thermostability firstly, especially in the level of amino acid sequence. In this research field, how to extract features from amino acid sequence and search an effective tool to recognize mesophilic and thermophilic proteins are what we should consider. Here, we focused on predicting whether an amino acid residue is thermophilic or mesophilic proteins. With the development of bioinformatics, there are many effectual methods to be introduced to this research. It is a two-type classification problem to discriminate protein thermostability based on bioinformatics. If an amino acid residue

---

* Corresponding author.

B.-L. Lu, L. Zhang, and J. Kwok (Eds.): ICONIP 2011, Part I, LNCS 7062, pp. 107–112, 2011.

was mesophilic proteins, it was labeled as '1', otherwise it was labeled as '0' [3]. The main steps of this work are as follows: (1) Select an appropriate data set; (2) extract important features; (3) use predictors to predict interaction sites; (4) determine the evaluation methods.

In this paper, we try to use new integrated feedforward artificial neural network which was optimized by particle swarm optimization (PSO-NN) to recognize the mesophilic and thermophilic proteins. Here, we adopted Genetic Algorithm based Selected Ensemble (GASEN) as our integration methods.

2   Materials and Methods

2.1   Dataset [14]

The training database came from reference [4]. It contained 3521 thermophilic protein sequences and 4895 mesophilic protein sequences which were derived from Swiss-Prot. The corresponding two testing datasets contained 859 proteins. The first one came from reference [4] which included 382 thermophilic protein sequences of Aquifex aeolicus [5], and 325 mesophilic protein sequences of Xylella fastidiosa [6]. The second one included 76 pairs set of thermophilic and mesophilic proteins which were also downloaded from Swiss-Prot for the reason of their non-redundancy. The protein of the second testing came from reference [7].

2.2   Feature Extraction

In this paper, four kinds of feature extraction method were used [8], they are the Amino acids models (AA), Chem-composition model (CC), the fusion of AA and CC and PseAA. Specific are as follows:

(1) Amino acids models (AA) [14]

Twenty kinds of common amino acid residues were calculated their frequency of occurrences in every protein sequences. Every dimension of characteristic vector can be expressed as:

$$X_i = N_i/N, i = 1, 2, 3......20 \tag{1}$$

In the formula, N_i indicates the numbers of the ith residue: N indicates the length of the sequence. This model contains twenty dimensions.

(2) Chem-Composition models(CC)

The 20 native amino acids are divided into three groups for their physicochemical properties, including hydrophobicity, polarity, polarizibility, charge, secondary structures and solvent accessibility. Take hydrophobicity attribute for example, all amino acids are divided into three groups: polar, neutral and hydrophobic. A protein sequence is then transformed into a sequence of hydrophobicity attribute. Therefore, the composition descriptor consists of three values: the global percent compositions of polar, neutral and hydrophobic residues in the new sequence. For seven types of attributes, the chem-composition model consists of twenty-one dimensions.

(3)Fusion

This is the process of combination of these existing features which were extracted by amino acids models and chem-composition models separately, that is to say, it contains forty-one dimensions.

(4) Pseudo-amino acid composition method (PseAA)

Pseudo-amino acid composition method (PseAA) was advanced by Chou [9]. According to the PseAA composition model, the protein sequence can be formulated as:

$$p = [p_1, p_2, \ldots, p_{20}, p_{20+1}, \ldots, p_{20+\lambda}]^T (\lambda < L) \tag{2}$$

Where the first 20 components are the same as those in the classic amino acid composition and $p_1, p_2, \ldots, p_{20}, p_{20+1}, \ldots, p_{20+\lambda}$ are related to λ which states different ranks of sequence order correlation factors. L is the length of protein sequence.

2.3   Discrimination Method [14]

Artificial neural network (ANN) [10] has the capabilities of learning, parallel processing, auto-organization, auto-organization, auto-adaptation, and fault tolerance. Therefore, ANN is well suited for a wide variety of engineering applications, such as function approximation, pattern recognition, classification, prediction, control, etc. In this paper, we choose ANN as our discrimination model to recognize thermophilic and mesophilic proteins, and particle swarm optimization (PSO) was used to optimize the parameters of ANN.

Particle swarm optimization (PSO) is a population-based stochastic optimization technique, which was originally designed by Kennedy and Eberhart [11]. It shares many similarities with evolutionary computation techniques such as GA. However, unlike GA, the PSO algorithm has no evolutionary operators, such as crossover and mutation. In the PSO algorithm, each single candidate solution can be considered a particle in the search space, each particle move through the problem space by following the current optimum particles. Particle i is represented as Xi, which represents a potential solution to a problem. Each particle keeps a memory of its previous best position Pbest, and a velocity along each dimension, represented as Vi. At each iteration, the position of the particle with the best fitness value in the search space, designated as Gbest, and the current particle are combined to adjust the velocity along each dimension, and that velocity is then used to compute a new position for the particle. The updating rules are as follows:

$$V_{k+1} = W * V_k + c1 * r1 * (Pbest - X_k) + c2 * r2 * (Gbest - X_k) \tag{3}$$

$$X_{k+1} = X_k + V_{k+1} \tag{4}$$

Where W is an inertia weight, it regulate the range of the solution space. $c1$ and $c2$ determine the relative influence of the social and cognition components (learning factors), while rand1 and rand2 denote two random numbers uniformly distributed in the interval [0,1]. In this paper, we choose W=0.5, $c1$=$c2$=2.

2.4   Integration

As shown in the previous research [12], integration always has good performance when the base classifiers have obvious differences. We used Genetic Algorithm based Selected Ensemble (GASEN) [13] as our integrated method. It is the extension of generalized ensemble method (GEM). GEM is calculated by the following formula:

$$f_{GEM} = \sum_{i=1}^{n} W_i * f_i(x) \tag{5}$$

Where W_i is an inertia weight,it is changed between 0 and 1. The sum of n*W_i is 1. In GASEN, the weights were optimized by Genetic Algorithm.

3   Experimental Results

First, we defined the followings [14]:

TP (true positives): it represents the number of mesophilic proteins that were predicted correctly.

TN (true negatives): it represents the number of thermophilic proteins that were predicted correctly.

FP (false positives): it represents the number of thermophilic proteins that were predicted as mesophilic proteins.

FN (false negatives): it represents the number of mesophilic proteins that were predicted as thermophilic proteins.

N: it represents the number of all protein molecule.

The following metrics were used to evaluate the prediction results:

sensitivity of the positive data:

$$sensitivity = TP/(TP + FN); \tag{6}$$

specificity of the negative data:

$$specificity = TN/(TN + FN); \tag{7}$$

accuracy of prediction:

$$accuracy = (TP + TN)/N; \tag{8}$$

In this paper, four kinds of feature extraction methods were used, they are the Amino acids models (AA), Chem-composition models (CC), the fusion of AA and CC and PseAA. So we need four PSO-NNs to train them. Every neutral network generated a result and the four groups of results were compared.Next, we used Genetic Algorithm based Selected Ensemble (GASEN) as our integrated method.The final experimental results were shown in Table 1:

The comparison of the results with prior research which also use the same database. It is shown in table 2:

Table 1. The final experimental results of database

method	*sensitivity*	*specificity*	*accuracy*
AA	93.27	82.42	87.71
CC	90.27	80.76	85.40
Fusion [14]	92.8	87.6	90.2
PSEAA	84.54	84.80	84.68
GASEN	95.52	89.88	92.70

Table 2. The comparison of the results with prior research

method	Test sample	*sensitivity*	*specificity*	*accuracy*
Our method(GASEN)	test1+test2	95.52	89.88	92.70
BPNN+Dipeptide[15]	test1	93.50	85.20	89.70
	test2	96.10	75.00	85.50
BPNN+AA[15]	test1	97.40	63.70	81.90
	test2	97.40	46.10	71.70
PSO-NN+fusion[14]	test1+test2	92.80	87.60	90.20
Adaboost[4]	test1+test2	82.97	92.27	87.31
Logitboost[4]	test1+test2	87.34	90.77	88.94

4   Conclusion

In this paper, Genetic Algorithm based Selected Ensemble (GASEN) as a kind of integrated method has good performance in the field of protein thermostability. The next work, we want to use more unique and critical features to predict protein thermostability. We also hope that more and more different and new methods about computational intelligence and biology can be applied to this subject.

Acknowledgments. This research was partially supported by the Natural Science Foundation of China (61070130, 60903176, 60873089), the Program for New Century Excellent Talents in University (NCET-10-0863), the Shandong Distinguished Middle-aged and Young Scientist Encourage and Reward Foundation of China (BS2009SW003), and the Shandong Provincial Key Laboratory of Network Based Intelligent Computing.

References

1. Xiangyu, W., Shouliang, C., Mingde, G.: General Biology (Version 2). Higher Education Press, Beijing (2005)
2. Marc Robinson, R., Adam, G.: Structural genomics of Thermotoga maritime proteins shows that contact order is a major determinant of protein thermostability. Structure 6, 857–860 (2005)

3. Changhui, Y., Vasant, H., Drena, D.: Predicting Protein-Protein Interaction Sites From Amino Acid Sequence. Technical report. Iowa State University (2002)
4. Zhang, G., Fang, B.: LogitBoost classifier for discriminating thermophilic and mesophilic proteins. Journal of Biotechnology 127(3), 417–424 (2007)
5. Fredj, Y., Edouard, D., Bernard: Amino acid composition of genomes, lifestyles of organisms, and evolutionary trends: a global picture with correspondence analysis. Gene. 297, 51–60 (2002)
6. David, J.L., Gregory, A.C., Donal, A.H.: Synonymous codon usage is subject to selection in thermophilic bacteria. Nucl. Acids Res. 30, 4272–4277 (2002)
7. Zhang, G., Fang, B.: Discrimination of thermophilic and mesophilic proteins via pattern recognition methods. Process Biochemistry 41(3), 552–556 (2006)
8. Chen, C., Chen, L., Zou, X., et al.: Predicting protein structural class based on multi-features fusion. Journal of Theoretical Biology 253, 388–392 (2008)
9. Chou, K.C.: Prediction of protein cellular attributes using pseudo-amino-acid-composition. PROTEINS: Struct. Func. Genetics 43, 246–255 (2001)
10. McCulloch, W.S., Pitts, W.A.: Logical Calculus of the Ideas Immanent in Nervous Activity. Bulletin of Mathematical Biophysics 5, 115–133 (1943)
11. Kennedy, J., Eberhart, R.C.: Particle swarm optimization. In: Proceedings of the 1995 IEEE International Conference on Neural Networks, Perth, vol. 4, pp. 1942–1948 (1995)
12. Hansen, I.K., Salamon, P.: Neural network ensembles. IEEE Trans. Pattern Anal. 12(10), 993–1001 (1990)
13. Zhihua, Z., Jianxin, W., Wei, T.: Ensembling neural networks: Many could be better than all. Artif. Intell. 137, 239–263 (2002)
14. Xu, J., Chen, Y.: Discrimination of thermophilic and mesophilic proteins via artificial neural network. In: Advances in Neural Networks, ISNN 2011 (in press, 2011)
15. Zhang, G., Fang, B.: Application of amino acid distribution along the sequence for discriminating mesophilic and thermophilic proteins. Process Biochemistry 41(8), 1792–1798 (2006)

Visual Analytics
of Clinical and Genetic Datasets
# of Acute Lymphoblastic Leukaemia

Quang Vinh Nguyen[1], Andrew Gleeson[1], Nicholas Ho[2], Mao Lin Huang[3],
Simeon Simoff[1], and Daniel Catchpoole[2]

[1] School of Computing and Mathematics, University of Western Sydney, Australia
[2] The Kids Research Institute, Children's Hospital at Westmead, Australia
[3] Faculty of Engineering & IT, University of Technology, Sydney, Australia

Abstract. This paper presents a novel visual analytics method that incorporates knowledge from the analysis domain so that it can extract
knowledge from complex genetic and clinical data and then visualizing
them in a meaningful and interpretable way. The domain experts that
are both contributors to formulating the requirements for the design of
the system and the actual user of the system include microbiologists,
biostatisticians, clinicians and computational biologists. A comprehensive prototype has been developed to support the visual analytics process. The system consists of multiple components enabling the complete
analysis process, including data mining, interactive visualization, analytical views, gene comparison. A visual highlighting method is also implemented to support the decision making process. The paper demonstrates
its effectiveness on a case study of childhood cancer patients.

Keywords: Visual analytics, Visualization, Microarray, Acute lymphoblastic leukamedia, Gene expression.

1   Introduction

The human genome contains a large amount of variations, and the simplest but
most abundant type of genetic variations among individuals is single nucleotide
polymorphisms (SNPs). Variations in SNPs can affect how humans develop diseases and respond to pathogens, chemicals, drugs, vaccines, and other agents.
For example, it is recognised that SNPs have high potential to be used as prognostic markers [1, 2]. As the activity of genes are co-regulated when one gene is
switched on, many other genes are also switched on or off, it is generally believed
that metabolic actions are driven by small changes in expression of a large number of genes rather than large changes in only a few. However, it is still unclear
exactly why particular genes are good indicators of patient outcome as it is not
known whether these genes directly cause the metabolic effects or whether they
are in the co-regulating ensemble.

Despite presenting with similar clinical features, ALL patients do not always
respond in a similar manner to the same treatment strategies. In other words,

B.-L. Lu, L. Zhang, and J. Kwok (Eds.): ICONIP 2011, Part I, LNCS 7062, pp. 113–120, 2011.

the underlying complexities of the disease are not always clearly reflected by the clinical presentation nor pathological results. As a result, understanding biological data (such as microarray and SNPs data) as well as clinical data (such as annotated clinical attributes, treatment details, domain ontologies and others) is crucial to improving the diagnosis and treatment. The analysis of the genetic data could bring the insight of information as well as the discovery of relationships, non-trivial structures and ir(regularity). Techniques that are acceptable to medical scientists should integrate rigorous automated analysis with the interactive interfaces whose visualizations are clearly interpretable, navigable and modifiable by medical analysts, so that in the decision making process they can interact with the automated analysis through these visualizations.

Recent biological data analysis techniques that provide reasonably advanced visualization were presented in [3–8]. Although these techniques provide somewhat effective ways for the analysis of data, they however do not fully provide a deep analysis mechanism and/or a complete platform so that medical analysts can interactively explore and manipulate the information. No matter how powerful the visualization or the visual analytics tools are, they are not useful if they cannot be interpreted by the domain experts in the field.

Presented work was developed with strong involvement from medical scientists at the Children's Hospital at Westmead, Sydney, Australia, who are experts in genetic research, especially related to ALL. The philosophy behind the work is that with this new way of looking at genetic and clinical details both from the perspectives of domain knowledge and technologies, visual analytics can provide a medium for the discovery of genetic and clinical problems leading to, potentially, improved ALL treatment strategies.

2   Data Set and Automated Analysis

The expression and genomic profiles of 100 paediatric B-cell ALL patients treated at the Children's Hospital at Westmead were generated using Affymetrix expression microarrays (U133A, U133A 2.0 and U133 Plus 2.0) and Illumina NS12 SNP microarrays respectively. Each Affymetrix expression microarray has over 22 thousands attributes whilst each Illumina SNP microarray has almost 14 thousands attributes. Each attribute is mapped to a gene and the value for each attribute corresponds to the expression levels or genotype for the gene. In addition, the details (annotations) for each gene are mapped on separate files. Expression microarrays were hybridised with diagnostic bone marrow samples and genomic microarrays with remission peripheral blood samples.

Using these datasets, we developed a model predictive of treatment outcome by identifying genes capable of differentiating patients that survived and those that did not. To achieve this, we applied an attribute deletion approach: identifying and removing genes that are almost certainly not involved in a biological phenomenon. We used the attribute importance ranking implemented in Random Forest [9] to identify these genes, called genes of interests (GOIs) [10]. The expression values were z-score normalised within each platform, concatenated and z-score normalised to minimise inter- and intra-platform biases. The expression and SNP values for the top 250 GOIs were used to create a 3-dimensional (3D) similarity space using Singular Value Decomposition (SVD) - a matrix decomposition technique that can be used to project highly multidimensional data into lower dimensional spaces [11].

3   The Visual Analytics Platform

The automated analysis produced a 3D similarity space of the patients based on their genetic properties. A mapping table of 250 genes of interests is also created during this process. In order to present these results in a form suitable for visual analysis by medical experts we have created an interactive visualisation which operates with the treated data. From the treated data, interactive visualization is applied to 1) present a meaningful display of the patient cohort 2) filter, explore and manipulate the information 3) interactively provide the details of both original and processed data on demands and 4) highlight patterns and/or abnormality in supporting the decision making process.

3.1   Overview of the Patients

A design of the prototype version of the new visual analytics system provides the set of views addressing the above described needs. It provides the global overview of the entire patient's structure, allowing to grasp the "big picture" before drilling into the details. Displaying the entire visual structure at once allows analysts to move rapidly to any location in the space so that the navigation across the large structures and exploration of specific parts of it is effortless.

The patients' positions are mapped into the 3D space according to their genetic similarity. The similarity was defined by applying the Random Forest technique on the 250 GOIs which was mentioned at section 2. Two patients are close together if their genes are similar and conversely, they are located far from each other if their genetic properties are different. Figure 1 shows the overall view of the entire patient population captured in the data set. This figure illustrates clearly two distinctive groups of patients, marked by the dash-line ellipses. The smaller group contains mostly the deceased patients (drawn with brighter and transparent outlines) whilst the other contains patients that responded well to the treatment and survived the decease. The fact that the deceased patients are closely located to each other in the gene space on the figure may support the hypothesis that genetic properties are essential to determine whether a patient is likely to respond well in the context of the clinical treatment.

Uniquely, with the incorporation of prior knowledge about the patient's clinical management and eventual outcome, it is anticipated that newly diagnosed and as yet untreated patients can then be directly compared to local neighbours on the basis that patients with similar genetic backgrounds and gene expression activities within their tumours will cluster together. By comparing how similar patients have previously been managed and their treatment outcomes, the analyst (or clinician) can glean specific information from the similar patients within

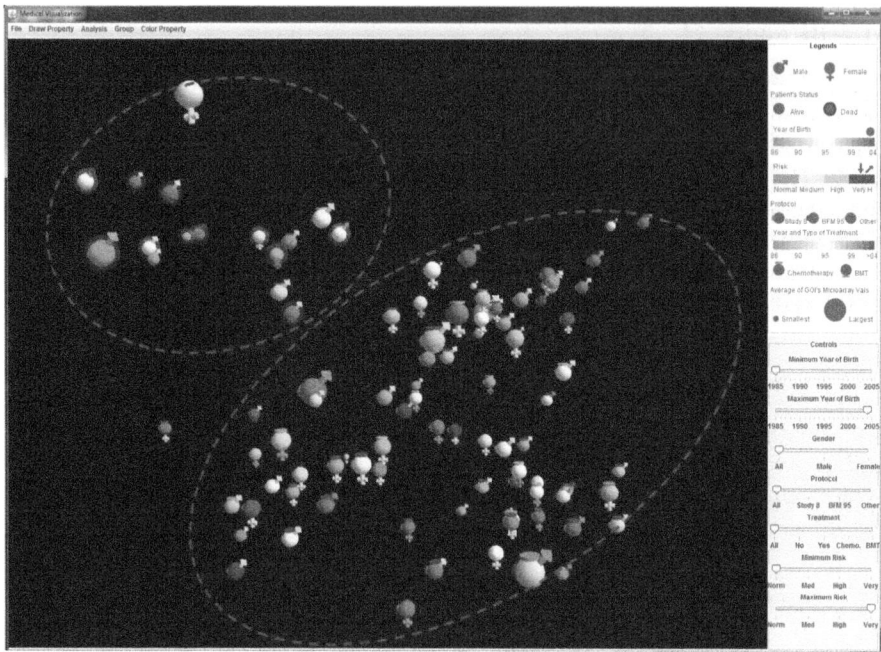

Fig. 1. An example showing 3D visualizations of the entire 100 patients

the cohort which will assist with their clinical decision making for the individual patient, thereby moving towards the development of more personalised treatment protocols. Rich graphical attributes are employed to provide the background and treatment properties of the patients. The attributes are carefully selected based on feedback from medical analysts. In addition, they can be easily adjusted, filtered and re-mapped via an interactive menu to suit the preferences of different users.

3.2   Interactive Navigation

From the visualization, users can interactively filter out uninteresting patients to enhance their visualization and the visualization of related data. The filtered patients are displayed dimly at the background using darker colours and transparency. Figure 2a indicates that all the patients born after the year 2000 and before the year 1995 have been filtered out and they are displayed transparently at the background. The treatment property and protocol figures have also been filtered out in this figure. Figure 2b illustrates another further stage where showing only high to very high risk patients. The interactive visualization interface allows the researcher/clinician (domain expert) to extract, picture and interrogate specific features for the patients.

From the overview of all patients, users can select one or multiple patients to analyse further. At first glance, the analytical view provides the full information of both genetic and clinical data of the selected patients. Layout of the panels is as follows: 1) the top-left panel displays information of Affymetrix expression microarrays; 2) the top-right panel displays information of Illumina SNPs and 3) the bottom panels displays all background and treatment information of the patient. For example, we select three patients highlighted at Figure 2b by the red-dash ellipses where the deceased patient ALL123 is at the top-left and the survived patients ALL26 and ALL302 are located close each other at the near bottom-right. Figure 3 shows the detail views of the three selected patients. The figure indicates that the gene expression values of the patient ALL123 is significantly different from the other two patients ALL26 and ALL302. This property confirms the result of generating the 3D similarity space of the patients based on their genetic properties.

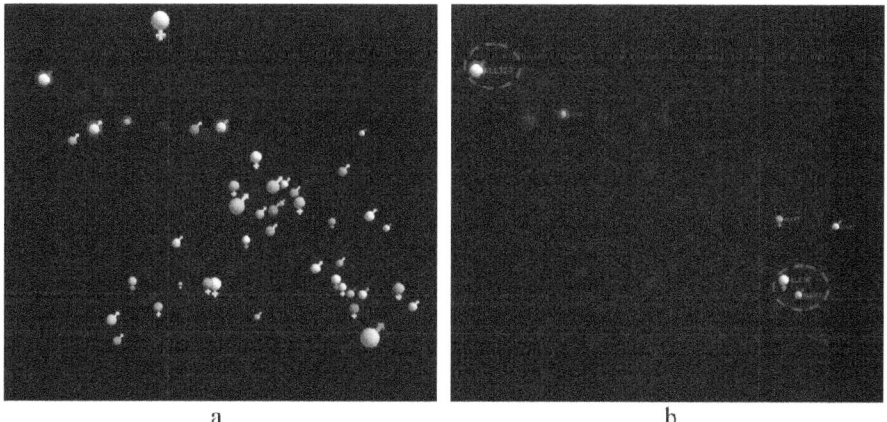

a   b

Fig. 2. An example of the visualizations at different navigational stages

3.3   Genes of Interests Analysis

The Gene Comparison Interactive Visualization is a tool we designed to allow the analyst to drill down further into the genetics and treatment data of patients identified as significant by the processes described in previous sections. This visualization figure implements several mechanisms whereby the similarity and differences between patients and groups of patients can be examined in greater detail.

An important feature of this interactive visualization is the use of active regions to indicate what the analyst is currently focused on. Knowing this makes it possible to infer what extra information the analyst might find useful, and to make sure that it is available. This interactive visualization consists of 5 separate visual components 1) the primary patient probeset heat-map component, 2) the gene zoom component, 3) the colour gradient and Gaussian curve component, 4) the gene ontology and probeset annotation component, and 5) the patient treatment data component.

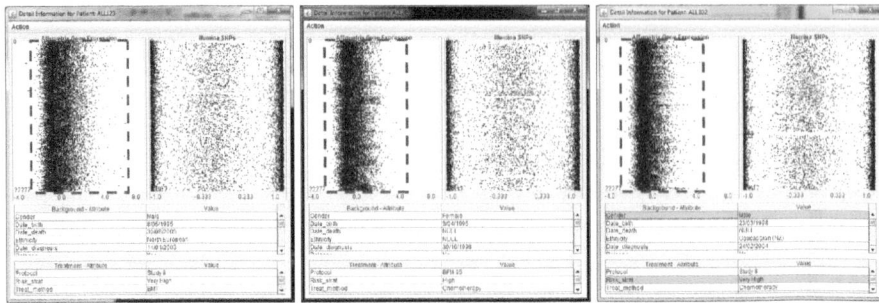

Fig. 3. An example of the comparison windows of the analytical views of three patients, namely ALL123, ALL26 and ALL302 respectively

Figure 4 shows an example of the gene comparison visualization of multiple patients. Patient's heat-maps are laid out one below the other in such a way to make sure that, for each patient, the cells representing particular genes and probesets line up vertically on the screen. The upper-layer represents the Illumina SNPs values whilst the lower-layer represents the Affymetrix expression microarrays. The significantly different probesets are also highlighted. This representation makes details like unusual measurements, and unusual variations of measurements between patients, stand out easily to the eye. For example, in this figure, it can be readily seen that the patient ALL144 might have significant difference in the expression values at the gene CYP1B1, HNMT and TTN.

4   Analytical Discussions

The above prototype has been used to explore and analyse the genetic and clinical details both from the perspectives of domain knowledge and technologies. In spite of the size limitation of the data set due to the expensive data collection, the discovery is quite encouraging in the case of ALL patients. The pilot results will enable further enhancement and verification on more comprehensive data. Details of the findings (and confirmation of the prognoses) are further described as following:

1. *There is little coherence between the genetic property and the background or the clinical property, such as age, gender, dead or survival rate.* In other words, similar patients in term of background and clinical information might have a significant difference in their genetic properties.

2. *Early treatments are more effective than late treatments.* Particularly, there is only one death out of 22 (95.5% chance of survival) if the treatment was started within a year after the birth. This is also a special case for very high risk patients who are likely dead because of the disease. If the treatment was started after 5 years from birth, the survival rate is 85% (3 dead cases out of 20). And if the treatment was used after 10 years from birth, the survival rate reduces significantly to 62.5% (3 dead cases out of 8).

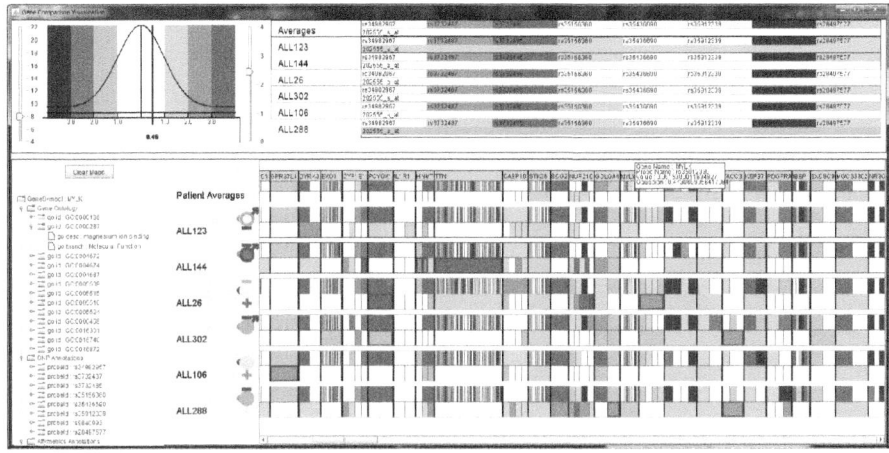

Fig. 4. Genes of Interests Comparison Visualization with multiple patients ALL123, ALL144, ALL26, ALL302, ALL106 and ALL288

3. Among the very high risk patients: ALL123, ALL143, ALL144 and ALL302, the patient ALL302 (located at the near bottom-right) was the only survivor whose genes were significantly different from the others (at the top-left). The further examination using the 250 Genes of Interests confirms the differences in the expression values, especially at the genes ABCA4, BRCA2 and ZNF267. The variation at the particular gene or the combination of the genes might contribute to the ability to survival or the improvement in treatment.

4. *Patients who were born at later years have a better chance of survival.* For example, the rate of survival for those patients born after 1995 is approximately 94% whilst the rate for those patients born before 1995 is approximately 83%. This property reflects the improvement in treatment technology, methodology and the living condition.

5. *Female patients tend to fare better than male patients in overall.* There is 92.5% chance of survival in females (3 dead cases out of 40) versus 88.3% chance of survival in males (7 dead cases out of 60).

6. *Study 8 protocol is dominant.* BFM 95 protocol is mostly applied to standard to medium risk patients who were born before year 2000. Study 8 protocol is a much more popular method and it was used for patients regardless their risk strategies and ages. There is only one case using Interfant 99 protocol.

7. *Chemotherapy is more effective in treatment.* Although nearly a half of all patients did not have any treatments, the untreated patients mostly have standard to medium risks. The BMT treatment method has little effect on the survival rate (only 1 out of 5 cases was survived). The most chosen treatment method is Chemotherapy whose survival rate is over 90%.

5   Conclusions

In this paper, we have presented a novel visual analytics approach that provides a comprehensive solution for analysing large and complex integrated genetic and clinical data. The set of interactive visualisations and the functionality of the visual data manipulation cover the needs defined by the above experts. It provides not only the overview of the importance of the patient's information but also the mechanism to zoom, to filter and to analyse further at each level of details, enabling by mapping the processed data to the original data and the clinical data to the genetic data. The pilot finding and the initial confirmation of the prognoses on the Acute Lymphoblastic Leukaemia patients are discussed comprehensively in this paper.

References

1. Goronzy, J.J., Matteson, E.L., Fulbright, J.W., et al.: Prognostic Markers of Radiographic Progression in Early Rheumatoid Arthritis. Arthritis & Rheumatism 50(1), 43–54 (2004)
2. Mei, R., Galipeau, P.C., Prass, C., et al.: Genome-wide Detection of Allelic Imbalance Using Human SNPs and High-density DNA Arrays. Genome Res. 10, 1126–1137 (2000)
3. Chao, S., Lihui, C.: Feature Dimension Reduction for Microarray Data Analysis Using Locally Linear Embedding. In: APBC, pp. 211–217 (2005)
4. Kaski, S., Venna, J.: Comparison of Visualization Methods for an Atlas of Gene Expression Data Sets. Information Visualization 6, 139–154 (2007)
5. Prasad, T.V., Ahson, S.I.: Visualization of Microarray Gene Expression Data. Bioinformation 1, 141–145 (2006)
6. Lex, A., Streit, M., Kruijff, E., Schmalstieg, D.: Caleydo: Design and Evaluation of a Visual Analysis Framework for Gene Expression Data in its Biological Context. In: 2010 IEEE Pacific Visualization Symposium, Taipeh, Taiwan, pp. 57–64 (2010)
7. Cvek, U., Rrutschl, M., Stone II, R., Syed, Z., Clifford, J.L., Sabichi, A.L.: Multidimensional Visualization Tools for Analysis of Expression Data. World Academy of Science, Engineering and Technology 54, 281–289 (2009)
8. Kilpinen, S., Autio, R., Ojala, K., et al.: Systematic Bioinformatics Analysis of Expression Levels of 17,330 Human Genes Across 9,783 Samples from 175 Types of Healthy and Pathological Tissues. Genome Biology 9(9), R139 (2008)
9. Breiman, L.: Radom Forests. Machine Learning 45, 5–32 (2001)
10. Ho, N., Morton, G., Skillicorn, D., Kennedy, P.J., Catchpoole, D.: Datamining gene expression and genomic microarray datasets of Acute Lymphoblastic Leukaemia. In: The Lowy Symposium: Discovering Cancer Therapeutics. UNSW, Sydney (2010)
11. Golub, G.H., Van Loan, C.F.: Matrix Computations. Johns Hopkins University Press, Baltimore (1996)

Complex Detection Based on Integrated Properties

Yang Yu[1], Lei Lin[2], Chengjie Sun[2], Xiaolong Wang[1,2], and Xuan Wang[1]

[1] Department of Computer Science and Technology, Harbin Institute of Technology Shenzhen Graduate School, Shenzhen, P.R. China
[2] School of Computer Science and Technology, Harbin Institute of Technology, P.R. China
{yuyang,linl,wangxl,cjsun,wangxuan}@insun.hit.edu.cn

Abstract. Most of current methods mainly focus on topological information and fail to consider the information from protein primary sequence which is of considerable importance for protein complex detection. Based on this observation, we propose a novel algorithm called CDIP (Complex Detection based on Integrated Properties) to discover protein complexes from the yeast PPI network. In our method, a simple feature representation from protein primary sequence is presented and become a novel part of feature properties. The algorithm can consider both topological and biological information (amino acid background frequency), which is helpful to detect protein complex more efficiently. The experiments conducted on two public datasets show that the proposed algorithm outperforms the two state-of-the-art protein complex detection algorithms.

Keywords: protein complex, PPI network, background frequency.

1 Introduction

Protein complexes are key molecular entities to perform important cellular functions in the PPI network. The detection about the components of these units from high-throughput experimental approaches on a proteome wide scales often suffers from high false positive and false negative rate [1] and some of them can not be detected under the given conditions. Protein complexes detection remains a challenge task in the present study by means of the computational methods.

Previous studies have provided alternative approaches for complex detection. MCODE [2] uses vertex weight to grow clusters from a starting vertex of high local weight by iteratively adding neighbor vertices with similar weights. It generates small complexes and some of complex often is two large. Cfinder [3] is a clique-finding algorithm to identify fully connected subgraphs of different minimum clique size and merges cliques based upon their percentage of shared members. CFinder results vary widely with each increment of minimum clique size. MCL [4] detects protein complexes by simulating random walks in the weighted or unweighted graph with iterative expansion and inflation operators. Amin et al. proposes a cluster periphery tracking algorithm (DPclus) to detect protein complexes by keeping tracking of the periphery of a detected cluster [5]. Limin et al. modifies the DPclus to detect protein complexes based on two topological constraints and reduce the number of parameters of the algorithm in IPCA [6]. King et al. proposes the restricted neighbors searching

B.-L. Lu, L. Zhang, and J. Kwok (Eds.): ICONIP 2011, Part I, LNCS 7062, pp. 121–128, 2011.

clustering (RNSC) to detect protein complexes based on both graph-theoretical and gene-ontological properties [7]. Leung at el. proposes the CORE algorithm based on a statistical framework to identify protein complex cores [8]. Wu at el. presents a core-attachment method which detects protein complex in two stages: core complex and attachment complex [9]. Qi et al. proposes a supervised graph clustering framework to predict protein complex with learning the topological and biological properties of the known complexes [10]. But the learned knowledge could be bias and affect the complex formation because of the limited training data. Although lots of clustering algorithms have been proposed to improve the performance of the protein complex detection, the limitation is that most of the methods are mainly based on observation of the topological structures and can not focus on the biological information within protein amino acid sequence.

To author's knowledge, there are few methods focusing on the primary sequence information for this task. Thus, in this paper, we present a novel graph clustering method, named complex detection based on integrated properties (CDIP), where the topological properties and protein amino acid background frequency are combined in the clustering process and it can help obtain insights into both the topological properties and functional organization of protein networks in cells [10]. Moreover, the similarity measure, namely cosine similarity, is introduced to locate protein complex from biological information. The topological properties are based on the fact that proteins are relatively connected densely in the complex [2] and protein amino acid background frequency is virtually axiomatic fact that "sequence specifies structure", which gives rise to an assumption that knowledge of the amino acid sequence might be sufficient to estimate the interacting property between two proteins for a specific biological function [11].

The remainder of this paper is organized as follows: Section 2 describes the proposed clustering method. Section 3 presents experimental results. Finally, Section 4 concludes this paper.

2   Method

In our framework, PPI network can be modeled as an undirected graph G= (V, E), in which V is the node set, corresponding with proteins and E is the edge set, representing interactions between pairs of proteins. Complex is presented as a subgraph in the whole PPI network. Features and the proposed method are described in detail for protein complex detection in the followings.

2.1   Features

In PPI network, complexes can be determined by multiple factors, such as node size, node degree distribution, substructure topology and biological properties. Thus, exploring features from subgraphs in the clustering process can measure the differences between complexes. 32 values are extracted as a vector to describe a complex. These properties include 12 values about subgraph topological structures and 20 biological values from protein amino acid background frequency. For cluster degree distribution part, mean, variance, median and maximum are used. Mean, variance and maximum properties are adopted in clustering coefficient and topological coefficient, respectively. Each is described in the following parts.

- Cluster size is the number of the nodes in a subgraph.
- Cluster density is defined in (1). $|V|$ is the number of vertexes and $|E|$ is the number of edges in a subgraph.

$$den(G) = \frac{2 * |E|}{(|V| * (|V| - 1))}$$ (1)

- Cluster degree distribution describes the degree of the nodes in a subgraph. Node degree is the sum of the direct neighbors of a node.
- Clustering coefficient [5] C_n for a node n is defined in (2):

$$C_n = 2e_n \Big/ (k_n(k_n - 1))$$ (2)

where k_n is the number of neighbors for the node n and e_n is the number of connected pairs between all neighbors of this node [2].

- Topological coefficient T_n for a node n with K_n neighbors is computed as (3):

$$T_n = average(J(n,m)) / K_n$$ (3)

$J(n,m)$ is the number of neighbors shared between n and m, plus one if there is a directed link between n and m. m is the node sharing at least one neighbor with n.

- As for biological properties, the frequency is defined in (4):

$$freq(C_i) = \frac{sum(C_i)}{\sum_{k=1}^{s} len(p_i)}$$ (4)

where C_i is a kind of amino acid among twenty amino acids, $sum(C_i)$ is the count of this amino acid C_i appearing in a subgraph, $len(p_i)$ is the amino acid sequence length of protein p_i in a subgraph, s is the size of subgraph.

2.2   The Proposed Algorithm

Our algorithm operates in four stages, including seed selection, cluster enlarge, enlarge judgment and filter process. Seed selection is based on the fact that protein complexes within PPI network form stable subgraph of interactions in same community [2]. In the part of enlarge judgement, two key parameters are applied to select the candidate subgraph. One is the topological constraint IN_{vk} [6], the other is the included angle cosine ($\cos\theta$) to measure the intrinsic similarity between interaction proteins. They are defined in (5) and in (6) respectively.

$$IN_{vk} = \frac{m_{vk}}{n_k} \qquad (5)$$

where m_{vk} is the number of edges between the vertex v and the subgraph k. n_k is the size of subgraph k.

$$\cos\theta = \frac{\sum\limits_{m=1}^{p} x_{im} x_{jm}}{\sqrt{\sum\limits_{m=1}^{p} x^2_{im} \sum\limits_{m=1}^{p} x^2_{jm}}} \qquad (6)$$

where p is the size of the vector for a protein (p=32), x_{im} and x_{jm} are the mth value of the vectors of protein i and protein j respectively.

Algorithm
1:input the PPI network G= (V, E);
2:initializing the set of selected clusters $DC = \varnothing$
3:calculating the degree of each node, ranking all nodes by non-increasing according to the degree and sorting the ranked nodes into Q;
4: while (Q! = ϕ) {select the first node v in Q and put it into the current cluster K. call Enlarge Cluster(K); call Filter Process(K); }
5: Output clusters in DC.
Enlarge cluster
1:Finding the neighbors of K and sorting neighbors to N by non-increasing according to degree;
2:For each node in N ; if a node satisfies the enlarge judgement condition in (7), add the node to the K; enlarge Cluster (K); else continue to select the nodes from other neighbors of K.
3: If no node is extended and the current cluster K can not be further extended, insert K to DC.
Filter process
1: $S = mostsimi\omega(D_i, K)$, $D_i \in DC$
2: if $\omega(S, K) < t$, insert K into DC, else if $den(K) * |V_k| \geq den(S) * |V_s|$, insert S into DC and delete K from DC else discard K; delete the nodes in K from Q.

The enlarge judgment is a key stage. When the algorithm locates a cluster, the cluster is restricted by $\cos\theta$ and IN_{vk}. If v satisfies the constraint condition in (7) at the same time, v will be added into a subgraph K. The types of related parameters $\delta = 0.7$ and $\mu = 0.8$ are obtained in our experiment for the best f-measure.

$$N_{vk} \geq \delta \text{ and } \cos\theta \geq \mu \quad \delta, \mu \in [0, 1] \qquad (7)$$

The filter process is to obtain meaningful complexes, in which S is the most similar graph in DC. Given two graphs, please see their judgement in equation (8) of the section 3. In step 2, t=0.2 is adopted.

3   Results and Discussion

3.1   Reference Datasets and Evaluation Matrices

Two reference datasets of protein complex with size no less than 4 are used. The first set comprises of hand-curated complexes MIPS [12] and the other set is generated from CYC 2008 catalog [13]. PPI dataset about yeast is from DIP [14].

Given a set of real complex $R=\{R_1, R_2, ..., R_n\}$ and a set of predicted clusters $C=\{C_1, C_2, ..., C_n\}$, the detected protein complex is matched with the real complex in the benchmark set by the evaluation score in (8):

$$\omega(C_i, R_j) = \frac{|C_i \cap R_j|^2}{|C_i| \times |R_j|} \tag{8}$$

Here, $|C_i \cap R_j|$ is the size of the interaction set between a detected complex C_i and a known complex R_j, $|C_i|$ is the size of detected complex and $|R_j|$ is the size of known complex. In [2, 5], a detected cluster is assumed to match a known complex if its overlapping score is at least 0.2, which is also adopted in this study. For comparison, recall, precision, f-measure and coverage rate are used [9] and they are defined as followings:

$$precision = \frac{|P_m|}{|C|}, P_m = \{C_i \mid C_i \in C \wedge \exists R_j \in R, R_j matchs C_i\} \tag{9}$$

$$recall = \frac{|K_m|}{|R|}, K_m = \{R_i \mid R_i \in R \wedge \exists C_j \in C, C_j matchs R_i\} \tag{10}$$

$$f-measure = \frac{2 * recall * precision}{recall + precision} \tag{11}$$

$$CR = \frac{\sum_{i=1}^{k} \max_j \{C_{ij}\}}{\sum_{i=1}^{k} N_i} \tag{12}$$

Where C_{ij} is the number of the common proteins between the ith real complex and the jth detected cluster N_i is the number of proteins in the ith real complex.

3.2   Comparison Results

In order to show the efficiency of CDIP, we conduct the experiments on two benchmark datasets in the DIP network. Since MCODE and CFinder predict different number of clusters with different parameters, the default parameters are utilized. Table 1 shows the comparison results of predictions among the three methods. It can be observed that CDIP performs consistently better than CFinder and MCODE on all evaluation matrices from two datasets, which may be due to the combination of biological and topological properties. The best f-measures are provided by the highest recall and precision, which illustrates that our method can match more real complexes and predict complex more accurately. Moreover, the highest coverage rates show that located complexes can cover the most proteins in the benchmark datasets.

Table 1. The comprehensive comparison results in DIP PPI network

methods	MIPS (complexes:109)			CYC2008(complexes:148)		
	CDIP	CFinder	MCode	CDIP	CFinder	MCode
$|C|$	181	112	40	181	112	40
$|P_m|$	65	26	14	91	42	16
$|K_m|$	52	28	19	73	41	23
recall	0.477	0.257	0.174	0.493	0.277	0.155
precision	0.359	0.232	0.35	0.503	0.375	0.4
f-measure	0.410	0.244	0.233	0.498	0.319	0.224
CR	0.289	0.212	0.250	0.368	0.229	0.309

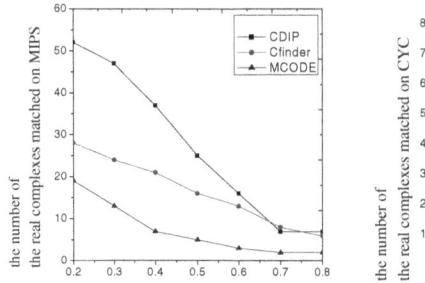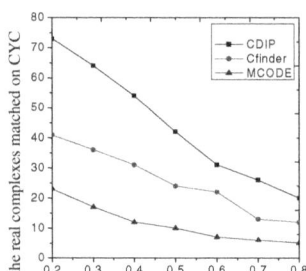

Fig. 1. The robustness to seven thresholds

In order show the robustness of the performance improvement of our method, three method are compared on seven thresholds t= {0.2, 0.3, 0.4, 0.5, 0.6, 0.7, 0.8} in terms of the number of real complexes matched on two individual datasets in Fig.1. CDIP can match more real complexes than other two methods, except t=0.7 on benchmark MIPS.

3.3   Functional Annotation of Network Modules

The basic hypothesis is that proteins in a reliable protein complex are shown to share the same function and the functional identities of proteins in the predicted complexes may be an alternative index to assess the reliability of predictions [15]. To validate this idea, the located modules were annotated and the comparison was made between the three methods based on the annotated functional of Yeast genes in MIPS Functional Catalog database. Fig. 2 shows the functional annotations ratios of the located modules of the three methods. Located modules were assigned benchmark dataset of funcat by analysis of constituent protein function. We have computed the ratio of the number of proteins belonging to the specific function over the located modules size. The modules with the ratio at least 0.5 are selected and the first three function class ratios are plotted. The column shows the proportion of the located modules of each method belongs to the first three specific function classes. We found that ratio of our methods is 86%, which is 10% and 32% higher than CFinder and MCODE, respectively. The reports illustrate that integrated properties are in favor of mining more functional modules. Moreover, it is helpful to infer protein function type. As shown in Fig. 3, the complex with size 14 is located by our method, 13 proteins belong to the splicing function and there is a new protein YKR002W.

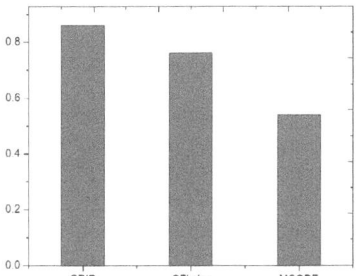 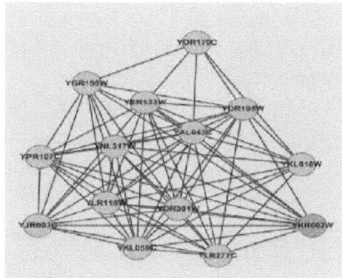

Fig. 2. Comparison of functional annotation ratios   **Fig. 3.** One example of predicted cluster

4   Conclusions and Future Work

We have proposed a new method for protein complex detection, which combines topological and biological properties to locate clusters. The feature based on protein primary sequence is proposed and become a novel part of features and both are investigated for a complex. The experimental results have shown that this algorithm clearly outperforms the two clustering algorithms. We validate the detected complexes using function analysis which shows that the CDIP is favor of locate more biological modules.

Acknowledgments. This work is supported by the National Natural Science Foundation of China (61073127 and 60973076) and Research Fund for the Doctoral Program of Higher Education of China (20102302120053) and the Fundamental Research Funds for the Central Universities (Grant No. HIT.KLOF.2010061) and the opening foundation of Shanghai Key Laboratory of Intelligent Information Processing (IIPL-2010-005).

References

1. von Mering, C., Krause, R., Snel, B., Cornell, M., Oliver, S.G., Fields, S., Bork, P.: Comparative assessment of large-scale data sets of protein-protein interactions. Nature 417(6887), 399–403 (2002)
2. Bader, G.D., Hogue, C.W.: An automated method for finding molecular complexes in large protein interaction networks. Bmc Bioinformatics 4 (2003)
3. Adamcsek, B., Palla, G., Farkas, I.J., Derenyi, I., Vicsek, T.: CFinder: locating cliques and overlapping modules in biological networks. Bioinformatics 22(8), 1021–1023 (2006)
4. Van Dongen, S.: Graph clustering by flow simulation. University fo Utrech (2000)
5. Altaf-Ul-Amin, M., Shinbo, Y., Mihara, K., Kurokawa, K., Kanaya, S.: Development and implementation of an algorithm for detection of protein complexes in large interaction networks. Bmc Bioinformatics 7 (2006)
6. Li, M., Chen, J.E., Wang, J.X., Hu, B., Chen, G.: Modifying the DPClus algorithm for identifying protein complexes based on new topological structures. Bmc Bioinformatics 9 (2008)
7. King, A.D., Przulj, N., Jurisica, I.: Protein complex prediction via cost-based clustering. Bioinformatics 20(17), 3013–3020 (2004)
8. Leung, H.C., Xiang, Q., Yiu, S.M., Chin, F.Y.: Predicting protein complexes from PPI data: a core-attachment approach. J. Comput. Biol. 16(2), 133–144 (2009)
9. Wu, M., Li, X.L., Kwoh, C.K., Ng, S.K.: A core-attachment based method to detect protein complexes in PPI networks. Bmc Bioinformatics 10 (2009)
10. Qi, Y., Balem, F., Faloutsos, C., Klein-Seetharaman, J., Bar-Joseph, Z.: Protein complex identification by supervised graph local clustering. Bioinformatics 24(13), i250–i258 (2008)
11. Shen, J., Zhang, J., Luo, X., Zhu, W., Yu, K., Chen, K., Li, Y., Jiang, H.: Predicting protein-protein interactions based only on sequences information. Proc. Natl. Acad. Sci. U S A 104(11), 4337–4341 (2007)
12. Mewes, H.W., Dietmann, S., Frishman, D., Gregory, R., Mannhaupt, G., Mayer, K.F.X., Munsterkotter, M., Ruepp, A., Spannagl, M., Stuempflen, V., et al.: MIPS: analysis and annotation of genome information in 2007. Nucleic Acids Research 36, D196–D201 (2008)
13. Pu, S.Y., Wong, J., Turner, B., Cho, E., Wodak, S.J.: Up-to-date catalogues of yeast protein complexes. Nucleic Acids Research 37(3), 825–831 (2009)
14. Xenarios, I., Salwinski, L., Duan, X.Q.J., Higney, P., Kim, S.M., Eisenberg, D.: DIP, the Database of Interacting Proteins: a research tool for studying cellular networks of protein interactions. Nucleic Acids Research 30(1), 303–305 (2002)
15. Maraziotis, I.A., Dimitrakopoulou, K., Bezerianos, A.: Growing functional modules from a seed protein via integration of protein interaction and gene expression data. Bmc Bioinformatics 8 (2007)

Exploring Associations between Changes in Ambient Temperature and Stroke Occurrence: Comparative Analysis Using Global and Personalised Modelling Approaches

Wen Liang, Yingjie Hu, Nikola Kasabov, and Valery Feigin

Knowledge Engineering and Discovery Research Institute,
Auckland University of Technology, New Zealand
{linda.liang,raphael.hu,nikola.kasabov,valery.feigin}@aut.ac.nz

Abstract. Stroke is a major cause of disability and mortality in most economically developed countries that increasing global importance. Up till now, there is uncertainty regarding the effect of weather conditions on stoke occurrence. This paper is offering a comparative study of exploring associations between changes in ambient temperature and stroke occurrence using global and personalised modelling methods. Our study has explored weather conditions have significant impact on stroke occurrence. In addition, our experimental results show that the personalised modelling approach outperforms the global modelling approach.

Keywords: weather, stroke occurrence, personalised modelling, global modelling, FaLK-SVM.

1 Introduction

Stroke is known as an acute cerebrovascular disease (CVD), it can cause neurological damages or even death (particularly in the elderly) by the reason of the blood supply suddenly disrupted or stopped to part of the brain. It is becoming a major public health concern and challenge in many countries.

Recently, there is increasing evidence linking weather conditions and stroke occurrence [1][2]. However, thus far, only few studies on exploring the effect of weather on stroke occurrence, and most of these studies examined stroke occurrence have also been inconsistent [3][4][5], which remains a matter of uncertainty and controversy. From early evidence, environmental triggers of different stroke subtypes are dependent to age, gender and climate characteristics. However, these data are selection bias (e.g. unclear CT/MRI verification of different stroke subtypes), or no reliable data exists in various population groups (e.g. by age, gender, and region).

Previous studies attempted using different techniques for studying complex stroke data. These techniques can be generally divided into two categories, statistical and machine learning methods (e.g. conventional statistical methods are more widely applied, in particular). However, in many cases, the conventional

B.-L. Lu, L. Zhang, and J. Kwok (Eds.): ICONIP 2011, Part I, LNCS 7062, pp. 129–137, 2011.

statistical methods have limitations in efficiency and improving the prediction accuracy compared to machine learning methods. Khosla et al. [6] presented an integrated machine learning approach that significantly outperformed the Cox proportional hazards model (one of the most popular used statistical methods in medical research) on the Cardiovascular Health Study (CHS) dataset for stroke risk prediction.

Personalised modelling is an emerging machine learning approach, where a model is created for every single new input vector of the problem space based on its nearest neighbours using transductive reasoning approach [7]. The basic philosophy behind this approach when applied to medicine is that every person is different from others, thus he/she needs and deserves a personalised model and treatment that best predicts possible outcomes for this person. Such characteristic makes personalised modelling an appropriate method for solving complex modelling problems.

This paper therefore presents a comparative analysis using global and personalised modelling methods to explore associations between changes in ambient temperature and stroke occurrence. This knowledge will contribute to the understanding of environmental triggers of stroke. In turn, this will help identify other new areas of research, such as physiological studies on weather-stroke associations or clinical trials, to test preventive strategies to reduce the hazardous effects of harmful weather conditions.

The remainder of this paper is organized as follows. Section 2 briefly reviews global and personalised modelling methods. Section 3 describes a recently developed personalised modelling method, Fast Local Kernel Support Vector Machines (FaLK-SVM). Section 4 provides the experimental results of the comparative study. Finally, section 5 gives the conclusion and future direction.

2   Background and Related Work

2.1   Global Modelling

A global model is created from the entire data set for the whole problem space based on the inductive inference method. It focuses on the whole problem space rather than individual vectors. This model is usually difficult to be adapted on new incoming input vectors.

Support vector machine (SVM) is one of popular global modelling algorithm, which has been widely used to deal with regression and classification problems. It is a powerful tool for separating a set of binary labeled data in a feature space by an optimal hyperplane. The two major types of SVM used far and wide, are linear SVM [8] and non-linear SVM [9].

2.2   Personalised Modelling

A personalised model is created for every single new input vector of the problem space based on its nearest neighbours using the transductive reasoning approach

[7]. It is more concerned with solving an individual given problem rather than solving a general problem across the whole population.

Personalised modelling has been successfully applied to deal with a variety of modelling problems such as personalised drug design for known diseases (e.g. cancer, diabetes, brain disease) as well as for other modelling problems in ecology, business, finance, crime prevention. Nowadays, personalised medicine is an emerging trend in the research areas of medicine, health care and life science. Ginsburg and McCarthy [10] present the objective of personalised medicine to determine a patient's disease at the molecular level, so the right therapies are able to be applied on the right people at the right time. Multiple examples have significantly proved that the traditional form of medicine is declining in favor of more accurate marker-assisted diagnosis and treatment.

The basic principle and framework of personalised modelling is summarized in Figure 1:

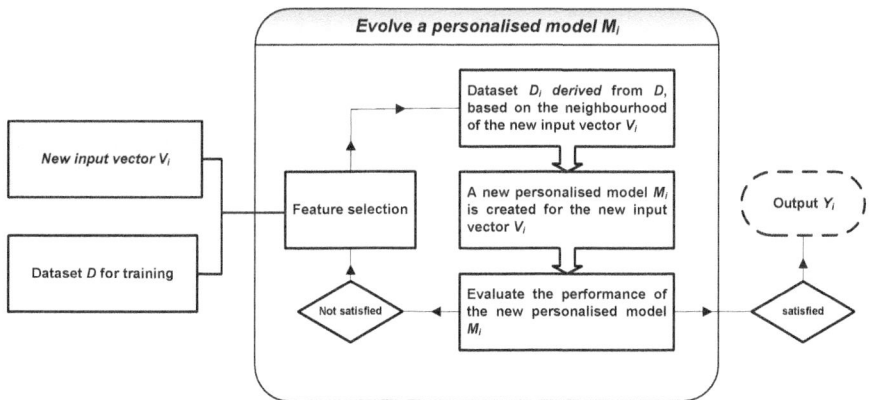

Fig. 1. Basic principle and framework of personalised modelling

K-nearest neighbour (KNN) is the simplest personalised modelling algorithm, was originally proposed by Fix and Hodges in 1951 [11]. It is a supervised learning algorithm that has been widely used for classifying sets of samples based on nearest training samples in a multi-dimensional feature space by using some suitable distance metric such as Euclidean distance or Manhattan distance.

Weighted K-Nearest Neighbour (WKNN) is a extension of KNN as developed by Dudani in 1976 [12]. In the WKNN algorithm, the output of a new input vector is calculated not only dependent upon its k-nearest neighbour vectors, but also upon the distance between the existing vectors and a new input vector (known as a weight vector w), this being the basic idea behind the WKNN algorithm.

Transductive Neural Fuzzy Inference System with Weighted Data Normalisation (TWNFI) is a dynamic neuro-fuzzy inference system with local generalization proposed by Song and Kasabov in 2006 [13]. In the TWNFI algorithm, the input variables are weighted based on their importance for the problem, derived through the back-propagation or an evolutionary optimisation algorithm.

3   Fast Local Kernel Support Vector Machines - A Recently Developed Personalised Algorithm

Fast Local Kernel Support Vector Machines (FaLK-SVM) is a fast and scalable local SVM algorithm [14]. In the FaLK-SVM algorithm, the cover tree data-structure [15] is implemented for fast retrieval of neighbourhoods in the feature space, and integrates the LibSVM for SVM training and prediction [16].

FaLK-SVM consists of training phase and testing phase. The training phase is designed to reduce the number of local models by pre-computing a set of local SVMs in the training set and assigns to each model all the points lying in the central neighbourhood of the k points on which it is trained. The testing phase is to apply to a query point that it is the nearest neighbour to the new vector in the training set.

Mathematically, the FaLK-SVM algorithm can be formulated with following equation, which derived from original formulation of KNNSVM:

$$FastLSVM(x) = sign(\sum_{i=1}^{k} \alpha_{rc(i)} y_{rc(i)} K(x_{rc(i)}, x) + b) \text{ with } c = cnt(x_{rx(l)}) \quad (1)$$

where $r_{c(i)}$ is a function to order the indexes of the training samples, $\alpha_{rc(i)}$, and b are two scalar values derived from the training of an SVM on the k-nearest neighbours of c in the feature space. c is the selected center of the local models. In this way, the total number of SVMs trained can be reduced.

$$cnt(x_i) = x_j \in C$$
$$\text{with } j = min(z \in \{1, ..., n\} | x_z \in C \text{ and } x_i \in X_{x_z}) \quad (2)$$

where $X_{x_z} = \{x_{rxz(h)} | h = 1, ..., k'\}$ With the cnt function, each training point is assigned to a unique corresponding center and thus to the SVM model trained on the center neighbourhood.

4   Experiment

4.1   Dataset and Experiment Setup

This study aims to explore the significant associations between ambient temperature and stroke occurrence using Weather and Stroke Occurrence dataset. This international collaborative study is carried out under the auspices of six population regions: Auckland (NZ), Perth and Melbourne (Australia), Oxfordshire (UK), Dijon (France), and Norrbotten and Vasterbotten counties (Northern Sweden).

The dataset consists of 11,453 samples (all with first-ever occurrence of stroke) and 9 features (4 *patient clinical features* - categorical data & 5 *weather features* - continuous data). *Patient clinical features* include information such as: age, gender, history of hypertension and smoking status. *Weather features* include information such as: temperature, humidity, wind speed, windchill and atmospheric pressure. All these weather parameters are measured for the date of stroke occurrence.

To our understanding, this work is the first study to use computational intelligent modelling techniques to investigate the associations between the weather and stroke occurrence. Many data samples are collected based on the interview and questionnaire of the patients, which leaves plenty of missing values. Since our work is a pilot study that focuses on this real world medical data, we use the data only from Auckland region and select 500 samples without missing values from the whole dataset.

We applied *case-crossover* design for the data pre-processing, because there is no "non-stroke" patients in the original dataset. We use the date of stroke occurrence (1 day lag) as the "stroke" group and 30 days before stroke occurrence (1 day lag) for the same participant as the "normal/control" group, assuming that weather parameters 30 days before the index stroke had no influence on the stroke occurrence 30 days later. This approach is known as *case-crossover* design. Mukamel and his colleagues [17] adopted case-crossover design for comparing the measures of weather and ambient air pollution on the day of presentation and control days for each patient.

We used the data by counting down 30 days from the date of stroke occurrence for creating the "normal/control" data samples. These samples are created in the following way: assuming we have a patient who firstly trigged stroke on i.e. *1 Jan 1981* - "stroke" sample. For creating the "normal/control" data sample, we count down 30 days from the date this patient trigged stroke, i.e. *1 Dec 1980*. If this patient did not have the record of stroke occurrence on *1 Dec 1980*, which means he/she was "normal/control" on that day. Then, we look for next following year stroke triggered, i.e. *1 Dec 1982* and use this measurement to create the "normal/control" data sample. This approach is based on the assumption that the weather does not have big changes for the same month but in different years.

Hence, using the *case-crossover* design approach for the experiments, the data consists of 1,000 samples (500 "normal/control" patients (class1) and 500 "stroke" patients (class 2)).

Our experiments are carried out in three steps: [(1) using only 4 patient clinical features to perform a comparative analysis of the global and personalised modelling approaches; (2) using all 9 features (4 patient clinical features and 5 weather features); (3) using 6 features (age and 5 weather features), as age is a continuous value and suggested to be used for our experiments by experts.] For all experiments in this study, *K-fold (k=3)* cross-validation is used to evaluate the performance obtained by global and personalised modelling methods.

4.2   Experiment Result and Discussion

In the first experiment, we applied all the modelling techniques on the data with 4 patient clinical features, and the best accuracy is manifested by the FaLK-SVM personalised model. Its classification accuracy is 51.69% (46.81% for class 1 - Normal, and 56.75% for class 2 - Stroke). FaLK-SVM outperforms all other methods in terms of classification accuracy. However, the accuracy obtained from FaLK-SVM with 4 patient clinical features is close to random, though the patient clinical variables such as age, gender, blood pressure and smoke are identified as very important stroke risk factors [18].

In the second experiment, we applied the same modelling techniques on the data with 9 features in order to explore whether the accuracy will be improved by taking weather features into account. The FaLK-SVM personalised model obtains the best classification accuracy as compared with all other methods, which is 65.00% (61.94% for class 1 - Normal, and 68.06% for class 2 - Stroke). It can be obviously seen that the classification accuracy is improved compared with the results achieved using only 4 patient clinical features. It is easy to elucidate that weather conditions have significant effect on stoke occurrence.

In the third experiment, we applied the same modelling techniques on the data with 6 features (age and 5 weather features), as age is a continuous value and suggested to be used for our experiments by experts. A comparison of classification performance using 6 features from SVM, KNN, WKNN, TWNFI and FaLK-SVM is summarized in Table 1.

Table 1. Experimental results in terms of model accuracy tested through 3-folds cross-validation method when using 6 features (age and 5 weather features) to perform a comparative analysis of global and personalised modelling approaches

Model	Global	Personalised			
	SVM	KNN	WKNN	TWNFI	FaLK-SVM
	(Poly kernel, $g=1$)	(k=51)	(k=51, thr=0.49)	(thr=0.5, Epochs=12)	(k=5, RBF kernel, $g=-0.8$, $c=115$)
Number of Features	6	6	6	6	6
Accuracy of Each Class (%) *Normal*	60.00	63.40	61.00	61.35	**69.90**
Accuracy of Each Class (%) *Stroke*	69.20	68.00	70.60	**72.73**	70.64
Overall Accuracy (%)	64.60	65.70	65.80	67.00	**70.27**

Note: g is gamma, thr is threshold, and c is a kernel parameter

Table 1 shows that the best classification accuracy is again achieved by FaLK-SVM. The overall classification accuracy is 70.27% (69.90% for class 1 - Normal, and 70.64% for class 2 - Stroke). The overall classification accuracy is approximately increased by 5% compared with the result obtained using 9 features, and class 1 and class 2 accuracy are significantly improved.

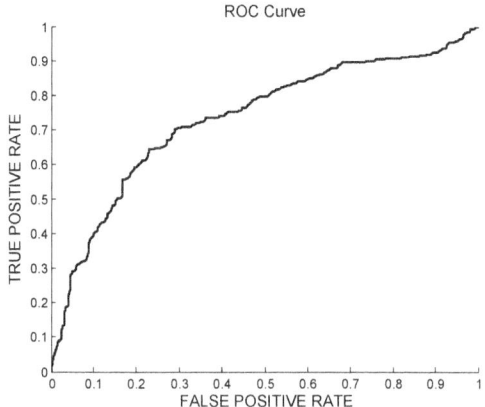

Fig. 2. Plot showing the area under the ROC curve of FaLK-SVM personalised model applied on the data with 6 Features

Figure 2 shows the classification performance obtained by FaLK-SVM personalised model on the weather-stroke data with 6 features. The classification performance is measured by Receiver Operating Characteristics (ROC) curve (area is .70). ROC graph is a widely used tool for measuring how well a parameter can distinguish between two groups (e.g. normal/diseased). Nowadays, they are becoming increasingly common in medical decision-making and diagnostic systems for evaluating medical tests [19] [20] [21].

As a general conclusion, the knowledge discovered through our experiments are: (1) weather conditions have significant impact on stroke occurrence. The overall classification accuracy is significantly improved due to taking weather features into account in the experiments. This knowledge will contribute to have good understanding of environmental triggers of stroke. In addition, this will help the health and medical experts to easily identify other new areas of research, such as physiological studies on weather-stroke associations, or clinical trials to test preventive strategies to reduce the hazardous effects of harmful weather conditions; (2) we have found out that all the models using same amount of features but different numbers of K produced different accuracy. So we can say that finding an appropriate size of testing samples' neighbors (the samples having similar data patterns) is a decisive factor to improve the accuracy, which worth to be further investigated in the future.

5   Conclusion and Future Direction

In this study, we have presented a comparative study of exploring associations between changes in ambient temperature and stroke occurrence using global and personalised modelling methods. Our study has explored weather conditions have significant impact on stroke occurrence. Our experimental results also show that the personalised modelling approach outperforms the global modelling approach on the weather-stroke data in terms of classification accuracy. The knowledge discovered from this study will bring outstanding contribution to the health and medical experts for better understanding the associations between weather and stroke. In turn, this will ensure the experts providing more accurate diagnosis and physiological treatment for individual patient.

However, this study only investigated the cases in the Auckland region and selected 500 samples out of 2850 samples as a preliminary study. Therefore, this work will be further extended to explore all samples in the Auckland region and also other five regions. Furthermore, in the future study, we will consider to explore the personalised risk for each individual patient, rather than only simply classify patients into normal or diseased group. Because accurately quantifying this risk can be helpful for medical decision support to ensure patients receive treatment that is optimal for their individual profile. In addition, the experimental results show that k-nearest neighbour could be an important factor to improve the accuracy. In this study, we manually selected neighbors and model parameters based on the training accuracy. In the future, we will develop new methods for personalised modelling in order to improve the robustness and generalisability of feature selection, neighborhood selection, model and its parameter selection for classification, diagnostic and prognostic problems. For instance, the evolutionary algorithm (EA) can be integrated with the personalised modelling approach for solving optimisation problems.

References

1. Chen, Z.Y., Chang, S.F., Su, C.L.: Weather and stroke in a subtropical area: Ilan, Taiwan. Stroke 26, 569–572 (1995)
2. Feigin, V.L., Nikitin, Y.P., Bots, M.L., Vinogradova, T.E., Grobbee, D.E.: A population-based study of the associations of stroke occurrence with weather parameters in Siberia, Russia (1982C92). Eur. J. Neurol. 7, 171–178 (2000)
3. Ricci, S., Celani, M.G., Vitali, R., La Rosa, F., Righetti, E., Duca, E.: Diurnal and seasonal variations in the occurrence of stroke: a community-based study. Neuroepidemiology 11, 59–64 (1992)
4. Biller, J., Jones, M.P., Bruno, A., Adams, H.P., Banwart, K.: Seasonal variation of stroke: does it exist? Neuroepidemiology 7, 89–98 (1988)
5. Nyquist, P.A., Brown, R.D., Wiebers, D.O., Crowson, C.S., OFallon, W.M.: Circadian and seasonal occurrence of subarachnoid and intracerebral hemorrhage. Neurology 56, 190–193 (2001)
6. Khosla, A., Cao, Y., Lin, C.Y., Chiu, H.K., Hu, J.L., Lee, H.: An itegrated machine learning approach to stroke prediction. In: 16th ACM SIGKDD Conference on Knowledge Discovery and Data Mining. ACM, Washington (2010)
7. Kasabov, N.: Evolving connectionist systems: the knowledge engineering approach. Springer, London (2007)
8. Vapnik, V., Lerner, A.: Pattern recognition using generalized portrait method. Automation and Remote Control 24, 774–780 (1963)
9. Aizerman, E.M., Braverman, L.R.: Theoretical foundations of the potential function method in pattern recognition learning. Automat. Remote Control 25, 821–837 (1964)

10. Ginsburg, G.S., McCarthy, J.J.: Personalized medicine: revolutionizing drug discovery and patient care. Trends in Biotechnology 19(12), 491–496 (2001)
11. Fix, E., Hodges, J.L.: Discriminatory analysis: Nonparametric discrimination: Consistency properties. Randolph Field, Texas (1951)
12. Dudani, S.A.: The distance-weighted k-nearest-neighbor rule. IEEE Transactions on System Man and Cybernetics, 325–327 (1976)
13. Song, Q., Kasabov, N.: TWNFI - a transductive neuro-fuzzy inference system with weighted data normalization for personliased modelling. Neural Networks 19, 1591–1596 (2006)
14. Segata, N., Blanzieri, E.: Fast and scalable local kernel machines. Journal of Machine Learning Research 11, 1883–1926 (2010)
15. Beygelzimer, A., Kakade, A., Langford, J.: Cover trees for nearest neighbour. In: 23rd International Conference on Machine Learning, Pittsburgh (2006)
16. Chang, C.C., Lin, C.J.: LIBSVM: A library for support vector machines (2001), http://www.csie.ntu.edu.tw/~cjlin/libsvm
17. Mukamal, K.J., Wellenius, G.A., Suh, H.H., Mittleman, M.A.: Weather and air pollution as triggers of severe headaches. Neurology 72, 922–927 (2009)
18. Stroke association, http://www.strokeassociation.org/STROKEORG/AboutStroke/TypesofStroke/Types-of-Stroke_UCM_308531_SubHomePage.jsp (n.d.)
19. Swets, J.: Measuring the accuracy of diagnostic systems. Science 240, 1285–1293 (1988)
20. Zou, K.H.: Receiver operating characteristic (ROC) literature research (2002), http://splweb.bwh.harvard.edu:8000/pages/ppl/zou/roc.html
21. Zweig, M.H., Campbell, G.: Receiver-operating characteristic (ROC) plots: a fundamental evaluation tool in clinical medicine. Clin. Chem. 39(4), 561–577 (1993)

Recognition of Human's Implicit Intention Based on an Eyeball Movement Pattern Analysis

Young-Min Jang[1], Sangil Lee[2], Rammohan Mallipeddi[1],
Ho-Wan Kwak[2], and Minho Lee[1]

[1] School of Electronics Engineering, and [2] Department of Psychology
Kyungpook National University 1370 Sankyuk-Dong, Puk-Gu, Daegu 702-701, Korea
ymjang@ee.knu.ac.kr, tarsys@nate.com, mallipeddi.ram@gmail.com,
{kwak,mholee}@knu.ac.kr

Abstract. We propose a new approach for a human's implicit intention recognition system based on an eyeball movement pattern analysis. In this paper, we present a comprehensive classification of human's implicit intention. Based on Bernard's research, we define the Human's implicit intention as informational and navigational intent. The intent for navigational searching is to locate a particular interesting object in an input scene. The intent for informational searching is to locate interesting area concerning a particular topic in order to obtain information from a specific location. In the proposed model, eyeball movement pattern analysis is considered for classifying the two different types of implicit intention. The experimental results show that the proposed model generates plausible recognition performance using a fixation length and counts with a simple nearest neighborhood classifier.

Keywords: eyeball movement, implicit intention, navigational intent, informational intent, human intention monitoring system, human computer interface & interaction.

1   Introduction

In recent, intention modeling and recognition have been important research issues for creating a new paradigm of human computer interface (HCI) and human robot interaction (HRI), which becomes a common issue to psychology and cognitive science [1]. Some researches of computer science and robotics have shown good results by cognitive science, HCI and psychology [2]. Human can inform the system of his explicit intention by facial expressions, explicit speech, and hand gesture as well as using a keyboard and a computer mouse. It is easy for the system to understand the explicitly represented intentions like "copy this file" in HCI, or "close the door" in HRI. Not only the explicit intention but also an implicit intention can be useful information for doing something related with human's intention by a computer or a robot. However, implicitly represented intentions might not be clear to the system. There have been many researches to handle this problem [3]. We focus on the implicit intention recognition based on eyeball movement pattern analysis.

B.-L. Lu, L. Zhang, and J. Kwok (Eds.): ICONIP 2011, Part I, LNCS 7062, pp. 138–145, 2011.
© Springer-Verlag Berlin Heidelberg 2011

Being "a window to the mind," the eye and its movements are tightly coupled with human cognitive processes. The possibility of taking advantage of information conveyed in eye-gaze has attracted many researchers in human-computer interaction [4-7]. In this research, we develop a methodology to classify human's implicit intention in real-world environment based on computer vision technology. We propose a new computational approach for a human's implicit intention recognition system based on an eyeball movement pattern analysis. Based on Bernard's research [8], we categorize human's implicit intention as navigational and informational intention. We propose a novel intent recognition model that can discriminate human's different implicit intent in the course of seeing real-world scenes with different intent in out-door environment as well as in-door environment. We analyze real human eyeball movement patterns measured by eye-track system (Tobii 1750) when human see visual scenes with different intent. As a reliable indicator to monitor human's implicit intention, we use a fixation length and account in human scan paths and a simple nearest neighborhood (NN) classifier is adopted to implement the human intention model. The proposed system shows reliable performance over 80% to recognize human's different implicit intent through the experiments using real in-door and out-door visual scenes.

This paper is organized as follows. Section 2 describes the proposed recognition of human intention based on eyeball movement. Section 3 includes experimental results for intention recognition of the proposed model in indoor and outdoor visual environments. Finally, conclusions and discussions are given in Section 4.

2   Recognition of Human Intention Based on Eyeball Movement

Figure 1 shows the block diagram of the proposed model for recognizing the human's implicit intention based on feature extraction of eyeball movement patterns. We use an eye tracking system to measure human's eyeball movement patterns according to the given visual stimuli.

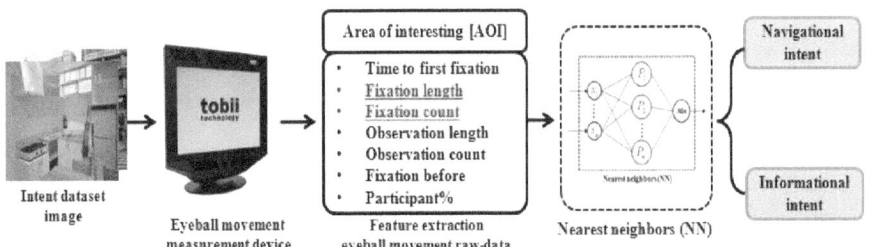

Fig. 1. Block diagram of the proposed system for recognizing human's implicit intention

We preset the areas of interest (AOI) in an image for visual stimuli, and analyze the eyeball movement pattern according to human's intention. There are several features related with intention driven eyeball movement pattern such as time to first fixation, fixation length and count, observation length and count as well as pupil size variation with eye blinking. In order to select the most dominant features among them, we have analyzed eyeball movement patterns in detail for real-time implementing a human's implicit intention recognition system in real world. A classifier is used to recognize whether the features are related to the navigational intention or informational intention. The detail processing in each part is described in the following sections.

2.1 Definition of User Intention

Several researchers have examined the elements of human intention in web searching using a variety of controlled studies, surveys, and direct observation. Carmel, Crawford, and Chen distinguished three types of categorization of user intent in web searching [9]: (1) search-oriented browsing which is the process finding information relevant to a fixed task, (2) review browsing which is the process of scanning to find interesting information, and (3) scan browsing which is the process of scanning to find information with no reviewing or integration involved. Marchionini articulated similar browsing patterns as directed browsing, semi-directed browsing, and undirected browsing [10].

In this research, we develop a new approach for a human's implicit intention recognition system based on an eyeball movement patterns during seeing real-world scenes. We redefine and present a comprehensive classification of human intention derived from the work of Rose and Levinson as follows [11]:

- Navigational intent: The intent of navigational searching is to locate a particular interesting object in an input scene.

- Informational intent: The intent of informational searching is to locate interesting area concerning a particular topic in order to obtain information from a specific location.

2.2 Eyeball Movement Measurement Device

For measuring human's visual scan path with each given visual stimulus, an eye tracking system, Tobii 1750 eye trackers, produced by Tobii Technology incorporation was utilized. Visual stimuli are displayed on the computer monitor located at 40cm from the human head. After visual stimulus is given, the eye tracking system measures the human's eyeball movement and obtains a fixation length and count, observation length and count, and visual scan path etc. of participant when seeing the given visual stimuli. The followings are several features that can be measured by the eye-track system as shown in figure 2 [12]:

- Fixation count: a number of fixations within an AOI.
- Fixation length: length of the fixations in seconds within an AOI.
- Observation count: a number of visits to an AOI.
- Observation length: total time in seconds for every time a person has looked within an AOI.
- Etc.

Element	Technical Specifications
TFT Display	• 17" TFT monitor, max. resolution 1280 x 1024 pixels
Freedom of head movement	• 30 x 16 x 20 cm at 63 cm from tracker.
Field of view of the camera	• 21 x 16 x 20 cm at 60 cm from tracker.
Accuracy / Spatial resolution	• 0.5 degrees / 0.25 degrees
Data output	• Gaze position relative to stimuli for each eye (X and Y) • Position in camera field of view of each eye (X and Y) • Distance from camera of each eye • Pupil size of each eye • Validity code of each eye

Fig. 2. Eye movement measurement device of Tobii 1750 (technical specifications)

2.3 Correlation between Eyeball Movement and Human's Implicit Intent

Eyeball movements are known as tightly being coupled with human cognitive processes and the eye-gaze contains richer and more complex information regarding a person's interest and intentions than what is used in pointing [7].

Therefore, it might be assumed that human generate specific eyeball movement patterns according to a different kind of human's implicit intention during visual searching. In accordance, the eyeball movement patterns are able to be considered as possible factors for recognizing the human's implicit intent by verifying the correlation between the eyeball movement patterns and specific human's implicit intent. Different implicit intent may cause different eyeball movement patterns when seeing or searching the same visual scene.

In this paper, we analyze the characteristics of eyeball movement patterns measured using the Tobii 1750 eye track system in order to verify the correlation between implicit intent and eyeball movement patterns. We have found that the fixation length and count are most dominant features to distinguish the navigational and informational intentions of a user through the experiment. When human search a visual scene with a navigational intent that is trying to localized user's interesting object, according to the analysis of eyeball movement patterns, fixation length and count features among characteristics of eyeball movement patterns are short and small, respectively. Instead, in the case of the informational intention, fixation length and count features are relatively long and large compared to those of navigation intention case. Therefore, it might be experimentally concluded that the fixation length and count features are highly correlated with the human's implied intention. By comparing the fixation length and count in each AOI of a given input image, we can construct a classifier to discriminate the navigation and information intent.

2.4 Discrimination between Navigational and Informational Intent

Based on the experimental verification of there being distinctive correlation between the fixation length/count and human's implicit intention, we considered the fixation length and count as input features of intention classifier. The NN classifier is applied for recognizing whether the current fixation length and count are related with the navigational intention or informational intention. Eyeball movement features such as fixation length and count show irregular characteristics during visual searching with navigational intention. Nevertheless, the simple classifier implemented by NN algorithm successfully discriminates the intentions as two classes. One class is for navigational intention and the other one is for informational intentions.

3   Experiments

We developed a methodology to classify user's intention in real-world environment. The indoor scene and outdoor scene images with five AOI areas were prepared. The AOI is a candidate fixation region by an eyeball movement with each given visual stimulus. We defined AOI as the significantly salient object in given visual stimulus. The set up AOI area is shown in figure. 3. Fig. 3 (a) is indoor image as a kitchen and Fig. 3 (b) is outdoor image as a downtown street.

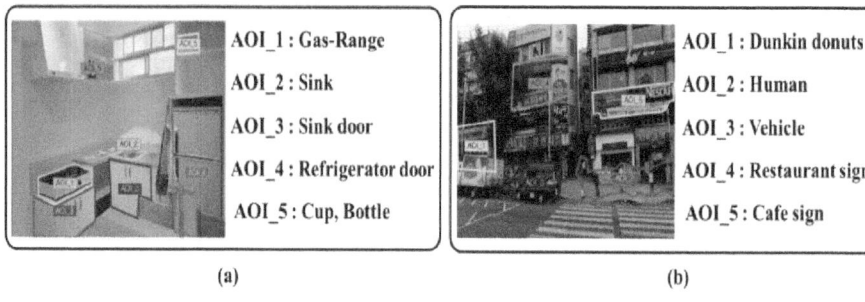

AOI_1 : Gas-Range
AOI_2 : Sink
AOI_3 : Sink door
AOI_4 : Refrigerator door
AOI_5 : Cup, Bottle

AOI_1 : Dunkin donuts
AOI_2 : Human
AOI_3 : Vehicle
AOI_4 : Restaurant sign
AOI_5 : Cafe sign

(a)   (b)

Fig. 3. Visual stimulus images and set up for AOI

Figure 4 shows that the proposed model for human's implicit intent recognition system in experimental sequence. Following is an actual working scenario of the proposed human intent recognition system based on an eyeball movement:

· Step 1) Command: searching is to locate a particular interesting object in visual stimuli images
· Step 2) Visual stimuli: Kitchen (Indoor image) / Downtown (Outdoor image)
· Step 3) Command: searching is to locate interesting area concerning a particular topic in order to obtain information.
    - Indoor: searching is to cup, bottle and refrigerator door.
    - Outdoor: searching is to humans and Dunkin donuts.
· Step 4/5) Visual stimuli: Kitchen (Indoor image) / Downtown (Outdoor image)

We obtain the essential experimental data for 40 people participated in human's intention recognition experiment using an eye tracking system. Using the raw data and their statistical analysis, we determine the baseline value as a threshold to distinguish the two implicit intents.

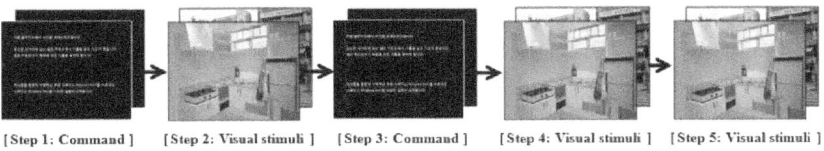

[Step 1: Command]   [Step 2: Visual stimuli]   [Step 3: Command]   [Step 4: Visual stimuli]   [Step 5: Visual stimuli]

Fig. 4. Human's implicit intention recognition system in experimental sequence

3.1   Experimental Results for an Eyeball Movement Analysis of Human's Intention

As a pre-step, in order to calibrate the eye tracker system (Tobii 1750), we demonstrate a simple test that is tracking an eyeball movement on the monitor. When visual stimulus is given, eye tracker system detects the pupils and measures the human's eye movement using an IR (infrared) camera. The whole processes detecting human's eyeball movement are shown in figure 5. After detecting the pupil using the IR camera, we get the pupil center data in an image as shown Figs. 5 (a) and (b). Fig. 5 (c) shows the points of the two pupils' center with depth information. Fig. 5 (d) is the calibrated test image. Finally, Fig. 5 (e) represents the positions of detected eyeball as color circles and green dot-circle is eye gaze point.

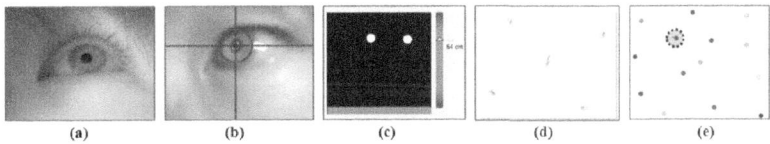

(a)   (b)   (c)   (d)   (e)

Fig. 5. The calibration process of eye tracker system to measures the eyeball movement

Figure 6 shows the heat maps representing the fixation length of the eyeballs for the given visual stimuli. Figs. 6 (a) and (c) are the heat maps of the navigational intent results for the indoor and outdoor scenes, respectively. Figs. 6 (b) and (d) are the heat maps of the informational intent results for the indoor and outdoor scenes, respectively.

(a)   (b)   (c)   (d)

Fig. 6. The heat maps of the human's eyeball movement analysis for visual stimuli

Figure 7 show the experimental results such as scan path, fixation length and count for raw-data analysis based on eyeball movement pattern. Fig. 7 (a) shows the results for indoor visual stimulus image. Fig. 7 (b) shows the results for outdoor visual stimulus image. In the Fig. 7, the gray bars are the data for the navigational intent and white bars are the data for the informational intent. We confirmed that the fixation length and count are dominant factors to discriminate the navigational and informational intent through the experiment.

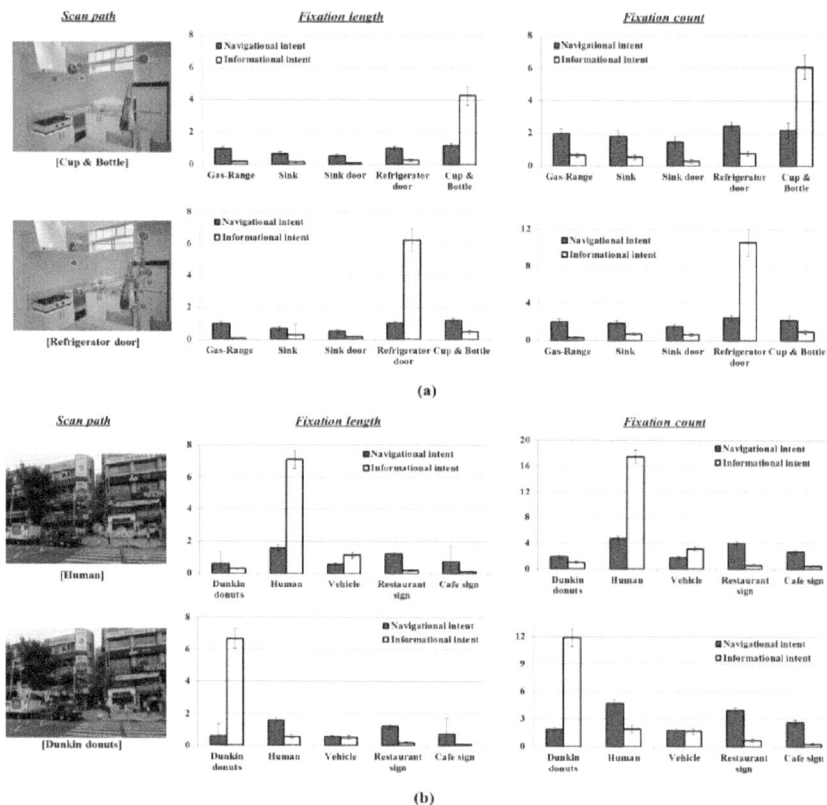

Fig. 7. The raw-data analysis experiments for recognition of human's implicit intents

3.2   Performance for Human's Implicit Intent Using the Nearest Neighbor

In order to classify the human's implicit intent based on eyeball movement pattern, we used a NN classifier. The data set of 40 people was used for training of the input images. Through the statistical analysis of data result in training process, we determined the baseline value as a reference of eyeball movement to distinguish the implicit intent. Test data set number is 180 which consist of 60 navigational intent data and 120 informational intent data. Table 1 shows the average recognition accuracy of NN classifier for human's implicit intents which consist of navigational and informational intent based on eyeball movement pattern.

Table 1. Human's intent recognition performance by the proposed model

Type of intent	Recognition rate		
	Indoor image	Outdoor image	Average
Navigational intent	85.98 (%)	86.40 (%)	86.19 (%)
Informational intent	79.01 (%)	81.82 (%)	80.42 (%)

4   Conclusions

In this paper, a new approach for a human's implicit intention recognition model based on an eyeball movement pattern analysis is proposed. We define and present a comprehensive classification of human's implicit intent. In order to recognize the human's implicit intent for given visual stimulus, an eye tracking system, Tobii 1750 was used. Through the experiments, we confirmed that the fixation length and counts are the main factors to discriminate the navigational and informational intent. In order to classify the human's implicit intent, we used a NN classifier. The experimental results show that the proposed method shows plausible performance.

We are now implementing a prototype for human intention monitoring system. For performance verification, however, more experiments using complex real scenes reflecting various situations are needed as a further work.

Acknowledgments. This research was supported by the Converging Research Center Program funded by the Ministry of Education, Science and Technology (2011K000659) (50%) and also the Original Technology Research Program for Brain Science through the National Research Foundation of Korea(NRF) funded by the Ministry of Education, Science and Technology (2011-0018292) (50%).

References

1. Cynthia, B.: Social interactions in HRI: The robot view. IEEE Trans. Systems, Man, and Cybernetics 52(6), 181–186 (2004)
2. Wong, F., Park, K.-H., Kim, D.-J., Jung, J.-W., Bien, Z.: Intention reading towards engineering applications for the elderly and people with disabilities. International Journal of ARM 7(3), 3–15 (2006)
3. Youn, S.-J., Oh, K.-W.: Intention Recognition using a Graph Representation. World Academy of Science, Engineering and Technology 25, 13–18 (2007)
4. Bolt, R.A.: Eyes at the interface. In: Proc. Human Factors in Computer Systems, pp. 360–362. ACM (1982)
5. Jacob, R.J.K.: What you look at is what you get: Eye movement-based interaction techniques. In: Proc. CHI, pp. 11–18. ACM (1990)
6. Ware, C., Mikaelian, H.H.: An evaluation of an eye tracker as a device for computer input. In: Proc. CHI+GI, pp. 183–188. ACM (1987)
7. Zhai, S., Morimoto, C., Ihde, S.: Manual and gaze input cascaded (MAGIC) pointing. In: Proc., pp. 246–253. ACM (1999)
8. Bernard, J.J., Danielle, L.B., Amanda, S.: Determining the informational, navigational, and transactional intent of Web queries. Information Processing and Management 44, 1251–1266 (2008)
9. Carmel, E., Crawford, S., Chen, H.: In Browsing in hypertext: A cognitive study. IEEE Transactions on Systems, Man and Cybernetics, 865–884 (1992)
10. Marchionini, G.: Information seeking in electronic environments. Cambridge University Press, Cambridge (1995)
11. Rose, D.E., Levinson, D.: In Understanding user goals in Web search. In: World Wide Web Conference, pp. 13–19 (2004)
12. Eye tracking system of Tobii Technology, http://www.tobii.com/

ECG Classification Using ICA Features and Support Vector Machines

Yang Wu and Liqing Zhang

Department of Computer Science and Engineering, Shanghai Jiaotong University,
Shanghai 200240, China
wuyang@sjtu.edu.cn,
zhang-lq@cs.sjtu.edu.cn

Abstract. Classification accuracy is vital in practical application of automatic ECG diagnostics. This paper aims at enhancing accuracy of ECG signals classification. We propose a statistical method to segment heartbeats from ECG signal as precisely as possible, and use the combination of independent component analysis (ICA) features and temporal feature to describe multi-lead ECG signals. To obtain the most discriminant features of different class, we introduce the minimal-redundancy-maximal-relevance feature selection method. Finally, we designed a vote strategy to make the decision from different classifiers. We test our method on the MIT-BIT Arrhythmia Database, achieving a high accuracy.

Keywords: ECG segmentation, ICA feature extraction, SVM.

1 Introduction

The electrocardiograic(ECG) signals is a key approach for heart disease diagnosis. The automatic ECG diagnostics has very high clinical value of modern medical diagnosis. Nowadays, many researches focus on feature extraction and pattern recognition of ECG signals.

Among those recently related works, we can find some typical techniques dealing with this issue. Wave detection of ECG signals is a key step of preprocessing. Previous work is mainly obtained from the shape characters [1], and more from frequency domain in recent works [2]. Many algorithms from pattern classification problems have been applied to the ECG signals classification. Zhao et al. used SVM and wavelet transform achieving a high accuracy [3]. E. Pasolli et al. applied active learning method to the ECG classification using some ECG morphology and temporal features [4]. Some other methods such as SOM, ANN, HMM have been proven to be efficient in automatic ECG Diagnosis [5]. To acquire high classification accuracy, efficient representation of ECG signals are needed. ECG feature extraction is important to diagnosing most of the cardiac disease. Various methods have proposed in different approaches for ECG features extraction [6],[7].

In order to realize the automatic diagnosis, to enhance the accuracy of automated classification of the ECG signals is an important task. This paper is to

B.-L. Lu, L. Zhang, and J. Kwok (Eds.): ICONIP 2011, Part I, LNCS 7062, pp. 146–154, 2011.

develop new feature extraction method to improve classification performance. Our system structure is illustrated in Fig.1, showing the general framework of the ECG data processing.

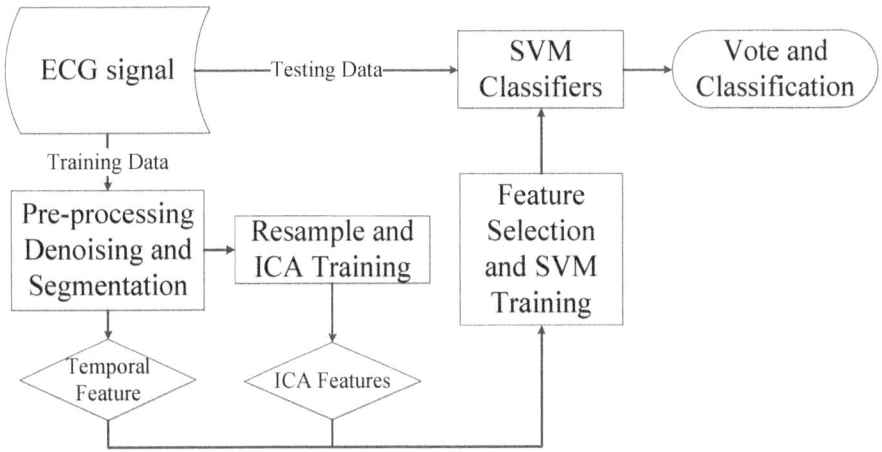

Fig. 1. Structure of out System

2   Feature Extraction

2.1   Independent Component Analysis Features

The ECG signals represent the current produced by myocardium in every heart-beat. ECG signals can be considered as a mixture of different source of ECG. Thus, the signals can be represented as the following linear combination of some unknown sources.

$$S_{ECG} = \sum_{i=1}^{N} a_i \mathbf{s}_i \tag{1}$$

where $\mathbf{s}_i$ is one source, and a_i is the coefficient.

The independent component analysis (ICA) is very closely related to the method called blind source separation (BSS) [8]. The linear combinations (1) above is the same as the basic model of ICA. Using a vector-matrix notation, the mixing model can be written as

$$\mathbf{x} = \mathbf{As} \tag{2}$$

where s is a $N \times 1$ column vector of the unknown source signals, A is a $N \times N$ mixing matrix we may want to know. Estimating the independent components equals finding the proper linear combination of the $\mathbf{s}_i$, i.e.,

$$\mathbf{s} = \mathbf{A}^{-1}\mathbf{x} \tag{3}$$

Our purpose is to find the independent components $\mathbf{s}_i$ with the observed data x [9]. There are several approaches have been proposed to solve this problem. We select the algorithm FastICA to estimate the ICs, which performs better in batch algorithm [10].

In this paper, we use ICA model to find efficient representation of ECG signals. Then, $\mathbf{s}_i$ is considered as the basis functions and a_i is the coefficient corresponding to the feature. ICA-based feature extraction method has been widely used in many feature extraction fields, such as image data, video data, audio data and hyperspectral data [9], and also used in ECG signals feature extraction and worked well.

We train the ICA basis functions using some random samples with zero mean selected from different heart disease. We chose twelve ICs as basis functions in computer simulations. The learned ICs are illustrated in Fig.2. The coefficients in the IC basis is considered as a features for disease classification. One fitting result using the IC basis shows in Fig.3.

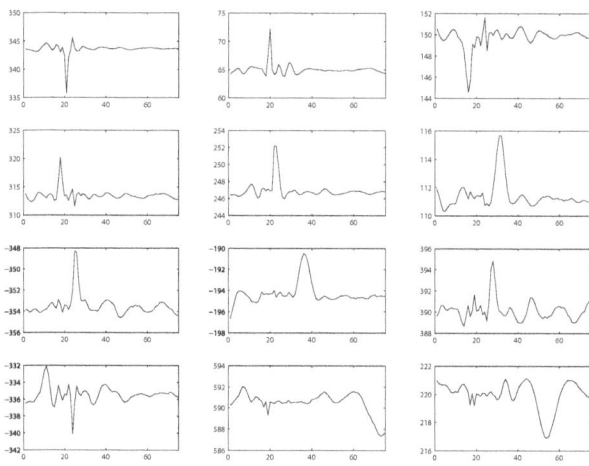

Fig. 2. A group of twelve ICA basis learned from random selected heartbeat data

2.2   Statistical Segmentation and Temporal Feature

But when we get features of ICA basis functions, we omit the temporal features casually. ICA described features can represent the morphology well, but lack of time domain ones. The dynamic feature is important for doctors to make clinic diagnosis. To overcome this shortcoming, we combine some time domain features with the ICA-based features.

Wave detection methods show good performance when the wave is regular. But there many varying morphologies of normal and abnormal complexes in the pathological signals. To obtain the length of abnormal ones, we record the wave temporal features statistically of the regular beats, and use those information to estimate the boundary of the abnormal records from the same person in the

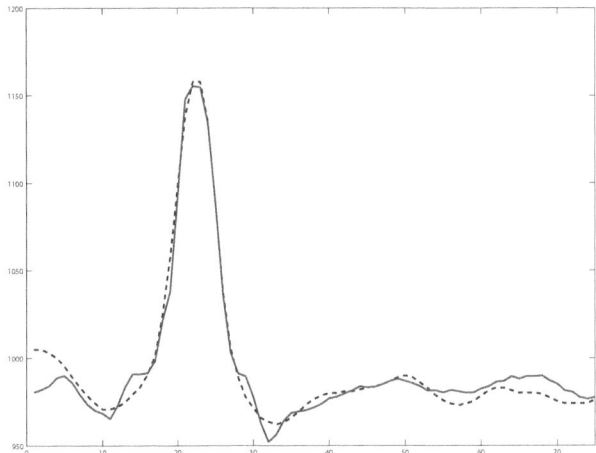

Fig. 3. One heartbeat fitting result with ICA basis. The blue dotted line express the original heartbeat, the red solid line express the fitting curve.

same time period. Meanwhile, we get the beat duration defined as the start of the P wave to the end of the T wave. Fig.4. shows 200 heartbeat segmentation length of both manual determined and statistical determined. The estimation results are very closed to the real ones. And we take this reliable length as a feature of time domain.

2.3   Feature Selection

Different types of heart disease may need different set of features to be described. So, it's important to select the most relevant features for classification purpose. Feature selection is a useful dimensionality reduction technique widely used in pattern recognition, machine learning and data mining research fields. Feature selection can also help us to find the distinctive features for the class. [11]

We chose a efficient heuristic feature selection method, minimal-redundancy maximal-relevance (mRMR) in our framework [12].

Relevance is described with the mean value of all mutual information values between individual feature $\mathbf{x}_i$ and class c:

$$D = \frac{1}{|S|} \sum_{x_i \in s} I(x_i; c) \tag{4}$$

Redundancy is defined as:

$$R = \frac{1}{|S|^2} \sum_{x_i, x_j \in s} I(x_i; x_j) \tag{5}$$

Therefore, the criterion of minimal-redundancy-maximal-relevance is define as the following form, and the objective is to optimize D and R simultaneously:

$$\max(D - R) \tag{6}$$

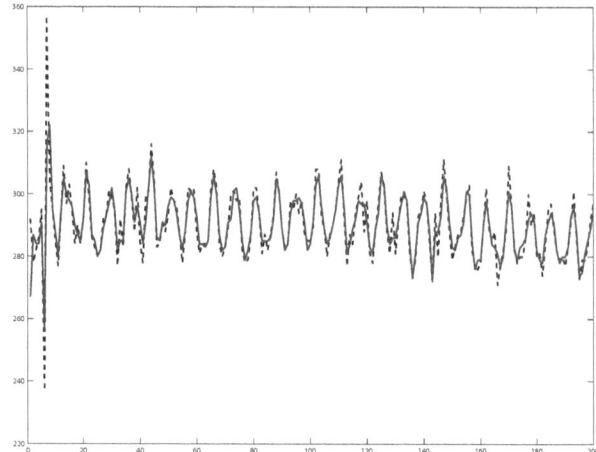

Fig. 4. Length comparison of 300 heartbeat between manual measured and statistical estimated. The blue dotted line express the former length, and the red solid line express the latter.

Fig.5 shows the ICA features distributions of two type of heart disease segments after feature selection. The first 10 small figures represent the 10 features selected by the mRMR approach. The last 5 small figures with no axis labels in the frame represent some of the less significant features for the two classes. We can clearly see that the selected features are more discriminant.

3   Classification

The Support Vector Machine (SVM) is an extensively adopted classification technique [13]. Given two classes of training vectors $x_i \in \Re^n, i = 1, \ldots, l$ from the d-dimensional feature space $\mathbf{X}$, and an indicator vector $y \in \Re^l$ such that $y_i \in \{+1, -1\}$. The linear SVM classification approach is trying to find a separation between the two classes by means of an optimal hyperplane that maximizes the separating margin, i.e. , to solve the following primal optimization problem.

$$\min_{\mathbf{w},b,\xi} \frac{1}{2}\mathbf{w}^T\mathbf{w} + C\sum_{i=l}^{l}\xi_i \qquad (7)$$

$$subject\ to \quad y_i(\mathbf{w}^T\phi(x_i) + b) \geq 1 - \xi_i, \xi_i \geq 0, i = 1, \ldots, l$$

Because of the possible high dimensionality of the vector w. The problem above can be reformulated by Lagrange functional. Then, we can solve the following equivalent dual problem.

$$\min_{\alpha} \frac{1}{2}\alpha^T y_i y_j K(x_i, x_j)\alpha - e^T\alpha \qquad (8)$$

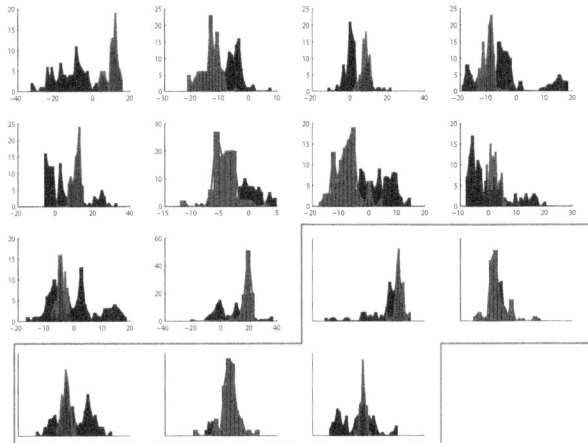

Fig. 5. Feature distributions of two type of heart disease selected by mRMR. The first 10 figures represent the most discriminant ones. The red bars and the blue bars represent two classes respectively.

$$subject\ to \quad \mathbf{y}^T\boldsymbol{\alpha} = 0, 0 \leq \boldsymbol{\alpha}_i \leq C, i = 1,\ldots,l,$$

In the nonlinear case, we can map the two classed in a higher dimensional feature space with a kernel method. The radial basis function (RBF) is introduce into this work.

SVM were originally designed for binary classification. When face to multiclass problem, we need to find an approaches for it. Now there are some types of approaches: "one-against-all", "one-against-one", DAGSVM and Binary Tree[14][15]. Hsu and Lin give a detailed comparison demonstrate that "one-against-one" is a competitive one [16]. In our framework, we designed some voting strategy rules based on probability estimated by SVM [17]. We trained different classifiers with optimum parameters by cross-validation (CV). When to predict the test data, we give different decision based on the probability given by the classifier. For example, we give bonus to the high certainty estimation but penalty to the uncertainty. After all, we give the most high score candidate as the predicted label. Through experiment, we find it shows significantly fair to unbalance training data. In other words, this strategy give much higher prediction accuracy to smaller scale training data, but a little bit lower to bigger ones.

4   Experiment and Result

In our experiment, we perform a number of computer experiments on the MIT-BIH Arrhythmia Database [18]. This database is an open ECG database with high quality expert labels. We chose all the 48 data with approximately 109,000 beats of following 13 types (Symbol in parenthesis): Normal beat (N), Left bundle branch

Table 1. The results of heartbeat classification accuracy of vote strategy and one-against-one. Our strategy performs well especially in small training data.

Beat Type	#Training Beats	Training Acc.	# Testing Beats	Test Acc. Vote Strategy	Test Acc. One-One
N	10000	97.71%	74931	97.90%	99.05%
L	1997	99.66%	8070	99.01%	99.30%
R	1738	99.03%	7254	95.88%	99.45%
A	570	98.21%	2541	89.85%	79.81%
a	53	97.46%	149	96.64%	77.85%
J	50	97.74%	83	90.36%	60.24%
V	1676	98.61%	7118	94.20%	97.53%
F	401	98.57%	802	94.51%	86.66%
!	133	98.73%	405	99.01%	99.01%
j	76	98.74%	229	99.56%	78.60%
E	50	99.71%	106	98.11%	97.17%
\	1729	99.27%	7020	98.56%	99.44%
f	242	99.33%	982	98.88%	94.91%

Table 2. Comparisons of the classification accuracy between our method and X.Jiang's

Beat Type	N	L	R	A	a	J	V
Our Result	97.90%	99.01%	95.88%	89.85%	96.64%	90.36%	94.20%
X.Jiang	98.54%	99.28%	99.14%	90.42%	70.13%	86.88%	97.44%

Beat Type	F	!	j	E	\	f
Our Result	94.51%	99.01%	99.56%	98.11%	98.56%	98.88%
X.Jiang	84.72%	77.43%	80.81%	91.84%	86.67%	94.53%

block beat (L), Right bundle branch block beat (R), Atrial premature beat (A), Aberrated atrial premature beat (a), Nodal (junctional) premature beat (J), Premature ventricular contraction (V), Fusion of ventricular and normal beat (F), Ventricular flutter wave (!), Nodal (junctional) escape beat (j), Ventricular escape beat (E), Paced beat (\), Fusion of paced and normal beat (f).

We use statistical segment method mentioned before to segment the original signals into single heartbeat segments [19]. The we center the R spike of each beat and resample them to the same length and keep the original beat length as a feature. Then we trained the ICA basis and extracted features use both leads in the data. The combination of the temporal feature and ICA features are 25 dimensions in all. After feature selection using mRMR, we get data to train the SVM classifiers [20]. Finally, the result in Table 1. shows high accuracy in classification.

We compared our work with other similar experiments published. The competitive one is X.Jiang et al. 's work [21]. They used similar feature extract method and got a high overall accuracy. But we get more stable classification accuracy, most of them are above 90%. The comparison results are given in Table 2. The other is the types of premature and escape beat are strongly exhibited in temporal feature and relied on the accuracy of the segmentation. It shows that our methods are greatly improved for many types.

Acknowledgement. The work was supported by the National Natural Science Foundation of China (Grant No. 90920014) and the NSFC-JSPS International Cooperation Program (Grant No. 61111140019).

References

1. Tan, K.F., Chan, K.L., Choi, K.: Detection of the QRS complex, P wave and T wave in electrocardiogram. In: First International Conference on Advances in Medical Signal and Information Processing, IEEE Conf. Publ. No. 476, pp. 41–47 (2000)
2. Pal, S., Mitra, M.: Detection of ECG characteristic points using Multiresolution Wavelet Analysis based Selective Coefficient Method. Measurement 43(2), 255–261 (2010)
3. Zhao, Q.B., Zhang, L.Q.: ECG Feature Extraction and Classification Using Wavelet Transform and Support Vector Machines. In: International Conference on Neural Networks and Brain, pp. 1089–1092 (2005)
4. Pasolli, E., Melgani, F.: Active learning methods for electrocardiographic signal classification. IEEE Transactions on Information Technology in Biomedicine 14(6), 1405–1416 (2010)
5. Gacek, A.: Preprocessing and analysis of ECG signals - A self-organizing maps approach. Expert Systems with Applications 38(7), 9008–9013 (2011)
6. Karpagachelvi, S., Arthanari, M., Sivakumar, M.: ECG Feature Extraction Techniques - A Survey Approach. International Journal of Computer Science and Information Security 8(1), 76–80 (2010)
7. Soria, L.M., Martínez, J.P.: Analysis of Multidomain Features for ECG Classification. Computers in Cardiology, 561–564 (2009)
8. Zhang, L.Q., Cichocki, A., Amari, S.I.: Self-adaptive blind source separation based on activation functions adaptation. IEEE Trans. Neural Networks 15(2), 233–244 (2004)
9. Hyvärinen, A., Karhunen, J., Oja, E.: Independent Component Analysis. Wiley Inter-science (2001)
10. Hyvärinen, A., Oja, E.: A Fast Fixed-Point Algorithm for Independent Component Analysis. Neural Computation 9, 1483–1492 (1997)
11. Dash, M., Liu, H.: Feature Selection for Classification. In: Intelligent Data Analysis, vol. 1, pp. 131–156 (1997)
12. Peng, H., Long, F., Ding, C.: Feature Selection Based on Mutual Information: Criteria of Max-Dependency, Max-Relevance, and Min-Redundancy. IEEE Trans. on Pattern Analysis and Machine Intelligence 27(8) (2005)
13. Hastie, T., Tibshirani, R., Friedman, J.: The Elements of Statistical Learning: Data Mining, Inference, and Prediction, 2nd edn. Springer, Heidelberg (2009)
14. Kressel, U.H.-G.: Pairwise classification and support vector machines. In: Advances in Kernel Methods, pp. 255–268. MIT Press, Cambridge (1999)

15. Cheong, S., Oh, S.H., Lee, S.Y.: Support Vector Machines with Binary Tree Architecture for Multi-Class Classification. Neural Information Processing 2(3), 47–51 (2004)
16. Hsu, C., Chang, C., Lin, C.-J.: A practical guide to support vector classification, Technical report, Department of Computer Science. National Taiwan University (2003)
17. Wu, T.F., Lin, C.J., Weng, R.C.: Probability Estimates for Multi-class Classification by Pairwise Coupling. Journal of Machine Learning Research 5, 975–1005 (2004)
18. Mark, R., Moody, G.: MIT-BIH Arrhythmia Database,
    `http://ecg.mit.edu/dbinfo.html`
19. Goldberger, A.L., Amaral, L.A.N., Glass, L., Hausdorff, J.M., Ivanov, P., Mark, R.G., Mietus, J.E., Moody, G.B., Peng, C.K., Stanley, H.E.: PhysioBank, PhysioToolkit, and PhysioNet: Components of a New Research Resource for Complex Physiologic Signals. Circulation 101(23), e215–e220 (2000)
20. Chang, C.C., Lin, C.J.: LIBSVM: a library for support vector machines. ACM Transactions on Intelligent Systems and Technology 2, 27:1–27:27 (2011)
21. Jiang, X., Zhang, L.Q., Zhao, Q.B., Albayrak, S.: ECG Arrhythmias Recognition System Based on Independent Component Analysis Feature Extraction. In: Conference Proceedings of IEEE Region 10 Conference (TENCON), pp. 464–471 (2006)

Feature Reduction Using a Topic Model for the Prediction of Type III Secreted Effectors

Sihui Qi, Yang Yang*, and Anjun Song

Department of Computer Science and Engineering, Information Engineering College, Shanghai Maritime University, 1550 Haigang Ave., Shanghai 201306, China
yangyang@shmtu.edu.cn

Abstract. The type III secretion system (T3SS) is a specialized protein delivery system that plays a key role in pathogenic bacteria. Until now, the secretion mechanism has not been fully understood yet. Recently, a lot of emphasis has been put on identifying type III secreted effectors (T3SE) in order to uncover the signal and principle that guide the secretion process. However, the amino acid sequences of T3SEs have great sequence diversity through fast evolution and many T3SEs have no homolog in the public databases at all. Therefore, it is notoriously challenging to recognize T3SEs. In this paper, we use amino acid sequence features to predict T3SEs, and conduct feature reduction using a topic model. The experimental results on *Pseudomonas syringae* data set demonstrate that the proposed method can effectively reduce the features and improve the prediction accuracy at the same time.

Keywords: type III secretion system, type III secreted effector, topic model, feature reduction.

1 Introduction

The type III secretion system (T3SS) is one of the six types of secretion systems that have been discovered in gram-negative bacteria. T3SS is an essential component for a large variety of pathogens, such as *Pseudomonas*, *Erwinia*, *Xanthomonas*, *Ralstonia*, *Salmonella*, *Yersinia*, *Shigella*, *Escherichia*, *etc* [1], which can cause devastating diseases on plants, animals and human beings. T3SS plays an important role in developing the diseases by injecting virulence proteins into the host cells.

Researchers have been exploring the working principle and mechanism of T3SS for over a decade. The detailed structure of T3SS has been identified, including a needle-like structure and bases embedded in the inner and outer bacterial membranes [1]. The virulence proteins, called type III secreted effectors (T3SEs), are secreted directly from the bacterial cell into the host cell through the needle. Although the structure of T3SS apparatus has been uncovered, the precise mechanism of the secretion process has not been fully understood. In recent years,

---

* Corresponding author.

B.-L. Lu, L. Zhang, and J. Kwok (Eds.): ICONIP 2011, Part I, LNCS 7062, pp. 155–163, 2011.

more and more effort has been put into the studies of T3SEs, because the characteristics determining what kind of proteins could be secreted have not been discovered yet. A lot of questions remain unresolved, such as how the T3SEs are recognized, and how they are transported into host cells. Once we know the answers, we know much better about how the T3SS works.

Although the structure of T3SS is conserved, T3SEs are highly variable even among different strains of the same bacterial species. This is because they evolve fast in order to adapt to different hosts and respond to the resistance from the host immune systems. Therefore, it is notoriously challenging to recognize T3SEs. Some wet-bench methods have been used to verify T3SEs, $e.g.$, functional screen and protein secretion assay [2]. These methods are time and labor consuming, and cannot deal with high-throughput screening, while computational tools can save the laborious work in wet-bench experiments and help biologists find the T3SE candidates more quickly. Therefore, bioinformatics approaches are in great demand for the study of T3SS.

There is very little domain knowledge could be used for identifying T3SEs. Actually, many T3SEs were hypothetical proteins before they were verified. As the sequencing techniques have gained breakthrough for the past decade, a large number of sequenced genomes for plant and animal pathogens became available, thus the genome sequences and amino acid sequences are widely used to discriminate effectors and non-effectors. Researchers have detected amino acid composition biases in T3SEs, especially in the N-termini. For example, Guttman $et\ al.$ [2] reported that the first 50 amino acids of $P.\ syringae$ effectors have a high proportion of Ser and a low proportion of Asp residues. A conserved regulatory motif on promoters was also found in some T3SEs [3]. However, these features are not accurate enough to identify new effectors because some effectors do not possess these features at all.

Recently, some machine learning methods have been proposed for the prediction of T3SEs. Arnold $et\ al.$ [4] used the frequencies of amino acids as well as the frequencies from two reduced alphabets, $i.e.$, they mapped amino acids to groups according to the amino acid properties. They also computed the frequencies of di- and tri-peptides from each of the alphabets. Löwer and Schneider [5] used sliding-window technique to extract features. The sliding window procedure divides a sequence in a number of overlapping segments. Each segment is encoded by a bit string containing $W \times 20$ bits (W is the size of the window). Yang $et\ al.$ used amino acid composition, K-mer composition, as well as SSE-ACC method (amino acid composition in terms of different secondary structures and solvent accessibility states) [6]. Wang $et\ al.$ [7] proposed a position-specific feature extraction. The position-specific occurrence time of each amino acid is recorded, and then the profile is analyzed to compose features.

These methods mainly utilize sequence features, like amino acid composition and position information, but they do not consider the most discriminating residues or peptides. In this paper, we regard the protein sequences as text written in a certain kind of biological language. The residues and peptides, $i.e.$, K-mers (K-tuple amino acid sequences), are the words composing the text.

Since the number of K-mers would be very large when K increases, we conduct feature reduction instead of using all the K-mers as features. In order to eliminate the noisy words effectively, we adopt a topic model, called HMM-LDA, which integrating the hidden Markov model (HMM) and latent Dirichlet allocation (LDA) model. The advantage of HMM-LDA over other LDA models is that it introduces both syntax states and topics. In our method, we keep the words which are assigned to some topics with high probability but do not pervasively assigned to many topics, *i.e.*, the typical words particularly to some topics, thus we get the condensed word set. We conduct a series of experiments to compare the new method with other approaches, including the methods using all di-mers/tri-mers, and the methods that reduce features by word frequency and tf-idf criteria. The experimental results show that the new method achieves better prediction accuracy with much less features.

2   Methods

In our previous studies [8,9], protein sequences were modeled as text, and segmented into words in a non-overlapping way. The words are predefined in a dictionary, which includes valuable words according to some criteria. This method has been proved to have good performance in protein subcellular localization [8] and protein family classification [9]. In this paper, the training data are processed by segmentation, and the dictionary is constructed using the criteria of word frequency and tf-idf value, respectively. Then the word set is further condensed by a topic model to eliminate noisy words.

The rest of this section consists of two parts. In the first part, we introduce the topic models, especially the HMM-LDA model used in our study. And the second part describes the details of our method.

2.1   Topic Model

Topic model is a kind of statistical model in the realm of machine learning and natural language processing. It is able to discover the implicit topic information in the document. Over the past decade, topic models have been developed very fast. Besides in text automatic classification, information retrieval and other related applications in natural language processing realm, topic models have been successfully applied in image searching and classification, social network analysis, *etc*. In the realm of bioinformatics, some researchers have also used topic models to process biological data. For example, in the study of protein remote homology detection, Liu *et al.* used latent semantic indexing model [10], and Yeh *et al.* used latent topic vector model [11], both of which achieved satisfying results.

Among the numerous topic models, the latent Dirichlet allocation (LDA) model [12] has been widely used in recently years for its excellent performance, in which each document is represented as a mixture of topics, and each topic is a multinomial distribution over words in a vocabulary. In this study, we regard each protein sequence as a document, and the K-mers as words. We adopted a variant of LDA, the HMM-LDA model [13] to conduct the feature selection. This model further extends the topic mixture model by separating syntactic words from content words, whose distributions depend primarily on local context and document topic, respectively. The major difference between LDA and HMM-LDA is that each word is generated independently in LDA, while there is local dependencies between nearby words in HMM-LDA. We have experimented both original LDA and HMM-LDA, and the later performs better. That is because the HMM-LDA model discovers both syntactic classes and semantic topics in the document, which may be more helpful to eliminate the noisy words.

2.2   Our Method

The flowchart of our method is shown in Fig. 1. Our method mainly consists of four steps: a) Construct a dictionary, *i.e.*, word set, using certain criteria; b) Segment the protein sequences by matching the words in the dictionary; c) Select typical words in order to get a condensed feature set; d) Use support vector machines to classify the feature vectors into effectors or non-effectors.

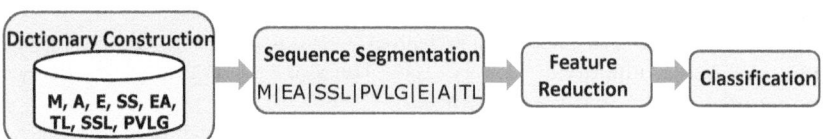

Fig. 1. Flowchart of the new method

For the first step, the dictionary can include both amino acids and K-mers ($K > 1$). Two criteria, word frequency and tf-idf value, are considered here to select words. They are defined as follows.

1) Frequency: We record the frequency for each K-mer appearing in the training set and keep a predefined proportion of the most frequent K-mers.

2) *tf-idf* value: According to its definition in text categorization, *tf-idf* is calculated for a term in a single document. The value is in proportion to the occurrence time of the term in the document, *i.e.*, the *tf* (term frequency) part; and in inverse proportion to the number of documents in which the term occurs at least once, *i.e.*, the *idf* (inverse document frequency) part. Here we redefine it as the following equation.

Let $f_{t,s}$ be the frequency of K-mer t in sequence s, N be the size of the training set, $w_{t,s}$ be the *tf-idf* value for a K-mer t in sequence s, and n_t be the number of sequences in which t appears.

$$w_{t,s} = f_{t,s} \times log\frac{N}{n_t}. \tag{1}$$

The weight of t, w_t, is defined as the maximum value of $w_{t,s}$ among all the sequences in data set $\mathcal{T}$. The words with high weight will be selected.

$$w_t = \max_{s \in \mathcal{T}} w_{t,s}, \tag{2}$$

In the second step, we adopt the segmentation method proposed in [8]. And in the third step, by using the HMM-LDA model to implement dimension reduction, we use the strategy shown in Algorithm 1 to extract the typical words. Let $n_{w,t}$ denote the number of times that word w has been assigned to topic t. There are two predefined thresholds α and β. α is used to eliminate the obscure words, and β is for selecting discriminating words by removing those words which occur nearly equally on multiple topics.

Algorithm 1. Feature Reduction

Input: Word set $\mathcal{W}$
Output: Reduced word set $\mathcal{W}'$
  Set $\mathcal{W}' = \phi$.
  for each word $w \in \mathcal{W}$ **do**
    if w is assigned to only one topic t **then**
      if $n_{w,t} > \alpha$ **then**
        Add w to $\mathcal{W}'$
      end if
    end if
    if w is assigned to multiple topics, $t_1, t_2, ..., t_m$ **then**
      Find j, where $n_{w,t_j} = \max\{n_{w,t_i}, 1 \leq i \leq m\}$
      $min_{diff} = \min\{(n_{w,t_j} - n_{w,t_i}), 1 \leq i \leq m, i \neq j\}$
      if $min_{diff} > \beta$ **then**
        Add w to $\mathcal{W}'$
      end if
    end if
  end for

---

3 Experimental Results

3.1 Data Set

Pseudomonas syringae, which has the biggest number of verified T3SEs, has been used as a model organism in the study of T3SS. Therefore, we collected data from this species. To our knowledge, there is a total of 283 confirmed effectors, from *P. syringae* pv. tomato strain DC3000, *P. syringae* pv. syringae strain B728a and *P. syringae* pv. phaseolicola strain 1448A. Considering that the redundancy of the data set would result in overestimation on the accuracy of the classifier, we eliminated the samples with sequence similarity over 60%. After redundancy removing, there are 108 positive samples.

    The negative data set was extracted from the genome of *P. syringae* pv. tomato strain DC3000. We excluded all the proteins related to T3SS, as well as the hypothetical proteins (Note that this set may still contain some unknown effectors). And then we selected randomly from the remaining samples to constitute a negative set, keeping the ratio of positive samples to negative samples as 1:7. The number of the negative samples is 760, thus there is a total of 868 samples.

3.2   Experimental Settings and Evaluation Criteria

In the experiments, we used HMM-LDA in the Matlab Topic Modeling Toolbox 1.4 [14]. Except the number of topics, all other parameters were set to be default values. The parameters α and β were set as 14 and 10, respectively. And we used the support vector machines (SVMs) as the classifier, which is widely used in bioinformatics. Our implementation of the SVM adopted LibSVM version 2.8 [15]. We chose the RBF kernel, and the kernel parameter gamma and C were set as 2^{-7} and 128, respectively.

Multiple measures were used to assess the performance of our proposed method, including sensitivity ($Sens$), specificity ($Spec$) and total accuracy (TA). TA is the ratio of the samples classified correctly compared to the total size of the data set. The sensitivity and specificity are defined in terms of the number of true positives (TP), the number of false positives (FP), and the number of false negatives (FN) as follows.

$$Sens = \frac{TP}{TP + FP}, Spec = \frac{TP}{TP + FN}. \tag{3}$$

3.3   Results and Discussions

We examined the performance of the feature reduction method by comparing the prediction accuracies of different feature vectors obtained by multiple methods. Table 1 lists the number of dimension, total accuracy (TA), sensitivity (Sens) and specificity (Spec) of five methods, respectively. The second and third methods perform feature reduction by word frequency/tf-idf only, and the fourth and fifth methods use all the di/tri-mers without feature reduction.

Table 1. Result Comparison

Method	Dimension	TA (%)	Sens (%)	Spec (%)
HMM-LDA	228	93.2	72.2	77.2
frequency	520	92.6	58.3	76.8
tf-idf	520	93.0	66.7	74.2
di-mer	400	93.2	65.7	76.3
tri-mer	8000	91.6	32.4	100

Table 1 clearly shows that the reduced feature set achieves good performance. The tri-mer method has the biggest number of dimensions, but its accuracy is the lowest. The reason why the tri-mer method has such high specificity and low sensitivity is that the number of false negative is zero while the number of false positive is very big. We also conducted experiments using K-mers with $K > 3$, and the accuracy is even worse. On the contrary, the HMM-LDA method has only 228 dimensions, but it has the best classification performance, especially for the sensitivity, which is nearly 14% higher than that of using frequency only as feature reduction method, and 6.5% higher than that of using all the dimers. We can also observe that an initial feature reduction using frequency or tf-idf has much better performance than that of tri-mer method, but has little advantage over di-mer method. That may be because the di-mers are better for the classification.

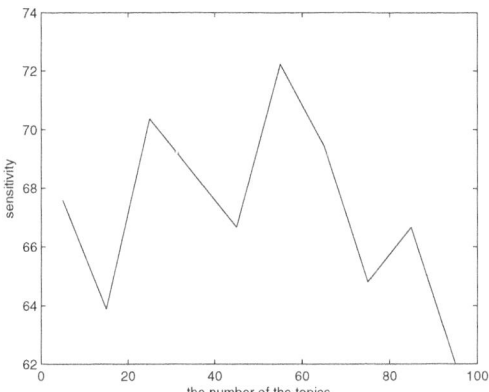

Fig. 2. Sensitivities obtained with different numbers of topics

In addition, we examined the impact of the number of topics on the prediction accuracy. The optimum number of topics was searched in the range from 5 to 95, including all the multiples of 5. The relationship between the number of topics and sensitivity is depicted in Figure 2. We found that the highest sensitivity (72.2%) was obtained when the number of topics is 55. And it could be observed from the figure that the number of topics has great influence on the final result. Even using the HMM-LDA model to reduce the dimension, if the number of topics is not appropriately selected, the sensitivity would be not higher than using frequency only. The reason is that we may remove some useful features in the process of dimension reduction.

On the whole, the experimental results demonstrate that using the HMM-LDA model for dimension reduction can improve the prediction accuracy, and it is helpful for predicting novel effectors.

4   Conclusion

In this paper, we use machine learning approaches to predict proteins secreted via the type III secretion system. We mainly focus on the sequence features, *i.e.*, the frequencies of amino acid subsequences. Instead of using all K-mers, we propose to use the HMM-LDA to eliminate noisy features. We compare the new method with the methods that use all di-mers/tri-mers, and that use only frequency/tf-idf as feature reduction method. The cross-validation tests show that our method achieves higher values on all of the accuracy measures.

This work is a preliminary study utilizing topic models to perform the feature reduction for protein sequence classification. For the future work, we will keep exploring better criteria in selecting the informative subsequences, as well as seeking the specific signals in the sequences that direct the secretion of effectors so as to advance our understanding on type III secretion mechanism.

Acknowledgments. This work was supported by the National Natural Science Foundation of China (Grant No. 61003093), and the Science & Technology Program of Shanghai Maritime University (Grant No. 20110009).

References

1. He, S.Y., Nomura, K., Whittam, T.S.: Type III protein secretion mechanism in mammalian and plant pathogens. BBA-Molecular Cell Research 1694(1-3), 181–206 (2004)
2. Guttman, D.S., Vinatzer, B.A., Sarkar, S.F., Ranall, M.V., Kettler, G., Greenberg, J.T.: A functional screen for the type III (Hrp) secretome of the plant pathogen Pseudomonas syringae. Science 295(5560), 1722–1726 (2002)
3. Vencato, M., Tian, F., Alfano, J.R., Buell, C.R., Cartinhour, S., DeClerck, G.A., Guttman, D.S., Stavrinides, J., Joardar, V., Lindeberg, M., et al.: Bioinformatics-enabled identification of the HrpL regulon and type III secretion system effector proteins of Pseudomonas syringae pv. phaseolicola 1448A. Molecular Plant-Microbe Interactions 19(11), 1193–1206 (2006)
4. Arnold, R., Brandmaier, S., Kleine, F., Tischler, P., Heinz, E., Behrens, S., Niinikoski, A., Mewes, H., Horn, M., Rattei, T.: Sequence-based prediction of type III secreted proteins. PLoS Pathogens 5(4), e1000376 (2009)
5. Löwer, M., Schneider, G.: Prediction of Type III Secretion Signals in Genomes of Gram-Negative Bacteria. PloS One 4(6), e5917 (2009)
6. Yang, Y., Zhao, J., Morgan, R., Ma, W., Jiang, T.: Computational prediction of type III secreted proteins from gram-negative bacteria. BMC Bioinformatics 11(suppl. 1), S47 (2010)
7. Wang, Y., Zhang, Q., Sun, M., Guo, D.: High-accuracy prediction of bacterial type iii secreted effectors based on position-specific amino acid composition profiles. Bioinformatics 27(6), 777–784 (2011)
8. Yang, Y., Lu, B.L.: Extracting features from protein sequences using Chinese segmentation techniques for subcellular localization. In: Proceedings of the 2005 IEEE Symposium on Computational Intelligence in Bioinformatics and Computational Biology, pp. 288–295 (2005)
9. Yang, Y., Lu, B., Yang, W.: Classification of protein sequences based on word segmentation methods. In: Proceedings of the 6th Asia-Pacific Bioinformatics Conference, vol. 6, pp. 177–186 (2008)
10. Liu, B., Wang, X., Lin, L., Dong, Q., Wang, X.: A discriminative method for protein remote homology detection and fold recognition combining top-n-grams and latent semantic analysis. BMC Bioinformatics 9(1), 510 (2008)

11. Yeh, J., Chen, C.: Protein remote homology detection based on latent topic vector model. In: Proceedings of 2010 International Conference on Networking and Information Technology, pp. 456–460 (2010)
12. Blei, D., Ng, A., Jordan, M.: Latent dirichlet allocation. The Journal of Machine Learning Research 3, 993–1022 (2003)
13. Griffiths, T., Steyvers, M., Blei, D., Tenenbaum, J.: Integrating topics and syntax. In: Advances in Neural Information Processing Systems, vol. 17, pp. 537–544 (2005)
14. Steyvers, M., Griffiths, T.: Matlab Topic Modeling Toolbox 1.4 (2011), Software available at `http://psiexp.ss.uci.edu/research/programs_data/toolbox.htm`
15. Chang, C.C., Lin, C.J.: LIBSVM: a library for support vector machines (2001), Software available at `http://www.csie.ntu.edu.tw/~cjlin/libsvm`

A Saliency Detection Model Based on Local and Global Kernel Density Estimation

Huiyun Jing, Xin He, Qi Han, and Xiamu Niu

Department of Computer Science and Technology, Harbin Institute of Technology,
No.92, West Da-Zhi Street, Harbin, China
xm.niu@hit.edu.cn

Abstract. Visual saliency is an important and indispensable part of visual attention. We present a novel saliency detection model using Bayes' theorem. The proposed model measures the pixel saliency by combining local kernel density estimation of features in center-surround region and global density estimation of features in the entire image. Based on the model, a saliency detection method is presented that extracts the intensity, color and local steering kernel features and employs feature level fusion method to obtain the integrated feature as the corresponding pixel feature. Experimental results show that our model outperforms the current state-of-the-art models on human visual fixation data.

Keywords: Visual attention, Saliency map, Bayes' theorem, Kernel density estimation.

1 Introduction

The Human visual system rapidly and automatically detects salient locations of images or videos to reduce the computational complexity. Saliency maps are topographical maps of the visually salient parts in static and dynamic scenes, which can be classified as bottom-up saliency map and top-down saliency map. The former are automatically-driven, while the latter are task-driven. The detection of salient image locations is important for applications like object detection[1], image browsing[2], and image/video compression[3].

Several computational models have been proposed to compute saliency maps from digital imagery. Some of these models[4] are biologically based, while other models [5–7] are partly based on biological models and partly on computational ones. Furthermore, the learning techniques training from human fixation data are recently introduced to compute bottom-up saliency map[8, 9].

The saliency model of Itti and Koch [4] is the earliest and the most influential model. Based on the feature integration theory, the model decomposes the input image into three channels (intensity, color, and orientation) and combines the multiscale center-surround excitation responses of feature maps in the three channels into a single saliency map. Gao et al. [10] proposed the discriminant center-surround saliency hypothesis, which is obtained by combining the center-surround hypothesis and the hypothesis that all saliency decisions are optimal in

B.-L. Lu, L. Zhang, and J. Kwok (Eds.): ICONIP 2011, Part I, LNCS 7062, pp. 164–171, 2011.
© Springer-Verlag Berlin Heidelberg 2011

a decision-theoretic sense. The saliency of each image location is equated to the discriminant power of a set of features observed in that location to distinguish between the region and its surround. Bruce and Tsotsos[5] modeled saliency at a location as the self-information of the location relative to its local surround or the entire image. Zhang et al.[6] proposed a Bayesian framework (SUN) from which the bottom-up saliency are computed as the self-information of local visual features. The underlying hypothesis of the framework is that the probability of local visual features is equal to saliency. Seo et al. [11] computed a saliency map through computing the self-resemblance of a feature matrix at a pixel with respect to its surrounding feature matrices. In [8], the authors learned optimal parameters for saliency detection model based on low-level, middle-level and high-level image features. Murray et al.[9] used a non-parametric low-level vision model to compute saliency, where the scale information is integrated by a simple inverse wavelet transform over the set of extended contrast sensitivity function (ECSF) outputs and the ad-hoc parameters are reduced by introducing training steps on both color appearance and eye-fixation data.

Considering that Zhang et al.[6] and Seo et al.[11] respectively utilize a part of Bayes' equation of saliency, we integrate the two parts of Bayes' equation and measure saliency as a function of local and global kernel density estimation using Bayes' theorem.

The rest of the paper is organized as follows. The proposed saliency detection model is described in Section 2. Section 3 presents the implementation of proposed saliency detection model. Experimental results and conclusions are given in Sections 4 and 5, respectively.

2 Proposed Saliency Detection Model

Motivated by the approach in [6], [10] and [11], we measure saliency of each pixel by Bayes' theorem. Firstly, representing saliency of each pixel i under the feature F_i and the location L_i as a binary random variable, we define binary random variables $\{y_i{}_{i=1}^M\}$ as follows

$$y_i = \begin{cases} 1, \text{ if pixel } i \text{ is salient,} \\ 0, \text{ otherwise.} \end{cases} \tag{1}$$

where $i = 1, \ldots, M$ and M is the total number of pixels in the image.

Thus, the saliency of a pixel i is defined as a posterior probability $Pr(y_i = 1|F, L)$ as follows

$$S_i = Pr(y_i = 1|F_i, L_i) \tag{2}$$

where $F_i = [f_i^1, f_i^2, \ldots, f_i^K]$ contains a set of features $\{f_i{}_{k=1}^K\}$ extracted from the local neighborhood of the corresponding pixel, K is the number of features in that neighborhood and L_i represents the pixel coordinates.

Eq. 2 can be rewritten using Bayes' rule:

$$S_i = Pr(y_i = 1|F_i, L_i) = \frac{p(F_i, L_i|y_i = 1)Pr(y_i = 1)}{p(F_i, L_i)} \tag{3}$$

We assume that 1) the feature and location are independent and conditionally independent given $y_i = 1$; and 2) under location prior, $Pr(y_i = 1|L)$ is equal to be salient. Then Eq. 2 is simplified as follows

$$
\begin{aligned}
S_i &= \frac{p(F_i, L_i|y_i = 1)Pr(y_i = 1)}{p(F_i, L_i)} \\
&= \frac{1}{p(F_i)}p(F_i|y_i = 1)Pr(y_i = 1|L_i) \\
&= \frac{1}{p(F_i)}p(F_i|y_i = 1)
\end{aligned}
\tag{4}
$$

$p(F_i)$ depends on the visual features and implies that the feature of less probability seems to have higher saliency. In Seo et al.[11], $p(F_i)$ is considered uniform over features. In Bruce et al.[5] and Zhang et al.[6], $p(F_i)$ is used to detect saliency, where F_i is the feature vector and the features are calculated as the responses to filters learned from natural images. Different from Bruce et al.[5] and Zhang et al.[6], we directly calculate $p(F_i)$ using normalization kernel density estimation for F_i. Then we obtain Eq. 5.

$$
\frac{1}{p(F_i)} = \frac{\sum_{i=1}^{M} \sum_{j=1}^{M} \kappa(F_i - F_j)}{\sum_{j=1}^{M} \kappa(F_i - F_j)}
\tag{5}
$$

where κ is the kernel density function and M is the total pixels number of the image.

In Zhang et al.[6], $p(F_i|y_i = 1)$ of Eq. 4 is considered with knowledge of the target and is not used when calculating saliency. However, Seo et al.[11] adopt local "self-resemblance" measure to calculate $p(F|y_i = 1)$ using nonparametric kernel density estimation. Similar to Seo et al.[11], we make a hypothesis that $y_i = 1$ of the center pixel i in the center-surround region. It means that F_i is the only sampled feature in the center-surround features' space. Under this hypothesis, we estimate all $F = [F_1, F_2, \ldots, F_N]$ including Fi using kernel density estimation in the center-surround region where F is a feature set containing features from the center and surrounding region and N is the number of pixels in the center-surround region. Then we normalize $p(Fi|y_i = 1)$ under the hypothesis of $y_i = 1$.

$$
\begin{aligned}
p(F_i|y_i = 1) &= \frac{\kappa(F_i - F_i)}{\sum_{j=1}^{N} \kappa(F_i - F_j)} \\
&= \frac{1}{\sum_{j=1}^{N} \kappa(F_i - F_j)}
\end{aligned}
\tag{6}
$$

Now we rewrite Eq. 4 using Eq. 5 and Eq. 6 and obtain the saliency formula of each pixel

$$S_i = \frac{\sum_{i=1}^{M} \sum_{j=1}^{M} \kappa(F_i - F_j)}{\sum_{j=1}^{M} \kappa(F_i - F_j)} \frac{1}{\sum_{j=1}^{N} \kappa(F_i - F_j)} \qquad (7)$$

Eq. 7 could be represented as follows

$$S_i = \frac{K_{local}(F_i)}{K_{global}(F_i)} \qquad (8)$$

where $K_{local}(F_i)$ represents normalization kernel density estimation in the local center-surround region and $K_{global}(F_i)$ represents normalization kernel density estimation in the entire image. Thus, we model pixel saliency as local and global kernel density estimation of features of the corresponding pixel.

3   Implementation

Due to the contributions of Takeda et al.[12, 13] to adaptive kernel regression, local steering kernel was proposed as a feature to compute $p(F|y_i = 1)$ in [11]. Though local steering kernel(LSK) is to robustly obtain the local structure of images by analyzing pixel value difference, it is unable to represent intensity or color information of the corresponding pixel, which are the important cues for computing saliency. Lacking intensity and color information, the LSK feature only represents a weighted relation which penalizes distance away from the local position where the approximation is centered[12]. So, it is not used appropriately to compute $p(F)$.

In our implementation, we extract the intensity, color and LSK features and employ feature-level fusion method to obtain the integrated feature of these features. Then, we compute the local and global kernel density estimation of the integrated feature to obtain the saliency value. The implementation could be detailed by the following steps:

1.We extract intensity and color features from the Lab color space. In the Lab color space, each pixel location is an $[L; a; b]^T$ vector. For the input image of M pixels, the normalized L component of each pixel composes the intensity features $F^L = [F_1^L, \ldots, F_i^L, \ldots, F_M^L]^T$; the normalized a and b components compose the color features $F^C = [F_1^C, \ldots, F_i^C, \ldots, F_M^C]^T$, where F_i^L and F_i^C are respectively the intensity feature and color feature of the ith pixel.

2.Local steering kernel(LSK) features are extracted. The local steering kernel [11] is represented as follows

$$K(\mathbf{x_j} - \mathbf{x_i}) = \frac{\sqrt{det(\mathbf{C_j})}}{h^2} \exp\{\frac{(\mathbf{x_j} - \mathbf{x_i})^T \mathbf{C_j}(\mathbf{x_j} - \mathbf{x_i})}{-2h^2}\} \qquad (9)$$

where $j \in \{1, \cdots, P\}$, P is the the number of pixels in a local window, h is a global smoothing parameter and the matrix $\mathbf{C_j}$ is a covariance matrix estimated from a collection of spatial gradient vectors within the local analysis window around a sampling position $\mathbf{x_j} = [x_1, x_2]_j^T$. We extract the normalized local steering kernels from each color channel L,a, b as $\overline{FL_i^S}$, $\overline{Fa_i^S}$, $\overline{Fb_i^S}$ and collect them to form LSK feature $\overline{F_i^S} = [\overline{FL_i^S}, \overline{Fa_i^S}, \overline{Fb_i^S}]$. These normalized local steering kernels $\overline{FL_i^S} = [\overline{fL_i^1}, \ldots, \overline{fL_i^j}, \ldots, \overline{fL_i^P}]$ are computed from the P number of pixels in a local window centered on the pixel i, where $\overline{fL_i^j}$ is computed as Eq. 10.

$$\overline{fL_i^j} = \frac{K(\mathbf{x_j} - \mathbf{x_i})}{\sum_{j=1}^{P} K(\mathbf{x_j} - \mathbf{x_i})}, i = 1, \cdots, M; j = 1, \cdots, P \tag{10}$$

The the normalized local steering kernels Fa_i^S and Fb_i^S can be obtained by the same process.

3. We obtain the integrated features. Because the LSK features are dense, PCA is applied to $\overline{F_i^S}$ for dimensionality reduction and retain only the largest d principal components. The lower dimensional features of local steering kernel features are obtained as $F^S = [F_1^S, \ldots, F_i^S, \ldots, F_M^S]$, where F_i^S is the lower dimensional LSK feature of the ith pixel. Then the F_i^L, F_i^C and F_i^S are concatenated to form the integrated features FC_i.

$$FC_i = [F_i^L, F_i^C, F_i^S]. \tag{11}$$

4.Calculating kernel density estimation of $p(FC_i)$ is usually time-consuming. In order to speed up the operation, we perform a kernel density estimation using a Gaussian kernel with with the rule-of-thumb bandwidth[1].

5.Because using a feature matrix consisting of a set of feature vectors provides more discriminative power than using a single feature vector, "Matrix Cosine Similarity" [11] is chosen to compute $p(FC_i|y_i = 1)$. The feature vectors $FC_i' = [FC_1, \ldots, FC_i, \ldots, FC_L]$ are used to replace FC_i to compute $p(FC_i|y_i = 1)$, where L is the number of features in a local neighborhood centered on the pixel i . Then we use the following equation to calculate $p(FC_i'|y_i = 1)$

$$p(FC_i'|y_i = 1) = \frac{1}{\sum_{k=1}^{N} \exp(\frac{-1+\rho(FC_i', FC_k')}{\sigma^2})}, k = 1, \cdots, N \tag{12}$$

where $\rho(FC_i', FC_k')$ is equal to trace ($\frac{FC_i'^T FC_k'}{\|FC_i'\|\|FC_k'\|}$) and σ is set to 0.07 as default value.

6.By step 4 and step 5, we obtain the value of $p(FC_i)$ and $p(FC_i|y_i = 1)$. Then saliency of each pixel is calculated using Eq. 4.

4   Experimental Results

We evaluated our method's performance with respect to predicting human visual fixation data from natural images. The dataset and the corresponding fixation

---

[1] We use the Kernel Density Estimation Toolbox for Matlab provided by Alexander Ihler (Available at http://www.ics.uci.edu/~ihler/code/kde.html)

Table 1. Performance of Three Kinds of Saliency Models

Model	KL(SE)	AUC(SE)
saliency model only using $p(FC_i)$	0.3792(0.0032)	0.6897(0.0008)
saliency model only using $p(FC_i \| y_i = 1)$	0.4088(0.0030)	0.6841(0.0007)
our model	0.4302(0.0031)	0.7022(0.0008)

Table 2. Experimental Results

Model	KL(SE)	AUC(SE)
Itti et al.[4]	0.1130(0.0011)	0.6146(0.0008)
Bruce and Tsotsos[5]	0.2029(0.0017)	0.6727(0.0008)
Gao et al.[10]	0.1535(0.0016)	0.6395(0.0007)
Zhang et al.[6]	0.2097(0.0016)	0.6570(0.0008)
Seo and Milanfar[11]	0.3432(0.0029)	0.6769(0.0008)
Our Method	0.4302(0.0031)	0.7022(0.0008)

data we used were collected by Bruce and Tsotsos[5] as the benchmark dataset for comparison. The dataset contains eye fixation data from 20 subjects for a total of 120 natural images of size 681×511.

In order to reduce the computation cost, the input images are down-sampled to an appropriate scale (64×64). We then set $P = 3 \times 3$ and $h = 0.2$ to extract the normalized LSK values in three color channels of size 3×3 to form the image local structure features F_i^S. The lower dimensional LSK feature F_i^S only contains three dimensions. The probability of $p(FC_i'|y_i = 1)$ is computed under $L = 7 \times 7$ and $N = 3 \times 3$.

The Kullback-Leibler(KL) divergence and the area under receiver operating characteristic(AUC) were computed as performance metrics. A high value of two metrics means better performance. In Zhang et al.[6], Zhang et al. noted that the original KL divergence and ROC area measurement are corrupted by an edge effect which yielding artificially high results. For eliminating border effects, we adopt the same procedure described by Zhang et al.[6] to measure KL divergence and ROC area. As we alluded to earlier, the proposed saliency detection model, combining local and global kernel density estimation using Bayes' theorem, provides more discriminative power than the saliency models only utilizing a part of Bayes' equation of saliency. Table 1 well demonstrate that.

We compared our method against state-of-the-art methods [4–6, 10, 11]. The mean and the standard error are reported in Table 2. Our method performs better than the current state-of-the-art models in KL-divergence and AUC metrics. Limited by space, we only present some examples of visual results of our method compared with Seo et al.[11] and Bruce et al.[5] in Fig. 1. Visually, our method also exceeds the other two models in term of accuracy.

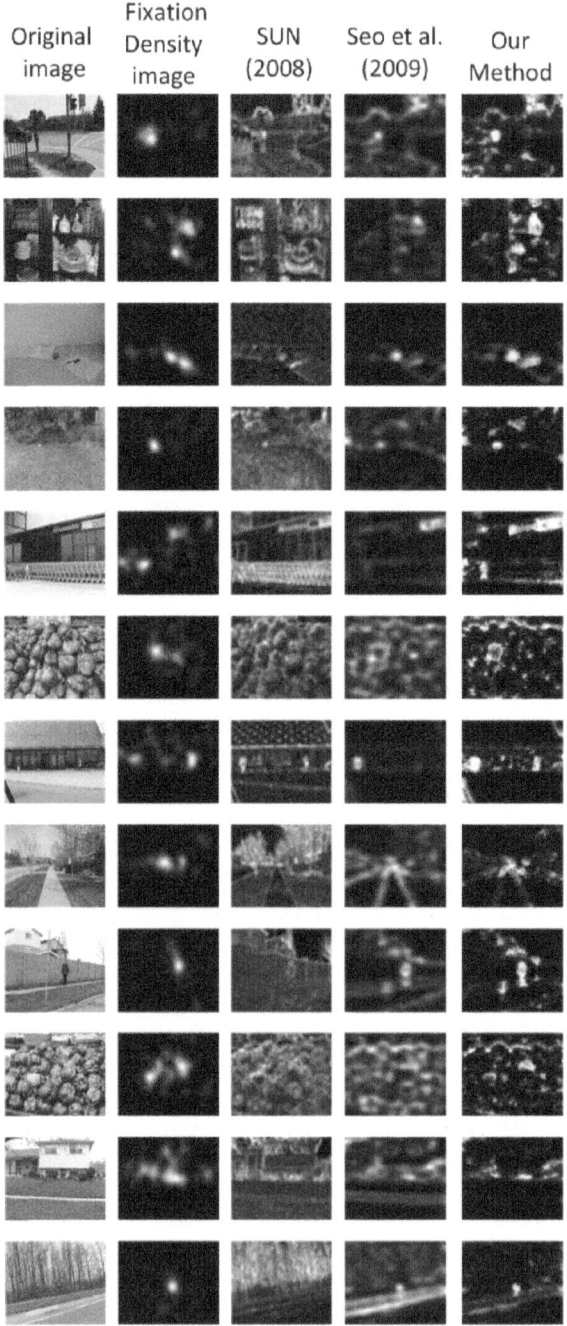

Fig. 1. Examples of saliency map on human visual fixation data

5 Conclusions and Future Work

In this paper, we presented a novel saliency detection model using Bayes' theorem. Because of the limitation of the common saliency models utilizing only a part of Bayes' equation of saliency, the proposed model integrates the two parts of Bayes' equation. Then the pixel saliency is modeled as the combination the global and local kernel density estimation of the integrated features. Experimental results demonstrate that the proposed model exceeds the current state-of-the-art models. In future work, we will investigate the saliency detection method for dynamic video.

Acknowledgments. This work is supported by the National Natural Science Foundation of China (Project Number: 60832010) and the Fundamental Research Funds for the Central Universities (Grant No. HIT. NSRIF. 2010046).

References

1. Papageorgiou, C., Poggio, T.: A trainable system for object detection. International Journal of Computer Vision 38, 15–33 (2000)
2. Rother, C., Bordeaux, L., Hamadi, Y., Blake, A.: AutoCollage. ACM Transactions on Graphics 25, 847–852 (2006)
3. Itti, L.: Automatic foveation for video compressing using a neurobiological model of visual attention. IEEE Transactions on Image Processing 13, 1304–1318 (2004)
4. Itti, L., Koch, C., Niebur, E.: A model of saliency-based visual attention for rapid scene analysis. IEEE Transactions on Pattern Analysis and Machine Intelligence 20, 1254–1259 (1998)
5. Bruce, N.D.B., Tsotsos, J.K.: Saliency based on information maximization. In: Advances in Neural Information Processing Systems, vol. 18, pp. 155–162 (2006)
6. Zhang, L., Tong, M.H., Marks, T.K., Shan, H., Cottrell, G.W.: SUN: a bayesian framework for saliency using natural statistics. Journal of Vision 8, 32–51 (2008)
7. Gopalakrishnan, V., Hu, Y.Q., Rajan, D.: Random Walks on Graphs to Model Saliency in Images. In: 2009 IEEE Conference on Computer Vision and Pattern Recognition, pp. 1698–1750. IEEE Press, New York (2009)
8. Judd, T., Ehinger, K., Durand, F., Torralba, A.: Learning to predict where humans look. In: 12th International Conference on Computer Vision, pp. 2106–2113. IEEE Press, New York (2009)
9. Murray, N., Vanrell, M., Otazu, X., Parraga, C.A.: Saliency Estimation Using a Non-Parametric Low-Level Vision Model. In: 2011 IEEE Conference on Computer Vision and Pattern Recognition. IEEE Press, New York (2011)
10. Gao, D., Mahadevan, V., Vasconcelos, N.: On the plausibility of the discriminant center-surround hypothesis for visual saliency. Journal of Vision 8, 13–31 (2008)
11. Seo, H.J., Milanfar, P.: Static and space-time visual saliency detection by self-resemblance. Journal of Vision 9, 15–41 (2009)
12. Takeda, H., Farsiu, S., Milanfar, P.: Kernel regression for image processing and reconstruction. IEEE Transactions on Image Processing 16, 349–366 (2007)
13. Takeda, H., Milanfar, P., Protter, M., Elad, M.: Super-Resolution Without Explicit Subpixel Motion Estimation. IEEE Transactions on Image Processing 18, 1958–1975 (2009)
14. Seo, H.J., Milanfar, P.: Training-free, generic object detection using locally adaptive regression kernels. IEEE Transactions on Pattern Analysis and Machine Intelligence 32, 1688–1704 (2010)

Saliency Detection Based on Scale Selectivity
of Human Visual System

Fang Fang[1,3], Laiyun Qing[2,3], Jun Miao[3], Xilin Chen[3], and Wen Gao[1,4]

[1] School of Computer Science and Technology,
Harbin Institute of Technology, Harbin 150001, China
[2] Graduate University of Chinese Academy of Science, Beijing 100049, China
[3] Key Laboratory of Intelligent Information Processing, Institute of Computing
Technology, Chinese Academy of Sciences, Beijing 100190, China
[4] Institute of Digital Media, Peking University, Beijing 100871, China
{ffang,lyqing,jmiao,xlchen,wgao}@jdl.ac.cn

Abstract. It is well known that visual attention and saliency mechanisms play an important role in human visual perception. This paper proposes a novel bottom-up saliency mechanism through scale space analysis. The research on human perception had shown that our ability to perceive a visual scene with different scales is described with the Contrast-Sensitivity Function (CSF). Motivated by this observation, we model the saliency as weighted average of the multi-scale analysis of the visual scene, where the weights of the middle spatial frequency bands are larger than others, following the CSF. This method is tested on natural images. The experimental results show that this approach is able to quickly extract salient regions which are consistent with human visual perception, both qualitatively and quantitatively.

Keywords: Saliency Detection, Human Visual System, Scale Selectivity.

1   Introduction

It is well known that the visual attention and saliency mechanism play an important role in human visual perception. In recent years, there have been increasing efforts to introduce computational models to explain the fundamental properties of biological visual saliency. It is generally agreed that visual attention of Human Vision System (HVS) is an interaction between bottom-up and top-down mechanisms. In this paper, we will focus on the bottom-up saliency detection.

Bottom-up saliency drives attention only by the properties of the stimuli in a visual scene and is independent of any high level visual tasks. Inspired by the early visual pathway in biological vision, the features used in saliency models include low-level simple visual attributes, such as intensity, color, orientation and motion. In one of the most popular models for bottom-up saliency [1], saliency is measured as the absolute difference between feature responses at a location and those in its neighborhood, in a center-surround fashion. This model has been shown to successfully replicate many

B.-L. Lu, L. Zhang, and J. Kwok (Eds.): ICONIP 2011, Part I, LNCS 7062, pp. 172–181, 2011.

observations from psychophysics [2]. Gao and Mahadevan [3] implemented the center-surround mechanisms as a discriminate process based on the distributions of local features centering and surrounding at a given point. In a recent proposal, Kienzle et al. [4] employed machine learning techniques to build a saliency model from human eye fixations on natural images, and showed that a center-surround receptive field emerged from the learned classifier. Some other recent works model the saliency in information theory and deriving saliency mechanisms as optimal implementations of generic computational principles, such as the maximization of self-information [5], local entropy [6] or 'surprise' [7]. Other methods are purely computational and are not based on biological vision principles [8, 9].

This paper proposes a saliency detection method based on the scale selectivity of HVS. In natural scenes, objects and patterns can appear at a wide variety of scales. In other words, a natural image includes signals of multiple scales or frequencies. However, the HVS is able to quickly select the frequency band that conveyed the most information to solve a given task and interpret the image, not conscious of the information in the other scales [10]. Researches on human perception had suggested that image understanding is based on a multi-scale, global to local analysis of the visual input [11, 12]. The global precedence hypothesis of image analysis implies that the low spatial frequency components dominate early visual processing. Physiological research has also shown that our ability to perceive the details of a visual scene is determined by the relative size and contrast of the detail present. The threshold contrast necessary for perception of the signal is found to be a function of its spatial frequency, described by the CSF [13, 14], in which contrast sensitivities are higher in the middle frequency bands than the other bands. These inspire us to detect saliency in the visual by using the low-middle frequency components and ignoring the other ones.

The rest of the paper is organized as follows. Some related works on scale selectivity of human vision system will be given in Section 2. The analysis of the eye tracking data and the details of the proposed saliency detection methods are described in Section 3. Experimental results are shown in Section 4. Finally the conclusions are given in Sections 5.

2   Scale Selectivity of Human Vision System

Scale selectivity is a visual processing property which passes different spatial frequencies. This behavior is characterized by a modulation-transfer function (MTF) which assigns an amplitude scale factor to different spatial frequency [15]. The amplitude scale factor ranges from 1.0 for spatial frequencies that are completely passed by the filter to 0.0 for spatial frequencies that are completely blocked. In certain situations, the MTF can be described by the CSF [13] which is the reciprocal of contrast sensitivity as a function of spatial frequency. This function describes the sensitivity of the human eye to sine-wave gratings at various frequencies.

The CSF tells us how sensitive we perceive different frequencies of visual stimuli. As illustrated in Fig.1 [15], the CSF curve is band pass in which spatial frequencies around 3-6 cycles/degree are best represented, while both lower and higher are poorly, and it is meaningless for that above 60 cycles/degree. That is to say, if the frequency of visual stimuli is too high, we will not be able to recognize the stimuli pattern any more. For example, the stripes in an image consisting of vertical black and white stripes are thin enough (i.e. a few thousand per millimeter), then we will not be able to see the individual stripes.

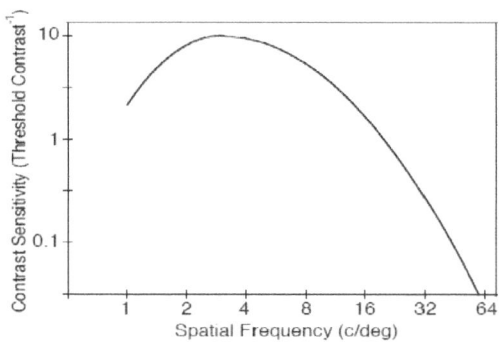

Fig. 1. Contrast-Sensitivity Function [15]

3   Saliency Detection Based on Scale Selectivity

Motivated by the observation in scale selectivity of human vision system, we propose a novel algorithm in this section. Although the form of CSF is known, the measurements of CSF only utilize sine-wave gratings, which is a simple stimulus [13]. Nevertheless, its application on complex stimuli, such as nature images, is dubious. Therefore we firstly analyze the sensitivity of the human eye at each spatial frequency according to eye tracking data over images.

3.1   Analysis of Eye Tracking Data

It is well known that any image can be represented equivalently in image space and frequency space. One can move back and forth between image space and frequency space via Fourier transformation and inverse Fourier transformation [17]. Fig.2 shows some natural images, their corresponding pictures in different frequency band and human saliency map, respectively. It is obvious that low-middle frequency components of image are closest to the human's perception when compare with the saliency map from the eye tracking data, which is consistent with previously described CSF.

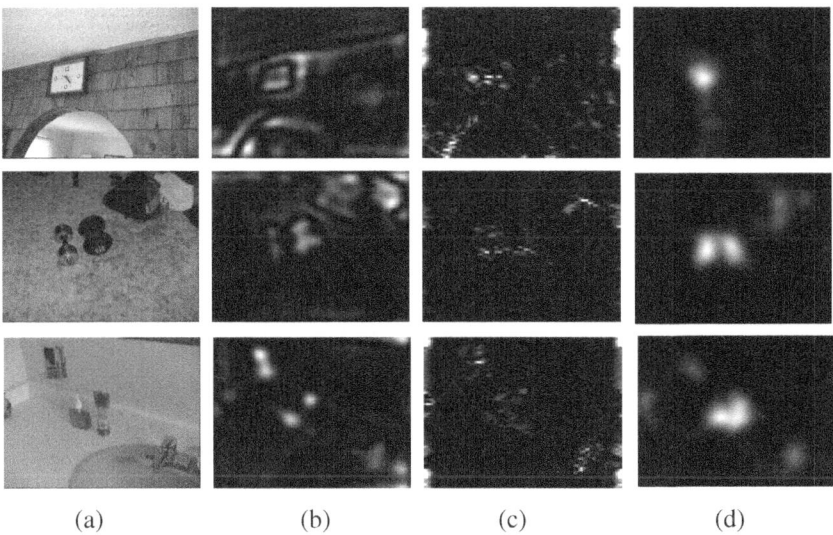

(a)   (b)   (c)   (d)

Fig. 2. Analysis of spatial frequencies. In each group, we present (a) original image, (b) decomposition of image into low-middle frequency components, (c) decomposition of image into high frequency bands, (d) saliency map from the eye tracking data.

As stated before, different frequencies which human perceive are quite distinct, namely, human have intense perception in some frequency but in others have a little or scarcely sensitive. We analyze the sensitivity of the human eye at each spatial frequency on two pubic datasets proposed by Bruce et al. [5] and Judd et al. [16]. The dataset of Bruce et al. [5] contains eye fixation records from 20 subjects on 120 images of size 511×681. The dataset of Judd et al. [16] contains 1003 natural images covering a wide range of situations, and the eye fixation data is collected from 15 subjects. All of images are down-sampled to the size of 64× 86.

We use the ROC metric for quantitative analysis human sensitive on each frequency. Under the criterion, we make comparisons between every frequency map from an image and eye fixation records from humans [5, 16]. By varying the threshold, we draw the ROC curve, and the area under the curve indicates how well the frequency map of any frequency can predict the ground truth fixations. More specifically, the larger area, the more sensitive human perceive, and vice versa. According to the value of the area, we obtain a set of coefficients about human sensitivity in every frequency. Fig.3(a)(b) and Fig.4(a)(b) are the sensitivity curves showing the changes in scales which are obtained from both the whole and part of the dataset, respectively. The red curves in these figures represent the sensitivity statistic of human fixation in different scales, while the blue ones are the 5-order polynomial fitting with respect to the red. As shown in these figures, we find that the sensitivity curve of different scenarios share similar trends.

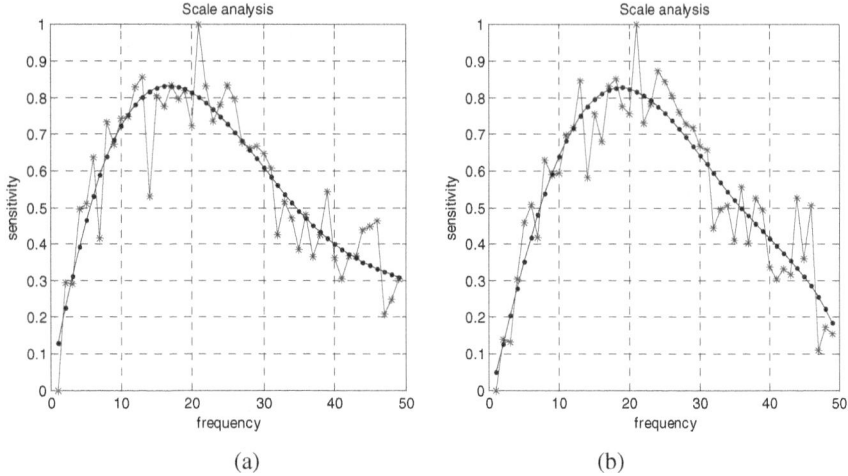

Fig. 3. Sensitivity curves on different frequency (a) over 60 images and (b) the whole images in Bruce et al. [5], where red curves represent the sensitivity statistic of human fixation in different scales, and the blue ones are the polynomial fitting with respect to the red ones

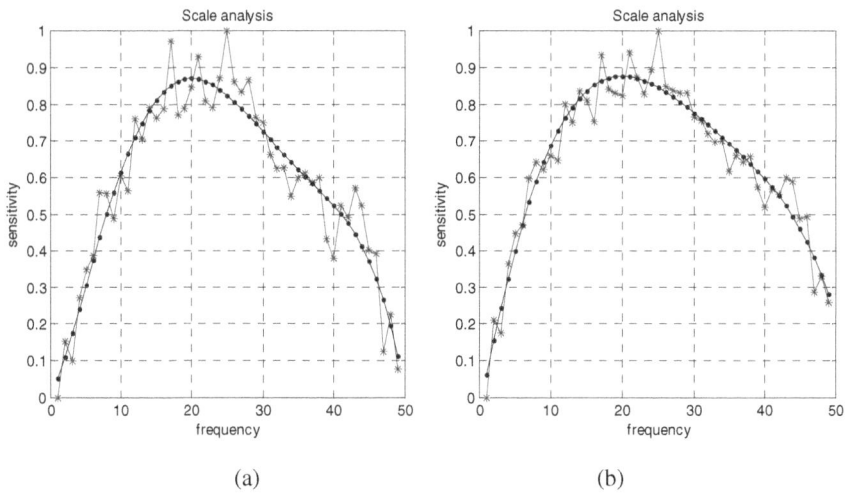

Fig. 4. Sensitivity curves on different frequency (a) over 300 images in Judd et al. and (b) the whole images in Judd et al. [16], where red curves represent the sensitivity statistic of human fixation in different scales, and the blue ones are the polynomial fitting with respect to the red ones

Table 1. Test results of frequency sensitivity by cross-validation

	Bruce dataset [5]	Judd dataset [16]
Bruce dataset [5]	0.71259	0.68612
Judd dataset [16]	0.71590	0.69438

According to the sensitivity coefficients from Fig.3 and Fig.4, we conduct four tests. In the first two tests, we utilize the sensitivity coefficients from part of the images as the weighted value of amplitude, and test on the other part of same dataset. In the later two, we carry out in a cross-dataset validation way, where training and testing are on different datasets. More specifically, we use each sensitivity coefficient from the whole dataset of Judd as the weighted value of amplitude, and test on Bruce dataset. Similarly, we conducted on the whole datasets of Bruce and test on Judd. Table 1 lists the four test results of cross-validation. For example, the value 0.71259 indicates the result training and testing both on Bruce dataset. As shown in Table 1, the best test result comes from training on Judd and testing on Bruce; while the worst result from training on Bruce and testing on Judd. This may due to the fact that the images in Judd dataset include more semantic objects which attract human eyes such as faces, cars, which can not explain by low level signal features only. However, difference between training results are little, which indicate that we can mimic the scale selectivity of HVS by designing a band-pass filter which passes the low-middle frequency components while suppresses the others.

3.2   Saliency Based on Scale Selection

Based on the scale selection properties of human vision system, we decompose the input image into the multi-scale bands and detect the saliency of the input scene by designing a band-pass filter which mimic the scale selectivity of human vision system.

We use Fourier transform to get a multi-scale representation of the input. The 2D formulation of Fourier function is:

$$F(u,v) = \sum_{(x,y)} I(x,y)e^{2\pi j(ux+vy)}$$ (1)

where (x, y) is the coordinate of the current pixel and $I(x,y)$ is the intensity function of input image. The variable u and v represent spatial frequency coordinate of natural image in horizontal and vertical directivity, respectively. The amplitude in each frequency band is represented as:

$$A(u,v) = |F(u,v)|.$$ (2)

Weighted value of each scale in image is equal to the corresponding coefficients. As a result, the amplitude $B(\omega)$ in each scale after weighting can be represented as:

$$B(\omega) = A(\omega) \cdot H(\omega).$$ (3)

The weighted amplitude map in different frequencies bands is shown in Fig.5(c). It can be seen that the center of the amplitude map is lighter, which further demonstrate the amplitude is more intense at low-middle frequency.

Then we can acquire saliency image in each spatial frequency by Inverse Fourier Transform. The value of the saliency map is obtained by Eq.4:

$$S(x, y) = g(x, y) * F^{-1}\left[B(\omega) \cdot \exp(i \cdot P(w))\right]^2,$$ (4)

where F^{-1} denotes Inverse Fourier Transform, $P(w)$ is the phase spectrum of the image, which is preserved during the process, and $g(x,y)$ indicates a 2D Gaussian filter to smooth the saliency map. An example of the saliency image is shown in Fig.5(d).

As stated before, human vision system are more sensitive to the low-middle frequency components. Based on the observation, we find that simply using the low-middle frequency components of image as the saliency map produces excellent result.

The low-middle frequency component is extracted using the following band-pass filter:

$$H(\omega) = \exp(-\omega^2 / 2\sigma_1^2) - \exp(-\omega^2 / 2\sigma_2^2) \,, \tag{5}$$

where w is the frequency, it is represented as: $w = \sqrt{u^2 + v^2}$, σ_1, σ_2 are the variances of the Gaussian function. The relationship between σ and cut-off frequency ω_0 is described as:

$$\sigma = \omega_0 \frac{1}{\sqrt{2\ln 2}} \cdot \tag{6}$$

According to perception ability that HVS on different spatial frequency visual signal, we indicate that the variances of σ are 15.2 and 90.5, respectively. The band-pass filter that we defined is shown in Fig.5(b). It preserves the low-middle frequency of the image for the detection of saliency.

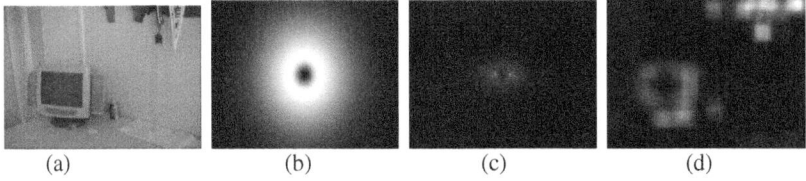

| (a) | (b) | (c) | (d) |

Fig. 5. (a) is the original image. The corresponding weighted amplitude map (c) is computed using the band-pass filter (b). And the saliency map is shown in (d).

4   Experimental Results

In this section, we evaluate the proposed method on natural images to demonstrate its effectiveness. In the experiments, we use the image dataset and its corresponding eye fixations collected by Bruce et al. [5] as the benchmark for comparison. We down-sample the images to the size of 64× 86 pixels. The results of our model are compared with two state-of-the-art methods: information maximization approach [5] and spectral residual approach [9], as shown in Fig.6 and Fig.7.

For qualitative analysis, we show two challenging saliency detection cases. The first case (Fig.6) includes images with a large amount of textured regions in the background. These textured regions are usually neglected by the human beings, whose saliency will be obviously inhibited. For such images, we expect that only the object's pixels will be identified as salient. In Bruce et al.'s method [5], the pixels on the objects are salient, but other pixels which are on the background are partially salient as well. As a consequence, Bruce et al.'s method is quite sensitive to textured regions. In Hou et al.'s method [9], which is somewhat better in this respect, however, many pixels on the salient objects are not detected as salient, e.g., the clock. Our method detects the pixels on the salient objects and is much less sensitive to background texture.

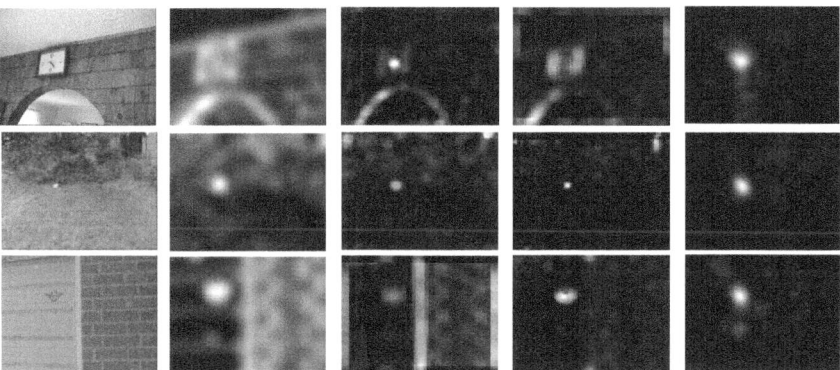

Fig. 6. Comparative saliency results on images with a large amount of textured regions in the background. The first image in each row is the original image (a), the rest saliency maps from left to right are produced by Bruce et al.[5] (b), Hou et al.[9] (c), our method (d) and human fixations (e), respectively.

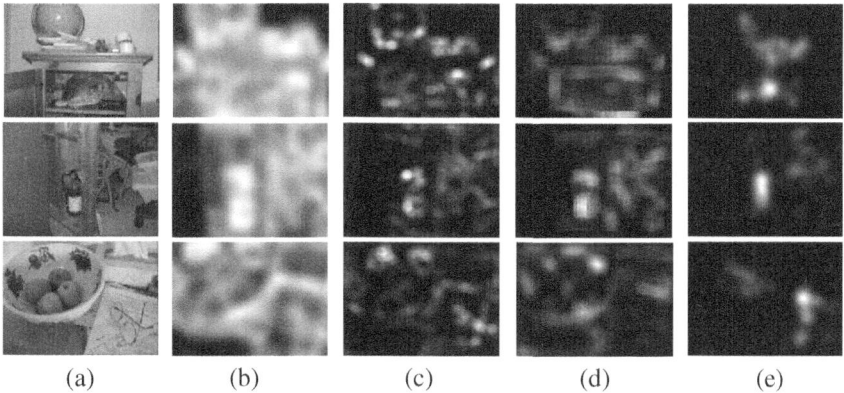

| (a) | (b) | (c) | (d) | (e) |

Fig. 7. Comparative saliency results on images of complex scenes. The first image in each line is the original image (a), the rest saliency maps from left to right are produced by Bruce et al.[5] (b), Hou et al.[9] (c), our method (d) and human fixations and (e), respectively.

The second case includes images of complex scenes. Fig.7 shows images of messy scene indoor. In this situation, the core objects in cluttered scene are expected as salient. It can be observed that our approach capture salient parts. For example both the globe and the table are detected in the first scene, and the hydrant is detected in the second scene. Taking the advantage of the property of middle spatial frequency, the proposed method achieves the best visual-evaluated performances among all comparative studies.

For quantitative evaluation, we exploit Receiver Operating Characteristic (ROC) curve. The fixation data collected by Bruce et al. [5] is compared as the benchmark. From Fig.8, we could see that our algorithm outperforms other methods. In addition, we computed the Area Under ROC Curve (AUC). The average values of ROC areas are calculated over all 120 test images. And the results are reported in Table 2, which further demonstrates the superiority of the proposed method.

Table 2. Performances on static image saliency

Method	Bruce et al.[5]	Hou et al.[9]	Our method
AUC	0.6919	0.7217	0.7265

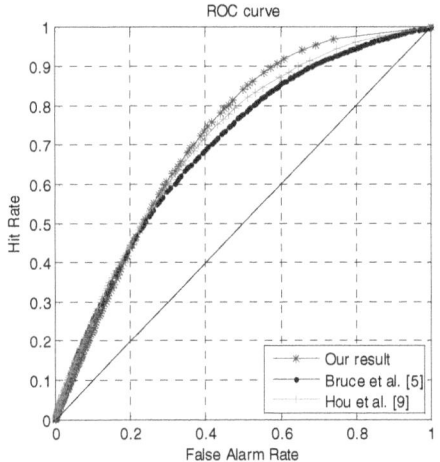

Fig. 8. ROC curves for different methods

5   Conclusion

In this paper, we propose a novel saliency detection method through scale space analysis. The saliency is based on the principle which is observed in the psychological literature: our ability to perceive a visual scene on different scales. This inspires us to detect saliency by using the low-middle frequency components and ignoring the others. Experiments on real world datasets demonstrated that our method achieves a high degree of accuracy and the computing cost is less. We would like to learn the benefits of our method in applications, such as image classification and image quality assessment in the future.

Acknowledgements. This research is partially sponsored by National Basic Research Program of China (No.2009CB320902), Beijing Natural Science Foundation (No. 4102013), Natural Science Foundation of China (Nos.60970087, 61070116 and 61070149) and President Fund of Graduate University of Chinese Academy of Sciences(No.085102HN00).

References

1. Itti, L., Koch, C., Niebur, E.: A Model of Saliency-Based Visual Attention for Rapid Scene Analysis. IEEE Trans. on Pattern Analysis and Machine Intelligence 20(11), 1254–1259 (1998)
2. Frintrop, S., Klodt, M., Rome, E.: A real-time visual attention system using integral images. In: Proceedings of the 5th International Conference on Computer Vision Systems, Biefeled (2007)
3. Gao, D.S., Mahadevan, V., Vasconcelos, N.: The Discriminant Center-Surround Hypothesis for Bottom-Up Saliency. In: Proceedings of Neural Information Processing Systems, pp. 497–504. MIT Press, Cambridge (2008)
4. Kienzle, W., Wichmann, F.A., Schölkopf, B., Franz, M.O.: A Nonparametric Approach to bottom-Up Visual Saliency. In: Proceedings of Neural Information Processing Systems, Canada, pp. 689–696 (2006)
5. Bruce, N., Tsotsos, J.: Saliency Based on Information Maximization. In: Proceedings of Neural Information Processing Systems, Vancouver, pp. 155–162 (2006)
6. Kadir, T., Brady, M.: Saliency, Scale and Image Description. Journal of Computer Vision 45(2), 83–105 (2001)
7. Itti, L., Baldi, P.F.: Bayesian Surprise Attracts Human Attention. In: Proceedings of Neural Information Processing Systems, pp. 547–554. MIT Press, Cambridge (2005)
8. Achanta, R., Estrada, F., Wils, P., Süsstrunk, S.: Salient Region Detection and Segmentation. In: Gasteratos, A., Vincze, M., Tsotsos, J.K. (eds.) ICVS 2008. LNCS, vol. 5008, pp. 66–75. Springer, Heidelberg (2008)
9. Hou, X.D., Zhang, L.Q.: Saliency detection: A Spectral Residual Approach. In: Proceedings of Computer Vision and Pattern Recognition, Minnesota, pp. 1–8 (2007)
10. Campbell, F.: The Transmission of Spatial Information Through the Visual System. In: Schmitt, F.O., Wonden, F.G. (eds.) The Neurosciences Third Study Program. M.I.T. Press, Cambridge (1974)
11. Schyns, P., Oliva, A.: From Blobs to Boundary Edges: Evidence for Time and Spatial Scale Dependent Scene Recognition. Journal of Psychological Science 5, 195–200 (1994)
12. Navon, D.: Forest before trees: The precedence of global features in visual perception. Journal of Cognitive Psychology 9, 353–383 (1977)
13. Van Nes, R.L., Bouman, M.A.: Spatial Modulation Transfer in the Human Eye. Journal of the Optical Society of America 57, 401–406 (1967)
14. Campbell, F., Robson, J.: Application of Fourier Analysis to the Visibility of Gratings. Journal of Physiology 197, 551–566 (1968)
15. Loftus, G.R., Harley, E.M.: Why Is It Easier to Identify Someone Close Than Far Away. Journal of Psychonomic 12, 43–65 (2005)
16. Judd, T., Ehinger, K., Durand, F., Torralba, A.: Learning to predict where humans look. In: Proceedings of International Conference on Computer Vision, Kyoto, pp. 1–8 (2009)
17. Bracewell, R.N.: The Fourier Transform and its Applications, 2nd edn. McGraw Hill, New York (1986)

Bio-inspired Visual Saliency Detection
and Its Application on Image Retargeting

Lijuan Duan[1], Chunpeng Wu[1], Haitao Qiao[1], Jili Gu[1], Jun Miao[2],
Laiyun Qing[3], and Zhen Yang[1]

[1] College of Computer Science and Technology, Beijing University of Technology,
Beijing 100124, China
[2] Key Laboratory of Intelligent Information Processing, Institute of Computing Technology,
Chinese Academy of Sciences, Beijing 100190, China
[3] School of Information Science and Engineering, Graudate University of the Chinese Academy
of Sciences, Beijing 100049, China
{ljduan,yangzhen}@bjut.edu.cn,
{wuchunpeng,qht,gujiligujili}@emails.bjut.edu.cn
jmiao@ict.ac.cn, lyqing@gucas.ac.cn

Abstract. In this paper, we present a saliency guided image retargeting method.
Our bio-inspired saliency measure integrates three factors: dissimilarity, spatial
distance and central bias, and these three factors are supported by research on
human vision system (HVS). To produce perceptual satisfactory retargeting
images, we use the saliency map as the importance map in the retargeting
method. We suppose that saliency maps can indicate informative regions, and
filter out background in images. Experimental results demonstrate that our
method outperforms previous retargeting method guided by the gray image on
distorting dominant objects less. And further comparison between various
saliency detection methods show that retargeting method using our saliency
measure maintains more parts of foreground.

Keywords: visual saliency, dissimilarity, spatial distance, central bias, image
retargeting.

1 Introduction

Human vision system is able to select salient information among mass visual input to
focus on. Via the ballistic saccades of the eyes, the limited resources of the visual
apparatus are directed to points of attentional awareness. Computationally modeling
such mechanism has become a popular research topic in recent years [1], [2], [3] and
the models has been applied to visual tasks such as image classification [4], image
segmentation [5] and object detection [6].

In this paper, a saliency guided image retargeting method is proposed. We use our
biologically inspired saliency measure proposed in [7], and this measure integrates
three factors: dissimilarity, spatial distance and central bias. The dissimilarity is
evaluated using a center-surround operator simulating the visual receptive field.
Visual neurons are typically more sensitive in a small region of the visual space (the

B.-L. Lu, L. Zhang, and J. Kwok (Eds.): ICONIP 2011, Part I, LNCS 7062, pp. 182–189, 2011.
© Springer-Verlag Berlin Heidelberg 2011

center) when stimuli are presented in a broader region around the center (the surround) [1], and this structure is a general computational principle in the retina and primary visual cortex [8]. The above second factor, i.e., spatial distance, is supported by the research [9] on foveation of HVS. HVS samples the visual field by using a variable resolution, and the resolution is highest at center (fovea) and drops rapidly toward the periphery [10]. Therefore with the increasing spatial distance between the current fixation and another image location, the influence of dissimilarity between them is decreased due to the decreased resolution. In addition, subjects tend to look at the center of images according to previous studies on the distribution of human fixations [11]. This fact, also known as central bias (the above third factor), has often been contributed to the experimental setup (e.g. experiments typically start in the center) but also reflects that photographer tent to center objects of interest [12].

To produce perceptual satisfactory retargeted images, we use the saliency map as the importance map in the retargeting method. Image retargeting aims at resizing an image by expanding or shrinking the non-informative regions. Therefore, an essential component of retargeting techniques is to estimate where the important regions of an image are located, and then these regions will be preserved in the final resized image [13]. The map recording the importance of all regions in an image is called an importance map. Various importance maps have been introduced, such as a Harris corners based map [14], a gradient-based map [15] or the corresponding gray image [16]. However, these importance maps are sensitive to strong edges appeared in background regions of an image due to the presence of noise. On the contrary, background can be filtered out by a saliency map, and informative regions are often salient in an image. Therefore we will use the saliency map as the importance map in our retargeting method.

The paper is organized as follows: Review of our saliency detection is in Section 2. In Section 3, we present our saliency guided retargeting method. In Section 4, we compare our retargeting method with previous method, and the performance of retargeting methods based on different saliency measures is also compared. The conclusions are given in Section 5.

2   Review of Our Saliency Measure

We use our saliency measure proposed in [7], and it is shown in Fig. 1. There are four main steps in our method: splitting image into patches, reducing dimensionality, evaluating global dissimilarity and weighting dissimilarity by distance to center. First non-overlapping patches are drawn from an image, and all of the color channels are stacked to represent each image patch as a feature vector of pixel values. All vectors are then mapped into a reduced dimensional space. The saliency of image patch p_i is calculated as

$$Saliency(p_i) = \omega_1(p_i) \cdot GD(p_i) \tag{1}$$

where $\omega_1(p_i)$ represents central bias and $GD(p_i)$ is the global dissimilarity. $\omega_1(p_i)$ and $GD(p_i)$ are computed as

$$\omega_1(p_i, p_j) = 1 - \frac{DistToCenter(p_i)}{D} \tag{2}$$

$$GD(p_i) = \sum_{j=1}^{L} \{\omega_2(p_i, p_j) \cdot Dissimilarity(p_i, p_j)\} \qquad (3)$$

In Eq. 2, $DistToCenter(p_i)$ is the spatial distance between patch p_i and center of the original image, and $D = \max_j \{DistToCenter(p_j)\}$ is a normalization factor. In Eq. 3, L is total number of patches, $\omega_2(p_i, p_j)$ is inverse of spatial distance, and $Dissimilarity(p_i, p_j)$ is dissimilarity of feature response between patches. $\omega_2(p_i, p_j)$ and $Dissimilarity(p_i, p_j)$ are computed as

$$\omega_2(p_i, p_j) = \frac{1}{1 + Dist(p_i, p_j)} \qquad (4)$$

$$Dissimilarity(p_i, p_j) = \| f_i - f_j \| \qquad (5)$$

In Eq. 4, $Dist(p_i, p_j)$ is the spatial distance between patch p_i and patch p_j in the image. In Eq. 5, feature vectors f_i and f_j correspond to patch p_i and patch p_j respectively. Finally, the saliency map is normalized and resized to the scale of the original image, and then is smoothed with a Gaussian filter ($\sigma = 3$).

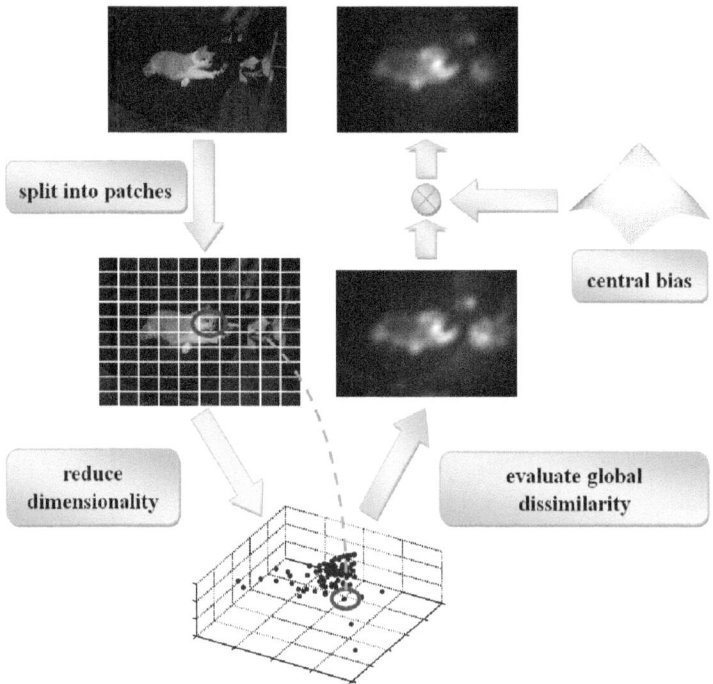

Fig. 1. Our saliency detection method [7]

3   Saliency Guided Image Retargeting

The proposed saliency guide image retargeting is demonstrated in Fig. 2. Rubinstein et al. [16]'s retargeting method using seam carving operator is exploited. However, as stated in Section 1, we will use saliency maps as importance maps to guide seams carving, instead of original gray image in Rubinstein's method. To introduce high-level semantic information, the saliency map is combined with the results of face detection [17], pedestrian detection [18] and vehicle detection [19]. In Fig. 2, the results of respective detectors are called face conspicuity map, pedestrian conspicuity map and vehicle conspicuity map. The conspicuity map is a binary map, and is the same size as original image. On the map, pixels belonging to the corresponding object (face, pedestrian or vehicle) are set to 1, and other pixels are set to 0. Then the "saliency map with semantic information" in Fig. 2 is calculated by linearly adding the saliency map, face conspicuity map, pedestrian conspicuity map and vehicle conspicuity map with respective weights ω_s, ω_f, ω_p and ω_v. Other object detectors can also be added into the above framework.

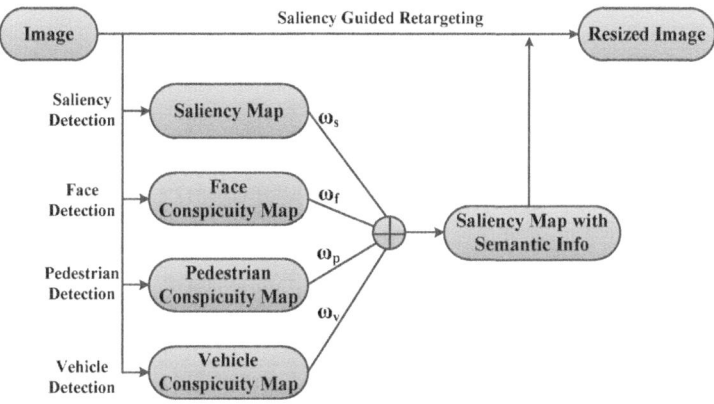

Fig. 2. Saliency guided image retargeting combined with object detectors

4   Experimental Validation

The saliency guided image retargeting method is applied on two public image datasets [20], [21] to evaluate its performance. We will compare the distortions induced by our method and Rubinstein's original method [16]. Performance of image retargeting based on different saliency detection methods including ours is also compared. The same parameters of our method will be used across all images.

4.1   Parameters Selection

According to our previous parameter settings [7], the color space is YCbCr, the size of image patch is 14x14 and the dimensions to which each feature vector reduced is

11. Our saliency detection method with above parameters outperforms some state-of-the-art [1], [20], [22], [23] on predicting human fixations, please see [7] for details. The weights ω_s, ω_f, ω_p and ω_v stated in Section 3.2 are set equally to each other , i.e., 1/4. We do not tune these four weights to specific image dataset.

4.2   Saliency Map vs. Gray Image

The performance of our saliency guided retargeting is compared with Rubinstein's original method based on gray image [16] in Fig. 3. On each input image, the foreground object is in the manually labeled regions using two red dashed lines, and these objects are supposed to be least distorted. In Fig. 3, car and people in images are less distorted by using our method. This is because these objects are dissimilar from surroundings (i.e., salient), and detectors of face, pedestrian and vehicle are used in our method. Other foreground objects are also less resized by our method such as rocks in the first input image. Another reason for our better performance is that foreground objects are often near the center of images due to photographers preferring placing objects of interest in the center [12], and this factor called central bias is integrated into our saliency detection method as stated in Section 3.

Fig. 3. Comparison between image retargeting based on saliency maps (the second row) and gray images (the third row)

4.3   Qualitative Comparison between Saliency Measures

In Fig. 4, we compare the retargeting results based on Itti et al. [1], Bruce et al. [20], Hou et al. [22], Harel et al. [23], and our saliency measure. Our method achieves better results on the first three images whose foreground objects are not pedestrians or vehicles, therefore the results mainly attribute to our original saliency maps without semantic information. As shown in Fig. 1, our saliency map is spread-out, therefore more parts of the foreground object on our map are detected as salient. According to Rubinstein's retargeting method [16], non-informative or non-salient pixels are removed first. Consequently, our method prevents more pixels belonging to the dominant object from being removed.

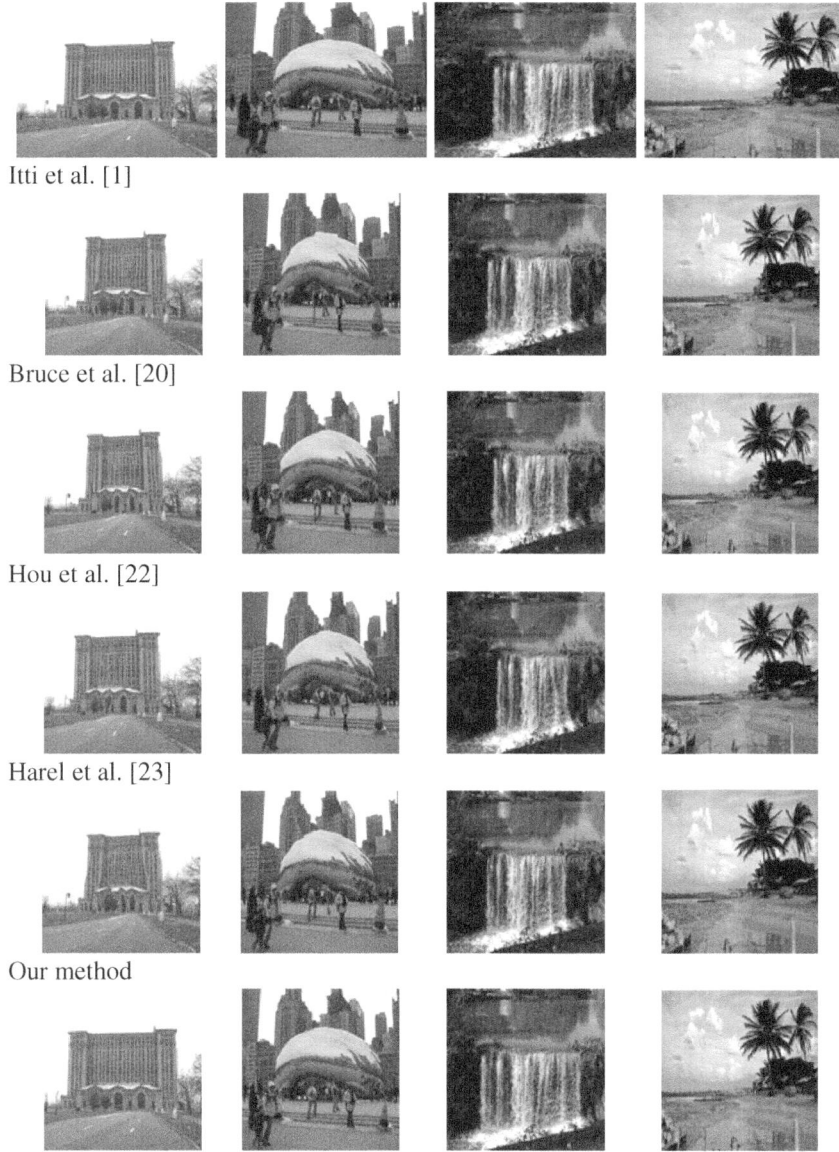

Fig. 4. Comparison of image retargeting based on Itti et al. [1], Bruce et al. [20], Hou et al. [22], Harel et al. [23] and our saliency measure

4.4   Quantitative Comparison between Saliency Measures

Ten students were asked to evaluate the retargeting results for each saliency measure according to the following criteria: almost perceptual satisfactory, moderate, and severely distorted. Therefore for each saliency measure, an image will be scored ten times. Fig. 5 shows the quantitative comparison based on Itti et al. [1], Bruce et al. [20], Hou et al. [22], Harel et al. [23] and our measure. Our method is more perceptual satisfactory and less distorted.

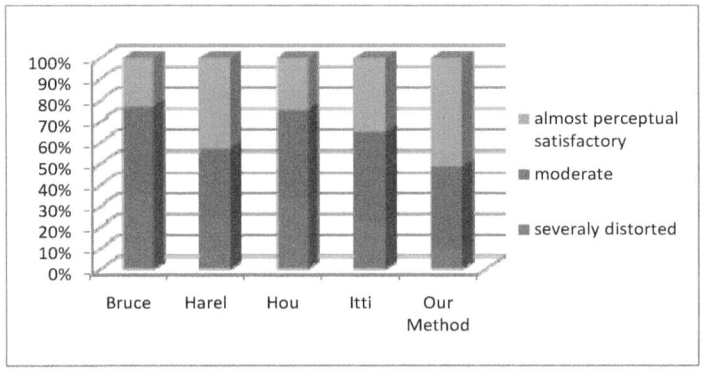

Fig. 5. Quantitative comparison of image retargeting based on Itti et al. [1], Bruce et al. [20], Hou et al. [22], Harel et al. [23] and our saliency measure

5   Conclusions

In this paper, we have presented a saliency guided image retargeting method. The saliency detection method is biologically inspired, and the three key factors of the method are supported by research on HVS. We use the saliency map as the importance map in the retargeting method. Experimental results show that dominant objects are less distorted by using our method than by previous gray image guided method. And comparing with some other saliency measures, our method prevents more pixels belonging to foreground from being removed. The performance of our method can be further improved by extracting more visual features and introducing the top-down control of visual attention.

Acknowledgments. This research is partially sponsored by National Basic Research Program of China (No.2009CB320902), Hi-Tech Research and Development Program of China (No.2006AA01Z122), Beijing Natural Science Foundation (Nos. 4102013 and 4072023), Natural Science Foundation of China (Nos.60702031, 60970087, 61070116 and 61070149) and President Fund of Graduate University of Chinese Academy of Sciences(No.085102HN00).

References

1. Itti, L., Koch, C., Niebur, E.: A Model of Saliency-Based Visual Attention for Rapid Scene Analysis. IEEE TPAMI 20, 1254–1259 (1998)
2. Gao, D., Vasconcelos, N.: Bottom-Up Saliency is a Discriminant Process. IEEE ICCV, 1–6 (2007)
3. Murray, N., Vanrell, M., Otazu, X., Parraga, C.A.: Saliency Estimation Using A Non-Parametric Low-Level Vision Model. IEEE CVPR, 433–440 (2011)
4. Kanan, C., Cottrell, G.: Robust Classification of Objects, Faces, and Flowers Using Natural Image Statistics. IEEE CVPR, 2472–2479 (2010)
5. Yu, H., Li, J., Tian, Y., Huang, H.: Automatic Interesting Object Extraction from Images Using Complementary Saliency Maps. ACM Multimedia, 891–894 (2010)
6. Navalpakkam, V., Itti, L.: An Intergrated Model of Top-Down and Bottom-Up Attention for Optimizing Detection Speed. IEEE CVPR, 2049–2056 (2006)
7. Duan, L., Wu, C., Miao, J., Qing, L., Fu, Y.: Visual Saliency Detection by Spatially Weighted Dissimilarity. IEEE CVPR, 473–480 (2011)
8. Levelthal, A.G.: The Neural Basis of Visual Function: Vision and Visual Dysfunction. CRC Press, Fla (1991)
9. Rajashekar, U., van der Linde, I., Bovik, A.C., Cormack, L.K.: Foveated Analysis of Image Features at Fixations. Vision Research 47, 3160–3172 (2007)
10. Wandell, B.A.: Foundations of vision. Sinauer Associates (1995)
11. Tatler, B.W.: The Central Fixation Bias in Scene Viewing: Selecting an Optimal Viewing Position Independently of Motor Biased and Image Feature Distributions. J. Vision 7(4), 1–17 (2007)
12. Zhao, Q., Koch, C.: Learning A Saliency Map Using Fixated Locations in Natural Scenes. J. Vision 11(9), 1–15 (2011)
13. Rubinstein, M., Gutierrez, D., Sorkine, O., Shamir, A.: A Comparative Study of Image Retargeting. ACM Trans. Graphics 29(160), 1–10 (2010)
14. Shamir, A., Sorkine, O.: Visual Media Retargeting. ACM SIGGRAPH Asia Courses (11), 1–11 (2009)
15. Wang, Y., Tai, C., Sorkine, O., Lee, T.: Optimized Scale-and-Stretch for Image Resizing. ACM Trans. Graphics 27(118), 1–8 (2008)
16. Rubinstein, M., Shamir, A., Avidan, S.: Improved Seam Carving for Video Retargeting. ACM Trans. Graphics 3(16), 1–9 (2008)
17. Viola, P., Jones, M.: Robust Real-Time Object Detection. IJCV 57, 137–154 (2001)
18. Dalal, N., Triggs, B.: Histograms of Oriented Gradients for Human Detection. IEEE CVPR, 886–893 (2005)
19. Felzenszwaklb, P., Girshick, R., McAllester, D., Ramanan, D.: Object Detection with Discriminatively Trained Part Based Models. IEEE TPAMI 32, 1627–1645 (2010)
20. Bruce, N.D.B., Tsotsos, J.K.: Saliency Based on Information Maximization. NIPS, 155–162 (2005)
21. Judd, T., Ehinger, K., Durand, F., Torralba, A.: Learning to Predict Where Humans Look. IEEE ICCV, 2106–2113 (2009)
22. Hou, X., Zhang, L.: Dynamic Visual Attention: Searching for Coding Length Increments. In: NIPS, pp. 681–688 (2008)
23. Harel, J., Koch, C., Perona, P.: Graph-Based Visual Saliency. In: NIPS, pp. 545–552 (2006)

An Approach to Distance Estimation with Stereo Vision Using Address-Event-Representation

M. Domínguez-Morales[1], A. Jimenez-Fernandez[1], R. Paz[1],
M.R. López-Torres[1], E. Cerezuela-Escudero[1], A. Linares-Barranco[1],
G. Jimenez-Moreno[1], and A. Morgado[2]

[1] Robotic and Technology of Computers Lab, University of Seville, Spain
[2] Electronic Technology Department, University of Cadiz, Spain
mdominguez@atc.us.es

Abstract. Image processing in digital computer systems usually considers the visual information as a sequence of frames. These frames are from cameras that capture reality for a short period of time. They are renewed and transmitted at a rate of 25-30 fps (typical real-time scenario). Digital video processing has to process each frame in order to obtain a result or detect a feature. In stereo vision, existing algorithms used for distance estimation use frames from two digital cameras and process them pixel by pixel to obtain similarities and differences from both frames; after that, depending on the scene and the features extracted, an estimate of the distance of the different objects of the scene is calculated. Spike-based processing is a relatively new approach that implements the processing by manipulating spikes one by one at the time they are transmitted, like a human brain. The mammal nervous system is able to solve much more complex problems, such as visual recognition by manipulating neuron spikes. The spike-based philosophy for visual information processing based on the neuro-inspired Address-Event-Representation (AER) is achieving nowadays very high performances. In this work we propose a two-DVS-retina system, composed of other elements in a chain, which allow us to obtain a distance estimation of the moving objects in a close environment. We will analyze each element of this chain and propose a Multi Hold&Fire algorithm that obtains the differences between both retinas.

Keywords: Stereo vision, distance calculation, address-event-representation, spike, retina, neuromorphic engineering, co-design, Hold&Fire, FPGA, VHDL.

1 Introduction

In recent years there have been numerous advances in the field of vision and image processing, because they can be applied for scientific and commercial purposes to numerous fields such as medicine, industry or entertainment.

As we all know, the images are two dimensional while the daily scene is three dimensional. Therefore, in the transition from the scene (reality) to the image, what we call the third dimension is lost. Nowadays, society has experienced a great advance in these aspects: 2D vision has given way to 3D viewing. Industry and

B.-L. Lu, L. Zhang, and J. Kwok (Eds.): ICONIP 2011, Part I, LNCS 7062, pp. 190–198, 2011.

research groups have started their further research in this field, obtaining some mechanisms for 3D representation using more than one camera [14]. Trying to simulate the vision of human beings, researchers have experimented with two-camera-based systems inspired in human vision ([12][13]). Following this, a new research line has been developed, focused on stereoscopic vision [1]. In this branch, researchers try to obtain three-dimensional scenes using two digital cameras. Thus, we try to get some information that could not be obtained with a single camera, i.e. the distance at which objects are.

By using digital cameras, researchers have made a breakthrough in this field, going up to create systems able to achieve the above. However, digital systems have some problems that, even today, have not been solved. A logical and important result in stereoscopic vision is the calculation of distances between the point of view and the object that we are focused on. This problem is still completely open to research and there are lots of research groups focusing on it. The problems related to this are the computational cost needed to obtain appropriate results and the errors obtained after distance calculation. There are lots of high-level algorithms used in digital stereo vision that solve the distance calculation problem, but this implies a computer intervention into the process and it is computationally expensive.

The required computational power and speed make it difficult to develop a real-time autonomous system. However, brains perform powerful and fast vision processing using millions of small and slow cells working in parallel in a totally different way. Primate brains are structured in layers of neurons, where the neurons of a layer connect to a very large number (~10^4) of neurons in the following one [2]. Most times the connectivity includes paths between non-consecutive layers, and even feedback connections are present.

Vision sensing and object recognition in brains are not processed frame by frame; they are processed in a continuous way, spike by spike, in the brain-cortex. The visual cortex is composed of a set of layers [2], starting from the retina. The processing starts when the retina captures the information. In recent years significant progress has been made in the study of the processing by the visual cortex. Many artificial systems that implement bio-inspired software models use biological-like processing that outperform more conventionally engineered machines ([3][4][5]). However, these systems generally run at extremely low speeds because the models are implemented as software programs. Direct hardware implementations of these models are required to obtain real-time solutions. A growing number of research groups around the world are implementing these computational principles onto real-time spiking hardware through the development and exploitation of the so-called AER (Address Event Representation) technology.

AER was proposed by the Mead lab in 1991 [8] for communicating between neuromorphic chips with spikes. Every time a cell on a sender device generates a spike, it transmits a digital word representing a code or address for that pixel, using an external inter-chip digital bus (the AER bus, as shown in figure 1). In the receiver the spikes are directed to the pixels whose code or address was on the bus. Thus, cells with the same address in the emitter and receiver chips are virtually connected by streams of spikes. Arbitration circuits ensure that cells do not access the bus simultaneously. Usually, AER circuits are built with self-timed asynchronous logic.

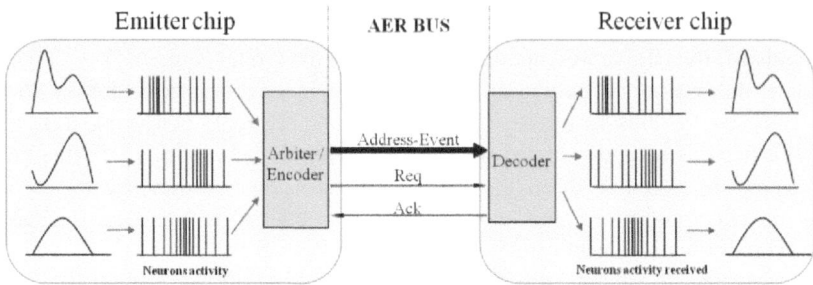

Fig. 1. Rate-coded AER inter-chip communication scheme

Several works are already present in the literature regarding spike-based visual processing filters. Serrano et al. presented a chip-processor able to implement image convolution filters based on spikes that work at very high performance parameters (~3GOPS for 32x32 kernel size) compared to traditional digital frame-based convolution processors (references [6],[7],[5]).

There is a community of AER protocol users for bio-inspired applications in vision and audition systems, as evidenced by the success in the last years of the AER group at the Neuromorphic Engineering Workshop series. One of the goals of this community is to build large multi-chip and multi-layer hierarchically structured systems capable of performing complicated array data processing in real time. The power of these systems can be used in computer based systems under co-processing.

First, we describe element by element of our processing chain until obtain the complete system. Then we propose an AER algorithm, which can be developed in AER systems using a FPGA to process the information; and that is able to obtain differences from both retinas and calculate a distance estimation of the object in movement. Finally, we present distance estimation results of the whole system, and compare it with the real distance.

2   System Description

In this section we will describe in detail the system used and each one of the components that form our system. We can see from figure 2 a block diagram of the whole system.

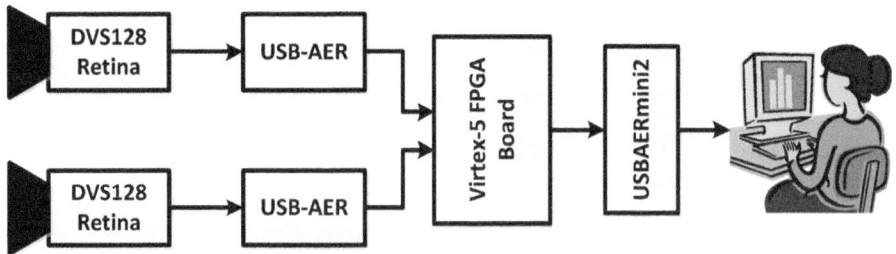

Fig. 2. Complete system with all the elements used

All the elements that compose our system are these (from left to right): two DVS128 retinas [10], two USB-AER, a Virtex-5 FPGA board, an USBAERmini2 [9] and a computer to watch the results with jAER software [11]. Next, we will talk about the USB-AER and the Virtex-5 FPGA board (Fig. 3).

USB-AER board was developed in our lab during the CAVIAR project, and it is based on a Spartan II FPGA with two megabytes of external RAM and a cygnal 8051 microcontroller.

To communicate with the external world, it has two parallel AER ports (IDE connector). One of them is used as input, and the other is the output. In our system we have used two USB-AER boards, one for each retina. In these boards we have synthetized in VHDL a filter called Background-Activity-Filter, which allows us to eliminate noise from the stream of spikes produced by each retina. This noise (or spurious) is due to the nature of analog chips and since we cannot do anything to avoid it in the retina, we are filtering it. So, at the output of the USB-AER we have the information filtered and ready to be processed.

Fig. 3. Left, USB-AER board; right, Virtex-5 FPGA board

The other board used is a Xilinx Virtex-5 board, developed by AVNET [17]. This board is based on a Virtex-5 FPGA and mainly has a big port composed of more than eighty GPIOs (General Purpose Inputs/Outputs ports). Using this port, we have connected an expansion/testing board, which has standard pins, and we have used them to connect two AER inputs and one output.

The Virtex-5 implements the whole processing program, which works with the spikes coming from each retina, processes them and obtains the differences between both retinas and the spikes rate of these differences. The whole program block diagram is shown in figure 4. The system behavior and its functionality are shown in the following sections.

3   Multi Hold and Fire Algorithm

Once we have all the elements of our chain, we can start thinking about the algorithms used. In this work we receive the traffic from both retinas and calculate the differences between them. To do that we have used the idea of the Hold&Fire building block [16] to obtain the difference of two signals. With this block working correctly, we have extrapolated this to a 128x128 signals system (one for each pixel of the retinas) and obtained a Multi Hold&Fire system that allows us to calculate the differences between both retinas' spikes streams.

The Hold&Fire subtracts two pulses, received from two different ports. When it receives an event, it waits a short fixed time for another event with the same address. If it does not receive a second event and the fixed time is over, it fires the pulse. If, otherwise, it receives another event with the same address, then, if the new event comes from the other retina, the event is cancelled and no event is transmitted, but if this second event comes from the same retina, the first event is dispatched and this second event takes the role of the first event and the system waits again the short fixed time. This Hold&Fire operation for subtracting or cancelling two streams of spikes is described in depth in the paper indicated before.

To summarize, if both events with the same address have the same polarization and come from different ports they are erased and the Hold&Fire block does not fire anything. In this case it fires only when both events have different polarizations. On the other hand, when both events come from the same port, then the opposite happens: it only fires when both events have the same polarization. There are other cases to complete the truth table.

Our algorithm is based on the Hold&Fire block, as it is said before, but it has one Hold&Fire block for each pixel of the retina. It treats each pixel separately and obtains the difference between this pixel in the left retina and the same pixel in the right retina. At the end, we have the difference of both retinas in our system output.

The complete VHDL system, which is completely implemented on the Virtex-5 FPGA board, consists of:

- One handshake block for each retina: these blocks establish the communication protocol with the retina. They are based on a state machine that waits for the retina request signal, receives the AER spike and returns the acknowledge signal.
- Two FIFOs: to storage a great amount of spikes and lose none.
- One Arbitrator: select spikes from both FIFOs depending on the occupation of them.
- Multi Hold&Fire module: applies the algorithm explained before to the stream of spikes received. To storage the information of the first pulse of each sequence, this block uses a dual-port RAM block.
- Distance estimation block: this block will be explained in the next section.
- Handshake out: this block establishes the communication between our processed information and the out world.

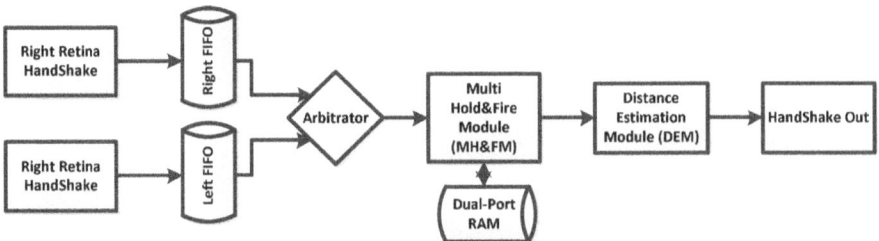

Fig. 4. VHDL block diagram inside Virtex-5 FPGA

4   Calculating Distances

In this section we will talk about the schematic block that was only named in the previously section: distance estimation module.

The algorithm explained is a first approximation to the distance calculation using spikes. Nowadays, there is no algorithm that can solve this problem using only spikes in a real-time environment. That is why we are trying to focus on this field.

Existing algorithms in digital systems extract features from both cameras, process them, and try to match objects from both cameras frame by frame [15]. This process involves high computational costs and does not work in real time.

We want to do this same thing in real time using AER. As a first step to achieve this goal we propose an algorithm based on the spikes rate of the Multi Hold&Fire output.

Theoretically we have to explain the meaning of what we are going to obtain and, after that, we will show the practical results and we will test if both match.

In our system, both retinas are calibrated with a certain angle to obtain a focus distance of 1 m. To do that, we have put our retinas in a base, separated 13'5 cm. We have obtained the system shown below (Fig. 5).

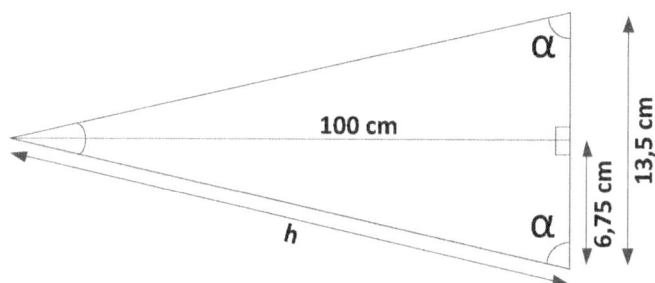

Fig. 5. Retinas situations and trigonometry

Applying Pythagoras and trigonometric rules, we can obtain:

$$h^2 = 6'75^2 + 100^2 \quad \rightarrow \quad h = 100'22755 \text{ cm}$$

$$\sin \alpha = 100 / 100'22755 = 0'99773$$

$$\arcsin 0'99773 = 86'1387°$$

So, our retinas are calibrated with an angle of 86'1387° to obtain a focal distance of one meter. After that, we have measured the spikes rates at the output of our Multi Hold&Fire algorithm using a recorded video about one object in movement. This video has been played many times at different distances from the retinas and we have annotated the number of spikes fired during this video. The resulting spike stream was recorded using jAER software [11]. After measurements, we have recorded all the results and have made a graph with all of them. This graph indicates the number of spikes versus distance.

It is logical to think that, at the central match point of the focal length of each retina, the Multi Hold&Fire acts like a perfect subtractor and do not fire any spike at all (except for the retinas' spurious), so the spike rate at this point is near zero. If we approach the retinas, the spike rate will be increased because the object becomes bigger and each retina sees it from a different point of view, so the subtractor will not act so perfectly.

Otherwise, if we put the video further from the retinas, spike rate will be slightly increased due to the subtraction result (different points of view of the retinas), but the object becomes smaller, so it compensates this failure: as the further the object is, the smaller it is and therefore, the lower the spike rate is; because less spikes are fired, but the subtraction acts worse and fires more spikes. That is why the second aspect is balanced with the first one.

So, in theory, we will see a graph where the number of spikes is greatly increased near the retinas and is slightly increased as we move away the object from the focal collision point. We can see the experimental results in the next figure. We have stimulated our system using two types of stimuli: an oscillating pendulum and a vibrating ruler. Measurements were taken from the starting point of 10 centimeters to 150 centimeters. They were taken every 10 centimeters.

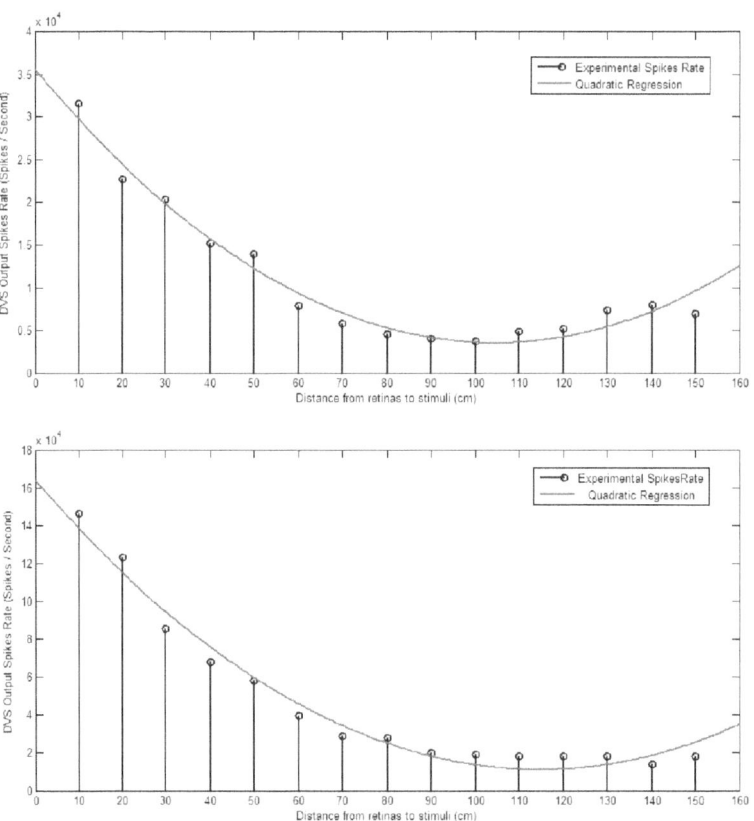

Fig. 6. Spike rate versus Distance. Up: Pendulum. Down: Ruler.

In figure 6 we can see the experimental results obtained. It is interesting to remark that, approximately, at a distance of 100 centimeters (focal collision of both retinas) we obtained the lowest spike rate. If we see measurements taken closer, it can be seen that spike rate increases, and far away from the focal collision point, the spike rate is increased a little. Finally, we obtained the results we expected. In the first case, with the pendulum, it can be seen better than the other one; but, what is true is that, in a short distance, we can estimate the distance of the object in movement with a quadratic regression (line in blue shown in figure 6).

It is very interesting to see that our system behaves similarly to human perception: a priori, without knowing the size of the object, we cannot give exact distances, only an approximation. This approximation depends on our experience, but the system proposed cannot learn. However, we can measure the distance qualitatively, and interacting with an object in a nearby environment.

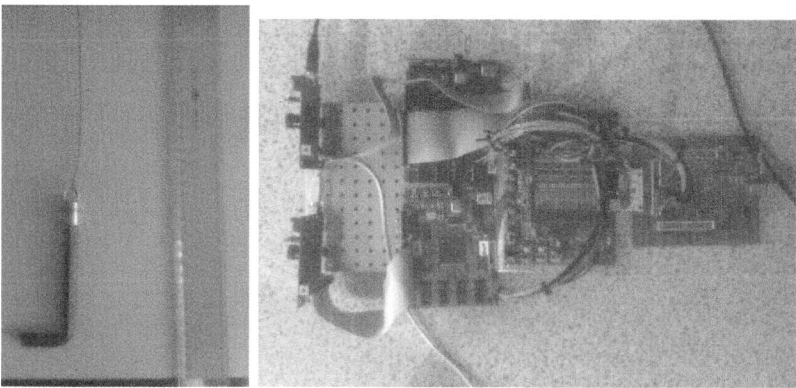

Fig. 7. Hardware used to test our system (right) and stimulus used (left)

5   Conclusions

The existing difficulties to calculate distances in digital systems have been shown. That is why a biological approach (Address-Event-Representation) to work with has been presented. We have introduced the Address-Event-Representation notation to communicate neuro-inspired chips as a new paradigm in Neuromorphic Engineering. We have evaluated the advantages of this method and explained why we work with it.

In this work we propose a first approximation to distance estimation using spikes in a close environment. To do that, a stereoscopic vision system with two DVS retinas has been used, working with VHDL over a Virtex-5 FPGA.

We have described and shown the whole system used and each one of its elements required to obtain the distance estimation. With the hardware system described, we have explained the algorithms used. The first algorithm uses a method to obtain differences between both retinas in real time and without sampling. With these differences the second algorithm was explained, which works with the spike rate obtained in our system after the differences calculation.

With the results of these two algorithms, we have been able to model the spike rate versus the distance of the object. The simulation results are very encouraging, because we can see in the graphs shown that there is a relationship between distance and the spike rate after our processing and that this system works quite similar to human perception.

Acknowledgements. First, we want to thank the contribution of Tobias Delbruck, whose retinas are used in this work; as well as Raphael Berner, whose USBAERmini2 was used too. Without their contributions this work could not have been done. At last, but not least, we want to thank the Spanish government, which has supported the project VULCANO (TEC2009-10639-C04-02).

References

1. Barnard, S.T., Fischler, M.A.: Computational Stereo. Journal ACM CSUR 14(4) (1982)
2. Shepherd, G.M.: The Synaptic Organization of the Brain, 3rd edn. Oxford University Press (1990)
3. Lee, J.: A Simple Speckle Smoothing Algorithm for Synthetic Aperture Radar Images. Man and Cybernetics SMC-13 (1981)
4. Crimmins, T.: Geometric Filter for Speckle Reduction. Applied Optics 24, 1438–1443 (1985)
5. Linares-Barranco, A., et al.: AER Convolution Processors for FPGA. In: ISCASS (2010)
6. Cope, B., et al.: Implementation of 2D Convolution on FPGA, GPU and CPU. Imperial College Report (2006)
7. Cope, B., et al.: Have GPUs made FPGAs redundant in the field of video processing? In: FPT (2005)
8. Sivilotti, M.: Wiring Considerations in analog VLSI Systems with Application to Field-Programmable Networks. Ph.D. Thesis, Caltech (1991)
9. Berner, R., Delbruck, T., Civit-Balcells, A., Linares-Barranco, A.: A 5 Meps $100 USB2.0 Address-Event Monitor-Sequencer Interface. In: ISCAS, New Orleans, pp. 2451–2454 (2007)
10. Lichtsteiner, P., Posh, C., Delbruck, T.: A 128×128 120dB 15 us Asynchronous Temporal Contrast Vision Sensor. IEEE Journal on Solid-State Circuits 43(2), 566–576 (2008)
11. jAER software: http://sourceforge.net/apps/trac/jaer/wiki
12. Benosman, R., Devars, J.: Panoramic stereo vision sensor. In: International Conference on Pattern Recognition, ICPR (1998)
13. Benosman, R., et al.: Real time omni-directional stereovision and planes detection. In: Mediterranean Electrotechnical Conference, MELECON (1996)
14. Douret, J., Benosman, R.: A multi-cameras 3D volumetric method for outdoor scenes: a road traffic monitoring application. In: International Conference on Pattern Recognition, ICPR (2004)
15. Dominguez-Morales, M., et al.: Image Matching Algorithms using Address-Event-Representation. In: SIGMAP (2011)
16. Jimenez-Fernandez, A., et al.: Building Blocks for Spike-based Signal Processing. In: IEEE International Joint Conference on Neural Networks, IJCNN (2010)
17. AVNET Virtex-5 FPGA board: http://www.em.avnet.com/drc

AER Spiking Neuron Computation on GPUs: The Frame-to-AER Generation

M.R. López-Torres, F. Diaz-del-Rio, M. Domínguez-Morales,
G. Jimenez-Moreno, and A. Linares-Barranco

Department of Architecture and Technology of Computers,
Av. Reina Mercedes s/n, 41012, University of Seville, Spain
rlopez@atc.us.es

Abstract. Neuro-inspired processing tries to imitate the nervous system and may resolve complex problems, such as visual recognition. The spike-based philosophy based on the Address-Event-Representation (AER) is a neuromorphic interchip communication protocol that allows for massive connectivity between neurons. Some of the AER-based systems can achieve very high performances in real-time applications. This philosophy is very different from standard image processing, which considers the visual information as a succession of frames. These frames need to be processed in order to extract a result. This usually requires very expensive operations and high computing resource consumption. Due to its relative youth, nowadays AER systems are short of cost-effective tools like emulators, simulators, testers, debuggers, etc. In this paper the first results of a CUDA-based tool focused on the functional processing of AER spikes is presented, with the aim of helping in the design and testing of filters and buses management of these systems.

Keywords: AER, neuromorphic, CUDA, GPUs, real-time vision, spiking systems.

1   Introduction

Standard digital vision systems process sequences of frames from video sources, like CCD cameras. For performing complex object recognition, sequences of computational operations must be performed for each frame. The computational power and speed required make it difficult to develop a real-time autonomous system. However, brains perform powerful and fast vision processing using millions of small and slow cells working in parallel in a totally different way. Vision sensing and object recognition in brains are not processed frame by frame; they are processed in a continuous way, spike by spike, in the brain-cortex. The visual cortex is composed of a set of layers [1], starting from the retina, which captures the information. In recent years, significant progress has been made in the study of the processing by the visual cortex. Many artificial systems that implement bio-inspired software models use biological-like processing that outperforms more conventionally engineered machines [2][3][4]. However, these systems generally run at extremely low speeds because the models are implemented as software programs in classical CPUs. Nowadays, a

B.-L. Lu, L. Zhang, and J. Kwok (Eds.): ICONIP 2011, Part I, LNCS 7062, pp. 199–208, 2011.

growing number of research groups are implementing some of these computational principles onto real-time spiking hardware through the so-called AER (Address Event Representation) technology, in order to achieve real-time processing.

AER was proposed by the Mead lab in 1991 [5][7] for communicating between neuromorphic chips with spikes. In AER, a sender device generates spikes. Each spike transmits a word representing a code or address for that pixel. In the receiver, spikes are processed and finally directed to the pixels whose code or address was on the bus. In this way, cells with the same address in the emitter and receiver chips are virtually connected by streams of spikes. Usually, these AER circuits are built using self-timed asynchronous logic [6]. Several works have implemented spike-based visual processing filters. Serrano et al. [10] presented a chip-processor that is able to implement image convolution filters based on spikes, which works with a very high performance (~3 GOPS for 32x32 kernel size) compared to traditional digital frame-based convolution processors [11]. Another approach for solving frame-based convolutions with very high performances are the ConvNets [12][13], based on cellular neural networks, that are able to achieve a theoretical sustained 4 GOPS for 7x7 kernel sizes.

One of the goals of the AER processing is a large multi-chip and multi-layer hierarchically structured system capable of performing complicated array data processing in real time. But this purpose strongly depends on the availability of robust and efficient AER interfaces and evaluation tools [9]. One such tool is a PCI-AER interface that allows not only reading an AER stream into a computer memory and displaying it on screen in real-time, but also the opposite: from images available in the computer's memory, generate a synthetic AER stream in a similar manner a dedicated VLSI AER emitter chip [8][9][4] would do. This PCI-AER interface is able to reach up to 10Mevents/sec bandwidth, which allows a frame-rate of 2.5 frames/s with an AER traffic load of 100% for 128x128 frames, and 25 frames/s with a typical 10% AER traffic load.

Nowadays computer-based systems are increasing their performances exploiting architectural concepts like multi-core and many-core. Multi-core is referred to those processors that can execute in parallel as many threads as cores are available on hardware. On the other hand, many-core computers consist of processors (that could be multi-core) plus a co-processing hardware composed of several processors, like the Graphical Processing Units (GPU). By joining a many-core system with the mentioned PCI-AER interface, a spike-based processor could be implemented.

In this work we focus on the evaluation of GPU to run in parallel an AER system, starting with the frame-to-AER software conversion and continuing with several usual AER streaming operations, and a first emulation of a silicon retina using CCD cameras. For this task monitoring internal GPU performance through the Compute Visual Profiler [17] for NVIDIA® technology was used. Compute Visual Profiler is a graphical user interface based profiling tool that can be used to measure performance. Through the use of this tool, we have found several key points to achieve maximum performance of AER processing emulation.

Next section briefly explains the pros and cons of Simulation and emulation of Spiking systems, when compared to real implementation. In section 3, a basic description of Nvidia CUDA (Compute Unified Device Architecture) architecture is shown, focusing on the relevant points that affect the performance of our tool. In section 4 the main parts of the proposed emulation/simulation tool are discussed. In section 5 the performance of the operations for different software organization is evaluated, and in section 6 the conclusions are presented.

2   Simulation and Emulation of Spiking Systems

Nowadays, there is a relatively high number of spiking processing systems that use FPGA or another dedicated integrated circuits. Needless to say that, while these implementations are time effective, they present many design difficulties: elevated time and cost of digital synthesis, tricky and cumbersome testing, possibility of wiring bugs (which often are difficult to detect), etc. Support tools are then very helpful and demanded by circuit designers. Usually there are several levels in the field of emulation and simulation tools: from the lower (electrical) level to the most functional one. There is no doubt that none of these tools can cover the whole "simulation spectrum". In this paper we present a tool focused on the functional processing of AER spikes, in order to help in the design and testing of filters and buses management of these systems.

CPU is the most simple and straightforward simulation, and there are already some of these simulators [25][26]. But simulation times are currently far from what one can wish for: a stream processing that lasts microseconds in a real system can extend for minutes on these simulators even when running in high performance clusters.

Another hurdle that is common in CPU emulators is their dependence on input (or traffic) load. In a previous work [14], the performance of several frame-to-AER software conversion methods for real-time video applications was evaluated, by measuring execution times in several processors. That work demonstrated that for low AER traffic loads any method in any multicore single CPU achieved real-time, but for high bandwidth AER traffics, it depends on which method and CPU are selected in order to obtain real-time. This problem can be mitigated when using GPGPUs (General Purpose GPUs).

A few years ago GPU software development was difficult and close to bizarre [23]. During the last years, with the expansion of GPGPUs, tools, function libraries, and hardware abstraction mechanisms that hide the GPU hardware from developers, have appeared. Nowadays there are reliable debugging and profiling tools like those from CUDA [18]. Two additional ones complement these advantages. First, the very low cost per processing unit, which is supposed to persist, since the PC graphics market subsidizes GPUs (only Nvidia has already sold 50 million CUDA-capable GPUs). Secondly, the annual growth performance ratio is predicted to stay very high: by 70% per year due to the continuing miniaturization. The same happens for Memory Bandwidth. For example, extant GPUs with 240 floating point arithmetic cores, realizing a performance of 1 TFLOPS with one chip. One of the consequences of this advantage to its competitors, is that the top leading supercomputer (November 2010) is a GPU based machine (www.top500.org).

Previous reasons have pushed the scientific community to incorporate GPUs in several disciplines. In video processing systems, GPUs application is obvious, therefore some very interesting AER emulation systems have begun to appear [15][16].

The last fact regards the need for spiking output cameras. These cameras (usually called silicon retinas) are currently expensive, rare and inaccurate (due to electrical mismatching between their cells [22]). Moreover, their actual resolution is very low, making it impossible to work with real elevated address streams at present. This hurdle is expected to be overcome in next years. However, current researchers cannot evaluate their spike processing systems for high resolutions. Therefore, a cheap CCD camera where a preprocessing module generates a spiking streaming is a very attractive alternative [20], even if the real silicon retina speed cannot be reached. And we believe GPUs may be the most well-placed platform to do it. Emulating a retina by a GPU has additional benefits: on the one hand, different address coding can be easily implemented [24]. On the other hand, different types of silicon retinas could be emulated over the same GPU platform, simply by changing the preprocessing filter executed before generating the spiking stream. Nowadays implemented retinas are mainly of three types [22]: gray level, spatial contrast (gradient) and dynamic vision sensor (diferential) retinas. The first one can be emulated by generating spikes for each pixel [8], the second option through a gradient and the last one, with a discrete time derivative.

In this work the balance between generality and efficiency of emulation is considered. Tuning simulator routines with a GPU to reach a good performance involves a loss of generality [21].

3   CUDA Architecture

A CUDA GPU [18] includes an array of streaming multiprocessors (SMs). Each SM consists of 8 floating-point Scalar Processors (SPs), a Special Function Unit, a multi-threaded instruction unit, and several pieces of memory. Each memory type is intended for a different use; in most cases they have to be managed by the programmer. This makes the programmer to be aware of their resources, achieving maximum performance. Each SM has a "warp scheduler" that selects at any cycle a group of threads for execution, in a round-robin fashion. A warp is simply a group of (currently) 32 hardware-managed threads. As the number and types of threads may be enormous, a five dimension organization is supported by CUDA, with two important levels: a grid contains blocks that must not be very coupled, while every block contains a relatively short number of threads, which can cooperate deeply.

If a thread in a warp issues a costly operation (like an external memory access), then the warp scheduler switches to a new warp, in order to hide the latency of the other thread. In order to use the GPU resources efficiently, each thread should operate on different scalar data, with a certain pattern. Due to these special CUDA features, some key points must be kept in mind to achieve a good performance of an AER system emulator. These are basically: It is a must to launch thousands or millions of very light threads; the pattern memory access is of vital importance (the CUDA manual [18] provides detailed algorithms to identify types of coalesced/uncoalesced memory accesses); there is a considerable amount of spatial locality in the image access pattern required to perform a convolution (GPU texture memory is used in this case). Any type of bifurcation (branch, loops, etc.) in the thread code should be avoided. The same for any type of thread synchronization, critical sections, barriers, atomic accesses, etc. (this means that each thread must be almost independent from the others); Nevertheless, because GPU do not usually have hardware-managed caches there will be no problem with false dependencies (as usual in multicore systems when every thread has to write in the same vector as the others).

4   Main Modules of the Emulation/Simulation Tool

An AER system emulation/simulation tool must contain at most the following parts:

- Images Input module. It can include a preprocessing filter to emulate different retinas.
- Synthetic AER spikes generation. It is an important part since the distribution of spikes throughout time must be similar to that produced by real retinas.
- Filters. Convolution kernels are the basic operation, but others like low pass filters, integrators, winner takes all, etc. may be necessary.
- Buses management. It includes buses splitters, merges, and so on.
- Result output module. It must collect the results in time order to send them to the CPU.
- AER Bus Probes. This module appears necessarily as discussed below.

The tool presented here is intended to emulate a spiking processing hardware system. As a result, the algorithms to be carried out in the tool are generally simple. In GPGPU terminology, this means that the "arithmetic intensity" [18] (which is defined as the number of operations performed per word of memory transferred) is going to be very low. Therefore, optimisation must focus on memory accesses and types. As a first consequence some restrictions on the number and size of the data objects were done. Besides, a second conclusion is presented: instead of simulating several AER filters in cascade (as usual in FPGA processing), it will be usually better to execute only a combined filter that fuses them, in order to save GPU DDRAM accesses or CPU-GPU transactions. This is to be discussed in next sections, according to the results. Finally the concept of AER Bus Probe is introduced here in order to only generate the intermediate AER values that are strictly necessary. Only when a probe is demanded by the user, a GPU to CPU transaction is inserted to collect AER spikes in the temporal order. Besides, some code adjustments have been introduced to avoid new data structures despite of adding more computation.

Synthetic AER generation is one of the key pieces of an AER tool. This is because it usually lasts a considerable time and because the spike distribution should have a considerable time uniformity to ensure that neuron information is correctly sent [8]. Besides, in this work two additional reasons have to be considered. Firstly, it has to be demonstrated that a high degree of parallelism can be obtained using CUDA, so that the more cores the GPU has, the less time the frame takes to be generated. And secondly, we have performed a comparison of execution times with those obtained for the multicore platforms used previously in [14].

In [14] these AER software methods were evaluated in several CPUs regarding the execution time. In all AER generation methods, results are saved in a shared AER spike vector. Actually, spike representation can be done in several forms (in [24] a wide codification spectrum is discussed). Taking into account the consideration of previous section, we have concluded that the AER spike vector (the one used in [14]) is very convenient when using GPUs.

One can think of many software algorithms to transform a bitmap image (stored in a computer's memory) into an AER stream of pixel addresses [8]. In all of them the frequency of appearance of the address of a given pixel must be proportional to the intensity of that pixel. Note that the precise location of the address pulses is not critical. The pulses can be slightly shifted from their nominal positions; the AER receivers will integrate them to recover the original pixel waveform.

Whatever algorithm is used, it will generate a vector of addresses that will be sent to an AER receiver chip via an AER bus. Let us call this vector the *"frame vector"*. The *frame vector* has a fixed number of time slots to be filled with event addresses. The number of time slots depends on the time assigned to a frame (for example *Tframe*=40 ms) and the time required to transmit a single event (for example *Tpulse*=10 ns). If we have an image of *N*×*M* pixels and each pixel can have a grey level value from *0* to *K*, one possibility is to place each pixel address in the *frame vector* as many times as the value of its intensity, and distribute it with equidistant positions. In the worst case (all pixels with maximum value *K*), the *frame vector* would be filled with *N*×*M*×*K* addresses. Note that this number should be less than the total number of time slots in the *frame vector*. Depending on the total intensity of the image there will be more or less empty slots in the *frame vector Tframe/Tpulse*.

Each algorithm would implement a particular way of distributing these address events, and will require a certain time. In [8] and [14] we discussed several algorithms that were convenient when using classical CPUs. But if GPUs are to be used, we have to discard those where the generation of each element of *frame vector* cannot be independent from the others. This clearly happens in those methods based on Linear Feedback Shift Registers (LFSR): as the method requires calling a random function that always depends on itself, the method cannot be divided in threads.

To sum up, the best-suited method for GPU processing is the so-called Exhaustive method. This algorithm divides the address event sequence into *K* slices of *N*×*M* positions for a frame of *N*×*M* pixels with a maximum gray level of *K*. For each slice (*k*), an event of pixel (*i,j*) is sent on time *t* if the following condition is asserted:

$$(k \cdot P_{i,j}) \bmod K + P_{i,j} \geq K \quad \text{and} \quad N \cdot M \cdot (k-1) + (i-1) \cdot M + j = t$$

where $P_{i,j}$ is the intensity value of the pixel (*i,j*).

The Exhaustive method tries distributing the events of each pixel in equidistant slices. In this method, there is a very important advantage when using CUDA: elements of *frame vector* can be sequentially processed, because the second condition above can be implemented using *t* as the counter of the frame vector (that is, the thread index in CUDA terminology). This means that several accesses (performed by different threads) can be coalesced to save DDRAM access time. The results section is based on this algorithm.

Nevertheless, other algorithms could be slightly transformed to adapt them to CUDA. The most favorable case is that of the Random-Square method. While this method requires the generation of pseudorandom numbers (which are generated by two LFSR of 8 and 14 bits), LFSR functions can be avoided if all the lists of numbers are stored in tables. Although this is possible, it will involve two additional accesses per element to long tables, which probably will reside in DDRAM. This will add a supplementary delay that is avoided with the exhaustive method.

The rest of methods are more difficult to fine-tune to CUDA (namely the Uniform, Random and Random-Hardware methods) because of the aforementioned reasons.

There are another group of software methods dedicated to manage the AER buses. Fortunately, these operations are intrinsically parallel, since in our case they basically consist of a processing of each of the frame vector elements (which plays the role of a complete AER stream).

The other operations involved in AER processing are those that play the role of image filtering and convolution kernels. Nowadays, the algorithms that execute them on AER based systems are intrinsically sequential: commonly for every spike that appears in the bus, the values of some counters related to this spike address are changed [15][10]. Therefore, these counters must be seen as critical sections when emulating this process through software. This makes impractical to emulate this operation in a GPU. On the contrary, the standard frame convolution operation can be easily parallelized [19] if the image output is placed in a memory zone different from that of image input. Execution of a standard convolution gives an enormous speedup when comparing to a CPU. Due to this and considering that the other group of operations does present a high degree of parallelism, in this paper convolutions are processed in a classical fashion. Nonetheless, this combination of AER-based and classical operations results in a good enough performance as seen in the following section.

5   Performance Study

In order to analyze the performance and scalability of CUDA simulation and emulation of AER systems, a series of operations have been coded and analyzed in two Nvidia GPUs. **Table 1** summarizes the main characteristics of the platforms tested. The second GPU have an important feature: it can concurrently copy and execute programs (while the first one cannot).

Table 1. Tested Nvidia GPUs

Characteristics	GeForce 9300 ION	GTX 285
Global memory	266010624 bytes	1073414144 bytes
Maximum number of threads per block	512 threads	512 threads
Multiprocessors x Cores/MP	2 x 8 = 16 Cores	30 x 8 = 240 Cores
Clock rate	1.10 GHz	1.48 GHz

Fig. 1 depicted a typical AER processing scenario. The first module 'ImageToAER' transforms a frame into an AER stream using the Exhaustive method. The 'Splitter' divides spikes into two buses according to an address mask that represents the little clear square in the almost black figure. The upper bus is then rotated 90 degrees simply by going through the 'Mapper' module, which changes each spike's address into another. Finally a merge between the original image and the upper bus gives a new emulated AER bus, which can be observed with a convenient AERToImage module (which, in a few words, makes a temporal integration). A consequence of the use of a rigid size frame vector composed of time slots is that a merge operation between two full buses cannot fuse perfectly the two images represented in the initial buses. In our implementation, if both buses have a valid spike in a certain time slot, the corresponding output bus slot is going to be filled only by one of the input buses (which is decided with a simple circuit). This aspect appears also in hardware AER implementations, where an arbiter must decide which input bus "looses", and then its spike does not appear in the merged bus [8][10].

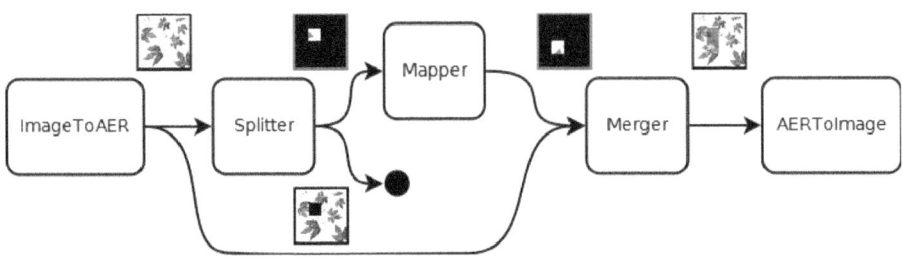

Fig. 1. Cascade AER operations benchmark

According to section 4, three kinds of benchmarks have been carried out. In a first group, an AER cascade operations were simulated: in the middle of two operations, (that is, in an AER bus) an intermediate result is collected by the CPU to check and verify the bus values. In the second group, operations are executed in cascade, but transitional values are not "downloaded" to the CPU, thus preserving GPU to CPU transactions. This implies that no AER Bus Probe modules are present, thus inhibiting inspection opportunities. Finally a third collection is not coded in previous modular fashion since the first four operations are grouped together in one CUDA kernel (the same thread executes all of them sequentially). This avoids several GPU DDRAM accesses, saving an enormous execution time in the end. The fifth operation that transforms an AER stream into an image frame cannot be easily parallelized for the same reasons described in previous section for convolution kernels. Timing comparison for these three groups is summarized in Table 2.

It is important to remark that the execution time for the third group almost coincides with the maximum execution time of all the operations in the first group. Another obvious but interesting fact is that transactional times are proportionally reduced when CPU-GPU transactions are eliminated. And speed-up between GTX285 and ION 9300 is also near to the ideal. One can conclude that scalability for our tool is good, which means for faster upcoming GPU a shorter execution time is expected.

Finally, contrast retina emulation has been carried out: for a frame, first a gradient convolution is done in order to extract image edges, and secondly, the AER frame vector is generated (in the same CUDA thread). A hopeful result is obtained: the mean execution time to process one frame is 313.3 μs, that is, almost 3200 frames per second. As the execution times are small we can suppose that these times could be overlapped with the transaction ones. The resulting fps ratio is very much higher (around 50x) than those obtained using multicore CPUs in previous studies of AER spikes generation methods [14].

Table 2. Benchmarking times for GPU GTX285 and for 9300 ION. Times in microseconds.

Measured Time			First Group GTX285	Second Group GTX285	Third Group GTX285	Third Group 9300
CPU	to	GPU	8649.2	360.3	125.5	5815.5
transactions						
CUDA		kernel	2068.6	1982.7	781.6	15336.8
execution						
GPU	to	CPU	11762.8	2508.9	2239.2	12740.8
transactions						
Total Time			22480.5	4851.9	3146.3	33893.0

Through these experiments we have demonstrated that major time is spent in these types of AER tools in external DDRAM GPU accesses, since data sizes are necessarily big while algorithms can be implemented with a few operations per CUDA thread. This conclusion gives us an opportunity to develop a completely functional AER simulator. A second consequence derived from this is that the size of the image can introduce a considerable increment of time emulation. In this work, the chosen size (128×128 pixels) results in a frame vector of 8 MB (4 Mevents $\times$ 2 bytes/event). However, a 512×512 pixel image will have 4x the size image, plus twice the bytes per address (if no compression is implemented). This means an 8x total size, and then an 8x access time. To sum up, eliminating the restrictions on the number and size of data objects can have an important impact on the tool performance.

6 Conclusions and Future Work

A CUDA-based tool focused on the functional processing of AER spikes and its first timing results are presented. It intends to emulate a spiking processing hardware system, using simple algorithms with a high level of parallelism. Through experiments, we demonstrated that major time is spent in DDRAM GPU accesses, so some restrictions on the number and size of the data objects have been done. A second result is presented: instead of simulating several AER filters in cascade (as usual in FPGA processing), it is better to execute only a combined filter that fuses them, in order to save GPU DDRAM accesses and CPU-GPU transactions. Due to the promising timing results, the immediate future work comprises a fully emulation of an AER retina using a classical video camera. Running our experiments on a multiGPU platform is another demanding extension because of the scalability of our tool.

Acknowledgments. This work was supported by the Spanish Science and Education Ministry Research Projects TEC2009-10639-C04-02 (VULCANO).

References

1. Drubach, D.: The Brain Explained. Prentice-Hall, New Jersey (2000)
2. Lee, J.: A Simple Speckle Smoothing Algorithm for Synthetic Aperture Radar Images. IEEE Trans. Systems, Man and Cybernetics SMC-13, 85–89 (1983)
3. Crimmins, T.: Geometric Filter for Speckle Reduction. Applied Optics 24 (1985)
4. Linares-Barranco, A., et al.: On the AER Convolution Processors for FPGA. In: ISCAS 2010, Paris, France (2010)

5. Sivilotti, M.: Wiring Considerations in analog VLSI Systems with Application to Field-Programmable Networks, Ph.D. Thesis, California Institute of Technology (1991)
6. Boahen, K.A.: Communicating Neuronal Ensembles between Neuromorphic Chips. In: Neuromorphic Systems. Kluwer Academic Publishers, Boston (1998)
7. Mahowald, M.: VLSI Analogs of Neuronal Visual Processing: A Synthesis of Form and Function. Ph.D. Thesis. California Institute of Technology Pasadena, California (1992)
8. Linares-Barranco, A., Jimenez-Moreno, G., Civit-Ballcels, A., Linares-Barranco, B.: On Algorithmic Rate-Coded AER Generation. IEEE Transaction on Neural Networks (2006)
9. Paz, R., Gomez-Rodriguez, F., Rodríguez, M.A., Linares-Barranco, A., Jimenez, G., Civit, A.: Test Infrastructure for Address-Event-Representation Communications. In: Cabestany, J., Prieto, A.G., Sandoval, F. (eds.) IWANN 2005. LNCS, vol. 3512, pp. 518–526. Springer, Heidelberg (2005)
10. Serrano, et al.: A Neuromorphic Cortical-Layer Microchip for Spike-Based Event Processing Vission Systems. IEEE Trans. on Circuits and Systems Part 1 53(12), 2548–2566 (2006)
11. Cope, B., Cheung, P.Y.K., Luk, W., Witt, S.: Have GPUs made FPGAs redundant in the field of video processing? In: IEEE International Conference on Field-Programmable Technology, pp. 111–118 (2005)
12. Farabet, C., et al.: CNP: An FPGA-based Processor for Convolutional Networks. In: International Conference on Field Programmable Logic and Applications (2009)
13. Farriga, N., et al.: Design of a Real-Time Face Detection Parallel Architecture Using High-Level Synthesis. EURASIP Journal on Embedded Systems (2008)
14. Domínguez-Morales, M., et al.: Performance study of synthetic AER generation on CPUs for Real-Time Video based on Spikes. In: SPECTS 2009, Istambul, Turkey (2009)
15. Nageswaran, J.M., Dutt, N., Wang, Y., Delbrueck, T.: Computing spike-based convolutions on GPUs. In: IEEE International Symposium on Circuits and Systems (ISCAS 2009), Taipei, Taiwan, pp. 1917–1920 (2009)
16. Goodman, D.: Code Generation: A Strategy for Neural Network Simulators. Neuroinformatics 8.3, 183–196 (2010), Issn: 1539-2791
17. Compute Visual Profiler User Guide, http://developer.nvidia.com/
18. NVIDIA CUDA Programming Guide, Version 2.1,
    http://developer.nvidia.com/
19. NVIDIA Corporation. CUDA SDK Software development kit,
    http://developer.nvidia.com/
20. Paz-Vicente, R., et al.: Synthetic retina for AER systems development. In: IEEE/ACS International Conference on Computer Systems and Applications, AICCSA, pp. 907–912 (2009)
21. Owens, J.D., Houston, M., Luebke, D., Green, S., Stone, J.E., Phillips, J.C.: GPU Computing. Proceedings of the IEEE 96(5) (May 2008)
22. Indiveri, G., et al.: Neuromorphic Silicon Neurons. Frontiers in Neuromorphic Engineering 5, 7 (2011)
23. Halfhill, T.R.: Parallel Processing With CUDA. Microprocessor The Insider's Guide To Microprocessor Hardware (2008)
24. Thorpe, S., et al.: Spike-based strategies for rapid processing. Neural Networks 14(6-7), 715–725 (2001)
25. Pérez-Carrasco, J.-A., Serrano-Gotarredona, C., Acha-Piñero, B., Serrano-Gotarredona, T., Linares-Barranco, B.: Advanced Vision Processing Systems: Spike-Based Simulation and Processing. In: Blanc-Talon, J., Philips, W., Popescu, D., Scheunders, P. (eds.) ACIVS 2009. LNCS, vol. 5807, pp. 640–651. Springer, Heidelberg (2009)
26. Montero-Gonzalez, R.J., Morgado-Estevez, A., Linares-Barranco, A., Linares-Barranco, B., Perez-Peña, F., Perez-Carrasco, J.A., Jimenez-Fernandez, A.: Performance Study of Software AER-Based Convolutions on a Parallel Supercomputer. In: Cabestany, J., Rojas, I., Joya, G. (eds.) IWANN 2011, Part I. LNCS, vol. 6691, pp. 141–148. Springer, Heidelberg (2011)

Skull-Closed Autonomous Development

Yuekai Wang[1,2], Xiaofeng Wu[1,2], and Juyang Weng[3,4,*]

[1] State Key Lab. of ASIC & System, Fudan University, Shanghai, 200433, China
[2] Department of Electronic Engineering, Fudan University, Shanghai, 200433, China
[3] School of Computer Science, Fudan University, Shanghai, 200433, China
[4] Dept. of Computer Sci. and Eng., Neuroscience Prog., Cognitive Science Prog.
Michigan State University, East Lansing, MI 48824, USA

Abstract. It seems how the brain develops its representations inside its closed skull throughout the lifetime, while the child incrementally learns one new task after another. By closed skull, we mean that the brain (or the Central Nervous System) inside the skull is off limit to the teachers in the external environment, except its sensory ends and the motor ends. We present Where-What Network (WWN) 6, which has realized our goal of a fully developmental network with closed skull, which means that the human programmer is not allowed to handcraft the internal representation for any concepts about extra-body concepts. We present how the developmental program (DP) of WWN-6 enables the network to learn and perform for attending and recognizing objects in complex backgrounds while the skull is closed.

Keywords: Development, neural nets, attention, object recognition.

1 Introduction

Symbolic methods [1,2] for object detection and recognition are "skull-open". By "skull-open", we mean that it is the human programmer who understands each given task (e.g., object recognition) that the machine is supposed to perform. The human programmer handpicks a set of symbols (e.g., "car", Gabor features, parts, whole) along with their meanings for the task. He then handcrafts a task specific representation (e.g., Hidden Markov Model (HMM) using a symbolic graph) plus a way to use the representation (e.g., search across the input image for a good fit to the HMM model). Effectively, he opens the "skull" and handcrafts the "open brain". Two problems exist with this paradigm of manual development. (1) Brittle systems: Many unpredictable events take place in the external environments. It is impractical for a human programmer to guarantee that all the conditions required by the handcrafted model (e.g., eyes are open) are met. This situation results in brittle systems. (2) No new tasks: The system is not able to go beyond the handcrafted representation so as to deal with

---

* This work was supported by the Fund of State Key Lab. of ASIC & System (11MS008) and the Fundamental Research Funds for the Central Universities to XW, and a Changjiang Visiting Scholar Fund of the Ministry of Education to JW.

B.-L. Lu, L. Zhang, and J. Kwok (Eds.): ICONIP 2011, Part I, LNCS 7062, pp. 209–216, 2011.

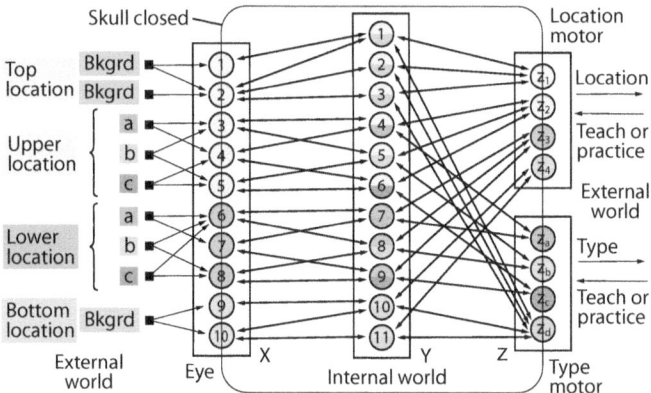

Fig. 1. The skull is closed, with X and Z exposed to the external environment. Operation, learning or performance, is fully autonomous inside the skull. The color of a Y neuron matches the sources in X and Z.

new tasks. In contrast, a child's brain seems to develop with the skull closed throughout the life (e.g., he can learn English even if all his ancestors did not).

Neural networks, whose internal representations autonomously emerge through interactions during training, have a potential to address the above two fundamental problems. However, the traditional neural networks face several intertwined problems. The most prominent one is a lack of emergent goals and emergent-goal directed search. The series of Where-What Networks (WWNs), from WWN-1 [3] to the latest WWN-5 [4] by Juyang Weng and his co-workers, was a series of brain-inspired networks that was meant to address these and other related problems. In a WWN, learned goals emerge as actions in motor areas, which in turn direct next internal search through top-down connections.

Although the prior WWNs are consistent with the idea of skull closure, they require, during learning only, a functioning "pulvinar". Pulvinar is an area inside the thalamus believed to play a role in attention [5]. The pulvinar module in WWN suppresses the neurons whose receptive fields are outside the current object being learned. The pulvinar module in a WWN is pre-programmed, assuming that it could be developed earlier. Without the pulvinar module, neurons responding to background patches can also fire, causing two major challenges. (a) Neuronal resource could be wasted for learning useless backgrounds. (b) With the skull closed, how can the network attribute the action currently supervised at the motor end to the correct object patch in the sensory image, instead of any other possible patches in the large background?

The major novelty of the work reported here is a biology-inspired computational mechanism for attention-based resource competition during autonomous development. Frequently attended objects recruit more neuronal resources than less frequently attended sensory components. Rarely firing neurons die to release resource. The brain has three areas, sensory X, internal Y (brain), and motor Z, with only X and Z open to the external environment, as illustrated in Fig. 1.

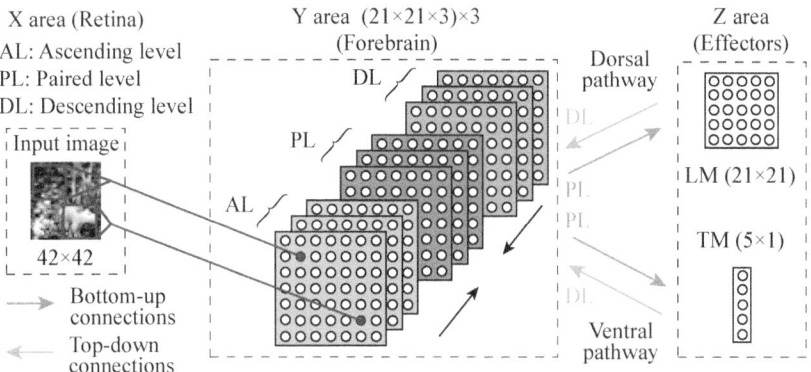

Fig. 2. The architecture and resource parameters of the new WWN-6. AL corresponds to layer L4 & L6, DL corresponds to layer L2/3 & L5 in the cerebral cortex.

Each Y neuron has a synaptic vector $(\mathbf{v}_x, \mathbf{v}_z)$ to match a local patch $\mathbf{x} \in S(X)$ and a pattern in motor Z, where $S(X)$ is a subspace in X by keeping only a subset (e.g., foreground pixels) of the components in X. Since Z is supervised with the action pattern $\mathbf{z}$ that corresponds to the learned concepts (e.g., location and type) of a particular object patch in $S(X)$, a neuron that happens to match an object patch $\mathbf{x}$ with its corresponding $\mathbf{z}$ concepts should win more often than other neurons, assuming that the actions of the network are related to some patches in X while the agent gets more and more mature. In this view, the brain gradually devotes its limited resources to the manifolds that have a high hit rate in the sensorimotor space $X \times Z$. This perspective enables us to realize the first, as far as we know, fully emergent "closed-skull" network that is able to detect and recognize general objects from complex backgrounds.

2   Overview of the Network Architecture

It has been shown [6], adding an IT (inferior temporal) and PP (posterior parietal) degrades the performance of WWNs. This seems to suggest that at least for single-frame recognition and single-frame actions, shallow structure is better. Therefore, the structure of WWN-6 is shown in Fig. 2.

The Y area uses a prescreening area for each source, before integration, resulting in three laminar levels: the ascending level that prescreenings the bottom-up input, the descending level that prescreenings the top-down input and paired level that takes pre-screened inputs. The neurons in each level are arranged in a 3-D plane whose scale is shown in the Fig. 2 corresponding to row × column × depth. The Z area has the type motor (TM) area and the location motor (LM) area. The number of neurons in each area in Fig. 2 can be considered as the limited resource of a "species", as an experimental example. The number is typically much larger in an biological animal. The number of firing neurons in LM and TM is not limited to one either. Depending on the complexity of the "manipulatory" language in LM and the "verbal" language in TM, each LM and TM can represent an astronomical number of concepts. Consider how the 26 English letters give rise to rich meanings in English.

In this part, some major characteristics of the WWNs are reviewed, which are consistent with finds about human visual cortex system in neuroscience, and also compared to the mechanisms of the human brain.

1. The WWN is inspired by the dorsal ("where" corresponding to LM) and ventral ("what" corresponding to TM) pathways in cortical visual processing and integrates both bottom-up (feed-forward) and top-down (feedback) connections.
2. The in-place learning algorithm in WWN is used to develop the internal representations, such that each neuron is responsible for the learning of its own signal processing characteristics within its connected network environment, through interactions with other neurons in the same level. This indicates that the signal processing of each neuron takes place independently which is consistent with the mechanism of the brain.
3. There is no symbolic representations in Y area (i.e., closed skull), which are human handcrafted concepts. The internal representations of each neuron in Y can be only learned from connective sensory end and motor end which are allowed to be interacted with the external environment during development.
4. Similar to the brain, WWN is not task-specific. Therefore, there is nothing hard-coded to bias the system to do a certain task (e.g., object recognition).

However, in the previous versions of WWNs, there still exist some problems.

1. The initialization of the bottom-up weights of the neurons in Y area uses the information of the foreground objects to be learned, which means that the foreground objects to be learned should be "told" to the agent (i.e., network) by the external environments (e.g., teacher) before learning. Apparently, this is not the way working in our brain.
2. There is a mechanism called "pulvinar" in the previous WWNs which allows the external environments to draw the agent's attention to the foreground object by forcing the corresponding neurons fired.

Here, we try to deal with the problems by introducing the new mechanisms called "firing threshold" and "cell regenesis" to realize a skull-closed WWN.

3   New Mechanisms for Skull Closure

3.1   Concepts of the New Mechanisms

Firing Threshold: Before describing this mechanism, two states of the neurons in Y area are defined: one is "memory state" (i.e., learning state), the other is "free state" (i.e., initial state). After the initialization of the network, all the neurons are in initial state. During the learning stage, once the stimulus to one neuron is strong enough, the state of this neuron will be changed into learning state. In our work, the intensity of the stimulus is measured by the pre-response of the neuron, and the firing threshold decides whether the neuron is

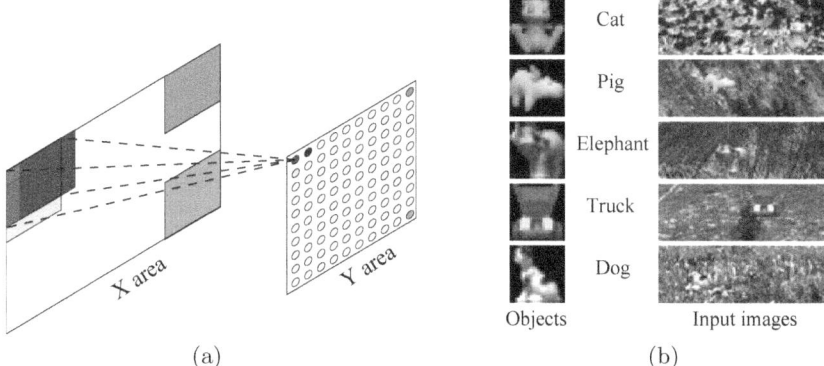

Fig. 3. (a) The illustration of the receptive fields of the neurons. (b) The training samples and input images.

fired (i.e., in learning state) with such stimulus. That is to say, if the pre-response of the neuron is over the threshold, it will be in learning state. Otherwise, it will remain the previous state. All the neurons in initial state and the top-k pre-responding neurons in learning state will do the Hebbian learning. After the Hebbian learning, the age of the neuron in initial state does not increase while the age of the neuron in learning state increases one for each time. When the age is smaller, it is easier for the neuron to learn something just like the child. Therefore, the neurons in initial state remain the high learning rate (high plasticity).

Cell regenesis: Neurons that do not fire sufficiently often are not attended enough to justify the dedication of the resource. We hypothesize that the internal cellular mechanisms turn them back to the initial state so that they are available for more appropriate roles.

3.2   Algorithm

In order to clearly illustrate all the algorithms to implement a skull-closed developmental WWN, all the following descriptions are from the viewpoint of a single neuron (e.g., neuron (i, j)) in the network.

Initialization of the network: WWN-6 initializes the bottom-up synaptic vectors of neurons randomly before the learning. Therefore, the foreground objects to be learned are not necessary to be known in advance any more and this makes the network more biologically plausible (internal neurons can not be initialized by using task information when the skull is closed).

Perception of the input image: Each neuron in Y area has a local receptive field from X area (i.e., sensory end), which perceives $a \times a$ region of the input image. The distance between two adjacent receptive field centers in horizontal and vertical directions is 1 pixel shown as Fig. 3 (a).

Computation of the pre-response: The pre-response of the neuron (i, j) is calculated as the following formula:

$$z_{i,j}^{p}(t) = \alpha z_{i,j}^{a}(t) + (1 - \alpha) z_{i,j}^{d}(t)$$

where $z_{i,j}^{p}(t)$ represents the pre-response of the neuron in the paired level called paired response, $z_{i,j}^{a}(t)$ represents the pre-response of the neuron in the ascending level called bottom-up response, and $z_{i,j}^{d}(t)$ represents the pre-response of the neuron in the descending level called top-down response. Here, $\alpha = 0.5$ which indicates that the bottom-up input accounts for 50% of "energy". It is important to note that, as indicated below, the dimension and contrast of the ascending and descending input have already normalized. The "genome" parameter α is assumed to be selected by evolution through many generations.

The bottom-up response is calculated as

$$z_{i,j}^{a}(t) = \frac{\mathbf{w}_{i,j}^{a}(t) \cdot \mathbf{x}_{i,j}^{a}(t)}{\|\mathbf{w}_{i,j}^{a}(t)\| \|\mathbf{x}_{i,j}^{a}(t)\|}$$

where $\mathbf{w}_{i,j}^{a}(t)$ and $\mathbf{x}_{i,j}^{a}(t)$ represent the ascending weight vector of and the input vector of the neuron; The top-down connections from the neurons in Z area consist of two parts, TM and LM. Accordingly

$$z_{i,j}^{d}(t) = \beta \frac{\mathbf{w}_{i,j}^{\mathrm{TM}}(t) \cdot \mathbf{x}_{i,j}^{\mathrm{TM}}(t)}{\|\mathbf{w}_{i,j}^{\mathrm{TM}}(t)\| \|\mathbf{x}_{i,j}^{\mathrm{TM}}(t)\|} + (1 - \beta) \frac{\mathbf{w}_{i,j}^{\mathrm{LM}}(t) \cdot \mathbf{x}_{i,j}^{\mathrm{LM}}(t)}{\|\mathbf{w}_{i,j}^{\mathrm{LM}}(t)\| \|\mathbf{x}_{i,j}^{\mathrm{LM}}(t)\|}$$

where $\mathbf{w}_{i,j}^{\mathrm{TM}}(t)$ and $\mathbf{x}_{i,j}^{\mathrm{TM}}(t)$ represent the weight vector and input vector corresponding to the connection with TM; $\mathbf{w}_{i,j}^{\mathrm{LM}}(t)$ and $\mathbf{x}_{i,j}^{\mathrm{LM}}(t)$ represent the weight vector and input vector corresponding to the connection with LM. Like α, $\beta = 0.5$ as a hypothesized "genome" parameter.

Top-k competition: Lateral inhibition among the neurons in the same level is used to obtain the best features of the training object. For the paired level in Y area, top-k competition is applied to imitate the lateral inhibition which effectively suppresses the weakly matched neurons (measured by the pre-responses). The response $z_{i,j}^{p}(t)$ after top-k competition is

$$z_{i,j}(t) = \begin{cases} z_{i,j}^{p}(t)(z_q - z_{k+1})/(z_1 - z_{k+1}) & \text{if } 1 \leq q \leq k \\ 0 & \text{otherwise} \end{cases}$$

where z_1, z_q and z_{k+1} denote the first, qth and $(k+1)$th neuron's paired-response respectively after being sorted in descending order. This means that only the top-k responding neurons can fire while all the other neurons are set to zero.

Firing threshold: If the paired pre-response (before top-k competition) of one neuron exceeds the firing threshold $(1 - \epsilon)$, its state will be changed into the learning stage. In the followings, the neurons in learning state are named as LSN, and the neurons in initial state are named as ISN.

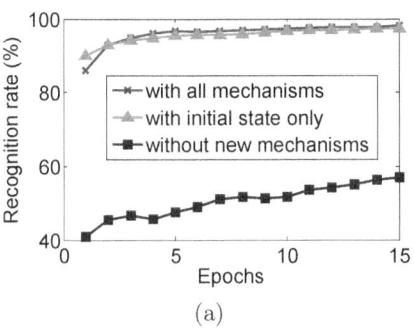
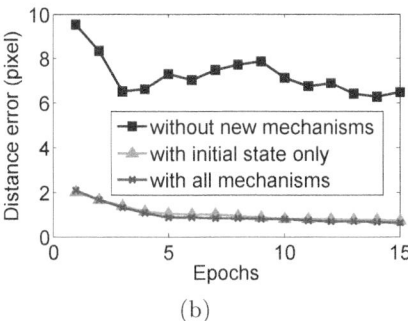

(a)   (b)

Fig. 4. Comparisons of the network performances with/without new mechanisms (i.e., initial state and cell regenesis) in 15 epoches. The data marked "with initial state only" is without cell regenesis. (a) Recognition rate. (b) Distance error.

Hebbian-like learning: Two types of neurons need to be updated by Hebbian-like learning: ISN and the firing LSN (i.e., the top-k responding LSN) whose age and weight vector will be updated. The age is updated as

$$n_{i,j}(t+1) = \begin{cases} n_{i,j}(t) & \text{if neuron } (i,j) \text{ is ISN} \\ n_{i,j}(t) + 1 & \text{if neuron } (i,j) \text{ is top-k LSN.} \end{cases}$$

The weight vector of the neuron is updated as

$$\mathbf{w}_{i,j}(t+1) = w_1(t)\mathbf{w}_{i,j}(t) + w_2(t)z_{i,j}(t)\mathbf{x}_{i,j}(t)$$

where $w_1(t)$ and $w_2(t)$ are determined by the following equations.

$$w_1(t) = 1 - w_2(t), \quad w_2(t) = \frac{1 + u(n_{i,j}(t))}{n_{i,j}(t)} \tag{1}$$

where $u(n_{i,j})$, the firing age of the neuron, is defined as

$$u(n_{i,j}(t)) = \begin{cases} 0 & \text{if } n_{i,j}(t) \leq t_1 \\ c(n_{i,j}(t) - t_1)/(t_2 - t_1) & \text{if } t_1 < n_{i,j}(t) \leq t_2 \\ c + (n_{i,j}(t) - t_2)/r & \text{if } t_2 < n_{i,j}(t) \end{cases} \tag{2}$$

where $t_1 = 20$, $t_2 = 200$, $c = 2$, $r = 10000$ in our experiment.

Cell regenesis: For the firing neuron in the learning, the age of its 6 neighbors (from 3 directions in 3-D space) are checked. The average age is defined:

$$\bar{n}(t+1) = w_1(t)\bar{n}(t) + w_2(t)n(t)$$

where $w_1(t)$ and $w_2(t)$ are described in equation (1) and (2), $\bar{n}(t)$ represents the average age and $n(t)$ represents the age of the neuron. Suppose that the neuron A fires, and one of its neighbors, neuron B, is checked. If

$$\bar{n}_A > t_1 \quad \text{and} \quad \bar{n}_A > 4\bar{n}_B$$

then neuron B needs to be turned back to initial state.

With above modifications, the new algorithm only allows external environment to interact with the sensory end (i.e., X area) and motor end (i.e., Z area), not the internal representation (i.e., Y area), which realizes the skull closure.

4   Experimental Results

In the experiment, background images were extracted from 13 natural images[1] and cropped into images of 42×42 pixels randomly. The objects of interest are selected from the MSU-25 image dataset [7]. In the training and testing sessions, different background patches were used for simulating a learned object present in a new unknown background. Foreground objects and some examples of the images that WWN-6 learns and recognizes are displayed in Fig. 3(b).

Shown as Fig. 4, it can be found that the network hardly works without the new mechanisms if the skull is closed. Without "pulvinar", the network cannot attend the foreground object and may learn the background. Moreover, without "initial state", the neurons may commit too soon. "Cell regenesis" can turn the neurons firing insufficiently back to initial state for more appropriate roles improving the network performance a little.

5   Conclusion

This is the first work, as far as we are aware, to establish a skull closed network which can be fully emergent (as opposed to programmed pulvinar) for general-purpose object detection and recognition from complex scenes. This framework needs to be tested on larger object sets in the future.

References

1. Tenenbaum, J.B., Kemp, C., Griffiths, T.L., Goodman, N.D.: How to grow a mind: Statistics, structure, and abstraction. Science 331, 1279–1285 (2011)
2. Tu, Z., Chen, X., Yuille, A.L., Zhu, S.C.: Image parsing: Unifying segmentation, detection, and recognition. Int'l J. of Computer Vision 63(2), 113–140 (2005)
3. Ji, Z., Weng, J., Prokhorov, D.: Where-what network 1: "Where" and "What" assist each other through top-down connections. In: Proc. IEEE Int'l Conference on Development and Learning, Monterey, CA, August 9-12, pp. 61–66 (2008)
4. Song, X., Zhang, W., Weng, J.: Where-what network 5: Dealing with scales for objects in complex backgrounds. In: Proc. Int'l Joint Conference on Neural Networks, San Jose, CA, July 31-August 5, pp. 1–8 (2011)
5. Olshausen, B.A., Anderson, C.H., Van Essen, D.C.: A neurobiological model of visual attention and invariant pattern recognition based on dynamic routing of information. Journal of Neuroscience 13(11), 4700–4719 (1993)
6. Wang, Y., Wu, X., Weng, J.: Synapse maintenance in the where-what network. In: Proc. Int'l Joint Conference on Neural Networks, San Jose, CA, July 31-August 5, pp. 1–8 (2011)
7. Luciw, M., Weng, J.: Top-down connections in self-organizing Hebbian networks: Topographic class grouping. IEEE Trans. Autonomous Mental Development 2(3), 248–261 (2010)

---

[1] Available from http://www.cis.hut.fi/projects/ica/imageica/

Enhanced Discrimination of Face Orientation Based on Gabor Filters

Hyun Ah Song[1], Sung-Do Choi[2], and Soo-Young Lee[1]

[1] Department of Electrical Engineering,
KAIST, Daejeon 305-701, Republic of Korea
{hyunahsong,sy-lee}@kaist.ac.kr
[2] Department of Bio and Brain Engineering,
KAIST, Daejeon 305-701, Republic of Korea
umumbe@gmail.com

Abstract. In general, a face analysis relies on the face orientation; therefore, face orientation discrimination is very important for interpreting the situation of people in an image. In this paper, we propose an enhanced approach that is robust to the unwanted variation of the image such as illumination, size of faces, and conditions of picture taken. In addition to the conventional algorithm (Principal Component Analysis and Independent Component Analysis), we imposed the Gabor kernels and Fourier Transform to improve the robustness of the proposed approach. The experimental results validate the effectiveness of the proposed algorithm for five kinds of face orientation (front, quarter left, perfect left, quarter right, and perfect right side of faces). In real application, the proposed algorithm will enable a Human-Computer Interface (HCI) system to understand the image better by extracting reliable information of face orientation.

Keywords: Face orientation discrimination, Gabor filter, Fourier transform, Principal independent analysis, Independent component analysis.

1 Introduction

The discrimination of face orientation plays an important role in applications such as car driver attention monitoring and Human-Computer Interface (HCI). Specifically, most face recognition and tracking systems are based on frontal face views [1]. To make it operate based on non-frontal face views, a large number of training samples are collected in different face orientation angles [2]. Another solution is to apply smart camera networks to the tasks [3]; however, its time complexity is too high.

  Alternatively, the orientation angle of the face is estimated from the captured 2D images. Methods for face orientation discrimination can be categorized into two main categories: local-feature based and global-feature based. Local-feature based approaches are based on geometrical relation among local facial feature (eyes, nostrils, and mouth) [4], [5]. Global-feature based approaches suppose the relationship between face orientation and certain properties of the facial image is unique [6]. The local-feature based approaches are accurate, but more complex and

B.-L. Lu, L. Zhang, and J. Kwok (Eds.): ICONIP 2011, Part I, LNCS 7062, pp. 217–224, 2011.

sensitive to noises (illumination, skin color, and glasses). On the contrary, the global-feature based approaches are simple and robust to the noises but are less accurate.

We propose a novel approach for face orientation discrimination based on Gabor kernels. Basically, the proposed approach is based on Principal Component Analysis (PCA) and Independent Component Analysis (ICA) to linearly separate datasets for feature discrimination. In addition, we applied Fast Fourier Transform (FFT) to remove shift variance of Gabor edge features. Experimental results show excellent performance rate for randomly selected dataset images. The results demonstrate the robustness of the proposed approach to any kind of variations such as illumination, skin color and occlusion (glasses) for five classes of face orientation (front, quarter left, perfect left, quarter right, and perfect right side of faces).

This paper consists of following. The proposed algorithm is described in Section 2. The experimental results are presented in Section 3. The paper concludes with Section 4.

2   Method

Our proposed algorithm for face orientation discrimination consists of Gabor filter, FFT, PCA, and ICA. Fig. 1 shows the flow chart of the proposed algorithm.

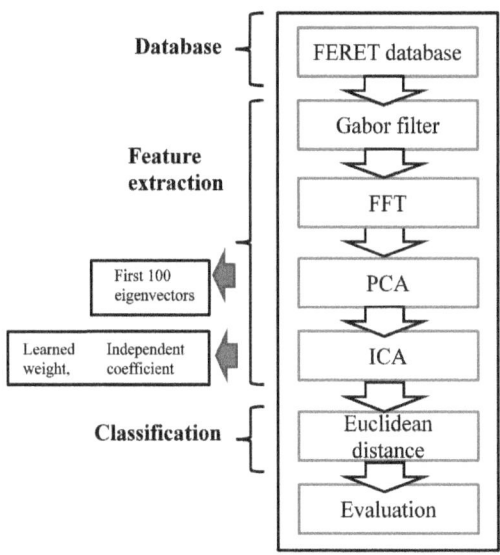

Fig. 1. Flow chart of proposed algorithm

2.1   Gabor Filters

Gabor filters are generally used for detecting edges over images. Gabor filters mimic receptive field of primary visual cortex, which shows various direction-orientation of edges with spatial locality [7]. Because of this biological relevance to visual processing, Gabor filtered face images "show strong characteristics of spatial locality, scale and orientation selectivity, which gives local features that are most suitable for face recognition" [7]. For its favorable characteristics, many studies have been conducted using Gabor filters for low level feature extraction and also as a pre-feature extraction method prior to ICA for general face recognition system [7].

Gabor filters are product of Gaussian envelope and complex sinusoidal waves. By combining two equations, we can adjust the width of kernels. By changing parameters in (1) described in Chengjun Liu's paper [7], Gabor filters are rotated to get desired edge directions:

$$B_1(x,y) = \exp\left[-\left(\frac{x^2}{\sigma_x^2}+\frac{y^2}{\sigma_y^2}\right)\right]\exp\left[-j(\omega_x x + \omega_y y)\right]. \tag{1}$$

For general face recognition, usually 40 different Gabor kernels are designed to detect 40 different edge features from face images [7]; kernels with five different scales and eight different rotational angles. To change the parameters efficiently, equation of the Gabor filters are modified to (2), which is adopted from Chengjun Liu's method [7]. In (2), μ is parameter for adjusting orientation and υ is for scale of Gabor filters. The wave vector $k_{\mu,\upsilon}$ is defined as $k_{\mu,\upsilon} = k_\upsilon e^{i\emptyset\mu}$, where $k_\mu = k_{max}/f^\upsilon$.

$$\varphi_{\mu,\upsilon}(z) = \left\|k_{\mu,\upsilon}\right\|^2/\sigma^2 e^{-(\|k_{\mu,\upsilon}\|^2\|z\|^2/2\sigma^2}\left[e^{ik_{\mu,\upsilon}z} - e^{-(\sigma^2/2)}\right]. \tag{2}$$

Each Gabor kernel is applied over an image to detect specific edge orientations corresponding to 40 different kernels using (3):

$$O_{\mu,\upsilon}(z) = I(z) * \varphi_{\mu,\upsilon}(z). \tag{3}$$

When utilizing Gabor edge features, only magnitude of convolution output, not the phase is used for further process [7].

2.2   FFT

In image processing, FFT is used to convert information from spatial domain to frequency domain. FFT represents information by sum of sinusoidal waves in frequency domain. Below, (4) shows basic FFT:

$$F(\mu,\upsilon) = \text{sum}\left\{f(x,y) * e^{(-\frac{2j\pi(\mu x + \upsilon y)}{N})}\right\}. \tag{4}$$

FFT result comprises of two parts: magnitude and phase. Magnitude tells us the amount of specific frequency component and phase tells us the location of the specific frequency. By extracting magnitude part only, one can remove the shift variance of pixel information of an image. Since the shift invariance is one of the biggest issues for image processing, FFT method is widely used for this purpose.

2.3   PCA

PCA is one of frequently used method in face recognition, along with ICA [8], [9]. Although PCA alone can be used as feature extraction method for face recognition, it is also used prior to ICA for dimension reduction [7], [8], [9].

PCA projects data linearly to two dimensional subspaces [9]. By linear projection onto second order dimensional subspaces, it reduces the dimension of data and decorrelates lower dimension information of data [7].

The basic equation for PCA is shown in (5), where $\mathbf{P}$ is eigen vector matrix, $\mathbf{X}$ is covariance matrix of face image data, and $\mathbf{Y}$ is the output of PCA, which is dimension-reduced face image data:

$$\mathbf{Y} = \mathbf{PX}. \tag{5}$$

By selecting a few eigen vectors with largest eigen values, information of weak independent coefficients are lost. This step reduces dimension of information while maintaining information of strong independent coefficients still [9].

2.4   ICA

The goal of ICA is to seek "nonaccidental and sparse feature codes and to form a representation in which these redundancies are reduced and the independent features and objects are represented explicitly" [7]. Learned filters of ICA show edge-oriented features, which mimics the receptive field properties of mammal visual cortex.

While Gabor filters extract local features, ICA reduces redundancy and represents independent features explicitly [7]. While PCA considers second order only, ICA considers higher order statistics, reduces statistical dependencies, and produces a sparse and independent code that is useful for subsequent pattern discrimination. Therefore, it enhances overall classification performance when utilized in combination of Gabor filters and PCA.

The basic equation of ICA is as follows in (6) and (7), where $\mathbf{X}$ is observed signals, $\mathbf{S}$ is unknown source signals, $\mathbf{A}$ is mixing matrix, $\mathbf{W}$ is separating matrix, and $\mathbf{U}$ is independent coefficients:

$$\mathbf{X} = \mathbf{AS}, \tag{6}$$

$$\mathbf{U} = \mathbf{WX}. \tag{7}$$

ICA assumes that all images are composed of independent components, $\mathbf{U}$ or $\mathbf{S}$, on mixing matrix of basis images, or axes, $\mathbf{A}$. ICA finds $\mathbf{W}$ matrix that best separates the image points. ICA works as a method for feature extraction as well as classification by finding $\mathbf{U}$ matrix that contains independent coefficient information, which is separated by $\mathbf{W}$ matrix.

3   Experimental Results

The database used for the experiment is FERET (FacE REcognition Technology) database, provided by NIST. We trained and tested our proposed method for discrimination of five-class problem for front, quarter left, perfect left, quarter right, and perfect right oriented faces. For each of five classes, 100 images were randomly selected and they were divided into 80 and 20 images each for training and testing dataset. Five cross validation tests were conducted for more reliable results. Initial images used are grayscale images with 384 x 256 resolutions. To make the problem global, we did not limit the conditions of the images; we did not discriminate images with different gender, races, clothing, date of image taken, facial sizes, facial expressions, or illuminations. Since dataset was collected randomly, each class may or may not include image of the same person.

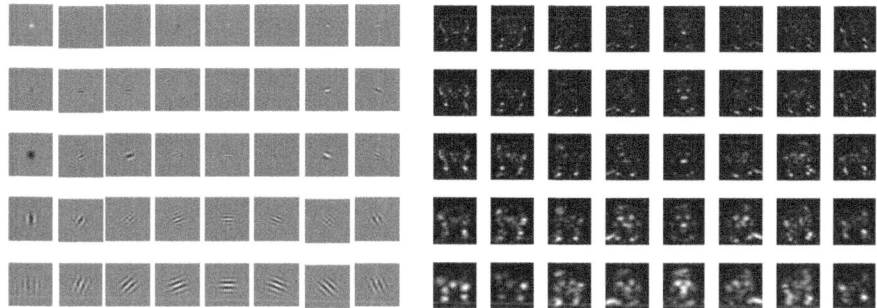

Fig. 2. Gabor kernels generated *(left)*. Five different scales and eight different rotational angles produce forty different Gabor kernels. To generate Gabor filters with five scales, υ is varied as $\upsilon=\{0,1,2,3,4\}$, and to make eight different angles, μ is varied as $\mu=\{0,1,2,3,4,5,6,7\}$. The parameters used are $\sigma=2\pi$, k_max=$\pi/2$, and f=$\sqrt{2}$. (Adopted from Chengjun Liu [7].) An image filtered with 40 different Gabor kernels *(right)*.

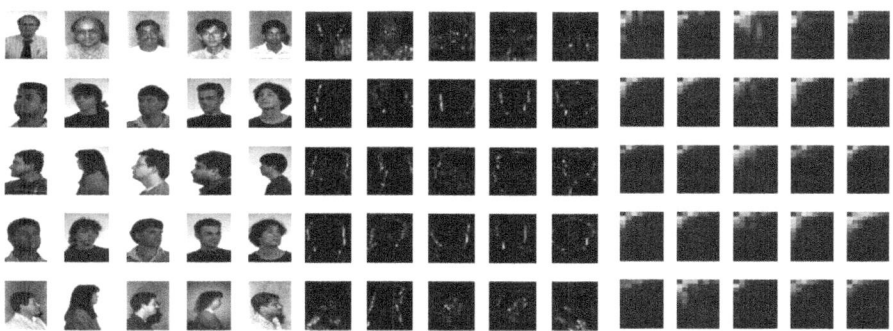

Fig. 3. Example of raw dataset images *(left)*, magnitude part of convolutional outputs of Gabor filtered images *(middle)*, and FFT2 result images *(right)*. The first Gabor kernel, which is located in the top-left of Fig. 2, is applied to each corresponding image shown on the left.

For Gabor kernels, we referred to the algorithm explained in Chengjun Liu's paper [7]. As shown and described in left image of Fig. 2, we generated 40 Gabor kernels in total for 5 different scales and 8 different rotational angles to extract 40 different kinds of edge features from images [7]. Right image of Fig. 2 shows convolutional output of 40 Gabor kernels applied to one face image. Fig. 3 shows the convolutional output of images in each of five classes with one Gabor kernel. Fig. 3 shows clear and distinct edge features for each class.

Gabor filtered images were down-sized by scale of 4x4 as known method [7], which resulted in 96x64 resolution for each Gabor images. Gabor filtered images were concatenated to form a column vector of 6144 rows for each Gabor image. Then, FFT was applied to Gabor images. We used 'fft2' function provided in Matlab. FFT result was then linearly separated by PCA and ICA. For PCA, we used first 100 eigenvectors as axes to be projected. For ICA, 'FastICA' algorithm was used for our experiment. For classification, Euclidean distance measure was used. Here, two kinds of measures were tested for better classification. In first measuring method, we treated each of Gabor image as one independent data source. We averaged column vectors of independent coefficient that correspond to the same Gabor kernel in each class. As a result, 40 averaged-independent-coefficient-columns for each Gabor images are obtained. In second measuring method, we averaged all independent coefficients of Gabor images in each class where one column vector of independent coefficient is generated as each class mean. The result showed that first measuring method provides better performance rate and we used the first measure for the result.

Table 1. Face orientation discrimination performance rate

Experiment	Front	Quarter left	Perfect left	Quarter right	Perfect right	Total
Proposed (Gabor + FFT + PCA +ICA)	96 %	97 %	88 %	95 %	90 %	93.2 %
FFT + PCA + ICA	65 %	75 %	80 %	80 %	80 %	76.0 %
Gabor + PCA + ICA	96 %	93 %	86 %	95 %	87 %	91.4 %
PCA + ICA	69 %	84 %	89 %	79 %	81 %	80.4 %
FFT + Gabor + PCA + ICA	75 %	60 %	55 %	85 %	90 %	73.0 %

The final classification result is as shown in Table 1. We conducted additional four tests to compare the performance rate with that of our proposed algorithm. Our proposed method showed 93.2% classification performance rate, the best performance for facial orientation discrimination among other four methods. Four additional tests are designed as follows: 1) FFT, PCA, and ICA, 2) Liu's algorithm: Gabor, PCA, and ICA, 3) conventional algorithm: PCA and ICA, 4) switched orders of proposed method: FFT, Gabor, PCA, and ICA. For FFT application in test 1) and 4), we tried two methods for extracting features from FFT2 image with several FFT-point resolution parameters. First method is to truncate leftmost corner of the enlarged FFT-point resolution of FFT2 image to extract essential low frequency information and second is to down-size the enlarged FFT-point resolution of FFT2 image by averaging. Truncation method showed no response towards increasing resolution parameter when already-down-sized image of 96x64 was introduced whereas down-size method showed increase in classification rate in response with increasing resolution parameter. Since we had to introduce down-sized image to FFT2 function due to computational problem, we utilized down-size method for extracting features from FFT2 images. The optimal resolution parameter for FFT2 we used is 8 where classification result of test 1) showed 76.0%; result showed saturation for parameter of 16 with result of test 1) of 72.0%. In Fig. 4, the FFT2 result images using averaging method are shown, for each FFT-point resolution parameter of 1, 4, 8, and 16. It is clear that as FFT-point resolution parameter increases, FFT images look more detailed.

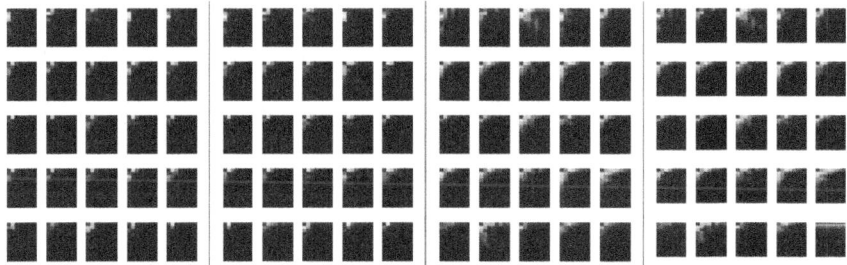

Fig. 4. FFT2 result images of averaging method for FFT-point resolution parameter of 1 *(first)*, 4 *(second)*, 8 *(third)*, 16 *(fourth)*. For observation, leftmost corner is magnified for images shown above. FFT2 result image shows more discriminancy as parameter increases until 8.

The result proved that our proposed algorithm works most successfully for facial orientation discrimination of random images, compared to several tests. By comparing result of test of FFT and test of Gabor only, we can analyze the reasons how our proposed algorithm enhances the performance compared to conventional method using PCA and ICA. The result for test with FFT shows that utilizing FFT with PCA and ICA does not contribute any benefits for facial orientation discrimination. On the other hand, result of Liu's method, test with Gabor with PCA and ICA, confirms that Liu's method enhances the performance rate when compared to conventional method of PCA and ICA, even for the case classification of orientation of random images, like it does for general recognition which is already known. The result for the proposed algorithm shows that when combining Gabor with FFT, performance is enhanced even more than adding Gabor alone to the conventional method. We can analyze the result as synergy effect of Gabor along with FFT; FFT removes the shift variance properties of Gabor filtered edge features and makes the classification performance more robust for pixel shifts in spatial location, which is a common problem encountered for conventional methods.

However, when we conducted experiment with dataset of images taken on same date, meaning all images are in the same condition, the classification performance turned out to be highest for Liu's Gabor, PCA, and ICA method with 92.0% whereas proposed method showed result of 78.0%. This may be because of no shift variations in images so that role of FFT is diminished. With this result, it is shown that our proposed method works the best in condition of random images.

4   Conclusions

In this paper, we proposed a novel algorithm for five-class face orientation discrimination that consists of four methods: Gabor filters, FFT, PCA, and ICA. The proposed algorithm demonstrated successful classification result of 93.2%. The result convinces that Gabor-based approach for five-class face orientation discrimination along with linear projection of simple edge features with shift invariant properties provides reliable feature extraction result for analyzing orientation of faces.

Our proposed algorithm proves that Gabor filter also contributes to excellent performance of face orientation discrimination as well, like its reputation for general face recognition introduced by Liu, and its performance is enhanced even more when working along with FFT as posterior step. We also suggest several useful comparisons on optimizing methods. With simple and straight-forward methods, the proposed algorithm will help analyze situation of a given image in fields of Human-Computer Interaction.

Acknowledgments. This research was supported by Basic Science Research Program through the National Research Foundation of Korea (NRF) funded by the Ministry of Education, Science and Technology (2009-0092812, 2010-0028722, and 2010-0018844). Also, we express our gratitude to CNSL members for valuable feedbacks.

References

1. Kurata, D., Nankaku, Y., Tokuda, K., Kitamura, T., Ghahramani, Z.: Face Recognition Based on Separable Lattice HMMs. In: 2006 IEEE International Conference on Acoustics, Speech and Signal Processing, pp. V–V. IEEE Press, New York (2006)
2. Darrell, T., Moghaddam, B., Pentland, A.P.: Active Face Tracking and Pose Estimation in an Interactive Room. In: 1996 IEEE Computer Society Conference on Computer Vision and Pattern Recognition, pp. 67–67. IEEE Press, New York (1996)
3. Chang, C.C., Aghajanet, H.: Collaborative Face Orientation Detection in Wireless Image Sensor Networks. In: International Workshop on Distributed Smart Cameras (2006)
4. Dervinis, D.: Head Orientation Estimation using Characteristic Points of Face. Electronics and Electrical Engineering 8(72), 61–64 (2006)
5. Kaminski, J.Y., Knaan, D., Shavit, A.: Single Image Face Orientation and Gaze Detection. Machine Vision and Applications 21(1), 85–98 (2009)
6. Rae, R., Ritter, H.J.: Recognition of Human Head Orientation based on Artificial Neural Networks. IEEE Transactions on Neural Networks 9(2), 257–265 (1998)
7. Liu, C.: Independent Component Analysis of Gabor Features for Face Recognition. IEEE Transactions on Neural Networks 14(4), 919 (2003)
8. Kailash, J., Karande, N.T., Sanjay, N.T.: Independent Component Analysis of Edge Information for Face Recognition. International Journal of Image Processing 3(3), 120–130 (2009)
9. Bartlett, M.S.: Face Recognition by Independent Component Analysis. IEEE Transactions on Neural Networks 13(6), 1450 (2002)
10. FacE REcognition Technology data base information, http://face.nist.gov/colorferet/

Visual Cortex Inspired Junction Detection

Shuzhi Sam Ge[1,3,*], Chengyao Shen[1,2], and Hongsheng He[1,3]

[1] Social Robotics Lab, Interactive Digital Media Institute
[2] NUS Graduate School for Integrative Science and Engineering
[3] Department of Electrical and Computer Engineering,
National University of Singapore
{samge,chengyao.shen,hongshenghe}@nus.edu.sg

Abstract. This paper proposes a visual cortex inspired framework to detect and discriminate junction points in a visual scene. Inspired by the biological research in primary visual cortex V1 and secondary visual cortex V2, banks of filters are invented to simulate the response of complex cells, horizontal lateral connections and angle detectors. Before junction detection, contour enhancement and texture suppression are performed in determining the sketch of a scene. The proposed algorithm can extract primal sketch of the scene and detect the response of junction points in the scene. Experiments prove the good performance on scenes with man-made structures.

Keywords: Visual Cortex, Primal Sketch, Junction Detection.

1 Introduction

With scale-invariant properties and semantic meanings, junctions constitute the "alphabet" in image grammar and could provide robust feature descriptors for feature matching in stereo vision and motion tracking. As junctions are points where two or more edges meet, they usually indicates occlusion, transparency, surface bending and other geometric properties in the scene [1, 2].

In the research of computer vision, the exploration on the detection of junctions has a long history. Traditional approaches usually detect junctions locally with a structure tensor like Harris operator or a corner template such as SU-SAN detector [3], and annotate junctions by a template with pre-defined parameters [4]. However, as local detectors are vulnerable to erroneous response in textured regions, multi-scale detection and contextual information obtained rich attention in recent literature on junction detection [5].

There is substantial evidence in psychophysiology, neurophysiology and computational modeling that secondary visual cortex (V2) is the place where two-dimensional features such as junctions and corners are encoded. Neurophysiological experiments on macaques [6, 7] show that neurons in V2 respond selectively to angle stimuli. Multi-layer self-organized model such as LISSOM [8] and

---

* To whom all correspondences should be addressed. Tel. (65) 6516-6821; Fax. (65) 6779-1103. samge@nus.edu.sg.

B.-L. Lu, L. Zhang, and J. Kwok (Eds.): ICONIP 2011, Part I, LNCS 7062, pp. 225–232, 2011.

Deep Belief Network [9] also present that preferred visual patterns of the second layer tend to combine two different orientations. Hence, one possible mechanism of junction detection in visual cortex is proposed that neurons in V2 form different feature maps which sample each combination of two orientation preference responses from primary visual cortex (V1) and generate sparse responses for each feature map [10].

On the basis of biological research in V2 and some existing computational model on V1 [2, 11], we propose a framework for junction detection. It is assumed that horizontal connections in V1 play an important role on contour enhancement and neurons in V2 combines inputs from two or more groups of neurons with different orientation preference in V1.The framework consists of two stages. At the first stage, a bank of Gabor filters are employed to generate contour responses, and horizontal connections in V1 are modeled for a contour representation which is perceptually evident. At the second stage, junction points are discriminated by annotating two or more largest orientation responses in a neighborhood region. In our framework, we only model the finest resolution in fovea and we choose to design receptive fields which lead to better results instead of totally simulating the receptive fields that was proposed in previous models. The main contributions of this paper are highlighted as follows:

(i) In computer vision, this paper proposes a framework for primal sketch and junction detection inspired from neural circuitry in V1 and V2;
(ii) In visual neuroscience, this paper presents a computational scheme to testify and predict the functions of feedforward and lateral connections in V1 and V2.

The rest of the paper is organized as follows. In Section 2, we present a model of primal sketch to get an enhanced contour map for junction detection. Methods in junction detection would be detailed in Section 3. In Section 4, experiments are conducted and a discussion on the model and experiment results is given. Section 5 concludes the article.

2   Primal Sketch

As junctions are usually intersection of edges, the extraction of edges is a necessary prepossessing step before junction detection. In this paper, we adopt the term "Primal Sketch", which is first proposed by David Marr in [12], to describe the process of contour extraction, contour enhancement and texture suppression at the first stage of our framework. In this section, the model of discovering primal sketch will be studied in detail.

2.1   Contour Extraction

To extract the contours in different orientations of a scene, a bank of oriented 2D quadrature Gabor filters is designed. The responses of 2D Gabor filter share

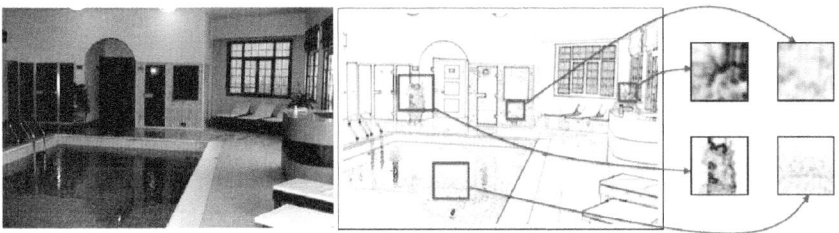

Fig. 1. Result of contour extraction after Gabor filtering and non-linear gain control and undesirable texture response that would interfere junction detection

many similarities with receptive fields of simple cells in V1 [13]. In spatial domain, a 2D Gabor filter can be denoted as the product of a Gaussian envelope times a complex sinusoid,

$$g(x, y, \theta; \lambda, \psi, \sigma, \gamma) = \exp(-\frac{x'^2 + \gamma^2 y'^2}{2\sigma^2}) \exp(i(2\pi\frac{x'}{\lambda} + \psi)) \tag{1}$$

with $x' = x\cos\theta + y\sin\theta$ and $y' = -x\sin\theta + y\cos\theta$, where θ represents the orientation preference, λ represents the wavelength of the sinusoidal factor, ψ is the phase offset, σ denotes the width of the Gaussian envelope and γ is the spatial aspect ratio which specifies the ellipticity of filter. In practice, the DC component in real part is eliminated in to ensure zero response in homogeneous area.

In contour extraction, a Gabor filter $g(x, y, \theta_k; 0.2\pi, 0, 1, 0.5)$ is applied with the orientation $\theta_k = k\pi/N$ $(k = 0, 1, ..., N - 1)$ and $N = 8$. The responses of even and odd simple cells, which correspond to the real and imaginary parts of $g(x, y, \theta)$, are obtained by a convolution of the input image I with Gabor filter. The response of complex cells are modeled by Euclidean norm of corresponding even and odd simple cells,

$$C'(\theta_k) = \|I * g(x, y, \theta_k)\| \tag{2}$$

After this, the responses of complex cells are normalized in contrast by non-linear gain control and thresholded by a piecewise linear approximation of sigmoid $F(\cdot)$ to constrain the activity in a range of $[0, 1]$:

$$C(\theta_k) = F(\frac{C'(\theta_k)}{l + C'(\theta_k)}) \tag{3}$$

The parameter $l > 0$ could be tuned to determine the degree of enhancement on weak response. Here we choose $l = 0.5$ and get a contour response as is shown in Fig. 1, which is very similar to a pencil sketch with noise from texture.

2.2   Contextual Moudulation

Contextual Modulation, which includes contour enhancement and texture suppression here, is originated from the fact that the result of contour extraction is not very 'clean'. As can be seen from Fig. 1, textured response that is undesirable in our primal sketch for contours also yields strong responses. This also cause interference in later stage of junction detection as textures are usually areas with random large response in multiple directions.

Biological evidence from the horizontal lateral connections in V1 also suggests such a mechanism [14]. Studies show that long-range projections of pyramidal cells in layer 2/3 project primarily to neurons with similar orientation preference and appear to be predominantly excitatory. However short-range projections appear to be largely isotropic and have been argued to be largely inhibitory.

The long-range projections here are modeled as a bank of collinear facilitation filters on contour responses. The receptive field of collinear facilitation filters are modeled as the real part of an elongated Gabor filter $g(x, y, \theta_k; 0.5\pi, 0, 2, 0.25)$[1] and the complex responses $C_i(\theta_k)$ are filtered to get the excitatory response $R_{ex}(\theta_k)$ of collinear structure.

$$R_{ex}(\theta_k) = [g_{real}(x, y, \theta_k; 0.5\pi, 0, 2, 0.25) * C(\theta_k)]^+ \qquad (4)$$

where $*$ denotes convolution, $[x]^+ = \max(x, 0)$ denotes half-wave-rectification.

The inhibitory short-range projections are modeled by filters which collect responses across different orientations. The shape of inhibitory receptive fields is modeled by the negative part of even response of $g(x, y, \theta_k; \pi, 0, 2, 1)$ across eight orientations with weights of 1D Gaussian distribution. The inhibitory response $R_{in}(\theta_k)$ can then be described as follows,

$$R_{in}(\theta_k) = \sum_{i=1}^{N} ([-g_{real}(x, y, \theta_k; \pi, 0, 2, 1)]^+ * C(\theta_i)) \cdot G_{1, \theta_k}(\theta_i). \qquad (5)$$

Here N=8 denotes eight orientations and 1D Gaussian distribution $G_{\sigma, \theta_k}(\theta_i)$ denotes that similar orientation response in the neighborhood would contribute more to the inhibitory response.

As the response of horizontal lateral connections only have modulatory effects on feed forward response, the response of long-range projections and short-range projections $P(\theta_k)$ are integrated with modulatory excitation and divisive inhibition as,

$$P(\theta_k) = \frac{C(\theta_k) \cdot (1 + a \cdot R_{ex}(\theta_k))}{b + R_{in}(\theta_k)} \qquad (6)$$

where $C(\theta_k)$, $R_{ex}(\theta_k)$, $R_{in}(\theta_k)$ are complex responses, excitatory responses and inhibitory response defined in Eq. (3), Eq. (4) and Eq. (5). Here $P(\theta_k)$ can be seen as the steady state of a firing rate model, b and a are the decay parameter and facilitatory parameter that could be tuned by readers.

The responses of the contextual modulation $P(\theta_k)$ are then fed into neighboring neurons by recurrent connections and integrated with the feedforward complex response $C'(\theta_k)$. After this procedure, a more sparse and accurate results of the orientation responses are yielded. The effects of the feedback connections are illustrated in Fig. 2 (b) and (c).

---

[1] Here Gabor filter is borrowed to model the shape of receptive field for the convenience of mathematical representation.

3   Junction Detection

As some neurons in V2 tend to form different feature maps by sampling each combination of two or more orientation preference responses from V1 [8–10], we propose a corresponding computational scheme in this section to locate and discriminate the type of junctions in a scene. The scheme uses combined orientation responses to get sparse junction candidate maps from primal sketch and responses of end-stopped cells to decide on the type of the junction.

3.1   Junction Locating

In primal sketch, a junction usually locates at the position where multiple lines end or intersect. So locating positions of intersections is a fast and economical way to locate junction candidates in a scene.

Intersection detectors $J(\theta_m, \theta_n)$ here are inspired from the neurons that respond selectively to angle stimuli in V2. They could be seen as an "AND" gate on even responses of two Gabor filters with different orientation preference $O(\theta_k)$ on the primal sketch $P(\theta_k)$:

$$O(\theta_k) = [g_{real}(x, y, \theta_k; 0.2\pi, 0, 1, 0.25) * P(\theta_k)]^+ \qquad (7)$$

$$J(\theta_m, \theta_n) = (O(\theta_m) \cdot O(\theta_n))^{0.5} \qquad (8)$$

The output of intersection detectors can be seen as sparse maps for junction candidates in different orientation preference (Each $J(\theta_m, \theta_n)$ here represents a sparse map with preferences in two orientations). The position of the junction candidates can then be found by calculating the local maxima of each $J(\theta_m, \theta_n)$.

3.2   Junction Classification

After fixating the location of candidate junctions, a bank of end-stopped filters, which is inspired from end-stopped cells found in visual cortex, are conducted on the primal sketch. Their receptive field could be approximated as Gabor filters in orthogonal orientations to orientation responses in primal sketch $P(\theta_k)$. Here we only model single-stopped filters as the odd part of $g(x, y, \theta_{k\perp}; \pi, 0, 1, 2)$ to find out the position of line ends,

$$R'_{es}(\theta_k) = [g_{imag}(x, y, \theta_{k\perp}; \pi, 0, 1, 2)] * P(\theta_k)]^+ \qquad (9)$$

where $\theta_{k\perp}$ denotes the orthogonal orientation to θ_k and $\theta_k = k\pi/N$ ($k = 0, 1, ..., 2N, N = 8$) denotes the single-stopped responses in both directions.

As there usually exist 'false responses' on lines and edges in $R'_{es}(\theta_k)$, an additional step of 'surround inhibition' [15] is implemented to eliminate these false responses. The end-stopped response $R_{es}(\theta_k)$ can then be expressed as,

$$R_{es}(\theta_k) = [R'_{es}(\theta_k) - gR_{surr}]^+ \qquad (10)$$

where $g = 0.5$ here and R_{surr} is detailed in Fig. 9 and Eq. (11) and (12) of [15].

After getting the end-stopped responses, the type of junction can be determined by the combined responses of $R_{es}(\theta_m)$, $R_{es}(\theta_n)$ and $J(\theta_m, \theta_n)$.

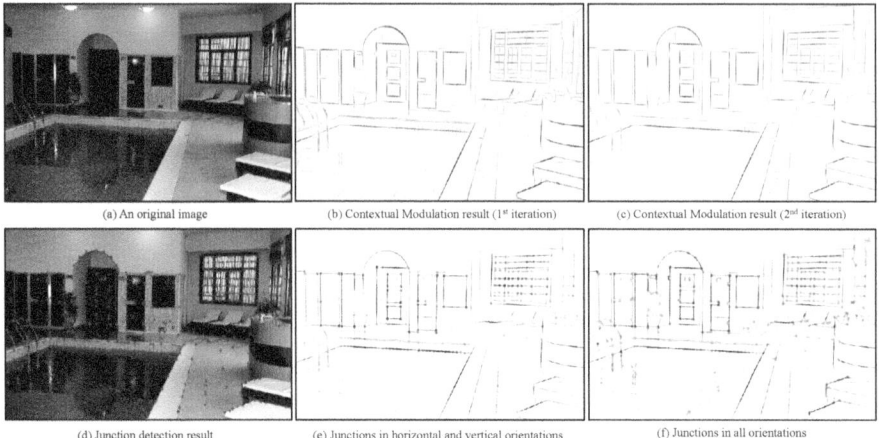

Fig. 2. Result of Primal Sketch and Junction Detection. (a) original image, (b-c) results of contextual modeling after 2 iterations, the primal sketch become more clean and sparse after two iterations, (d) junction points in the image after setting a threshold and get the local maxima on (f), (e) junction points where horizontal and vertical lines intersect. (f) junction points in the scene which get strong responses in two or more directions.

Fig. 3. Result of Primal Sketch and Junction Detection. Left: original image, Middle: junction points detected in the scene, Right: primal sketch and junction map.

4   Experiments and Discussions

In the experiments, the proposed model is applied to compute primal sketches and junctions in a variety of scenes with man-made structures as shown in Fig. 2 and Fig. 3. It can be seen from Fig. 2 (b) and (c) that texture areas are largely suppressed and contour regions are enhanced after 2 iterations of contextual modulation. The results of junction detection in Fig. 2 and Fig. 3 show that the detected junction generally captures the structure of the scene, where junctions capture the corner, occlusion, and surface intersection in the scene.

Another conclusion that can be drawn from the experiments is that, with well-tuned parameters and proper receptive size, filters similar receptive field of different neuron groups in visual cortex could to some extent realize some function of visual processing better than traditional computational algorithms. The filters used in this model simulate a subset of the neurons with different sizes and functional connections in brain, and we believe that brain has a similar mechanism to optimally choose filter banks of proper sizes and connection weights to adapt the environment.

5   Conclusion

In this paper, we have presented a brain inspired framework for detecting and recognizing junctions in low-middle level vision. We identified the primal sketch of the scene with filter banks inspired from neuron receptive fields and horizontal connections in V1. We then conducted junction detection with based on end-stopped filters and intersection detectors. The experiments show that the algorithm can simultaneously obtain good results in junction detection and the primal sketch which generalizes the main structure of the scene.

References

1. Wu, T., Xia, G., Zhu, S.: Compositional boosting for computing hierarchical image structures. In: 2007 IEEE Conference on Computer Vision and Pattern Recognition, pp. 1–8. IEEE (2007)
2. Rodrigues, J., du Buf, J.M.H.: Visual Cortex Frontend: Integrating Lines, Edges, Keypoints, and Disparity. In: Campilho, A.C., Kamel, M.S. (eds.) ICIAR 2004. LNCS, vol. 3211, pp. 664–671. Springer, Heidelberg (2004)
3. Parida, L., Geiger, D., Hummel, R.: Junctions: Detection, classification, and reconstruction. IEEE Transactions on Pattern Analysis and Machine Intelligence 20(7), 687–698 (1998)
4. Cazorla, M., Escolano, F., Gallardo, D., Rizo, R.: Junction detection and grouping with probabilistic edge models and bayesian a*. Pattern Recognition 35(9), 1869–1881 (2002)
5. Maire, M., Arbeláez, P., Fowlkes, C., Malik, J.: Using contours to detect and localize junctions in natural images. In: IEEE Conference on Computer Vision and Pattern Recognition, pp. 1–8. IEEE (2008)
6. Hegdé, J., Van Essen, D.: Selectivity for complex shapes in primate visual area v2. Journal of Neuroscience 20(5), 61–61 (2000)
7. Ito, M., Komatsu, H.: Representation of angles embedded within contour stimuli in area v2 of macaque monkeys. The Journal of Neuroscience 24(13), 3313 (2004)
8. Sit, Y., Miikkulainen, R.: Computational predictions on the receptive fields and organization of v2 for shape processing. Neural Computation 21(3), 762–785 (2009)
9. Lee, H., Ekanadham, C., Ng, A.: Sparse deep belief net model for visual area v2. In: Advances in Neural Information Processing Systems, vol. 19 (2007)

10. Loffler, G.: Perception of contours and shapes: Low and intermediate stage mechanisms. Vision Research 48(20), 2106–2127 (2008)
11. Hansen, T., Neumann, H.: A recurrent model of contour integration in primary visual cortex. Journal of Vision 8(8) (2008)
12. Marr, D.: Vision: A computational investigation into the human representation and processing of visual information (1982)
13. Movellan, J.: Tutorial on gabor filters. Open Source Document (2002)
14. Tucker, T., Fitzpatrick, D.: Contributions of Vertical and Horizontal Circuits to the Response Properties of Neurons in Primary Visual Cortex, vol. 1. MIT press, Cambridge (2004)
15. Heitger, F., Rosenthaler, L., Von Der Heydt, R., Peterhans, E., Kubler, O.: Simulation of neural contour mechanisms: from simple to end-stopped cells. Vision Research 32(5), 963–981 (1992)

A Quasi-linear Approach for Microarray Missing Value Imputation

Yu Cheng, Lan Wang, and Jinglu Hu

Graduate School of Information, Production and Systems, Waseda University
Hibikino 2-7, Wakamatsu-ku, Kitakyushu-shi, Fukuoka, Japan

Abstract. Missing value imputation for microarray data is important for gene expression analysis algorithms, such as clustering, classification and network design. A number of algorithms have been proposed to solve this problem, but most of them are only limited in linear analysis methods, such as including the estimation in the linear combination of other no-missing-value genes. It may result from the fact that microarray data often comprises of huge size of genes with only a small number of observations, and nonlinear regression techniques are prone to overfitting. In this paper, a quasi-linear SVR model is proposed to improve the linear approaches, and it can be explained in a piecewise linear interpolation way. Two real datasets are tested and experimental results show that the quasi-linear approach for missing value imputation outperforms both the linear and nonlinear approaches.

Keywords: microarray data, missing value imputation, quasi-linear, SVR.

1 Introduction

Microarray gene expression data generally suffers from missing value problem resulted from a variety of experimental reasons [1], while it has become a useful biological resource in recent years. In DNA microarray experiments, a large mount of genes are monitored under various conditions, and recovering missing values by repeating the experiment is often costly or time consuming. Therefore, computational methods are desired to achieve accurate result for imputation of microarray data.

Recently, linear analysis based approaches have been widely studied, among which, an imputation method based on local least squares (LLS) [2] has shown highly competitive performance compared with others, like K-nearest neighbors (KNN) imputation [3] and Bayesian principal component analysis (BPCA) imputation [4]. Missing values in a certain gene are represented as a linear combination of similar genes, and elimination of less similar genes is helpful to improve performance. Nevertheless, in most cases, hundreds of similar genes are selected for estimating missing values in one gene, and the linear approximation may be insufficient to capture nonlinearity in nature of the gene expression data. On the other hand, since the microarray dataset often comprises of huge

B.-L. Lu, L. Zhang, and J. Kwok (Eds.): ICONIP 2011, Part I, LNCS 7062, pp. 233–240, 2011.

size of genes with only a small number of observations [5], nonlinear regression techniques successfully apply in machine learning, such as neural networks, are prone to overfitting and generate models with little interpretability [6]. To fill the gap between the existed linear and nonlinear methods, a model with adjustable complexity is proposed for the maximum imputation performance.

In this paper, quasi-linear support vector regression (SVR) model [7] is proposed to improve linear approaches with proper nonlinearity (or flexibility) contained. It can be explained in a piecewise linear interpolation way to solve nonlinear regression problem with characteristics of high noise and small number of training examples for high input dimensional data. To generate appropriate partition in the input space, clustering technique is used according to the distribution of data, and flexibility of the model is determined by the number of partitions. For instance, only two or three subregions are enough in the case of microarray data missing value estimation, since the size of samples is limited within a small number, and unnecessary separating may deteriorate generalization. In the following, SVR is applied for estimation by a quasi-linear kernel, which includes piecewise linear information. Details of quasi-linear SVR model will be elaborated in Section 2.

The proposed imputation method falls in the category of local approaches [1], and gene selection algorithm from [2] is also applied for the gene-wise local scheme. In order to show the effectiveness of the proposed local SVR with quasi-linear kernel (LSVR/quasi-linear) imputation method, results from LLS, local SVR with linear kernel (LSVR/linear) and local SVR with nonlinear kernel (LSVR/nonlinear) will be compared. The paper is organized as following: Section 2 formulates quasi-linear SVR model. Section 3 gives the implementation of microarray data imputation in detail. Section 4 provides simulations based on two different data sets, and the results will be compared with others. Conclusions are provided in Section 5.

2   Quasi-Linear SVR Model

2.1   Quasi-Linear Formulation

Linear models often outperform nonlinear models in microarray missing value imputation problem due to the high noise and a small number of observations. However, linear models are difficult to capture nonlinear dynamics of the nature in microarray data. Therefore, a quasi-linear SVR model is proposed with proper flexibility contained. Let $\mathbf{x}_n = [x_{1,n}, x_{2,n}, \cdots, x_{k,n}]^T, n = 1, \cdots, N$ denotes the selected k genes in the n-th observation for estimating missing values of gene $\mathbf{y}$, then quasi-linear formulation [8] could be given as

$$y_n = \mathbf{x}_n^T \mathbf{w}(\mathbf{x}_n) + e_n \tag{1}$$

$$\mathbf{w}(\mathbf{x}_n) = \mathbf{\Omega}_0 + \sum_{j=1}^{M} \mathbf{\Omega}_j \mathcal{N}(\mathbf{p}_j, \mathbf{x}_n) \tag{2}$$

where $\mathbf{w}(\mathbf{x}_n)$ is a state-dependant parameter vector of linear structure in Eq.(1), which can be represented by a basis function network such as radial basis function (RBF) network in Eq.(2), and $\mathcal{N}(\mathbf{p}_j, \mathbf{x}_n)$ denotes the j-th basis function constructed by parameters $\mathbf{p}_j$. $\mathbf{\Omega}_j = [\omega_{0j}, \cdots, \omega_{kj}]^T (j = 1, \cdots, M)$ is the connection matrix between input genes and the corresponding basis functions, and M represents the size of basis functions.

The quasi-linear formulation in Eq.(1) can be explained in a piecewise linear interpolation way as

$$y_n = \sum_{j=0}^{M} \mathbf{x}_n^T \mathbf{\Omega}_j \mathcal{N}_{j,n} + e_n \tag{3}$$

in which $\mathcal{N}_{j,n}$ means interpolation value of the j-th piecewise linear part, which is the shortening of $\mathcal{N}(\mathbf{p}_j, \mathbf{x}_n)$ for simplicity, and it is defined $\mathcal{N}_{0,n} = 1$. M is the number of partitions for piecewise linear separating, and e_n is error of the model.

2.2   SVR with Quasi-Linear Kernel

SVR has demonstrated impressive generalization performance because it concentrates on minimizing the structural risk instead of the empirical risk. However, the kernel functions could be used are few, which can be categorized into linear kernel and nonlinear kernel (such as polynomial kernel, Gaussian kernel and Sigmoid kernel) [9]. Since quasi-linear formulation can be explained into a piecewise linear interpolation way, a SVR model is built with piecewise linear information contained into the kernel function.

Firstly, Eq.(3) can be written in a linear-in-parameter way as

$$y_n = \Phi_n^T \Theta + e_n \tag{4}$$

where

$$\Phi_n = [\mathbf{x}_n^T, \mathcal{N}_{1,n}^T \mathbf{x}_n^T, \cdots, \mathcal{N}_{M,n}^T \mathbf{x}_n^T]^T \tag{5}$$

$$\Theta = [\mathbf{\Omega}_0^T, \mathbf{\Omega}_1^T, \cdots, \mathbf{\Omega}_M^T]^T. \tag{6}$$

In the following, Structural Risk Minimization principal is introduced, and Eq.(4) can be solved as the QP optimization problem

$$\min \mathcal{J} = \frac{1}{2} \Theta^T \Theta + c \sum_{n=1}^{N} (\xi_n + \xi_n^*)$$

subject to:

$$\begin{cases} y_n - \Phi_n^T \Theta - b_0 \leq \epsilon + \xi_n \\ -y_n + \Phi_n^T \Theta + b_0 \leq \epsilon + \xi_n, \end{cases}$$

where N is the size of observations, and ξ_n, $\xi_n^* \geq 0$ are slack variables. C is a non-negative weight to determine how much prediction errors are penalized, which exceed the threshold value ϵ. The solution can be turned to find a saddle point of the associated Lagrange function

$$
\mathcal{L}(\Theta, \xi_n, \xi_n^*, \alpha_n, \alpha_n^*, \beta_n, \beta_n^*)
$$
$$
= \frac{1}{2}\Theta^T\Theta + c\sum_{n=1}^{N}(\xi_n + \xi_n^*) + \sum_{n=1}^{N}\alpha_n(y_n - \Phi_n^T\Theta - b_0 - \epsilon - \xi_n)
$$
$$
+ \sum_{n=1}^{N}\alpha_n^*(-y_n + \Phi_n^T\Theta + b_0 - \epsilon - \xi_n^*) - \sum_{n=1}^{N}(\beta_n\xi_n + \beta_n^*\xi_n^*). \tag{7}
$$

Then the saddle point could be acquired by $\frac{\partial\mathcal{L}}{\partial\Theta}$, thus it leads to

$$
\Theta = \sum_{n=1}^{N}(\alpha_n - \alpha_n^*)\Phi_n. \tag{8}
$$

Substitute Θ in Eq.(7) with (8), the dual cost can be written as

$$
\max \mathcal{W}(\alpha_n, \alpha_n^*) = -\frac{1}{2}\sum_{n,p=1}^{N}(\alpha_n - \alpha_n^*)(\alpha_p - \alpha_p^*)\Phi_n^T\Phi_p
$$
$$
+ \sum_{n=1}^{N}(\alpha_n - \alpha_n^*)(y_n - b_0) - \epsilon\sum_{n=1}^{N}(\alpha_n + \alpha_n^*). \tag{9}
$$

With solutions of α_n and α_n^* optimized from Eq.(9), the quasi-linear SVR model can be used as

$$
y_l = \sum_{n=1}^{N}(\alpha_n - \alpha_n^*)\mathcal{K}(l, n) \tag{10}
$$

in which y_l is the sample with missing value in the target gene, n is the sample points without missing value. In Eq.(10) the quasi-linear kernel is defined as

$$
\mathcal{K}(l, n) = \Phi_l^T\Phi_n = (\sum_{i=1}^{M}\mathcal{N}_{i,l}\mathcal{N}_{i,n} + 1)\mathcal{K}_L(l, n) \tag{11}
$$

and $\mathcal{K}_L(l, n) = \mathbf{x}_l^T\mathbf{x}_n$ is the expression of linear kernel. From Eq.(11) it is found that quasi-linear kernel contains the piecewise linear interpolation information, thus the flexibility of model can be adjusted by using different number of partition M, which is determined according to the knowledge from data.

3   Implementation of Quasi-Linear SVR for Microarray Missing Value Imputation

There are three steps in quasi-linear SVR based imputation method. The first step is to select k genes which are similar with the target gene. We use selection scheme which is the same with the one in LLS for convenience of comparison.

In the next step, quasi-linear formulation is realized in a piecewise linear interpolation way. The partitions are obtained by using affinity propagation (AP) clustering algorithm [10], which is recently introduced for exemplar-based clustering. The main advantage of this algorithm is that it can find good partitions of data and associate each partition with its most prototypical data point (exemplar) such that the similarity between points to their exemplar is maximized, and the overall cost associated with making a point an exemplar is minimized. Therefore, reasonable partitions can be generated automatically without pre-determining the number of clusters. In microarray missing value imputation problem, due to the size of samples is limited in a small number, two or three clusters are acquired each time. Then RBF is introduced as basis function to cover each subregion, which can be explained as interpolation. Therefore, $\mathcal{N}_i(t)$ is defined as

$$\mathcal{N}_{i,n} = \exp(-\frac{\mathbf{x}_n - \mathbf{m}_i}{\lambda \mathbf{d}_i}) \tag{12}$$

where $\mathbf{m}_i$ and $\mathbf{d}_i$ are the center vectors for the i-th subregion, λ is a scale parameter.

At last, the quasi-linear SVR model obtained from Eq.(7)-(9) is used to estimate the missing value y_l as Eq.(10) shown.

4   Experiments and Results

4.1   Experiment Setting

To show effectiveness of the proposed quasi-linear SVR model microarray missing value imputation, the results will be compared with LLS, LSVR/linear and LSVR/nonlinear. In LSVR/nonlinear method Gaussian kernel function is applied. For the reason of fairness, missing values are set to zero when they are used for estimation of other missing ones in all the four methods. Both experiments are implemented by Matlab 7.6, and Lib SVM toolbox version 2.91 [11] is applied for SVR implementation, and ν-SVR is used with default parameter setting.

4.2   Datasets

Two real datasets have been used in our experiments. The first one is from 784 cell-cycle-regulated genes, and 474 genes and 14 experiments (SP.CYCLE) are obtained after removing all gene rows that have missing values. It was designed to test how much an imputing method can take advantage of strongly correlated genes in estimating missing values. The second dataset is the cDNA microarray data relevant to human colorectal cancer (TA.CRC) and contains 758 genes in 50 samples. These two datasets have been used in studies of BPCA [4] and LLS [2] imputation.

Since missing values used in our experiments are introduced artificially, the performance of imputation methods are evaluated by comparing the estimated values with the corresponding real values. The matric to assess the accuracy of estimation is normalized root mean squared error (NRMSE):

$$\text{NRMSE} = \sqrt{\frac{\text{mean}[(y_{guess} - y_{ans})^2]}{\text{variance}[y_{ans}]}}$$

where y_{guess} and y_{ans} are estimated and real vectors for the artificial missing values.

4.3   Experimental Results

AP clustering algorithm is implemented, and two clusters are obtained in both the two cases. In Fig.1, the experimental results are shown for SP.CYCLE and TA.CRC dataset with missing data rate of 1%. In both cases, the NRMSE values for all the methods are tested with various values of k, and the best result of each method that do not depend on the number of genes are shown on the y-axis. It is found that linear imputation methods (LLS and LSVR/linear) outperform nonlinear one (LSVR/nonlinear) as k-value increases. And the quasi-linear imputation approach shows the best performance among all the methods compared. The results are given in Tab.1.

What's more, as the percentage of missing value increased, the NRMSE values of both two dataset with the missing rate of 10% and 20% are also shown in Tab.1. It is found when the missing rate is high, the results from LSVR/quasi-linear are competitive with LLS, LSVR/linear method, and better than LSVR/nonlinear method on both two datasets.

4.4   Discussions

As the results shown in above experiments, the quasi-linear approach can provide higher estimation accuracy compared with linear and nonlinear methods especially when the missing rate is not high. It owns to the proper flexibility of the model, which can be explained in a piecewise linear interpolation way. However, when the missing rate becomes to be high, inaccurate knowledge, such as data distribution for partition, is also incorporated into quasi-linear SVR model, hence the performance is only competitive with linear methods.

Table 1. NRMSE results with missing percentage of 1%, 10% and 20%

	SP.CYCLE			TA.CRC		
missing rate	1%	10%	20%	1%	10%	20%
LLS	0.523	**0.637**	**0.638**	0.389	0.393	0.503
LSVR/linear	0.513	0.639	0.649	0.385	**0.390**	0.498
LSVR/quasi-linear	**0.500**	**0.637**	0.643	**0.379**	**0.390**	**0.497**
LSVR/nonlinear	0.632	0.686	0.730	0.424	0.408	0.524

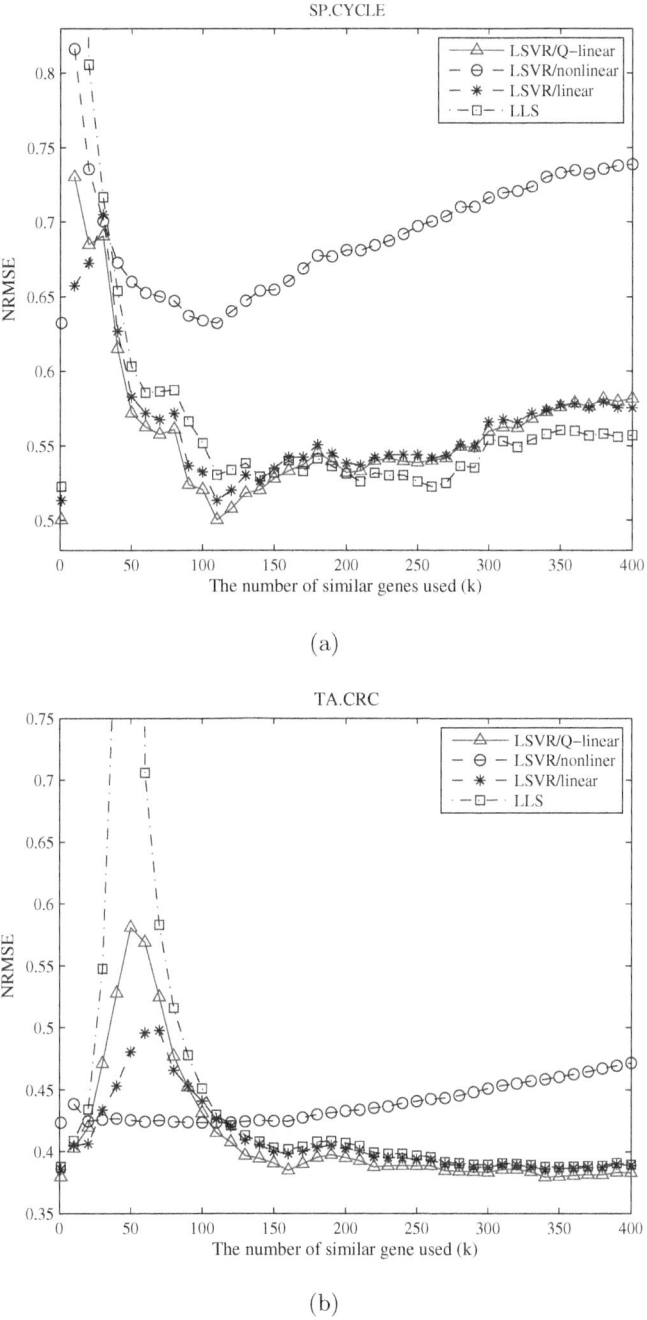

(a)

(b)

Fig. 1. Comparison of the NRMSEs of four methods and effect of the k-value on SP.CYCLE and TA.CRC dataset with 1% entries of each dataset

5 Conclusions

A quasi-linear SVR model is proposed to estimate the missing values in microarray data. It can be explained in a piecewise linear interpolation way to improve the linear method, and it also performs better than nonlinear methods because overfitting caused by the lack of observations and high noise of the datasets. The experimental results show the proposed method outperforms LLS imputation method, and SVR approaches with linear and RBF kernel functions as well.

References

1. Liew, A.W.C., Law, N.F., Yan, H.: Missing value imputation for gene expression data:computational techniques to recover missing data from available information. Briefings in Bioinformatics 12(3), 1–16 (2010)
2. Kim, H., Golub, G.H., Park, H.: Missing value estimation for dna microarray gene expression data: local least squares imputation. Bioinformatics 21(2), 187–198 (2005)
3. Troyanskaya, O., Cantor, M., Sherlock, G., Brown, P., Hastie, T., Tibshirani, R., Botstein, D., Altman, R.: Missing value estimation methods for dna microarrays. Bioinformatics 17(6), 520–525 (2001)
4. Oba, S., Sato, M.A., Takemasa, I., Monden, M., Matsubara, K.I., Ishii, S.: A bayesian missing value estimation method for gene expression profile data. Bioinformatics 19(16), 2088–2096 (2003)
5. Tarca, A.L., Romero, R., Draghici, S.: Analysis of microarray experiments of gene expression profiling. American Journal of Obstetrics and Gynaecology 195(2), 373–388 (2006)
6. Sahu, M.A., Swarnkar, M.T., Das, M.K.: Estimation methods for microarray data with missing values: a review. International Journal of Computer Science and Information Technologies 2(2), 614–620 (2011)
7. Cheng, Y., Wang, L., Hu, J.: Quasi-ARX wavelet network for SVR based nonlinear system identification. Nonlinear Theory and its Applications (NOLTA), IEICE 2(2), 165–179 (2011)
8. Hu, J., Kumamaru, K., Inoue, K., Hirasawa, K.: A hybrid Quasi-ARMAX modeling scheme for identification of nonlinear systems. Transections of the Society of Instrument and Control Engineers 34(8), 997–985 (1998)
9. Vapnik, V.: The Nature of Statistical Learning Theory. Springer, Berlin (1999)
10. Frey, B.J., Dueck, D.: Clustering by passing messages between data points. Science 315(5814), 972–976 (2007)
11. Chang, C.-C., Lin, C.-J.: LIBSVM: a library for support vector machines (2001), software available at http://www.csie.ntu.edu.tw/~cjlin/libsvm

Knowledge-Based Segmentation of Spine and Ribs from Bone Scintigraphy

Qiang Wang[1], Qingqing Chang[1], Yu Qiao[1], Yuyuan Zhu[2],
Gang Huang[2], and Jie Yang[1,*]

[1] School of Electronic, Information and Electrical Engineering,
Shanghai Jiao Tong University, Shanghai 200240, China
`{cnsd,changqing,qiaoyu,jieyang}@sjtu.edu.cn`
[2] Department of Nuclear Medicine, Renji Hospital, Shanghai Jiao Tong University,
Shanghai 200127, China
`hezicase@gmail.com, huang2802@163.com`

Abstract. We propose a novel method for the segmentation of spine and ribs from posterior whole-body bone scintigraphy images. The knowledge-based method is first applied to determine the thoracic region. An adaptive thresholding method is then used to extract the thoracic spine from it. The rib segmentation algorithm is carried out in two steps. First, the rib skeleton is extracted based on standard template and image information. The skeleton is then used to locate the accurate boundary of the respective ribs. The introduction of standard template can deal with significant variations among different patients well, while the skeleton-based method is robust against the low contrast between the ribs and the adjacent intervals. The experiments show that our method is robust and accurate compared to existing methods.

Keywords: Bone scintigraphy, Spine segmentation, Rib segmentation, Knowledge-based segmentation.

1 Introduction

Bone scintigraphy is a useful tool for diagnosing diseases such as bone tumors in nuclear medicine [1,2,3,4]. The accurate segmentation of the spine and ribs can be helpful for the analysis of the bone and location of the lesions. However, the automatic segmentation and analysis of the images have been a challenging task, due to the poor quality of the bone scan images and the significant variations of bone structures among different patients. The low signal noise ratio (SNR) and weak boundary contrast are the main difficulty for delineating the specific bone boundary. Traditional image segmentation algorithms like thresholding, region growing or level set may not work well due to these difficulties .

Previous works mainly focus on the region partition of the whole-body image [1, 3, 4, 5]. Only few works have been done on the accurate segmentation of the specific bones. Sajn [2] intend to separate body scan images into main skeletal regions and intend to segment individual bones. However, many simplifications are made in his work: the bones are represented as simple geometric shapes, for example, the spine

B.-L. Lu, L. Zhang, and J. Kwok (Eds.): ICONIP 2011, Part I, LNCS 7062, pp. 241–248, 2011.

area is seen as a rectangle, and not all ribs are segmented. The boundary of the bones cannot be accurately delineated, either.

Spine is the most important part of the thoracic region. The accurate segmentation of spine can not only be useful for hotspot detection in both spine and ribs area but also be important for rib segmentation. The rib segmentation is the most challenging work in the segmentation of bone scan images [2]. The contrast between rib and rib interval is too low, which makes the accurate delineation of the boundary really difficult. Few methods have been proposed for rib segmentation.

In this paper, the knowledge-based method is first applied to extract the thoracic region. An adaptive thresholding method is used for spine and vertebra segmentation. Then rib skeleton is extracted based on the standard template via a novel curve following algorithm. Finally, the rib skeleton is used to locate the accurate boundary of ribs combined with local information. The implementation of our algorithm shows that it can generate satisfactory results.

2   Materials and Methods

The bone scintigram images we used are acquired from department of nuclear medicine, Shanghai Renji Hospital. Each image has a resolution of 1024×512 and each pixel has 16-bit grayscale depth.

The segmentation algorithm consists 3 steps: thoracic region extraction, spinal column segmentation and rib segmentation. All input images first go through a Gaussian filter with size of 5×5 to remove additional noise.

2.1   Thoracic Area Extraction

After consulting with the physicians, our region of interest (ROI) is defined as the thoracic area between the shoulder points and the pelvis top points. We locate left point and the right landmark points separately by looking for the maximum width changes along the vertical direction within given ranges.

Image binarization is first performed by setting all pixels with value less than T_1 (which is set as 6 in our experiments) to 0. The medium filter is applied to remove the additional noise introduced during the binarization process. The patient height is then estimated by detecting the head and feet position on the binarized image.

According to statistics and anatomical knowledge, the shoulder points are generally located within 0.17-0.25 patient height down from the head line, while the pelvis points are generally located within 0.38-0.45 patient height down from the head line. The line with maximum width change (|current width-previous width|) in each range is considered as the corresponding shoulder/pelvis line; the landmark point is then located as the first non-zero point from outside to inside along the horizontal line. The thoracic region is set to be the minimum enclosing rectangle of the detected landmark points.

Fig. 1 demonstrates the thoracic region extraction process. (a) shows the original image, (b) shows the binarized image, (c) shows landmarks points and (d) shows extraction result.

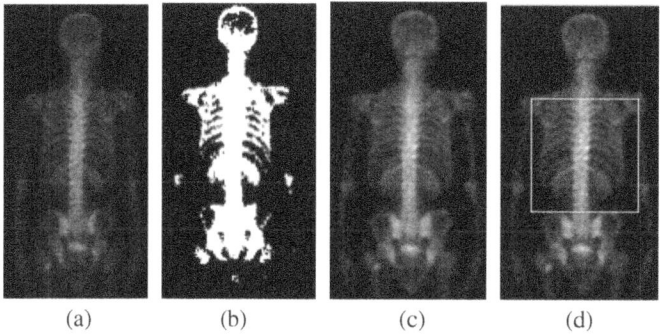

(a)   (b)   (c)   (d)

Fig. 1. Procedure of thoracic area extraction

2.2   Spinal Column Segmentation

Spine Area Segmentation

An optimal threshold T_2 is first computed by the OTSU algorithm [8] to segment spine from thoracic area. The OTSU algorithm considers the segmentation problem as a two-class classification problem and intends to minimize the intra-class variance. In our case, one class usually includes spine, kidney, and hotspot, which have relatively high intensity values, while the other one mainly includes the rib area.

Topological operations are then applied to remove kidney, hotspots or scapula from threshold results. For few cases in which some ribs or kidney may still be connected to the spine, the width of the spine area is computed for each horizontal line. If the width change surpasses certain predefined value, the corresponding boundary is adjusted according to its neighboring lines. The typical result of spine segmentation is shown in Fig. 2 (b).

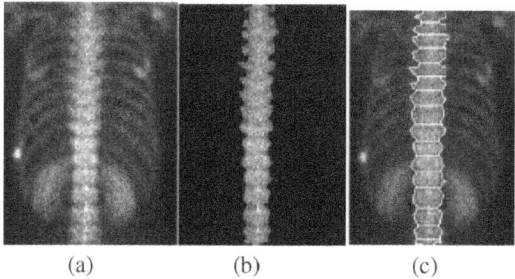

(a)   (b)   (c)

Fig. 2. The segmentation and partition of spinal column

Spinal Column Partition

It is not easy to separate spine into vertebras merely by the width or intensity sum distribution of each horizontal line (Fig. 3(a)). Considering the anatomical structure and intensity distribution of the spine area, we propose a new term $S(i)$ for vertebra segmentation, which is defined by:

$$S(i) = \sum_{m=-3}^{3} x(i, left(i) + m) + \sum_{n=-3}^{3} x(i, right(i) + n) \tag{1}$$

in which i denotes the row number of the thoracic area, $x(i, j)$ denotes the pixel value of the image at the position of (i, j), $left(i)$ and $right(i)$ respectively denote the column coordinates of left and right boundary of the spine.

The formula computes the intensity sum around the spine boundary, which shows significant periodic change along the vertical line, as is shown in Fig 3(b). The intervertebral discs correspond to the local minima of the distribution curve, which is used to partition vertebral bodies. The typical vertebra segmentation result is shown in Fig. 2(c).

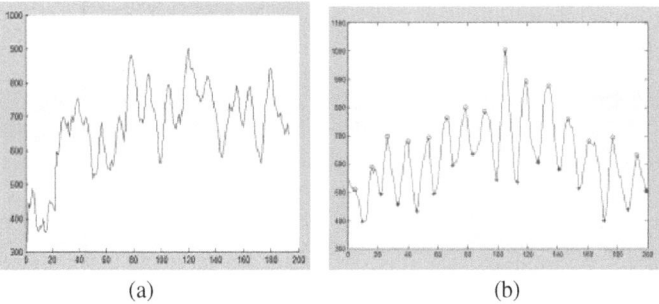

(a)   (b)

Fig. 3. The distribution of intensity sum and S(i) along the vertical line

2.3   Rib Segmentation

To include as much prior knowledge about ribs as possible, we introduce a standard template manually delineated by an experience physician from Shanghai Renji Hospital, which is shown in Fig. 4(a). We propose a skeleton-based algorithm for rib segmentation, which proves to be more robust and accurate.

Rib Skeleton Extraction
Rib skeleton is extracted via a novel curve following algorithm based on standard template. To control the following process, prior knowledge about the rib shape and local information are combined to obtain the best results. The detailed algorithm for left rib skeleton extraction is summarized as follows.

1. The starting points of the ribs care obtained from $S(i)$ that we got in section 2.2. As ribs are connected to corresponding thoracic vertebras from posterior view, the coordinates of local maximum points of $S(i)$ can then serve as the starting points of ribs.

2. For each rib skeleton point (a_k, b_k), we can predict the range of the coordinate of the neighboring point by:

$$a_{k+1} = a_k + \delta \ , b_{k+1} = b_k - 1 \quad \delta \in [\delta_{min}, \delta_{max}] \tag{2}$$

in which (a_k, b_k) is the coordinate of current rib skeleton point with line number of b_k. The range of δ can be obtained by manually measuring the standard template and storing $\delta_{min}, \delta_{max}$ for different rib length in a look-up-table (LUT).

3. The optimal δ_{op} can be obtained by solving the following optimization problem:

$$\delta_{op} = \arg \max \sum_{m=-2}^{2} x(a_k + \delta + m, b_k - 1) \quad \delta \in [\delta_{min}, \delta_{max}] \tag{3}$$

in which $x(a_k, b_k)$ is the corresponding intensity value; the coordinate of the current rib skeleton point is then $(a_k + \delta_{op}, b_{k+1})$.

4. Repeat step 2 and step 3 until all ribs reach the end or predefined maximum length.
5. Cubic spline fitting is applied for smoothing the obtained curve. The result of curve fitting is considered as rib skeleton.

The process of rib skeleton extraction is presented in Fig. 4, in which (b) shows the starting points of rib skeleton, (c) shows the curve following result and (d) shows the rib skeleton after curve fitting.

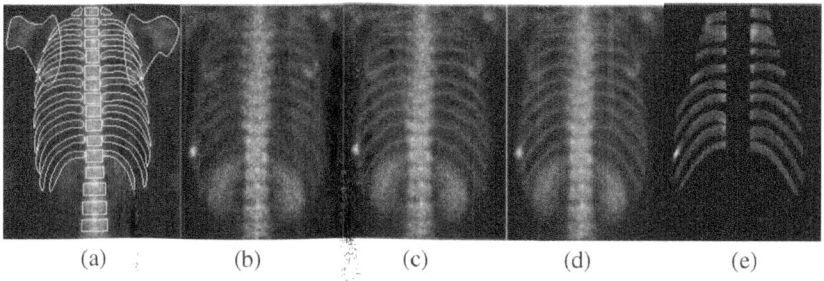

(a)   (b)   (c)   (d)   (e)

Fig. 4. The template we employed and the extraction of rib skeleton

Rib Boundary Delineation

To accurately delineate the boundary of a specific rib, we shift the rib skeleton up and down in predefined ranges. The sum of intensity values along the shifted curve is then calculated, the peak of each curve corresponds to the rib middle line while the valley corresponds to the rib interval. We define the rib boundary as the line with steepest intensity change. The rib segmentation result of our algorithm is shown in Fig. 4 (e).

3   Experimental Results

3.1   The Thoracic Region Extraction

To compare our thoracic region segmentation with existing methods, we ask physicians from Shanghai Renji Hospital to manually inspect the results. The comparison of accuracy between our method and Huang's method in our test set is shown in table 1. The width-change based method show better robustness than intensity-based method. And by locating the left and right landmark points separately, the method can resist noise or patient position. Fig. 5 (a) (c) show two cases in which Huang's algorithm fail due to noise and asymmetric landmark points. Our method, on the other hand, can deal with the problem well, (b) (d).

Table 1. Comparison of accuracy between landmark points detection methods

	Our method	Huang's method
Shoulder point detection	247/250(98.8%)	222/250(88.8%)
Pelvis point detection	166/175(94.8%)	148/175(84.6%)

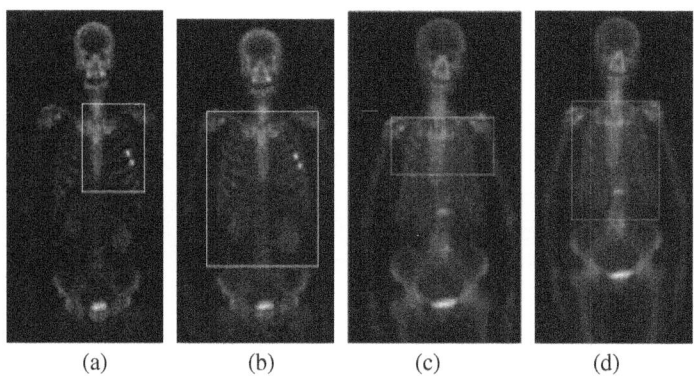

(a)   (b)   (c)   (d)

Fig. 5. Examples of thoracic region segmentation results

3.2   Spine and Vertebrae Segmentation

Fig. 6 shows some of our spine segmentation results, the knowledge-based algorithm can be deal with the variations among different patients well. Our algorithm can also show good results for scoliosis patients, as is shown in Fig. 6(c)(d). Knowledge-based method can remove ribs or kidney connected to the spine as well. The automatic segmentation and partition of vertebras is not only useful for the hotspot detection and location in posterior images but also can be used for image registration in later studies.

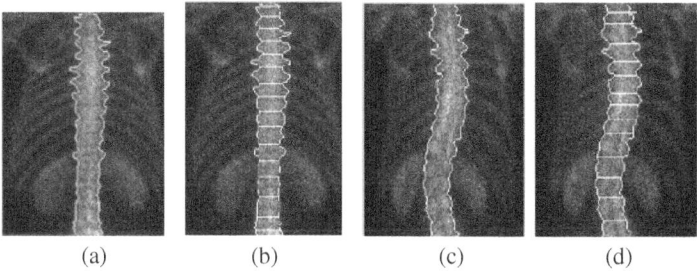

Fig. 6. The results of spine segmentation and partition

3.3   Rib Segmentation

Fig. 7 shows the rib segmentation results. (a)(e) show the original images; (b)(f) show the rib skeleton extracted by our algorithm, (c)(g) show the final rib segmentation of our method, (d)(h) show the results obtained by the adaptive region growing in [9], the spine and kidney are removed by our method. We can see that it is difficult to find the boundary of ribs merely by intensity information due to the low SNR. However, the skeleton-based algorithm can overcome the problem since it does not depend on the single pixel value. Another advantage of our algorithm over traditional methods like level set or graph cut is that each rib can be segmented, which can be analyzed individually in later studies.

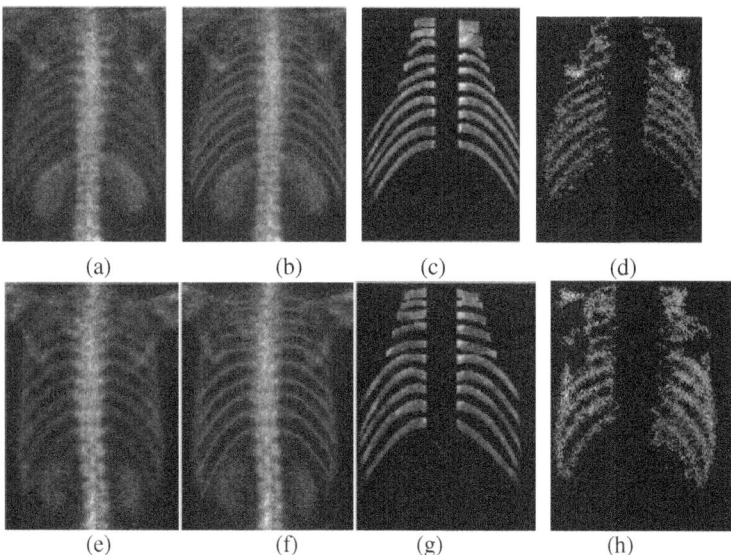

Fig. 7. The results of rib segmentation

The main difficulty with rib segmentation is the 1st to 5th ribs, which are quite unexpressive in the images. In few cases, the algorithm may fail due to poor quality.

This can be manually corrected by the physicians. The result of spine and rib segmentation is also applied in our CAD system for the interpretation of bone scintigraphy images and shows good clinical potential.

4   Conclusions

An automatic method for spine and rib segmentation from posterior bone scintigraphy images is proposed in this paper. Combined with statistic information and standard template, our algorithm can deal with the shape and intensity variety among different patients well. The knowledge-based method also shows good results delineating the boundary of the rib in spite of the low SNR and weak boundary contrast. The experiments show that our algorithm is robust and accurate.

References

1. Yin, T.K., Chiu, N.T.: A computer-aided diagnosis for locating abnormalities in bone scintigraphy by a fuzzy system with a three-step minimization approach. IEEE. Trans. Med. Imag. 23, 639–654 (2004)
2. Sajn, L., Kukar, M., Kononenko, I., Milcinski, M.: Computerized segmentation of whole-body bone scintigraphy and its use in automated diagnostics. Comput. Methods Programs Biomed. 80, 47–55 (2005)
3. Sadik, M., Jakobsson, D., Olofsson, F., Ohlsson, M., Suurkula, M., Edenbrandt, L.: A new computer-based decision-support system for the interpretation of bone scans. J. Nucl. Med. Commun. 27(5), 417–423 (2006)
4. Sadik, M., Hamadeh, I., Nordblom, P.: Computer-assisted diagnosis of planar whole-body bone scans. J. Nucl. Med. 49, 1958–1965 (2008)
5. Huang, J.Y., Kao, P.F., Chen, Y.S.: A Set of Image Processing Algorithms for Computer-Aided Diagnosis in Nuclear Medicine Whole Body Bone Scan Images. IEEE. Trans. Nucl. Sci. 54(3) (2007)
6. Ohlsson, M., Kaboteh, R., Sadik, M., Suurkula, M., Lomsky, M., Gjertsson, P., Sjostrand, K., Richter, J., Edenbrandt, L.: Automated decision support for bone scintigraphy. In: 22nd IEEE International Symposium on Computer Based Medical Systems, pp. 1–6 (2009)
7. Sadik, M., Suurkula, M., Höglund, P., Järund, A., Edenbrandt, L.: Improved Classifications of Planar Whole-Body Bone Scans Using a Computer-Assisted Diagnosis System: A Multicenter, Multiple-Reader, Multiple-Case Study. J. Nucl. Med. 50, 368–375 (2009)
8. Nobuyuki, O.: A threshold selection method from gray-level histograms. IEEE. Trans. Sys. Man Cyber. 9(1), 62–66 (1979)
9. Regina, P., Klaus, D.: Segmentation of medical images using adaptive region growing. In: Proc. SPIE, vol. 4322, p. 1337 (2001)

Adaptive Region Growing Based on Boundary Measures

Yu Qiao and Jie Yang

Institute of Image Processing and Pattern Recognition,
Shanghai Jiao Tong University, 800 Dongchuan Road, Shanghai 200240, P.R. China
qiaoyu@sjtu.edu.cn,jieyang@sjtu.edu.cn
http://www.pami.sjtu.edu.cn

Abstract. Weak boundary contrast, inhomogeneous background and overlapped intensity distributions of the object and background are main causes that may lead to failure of boundary detection for many image segmentation methods. An adaptive region growing method based on multiple boundary measures is presented. It consists of region expansion and boundary selection processes. During the region expansion process the region grows from a seed point. The background points adjacent to the current region are examined with local boundary measures. The region is expanded by iteratively growing the most qualified points. In the boundary selection process, the object boundary is determined with the global boundary measure that evaluates the boundary completeness. Experimental results demonstrate that our algorithm is robust against weak boundary contrast, inhomogeneous background and overlapped intensity distributions.

Keywords: Segmentation, Region growing, Local boundary measures, Global boundary measure, Growing cost.

1 Introduction

Image segmentation is one of the most critical tasks for automatic image analysis. Weak boundary contrast occurs frequently in medical images, in which the boundary contrast of some boundary points is too weak to distinguish the object and the background. These boundary points are called boundary openings because the boundary looks discontinuous around them. In some intensity images, the objects may be surrounded by both darker and brighter background structures, which are called inhomogeneous background. In medical images the intensity ranges of different medical structures are often overlapped, which are called overlapped intensity distributions of the object and background. All of three problems are main causes that may lead to failure of boundary detection for many image segmentation methods. Robust segmentation methods are required to deal with these three problems: 1) weak boundary contrast; 2) inhomogeneous background; and 3) overlapped intensity distributions.

Thresholding is a basic segmentation technique based on intensity histogram [1], [2]. But they may fail when the intensity ranges of object and background

B.-L. Lu, L. Zhang, and J. Kwok (Eds.): ICONIP 2011, Part I, LNCS 7062, pp. 249–256, 2011.
© Springer-Verlag Berlin Heidelberg 2011

overlap. Edge detection is one of the most common approaches for detecting meaningful intensity discontinuity in an intensity image, such as Canny and Sobel edge detectors. They may be badly influenced by weak boundary contrast. Seeded region growing is another basic technique for object. But some seeded region growing approaches may not identify the object from the inhomogeneous background [3]. Various deformable models have been widely used in medical image segmentation. However, snakes [4], [5] and level set methods [6], [7] may fail if the initial models are not in the vicinity of the objects.

In this paper, an adaptive region growing algorithm for object segmentation is developed. It consists of two processes, region expansion and boundary selection. In the region expansion process, current region is expanded by growing the most qualified point extracted with local boundary measures. The region expansion process continues until there is no contour point available to join the region. In the boundary selection process, the object boundary is selected with a global boundary measure. The application of the proposed method to real images shows that it is effective in dealing with weak boundary contrast, inhomogeneous background and overlapped intensity distributions.

2   Region Expansion

In order to avoid the unwanted influences of inhomogeneous background and overlapped intensity distributions, the information of local intensity distribution is used to guide the region expansion process. In an intensity image, the location and shape of a region can be defined by a set of background points adjacent to the region. For convenience, this set of background points is called the contour of the region in this paper.

Let Ω represent the region of an original image. Let $f(\mathbf{x})$ be the intensity function of the image point $\mathbf{x} = (x, y) \in \Omega$. The region of object to be extracted is called real object R, while $(\Omega - R)$ is called real background B. The contour of the real object R is denoted as object boundary L. To differ from the real object, the region examined at step i is called step region $R(i)$, while $(\Omega - R(i))$ is called step background $B(i)$. The contour of the step region $R(i)$ is called step contour $L(i)$.

In the process of region expansion, each contour point is examined within its neighborhood, which is a square window centered at the examined contour point. The size of neighborhood is application dependent. Let $N(\mathbf{x}_e)$ be the local neighborhood centered at the contour point $\mathbf{x}_e$. Let $R(\mathbf{x}_e)$ be a subset of step region points in $N(\mathbf{x}_e)$ and $B(\mathbf{x}_e)$ be a subset of step background points, i.e. $N(\mathbf{x}_e) = R(\mathbf{x}_e) \bigcup B(\mathbf{x}_e)$. Fig. 1 shows an example of a 5×5 neighborhood centered at a contour point.

The local contour contrast is defined by $d_R = |f(\mathbf{x}_e) - m_R|$, which is the contrast between intensity of the contour point $\mathbf{x}_e$ and the mean intensity m_R of $R(\mathbf{x}_e)$. The local background intensity variation is $d_B = |f(\mathbf{x}_e) - m_B|$, which is the contrast between intensity of $\mathbf{x}_e$ and the mean intensity m_B of $B(\mathbf{x}_e)$.

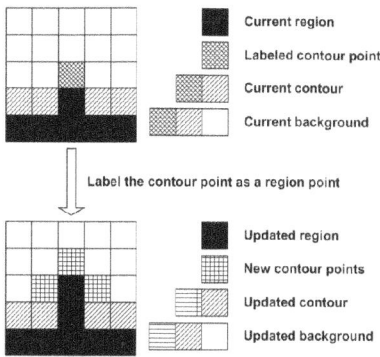

Fig. 1. The variations in the 5×5 neighborhood of the labeled contour point during labeling process

2.1 Local Boundary Measures

In the region expansion process, all of current contour points are examined with local boundary measures and are classified into two states: region and boundary candidates. The state of a contour point $\mathbf{x}_e$ is classified as region candidate if its intensity is similar to that of $R(\mathbf{x}_e)$. Otherwise it is classified as boundary candidate.

A local boundary measure called the local region competition measure, is introduced to classify the states of contour points. It evaluates the competition between local region $R(\mathbf{x}_e)$ and background $B(\mathbf{x}_e)$ on attraction to the contour point $\mathbf{x}_e$ in terms of intensity. This measure is defined by,

$$\upsilon_0 = u(d_R - d_B) = \begin{cases} 0, & \text{if } d_R \leq d_B \\ 1, & \text{if } d_R > d_B \end{cases}, \tag{1}$$

where $u(\cdot)$ is a unit step function. If $\upsilon_0 = 0$, $\mathbf{x}_e$ is classified as a region candidate because its intensity is more similar to that of $R(\mathbf{x}_e)$. Otherwise $\mathbf{x}_e$ is regarded as a boundary candidate with this measure. The contour points of real object R with significant boundary contrast are very likely classified as boundary candidates by this measure. This local region competition measure is application independent.

Two application dependent local boundary measures are introduced to assist the local region competition measure. The first application dependent measure examines the relative intensity range of local region $R(\mathbf{x}_e)$,

$$\upsilon_1 = u(d_R - \lambda\sigma_R) = \begin{cases} 0, & \text{if } d_R \leq \lambda\sigma_R \\ 1, & \text{if } d_R > \lambda\sigma_R \end{cases}, \tag{2}$$

where σ_R is the standard deviation of $R(\mathbf{x}_e)$. λ is the region intensity range parameter. $\upsilon_1 = 0$ implies that the intensity of $\mathbf{x}_e$ is within the relative intensity range of $R(\mathbf{x}_e)$, i.e. $f(\mathbf{x}_e) \in [m_R - \lambda\sigma_R, \; m_R + \lambda\sigma_R]$. $\mathbf{x}_e$ is therefore classified as a region candidate. Otherwise, $\mathbf{x}_e$ is treated as boundary candidate if $\upsilon_1 = 1$.

The second application dependent measure analyzes the absolute intensity range of local region $R(\mathbf{x}_e)$,

$$v_2 = u(d_R - \kappa) = \begin{cases} 0, & \text{if } d_R \leq \kappa \\ 1, & \text{if } d_R > \kappa \end{cases}, \tag{3}$$

where κ is the region intensity threshold. $v_2 = 0$ shows that the intensity of $\mathbf{x}_e$ is within the absolute intensity range of $R(\mathbf{x}_e)$, i.e. $f(\mathbf{x}_e) \in [m_R - \kappa, m_R + \kappa]$. $\mathbf{x}_e$ is hence classified as a region candidate. Otherwise, $\mathbf{x}_e$ is regarded as boundary candidate if $v_2 = 1$.

Many points inside the real object may be misclassified as boundary candidates with the local region competition measure during the early stage of region expansion process due to inhomogeneous local intensity variation inside the object. They may well be re-classified as region candidates with these two application dependent measures.

2.2   Local Classification Function

A classification function is introduced to combine three local boundary measures for the state classification of contour points, $h(\mathbf{x}_e) = v_0 \times v_1 \times v_2$. The function $h(\mathbf{x}_e) = 1$ indicates that the contour point $\mathbf{x}_e$ satisfies all three conditions: (1) $d_R > d_B$; (2) $d_R > \lambda \sigma_R$; (3) $d_R > \kappa$. It shows that there is significant intensity contrast between $\mathbf{x}_e$ and current local region $R(\mathbf{x}_e)$. Hence $\mathbf{x}_e$ is regarded as a boundary candidate. On the other hand, $h(\mathbf{x}_e) = 0$ implies that the intensity of $\mathbf{x}_e$ is similar to that of $R(\mathbf{x}_e)$. Therefore $\mathbf{x}_e$ is classified as a region candidate. After the state classification of all contour points, the current step contour $L(i)$ is divided into region candidate set $G(i)$ and boundary candidate set $V(i)$, i.e. $L(i) = G(i) \cup V(i)$.

2.3   Region Expansion Algorithm

In the region expansion process, some contour points will be selected to join current step region at each step. A measure called growing cost is introduced to evaluate the priority of contour points for step region expansion. The growing cost function is defined by the local contour contrast d_R of the evaluated contour point $\mathbf{x}_e$, $c_i(\mathbf{x}_e) = d_R(\mathbf{x}_e)$, where i indicates that $\mathbf{x}_e$ is located in the step contour $L(i)$. The contour point with minimum growing cost is treated as current most qualified point for region growing. It is labeled as the new region point. Obviously the step region grows along the path of minimum growing cost. The region expansion process continues until there is no contour point available to join the region.

3   Boundary Selection

The object boundary may look like discontinued at its boundary openings due to weak boundary contrast, which may lead to over-growing into the background through these boundary openings. However, compared with pseudo-boundaries, most of the object boundaries look salient because they have much more boundary points with significant boundary contrast, which are very likely classified as boundary candidates with local boundary measures.

Qiao and Ong [8], [9] developed an effective method to deal with boundary discontinuity for circle and ellipse fitting by evaluating the boundary completeness of discontinued arcs with their subtended angles. Here the idea of evaluating the boundary completeness is shared to measure the detected step contours. A global boundary measure is introduced to estimate the boundary completeness of step contour $L(i)$,

$$\rho(i) = \frac{n_V(i)}{n_L(i)}, \tag{4}$$

where $n_V(i)$ is the number of points in boundary candidate set $V(i)$ and $n_L(i)$ is the amount of contour points in $L(i)$. The global boundary measure $\rho(i)$ is the percentage of boundary candidates in the step contour $L(i)$. Obviously, the step contour with maximum global boundary measure looks most salient and complete because it has the highest percentage of boundary points with significant boundary contrast. It is therefore selected as the object boundary L. With global boundary evaluation, the object boundary can be identified even if there may be some boundary openings in it.

4   Experimental Results

The proposed approach is applied to the real medical images for evaluating its performance. The neighborhood is a 3×3 square centered at the contour point. The point adjacency is 4-connectedness.

Fig. 2(a) is an original magnetic resonance (MR) image of size 166×177. There are several brain structures in the MR image such as ventricles, caudate nucleus and putamen. In Fig. 2(b) the contours of both caudate nucleus and putamen are presented. The right structure adjacent to the ventricle is caudate nucleus while the left one is putamen. Fig. 2(c) and (d) are the results of applying canny and sobel edge detector on the MR image respectively. The background of the caudate nucleus is inhomogeneous because the ventricle is darker than it and the white matter is brighter. The boundary contrast of the putamen is very weak. The results show that both sobel and canny edge detector fail to detect the complete contours of putamen and caudate nucleus.

Fig. 3 illustrates the region expansion process for caudate nucleus segmentation with the intensity range parameter $\lambda = 3$ and the region intensity threshold $\kappa = 3$. Fig. 3(a) presents the initial step region. The step region at step 30 is described in Fig. 3(b), which is inside the caudate nucleus. Fig. 3(c) shows that the step region is over-expanded to the white matter through the bottom right corner of the caudate nucleus. In the boundary selection process, the step contour obtained at step 76 is selected as the object contour due to its maximum global boundary measure. Fig. 3(d) demonstrates the result of caudate nucleus extraction at step 76.

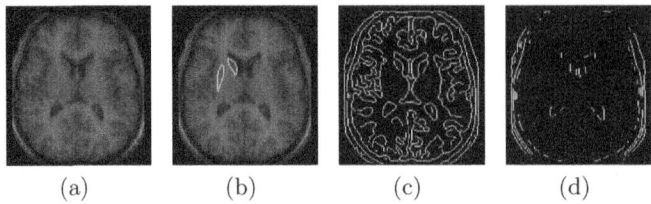

(a)   (b)   (c)   (d)

Fig. 2. MR Image: (a) original image; (b) the contour of caudate nucleus and putamen; (c) Canny edge detection; (d) Sobel edge detection

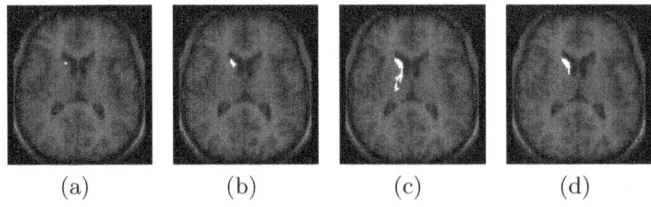

(a)   (b)   (c)   (d)

Fig. 3. Caudate nucleus segmentation: (a) initial step region obtained at step 1; (b) step region obtained at step 30; (c) step region obtained at step 150; (d) caudate nucleus extracted at step 76 with maximum global boundary measure.

Fig. 4 illustrates the examples of caudate nucleus segmentation with GVF-snake [5] and level set [6] methods. An initial model for snake method is located in Fig. 4(a), which is determined by the brain atlas registration. Fig. 4(b) shows the caudate nucleus detected by the snake method using gradient vector flow as its external force. Some parts of ventricle are misclassified as caudate nucleus. It indicates that the snake method is sensitive to the initial model. The seed point for level set method is determined in Fig. 4(c). It is actually the center of the initial model in Fig. 4(a) and is very close to the seed point in Fig. 3(a). Fig. 4(d) presents caudate nucleus segmentation obtained by applying the level set method in ITK-Snap software. The region is over-expanded to white matter through the bottom right corner of caudate nucleus at which the contrast between caudate nucleus and white matter is quite weak. It shows that the level set method may fail to prevent from region over-growing due to weak boundary contrast.

Fig. 5 presents the region growing process for putamen segmentation with the intensity range parameter $\lambda = 2$ and the region intensity threshold $\kappa = 2$. The initial step region of the seed point is located in fig. 5(a). The step region defined at step 50 is described in Fig. 5(b), which is a subset of putamen. Fig. 5(c) demonstrates that the region is over-expanded into background at step 400. Fig. 5(d) shows that the putamen is well extracted at step 109 when the global boundary measure of the corresponding step contour is maximum.

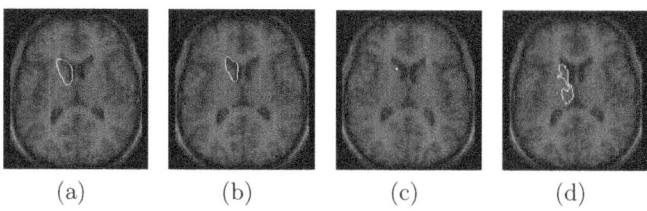

(a)   (b)   (c)   (d)

Fig. 4. Caudate nucleus segmentation with snake and level set methods: (a) the initial boundary for snake method; (b) the boundary of caudate nucleus obtained by snake method; (c) the seed point for level set method; (d) the boundary of caudate nucleus obtained by level set method

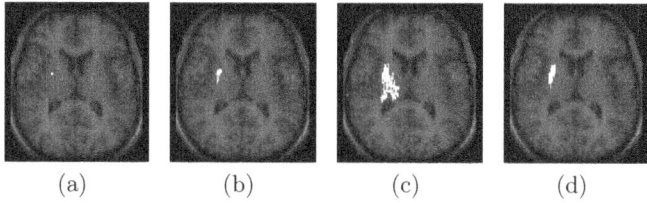

(a)   (b)   (c)   (d)

Fig. 5. Putamen segmentation: (a) initial step region obtained at step 1; (b) step region obtained at step 50; (c) step region obtained at step 400; (d) putamen extracted at step 109 with maximum global boundary measure

5   Discussion

5.1   Global Boundary Measure

The existence of boundary openings is a difficult problem for boundary detection based on the image gradient or its local intensity variation (Fig. 2(c) and (d)). Fig. 4(d) shows that the level set method may not be robust against boundary openings. The caudate nucleus is over expended into white matter through boundary openings at its right bottom corner.

The global boundary measure can well deal with boundary openings. It enables the proposed algorithm to tolerate the existence of boundary openings because it seeks the object boundary in sense of boundary completeness. The well extraction of medical structures (Fig. 3 and 5) demonstrate the the strategy of utilizing the global boundary measure for evaluating step contours is robust against boundary openings.

5.2   Local State Classification

In the region expansion process, the states of the contour points are examined with local boundary measures based on the intensity distribution of its neighborhood. Therefore the intensity information of other points outside its neighborhood have no influence on the classification of its state. The strategy of state classification based on local boundary measures may provide an effective way to deal with inhomogeneous background and overlapped intensity distribution of object and background. Besides, the state classification within the neighborhood simplifies the computational complexity of the region expansion process.

In some applications the background surrounding the object is inhomogeneous. Caudate nucleus in the real MR image is an example (Fig. 2(a)). The ventricle is darker than the caudate nucleus while the white matter is brighter. This kind of background may lead to failure of thresholding techniques [1], [2] and some seeded region growing methods [3]. In comparison, the extraction of caudate nucleus (Fig. 3) shows that state classification based on local boundary measures are robust against inhomogeneous background.

6 Conclusion

We have presented a novel region growing method characterized by the local point classification and global boundary evaluation. Its basic idea is to classify the state of contour points with local boundary measures, select most qualified point with minimum growing cost for current region expansion, and evaluate step contours with the global boundary measure. It has two novel features:

1. The local boundary measures enable the state classification of contour points to be robust against inhomogeneous background and overlapped intensity distributions.
2. The global boundary measure frees boundary selection from the influences of the weak boundary contrast and boundary openings.

Experimental results show that the algorithm can well deal with weak boundary contrast, inhomogeneous background and overlapped intensity distributions.

References

1. Kittler, J., Illingworth, J.: Minimum error thresholding. Pattern Recognit. 19(1), 41–47 (1986)
2. Qiao, Y., Hu, Q., Qian, G., Luo, S., Nowinski, W.L.: Thresholding based on variance and intensity contrast. Pattern Recognit. 40(42), 596–608 (2007)
3. Hojjatoleslami, S.A., Kittler, J.: Region Growing: A New Approach. IEEE Trans. Image Process. 7(7), 1079–1084 (1998)
4. Kass, M., Witkin, A., Terzopoulos, D.: Snakes: Active contour models. Int. J. Comput. Vis. 1(4), 321–331 (1987)
5. Xu, C., Prince, J.L.: Snakes, Shapes, and Gradient Vector Flow. IEEE Trans. Image Process. 7(3), 359–369 (1998)
6. Malladi, R., Sethian, J.A., Vemuri, B.C.: Shape Modeling with Front Propagation: A Level Set Approach. IEEE Trans. Pattern Anal. Mach. Intell. 17(2), 158–175 (1995)
7. Yan, P., Kassim, A.A.: Segmentation of volumetric MRA images by using capillary active contour. Med. Image Anal. 10(3), 317–329 (2006)
8. Qiao, Y., Ong, S.H.: Connectivity-based multiple-circle fitting. Pattern Recognit. 37(4), 755–765 (2004)
9. Qiao, Y., Ong, S.H.: Arc-based evaluation and detection of ellipses. Pattern Recognit. 40(7), 1990–2003 (2007)

Adaptive Detection of Hotspots in Thoracic Spine from Bone Scintigraphy

Qingqing Chang[1], Qiang Wang[1], Yu Qiao[1], Yuyuan Zhu[2], Gang Huang[2], and Jie Yang[1]

[1] School of Electronic, Information and Electrical Engineering, Shanghai Jiao Tong University, Shanghai 200240, China
{changqing,cnsd,qiaoyu,jieyang}@sjtu.edu.cn
[2] Department of Nuclear Medicine, Renji Hospital, Shanghai Jiao Tong University, Shanghai 200127, China
hezicase@gmail.com, huang2802@163.com

Abstract. In this paper, we propose an adaptive algorithm for the detection of hotspots in thoracic spine from bone scintigraphy. The intensity distribution of spine is firstly analyzed. The Gaussian fitting curve for the intensity distribution of thoracic spine is estimated, in which the influence of hotspots is eliminated. The accurate boundary of hotspot is delineated via adaptive region growing algorithm. Finally, a new deviation operator is proposed to train the Bayes classifier. The experiment results show that the algorithm achieve high sensitivity (97.04%) with 1.119 false detections per image for hotspot detection in thoracic spine.

Keywords: Bone scintigraphy, Hotspot detection, Gaussian fitting, Image classification.

1 Introduction

Bone scintigraphy is a useful tool in diagnosing bone diseases such as bone tumors, metabolic bone disease[1][2][3]. In clinical routine bone scanning, differences between operations, patients, and subjectivities of physicians limit the accuracy of diagnosis. It is urgent to develop an automatic recognition system, which will provide assistance for diagnosis and treatment of bone metastasis.

Yin et al[1] proposed a computer-aided diagnosis system, which chose Asymmetry and brightness as the inputs to the characteristic-point-based fuzzy inference system according to radiologists' knowledge. The sensitivity was 91.5% (227/248). Sadik [2][3][4] developed an automatic bone scan image analysis software. They extracted feature vectors from suspicious lesions and fed the vectors into an artificial neural network for classification. Huang[5] proposed a method to compute region threshold by least squares estimation of the mean and standard deviation for hotspot detection. Ohlsson [6] developed an automatic decision support system for whole-body bone scans using image analysis and artificial neural networks. The system sensitivity and specificity were respectively 95% and 64%.

B.-L. Lu, L. Zhang, and J. Kwok (Eds.): ICONIP 2011, Part I, LNCS 7062, pp. 257–264, 2011.
© Springer-Verlag Berlin Heidelberg 2011

In this paper, a new adaptive algorithm is proposed for hotspot detection in spine. First, a hotspot detection algorithm based on the intensity distribution is used to check whether hotspots exist or not. Then an edge detection algorithm which accurately delineates the boundary of hotspots is introduced. Finally, a deviation operator is calculated as the input of Bayes classifier, which can provide the probability of the bone metastasis in spine.

2   Materials and Methods

Bone scan images are collected from the Department of Nuclear Medicine, Shanghai Renji Hospital and obtained approximately $3\sim4$ hours after intravenous injection of ^{99m}Tc-MDP 20~25mCi . Each image has a resolution of 1024 *512, which each pixel 16-bit value in intensity. The clinical data are used for academic research only.

2.1   Hotspot Detection

Based on previous work[7], the spine is extracted from whole-body bone scintigraphy as Fig.1(b). And the hotspot detection method proposed in this paper is based on intensity probability distribution function(IPDF) which is found approximate to Gaussian distribution as shown in Fig.1(c). The detection proceeds in three steps. First, the intensity distribution normalization is applied to bring the various spine intensities into a reference level. Second, an optimal Gaussian fitting is employed to compute the confidence interval which can be used to obtain hotspot candidates. Finally, the boundaries of hotspots are delineated by an region growing algorithm.

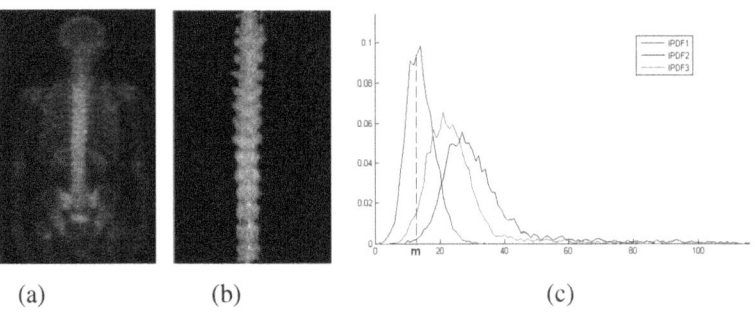

Fig. 1. (a)whole-body bone scintigraphy (b) extracted spine   (c) IPDFs of different patients

IPDF Normalization
The statistical data shows that the intensities of normal spine are relatively low, most of which range from 20 to 40. In order to eliminate the effect of hotspots, only n the normal intensity interval $[0, m]$ of the distribution is selected to obtain the mapping function as shown in (1).

$$x_1 = x_0 \cdot \frac{\mu}{m} \qquad where \; \sum_{x=0}^{m} f_0(x_0) = 0.5 \tag{1}$$

where x_0 and x_1 are respectively the original and mapped intensity, $f_0(x_0)$ is the probability of intensity x_0, and μ is the reference center of the normalized distribution , which is set to 30 according to our experiment.

Relatively, the probability mapping function is defined as:

$$f_1(x_1) = \begin{cases} f_0(x_1 \cdot \frac{m}{\mu}) \frac{m}{\mu} & m \geq \mu \\ \frac{1}{3} \sum_{k=-1}^{1} f_0((x_1 + k) \cdot \frac{m}{\mu}) \frac{m}{\mu} & m < \mu \end{cases} \tag{2}$$

Various IPDFs of different patients (see Fig.2(a)) are mapped into an approximate interval after normalization as shown in Fig.2(b). We notice that one remarkable feature of abnormal IPDF compared with normal one, is the long tail which is marked by an ellipse in the figures.

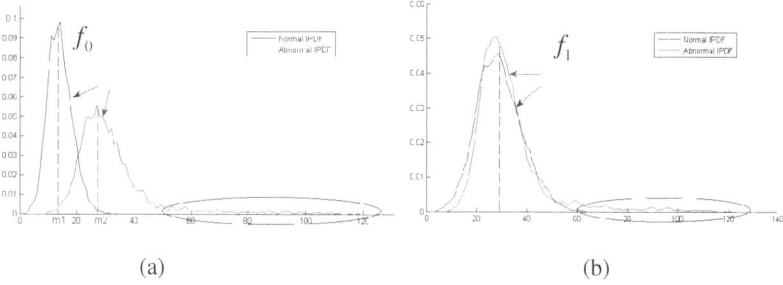

(a)   (b)

Fig. 2. Original IPDF and mapped IPDF

Hotspot Detection

Hotspot detection is based on the confidence interval which is obtained via optimal Gaussian fitting of normal IPDF model. while the pixels which are out of the confidence interval are considered as hotspot candidates. The detailed algorithm is implemented as follows.

1. Given mapped IPDF f_1, the predicted normal IPDF model f_2 is computed by:

$$f_2(x) = \begin{cases} f_1(x) & x \in (0, \mu] \\ f_1(2\mu - x) & x \in (\mu, 2\mu] \\ 0 & x \in (2\mu, x_{max}] \end{cases} \tag{3}$$

in which the part of $f_2(x)$ in the interval $[\mu, 2\mu]$ is mirrored by the part in $[0, \mu]$ to eliminate the effect of hotspots, and x_{max} is the maximum value of mapped intensity. At the same time , the deviation σ_2 of f_2 within $[0, 2\mu]$ is computed by:

$$\sigma_2 = \sqrt{\sum_{x=1}^{2\mu} f_2(x)(x-\mu)^2}$$

(4)

2. Compute the optimal Gaussian function f_{op}, which minimizes the mean square error between f_{op} and f_2.

$$f_{op} = \frac{1}{2\sigma_{op}\sqrt{\pi}} e^{\frac{-(x-\mu)^2}{4\sigma_{op}^2}}$$

(5)

$$where \quad \sigma_{op} = \arg\min \sum_{x=1}^{2\mu} (f_{op}(x) - f_2(x))^2 \quad s.t. \quad \frac{\sigma_2}{2} \leq \sigma_{op} \leq 2\sigma_2$$

(6)

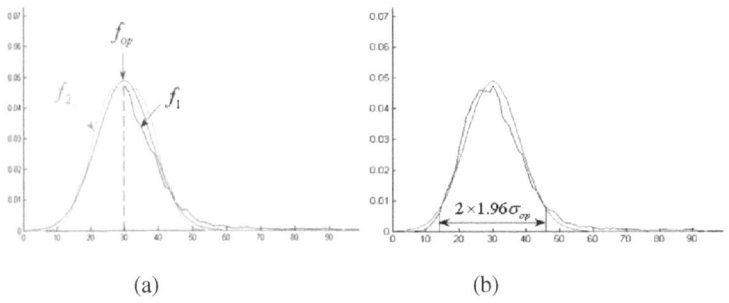

(a)   (b)

Fig. 3. The optimal Gaussian fitting of IPDF and confidence interval

3. Compute confidence interval of f_{op} with confidence level of 95%, as shown in Fig.3(b). The pixels out of this interval are considered as hotspot candidates.
4. Given the effect of noise and the anatomical knowledge of spine, connected region with area less than 10 pixels are excluded.

Hotspot Edge Detection
The edge detection of hotspots is essential for the quantification analysis. We adopt the adaptive region growing algorithm proposed by S.A.Hojjatoleslami [8]. In our implementation, the initial seed is the pixel with maximum intensity in each connected area. The maximum iteration time is set as two times the original area.

The hotspots are not fully detected as shown in Fig.4(a), since no local information is used in the detection algorithm. The adaptive region growing method algorithm is able to solve the problem by searching the maximum peripheral contrast(PC) and average contrast(AC) (Fig. 4(c)). The result of edge detection is shown in Fig. 4(b), in which the missing part is found again.

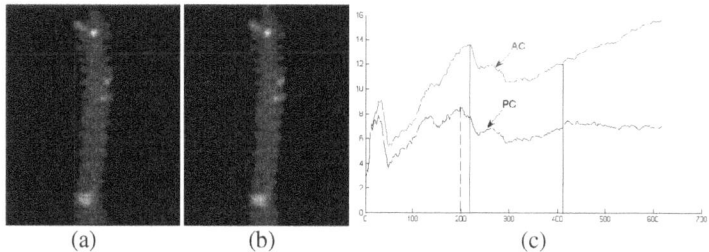

(a)   (b)   (c)

Fig. 4. The schematic drawing of hotspot detection

2.2 Classification of Spine Images

Deviation Operator
A deviation operator is proposed to measure the difference between the mapped IPDF and the optimal Gaussian fitting model in Fig.3. The deviation value is then used as the input of the Bayes classifier.

Deviation operator is defined by:

$$Dev = \ln(\sum_{x=\mu+1.96\sigma_{op}}^{x_{max}}(f_1(x)-f_{op}(x)(x-\mu)) + \ln(\sqrt{\sum_{x=\mu}^{x_{max}}f_1(x)(x-\mu)^2} - \frac{\sqrt{2}}{2}\sigma_2) \qquad (7)$$

Where $\displaystyle\sum_{x=\mu+1.96\sigma_{op}}^{x_{max}}(f_1(x)-f_{op}(x))$ is the difference between mapped IPDF $f_1(x)$

and optimal Gaussian fitting $f_{op}(x)$ out of the confidence interval; $(x-\mu)$ serves as a weight, since the further x is away from centre μ, which is of a higher probability

being pathological. Besides, $\dfrac{\sqrt{2}}{2}\sigma_2$ and $\sqrt{\displaystyle\sum_{x=\mu}^{x_{max}}f_1(x)(x-\mu)^2}$ are respectively the

deviations of left and right half of $f_1(x)$, which are constants for a given spine image. For abnormal IPDF, the right deviation is usually much larger than the left one because of the presence of hotspots. The logarithmic function reduces wide-ranging deviation values to smaller scopes.

Bayes Classifier
225 spine images, including 110 normal images and 115 abnormal ones, are used to train the Bayes classifier. The deviations of images are calculated and plotted in Fig.5(a). The dots 'o' denote deviations of normal images, and red dots '*' denote abnormal ones. The distribution curves of deviations are shown in Fig.5(b).

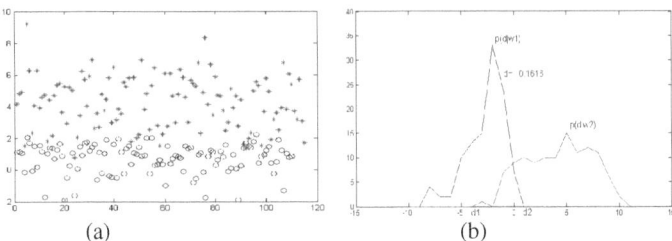

(a)   (b)

Fig. 5. The distribution of Dev

When deviation operators of newly input images are computed, the Bayes classifier divides them into two classes (normal ω_1 and pathological ω_2). The probability of the spine with deviation operator d belonging to class ω_2 is calculated as:

$$p(d\,|\,\omega_2) = \begin{cases} 0 & d \in (-\infty, d_1] \\ \dfrac{h_1(d)}{h_1(d) + h_0(d)} & d \in (d_1, d_2] \\ 1 & d \in (d_2, +\infty) \end{cases} \tag{8}$$

Where $h_0(d)$ and $h_1(d)$ are respectively distribution curves shown in Fig.5(b).

3   Results and Discussion

3.1   Results of Hotspot Detection

Our hotspot detection algorithm is tested on 327 images, 217 of which show the presence of a total 643 hotspots. 202 of the 217 patients are correctly detected, showing a sensitivity of 93.1%(202/217) in the patient level. And a total of 624 hotspots are detected with an average of 3.09 hotspots per image, which show a sensitivity of 97.04%(624/643) in the hotspot level. There are 243 false detections, which means 1.119 average false detections per image. Some typical hotspot detection results are shown in Fig. 6.

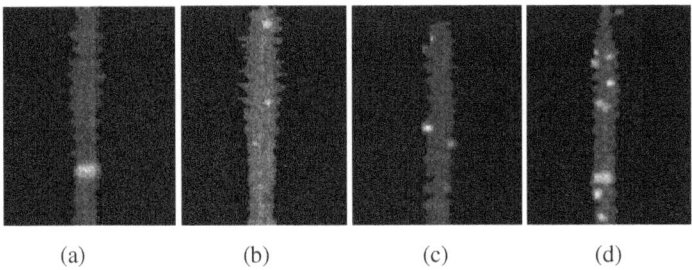

(a)   (b)   (c)   (d)

Fig. 6. Some results of hotspot detection

Fig.6(a) show the spines with single hotspot. The hotspots in Fig.6(b) have lower contrasts against their neighboring normal intensities, compared with the one in Fig.6(a). The correct detection in such sample is essential to help physicians diagnose the patients who have early-stage bone metastasis.

Fig.6(c) and Fig.6(d) show the spines with multiple hotspots, and although there are fewer hotspots in Fig.6(c), the uptakes of hotspots are different, which make it more difficult to detect. By applying the histogram-based adaptive detection algorithm, the hotspots with lower uptakes also are detected. The implementation of region growing method on each hotspot can delineate the boundary more accurately.

3.2 Comparison with Existing Methods

In this section, we compare our hotspot detection algorithm with two existing methods by Yin[2] and Huang[6]. Sensitivity (both patient level and hotspot level) and false positive are selected to evaluate the performances of different methods. The results are summarized in Table 1. We can see our method reduces the false detections while maintaining high sensitivity. Fig.7a(1) show the result of our method on one patient with single low uptake hotspot, but Huang's method and Yin's method fail to detect the hotspot due to the low contrast of the hotspot against it neighboring area as shown in Fig.7a(2). Fig7b(1-3) respectively show the results for patients with multiple hotspots by our method, Huang's method and Yin's method. Our method delineate the boundary of the hotspots more accurately for such patients.

Table 1. The comparison of hotspot detection accuracy

	Our method	Huang JY's method	Yin T K's method
Sensitivity on patient level	93.1%(202/217)	89.4%(194/217)	91.2%(198/217)
Sensitivity on hotspot level	97.1%(624/643)	90.4%(581/643)	92.4%(594/643)
False detections	243	265	271
False detections per image	1.119(243/217)	1.223(265/217)	1.249(271/217)

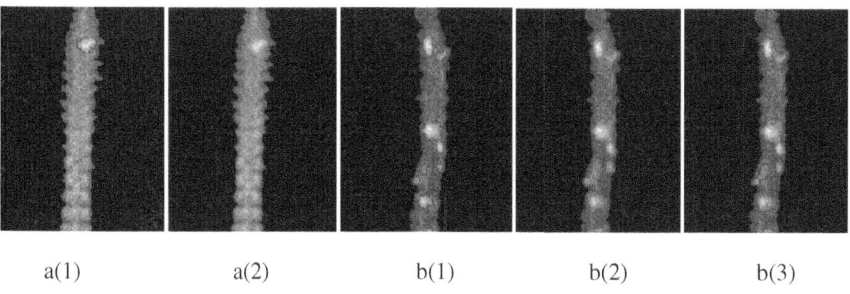

a(1)   a(2)   b(1)   b(2)   b(3)

Fig. 7. The comparison with existing methods

4 Conclusion

In this paper, an adaptive detection algorithm for bone scan spine image is developed. By applying this algorithm, we can obtain a sensitivity of 97.04% in hotspot level with an acceptable 1.119 false positive detections per image. Moreover, hotspots can be delineated accurately, which can provide a basis for quantitative analysis. Our algorithm shows better robustness and higher accuracy, and has the potential to be applied to whole body bone scan images.

References

1. Yin, T.K., Chiu, N.T.: A computer-aided diagnosis for locating abnormalities in bone scintigraphy by a fuzzy system with a three-step minimization approach. IEEE. Trans. Med. Imag. 23, 639–654 (2004)
2. Sadik, M., Jakobsson, D., Olofsson, F., Ohlsson, M., Suurkula, M., Edenbrandt, L.: A new computer-based decision-support system for the interpretation of bone scans. J. Nucl. Med. Commun. 27(5), 417–423 (2006)
3. Sadik, M., Hamadeh, I., Nordblom, P.: Computer-assisted diagnosis of planar whole-body bone scans. J. Nucl. Med. 49, 1958–1965 (2008)
4. Sadik, M., Suurkula, M., Höglund, P., Järund, A., Edenbrandt, L.: Improved Classifications of Planar Whole-Body Bone Scans Using a Computer-Assisted Diagnosis System: A Multicenter, Multiple-Reader, Multiple-Case Study. J. Nucl. Med. 50, 368–375 (2009)
5. Huang, J.Y., Kao, P.F., Chen, Y.S.: A Set of Image Processing Algorithms for Computer-Aided Diagnosis in Nuclear Medicine Whole Body Bone Scan Images. IEEE. Trans. Nucl. Sci. 54(3) (2007)
6. Ohlsson, M., Kaboteh, R., Sadik, M., Suurkula, M., Lomsky, M., Gjertsson, P., Sjostrand, K., Richter, J., Edenbrandt, L.: Automated decision support for bone scintigraphy. In: 22nd IEEE International Symposium on Computer Based Medical Systems, pp. 1–6 (2009)
7. Wang, Q., Chang, Q., Qiao, Y., Zhu, Y., Huang, G., Yang, J.: Knowledge-based Segmentation of Spine and Ribs from Bone Scintigraphy. In: Lu, B.-L., Zhang, L., Kwok, J. (eds.) ICONIP 2011, Part I. LNCS, vol. 7062, pp. 241–248. Springer, Heidelberg (2011)
8. Hojjatoleslami, S.A., Kittler, J.: Region Growing: A New Approach. IEEE Trans. Image Processing 7, 1079–1084 (1998)

ICA-Based Automatic Classification of PET Images from ADNI Database

Yang Wenlu[1,*], He Fangyu[1], Chen Xinyun[1], and Huang Xudong[2]

[1] College of Information Engineering, Shanghai Maritime University,
Shanghai, 200135, China
wenluyang@online.sh.cn
[2] Department of Radiology, Brigham and Women's Hospital and
Harvard Medical School,Boston, MA, USA
xhuang3@partners.org

Abstract. Due to the unknown pathogenesis and pathologies of Alzheimer's Disease(AD) that it brings about the serious of social problems, it is urgent to find appropriate technology for early detection of the Alzheimer's Disease. As a kind of function imaging, FDG-PET images can display lesions distribution of AD through the glucose metabolism in brain, directly reflect lesions of specific areas and the metabolic features, to diagnose and identify AD. In the paper, we propose a novel method combining Independent Component Analysis(ICA) and voxel of interest in PET images for automatic classification of AD vs healthy controls(HC) in ADNI database. The method includes four steps: preprocessing, feature extraction using ICA, selection of voxel of interest, and classification of AD vs healthy controls using Support Vector Machine(SVM). The experimental results show that the proposed method based on ICA is able to obtain the averaged accuracy of 86.78%. In addition, we selected different number of independent component for classification, achieving the average accuracy of classification results with the biggest difference only 1.47%. According to the experimental results, we can see that this method can successfully distinguish AD from healthy controls, so it is suitable for automatic classification of PET images.

Keywords: Alzheimer's Disease, Positron Emission Tomography, Independent Component Analysis, Support Vector Machine.

1 Introduction

Alzheimer's Disease(AD), also known as senile dementia, is a kind of old neurodegenerative diseases. It has pathologically and clinically unique characteristics. Post-mortem studies of AD have showed three typical lesions in AD brains: intraneuronal neurofibrillary tangles(NFTs), extracellular deposits of A amyloid plaques, and the loss of neurons[1]. Recent epidemiological data shows, until 2011, the number of global of AD reaches 35 million , increased 10 percent than

---

* To whom correspondence should be addressed.

B.-L. Lu, L. Zhang, and J. Kwok (Eds.): ICONIP 2011, Part I, LNCS 7062, pp. 265–272, 2011.
© Springer-Verlag Berlin Heidelberg 2011

the number of global of AD in 2005[2]. Whether developed or developing countries, the morbidity and mortality of dementia are growing fast. In China, for example, at present it has over eight million cases of AD. AD causes the economic burden and social issues which has been increasingly more serious, so it becomes humanity's great challenge in the 21st century.

Up to now, pathogenesis and pathologies of AD is unknown. In addition, it still has no special treatment for AD, hence early diagnosis becomes an important way to improve AD survival rate. To clinically diagnose AD patients at an early stage, many biomedical imaging techniques have been used, such as structural and functional magnetic resonance imaging(sMRI)[3][4],and positron emission tomography(PET)[5], etc. However, MR imaging is performed in the evaluation of patients who have suspected early AD, but the imaging study is neither sensitive nor specific for the diagnosis, also SPECT imaging has lower sensitivity than PET imaging, whereas, PET imaging of ^{18}F-2-fluoro-2-deoxy-d-glucose (FDG) is demonstrated to be accurate and specific in the early detection of AD[6]. As a kind of function imaging, FDG-PET images can display lesions distribution of AD through the glucose metabolism in brain, directly reflect lesions of specific areas and the metabolic features, to diagnose and identify AD. And PET images are non-invasive observation tools to assist the diagnosis, commonly used to explore the emissive nuclide in body. Relative to other imaging, PET imaging has unique advantages in early diagnosis of AD. However, how to deal with PET images, how to extract rich information containing in PET images, how to classify the useful information extracting from PET images, have become a focus of attention.

Despite these useful imaging techniques, early treatment of AD still remains a challenge because valuation of these images normally depends on manual reorientation, visual reading and semiquantitative analysis[7]. So several methods have been proposed in the literature aiming at providing an automatic tool that guides the clinician in the AD diagnosis process[8][9][10]. These methods can be classified into two categories: mono-variate and multivariate methods. Statistical parametric mapping(SPM)[11] is a mono-variate method, consisting of doing a voexl-wise statistical test and inference, with comparing the values of the image under study to the mean values of the group of normal images. This method suffers the main problem that the well-known small sample size problem, that is, the number of available samples is much lower than the number of features used in the training step. By contrast, independent component analysis(ICA) is a multivariate analysis method, an significant kind of blind signal separation, and has already applied to functional brain images[12][13][14]. It is also able to probe into PET datasets to provide useful information about the relationships among voexls. In order to distinguish AD and health controls, support vector machine(SVM), a kind of machine learning techniques, has received more attention [15][16]. In the current study, we propose a novel approach for automatic classification of PET images, which includes four steps: preprocessing using SPM, extracting features by ICA, selecting voxels of interest, and classification of AD vs healthy controls using SVM.

2 Materials and Methods

2.1 Database

In the study, data used in the preparation of this article were obtained from ADNI database(www.loni.ucla.edu/ADNI). The characteristics of the data set is showed in table 1.

Table 1. The characteristics of the data set

Group	AD	HC	Group	AD	HC
No. of subjects	80	80	CDR	1	0
Female/Male	31/49	34/46	MMSE	21.80 ± 4.21	29.01 ± 1.01
Age	76.21 ± 7.22	77.20 ± 5.26	FAQ	15.81 ± 8.32	0.55 ± 1.74

FAQ: Function Assessment Questionnaire.

2.2 Preprocessing

The original PET images were first preprocessed using Matlab toolbox SPM8. First, all PET images were realigned to the first image, making sure that all of them were consistent with each other in the spatial atlas. Then those output realigned images were normalized to a standard template PET image in SPM8. In this specific process, each image was resliced and voxel volume was set to be $2 \times 2 \times 2 mm^3$. At last, the normalized images were separated into 2 different groups and were named as group AD which included all the PET images of AD patients and group HC of healthy control.

2.3 Independent Component Analysis

To search for the source images that implied the underlying features of PET images, obviously a blind source separation problem, independent component analysis[18][19][20][21] was then come into use. ICA can be stated as follows: let X be an observed random vector and A an unknown full rank mixing matrix such that: $X=AS$, where the source signals S denote latent sources. Here since images of different subjects were already divided into the specific group, we might as well take each group as a session, a term broadly used in medical images acquisition, and another toolbox GroupICA [19] could be used. GroupICA was actually used to find out the independent components that can significantly differentiate the subjects belonging to a certain group. Using the program package GIFT finds the diversity between groups, including three steps: Constructing grouping model and setting the analysis parameters, executing independent component analysis, expanding independent component analysis results. After the whole analysis, some significant independent components were also shown in Fig 1.

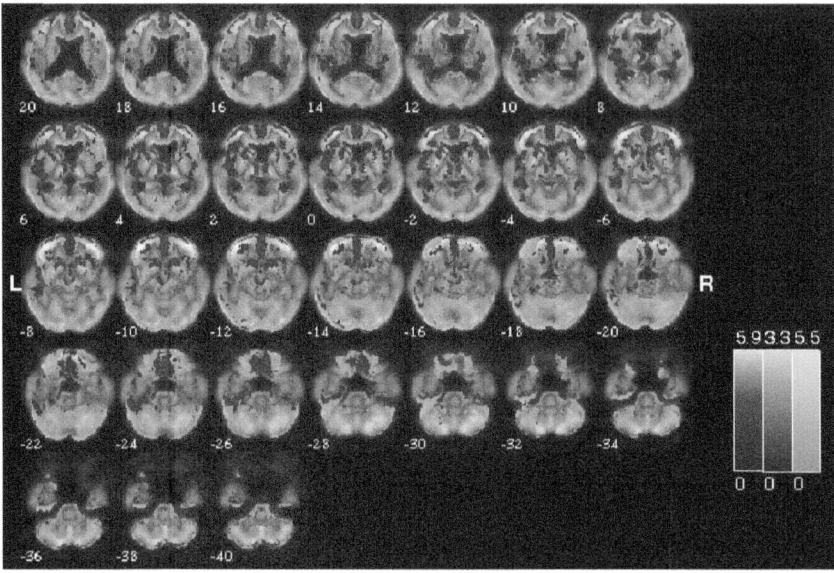

Fig. 1. The visualization of significant ICs

2.4 Feature Extraction

From the visualization of those independent components, it's not hard to see that each independent component occupied some space in the brain and different components corresponded to different areas. In order to extract features of each group of subjects, we thought about considering the voxel information in each source area. A common sense is that AD patients suffer from much worse brain structures atrophy compared to healthy control with similar background information. So the decision was made that the space area of each independent component would be mapped into every subjects, then count the numbers of the voxels whose value indicated as brain structure parts. If the number of independent components analysed to be K, each subject would correspondingly achieve K voxel numbers. So all subjects could together assemble a K-column feature matrix.

2.5 Classification

The goal of this research was set to better classify subjects from different groups from each other with a acceptable feature matrix. Since we have constructed a feature matrix in the foregoing procedures, we next would focus on classification of subjects involved in the research. we assigned 1 as the label value for the group AD components and -1 for the group HC ones. In considering of the situation that features in greater numeric ranges may dominate the classification, we scaled the features by column into the range of from -1 to 1. Support Vector Machine has showed great capacity in the field of classification and we chose LIBSVM[22] as the analysis method which was developed by Chih-JenLin of the National Taiwan University. All the subjects was halved into 2 parts, one half for training and the other half for testing. Taking the randomness of each classification trial into consideration, the classification procedure was carried out 100 times, and the final classification results including the accuracy, sensitivity and specificity were statistically calculated and demonstrated.

3   Results

3.1   Extraction of Independent Components

The PET images of this experiment are 3D images of size M=$95 \times 69 \times 79$ voxels. Each voxel represents a brain volume of $2 \times 2 \times 2 mm^3$. Every row of the source matrix was reshaped in to a 3D image called map[24]. Using GroupICA algorithm to decompose the brain images, we can obtain independent component(IC). Fig. 1 shows the three significant ICs obtained by the toolbox GroupICA. And the different color blocks represent different IC in this figure. To get more in-depth biochemical information of these sources, we transformed the coordinates of these two significant sources to the coordinates of the standard space of Talairach and Tournoux[25] by help of the Matlab conversion program developed by Mathew Brett. And the output Talairach and Tournoux coordinates of voxels were entered into TD client[26] which was created and developed by Jack Lancaster and Peter Fox. It is a high-speed database server for querying and retrieving data about human brain structure over the internet(www.talairach.org). For example, the output location information of those specific sources labeled of component 1 is listed in the following table 2. The left column of table 2 are source areas of the component while the right column the corresponding brodmann area.

Table 2. Talairach Lables of IC1

Source 1 areas	Brodmann area	Source 1 areas	Brodmann area
Superior Occipital Gyrus	19,39	Precuneus	7,19,39
Superior Parietal Gyrus	7	Cuneus	7,17,18,19
Middle Temporal Gyrus	19,39	Angular Gyrus	39
Middle Occipital Gyrus	18,19	Lingual Gyrus	17,18
Middle Frontal Gyrus	10,11,46,47	Insula	13,41
Superior Frontal Gyrus	10,11	Extra-Nuclear	CC
Inferior Frontal Gyrus	10,46	Anterior Cingulate	24,25,32
Inferior Occipital Gyrus	18	Subcallosal Gyrus	11,13,25,47
Inferior Parietal Gyrus	7,39	Caudate	CH
Transverse Temporal Gyrus	41	Lateral Ventricle	P
Superior Temporal Gyrus	13,41		

CC = Corpus Callosum; CH = Caudate Head; P = Puamen.

3.2   Classification

In this section, we randomly selected 50% of the samples to do training after we gain the feature matrix. For the randomness of the training samples, we get different results of the experiment. Therefore, we considered to classify the samples in many times to obtain the statistically averaged values. In the experiment, we repeated 100 times in classification and obtained the corresponding statistically averaged values.

Table 3 shows the classification results with the different number of IC. ACC means the accuracy of this classification, MaxA is the highest accuracy among all those 100 times results and MinA is the lowest one. Sen means the sensitivity of this classification, and Spe is the specificity of the classification. Three parameters that measure the results of the classification are: the Accuracy, the sensitivity and the Specificity, which can be described as $Accuracy = \frac{Tp+Tn}{Tp+Tn+Fp+Fn}, Sensitivity = \frac{Tp}{Tp+Fn}$, and $Specificity = \frac{Tn}{Tn+Fp}$. where Tp, Tn, Fp and Fn indicate true positives, true negatives, false positives and false negatives, respectively.

Table 3. The classification results (%)

No. of ICs	ACC	MaxA	MinA	Sen	Spe
3	85.31	93.75	75.00	71.81	98.81
4	86.23	93.75	76.56	72.75	99.71
5	86.62	98.43	73.43	74.03	99.21
6	86.78	93.75	76.56	73.88	99.69
7	85.81	92.18	76.56	71.87	99.75

4   Discussion and Conclusion

In table 3, the sensitivity of classification is not very ideal. It is possible that differences in the images due to factors not related to AD are an important source of variability that may affect classification, as scanner differences or formats. On the other hand, ADNI patient diagnostic are not pathologically confirmed, introducing some uncertainly on the subject's labels. Probably, trying to make PET images unification, which will improve the sensitivity of classification.

Our results is also comparable with other related study. Illan et al. [27] proposed an approach including three steps: training, cross-validation by means of the leave-one-out method, and test. They applied SVM based on RBF kernel to obtain classification accuracy of 88.24% by extracting features with PCA and 87.06% by ICA.

In the paper, an efficient classifier for distinguishing AD from HC has been presented. All the proposal aim at reducing the dimension of the original images, which after the reconstruction count with more than 300 000 voxels, to a number of features comparable to the number of samples, solving the so-called small sample size. The scheme combining feature extraction, feature selection and classification techniques has been deeply studied and tested on real PET database with promising results, and it can also be extended to other functional imaging modalities as SPECT or fMRI.

Acknowledgments. The authors would like to express their gratitude for the support from the Shanghai Maritime University Foundation(Grant No. 20090175), the national natural science foundation of China (Grant No. 60905065), and the research funds from BWH Radiology Department.

References

1. Gomez-ISLA, T., Spires, T., De Calignon, A., Hyman, B.T.: Neuropathology of Alzheimer's Disease. In: Hankbook of Clinical Neurology, Dementias, vol. 89, pp. 234–243 (2008)
2. Dartigues, J.F.: Altheimer's disease:a global challenge for the 21st century. Lancel Neurol. 8, 1023–1083 (2009)
3. Jack Jr., C.R., Bernstein, M.A., Fox, N.C., Thompson, P., Alexander, G., Harvey, D., Borowski, B., Britson, P.J., Whitwell, J.L., Ward, C., Dale, A.M., Felmlee, J.P., Gunter, J.L., Hill, D.L., Killiany, R., Schuff, N., Fox-Bosetti, S., Lin, C., Studholme, C., DeCarli, C.S., Krueger, G., Ward, H.A., Metzger, G.J., Scott, K.T., Mallozzi, R., Blezek, D., Levy, J., Debbins, J.P., Fleisher, A.S., Albert, M., Green, R., Bartzokis, G., Glover, G., Mugler, J., Weiner, M.W.: The Alzheimer's Disease Neuroimaging Initiative (ADNI): MRI methods. J. Magn. Reson. Imaging 27, 685–691 (2008)
4. Sperling, R.: Functional MRI studies of associative encoding in normal aging, mild cognitive impairment, and Alzheimer's disease. Imaging and the Aging Brain 1097, 146–155 (2007)
5. Walhovd, K.B., Fjell, A.M., Brewer, J., McEvoy, L.K., Fennema-Notestine, C., Hagler Jr., D.J., Jennings, R.G., Karow, D., Dale, A.M.: Combining MR Imaging, Positron-Emission Tomography, and CSF Biomarkers in the Diagnosis and Prognosis of Alzheimer Disease. AJNR Am. J. Neuroradiol. (2010)
6. Mosconi, L.: Brain glucose metabolism in the early and specific diagnosis of Alzheimer's disease. Eur. J. Nucl. Med. Mol. Imaging 32, 486–510 (2005)
7. Lopez, M., Ramirez, J., Gorriz, J.M., Alvarez, I., Salas-Gonzalez, D., Segovia, F., Chaves, R., Padilla, P., Gomez-Rio, M.: The Alzheimer's Disease Neuroimaging Initiative. Principal Component Analysis-Based Techniques and Supervised Classification Schemes for the Early Detection of Alzheimer's Disease 74, 1260–1271 (2011)
8. Gorriz, J.M., Ramirez, J., Lassl, A., Salas-Gonzalez, D., Lang, E.W., Puntonet, C.G., Alvarez, I., Lopez, M., Gomez-Rio, M.: Automatic computer aided diagnosis tool using component-based SVM. In: IEEE Nuclear Science Symposium Conference Record, Medical Imaging Conference, pp. 4392–4395 (2008)
9. Salas-Gonzalez, D., Gorriz, J.M., Ramirez, J., Lopez, M., Illan, I.A., Puntonet, C.G., Gomez-Rio, M.: Analysis of SPECT brain images for the diagnosis of Alzheimer's disease using moments and support vector machines. Neuroscience Letters 461, 60–64 (2009)
10. Ramirez, J., Gorriz, J., Salas-Gonzalez, D., Romero, A., Lopez, M., Alvarez, I., Gomez-Rio, M.: Computer-aided diagnosis of alzheimer's type dementia combining support vector machines and discriminant set of features. Information Sciences 05, 012 (2009)

11. Friston, K.J., Ashburner, J., Kiebel, S.J., Nichols, T.E., Penny, W.D.: Statistical Parametric Mapping: The Analysis of Functional Brain Images. Academic Press (2007)
12. Xu, L., Groth, K.M., et al.: Source-based morphometry: the use of independent component analysis to identify gray matter differences with application to schizophrenia. Hum. Brain Mapp. 30(3), 711–724 (2009)
13. Fink, F., Worle, K., Guber, P., Tome, A.M., Gorriz, J.M., Puntonet, C.G., Lang, E.W.: ICA analysis of retina images for glaucoma classification. In: 30th Annual International Conference of the IEEE Engineering in Medicine and Biology Society, pp. 4664–4667 (2008)
14. Górriz, J.M., Puntonet, C.G., Salmerón, M., Rojas Ruiz, F.: Hybridizing Genetic Algorithms with ICA in Higher Dimension. In: Puntonet, C.G., Prieto, A.G. (eds.) ICA 2004. LNCS, vol. 3195, pp. 414–421. Springer, Heidelberg (2004)
15. Magnin, B., Mesrob, L., et al.: Support vector machine-based classification of Alzheimer's disease from whole-brain anatomical MRI. Neuroradiology 51(2), 73–83 (2009)
16. Kloppel, S., Stonnington, C.M., et al.: Automatic classification of MR scans in Alzheimer's disease. Brain 131(Pt 3), 681–689 (2008)
17. Ashburner, J.: Computational anatomy with the SPM software. Magnetic Resonance Imaging 27, 1163–1174 (2009)
18. Calhoun, V., Pekar, J.: When and where are components independent? On the applicability of spatial and temporal ICA to functional MRI data. Hum. Brain Mapp. (2000)
19. Calhoun, V., et al.: A Method for Making Group Inferences Using Independent Component Analysis of Functional MRI Data: Exploring the Visual System (2001)
20. Calhoun, V.D., et al.: ICA of Functional MRI Data: An Overview. In: 4th International Symposium on Independent Component Analysis and Blind Signal Seperation (ICA 2003) (2003)
21. Li, Y.-O., Adal, T.l., Calhoun, V.D.: Estimating the Number of Independent Components for Functional Magnetic Resonance Imaging Data. Hum. Brain Mapp. 28, 1251–1266 (2007)
22. Hsu, C.-W., Chang, C.-C., Lin, C.-J.: A Practical Guide to Support Vector Classification (2009)
23. Yuan, Y., Gu, Z.X., Wei, W.S.: Fluorodeoxyglucose-Positron-Emission Tomography, Single-Photon Emission Tomography, and Structural MR Imaging for Prediction of Rapid Conversion to Alzheimer Disease in Patients with Mild Cognitive Impairment: A Meta-Analysis. American Journal of Neuroradiology 30(2), 404–410 (2009)
24. Calhoun, V., Adali, T., Pearlson, G., Pekar, J.: Group ICA of Functional MRI Data: Separability, Stationarity, and Inference (2007)
25. Lancaster, J.L., Rainey, L.H., Summerlin, J.L., Freitas, C.S., Fox, P.T., Evans, A.C., Toga, A.W., Mazziotta, J.C.: Automated labeling of the human brain: A preliminary report on the development and evaluation of a forwardtransform method. Human Brain Mapping 5(4), 238–242 (1997)
26. Lancaster, J.L., Woldorff, M.G., Parsons, L.M., Liotti, M., Freitas, C.S., Rainey, L., Kochunov, P.V., Nickerson, D., Mikiten, S.A., Fox, P.T.: Automated Talairach atlas labels for functional brain mapping. Human Brain Mapping 10, 120–131 (2000)
27. Illan, I.A., Gorriz, J.M., Ramrez, J., Salas-Gonzalez, D., Lopez, M.M., Segovia, F., et al.: 18F-FDG PET imaging analysis for computer aided Alzheimer's diagnosis. Information Sciences 181(4), 903–916 (2011)

A Novel Combination of Time Phase and EEG Frequency Components for SSVEP-Based BCI

Jing Jin[1], Yu Zhang[1,2], and Xingyu Wang[1]

[1] School of Information Science and Engineering,
East China University of Science and Technology, Shanghai, China
[2] Laboratory for Advanced Brain Signal Processing,
RIKEN, Brain Science Institute, Wako-shi, Saitama, Japan
`jinjing@ecust.edu.cn, yuzhang@brain.riken.jp, xywang@ecust.edu.cn`

Abstract. The steady-state visual evoked potential (SSVEP) has been widely applied in brain-computer interfaces (BCIs), such as letter or icon selection and device control. Most of these BCIs used different flickering frequencies to evoke SSVEP with different frequency components that were used as control commands. In this paper, a novel method combining the time phase and EEG frequency components is presented and validated with nine healthy subjects. In this method, four different frequency components of EEG were classified out from four time phases. When the SSVEP is evoked and what is the frequency of the SSVEP is determined by the linear discriminant analysis (LDA) classifier in the same time to locate the target image. The results from offline analysis show that this method yields good performance both in classification accuracy and information transfer rate (ITR).

Keywords: Brain-computer interfaces (BCIs), Electroencephalogram (EEG), Steady-state visual evoked potential (SSVEP), Time phase.

1 Introduction

Brain-computer interface (BCI) is a kind of communication system that allows a connection between the human brain and a computer [1]. According to the different types of brain activities used as command, many kinds of BCIs are available. Most of BCIs are based on four types of EEG potentials: (1) slow cortical potentials (SCP). (2) P300 potentials. (3) event related desynchronization/event related synchronization (ERD/ERS, motor imagery). (4) steady-state visual evoked potential (SSVEP).

In recent years, SSVEP-based BCIs have been increasingly studied and applied in many aspects, such as letter or icon selection, cursor movement and device control, due to the shorter calibration time and higher information transfer rate (ITR) than other types of BCIs [2], [3]. SSVEP evoked by repetition flicker stimulation has the same fundamental frequency as the stimulus and may also include higher harmonics [4]. The target stimulus on which the subject is focusing can be located by classifying the frequency components of recorded

B.-L. Lu, L. Zhang, and J. Kwok (Eds.): ICONIP 2011, Part I, LNCS 7062, pp. 273–278, 2011.
© Springer-Verlag Berlin Heidelberg 2011

EEG [5]. A 48-target environment controller based on the SSVEPs evoked by 48 flickering LEDs respectively was presented by Gao et al. [6]. The 48 LEDs were installed on a circuit board in a matrix way and flickered simultaneously. The average classification accuracy is 87.5 % and the information transfer rate is up to 68 bits/min. Although the exhibited transfer rate is relatively high when compared to the state of the other results [4], [5], such performance is not typical for most users in real world settings [7]. A simultaneous classification for so many targets requires a considerable number of stimulus frequencies and also may be difficult to achieve high classification accuracy. It is because that the user may be affected by the neighbor LEDs that are flickering simultaneously with other frequencies. Furthermore, the classification accuracy usually appears to decrease rapidly as the number of classes increase in a multi-classification problem [8]. Recently, Friman et al. [5] reported a spelling (letter selection) paradigm with the SSVEPs evoked by ten LEDs located on the top of rows and columns. A 55 row-column (i.e., a matrix way) paradigm similar with a traditional P300 spelling paradigm was used in their research. Two subjects can hardly complete any assignment for their weak SSVEP responses. The paradigm of simultaneous flickers for all stimulus used in [5], [6], [9] should be cautiously considered since it is difficult to prevent subjects from the mutual interferences among different stimuli installed with a limited distance. In order to avoid excessive requirement for the number of stimulus frequencies and the mutual interferences among different stimulus flickers, we present a time phase-based flicker paradigm with four different flickering frequencies. In this method, the four different frequency components of EEG are classified out from the four time phases (TPs).

2   Material and Methods

2.1   Experimental Paradigm and EEG Acquisition

In this study, a new stimulus layout design (see Fig. 1(a)) is presented. 16 icons are assigned to four groups and located in four sides of the monitor window respectively, which is different from the traditional matrix layout [5], [9], [10]. The icons show an Internet Explore, a Folder, a Email and other commonly used computer tools. SSVEP will be evoked when user is focusing on the red flicker beside the target image. The distance among the four flickering squares in each TP are large enough to avoid the interferences. An explanation for this layout is depicted in Fig. 1(b). A flicker cycle of our paradigm contains four time phases, namely TP1~TP4 (see Fig. 2), and the duration of each TP is 7 seconds. The white circles show the flickering positions in the each TP. The flicker frequency is same in the same side. The flickering frequency of the different sides are different as "f_1: 6.9 Hz (Up side), f_2: 8.9 Hz (Right side), f_3: 10.8 Hz (Down side) and f_4: 12.8 Hz (Left side)". The subject would be required to focus on the flickering square beside the target image.

Nine healthy right-handed volunteers (aged 21~28 years old, six males, and three females) with normal or corrected to normal vision participated in our study. During the whole experiment, all subjects were seated in a comfortable

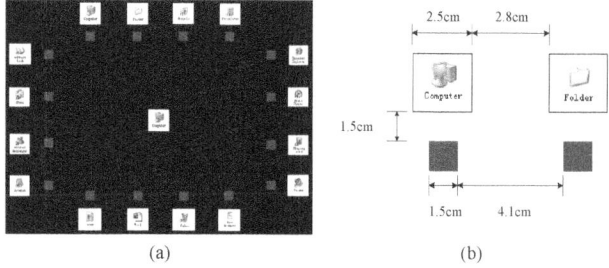

Fig. 1. Experimental icons and the distance and the size of the icons

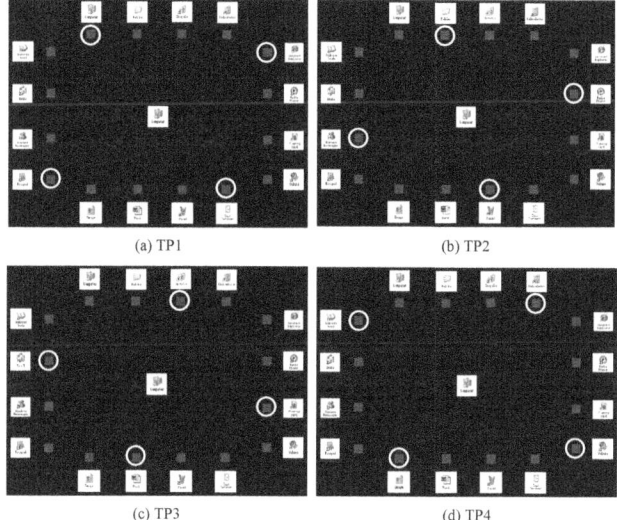

Fig. 2. Experimental flicker paradigm. TP1~TP4 respectively corresponds to the four flicker sequences in a flicker cycle.

chair with 50 cm distance from a standard 17 inch CRT monitor in a shield room. Each subject completed four recording sessions. The time required to select an image was called a run, and there were 32 runs for two selections of each of the 16 icons. Each run began with a target cue of 1 s, and then the flicker would start according to the TP paradigm in Fig. 2.

EEG was recorded at 250 Hz ($f_s = 250$). Twelve Ag/AgCl electrodes were placed on Fz, C3, Cz, C4, T5, P3, Pz, P4, T6, O1, Oz and O2 according to the standard positions of the 10-20 international system. The average signal from two mastoid electrodes was used for referencing and grand was at the forehead. A Nuamps (NuAmp, Neuroscan, Inc.) amplifier 100uv was used. The EEG was filtered with a band-pass of 5~35 Hz.

2.2   Feature Extraction and Classification

Each TP is divided into seven time windows (TWs) (1 s∼7 s). The frequency components are extracted by estimating the power spectra density (PSD) ($4 \times f_s$ points fast Fourier transform (FFT)) of the EEG signals. For each TW, the sum of the normalized fundamental stimulus frequency and its second harmonic components with a bandwidth of 0.5 Hz is calculated from each channel, and then would be taken as features for detect SSVEP. The classes for the four different stimulus frequencies are labeled as 1∼4 respectively and non-target class is labeled as class 5. For each subject, 32 target trials (8 trials for each class) and 96 non-target trials are obtained in each session. The train data contain 128 target trials and 384 non-target trials. Here the one-against-rest method is used to classify. There are four sub-classifiers to detect the four different frequency components respectively. The output of the classifiers is "0" or "1". If the corresponding frequency component is detected, the output would be "1", and if not, the output would be "0" (see Fig. 3). Four-fold cross-validation is used to estimate the average classification accuracy of each subject. The information transfer rate (ITR) which is computed by the definition of Wolpaw et al. [1] is used to evaluate the communication performance of the BCI system.

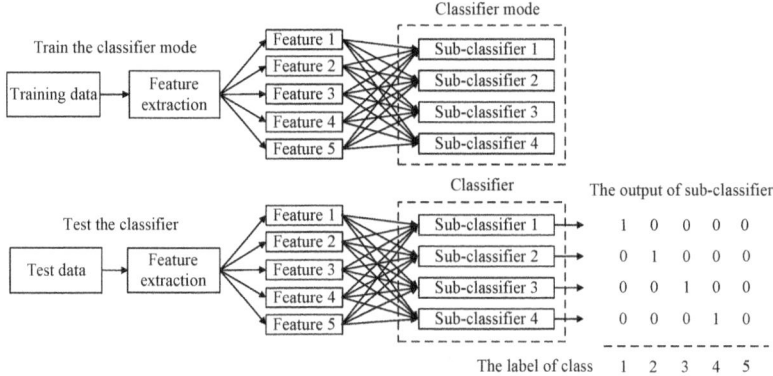

Fig. 3. The method of training and testing the classifier

3   Results

The target selecting accuracy averaged over sessions and the corresponding ITR are shown in Fig. 4. All of the subjects, except for subject 5, achieved an accuracy which is not less than 85 %. Subject 4 is the best, with the average accuracy of 92.5 % and the ITR of 35.1 bits/min. Subject 5 is the worst, with the average accuracy of 68.1 % and the ITR of 9.4 bits/min. The highest ITRs of all subjects with corresponding accuracies and TWs are listed in Table 1. A averaged performance, accuracy of 86.1 % and ITR of 26.7 bits/min within time window length of 3.4 s, is obtained by the proposed method.

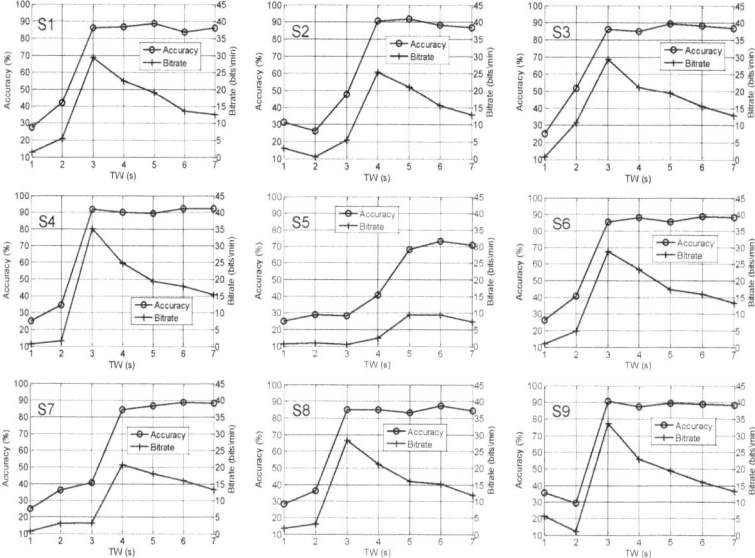

Fig. 4. Classification accuracy and information transfer rate across different time window lengths (TWs)

Table 1. The highest information transfer rate (ITR) (bits/min) with the corresponding accuracy (Acc) (%) and corresponding time window length (TW) (s), averaged on the four sessions

	S1	S2	S3	S4	S5	S6	S7	S8	S9	Av.
Acc	86.3	91.9	86.3	92.5	68.1	85.6	88.8	85.0	90.6	86.1±7.3
ITR	29.4	25.3	29.4	35.1	9.4	28.8	20.8	28.2	33.7	26.7±7.7
TW	3	4	3	3	5	3	4	3	3	3.4±0.7

4   Discussion and Conclusion

In this study, a novel method combining the time phase and EEG frequency components was presented. Compared with the matrix flickering paradigm used in the multi-command SSVEP-based BCIs [5], [6], [9], this new design was valuable in reducing the number of stimulus frequencies and the interferences among different stimulus frequencies. From the results, all of the subjects, except for subject 5, the target selecting accuracy were higher than 85 % and the ITR was over 20 bits/min. As shown in table 1, the average highest accuracy of 86.1 % and ITR of 26.7 bits/min within 3.4 s were achieved by our method. It is promising to apply the method of this paper in online BCIs. In online system, the system will continuously repeat the flickering cycle described in section 2.2. When one of the four frequency components was detected, the TP and the frequency were determined based on the classifier output and the target icon would be located. To locate the next target icon, if the target's TP was behind the previous target's TP, the user would not need to wait for the next flickering cycle and could continue to focus on the next target, then the target would be located in the same flicker cycle. In this way, the information transfer rate would be improved further. In the future work, an online system based on this method will be developed and this method should be validated with real world settings.

Acknowledgments. This study is supported by Nation Nature Science Foundation of China 61074113, Shanghai Leading Academic Discipline Project B504, and Fundamental Research Funds for the Central Universities WH0914028.

References

1. Wolpaw, J.R., Birbaumer, N., McFarland, D.J., Pfurtscheller, G., Vaughan, T.M.: Brain-computer interfaces for communication and control. Clin. Neurophysiol. 113, 767–791 (2002)
2. Zhu, D.H., Bieger, J., Molina, G.G., Aarts, R.M.: A survey of stimulation methods used in SSVEP-based BCIs. Comput. Intell. Neurosci. (2010), doi: 10.1155/2010/702357
3. Zhang, Y., Jin, J., Qing, X., Wang, B., Wang, X.: LASSO based stimulus frequency recognition model for SSVEP BCIs. Biomed. Signal Process. Control (2011), doi:10.1016/j.bspc.2011.02.002
4. Müller-Putz, G.R., Scherer, R., Brauneis, C., Pfurtscheller, G.: Steady-state visual evoked potential (SSVEP)-based communication: impact of harmonic frequency components. J. Neural Eng. 2, 123–130 (2005)
5. Friman, O., Lüth, T., Volosyak, I., Gräser, A.: Spelling with steady-state visual evoked potentials. In: Proceedings of the 3rd International IEEE/EMBS Conference on Neural Engineering (CNE 2007), Hawaii, May 2-5, pp. 510–523 (2007)
6. Gao, X., Xu, D., Cheng, M., Gao, S.: A BCI-based environmental controller for the motion-disabled. IEEE Trans. Neural Syst. Rehabi. Eng. 11, 137–140 (2003)
7. Graimann, B., Allison, B., Mandel, C., Lüth, T., Valbuena, D., Gräser, A.: Non-invasive brain-computer interfaces for semi-autonomous assistive devices. In: Schuster, A. (ed.) Robust Intelligent System, pp. 113–138. Springer, London (2008)
8. Li, T., Zhang, C., Ogihara, M.: A comparative study of feature selection and multiclass classification methods for tissue classification based on gene expression. Bioinformatics 20, 2429–2437 (2004)
9. Cheng, M., Gao, X., Gao, S., Xu, D.: Design and implementation of a brain-computer interface with high transfer rates. IEEE Trans. Biomed. Eng. 49, 1181–1186 (2002)
10. Wang, Y., Wang, R., Gao, X., Hong, B., Gao, S.: A practical VEP-based brain-computer interface. IEEE Trans. Neural Syst. Rehab. Eng. 14, 234–239 (2006)

A Novel Oddball Paradigm for Affective BCIs Using Emotional Faces as Stimuli

Qibin Zhao[1], Akinari Onishi[1], Yu Zhang[1,2], Jianting Cao[3], Liqing Zhang[4], and Andrzej Cichocki[1]

[1] Laboratory for Advanced Brain Signal Processing, Brain Science Institute, RIKEN, Saitama, Japan
[2] School of Control Science and Engineering, East China of Science and Technology, Shanghai, China
[3] Saitama Institute of Technology, Saitama, Japan
[4] MOE-Microsoft Laboratory for Intelligent Computing and Intelligent Systems, Department of Computer Science and Engineering, Shanghai Jiao Tong University, Shanghai, China

Abstract. The studies of P300-based brain computer interfaces (BCIs) have demonstrated that visual attention to an oddball event can enhance the event-related potential (ERP) time-locked to this event. However, it was unclear that whether the more sophisticated face-evoked potentials can also be modulated by related mental tasks. This study examined ERP responses to objects, faces, and emotional faces when subjects performs visual attention, face recognition and categorization of emotional facial expressions respectively in an oddball paradigm. The results revealed the significant difference between target and non-target ERPs for each paradigm. Furthermore, the significant difference among three mental tasks was observed for vertex-positive potential (VPP) ($p < 0.01$), late positive potential (LPP) / P3b ($p < 0.05$) at the centro-parietal regions and N250 ($p < 0.003$) at the occipito-temporal regions. The high classification performance for single-trial emotional face-related ERP demonstrated facial emotion processing can be used as a novel oddball paradigm for the affective BCIs.

Keywords: Brain Computer Interface (BCI), Event-Related Potential (ERP), P300.

1 Introduction

Brain computer interfaces (BCIs) are communication systems that enable the direct communication between human and computers through decoding of brain activity [1], which can be used to assist patients who have disabled motor functions. The P300 speller is one of the most popular BCI paradigm first introduced by Farwell and Donchin [2]. The P300 ERP elicited when users attend to an oddball stimulus, i.e., a random series of stimulus events that contains an infrequently presented target, is a positive deflection occurring at 300-500 millisecond (ms) post-stimulus over parietal cortex. This is usually done by performing a

B.-L. Lu, L. Zhang, and J. Kwok (Eds.): ICONIP 2011, Part I, LNCS 7062, pp. 279–286, 2011.
© Springer-Verlag Berlin Heidelberg 2011

mental count of the number of times the target character is highlighted, implying the fact that the neural processing of a stimulus can be modulated by attention [3]. So far, a number of variations of P300 speller have been explored such as an apparent motion and color onset paradigm [4], the checkerboard paradigm [5] and the auditory oddball ERP [6]. Although the speed and accuracy of P300 BCIs have been significantly improved by various signal processing methods [7,8], the single-trial classification of P300 ERP remains a challenging problem. Recent studies in neuroscience showed that larger N170 ERP is elicited in response to facial stimuli than non-face objects and scrambled faces [9], face-selective N250r is elicited by immediate repetitions of faces [10,11,12]. Emotional face type and anxiety modulated ERP responses are also investigated and divided into three stages around 200 ms, 250 ms, and 320 ms [13,14]. The neural processes involved in switching associations formed with angry and happy faces diverged 375 ms after stimulus onset [15]. The early posterior negativity (EPN) and late positive potentials (LPP) related to emotional processing were enhanced when the subjects seeing a fearful face compared to a neutral face [16].

In contrast to highlighting letters in the classical P300 speller, the present study investigates the three oddball BCI paradigms utilizing randomly flashed images of objects, faces and emotional faces [17]. The subjects were requested to perform three different mental tasks, i.e., visual attention, face recognition (identification), emotion discrimination, corresponding to three types of images. The main objective was to find the ERP waveforms elicited by oddball faces or emotional faces stimuli and whether it is feasible to apply face-related ERPs for BCI paradigm. Furthermore, the amplitude and latency of ERPs under these three paradigms as well as classification performance were compared.

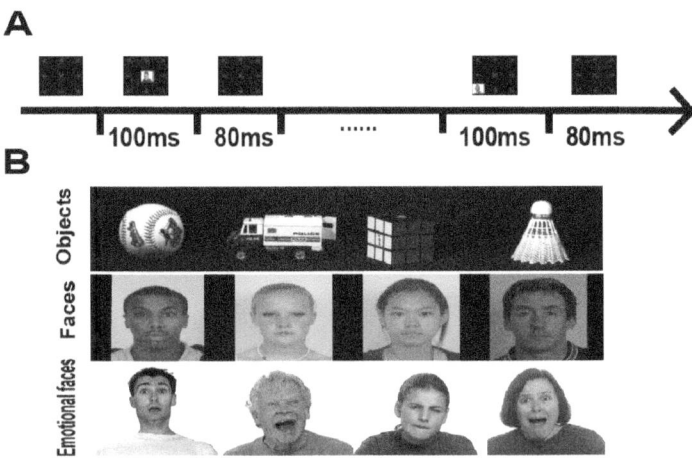

Fig. 1. A. The procedure of BCI paradigm. The stimuli were shown for 100 ms each with an ISI of 80 ms. B. Three groups of images were used as stimuli corresponding to three different experimental conditions.

2   Methods

Subjects and data collection: Five male subjects aged 25-31 years participated in the study. All participants were healthy, right-handed, and had normal or corrected to normal vision. We recorded the EEG from 16 electrodes (F5, Fz, F6, T7, C5, Cz, C6, T8, P7, P5, Pz, P6, P8, PO7, Oz, PO8) using an 16-channel amplifier (g.tec, Guger Technology, Austria). The left mastoid and Fpz served as reference and ground, respectively. The EEG signals were sampled at 256Hz and band-pass filtered to 0.1-100 Hz with a notch filter of 50 Hz.

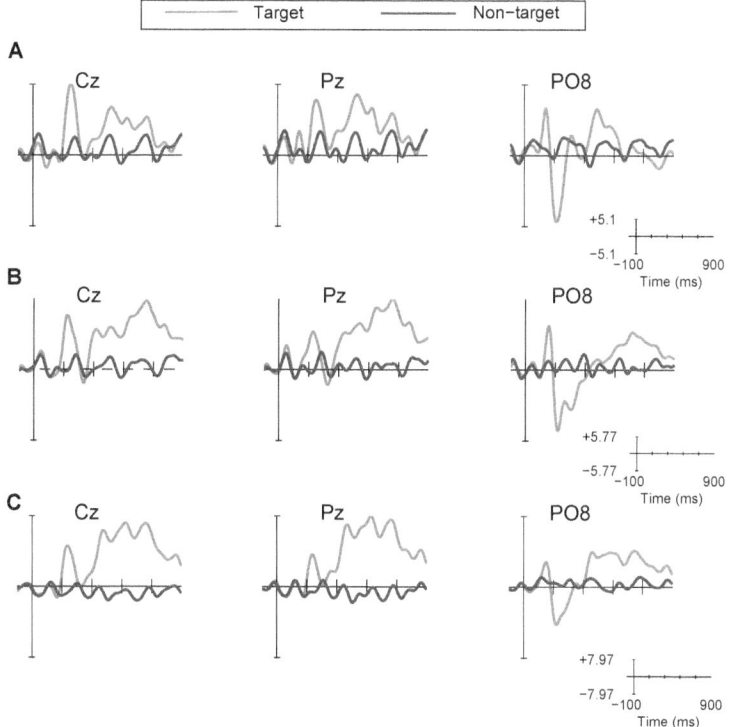

Fig. 2. Grand averaged ERPs at Cz over fronto-central region, Pz over centro-parietal region and PO8 over occipito-temporal region to target (green) and non-target (red) objects stimuli. ERPs using objects, faces, emotional faces are shown in panel (A), (B) and (C), respectively.

Procedure: Subjects were seated in a comfortable chair and the screen presented a 3 × 3 matrix of 9 arrows with gray color and black background (see Fig. 1A), corresponding to the 9 commands for the concrete BCI application. We collected data under three experimental conditions. In condition 1, the subjects were asked to focus on the target item and count the number of flashing, instead of highlighting the target arrow, the images from objects group were

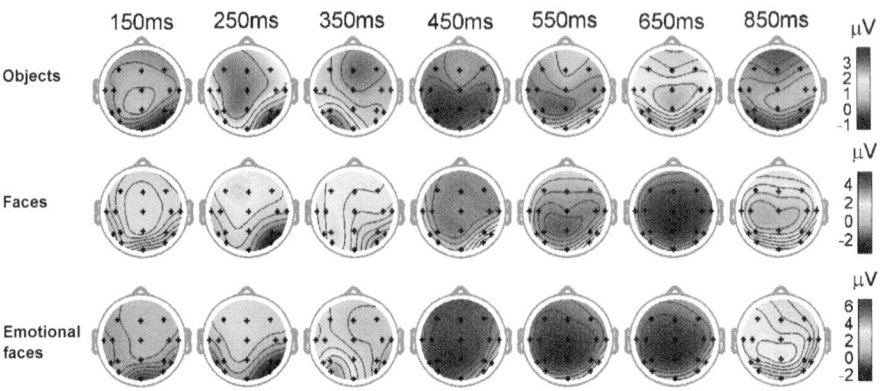

Fig. 3. Topography of target ERPs for specific time points using three different stimuli and mental tasks

shown randomly at each of 9 positions; In condition 2, the images from faces group were utilized for flashed targets and the subjects were asked to perform the face recognition tasks; In condition 3, the images from emotional faces group were presented as flashed targets and the subjects were asked to perform emotion discrimination tasks whenever the desired target is intensified. The procedure and the images groups are shown in Fig. 1. The subjects performed two sessions for each experimental condition. Each session consists of 5 runs and each run presented 9 different target items in succession using only 2 flashes for all items in random order.

Feature extraction: We first filtered the EEG signals between 0.1 and 20 Hz, then rejected bad trials by utilizing a threshold for amplitude detection. The one second (s) time window of EEG after each stimulus onset was extracted as a trial and the baseline of each EEG trial was corrected using 100 ms pre-stimulus interval. After moving average and downsampled to 20 Hz, the feature vector consisted of 16 spatial features × 20 temporal features = 320 spatio-temporal features.

Channels selection: The number of channels applied in BCI classification can be reduced by the appropriate channels selection method. By employing the channels selection method, we can faciliate the installation of EEG scalp and, in the meantime, the number of features could be reduced, indicating that smaller training samples is sufficient to build a classifier. In this study, we utilized the point-biserial correlation coefficient r^2-value which can be used for evaluating the discriminative ability of spatio-temporal features for channels selection.

Classification: The support vector machine (SVM), which has been widely used for ERP classifications, was adopted in this study to perform target detection. The principle of the SVM is to seek the maximal margin between two classes, to form the hyper-plan with the best generalization capabilities. In our study, we employed a linear kernel SVM, and chose the best parameters for each subject by 5-fold cross-validation from the offline data. Given a set of ERPs

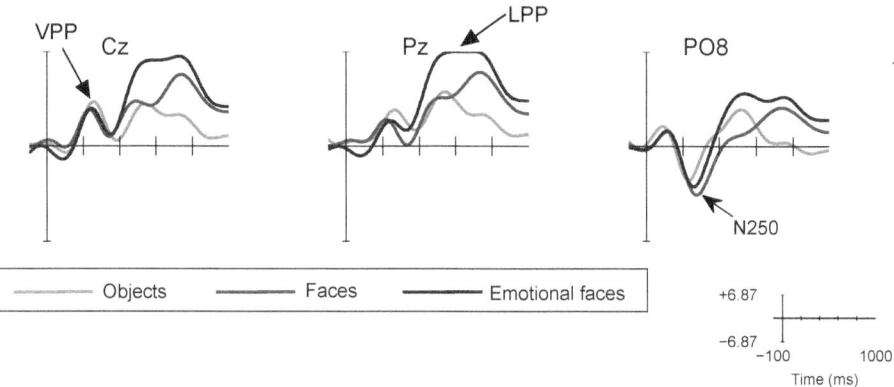

Fig. 4. Grand averaged ERPs at Cz, Pz over centro-parietal region and PO8 over occipito-temporal region to objects, faces, and emotional faces stimuli

data $(\mathbf{x}_i | i = 1, \ldots, n)$ and corresponding labels y_i, e.g., target $(+1)$ or nontarget (-1), the objective function of linear SVM is to maximize the margin, which is formulated as

$$\max_{\mathbf{w}, w_0} \frac{1}{2} \|\mathbf{w}\|^2$$
$$s.t. \ y_i(\mathbf{w} \cdot \mathbf{x}_i + w_0) \geq 1, i = 1, \ldots, n. \tag{1}$$

3   Results

For each experimental condition, grand-averaged ERPs were calculated separately for target and non-target events. We focus on key components of ERPs elicited by faces such as the face-specific N170 (150-190 ms), VPP (140-200 ms), N250 (240-280 ms), P300 (250-350 ms), P3b/LPP (400-800 ms). Early components are thought to reflect basic structural encoding of faces, whereas later components may reflect categorization and attention to motivationally relevant information, including emotion, gender, or identity. Thus, different experiment conditions may evoke different responses.

Attention task: The Fig. 2A shows the grand-averaged ERPs for target and non-target stimulus while Fig. 3 shows the topography of corresponding ERPs. VPP at Cz is clearly different between target and non-target $(F(1,18)= 8.08, p < 0.01)$, P300 at Cz is also significant $(F(1,18)=13.41, p < 0.002)$. LPP at Cz for target is larger than non-target $(F(1,18)=5.41, p < 0.032)$. The main effect of attention tasks is also significant for VPP at Pz $(F(1,18)= 5.47, p < 0.03)$, P300 $(F(1,18)= 17.4, p < 0.0006)$ and LPP $(F(1,18)=9.76, p < 0.006)$.

Face identification task: The ERPs elicited by faces stimuli are shown in Fig. 2B while topography map of ERPs are shown in Fig. 3. We observed the significant VPP at Cz $(F(1,18)= 14.15, p < 0.02)$, P300 at Cz $(F(1,18)= 24.29, p < 0.0001)$ and LPP at Cz $(F(1,18)= 7.98, p < 0.012)$, indicating the effects of face identification task for oddball paradigm. The similar difference are also clear at Pz such as VPP (F(1,18)= 13.71, $p < 0.02$), P300 (F(1,18)= 12.84, $p < 0.0022$) and LPP (F(1,18)= 17.2, $p < 0.0006$). Due to the faces stimulus, the N170 at PO8 is clearly observed (F(1,18)= 6.37, $p < 0.02$).

Emotional face discrimination task: The ERPs elicited by emotional faces stimuli are shown in Fig. 2C while topography map of ERPs are shown in Fig. 3. Analysis of VPP amplitude revealed a significant effect of emotion information processing at Cz (F(1,18)= 12.13, $p < 0.003$) and Pz (200 ms) (F(1,18)= 16.09, $p < 0.0008$). The LPP are clearly larger for target compared to non-target at Cz (F(1,18)= 27.97, $p < 0.0004$) and Pz (F(1,18)=19.99, $p < 0.0003$). Additionally, LPP at PO8 is also significant (F(1,18)= 8.08, $p < 0.011$).

Comparison among three conditions: Further analyses have been performed to explore the differences of ERPs among the three mental tasks and to find the best oddball paradigm for BCI application. The VPP at Pz revealed the significant difference between objects (task 1) and faces (task 2, 3) stimuli (F(2,27)= 5.9, $p < 0.01$), but there is no significant difference between faces and emotional faces. The main effects of emotional faces at Pz and Cz indicating that significant larger LPP (Cz: F(2,27)= 3.94, $p < 0.032$, Pz: F(2,27)= 3.45, $p < 0.05$) compared to faces and objects stimuli and N250 at PO8 also revealed larger negative potentials for faces and emotional faces compared to objects stimuli (F(2,27)= 7.48, $p < 0.003$) (Fig. 4).

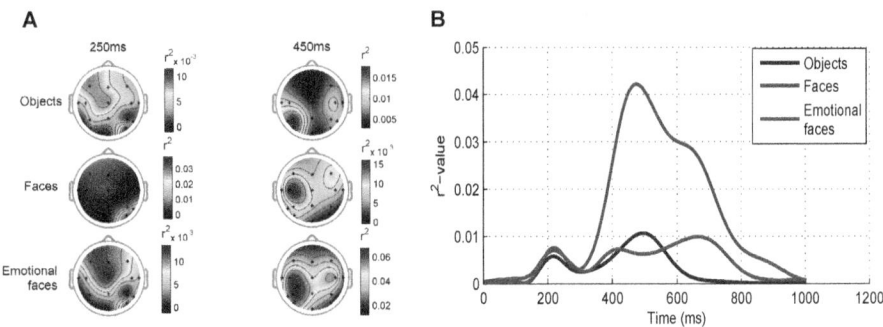

Fig. 5. Spatial and temporal distribution of discriminative information. (A) Topographic map of r^2-value at 250 ms and 450 ms, (B) r^2-value for temporal features.

Feature selection: The Fig. 5 depicted that the most discriminative information consisted of two parts: 1) VPP and N250 around 200 ms and 2) LPP at the time window (400 - 800 ms). We observed that LPP are more pronounced for emotional faces compared to the other two types of stimuli.

Classification performance: To compare the performance of three oddball paradigms, we performed 5-fold cross-validation procedure on single trial data sets by using various time windows with the trial lengths changed from 100 ms to 800 ms after stimulus onset, as shown in Fig. 6A. It is clear that using

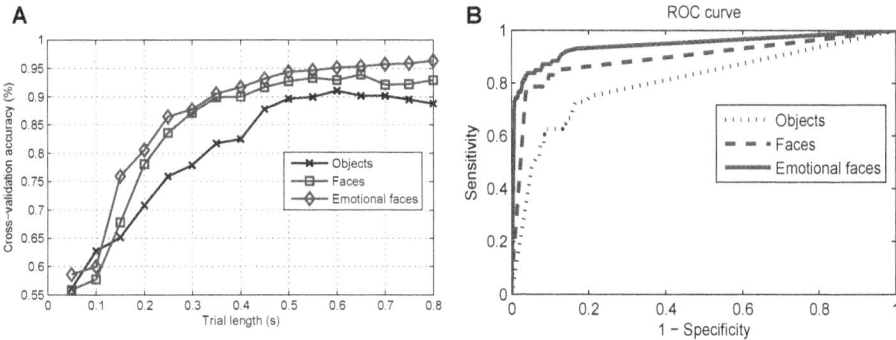

Fig. 6. (A) Cross-validation accuracy for single trial ERPs by increasing the trial length. (B) ROC curve for three different oddball paradigms.

emotional faces with emotion discrimination tasks outperform both of objects and faces paradigms, especially when the trial length is longer than 400 ms. The ROC curve is shown in Fig. 6B indicating that both emotional faces and faces paradigm are superior than objects paradigm, in particular, emotional faces paradigm greatly improved the performance for single trial EEG.

4 Conclusions

This study demonstrates that the components of ERPs, e.g. VPP, N250, LPP, can be modulated by the processing of emotional facial expressions. We compared the three oddball BCI paradigms by using objects, faces and emotional faces as stimuli and corresponding mental tasks, respectively. The results revealed VPP and LPP effects for emotional faces and N250 effects for face stimuli. The classification performance for single trial data demonstrated the superiority of emotional faces-related ERPs applied in the oddball BCI paradigm. These finding provide the evidence for the emotional-faces evoked ERPs in oddball paradigm, suggesting a novel affective BCI paradigm with improved reliability and flexibilities.

Acknowledgments. The work was supported in part by the national natural science foundation of China under grant number 90920014, NSFC international cooperation program under grant number 61111140019 and JSPS under the Japan - China Scientific Cooperation Program.

References

1. Wolpaw, J., Birbaumer, N., McFarland, D., Pfurtscheller, G., Vaughan, T.: Brain-computer interfaces for communication and control. Clinical Neurophysiology 113(6), 767–791 (2002)
2. Farwell, L., Donchin, E.: Talking off the top of your head: toward a mental prosthesis utilizing event-related brain potentials. Electroencephalography and Clinical Neurophysiology 70(6), 510–523 (1988)

3. Treder, M., Blankertz, B. (C) overt attention and visual speller design in an ERP-based brain-computer interface. Behavioral and Brain Functions 6(1), 28 (2010)
4. Martens, S., Hill, N., Farquhar, J., Schölkopf, B.: Overlap and refractory effects in a brain-computer interface speller based on the visual P300 event-related potential. Journal of Neural Engineering 6 (2009)
5. Townsend, G., LaPallo, B., Boulay, C., Krusienski, D., Frye, G., Hauser, C., Schwartz, N., Vaughan, T., Wolpaw, J., Sellers, E.: A novel P300-based brain-computer interface stimulus presentation paradigm: Moving beyond rows and columns. Clinical Neurophysiology 121(7), 1109–1120 (2010)
6. Furdea, A., Halder, S., Krusienski, D., Bross, D., Nijboer, F., Birbaumer, N., Kübler, A.: An auditory oddball (P300) spelling system for brain-computer interfaces. Psychophysiology 46(3), 617–625 (2009)
7. Lenhardt, A., Kaper, M., Ritter, H.: An adaptive P300-based online brain–computer interface. IEEE Transactions on Neural Systems and Rehabilitation Engineering 16(2), 121–130 (2008)
8. Xu, P., Yang, P., Lei, X., Yao, D.: An enhanced probabilistic LDA for multi-class brain computer interface. PloS One 6(1), e14634 (2011)
9. Sadeh, B., Zhdanov, A., Podlipsky, I., Hendler, T., Yovel, G.: The validity of the face-selective ERP N170 component during simultaneous recording with functional MRI. Neuroimage 42(2), 778–786 (2008)
10. Schweinberger, S., Huddy, V., Burton, A.: N250r: A face-selective brain response to stimulus repetitions. Neuro Report 15(9), 1501 (2004)
11. Nasr, S., Esteky, H.: A study of N250 event-related brain potential during face and non-face detection tasks. Journal of Vision 9(5), 5 (2009)
12. Neumann, M.F., Mohamed, T.N., Schweinberger, S.R.: Face and object encoding under perceptual load: ERP evidence. NeuroImage 54(4), 3021–3027 (2011)
13. Dennis, T., Chen, C.: Emotional face processing and attention performance in three domains: Neurophysiological mechanisms and moderating effects of trait anxiety. International Journal of Psychophysiology 65(1), 10–19 (2007)
14. Luo, W., Feng, W., He, W., Wang, N., Luo, Y.: Three stages of facial expression processing: ERP study with rapid serial visual presentation. NeuroImage 49(2), 1857–1867 (2010)
15. Willis, M., Palermo, R., Burke, D., Atkinson, C., McArthur, G.: Switching associations between facial identity and emotional expression: A behavioural and ERP study. Neuroimage 50(1), 329–339 (2010)
16. Lee, K., Lee, T., Yoon, S., Cho, Y., Choi, J., Kim, H.: Neural correlates of top-down processing in emotion perception: An ERP study of emotional faces in white noise versus noise-alone stimuli. Brain Research 1337, 56–63 (2010)
17. Zhao, Q., Onishi, A., Zhang, Y., Cichocki, A.: An affective BCI using multiple ERP components associated to facial emotion processing. State-of-the-Art in BCI Research: BCI Award (in print, 2011), http://www.bci-award.com/

Multiway Canonical Correlation Analysis for Frequency Components Recognition in SSVEP-Based BCIs

Yu Zhang[1,2], Guoxu Zhou[1], Qibin Zhao[1], Akinari Onishi[1,3], Jing Jin[2], Xingyu Wang[2], and Andrzej Cichocki[1]

[1] Laboratory for Advanced Brain Signal Processing,
RIKEN Brain Science Institute, Wako-shi, Saitama, Japan
[2] School of Information Science and Engineering,
East China University of Science and Technology, Shanghai, China
[3] Department of Brain Science and Engineering,
Kyushu Institute of Technology, Fukuoka, Japan
yuzhang@brain.riken.jp

Abstract. Steady-state visual evoked potential (SSVEP)-based brain computer-interface (BCI) is one of the most popular BCI systems. An efficient SSVEP-based BCI system in shorter time with higher accuracy in recognizing SSVEP has been pursued by many studies. This paper introduces a novel multiway canonical correlation analysis (Multiway CCA) approach to recognize SSVEP. This approach is based on tensor CCA and focuses on multiway data arrays. Multiple CCAs are used to find appropriate reference signals for SSVEP recognition from different data arrays. SSVEP is then recognized by implementing multiple linear regression (MLR) between EEG and optimized reference signals. The proposed Multiway CCA is verified by comparing to the standard CCA and power spectral density analysis (PSDA). Results showed that the Multiway CCA achieved higher recognition accuracy within shorter time than that of the CCA and PSDA.

Keywords: Brain-computer interface (BCI), Canonical Correlation Analysis (CCA), Electroencephalogram (EEG), Steady-State Visual Evoked Potential (SSVEP), Tensor Decomposition.

1 Introduction

SSVEP is evoked over occipital scalp areas with the same frequency as the visual stimulus and may also include its harmonics when subject focuses on the repetitive flicker of a visual stimulus [1]. According to this mechanism, a SSVEP-based BCI can be designed to recognize the frequency components of EEG signals. In recent years, SSVEP-based BCI has been increasingly studied and has demonstrated strength including shorter calibration time and higher information transfer rate (ITR) than other types of BCIs [2]. Although SSVEP provides aforementioned advantages for BCI systems, it may be contaminated

B.-L. Lu, L. Zhang, and J. Kwok (Eds.): ICONIP 2011, Part I, LNCS 7062, pp. 287–295, 2011.

by spontaneous EEG or noise and it is still a challenge to detect it with a high accuracy, especially at a short time window (TW) [3]. Hence, how to recognize SSVEP with higher accuracy and for shorter TW is a considerably important issue for obtaining an improved SSVEP-based BCI.

A traditional method for SSVEP recognition is power spectral density analysis (PSDA). PSD is estimated from the EEG signals within a TW typically by Fast Fourier Transform (FFT), and its peak is detected to recognize the target stimulus [4]. Instead of recognizing SSVEP by directly detecting the peak of PSD, some studies also took the PSDs as features and applied classification algorithm, such as linear discriminant analysis (LDA), to classify the target frequency [1]. A TW longer than 3 seconds (s) is usually required for estimating spectrum with sufficient frequency resolution when using the PSDA [4]. Such duration may limit the real-time performance of SSVEP-based BCIs. Lin et al. [5] proposed a promising and increasingly used method based on canonical correlation analysis (CCA) to recognize SSVEP. In their work, CCA was used to find the correlations between the EEG signals of multiple channels and reference signals of sine-cosine with different stimulus frequencies. Then, the target stimulus is recognized through maximizing these correlations. The use of CCA seems to provide better recognition performance than that of the PSDA since it delivers an optimization for the combination of multiple channels and improves the signal-to-noise ratio (SNR). A further comparison between the CCA and PSDA was done by Hakvoort et al. [6]. They also adopted the sine-cosine waves as reference signals used in the CCA for SSVEP recognition.

Although the CCA works quite well in SSVEP-based BCIs, we consider that the commonly used reference signals of sine-cosine may be not optimal for SSVEP recognition due to the inter-subject variability of SSVEP and effects of ongoing EEG and noises. Hence, our goal in this study is to find more efficient reference signals used in correlation analysis for SSVEP recognition. Tensor CCA proposed by Kim et al. [7] is a extension of the standard CCA and focuses on two multiway data arrays. Inspired by their work, We propose a Multiway CCA approach to discover the optimal reference signals from different modes (space and trial modes) of multidimensional EEG data and recognize SSVEP. The proposed method is verified with the EEG data of three healthy subjects and compared with the standard CCA, PSDA and the combination of PSDA and LDA (PSDA+LDA).

2   Experiment and EEG Acquisition

Three healthy volunteers (all males, aged 25, 31 and 34) participated in the experiments. The subjects were seated in a comfortable chair 50 cm from a LCD monitor (60 Hz refresh rate) in a shielded room. Four white squares, as stimuli, were flickered at four different frequencies: 8.5 Hz, 10 Hz, 12 Hz and 15Hz, respectively, on the black screen. In the experiment, each subject completed five runs with 5 ∼ 10 min rest after each of them. In each run, the subject was asked to focus on each of the four white squares for five times with a duration of 2 s for each time, respectively, with each target cue duration of 1 s. That is, each run contains 20 trials and totally 100 trials were completed for each subjects. EEG signals were recorded by a Biosemi Active Two amplifier at 256 Hz sampling rate ($f_s = 256$ Hz) from eight channels PO3, POz, PO4, PO7, PO8, O1, Oz and O2 placed on the standard position of the 10-20 international system. The average of them was used as reference. The EEG signals were bandpass filtered between 5 and 50 Hz.

3 Method

3.1 CCA and SSVEP Recognition

CCA is a multivariable statistical method to reveal the underlying correlation between two sets of data [8]. Consider two sets of random variables $\mathbf{X} \in \mathbb{R}^{I_1 \times J}$, $\mathbf{Y} \in \mathbb{R}^{I_2 \times J}$ and their linear combinations $\tilde{\mathbf{x}} = \mathbf{w}^T \mathbf{X}$ and $\tilde{\mathbf{y}} = \mathbf{v}^T \mathbf{Y}$, CCA tries to find a pair of linear transform $\mathbf{w} \in \mathbb{R}^{I_1 \times 1}$ and $\mathbf{v} \in \mathbb{R}^{I_2 \times 1}$ to maximize the correlation between $\tilde{\mathbf{x}}$ and $\tilde{\mathbf{y}}$, through solving the following optimization problem:

$$\rho = \max_{\mathbf{w},\mathbf{v}} \frac{E\left[\tilde{\mathbf{x}}\tilde{\mathbf{y}}\right]}{\sqrt{E\left[\tilde{\mathbf{x}}^2\right] E\left[\tilde{\mathbf{y}}^2\right]}} = \frac{\mathbf{w}^T \mathbf{X} \mathbf{Y}^T \mathbf{v}}{\sqrt{\mathbf{w}^T \mathbf{X} \mathbf{X}^T \mathbf{w} \mathbf{v}^T \mathbf{Y} \mathbf{Y}^T \mathbf{v}}}. \tag{1}$$

The maximum of ρ corresponds to the maximum canonical correlation between the canonical variates $\tilde{\mathbf{x}}$ and $\tilde{\mathbf{y}}$.

Lin et al. [5] introduced the CCA to recognize SSVEP for the first time. Assume there are M stimulus frequencies need to be recognized. $\mathbf{X}$ consists of EEG signals from I_1 channels, and $\mathbf{Y}_m$, as a reference signals set, is constructed by sine-cosine waves at the mth stimulus frequency f_m ($m = 1, 2, \ldots, M$):

$$\mathbf{Y}_m = \begin{pmatrix} \sin\left(2\pi f_m 1/f_s\right) & \cdots & \sin\left(2\pi f_m J/f_s\right) \\ \cos\left(2\pi f_m 1/f_s\right) & \cdots & \cos\left(2\pi f_m J/f_s\right) \\ \vdots & \vdots & \vdots \\ \sin\left(2\pi H f_m 1/f_s\right) & \cdots & \sin\left(2\pi H f_m J/f_s\right) \\ \cos\left(2\pi H f_m 1/f_s\right) & \cdots & \cos\left(2\pi H f_m J/f_s\right) \end{pmatrix}, \tag{2}$$

where H denotes the number of used harmonics (i.e., $I_2 = 2H$), J is the number of sampling points and f_s represents the sampling rate. We apply the optimization of Eq.(1) to solve the canonical correlations $\rho_1, \rho_2, \ldots, \rho_M$ corresponding to the M reference signals, respectively. Then the target stimulus frequency f_{target} is recognized as:

$$f_{\text{target}} = \max_{f_m} \rho_m, \quad m = 1, 2, \ldots, M. \tag{3}$$

Although the CCA works quite well for SSVEP recognition, sine-cosine waves may be not the optimal reference signals in using correlation analysis since they do not contain any information about the inter-subject variability and trial-to-trial variability. We consider that the recognition accuracy may be further improved by optimizing the reference signals. We will show how to find more efficient reference signals by a novel Multiway CCA approach from experimental multidimensional EEG data in the next section.

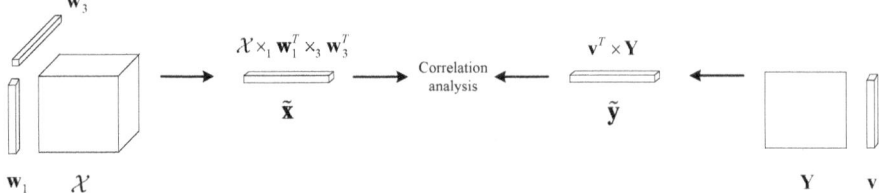

Fig. 1. Illustration of Multiway CCA mode

3.2   Multiway CCA and SSVEP Recognition

A tensor is a multiway array of data and the order of the tensor is the number of dimensions, also known as ways or models [9]. A first-order tensor is a vector and a second-order tensor is a matrix. A Nth-order tensor is denoted by $\mathcal{X} = (\mathcal{X})_{i_1 i_2 \ldots i_N} \in \mathbb{R}^{I_1 \times I_2 \times \ldots \times I_N}$. The n-mode product of the tensor with a vector $\mathbf{w} \in \mathbb{R}^{I_n \times 1}$ is

$$\left(\mathcal{X} \times_n \mathbf{w}^T \right)_{i_1 \ldots i_{n-1} i_{n+1} \ldots i_N} = \sum_{i_n=1}^{I_n} x_{i_1 i_2 \ldots i_N} w_{i_n}. \tag{4}$$

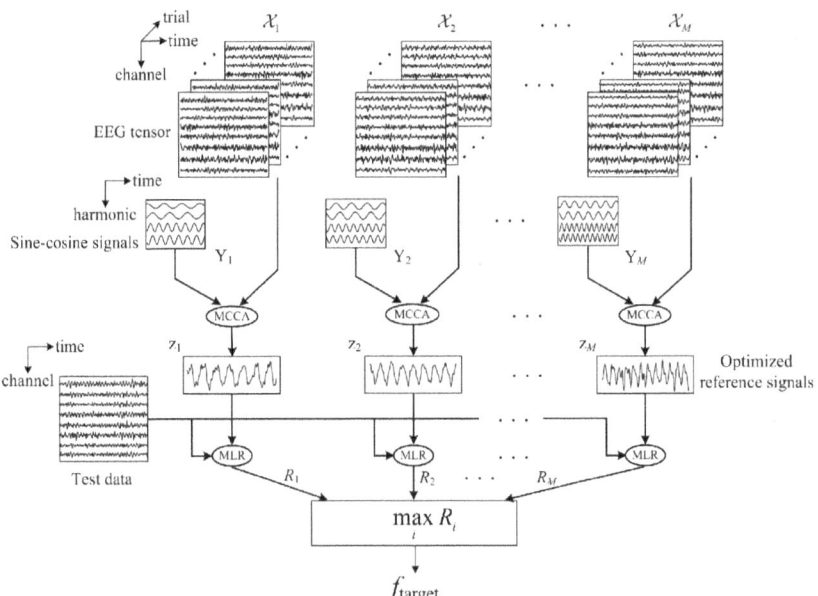

Fig. 2. Illustration of Multiway CCA approach for SSVEP recognition. Here, MCCA denotes the Multiway CCA.

Multiway CCA mode. Tensor CCA is an extension of the standard CCA, which focuses on inspecting the correlation between two multiway data arrays, instead of two sets of vector-based variables [7]. Drawing on the idea of Tensor CCA, we introduce a Multiway CCA which maximizes the correlation between multiway data (tensor) and two-way data (matrix) to optimize the reference signals used in correlation analysis for SSVEP recognition. We consider that EEG data from the trials with a specific stimulus frequency form a third-order (three-way) data tensor $\mathcal{X} \in \mathbb{R}^{I \times J \times K}$ (channel $\times$ time $\times$ trial) and a original reference signal matrix $\mathbf{Y} \in \mathbb{R}^{2H \times J}$ (harmonic $\times$ time) is constructed by the sine-cosine signals with frequencies as the stimulus frequency and its higher harmonics. Our aim is to find more efficient reference signals for SSVEP recognition from time domain (i.e. optimizing the channels and trials data ways) based on the original reference signals of sine-cosine. Then, the canonical correlation between mode-2 of $\mathcal{X}$ and $\mathbf{Y}$ is considered. The proposed Multiway CCA finds linear transforms $\mathbf{w}_1 \in \mathbb{R}^{I \times 1}$, $\mathbf{w}_3 \in \mathbb{R}^{K \times 1}$ and $\mathbf{v} \in \mathbb{R}^{2H \times 1}$ such that

$$\rho = \max_{\mathbf{w}_1, \mathbf{w}_3, \mathbf{v}} \frac{E\left[\tilde{\mathbf{x}}\tilde{\mathbf{y}}\right]}{\sqrt{E\left[\tilde{\mathbf{x}}^2\right] E\left[\tilde{\mathbf{y}}^2\right]}} \tag{5}$$

is maximized, where $\tilde{\mathbf{x}} = \mathcal{X} \times_1 \mathbf{w}_1^T \times_3 \mathbf{w}_3^T$ and $\tilde{\mathbf{y}} = \mathbf{v}^T \times \mathbf{Y}$. Fig. 1 illustrates the Multiway CCA mode. For solving this problem, we adopt an alternating algorithm which fixes $\mathbf{w}_3$ to solve $\mathbf{w}_1$ and $\mathbf{v}$, then fixes $\mathbf{w}_1$ and $\mathbf{v}$ to solve $\mathbf{w}_3$, and repeats this procedure until convergence criterion is satisfied. Then, the optimal reference signal denoted by $\mathbf{z} \in \mathbb{R}^{1 \times J}$ can be obtained by:

$$\mathbf{z} = \mathcal{X} \times_1 \mathbf{w}_1^T \times_3 \mathbf{w}_3^T. \tag{6}$$

When compared with the standard sine-cosine signals, the optimized reference signal contains not only the ideal SSVEP frequency components but also the information of inter-subject variability and trial-to-trial variability.

Multiway CCA based SSVEP recognition. We represent the experimental EEG data corresponding to the mth stimulus frequency f_m $(m = 1, 2, \ldots, M)$ as a third-order tensor $\mathcal{X}_m \in \mathbb{R}^{I \times J \times K}$ (channel$\times$time$\times$trial). The Multiway CCA is implemented to maximize the correlation between the EEG data tensor and the corresponding sine-cosine signals $\mathbf{Y}_m \in \mathbb{R}^{2H \times J}$ (harmonic$\times$time) defined by Eq.(2), and find the optimal reference signals denoted by $\mathbf{z}_m \in \mathbb{R}^{1 \times J}$, which will be used to replace the original sine-cosine based reference signals. Then, with the new reference signals, multiple linear regression (MLR) [11], a correlation technique focusing on a variable and a set of variables, is utilized to recognize target stimulus frequency. We consider the relationship between the test EEG data $\mathbf{X}_{\text{test}} \in \mathbb{R}^{I \times J}$ and optimized reference signal $\mathbf{z}_m$ as a multiple regression model, i.e.,:

$$\mathbf{z}_m = \mathbf{b}_m^T \mathbf{X}_{\text{test}} + \mathbf{e}_m, \tag{7}$$

where $\mathbf{b}_m \in \mathbb{R}^{I \times 1}$ is a coefficient vector to be estimated and $\mathbf{e}_m \in \mathbb{R}^{1 \times J}$ is a noise vector with zero mean and constant variance. With least square method,

292   Y. Zhang et al.

Algorithm 1. Multiway CCA algorithm for SSVEP Recognition

Input: M EEG tensor data $\mathcal{X}_1, \mathcal{X}_2, \ldots, \mathcal{X}_M \in \mathbb{R}^{I \times J \times K}$ and sine-cosine signals $\mathbf{Y}_1, \mathbf{Y}_2, \ldots, \mathbf{Y}_M \in \mathbb{R}^{2H \times J}$ corresponding to M stimulus frequencies, respectively. A test EEG data $\mathbf{X}_{\text{test}} \in \mathbb{R}^{I \times J}$.

Output: Recognition result f_{target}.

for $m = 1$ *to* M **do**

  Random initialization for $\mathbf{w}_{m,3}$ and do $\tilde{\mathbf{X}}_m \leftarrow \mathcal{X}_m \times_3 \mathbf{w}_{m,3}^T$.

  repeat

    Find $\mathbf{w}_{m,1}, \mathbf{v}_m$ which maximize the correlation between $\tilde{\mathbf{X}}_m$ and $\mathbf{Y}_m$ by the CCA. Do $\tilde{\mathbf{X}}_m \leftarrow \mathcal{X}_m \times_1 \mathbf{w}_{m,1}^T$, $\tilde{\mathbf{y}}_m \leftarrow \mathbf{v}_m^T \times \mathbf{Y}_m$.

    Find $\mathbf{w}_{m,3}$ which maximizes the correlation between $\tilde{\mathbf{X}}_m$ and $\tilde{\mathbf{y}}_m$ by the CCA. Do $\tilde{\mathbf{X}}_m \leftarrow \mathcal{X}_m \times_3 \mathbf{w}_{m,3}^T$.

  until *the maximum number of iterations is reached* ;

  Compute the optimized reference signal $\mathbf{z}_m \leftarrow \mathcal{X}_m \times_1 \mathbf{w}_{m,1}^T \times_3 \mathbf{w}_{m,3}^T$.

end

for $m = 1$ *to* M **do**

  Implement MLR between $\mathbf{X}_{\text{test}}$ and $\mathbf{z}_m$ to obtain the correlation R_m.

end

Recognize target stimulus frequency as $f_{\text{target}} = \max\limits_{f_m} R_m, (m = 1, 2, \ldots, M)$.

---

the esimation of $\mathbf{b}_m$ is solved as:

$$\hat{\mathbf{b}}_m = \left(\mathbf{X}_{\text{test}}\mathbf{X}_{\text{test}}^T\right)^{-1}\mathbf{X}_{\text{test}}\mathbf{z}_m^T, \tag{8}$$

and the estimated vector of fitting values $\hat{\mathbf{z}}_m$ is computed as:

$$\hat{\mathbf{z}}_m = \hat{\mathbf{b}}_m^T\mathbf{X}_{\text{test}} = \mathbf{z}_m\mathbf{X}_{\text{test}}^T\left(\mathbf{X}_{\text{test}}\mathbf{X}_{\text{test}}^T\right)^{-1}\mathbf{X}_{\text{test}}, \tag{9}$$

Then, the correlation coefficient R_m which reflects the relationship between $\mathbf{X}_{\text{test}}$ and $\mathbf{z}_m$ is calculated as:

$$R_m = \sqrt{1 - \frac{\|\mathbf{z}_m - \hat{\mathbf{z}}_m\|_2^2}{\|\mathbf{z}_m - E\left[\mathbf{z}_m\right]\|_2^2}}, \tag{10}$$

where $\|\cdot\|_2$ denotes l_2-norm. Larger R_m implies more significant relationship between $\mathbf{X}_{\text{test}}$ and $\mathbf{z}_m$. Then, the target stimulus frequency is recognized as:

$$f_{\text{target}} = \max\limits_{f_m} R_m, (m = 1, 2, \ldots, M). \tag{11}$$

The algorithm of the proposed Multiway CCA for SSVEP recognition is summarized in Algorithm 1. Fig. 2 illustrates SSVEP recognition based on the Multiway CCA. For each subject, five-fold cross-validation is used to estimated average classification accuracy. More specifically, a procedure, in which the EEG data from four runs (80 trials) are used to optimize the reference signals and that from the left-out run (20 trials) is used for SSVEP recognition, is repeated five times so that each run served once for SSVEP recognition validation.

4   Results

The proposed Multiway CCA was compared with the standard CCA, PSDA and PSDA+LDA. EEG data from all eight channels were used as the inputs for the standard CCA and Multiway CCA. For the PSDA, PSDs were estimated by $4f_s$-point-FFT (i.e., the frequency resolution is 0.25 Hz) from the EEG data with a bandwidth of 0.5 Hz, averaged on the channels O1, Oz and O2. For the PSDA+LDA, we took the PSDs as features and applied a 4-class classifier built by combining six single LDAs to classify the target frequency. The average accuracy was estimated by a five-fold cross-validation. Fig. 3 shows the recognition accuracy of the four methods for different subjects and harmonic combinations. While the standard CCA performed better than the PSDA and PSDA+LDA, the proposed Multiway CCA yielded higher recognition accuracies than the standard CCA for most time window (TW) lengths. There was no big difference between the accuracy of the PSDA and PSDA+LDA. For most of the four methods, the performance in using more harmonics was slightly better than that in using fewer harmonics.

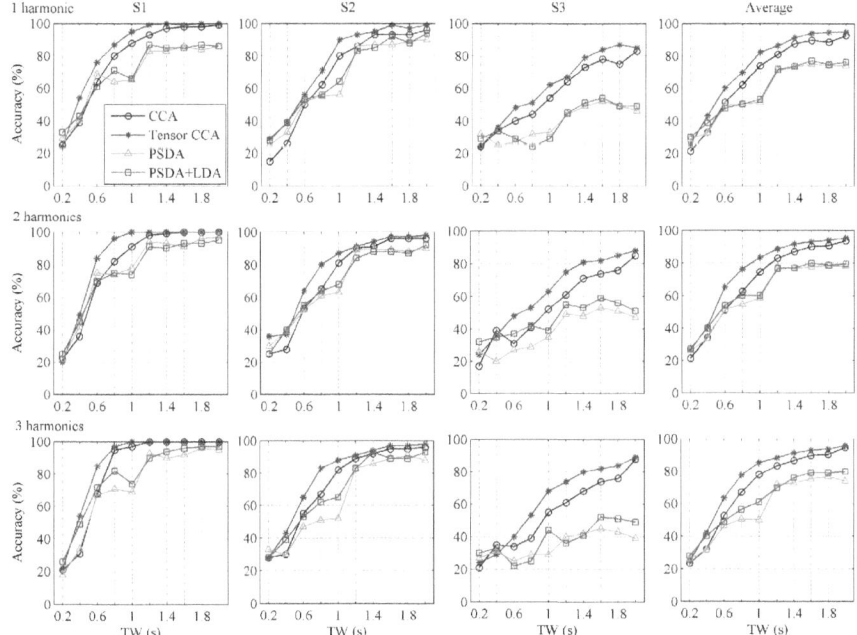

Fig. 3. SSVEP recognition accuracies obtained by the standard CCA, Multiway CCA, PSDA and the combination of PSDA and LDA, across different time window (TW) lengths, in using different harmonic combinations, for three subjects. The rightmost column shows the accuracies averaged on all subjects. 1 harmonic: fundamental frequency only, 2 harmonics: fundamental frequency and second harmonic, 3 harmonics: fundamental frequency, second and third harmonics.

Information transfer rate (ITR) [10] was also used to evaluate the performance of the CCA and Multiway CCA further. The ITR can be computed as:

$$B = \log_2 N + Acc \times \log_2 Acc + (1 - Acc) \times \log_2 \left[(1 - Acc)/(N - 1) \right], \quad (12)$$

$$\text{ITR} = B \times 60/T, \quad (13)$$

where the bit rate or bits/trial is denoted by B, N is the number of stimulus frequency, Acc is the recognition accuracy and T represents the duration per trial. Table 1 presents the recognition accuracies and ITRs of the CCA and Multiway CCA in using different channel combinations and TW lengths. Here, we focus on the TW lengths in range of 0.8 s $\sim$ 1.4 s to compromise between recognition accuracy and speed. For all of the four TW lengths, the Multiway CCA yielded higher recognition accuracies and higher ITRs than that of the CCA for various channel combinations. Furthermore, if fewer number of channels were used, bigger advantages over the CCA seemed to be achieved by the Multiway CCA. For both methods, the combination of more channels used yielded better performance than that in using fewer channels.

Table 1. Accuracy (Acc) (%) and information transfer rate (ITR) (bits/min) of the standard CCA and Multiway CCA (MCCA) in using different channel combinations and within different time window lengths (TW) (s), averaged on all subjects

TW	Channel	CCA		MCCA	
		Acc	ITR	Acc	ITR
0.8					
	8 channels	67.0	25.2	77.7	34.3
	6 channels	68.0	25.7	74.3	32.4
	3 channels	60.7	17.8	70.7	27.8
1.0					
	8 channels	78.0	30.9	85.3	38.7
	6 channels	73.7	29.8	81.3	36.3
	3 channels	67.7	22.9	78.0	32.6
1.2					
	8 channels	83.3	34.0	88.3	38.0
	6 channels	80.3	32.0	85.7	37.0
	3 channels	72.7	26.6	82.3	33.2
1.4					
	8 channels	86.7	33.8	91.3	37.8
	6 channels	85.0	33.3	88.0	35.4
	3 channels	78.7	28.9	83.3	33.1

Note: 8 channels: PO3, POz, PO4, PO7, PO8, O1, Oz and O2. 6 channels: PO3, POz, PO4, O1, Oz, O2. 3 channels: O1, Oz, O2.

5   Discussion and Conclusion

In this study, a Multiway CCA approach was proposed to recognize the stimulus frequency for SSVEP-based BCI. In this method, multiple CCAs were implemented between the EEG tensor data and sine-cosine signals to find appropriate reference signals used in correlation analysis for SSVEP recognition. After that, multiple linear regression was applied to inspect the correlation between the test EEG data and optimized reference signals for SSVEP recognition. From the results, the Multiway CCA achieved higher accuracy than that of the standard CCA, PSDA and the combination of PSDA and LDA, within shorter TW length. This shows the proposed method is promising for enhancing the real-time performance of SSVEP-based BCIs. Also, the better performance of the Multiway CCA confirmed that the reference signals optimized from space and trial data modes were more efficient than the commonly used sine-cosine signals for SSVEP recognition, since they might contain some information of subject-specific and trial-to-trial variability. It is possible to develop an online learning algorithm which gives real-time updates to the reference signals so that an adaptive SSVEP-based BCI can be established, which will be our future study.

Acknowledgments. This study is supported by Nation Nature Science Foundation of China 61074113, Shanghai Leading Academic Discipline Project B504, and Fundamental Research Funds for the Central Universities WH0914028.

References

1. Müller-Putz, G.R., Scherer, R., Brauneis, C., Pfurtscheller, G.: Steady-state visual evoked potential (SSVEP)-based communication: impact of harmonic frequency components. J. Neural Eng. 2, 123–130 (2005)
2. Zhu, D., Bieger, J., Molina, G., Aarts R.M.: A survey of stimulation methods used in SSVEP-based BCIs. Comput. Intell. Neurosci. (2010), doi: 10.1155/2010/702357
3. Zhang, Y., Jin, J., Qing, X., Wang, B., Wang, X.: LASSO based stimulus frequency recognition model for SSVEP BCIs. Biomed. Signal Process. Control (2011), doi:10.1016/j.bspc.2011.02.002
4. Cheng, M., Gao, X., Gao, S., Xu, D.: Design and implementation of a brain-computer interdace with high transfer rates. IEEE Trans. Biomed. Eng. 49, 1181–1186 (2002)
5. Lin, Z., Zhang, C., Wu, W., Gao, X.: Frequency recognition based on canonical correlation analysis for SSVEP-based BCIs. IEEE Trans. Biomed. Eng. 53, 2610–2614 (2006)
6. Hakvoort, G., Reuderink, B., Obbink, M.: Comparison of PSDA and CCA detection methods in a SSVEP-based BCI-system. Technical Report TR-CTIT-11-03. EEMCS (2011), ISSN 1381-3625
7. Kim, T.K., Cipolla, R.: Canonical correlation analysis of video volume tensor for action categorization and detection. IEEE Trans. PAMI 31, 1415–1428 (2009)
8. Hotelling, H.: Relations between two sets of variates. Biometrika 28, 321–377 (1936)
9. Cichocki, A., Zdunek, R., Phan, A., Amari, S.: Nonnegative matrix and tensor factorizations: Applications to exploratory multi-way data analysis and blind source separation. Wiley, New York (2009)
10. Wolpaw, J.R., Birbaumer, N., McFarland, D.J., Pfurtscheller, G., Vaughan, T.M.: Brain-computer interface for communication and control. Clin. Neurophysiol. 113, 767–791 (2002)
11. Chatterjee, S., Hadi, A.S.: Influential observations, high leverage points, and outliers in linear regression. Statistical Science 1, 379–416 (1986)

An Emotional Face Evoked EEG Signal Recognition Method Based on Optimal EEG Feature and Electrodes Selection

Lijuan Duan[1], Xuebin Wang[1], Zhen Yang[1], Haiyan Zhou[1], Chunpeng Wu[1], Qi Zhang[1], and Jun Miao[2]

[1] College of Computer Science and Technology, Beijing University of Technology, Beijing 100124, China
[2] Key Laboratory of Intelligent Information Processing, Institute of Computing Technology, Chinese Academy of Sciences, Beijing 100190, China
jmiao@ict.ac.cn, {ljduan,yangzhen,zhouhaiyan}@bjut.edu.cn,
{wangxuebin0216,wuchunpeng,yunuo014212}@emails.bjut.edu.cn

Abstract. In this work, we proposed an emotional face evoked EEG signal recognition framework, within this framework the optimal statistic features were extracted from original signals according to time and space, i.e., the span and electrodes. First, the EEG signals were collected using noise suppression methods, and principal component analysis (PCA) was used to reduce dimension and information redundant of data. Then the optimal statistic features were selected and combined from different electrodes based on the classification performance. We also discussed the contribution of each time span of EEG signals in the same electrodes. Finally, experiments using Fisher, Bayes and SVM classifiers show that our methods offer the better chance for reliable classification of the EEG signal. Moreover, the conclusion is supported by physiological evidence as follows: a) the selected electrodes mainly concentrate in temporal cortex of the right hemisphere, which relates with visual according to previous psychological research; b) the selected time span shows that consciousness of the face picture has a trend from posterior brain regions to anterior brain regions.

Keywords: EEG, expression classification, electrodes selection, principal component analysis.

1 Introduction

The EEG can directly reflect the brain's activities, as is well known; it is the comprehensive reflection of millions of nerve cells in the brain. Scientists are trying to find the coding mechanism of the brain from EEG signals. Brain computer interface is a new way to investigate brain processes involved in various stimuli [1][2][3]. In this paper, we introduce an EEG signal classification method to recognize facial expressions.

B.-L. Lu, L. Zhang, and J. Kwok (Eds.): ICONIP 2011, Part I, LNCS 7062, pp. 296–305, 2011.
© Springer-Verlag Berlin Heidelberg 2011

How the brain distinguishes different facial expressions of emotion has been the subject of a number of recent studies. At present, most of researchers determine facial expression emotion by image processing and pattern recognition, which is based on stimulus of facial image. Analyzing the physiological EEG signal is another way to recognize emotion, which is a more intuitive and effective means of emotion recognition. However, few researchers consider using EEG for emotion recognition [4][5][6].

Reference related to the classification of EEG emotion mainly consists of two aspects. One directly uses the overall EEG signal as a feature vector; Murugappan et al. [7] extracted wavelet features from EEG signals for classifying human emotions. It used three different wavelet functions on 5 different emotions (happy, surprise, fear, and disgust, neutral) for feature extraction, and then selected KNN algorithm and LDA classification, test results show that KNN algorithm is superior to LDA; the highest average recognition rate reaches 79.174%. However, the other uses the event related potentials (ERPs) extracted from the whole EEG signals. Haihong et al. [8] classify the ERP signals P300 by support vector machine classifier SVM. P300 is a task-related ERP, and the peak appears about 300ms after the stimulus occurs. The EEG feature selection in [8] is from the perspective of cognitive psychology and combined with the test-related ERP analysis. In summary, the present studies select different features in order to improve the performance of emotional facial expression classification.

In this paper, we proposed an emotional face evoked EEG signal recognition framework, within this framework the optimal statistic features were extracted from original signals according to time and space, i.e., the span and electrodes. First, the EEG signals were collected using noise suppression methods, and principal component analysis (PCA) was used to reduce dimension and information redundant of data. Then the optimal statistic features were selected and combined from different electrodes based on the classification performance. We also discussed the contribution of each time span of EEG signals in the same electrodes. Finally, experiments using Fisher, Bayes and SVM classifiers show that our methods offer the better chance for reliable classification of the EEG signal. There are two advantages in our method. The first is that we specifically investigate the contribution of each time span extracted from an EEG signal to this classification problem, while others directly use the whole EEG signal as an input. The second is that our model is purely computational, however, the result is supported by biological evidence as follows: a) The electrodes selected by our method mainly concentrate in temporal cortex of the right hemisphere which is related to vision according to previous psychological research. b) The results based on time sequence can show that the consciousness of the face picture has a trend from posterior brain regions to anterior brain regions.

In the remainder of this paper, we explain the framework of facial expression evoked EEG signal recognition method and EEG acquisition. Subsequently, we describe the method of EEG classification. Then, it shows the results of each experiment. Finally, we conclude and prospect the future in the end of the part.

2   The Framework of Emotional Face Evoked EEG Signal Recognition Method

2.1   Framework

The framework of our method includes two main parts: EEG acquisition and EEG processing as shown in Fig.1. As with other EEG signal recognition systems, we first designed a psychological experiment to collect facial expression evoked EEG signals. The left part of fig.1 shows the EEG acquisition step. In order to eliminate the effects of age, race and gender elements to facial expression processing, we use schematic facial expressions (happy, neutral and sad) to get the clear facial expression evoked EEG signal. The right part of fig.1 shows the EEG processing steps. First, we preprocess the collected EEG using a noise reduction method, then extract features through feature selection and electrode combination. Finally, classification is operated on the optimal signal by combining the selected electrodes. In the course of EEG signal processing, we adopt many mechanisms to get optimal features and electrodes, which mainly involve selection of time series, the selection of electrode and the combination of electrode. In order to get the optimal classification performance, we compare three classifiers: liner discriminant function classifier (Fisher), support vector machine classifier (SVM) and Bayes classifier (Bayes).

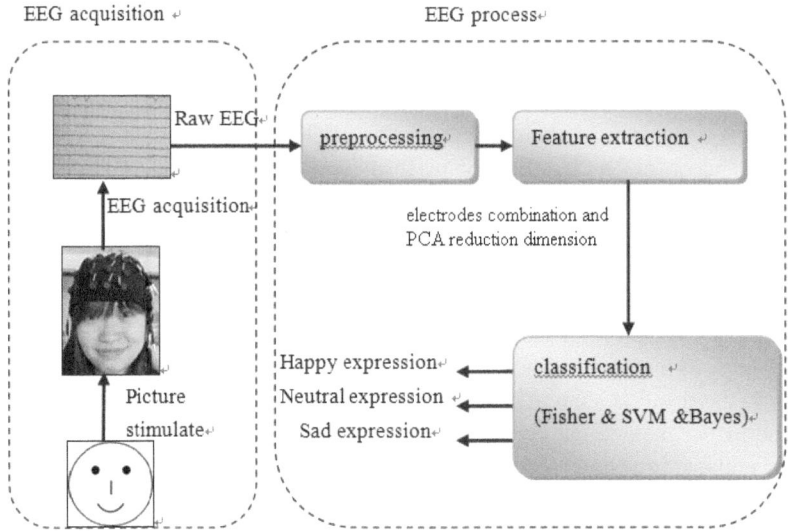

Fig. 1. The framework of emotional face evoked EEG signal recognition method

2.2 EEG Acquisition

Twelve subjects aged 20-30 years participate in the study. All the participants report normal vision and have no history of current or past neurological or psychiatric illness and take no medications known to affect the central nervous system. Participants are tested in a dimly lit room. They sit in a comfortable chair, and then they are instructed in how to complete the experiment. The stimuli are three types schematic facial expressions (happy, neutral, sadness). Each type of expressions has 18 samples of facial expressions. The stimuli are presented one at a time. The order of trials is randomized. Stimulus exposure time is 1000ms. These stimuli are used to form 2 stimulus conditions: (a) upright faces, (b) inverted faces. EEG is recorded continually by electrodes placed according to the 64 system. EEG is sampled at 500Hz using a Neuroscan Nuamps digital amplifiers system. EOG artifacts are corrected using a correlation method. The EEG is segmented in epochs of 1000 ms beginning 200 ms prior to stimulus onset and averaged separately. Each subject has done 408 trials, 136 trials for each type schematic facial expressions. Fig. 2 illustrates the EEG data acquisition processing using visual stimulus.

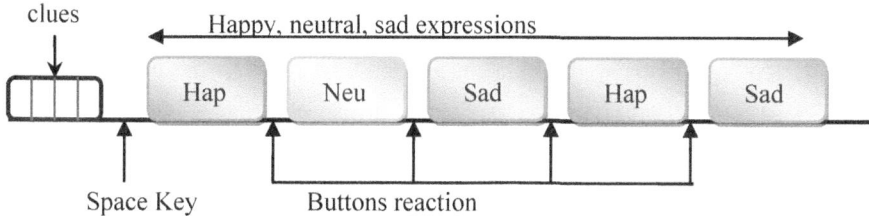

Fig. 2. EEG data acquisition processing using visual stimulus

3   Analysis of the Reorganization of EEG

Suppose the EEG signals are decomposed into sub-bands according to different time span and electrodes. E_{ij} represents the sub-band of the jth time span at the ith electrode. Then the optimal statistic features X can be written as follows:

$$X = \sum_i \sum_j \alpha_i \beta_j E_{ij} ,$$

(1)

Where $i \in \{1,...,64\}$, $j \in \{1,...,5\}$,

$$\alpha_i = \begin{cases} 1 & i^{th} \ electrode \ EEG \ signals \\ 0 & otherwise \end{cases}, \quad \beta_j = \begin{cases} 1 & j^{th} time \ range \ EEG \ signals \\ 0 & otherwise \end{cases}.$$

The first method is based on single electrode signals.

a) To begin with, we use 64 single electrodes ($i \in \{1, 2, ..., 64\}$) as a brain signal feature to classify and get 64 classification performances.

b) Subsequently, our purely computational model automatic selects the first three single electrodes to restructure. We used combination optimal single electrode EEG signals as a new EEG signals feature and PCA dimension reduction. In the experiment, $\alpha_i \in \{0,1\}$, $\beta_j = 1$ and according Experiment II result, it shows E_{CB2}, E_{PO8} and E_{P8} as the first three optimal classification electrodes. Fig. 3 illustrates the process of optimal single electrodes EEG signals feature selection and reorganization.

c) Finally, we chose three classifiers (Fisher, SVM and Bayes) to compare the classification performances. The results show that the fisher classifier performs well, as is shown in Experiment II.

The second method is based on time spans signals.

a) To begin with, we took 200ms as each time span. As the EEG signals hold 1000ms. There are five time spans in one electrode (-200-0ms, 0-200ms, 200-400ms, 400-600ms, 600-800ms). Therefore $j \in \{1, 2, 3, 4, 5\}$. Meanwhile, we chose E_{ij} as an initial signal feature to specifically investigate the contribution of each time span extracted from an EEG signal.

b) Subsequently, when i^{th} electrode j^{th} time span EEG signal got the best classification result as MAX, and we artificially set the threshold range as [MAX-5% MAX]. In this way, our model automatic selects these reasonable EEG signals based on each time span extracted from an EEG signal and random combination, and then PCA dimension reduction. Such as Fig.4 shows the process of time spans optimal electrode EEG signals combination.

c) Finally, based on a comparison of the results of the three classifiers, we chose Fisher as the classifier and this classification performance compares with the results of the first method. The results further show that EEG signal features based on time span perform well. As shown in Experiment IV.

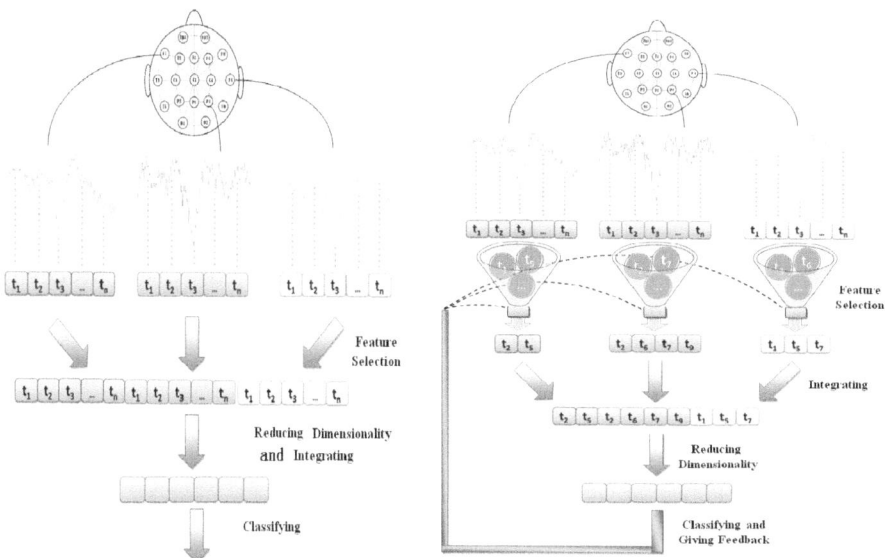

Fig. 3. Optimal single electrodes EEG signals feature selection and reorganization scheme

Fig. 4. Optimal time spans EEG signals feature selection and reorganization scheme

4   Experimental Results and Analysis

This section contains two parts. On the one hand, it describes PCA dimension reduction parameters. On the other hand, we list 4 experiments. Experiment I illustrates that the electrodes selected by our method mainly concentrate in temporal cortex of the right hemisphere which is related to vision. Experiment II shows the Fisher classifier is better than SVM and Bayes. Experiment III illustrates that the consciousness of the face expression trends from posterior brain regions to anterior brain regions. And experiment VI shows that EEG classification performances based on time spans are better than those based on electrodes.

4.1   PCA Parameter Setting Description

Since the combination EEG signals are typical high-dimension, it is necessary to reduce EEG signals dimensional. The PCA method was chosen because it can guarantee a minimum loss of information when it makes high-dimensional variable space dimension reduction. In the course of PCA dimension reduction, different dimension parameter of 400, 300, 200 and 100 were selected. We can get every single electrode's classification results in different dimensions through taking closed-end testing, which clearly shows the differences in the classification of different dimensions, the result shows the EEG's classification was best when it was reduced to 400 dimensions. Therefore, an EEG signal feature selection will be reduced to 400 dimensions.

4.2   Experiment I : Feature Selection – A Single Electrode

In the Experiment I, the single electrode is viewed as the EEG feature; we choose the Fisher classifier to do closed-end testing. The result shows that the electrode CB2 classification accuracy reached 90.98%. At the same time, the PO8, P8, CP4, P6, PO6, P4 classification accuracy rate reached 89.51%, 88.97%, 88.02%, 88.8%, 85.4%, and 84.36% respectively. Based on the classification results, we marked the optimal first six electrodes on brain mapping. Fig.5 shows they are mainly in the temporal occipital region [9] and right hemispheric dominance [10], while the temporal occipital region is associated with the visual region according to previous psychological research.

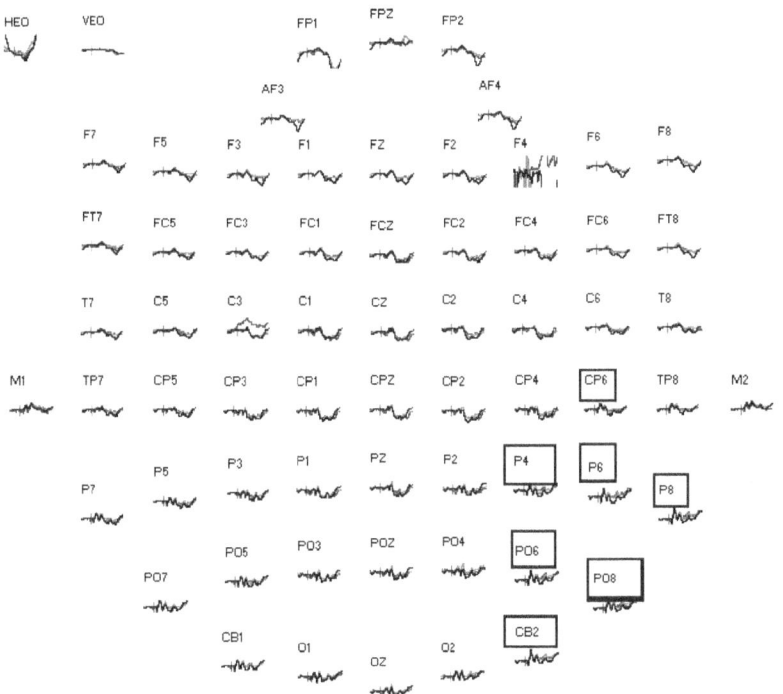

Fig. 5. The brain mapping (first six electrodes)

4.3   Experiment II: Feature Selection – A Combination of Optimal Single Electrodes

The purpose of experiment II is improving EEG classification through combination of optimal single electrodes. Thus according to experiment I's classification results, firstly, the first three electrodes in the temporal occipital (CB2, PO8, P8) are combined. Secondly, the combination EEG features are reduced dimensionally. Finally, we chose Fisher, SVM and Bayes to classify, and then used a 10 times 10 fold cross validation method to test results. The results shown as Tab.1: The Fisher classification accuracy rate was 88.51%. Therefore, the Fisher classifier had good classification results in the EEG emotional classification.

Table 1. The results of classification of emotion EEG in different ways

Classification of emotion EEG			
Classification method	PCA + FISHER	PCA + Bayes	PCA + SVM
Recognition rate	88.51	82.30	69.98

4.4   Experiment III: Feature Selection – Each Time Span of Each Single Electrode

This EEG signals in -200ms-0ms are omitted due to no ERP existence. Therefore, we choose EEG signals in 0-800ms, and use each time span (200ms) EEG signals as feature to classify them. Fig.6 illustrates the optimal electrodes based on different

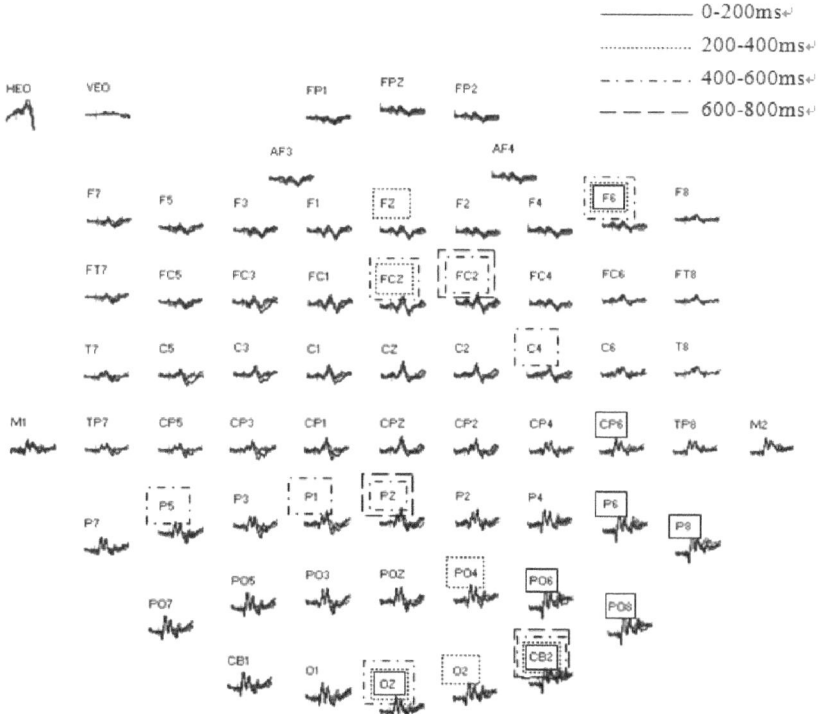

Fig. 6. The optimal electrodes based on different time range EEG signals classification

time range's EEG signals classification distributions for different electrodes. Such as PO8 is an optimal electrode in 0-200ms, PO4 is an optimal electrode in 200-400ms, C4 is an optimal electrode in 400-600ms, and FC2 is an optimal electrode in 600-800ms. According to this experiment, we not only found the emotional classification is predominantly on the right hemispheric, but also the most important is that we found the perception of pictures of facial expressions trends from posterior brain regions to anterior brain regions. This finding is the same as the reference [11].

4.5   Experiment IV: Feature Selection – A Combination of Optimal Time Spans

On the basis of the experiment III, we made the advantages of a single electrode in the scope of [MAX-5% MAX] random combination. The result is following.

Table 2. Optimal time series electrodes combination

The fourth experiment electrode combination		
0ms-200ms PO6	PO8	P8
200ms-400ms CB2	FZ	F6
400ms-600ms CB2	P1	PZ
600ms-800ms PZ	FC2	CB2

Compared with experiment II, the experiment IV's electrode combinations based on time series reached an accuracy rate of 98.38%. While in the experiment II, the accuracy rate reached 88.51%. Hence, selecting EEG based on time series can greatly improve the performance of emotional classification through reduction and recombination of EEG.

5   Conclusions and Outlook

This paper proposes an emotional facial expression EEG signal classification method by dimension reduction and reorganization of electrodes. In the EEG acquisition step, we get 500 dimension data for each electrode. In order to eliminate the redundant information, we use PCA to decrease the dimensions to 400 for single electrode and combined electrodes. We select 3~10 optimal electrodes for classification, while we use 64 electrodes for EEG acquisition. In order to get high classification performance, we compare Fisher, SVM and Bayes, and find that Fisher obtains better classification results. Meanwhile, it has also confirmed the psychology theory. In the future, we will consider ERP for the emotional expression classification, and classify all electrodes for each time point and analysis the continuity of the same electrodes in each time point to find the valuable ERP.

Acknowledgements. This research is partially sponsored by National Basic Research Program of China (No.2009CB320902), Beijing Natural Science Foundation (Nos. 4102013, 4102012), Natural Science Foundation of China (Nos.61175115, 60702031, 60970087, 61070116, 61070149, 61001178), and President Fund of Graduate University of Chinese Academy of Sciences (No.085102HN00).

References

1. Schaaff, K., Schultz, T.: Towards an EEG-Based Emotion Recognizer for Humanoid Robots. In: The IEEE Inter. Symposium on Robot and Human Interactive Communication Toyama, pp. 792–796 (2009)
2. Furui, S., Hwang, J.: Automatic Recognition and Understanding of Spoken Languages- A First Step Toward Natural Human Machine Communication. Proc. of the IEEE 88, 1142–1165 (2000)
3. Cowie, R., Douglas, E., Tsapatsoulis, N., Votsis, G., Kollias, G., Fellenz, W., Taylor, J.G.: Emotion Recognition in Human Computer Interaction. IEEE Trans. Signal Processing 18, 3773–3776 (2001)
4. Petrantonakis, P., Hadjileontiadis, L.: Emotion Recognition from EEG Using Higher Order Crossings. IEEE Trans. Information Techonology in Biomedicine 14(2), 186–197 (2010)
5. Meeren, H., Heijnsbergen, C., Gelder, B.: Rapid Perceptual Integration of Facial Expression and Emotional Body Language. PNAS 102(45), 16518–16523 (2005)
6. Chanel, G., Karim, A., Pun, T.: Valence-Arousal Evaluation Using Physiological Signals in an Emotion Recall Paradigm. In: IEEE Inter. Conference on Systems, Man and Cybernetics, pp. 530–537 (2007)
7. Murugappan, M., Nagarajan, R., Yaacob, S.: Comparison of Different Wavelet Features from EEG Signals for Classifying Human Emotions. In: IEEE Symposium on Industrial Electronics & Applications, vol. 2, pp. 836–841 (2009)
8. Haihong, Z., Cuntai, G., Chuanchu, W.: Asynchronous P300-Based Brain-Computer Interfaces: A Computational Approach with Statistical Models. IEEE Trans. on Biomedical Engineering 55(6), 1754–1763 (2008)
9. Wenbo, L.: The Neural Mechanism of Facial Processing in Rapid Serial Visual Presentation (2009)
10. Robinson, A., Morris, J.: Is the N170 ERP Component Face Specific? Poster presented at the Meeting of the Psychonomic Society (2002)
11. Li, P.: An Event-Related Potential Study of Facial Expression Perception. master thesis. Hunan Normal University (2007)

Functional Connectivity Analysis
with Voxel-Based Morphometry
for Diagnosis of Mild Cognitive Impairment

JungHoe Kim and Jong-Hwan Lee

Department of Brain and Cognitive Engineering, Korea University
Seoul 136-713, Republic of Korea
{kjh948,jonghwan_lee}@korea.ac.kr

Abstract. The cortical atrophy measured from the magnetic resonance imaging (MRI) data along with aberrant neuronal activation patterns from the functional MRI data have been implicated in the mild cognitive impairment (MCI), which is a potential early form of a dementia. The association between the level of cortical atrophy in the gray matter (GM) and corresponding degree of neuronal connectivity, however, has not systematically been presented. In this study, we aimed to provide anecdotal evidence that there would be a close link between the anatomical abnormality and corresponding functional aberrance associated with the neuropsychiatric condition (*i.e.* MCI). Firstly, the voxel-based morphometry (VBM) analysis identified the medial temporal lobe and inferior parietal lobule as the regions with substantially decreased (*i.e.* atrophy) and increased GM concentrations, respectively. In the subsequent functional connectivity (FC) analysis via Pearson's correlation coefficients, the FC patterns using the regions with a decreased GM concentration showed increased FC patterns (*i.e.* hyper-connectivity) associated with the MCI. On the other hand, the FC patterns using the seed regions with an increased GM concentration have shown decreased FC (*i.e.* hypo-connectivity) with the MCI in the task anti-correlated regions including superior frontal gyrus (*i.e.* task-negative networks or default-mode networks). These results provide a supplemental information that there may be an compensatory mechanism in the human brain function, which potentially allow to diagnose early phase of the neuropsychiatric illnesses including the Alzheimer's diseases (AD).

Keywords: Functional magnetic resonance imaging, mild cognitive impairment, voxel-based morphometry, functional connectivity, dementia.

1 Introduction

Using non-invasive magnetic resonance imaging (MRI) modalities, a number of studies have explored the characteristic traits of various neurodegenerative diseases such as the AD using structural MRI (sMRI) and functional MRI (fMRI) data [1,2]. These include the abnormalities of the regional volumes of the human brain, GM concentration, level of neuronal activity, and functional connectivity between multiple regions [2,3,4,5].

B.-L. Lu, L. Zhang, and J. Kwok (Eds.): ICONIP 2011, Part I, LNCS 7062, pp. 306–313, 2011.
© Springer-Verlag Berlin Heidelberg 2011

In recent studies, the VBM has been useful to characterize a potential biomarker estimated from the sMRI data to study GM degeneration [6]. The advantages of VBM include greater sensitivity for localizing regional characteristics of the GM in voxel-wise compared to the volume information from the volumetric analysis [7]. The functional abnormality has also been useful to diagnose the AD [4,5]. For example, Gili and colleagues [8] investigated GM concentration and functional connectivity from the MCI and AD group using VBM and independent component analysis (ICA), respectively. However, the association between the structural atrophy and functional connectivity abnormality toward the identification of the symptomatic traits of AD pathophysiology has not been extensively appreciated. The goal of the present study is to provide anecdotal evidence that there would be a close link between the anatomical abnormality and the corresponding functional aberrance associated with the neuropsychiatric condition (*i.e.* MCI). To do so, the symptomatic traits of the MCI were identified using sMRI and fMRI data via statistical comparison of the GM concentration via VBM and functional connectivity patterns using the seed voxels derived from the VBM analysis.

2   Method

Figure 1 illustrates the overall flow diagram of functional connectivity analysis together with VBM. The detailed information of each step is elaborated in the later sections. Both sMRI and fMRI data were obtained from the freely available repository of the fMRI data center (www.fmridc.org, #2-2000-1118W; n = 41; 1.5 T GE Signa LX scanner; [3]). Data from two groups of elderly subjects (*i.e.* healthy controls, or HC, $n = 14$; subjects with an MCI, $n = 13$) were adopted. T_1-weighted structural MRI data (TR/TE = 9.7/4 msec; FA=10°; $1 \times 1 \times 1.25$ mm^3 voxel-size) were acquired in a series of three to four separate T_1-weighted MP-RAGE anatomic images and were used for VBM. T_2^*-weighted fast spin-echo echo-planar-imaging (EPI) volumes were acquired while a subject performed sensory-motor task trials ($n = 3$ or 4; TR/TE = 2.68/0.096 sec; 128 volumes per run; 64×64 in-plane voxels; 16 axial slices; 3mm thickness; $3.75 \times 3.75 \times 8$ mm^3 voxel-size).

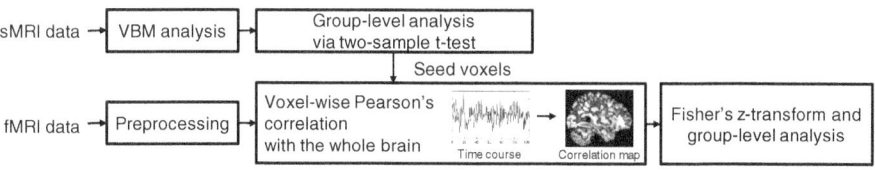

Fig. 1. Overall flow diagram of functional connectivity analysis together with the VBM

2.1   sMRI Analysis via Voxel-Based Morphometry

As shown in Figure 2, the sMRI data of T_1 images of each subject were preprocessed using the steps of VBM in SPM8 (Statistical Parametric Mapping software toolbox;

the Wellcome Trust Centre for Neuroimaging, London, UK). In detail, firstly, T_1 structural images were initially normalized using a 12-parameter affine model to the Montreal Neurological Institute (MNI) template that was derived from 550 healthy control subjects from the IXI-database (www.braindevelopment.org). Normalized images were interpolated with voxel dimension of $1.5 \times 1.5 \times 1.5$ mm^3 and segmented into gray, white, and cerebrospinal fluid (CSF) compartments using a modified mixture model based cluster analysis technique [6]. Only non-linear deformation was used to align with the template, which allows comparing the absolute amount of tissue corrected for individual brain sizes. This option is equivalent to non-linear deformation combined with affine transformation, in which the total intracranial volume is used as a covariate for the step of group-level analysis. The GM images were then smoothed using 8 mm full-width-at-half-maximum (FWHM) Gaussian kernel. Each voxel in a smoothed image contains the averaged GM concentration. In the group-level, a random-effect (RFX) model was administered to evaluate the group-level difference between the HC and MCI groups, in which a two-sample t-test was applied to the smoothed GM images. The clusters that showed significantly different GM concentrations between two groups were obtained. Then, the foci with maximum t-score of each of these clusters were defined as seed voxels and used for subsequent functional connectivity analysis.

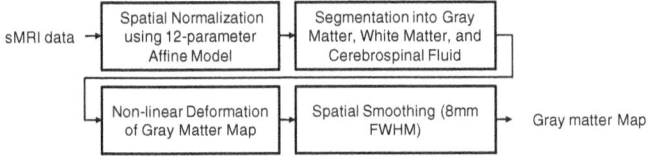

Fig. 2. Overall steps of VBM to analyze the sMRI data

2.2   fMRI Analysis via Functional Connectivity Patterns

The EPI volumes were preprocessed using SPM2 (*i.e.* slice timing correction, head motion correction, normalization to the MNI coordinate with 3mm isotropic voxel, and spatial smoothing using 8mm isotropic FWHM Gaussian kernel in order) as shown in Figure 3. For each run, the first four EPI volumes were removed to allow for T_1-related signals to reach steady-state equilibrium.

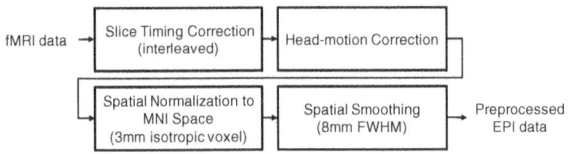

Fig. 3. Overall steps of preprocessing to analyze the fMRI data

In the functional connectivity analysis, the identified foci where GM concentration was significantly decreased or increased from the results of VBM analysis were used as seed voxels. The average of the fMRI time series (*i.e.* blood-oxygenation-level-dependent, or BOLD signals) from neighboring 27 voxels centered at a seed voxel (*i.e.* $3 \times 3 \times 3$ voxels) were regarded as a representative BOLD signal of the cluster and used as a reference BOLD signal of the region. The linear trend of the BOLD signals from each of the 27 neighboring voxels including the seed voxel was corrected before averaging to remove the potential confounding artifact of the low-frequency drift noise originated from hardware imperfection [9].

The level of functional connectivity between the seed region and the rest of the voxels within a brain region of each subject in each run was evaluated using the Pearson's correlation coefficient. The Pearson's correlation coefficient of the two linearly detrended BOLD signals (*i.e.*, $\mathbf{x}_{seed}$ and $\mathbf{x}_v$ from the seed voxel and each voxel within brain mask, respectively) was calculated as follows:

$$r_v = \frac{\sigma^2_{seed.v}}{\sigma_{seed}\sigma_v} \tag{1}$$

where σ_{seed} and σ_v are the standard deviation of $\mathbf{x}_{seed}$ and $\mathbf{x}_v$, respectively, and $\sigma^2_{seed.v}$ is the covariance of $\mathbf{x}_{seed}$ and $\mathbf{x}_v$. Subsequently, each correlation map was converted to normally distributed z-values using Fisher's r to z transformation. A fixed-effect (FFX) model was adopted by averaging the resulting z-scored correlation map across the three or four runs of each subject. In the group-level, a RFX model was administered to compare the group-level FC between the HC and MCI groups using two-sample *t*-test applied to the z-scored correlation maps from two groups.

3   Experimental Results

3.1   Group-Level Inference of VBM Results

From the group-level comparison using RFX to the VBM results, a total of 8 foci were identified from the two-sample *t*-test as shown in Fig. 4. The decreased GM concentrations from the MCI group were detected in five regions including bilateral fusiform gyrus, left parahippocampus, left hippocampus, and left anterior cingulate cortex ($p < 10^{-3}$ uncorrected, with a minimum of 20 connected voxels), in which the atrophy of these regions responsible for cognitive ability are believed to be due to the MCI status. GM concentration within right hippocampus was shown a moderate statistical significance between the MCI and HC group (*i.e.* $p = 0.004$ uncorrected).

Meanwhile, three cortical regions including right medial frontal gyrus, left inferior parietal lobule, and right paracentral lobule were identified as the regions with significantly increased GM concentrations from the MCI group ($p < 10^{-3}$ uncorrected, with a minimum of 20 connected voxels). Table 1 summarizes the results of the group-level analysis from the VBM. Subsequently, the eight foci from these identified clusters were used as seed regions for the subsequent functional connectivity analysis.

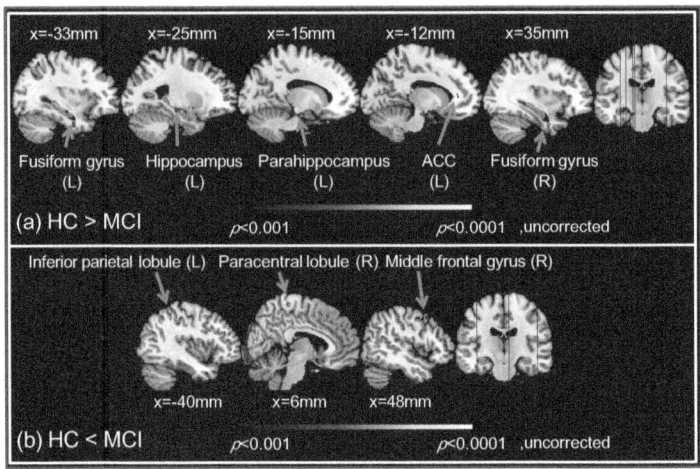

Fig. 4. Two-sample *t*-test results indicate that the GM concentrations were (a) significantly decreased from the MCI group in the bilateral fusiform gyrus, left parahippocampus, left hippocampus, and left anterior cingulate cortex, and (b) significantly increased from the MCI group including the bilateral right medial frontal gyrus, left inferior parietal lobule, and right paracentral lobule ($p < 10^{-3}$ uncorrected, with a minimum of 20 connected voxels)

Table 1. The identified foci from two-sample *t*-test applied to the GM intensities from the subjects in each of the two groups (*i.e.* cluster size as a number of voxels; x,y,z mm in MNI coordinate)

Contrast	Region	Side	Cluster size	x	y	z	Peak *t*-score	*p*-value
MCI < HC	FG Hippocampus	L	57	-33.0	-4.5	-31.5	3.93	3.0×10^{-4}
	FG	R	28	34.5	-1.5	-33.0	4.25	1.3×10^{-4}
	Parahippocampus Hippocampus	L	110	-15.0	-13.5	-21.0	4.00	2.5×10^{-4}
	Hippocampus	L	131	-25.5	-21.0	-10.5	3.87	3.5×10^{-4}
	ACC	L	21	-12.0	36.0	7.5	4.14	1.7×10^{-4}
MCI > HC	MFG	R	57	48.0	13.5	48.0	3.68	5.6×10^{-4}
	IPL	L	28	-40.5	-45.0	55.5	4.20	1.5×10^{-4}
	Paracentral lobule	R	110	6.0	-33.0	64.5	4.01	2.4×10^{-4}

FG: fusiform gyrus; MFG: middle frontal gyrus; ACC: anterior cingulate cortex; ITG: inferior temporal gyrus; IPL: inferior parietal lobule

3.2 Functional Connectivity Results Integrated with VBM

Figure 5 and Table 2 summarize the results from the functional connectivity analysis. Using the seed regions with reduced GM concentration (*i.e.* atrophy) from the sMRI data, functional connectivity patterns were significantly increased ($p < 10^{-3}$ uncorrected; a minimum of 20 connected voxels) from the MCI group in the regions including the right superior frontal gyrus, bilateral middle/posterior cingulate cortex, and right inferior parietal lobule. On the other hand, using the seed regions with increased GM concentration, functional connectivity patterns were significantly decreased from the MCI group within the anterior cingulate cortex and superior frontal gyrus ($p < 10^{-3}$ uncorrected; a minimum of 20 connected voxels). Functional connectivity patterns from the seed regions of bilateral fusiform gyrus and the paracentral lobule were not significantly different between the MCI and HC group ($p > 10^{-3}$ uncorrected).

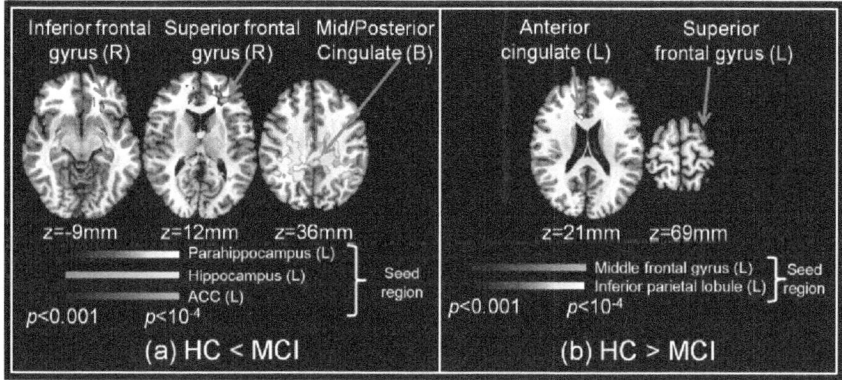

Fig. 5. Functional connectivity patterns via two-sample t-test was (a) significantly increased (*i.e.*, hyper-connectivity) from the MCI group within the right superior frontal gyrus, bilateral middle/posterior cingulate cortex, and right inferior parietal lobule, and (b) significantly decreased from the MCI group (*i.e.*, hypo-connectivity) within the anterior cingulate cortex and superior frontal gyrus ($p < 10^{-3}$ uncorrected; a minimum of 20 connected voxels).

Table 2. Summary of group-level differences of functional connectivity patterns via two-sample t-tests, in which the regions with significantly different GM intensity were used as seed regions (*i.e.* cluster size as a number of voxels; x,y,z mm in MNI coordinate)

GM intensity	Seed region (L/R)	Functional connectivity	Regions	x	y	z	Peak t-score	Cluster size
MCI < HC	Parahippocampus (L)	MCI > HC	SFG (R)	18	45	12	4.83	58
	Hippocampus (L)		MCC/PCC (B)	-21	-54	36	7.14	1119
	ACC (L)		IFG (R)	24	27	-9	4.80	21
			SFG (R)	27	36	12	5.71	58
MCI > HC	MFG (R)	MCI < HC	ACC (L)	-9	21	21	5.15	32
	IPL (L)		SFG (R)	27	0	69	4.30	21

SFG, MFG, & IFG: superior, middle, & inferior frontal gyrus; ACC, MCC, & PCC: anterior, middle, & posterior cingulate cortex; IPL: inferior parietal lobule

4   Discussion

In this study, we successfully demonstrated that the intensity changes of the cortical GM observed from the structural MRI data are closely related to the aberrant functional connectivity from the affected regions to the rest of the brain. In detail, the hyper-and hypo-connectivity patterns from the MCI subjects compared to the healthy subjects were observed from the seed regions with decreased and increased GM concentration, respectively. These changes of the functional connectivity are particularly interesting since this may be indicating the aberrant human brain functions compensating the structural abnormality, which may be evident much earlier than the progression of the structural detriment. Based on the neurophysiological evidence, a hyper-connectivity in early AD may be interpreted as a compensatory mechanism due to axonal sprouting and synaptic plasticity, so that the region with atrophy can maintain the level of functional integration within the affected region along the progression of the structural damage [10].

Further study would be warranted to investigate an effective connectivity (*e.g.*, Granger causality analysis; [11]) to identify the abnormal causal flow between the brain regions associated with the anatomical abnormality. Since all the MCI subjects recruited may not subsequently be developed to the AD, the future investigation using the longitudinally obtained data sets from MCI subjects would be greatly appreciated toward the potential utility for an early diagnosis of the illnesses.

5   Conclusion

In this study, we presented that the structural and functional abnormalities were tightly coupled. The analytical method of this study may provide a potentially valuable option toward an early diagnosis of various neuropsychiatric illnesses including the AD providing an early sign of neuropsychiatric condition estimated from the functional abnormality, which may be observed much earlier than the more evident structural damage.

Acknowledgments. This work was supported by WCU (World Class University) program through the National Research Foundation (NRF) of Korea funded by the Ministry of Education, Science and Technology (R31-10008) and Basic Science Research Program, NRF grant of Korea (2011-0004794).

References

1. Dickerson, B.C., Sperling, R.A.: Functional Abnormalities of the Medial Temporal Lobe Memory System in Mild Cognitive Impairment and Alzheimer's Disease: Insights from Functional MRI Studies. Neuropsychologia 46(6), 1624–1635 (2008)
2. Fan, Y., Resnick, S.M., Wu, X., Davatzikos, C.: Structural and Functional Biomarkers of Prodromal Alzheimer's Disease: a High-dimensional Pattern Classification Study. Neuroimage 41(2), 277–285 (2008)
3. Buckner, R.L., Snyder, A.Z., Sanders, A.L., Raichle, M.E., Morris, J.C.: Functional Brain Imaging of Young, Nondemented, and Demented Older Adults. J. Cogn. Neurosci. 12(suppl.2), 24–34 (2000)

4. Greicius, M.D., Srivastava, G., Reiss, A.L., Menon, V.: Default-mode Network Activity Distinguishes Alzheimer's Disease from Healthy Aging: Evidence from Functional MRI. Proc. Natl. Acad. Sci. U S A 101(13), 4637–4642 (2004)
5. Wang, K., Liang, M., Wang, L., Tian, L., Zhang, X., Li, K., Jiang, T.: Altered Functional Connectivity in Early Alzheimer's Disease: a Resting-state fMRI Study. Hum. Brain Mapp. 28(10), 967–978 (2007)
6. Ashburner, J., Friston, K.: Voxel-based Morphometry—the Methods. Neuroimage 11, 805–821 (2000)
7. Mak, H.K., Zhang, Z., Yau, K.K., Zhang, L., Chan, Q., Chu, L.W.: Efficacy of Voxel-based Morphometry with DARTEL and Standard Registration as Imaging Biomarkers in Alzheimer's Disease Patients and Cognitively Normal Older Adults at 3.0 Tesla MR Imaging. J. Alzheimers Dis. 23(4), 655–664 (2011)
8. Gili, T., Cercignani, M., Serra, L., Perri, R., Giove, F., Maraviglia, B., Caltagirone, C., Bozzali, M.: Regional Brain Atrophy and Functional Disconnection across Alzheimer's Disease Evolution. J. Neurol. Neurosurg. Psychiatry 82(1), 58–66 (2011)
9. Huettel, S.A., Song, A.W., McCarthy, G.: Functional Magnetic Resonance Imaging, 2nd edn. Sinauer Associates, Inc., Sunderland (2009)
10. Rytsar, R., Fornari, E., Frackowiak, R.S., Ghika, J.A., Knyazeva, M.G.: Inhibition in Early Alzheimer's Disease: An fMRI-based Study of Effective Connectivity. Neuroimage (2011) Epub ahead of print
11. Seth, A.K.: A MATLAB Toolbox for Granger Causal Connectivity Analysis. J. Neurosci. Methods 186, 262–273 (2010)

An Application of Translation Error to Brain Death Diagnosis

Gen Hori[1,4] and Jianting Cao[2,3,4]

[1] Department of Business Administration, Asia University, Tokyo 180-8629, Japan
[2] Department of Information System, Saitama Institute of Technology, Saitama 369-0293, Japan
[3] East China University of Science and Technology, Shanghai 200237, China
[4] Lab. for Advanced Brain Signal Processing, BSI, RIKEN, Saitama 351-0198, Japan
hori@brain.riken.jp

Abstract. The present study aims at the use of the translation errors of the EEG signals as criteria for brain death diagnosis. Since the EEG signals of the patients in coma or brain death contain several kinds of sources that differ from the viewpoint of determinism, we can exploit the difference of the translation errors for brain death diagnosis. We also show that the translation errors of the post-ICA EEG signals are more reliable than the ones of the pre-ICA EEG signals.

Keywords: Brain death diagnosis, Translation error, Embedding dimension, Independent component analysis.

1 Introduction

Brain death is defined as the complete, irreversible and permanent loss of all brain and brain stem functions. In many countries, the EEG signals of the patients are used to inspect the absence of cerebral cortex function for brain death diagnosis. Because the EEG recordings are usually corrupted by various kinds of artifacts, extracting informative features from noisy EEG signals and evaluating their significance is essential in brain death diagnosis. In 2000, Hori et al.[4][5] proposed the use of independent component analysis (ICA) for the flat EEG examination in brain death diagnosis. Cao[7], Cao and Chen[8] and Chen et al.[9] proposed the use of frequency-based and complexity-based statistics for quantitative EEG analysis in brain death diagnosis. Moreover, Hori and Cao[6] proposed the use of the translation error for selecting EEG components in brain death diagnosis. The purpose of the present study is to introduce the use of the translation errors of the EEG signals as criteria for brain death diagnosis. Based on the observation that the signals from the brain activities are deterministic while the environmental noises are stochastic, we exploit the difference of the translation errors for the EEG signals of the patients in coma or brain death. We also show that the translation errors of the post-ICA EEG signals are more reliable than the ones of the pre-ICA EEG signals.

B.-L. Lu, L. Zhang, and J. Kwok (Eds.): ICONIP 2011, Part I, LNCS 7062, pp. 314–321, 2011.

The rest of the paper is organized as follows. Section 2 presents the ICA algorithm applied to EEG data in Section 4. Section 3 explains the calculation procedure of the translation error. Section 4 applies the translation error together with the ICA algorithm to EEG data for diagnosis of brain death. Section 5 includes concluding remarks.

2   Independent Component Analysis

Independent component analysis (ICA) is a powerful method of signal processing for blind separation of statistically independent source signals. Several standard ICA algorithms have been developed and widely used for EEG analysis such as elimination of noise and extraction of components related to the brain activities. This section presents the ICA algorithm we apply to EEG data in Section 4.

Since the source of EEG signals propagate rapidly to the electrodes on the scalp, EEG observation is modeled by an instantaneous mixing model

$$x(t) = As(t)$$

where $s(t)$, $x(t)$ and A denote n-channel source signals, m-channel observed signals, and an $m \times n$ mixing matrix respectively. The source signals are all supposed to be zero-mean and statistically independent to each other. Corresponding demixing model is given as

$$y(t) = Wx(t)$$

where $y(t)$ and W denote n-channel demixed signals and an $n \times m$ demixing matrix respectively. ICA algorithms set the demixing matrix W to some initial value and update it iteratively so that the statistical independence of the demixed signals is maximized. In convergence, the demixed signals are statistically independent to each other and the matrix WA is a permutation matrix with amplitudes. In Section 4, we apply the natural gradient ICA algorithm[12] with an automatic detection of sub- and super-Gaussian sources[13],

$$\dot{W} = (I - K\tanh(y)y^T - yy^T)W, \tag{1}$$

where

$$\tanh(y) = (\tanh(y_1), \ldots, \tanh(y_n))^T,$$
$$K = \mathrm{diag}(k_1, \ldots, k_n),$$
$$k_i = \mathrm{sign}(E[\mathrm{sech}^2(y_i)]E[y_i{}^2] - E[\tanh(y_i)y_i]),$$

to the EEG data of patients in coma or brain death.

In our application to the EEG data of the patients in coma or brain death, the source signals are either the signals from the brain activities, the contamination from the power supply or the environmental noise. Applying the ICA algorithm to the EEG data, we obtain those as the demixed signals. As explained in

Section 4, six electrodes (Fp1, Fp2, F3, F4, C3 and C4) were placed for recording EEG data to produce six source signals, to which we apply the ICA algorithm to obtain six demixed signals. Following terminology in the field of ICA, we refer to pre-ICA EEG signals as "EEG channels" and post-ICA EEG signals "EEG components".

3   Translation Error

The translation error[14] provides a quantitative measure of determinism of single channel time series data from the viewpoint of determinism versus stochasticity. As the EEG signals of the patients in coma or brain death contain several kinds of sources that differ from the viewpoint, we can exploit the translation error of the EEG signals in brain death diagnosis. For the calculation of the translation error, a given single channel time series data is considered to be generated from a higher dimensional attractor. The attractor is reconstructed based on Takens' embedding theorem[15] and the translation error is calculated from the reconstructed attractor as follows.

Given a single channel time series data $z(t)$, let

$$\xi(t) = (z(t), z(t+\tau), \ldots, z(t+(D-1)\tau))$$

denote a point on the reconstructed attractor in D-dimensional state space where τ is a delay time. We choose a point $\xi_0(t)$ randomly from the reconstructed attractor and consider its neighborhood on the attractor $\{\xi_k(t)|k = 1, \ldots, K\}$ of the K nearest points to $\xi_0(t)$ and how the neighborhood displace along the time evolution of T steps,

$$v_k(t) = \xi_k(t+T) - \xi_k(t) \quad (k = 0, \ldots, K).$$

The translation error is defined as the variance of such displacements,

$$E_{trans} = \frac{1}{K+1} \sum_{k=0}^{K} \frac{|v_k(t) - \langle v(t) \rangle|^2}{|\langle v(t) \rangle|^2},$$

$$\langle v(t) \rangle = \frac{1}{K+1} \sum_{k=0}^{K} v_k(t).$$

For robust estimation of the translation error, we repeat the above calculation M times using randomly sampled $\xi_0(t)$ and use the median as the result. For a deterministic time series data, the displacements are expected to be concentrated on their center to make the variance small. The translation error E_{trans} will decrease as the embedding dimension D approaches the dimension of the attractor. On the other hand, for a stochastic time series data, the displacements will be scattered to make the variance large. The translation error E_{trans} stays at larger values not depending on the embedding dimension D.

As the brain activities reside in the ensembles of synchronous firings of neurons, the signals from the brain activities are considered to be deterministic.

Ikeguchi et al.[16] and Gallez and Babloyantz[17] calculated the Lyapunov exponent of the EEG signals and suggested that the signals from the brain activities are deterministic. The EEG signals of the patients in coma are mainly the signals from the brain activities for which the translation error is relatively small. On the other hand, the EEG signals of the patients in brain death are mainly stochastic environmental noises for which the translation errors keep larger values. In the following section, we exploit the difference of the translation errors for diagnosis of brain death.

4   Experiments

In the present study, the EEG recordings were carried out in the Shanghai Huashan Hospital in affiliation with the Fudan University (Shanghai, China). The EEG recording instrument was a portable NEUROSCAN ESI-32 amplifier associated with a laptop computer. During the EEG recordings, nine electrodes were placed on the forehead of the patient lying on the bed. Specifically, six electrodes were placed at Fp1, Fp2, F3, F4, C3 and C4 for recording, two at both ears for reference, and one was used as the ground. From June 2004 to March 2006, the EEG recordings were carried out for a total of 36 patients in coma or brain death, from which we analyze the following:

- Patient A : 18-year-old male patient in coma
- Patient B : 19-year-old female patient in brain death
- Patient C1 : 48-year-old male patient in coma
- Patient C2 : Patient C1 transitioned to brain death
- Patient D : 56-year-old female patient in brain death
- Patient E : 26-year-old female patient in brain death
- Patient F : 22-year-old male patient in brain death

In our analysis, we first apply the ICA algorithm (1) to separate the EEG channels of all the above EEG data to the EEG components. After that, we calculate the translation error for all the EEG channels and the EEG components for the embedding dimensions $D = 2$ through $D = 10$. In the calculation of the translation error, we set $K = 4$, $\tau = 50$, $T = 10$ and $M = 30$, that is, the number of neighborhood points is 4, the time delay for the embedding is 50msec, time evolution of 10msec is used to calculate displacements, and the median is taken through the 30 times repetition of the calculation procedure.

Fig.1 and Fig.2 display the translation errors of pre-ICA EEG signals (the EEG channels) and post-ICA EEG signals (the EEG components) respectively, for the embedding dimensions $D = 2$ through $D = 10$. The results of all the EEG channels or the EEG components for all the patients are put together in Fig.1 or Fig.2, where the open dots (∘) indicate the results for the patients in coma (A and C1) and the crosses (×) for the patients in brain death (B, C1, D, E and F). In Fig.2, we exclude the results for the EEG components that are similar to the sinusoid and are apparently the contaminations from the power supply. The translation errors of such components are very small because the

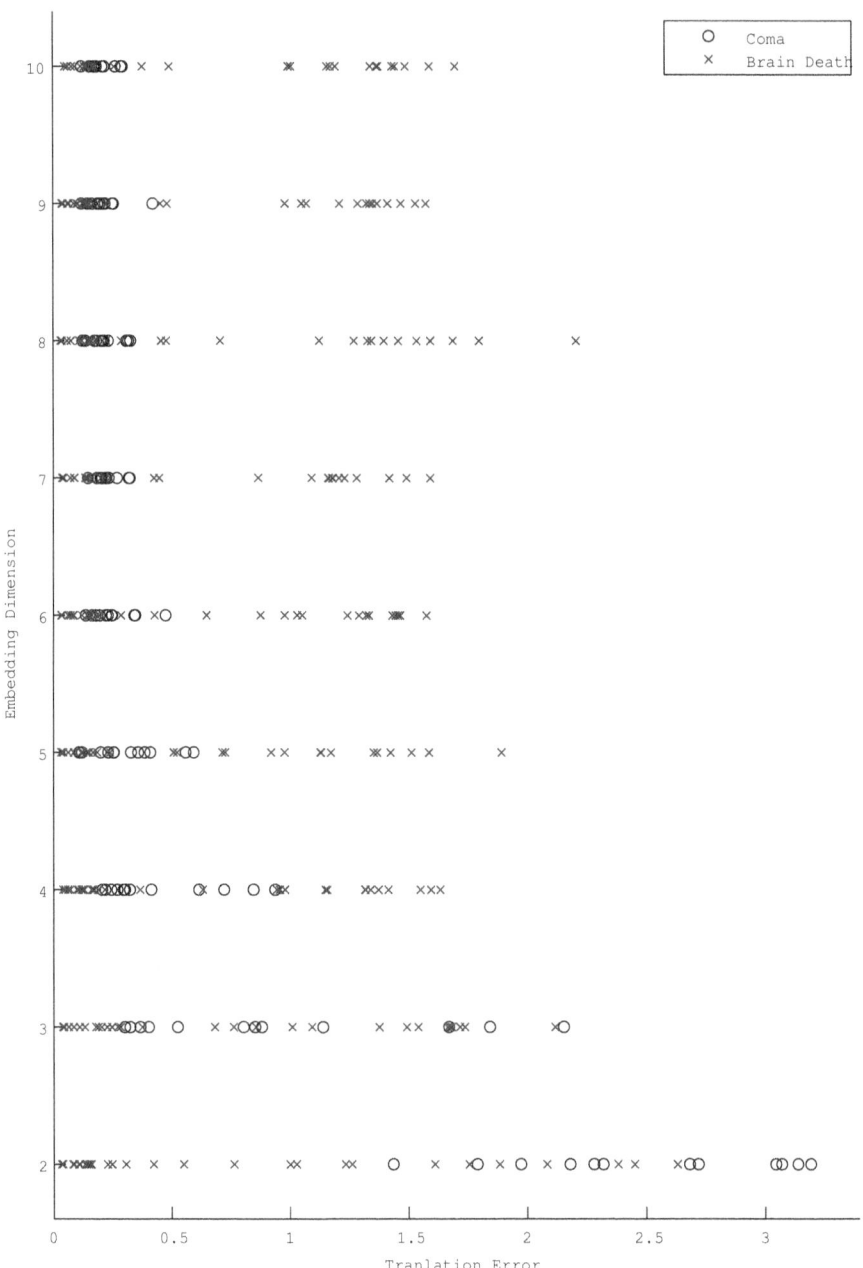

Fig. 1. Translation error of pre-ICA EEG signals (EEG channels)

Fig. 2. Translation error of post-ICA EEG signals (EEG components)

sinusoid is deterministic and can be embedded in a lower dimensional state space. In both figures, the open dots and the crosses are mixed at lower embedding dimensions ($D \leq 4$) while they tend to separate and form two clusters at higher embedding dimensions ($D \geq 7$), where the open dots (the patients in coma) forms the cluster for smaller values and the crosses (the patients in brain death) for larger values. In Fig.1, however, some crosses take smaller values than the cluster of open dots even at higher embedding dimensions. Those crosses correspond to the EEG channels contaminated by the sinusoid from the power supply which makes the translation error of the channels much smaller. On the other hand, in Fig.2, the open dots and the crosses separate clearly to form two clusters at higher embedding dimensions.

5   Concluding Remarks

Regarding the translation errors of the post-ICA EEG signals (the EEG components) at higher embedding dimensions ($D \geq 7$), the patients in coma forms the cluster for smaller values and the patients in brain death for larger values. This is explained by supposing that all the EEG components of patients in coma are related to the brain activities and the signals from the brain activities are deterministic whereas most EEG components of patients in brain death are related to stochastic environmental noises. On the other hand, the translation errors of the pre-ICA EEG signals (the EEG channels) do not show such clear separation mainly because the sinusoidal contaminations from the power supply make the translation errors of the EEG channels smaller for patients in brain death. From the observation so far, we conclude that the translation errors of the post-ICA EEG signals at higher embedding dimensions can be used as criteria for brain death diagnosis. Also the translation errors of the EEG signals can be used for verifying that certain components are stochastic noises and are not signals from the brain activities[5] as well as selecting the EEG component in brain death diagnosis[6].

Acknowledgments. The authors are indebted to Prof. Fanji Gu and Prof. Yang Cao of the Fudan University and Dr. Yue Zhang, Dr. Guoxian Zhu and Dr. Zhen Hong of the Shanghai Huashan Hospital for the EEG recordings. The present work was partially supported by KAKENHI (22560425).

References

1. Wijdicks, E.F.M.: Determining brain death in adults. Neurology 45, 1003–1011 (1995)
2. Wijdicks, E.F.M.: Brain death worldwide: accepted fact but no global consensus in diagnostic criteria. Neurology 58, 20–25 (2002)
3. Taylor, R.M.: Reexamining the definition and criteria of death. Seminars in Neurology 17(3), 265–270 (1997)

4. Hori, G., Amari, S., Cichocki, A., Mizuno, Y., Okuma, Y., Aihara, K.: Using ICA of EEG for judgment of brain death. In: Proc. Intl. Conf. on Neural Information Processing, pp. 1216–1219 (2000)
5. Hori, G., Aihara, K., Mizuno, Y., Okuma, Y.: Blind source separation and chaotic analysis of EEG for judgment of brain death. Artificial Life and Robotics 5(1), 10–14 (2001)
6. Hori, G., Cao, J.: Selecting EEG components using time series analysis in brain death diagnosis. Cogn. Neurodyn. 5 (2011), doi:10.1007/s11571-010-9149-2
7. Cao, J.: Analysis of the Quasi-Brain-Death EEG Data Based on a Robust ICA Approach. In: Gabrys, B., Howlett, R.J., Jain, L.C. (eds.) KES 2006. LNCS (LNAI), vol. 4253, pp. 1240–1247. Springer, Heidelberg (2006)
8. Cao, J., Chen, Z.: Advanced EEG signal processing in brain death diagnosis. In: Signal Processing Techniques for Knowledge Extraction and Information Fusion, pp. 275–298. Springer, Heidelberg (2008)
9. Chen, Z., Cao, J., Cao, Y., Zhang, Y., Gu, F., Zhu, G., Hong, Z., Wang, B., Cichocki, A.: An empirical EEG analysis in brain death diagnosis for adults. Cogn. Neurodyn. 2(3), 257–271 (2008)
10. Shi, Q.-W., Yang, J.-H., Cao, J.-T., Tanaka, T., Rutkowski, T.M., Wang, R.-B., Zhu, H.-L.: EMD Based Power Spectral Pattern Analysis for Quasi-Brain-Death EEG. In: Huang, D.-S., Jo, K.-H., Lee, H.-H., Kang, H.-J., Bevilacqua, V. (eds.) ICIC 2009. LNCS, vol. 5755, pp. 814–823. Springer, Heidelberg (2009)
11. Shi, Q., Cao, J., Zhou, W., Tanaka, T., Wang, R.: Dynamic Extension of Approximate Entropy Measure for Brain-Death EEG. In: Zhang, L., Lu, B.-L., Kwok, J. (eds.) ISNN 2010. LNCS, vol. 6064, pp. 353–359. Springer, Heidelberg (2010)
12. Amari, S., Cichocki, A., Yang, H.H.: A new learning algorithm for blind signal separation. In: Advances in Neural Information Processing Systems, vol. 8, pp. 757–763. The MIT press (1996)
13. Lee, T.W., Girolami, M., Sejnowski, T.J.: Independent component analysis using an extended infomax algorithm for mixed sub-Gaussian and super-Gaussian sources. Neural Computation 11(2), 417–441 (1999)
14. Wayland, R., Bromley, D., Pickett, D., Passamante, A.: Recognizing determinism in a time series. Phys. Rev. Lett. 70(5), 580–582 (1993)
15. Takens, F.: Detecting strange attractors in turbulence. In: Dynamical System and Turbulence. Lecture Notes in Mathematics, vol. 898, pp. 366–381. Springer, Heidelberg (1981)
16. Ikeguchi, T., Aihara, K., Itoh, S., Utsunomiya, T.: An analysis on the Lyapunov spectrum of electroencephalographic (EEG) potentials. Trans. IEICE E73(6), 842–847 (1990)
17. Gallez, D., Babloyantz, A.: Predictability of human EEG: a dynamic approach. Biol. Cybern. 64(5), 381–391 (1991)

Research on Relationship between Saccadic Eye Movements and EEG Signals in the Case of Free Movements and Cued Movements

Arao Funase[1,2], Andrzej Cichocki[2], and Ichi Takumi[1]

[1] Graduate School of Engineering, Nagoya Institute of Technology, Gokiso-cho,
Showa-ku, Nagoya, 466-8555, Japan
[2] Brain Science Institute, RIKEN, 2-1, Hirosawa, Wako, 351-0198, Japan

Abstract. Our final goal is to develop an inattentive driving alert system by EEG signals. To develop this system, it is important to predict saccadic eye moements by EEG signals.

In my previous studies, we found a sharp change of saccade-related EEG signals in P3 and P4 positions before the saccade. However, there are two problems in previous studies. As the first problem, we did not focus on slow cortical potentials that are important EEG signals like the P300 and the movement-related cortical potential. As the second problem, we did not observe EEG signals in the case of only cued saccade.

In this study, our purpose is to find differences of saccade-related cortical potentials between in cued movements and in free movements.

1 Introduction

Recently, many researcher have been developed the interface used by bio-signals to apply to physically challenged patients and to apply to quantitative evaluation of human statement. Especially, interfaces used by bio-signals generated from a brain make rapid progress. Brain Computer Interface (BCI) is interface used by bio-signals recorded on a brain non-invasively and Brain Machine Interface (BMI) is interface used by bio-signals recorded in a brain invasively. In many Brain computer interface, brain computer interface used by electroencephalography (EEG) is most popular.

EEG related to saccadic eye movements have been studied by our group toward developing a BCI eye-tracking system [1]. In previous research, saccade-related EEG signals were found by using the ensemble averaging method [1] and we can detect saccade-related EEG signals by using the single-trial method [2].

In the ensemble averaging method, the EEG signals were recorded during saccadic eye movements toward a visual or auditory stimulus and the experimental results showed the amplitude of the EEG declined just before to saccadic eye movements at one side of the occipital cerebrum. Moreover, the direction of the saccade coincides with the active side at the occipital cerebrum.

B.-L. Lu, L. Zhang, and J. Kwok (Eds.): ICONIP 2011, Part I, LNCS 7062, pp. 322–328, 2011.

2   Previous Results and This Problems

2.1   Experimental Settings

There were two tasks in this study (See Fig.1). The first task was to record the EEG signals during a saccade to a visual target when a subject moves its eyes to a visual stimulus that is on his right or left side. The second task was to record the EEG signals as a control condition when a subject dose not perform a saccade even though a stimulus has been displayed. Each experiment was comprised of 50 trials in total: 25 on the right side and 25 on the left side.

The experiments were performed in an electromagnetically shielded dark room to reduce the effect of electromagnetic noise and any visual stimuli in the environment. The visual targets were three LEDs placed in a line before the subject. One was located 30 [cm] away from the nasion of the subject (Fig. 2). The other two LEDs were placed to the right and left of the center LED, each separated by 25 degrees from the nasion. They were illuminated randomly to prevent the subjects from trying to guess which direction the next stimulus would be coming form next.

In order to record saccade-related EEG signals, we performed visually guided saccade task. This task was to record the EEG signals during a saccade to a visual target that is either his/her right side or left side. This experiment was comprised of 50 trials in total: 25 on the right side and 25 on the left side. The number of subjects is 5. Their age is from 22 to 24 years old.

The EEG signals were recorded through 19 electrodes (Ag-AgCl), which were placed on the subject's head in accord with the international 10-20 electrode position system (see Fig.3). The Electrooculogram (EOG) signals were simultaneously recorded through two pairs of electrodes (Ag-AgCl) attached to the top-bottom side and right-left side of the right eye.

All data were sampled at 1000 [Hz], and stored on a hard disk for off-line data processing after post-amplification. The raw EEG data was filtered by a high-pass filter (cut-off 0.53 [Hz]) and a low-pass filter (cut-off 120 [Hz]). The EOG data was recorded through a high-pass filter (cut-off 0.1 [Hz]) and a low-pass filter (cut-off 15 [Hz]).

2.2   Experimental Results

Fig. 4 shows electrical potential in a visually guided saccade task. Each graph are recorded on Fp1, Fp2, T3, T4, O1, O2 according to international 10-20 electrode position classification system [1]. The black lines show electrical potentials in right side movements and the gray lines show electrical potentials in left side movements. The horizontal axes indicate the time span and 0 [ms] is starting point of saccadic eye movements. The vertical axes indicate the electrical potentials.

From this figure, the amplitude of the EEG declined just before to saccadic eye movements at one side of the occipital cerebrum. Moreover, the direction of the saccade coincides with the active side at the occipital cerebrum.

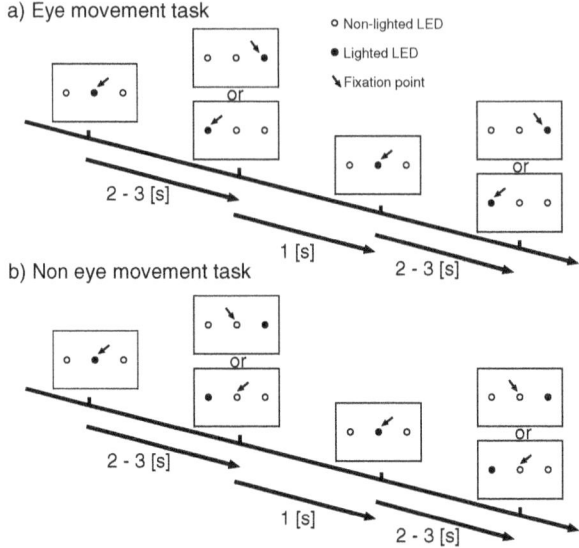

Fig. 1. Experimental tasks

These features are observed in the auditory guided saccade task. Therefore, these feature are dependent on visual and auditory stimuli and are related to only saccadic eye movements.

2.3   Problems in Previous Results

These features has a sharp change just before saccadic eye movements and a sharp change are changed in some dozens milliseconds before saccadic eye movements and Peak time of a sharp change is several milliseconds before saccade.

If we use these feature for input signals of BCI, It is short time between observing these feature and generating saccadic eye movement. Therefore, processing time for these feature is very short.

In this problem, we analyze anther electrical potentials before saccadic eye movements.

In previous experiments, we use a high-pass filter to extract sharp change of electrical potentials. The cut-off frequency of high-pass filter is 4 [Hz]. However, 0-4[Hz] include slow cortical potential components related to P300.

In this paper, we do not use the high-pass filter and focus on slow cortical potentials before saccadic eye movements.

3   Cued-Movement Experiments

3.1   Experimental Results

Fig. 5 is a result of a visual guided saccade task and subject-A moves his/her eyes to a right side target. This potential are recorded on P4 position. Horizontal

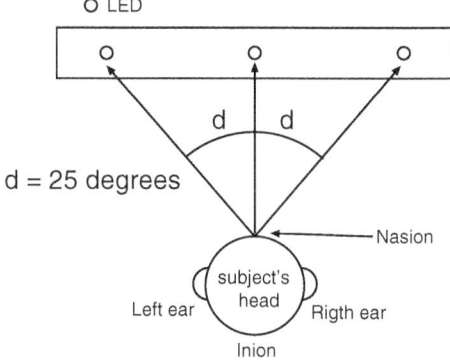

Fig. 2. Placement of LEDs

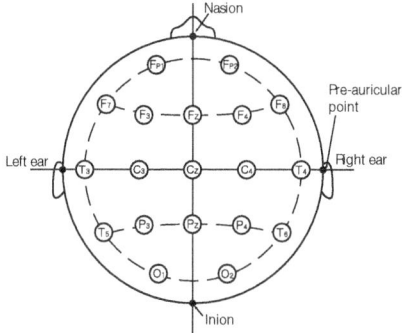

Fig. 3. International 10-20 electrode position classification system

axis indicates an electrical potential. Vertical axis indicated an time span and 0 [ms] is a starting point of eye movements. Averaging latency between showing visual stimuli and starting saccadic eye movements is 304 [ms].

A slow cortical potential are observed in this figure and a slow cortical potential decrease in about 800[ms] on P4, O2 position. When subjects moved to a left side target, a slow cortical potential decrease in about 800[ms] on P3, O1 position.

3.2   Discussion

From this results, we observed a slow cortical potential before saccadic eye movements. This slow cortical potential is not observed in no movements. This slow cortical potential is related to this experimental task. The position where a slow cortical potential are generated is related to direction of saccadic eye movements. From this result, this slow cortical potential is related to cued-movement based on visual stimuli. We do not decide whether this feature is related to processing of stimuli or to processing of movements. Therefore, in next step, we perform experiments in the case of free-movements.

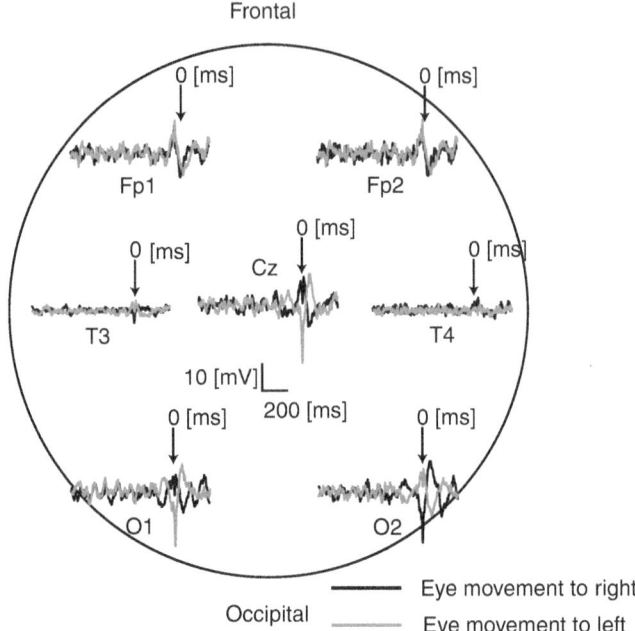

Fig. 4. Experimental results in previous results

4   Free-Movement Experiments

4.1   Experimental Setting

We performed Free-movement task in saccadic eye movements. Subjects watched center LED during 3-5[sec]. After 3-5[sec], subjects move their eyes to right or left side by subjects' free will and subjects watch right or left side LED during 1[sec]. After 1[sec], subjects return their eyes to center LED. This procedure is one trial. Each experiment was comprised of 50 trials in total: 25 on the right side and 25 on the left side. The number of subjects is 2 and they have normal vision.

4.2   Experimental Results

Fig. 6 is a result in free-movement of saccadic eye movements and subject-A moves his/her eyes to a right side target. This potential are recorded on P4 position. Horizontal axis indicates an electrical potential. Vertical axis indicated an time span and 0 [ms] is a starting point of eye movements.

We observed two components in this figure. First component has a slow cortical potential and has trend toward increase of potential from -1000[ms] to -500[ms]. Second component has a slow cortical potential and has trend toward decrease of potential from -500 [ms] to 0[ms].

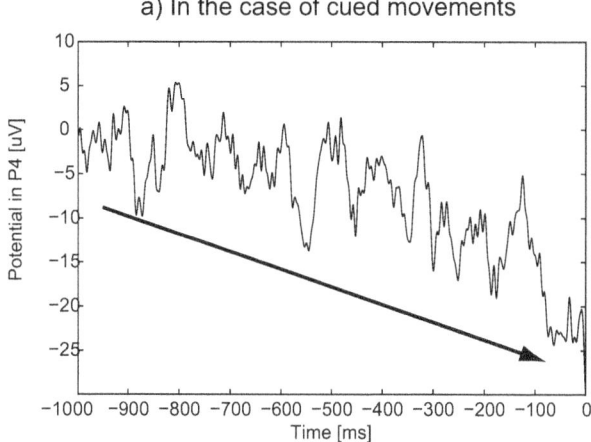

Fig. 5. Results in visually guided saccade task by ensemble averaging. (P4,Cued movement).

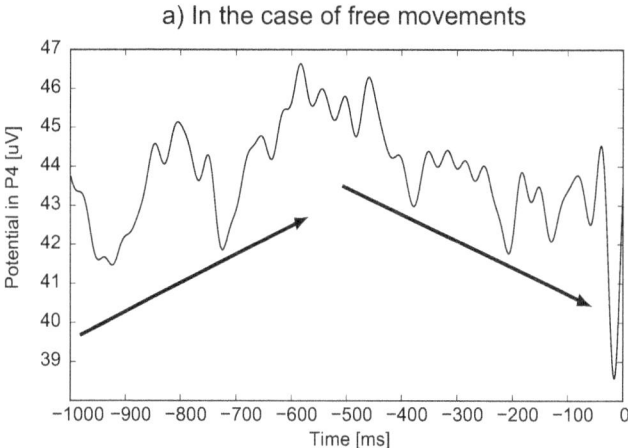

Fig. 6. Results in visually guided saccade task by ensemble averaging. (P4,Free movement).

These component are recorded in 2 subjects. When subjects move to left side target, these components are observed in P3 position.

4.3   Discussion

By compared with Fig. 5 and Fig. 6, we discussed features of EEG signals in saccadic eye movements.

First, in cued movement, a feature has only trend toward decrease. In the other hand, in free movements, one feature has a trend toward decrease from -500[ms] to 0[ms]. From this results, components from -500 to 0[ms] in two experiments

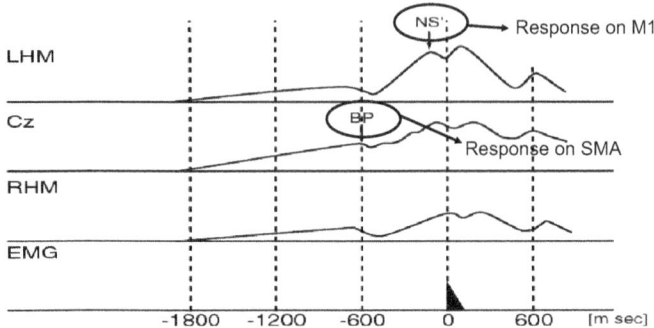

Fig. 7. Movement-related cortical potential

are the same components possibly. If this component is the same, this component is related to decision of saccadic eye movements.

Next, in free movements, one feature has a trend toward increase from -1000[ms] to -500[ms]. In the other hand, we do not observed this feature in cued movements. From this results, a brain function is not same from -1000[ms] to -500[ms] and it is difficult how brain function is related to this components from -1000[ms] to -500[ms]. However, this component has same tendency in BP components of movement related cortical potentials [3] (See Fig. 7). In the future, we must perfume experiments to estimate whether this component is related to BP components of movement related cortical potential.

5 Conclusion

In this paper, we focus on another components related to saccadic eye movements except for previous results. We performed experiments in cued-movement and free movements by not using high-pass filter. From these results, we observed a slow cortical potential in occipital lobe and this component has a trend toward decrease of potentials. This components are related to decision of saccadic eye movements possibly. In the future, we will confirm whether this components are related to decision of saccadic eye movements.

References

1. Funase, A., Yagi, T., Kuno, Y., Uchikawa, Y.: A study on electro-encephalo-gram (EEG) in eye movement. Studies in Applied Electromagnetics and Mechanics 18, 709–712 (2000)
2. Funase, A., Hashimoto, T., Yagi, T., Barros, A.K., Cichocki, A., Takumi, I.: Research for estimating direction of saccadic eye movements by single trial processing. In: Proc. of 29th Annual International Conference of the IEEE EMBS, pp. 4723–4726 (2007)
3. Deecke, L., et al.: Distribution of readiness potential, premotion positivity, and motor potential of the human cerebral cortex preceding voluntary finger movement. Exp. Brain Res. 7, 158–168 (1969)

A Probabilistic Model for Discovering High Level Brain Activities from fMRI

Jun Li and Dacheng Tao

Center for Quantum Computation & Intelligent Systems
University of Technology, Sydney

Abstract. Functional magnetic resonance imaging (fMRI) has provided an invaluable method of investing real time neuron activities. Statistical tools have been developed to recognise the mental state from a batch of fMRI observations over a period.

However, an interesting question is whether it is possible to estimate the real time mental states at each moment during the fMRI observation. In this paper, we address this problem by building a probabilistic model of the brain activity. We model the tempo-spatial relations among the hidden high-level mental states and observable low-level neuron activities. We verify our model by experiments on practical fMRI data. The model also implies interesting clues on the task-responsible regions in the brain.

Keywords: fMRI, Conditional Random Fields.

1 Introduction

Functional magnetic resonance imaging (fMRI) employs paramagnetic deoxy-haemoglobin in venous blood as the contrast agent for MRI and measures the haemodynamic response to neuron activities in the brain [1]. This makes non-invasive examination of the human brain possible, where the neuron activities can be observed with high-accuracy and in real time. Functional MRI investigates how the brain responds to a stimulus by measuring neuron-activity-related signals[1]. A typical fMRI test yields high-definition measurement of the brain. The neuron activities are measured at a spatial resolution as fine as several millimeters, and the measurement is made at a time interval of tenths of a second. The large volume of data requires specialised tools to aid human experts for analysis.

Early research has focused on establishing connection between fMRI observations and the neuron activations. On one side, statistical methods have been developed to discover which neurons are responsible for certain cognitive tasks by using fMRI measurements [2,3]. These methods consider the stimulus as a binary temporal signal. The neuron response to the stimulus is modelled by using a

---

[1] In particular, the neuron activities consume oxygen, and generates *blood oxygen level-dependent* (BOLD) signals, which is measurable by MRI.

B.-L. Lu, L. Zhang, and J. Kwok (Eds.): ICONIP 2011, Part I, LNCS 7062, pp. 329–336, 2011.

haemodynamic response function (HRF). Classical statistic hypothesis tests are then applied to examine whether individual neurons are related to a stimulus. To account for the tempo-spatial structure in the fMRI data, Bayesian and random field approach has been introduced in the model [4,5]. The hypothesis tests have proven useful to determine neuron activations for some cognitive tasks. However, the method depends the reliability of the underlying assumption. Ie the system may give error results if the employed HRF model is not consistent with the actual response of a neuron to a stimulus. Thus the HRF model must be designed carefully by experts; and it is not completely clear how to verify the model.

On the other side, fully automatic approaches have been developed based on data mining techniques, where no explicit neural response model is assumed [6,7]. Individual neurons are considered as observed objects, and the recorded activities are considered as attributes of the objects. Then the neurons are grouped based on the fMRI observations. One particular advantage of the structure-free methods is that they can be used to discover the hidden states beyond the individual neurons, for which explicit response function cannot be manually built.

Mitchell et al. [8] has proposed to use fMRI records to identify the mental states of the brain. They collected fMRI data of subjects during several trials, in which the subjects were required to perform a series of cognitive tasks, including reading, seeing and making judgements. A classifier is fitted to the observed fMRI data and the associated task tags. Then the fitted classifier is used of predict the cognitive states associated to testing fMRI data. The research shows that it is possible to recognise high level brain states from fMRI. However, despite the success of recognising the entire cognitive status during a period, structure-free methods discard the temporal relations in fMRI data, which may reveal interesting brain activities.

In this paper, we propose to use a probabilistic model to relate the brain states and the tempo-spatial structure of the neuron activities observed in fMRI data. For fMRI data analysis, the proposed approach is developed from the previous work of [8] by providing a dynamic analysis of the brain states. For technical development, the proposed approach takes advantage of both paradigm of fMRI analysis: (1) it models the temporal correlations in fMRI data; (2) it does not need assumption about the specific relation. Moreover, the model contains spatial correlations between the neurons, because it is reasonable to believe adjacent neurons behave similarly in performing cognitive tasks. Specifically, we employ a *conditional random field* (CRF, [9]) model to represent the dynamic relations of the fMRI records and the corresponding cognitive status. Learning the CRF model reveals the interested relations as conditional distributions of the cognitive activities given the fMRI observations. Moreover, the learned model may imply interesting clues of how different part of the brain functions in different cognitive activities. We have verified the model by applying it to the real fMRI data reported in [8]. Compared to the setting in [8], recognising the mental states at each moment is a more challenging task. We also show that the learned model parameters reveals interesting links between anatomical regions of the brain and neuron activities in particular tasks.

In the next section, we introduce the CRF framework and tailor the model for analysing fMRI data. We then report experimental results of recognising the cognitive status, and discuss some implications of the learnt model. Section 4 concludes the paper.

2   A Model for Dynamic Cognitive Activities

2.1   Conditional Random Field

In an fMRI test, a subject conducts a set of cognitive tasks, and his/her brain images are taken by MRI device. We consider the unknown mental states of the subject as *a hidden variable* and the observed brain image as a set of *input variables*, where each voxel in the image corresponds to one input variable. We will use X for input variables and Y for hidden variables. We are interested in the conditional distribution of Y given X, which reveals the dependencies of interest between the fMRI record and the cognitive activity. In a CRF model, the conditional distribution is as follows

$$p(\mathbf{y}|\mathbf{x}) = \frac{1}{Z(\mathbf{x})} \exp\left\{ \sum_{k=1}^{K} w_k f_k(\mathbf{y}_k, \mathbf{x}_k) \right\}. \tag{1}$$

In the distribution (1), the conditional density is affected by K possible relations between the variables, $f_1, \ldots, f_K$, which are called *feature functions*. Each feature function is affected by a subset of the variables. In (1), these input variables of the k-th feature function are denoted as $\mathbf{x}_k$ and $\mathbf{y}_k$. $Z()$ is constant with respect to the hidden variables. Given the values of the input variables, $Z(\mathbf{x})$ normalises (1) to be a proper probability distribution (unit sum),

$$Z(\mathbf{x}) = \sum_{\mathbf{y}} \exp\left\{ \sum_{k=1}^{K} w_k f_k(\mathbf{y}_k, \mathbf{x}_k) \right\}. \tag{2}$$

The CRF model in (1) is for general case. If the data has temporal characteristic, it is suitable to use a set of feature functions that fits sequential signal. In practice, it is often found useful to assume that the temporal variance of the signal is underlay by the hidden variables. Thus a pair of hidden states that correspond to adjacent moments are correlated by introducing a feature function defined on the two hidden variables and the input variables. The sequential CRF model is as follows

$$p(\mathbf{y}|\mathbf{x}) = \frac{1}{Z(\mathbf{x})} \exp\left\{ \sum_{k=1}^{K} w_k f_k(y_i, y_{i+1}, \mathbf{x}, i) \right\}. \tag{3}$$

Note that the feature index k and the variable index i are running over difference ranges respectively.

For the fMRI data, the fMRI observations are determined by hidden mental states. However, the response of fMRI measurement to neuron activities is not

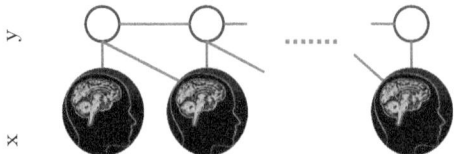

Fig. 1. Diagrams of CRF models

CRF for fMRI consider the relation between y_i and $\mathbf{x}_{i,...,i+L}$, in order to analyse the fMRI records after the inception of a latent cognitive activity. Circles represent hidden variables Y; and the brain fMRI illustrates represent observed variables X.

immediate. The interval between the change of mental state and reaction in fMRI can be up to several seconds [8]. Thus the model needs to consider the correlation between a hidden state at a moment and the input variables at several moments later. We introduce a set of feature functions for these correlations

$$p(\mathbf{y}|\mathbf{x}) = \frac{1}{Z(\mathbf{x})} \exp\Big\{ \sum_i w_{i,0} f(y_i, y_{i+1}, \mathbf{x}_i)$$
$$+ \sum_i w_{i,1} f(y_i, y_{i+1}, \mathbf{x}_{i+1}) + \dots$$
$$+ \sum_i w_{i,L} f(y_i, y_{i+1}, \mathbf{x}_{i+L}) \Big\}, \tag{4}$$

where L corresponds the delay of fMRI with respect to the change of hidden mental state. In (4), we use normal typeface for y_i, indicating they are scalars. The model structure is shown in Fig. 1. The figure shows the relationship between the fMRI observations and the mental states modelled by the CRF. The hidden variables (circles in the diagram) are connected to the observations (brain illustrations) that are moments later.

2.2 Model Learning by Maximum Likelihood

The probability of an observed sequence as a function of the model parameters w_k is the likelihood. The objective of model learning is to maximise the likelihood of all observed sequences over all possible model parameters. For the CRF model, the log-likelihood function is as follows

$$L(\{w_k\}) = \sum_m \Big\{ - \log Z(\mathbf{x}^{(m)}) + \sum_k \{ w_k f_k(\mathbf{y}_k^{(m)}, \mathbf{x}_k^{(m)}) \} \Big\}, \tag{5}$$

where m represents the data items. To find parameters $\{w\}$ that maximise the likelihood in (5), we consider the first order derivatives of L with respect to w,

$$\frac{\partial L}{\partial w_j} = \sum_m \Big\{ - \frac{1}{Z(\mathbf{x}^{(m)})} \frac{\partial Z(\mathbf{x}^{(m)})}{\partial w_j} + \sum_k f_k(\mathbf{y}_k^{(m)}, \mathbf{x}_k^{(m)}) \Big\} \tag{6}$$

$$= \langle f_j(Y_{A_j}, X_{A_j}) \rangle_{\tilde{p}(Y,X)} - \sum_m \langle f_j(Y_{A_j}, \mathbf{x}_{A_j}^{(m)}) \rangle_{p(Y|\mathbf{x}^{(m)}, \{w_k\})},$$

where $\langle \cdot \rangle_p$ represents taking expectation over a distribution p. The empirical estimate of the joint density is represented by $\tilde{p}(Y, X)$. Given the m-th sample, the conditional density over the hidden variables is represented by $p(Y|\mathbf{x}^{(m)}, \{w_k\})$. At a stationary point of (6) the expectation of each feature function over the empirical distribution is the same as the expectation over the model distribution.

2.3 Model Inference

Computing the empirical expectation (the first term on the r.h.s. of (6)) is straightforward. The second term on the r.h.s. of (6) takes expectation over the posterior of Y given $\mathbf{x}^{(m)}$ and the model parameters $\{w_k\}$

$$\langle f_k(Y, \mathbf{x}^{(m)}) \rangle_{p(Y|\mathbf{x}^{(m)}, \{w_k\})} = \sum_{y_1} \cdots \sum_{y_T} p(\{y_k\}|\mathbf{x}^{(m)}, \{w_k\}) f_k(\{y_k\}|\mathbf{x}^{(m)})$$

$$= \sum_{y_1} \cdots \sum_{y_T} \prod_i p(y_i, y_{i+1}|\mathbf{x}^{(m)}, \{w_k\}) f_k(\{y_k\}|\mathbf{x}^{(m)}). \tag{7}$$

We can exploit the chain structure of the CRF; and (7) is then amenable to belief propagation algorithm [9].

3 Experiment

We conduct experiments on modelling the relations between fMRI records and cognitive process. This section briefly report the experimental results and compare to the classical classifiers.

3.1 Brain Images and Task Settings

The fMRI data we used in our experiment is from [8]. The data consists of fMRI records of 6 subjects taking a cognitive test. The brain images are taken every 0.5 second during the test. In the test, each subject attends 40 trails. The protocol of carrying out a trail is as follows. Two types of stimuli are used.

- A pictorial stimulus is an arrangement of two symbols chosen from *(star), $(dollar) or +(plus). The symbols are put one over another.
- A semantic stimulus is a statement about the arrangement of the symbols, eg "It is true that the plus is over the star."

In each trial, there are two stimuli from each category. The first stimulus is shown for 4 seconds, following by 4 seconds rest. Then the second stimulus is shown, and the subject is asked to answer whether the statement about the arrangement of the symbols is valid. There are 15 seconds rest before the next trial.

The task is to identify five mental states that a subject may experience during a trial: (1) reading a sentence after a rest (2) seeing a picture after a rest (3) reading a sentence after seeing a picture (judge making) (4) seeing a picture

Table 1. Dynamic cognitive activity recognition rates

	S1	S2	S3	S4	S5	S6	Avg.
CRF	59.2	**77.6**	**81.1**	**76.6**	**68.1**	**79.6**	**73.6**
HMM	**60.0**	65.3	72.9	73.3	65.7	72.4	68.3
SVM	59.0	70.2	71.6	67.4	64.4	69.8	67.1
NB	52.6	54.5	60.0	38.2	50.4	52.7	54.6
1-NN	35.8	47.9	50.6	48.2	42.8	43.1	44.6

after reading a sentence (judge making) (5) resting For the training sequences, we tag the first 8 seconds as state (1)/(2), the next 8 seconds as state (3)/(4) and the rest of a trial as state (5). The mental state at each moment in a test sequence is predicted by the trained CRF.

The voxels in the fMRI scan have anatomical tags. Thus the voxels are segmented into regions according to brain anatomy. The signs of the average voxel values in each region are used as the input variables of the CRF.

For the temporal relations, we set $L = 8$ in (4). Since the time interval between two fMRI frames is 0.5 second, this value corresponds to consider a 4 second delay between change of cognitive state and the response in BOLD signal.

3.2   Recognising Dynamic Cognitive Activities

For each subject, training and test data consist of 50% trials, respectively. We have compared the proposed CRF based classifier to three widely used classifiers: nearest neighbour (NN), support vector machine (SVM, [10]), and naïve Bayes (NB). We also include *hidden Markov model* (HMM) in our comparison. This can be implemented within the proposed CRF framework. We can ignore the temporal relations between the fMRI inputs at one moment and the hidden states of other moments (let $L = 0$ in (4)), then the CRF degrades to an HMM.

The performances of different methods in the test of mental state recognition is shown in Table 1. In general, the temporal information allows HMM and CRF to outperform the classical classifiers, which predict the mental state for each moment independently. CRF outperform HMM in most tests because the model better fits the brain operations. Note that the performance on Subject 1 is inferior for all classifiers, which indicates that the data may be corrupted by bad alignment or noises.

In Fig. 2 (last page), we illustrate the weights uncovered by the CRF model. The colours of the ROIs indicates the contribution of the that area to the cognitive task (reading a sentence *without* recalling a picture in the shown example) in a period after the snapshot of interest (t_0). Though it is premature for making a tight conclusion, the model gives interesting clues of the function mapping of the brain from a statistical perspective. For example, when reading a sentence, the weight of the *right-dorsolateral prefrontal cortex* (RDLPFC, white boxed in the figure) peaks after about 3 seconds from t_0. The postulated function of the dorsolateral prefrontal cortex includes associating concepts, e.g. words [11]. The

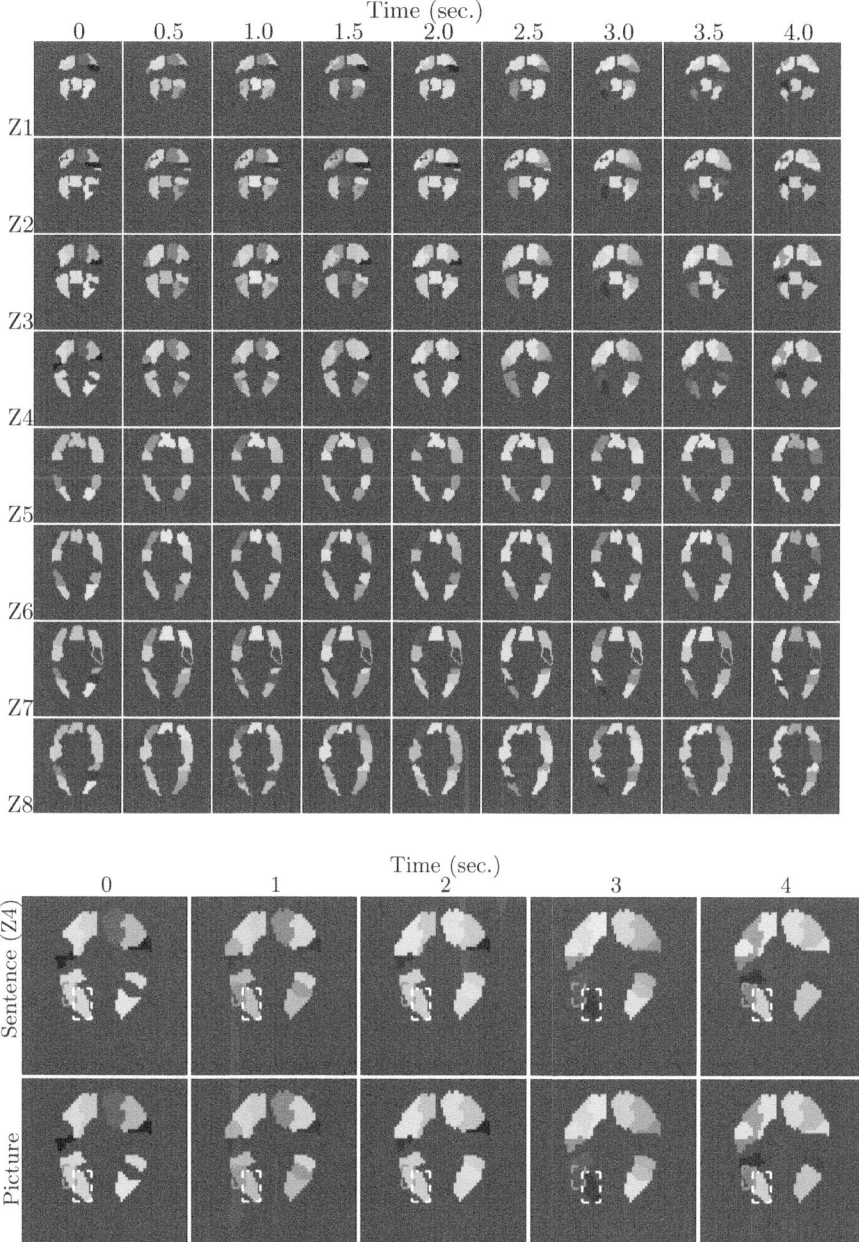

Fig. 2. Weights of ROI for cognitive activities over time
The figure shows the learnt weights of the feature functions. Each feature function corresponds to an *ROI* after a *latency* from the time of interest t_0 for a specific *cognitive activity*. Red colours represent large weights and blue colours for small weights. **Upper**: weights for reading a sentence, ROIs are shown with 8 slices along the Z-axis (vertical) of the brain; **Lower**: comparison between weights for reading a sentence and those for seeing a picture at slice $Z4$.

area's response in seeing a picture is less obvious. In contrast, when seeing a picture, the weight of the *right-frontal eye field* (RFEF, golden boxed) rises at t_0. Frontal eye field is reported responsible for eye movement [12].

4   Conclusion

In this paper, we propose to use CRF to model the relations between high-level mental states and the fMRI brain images. The structure CRF allows the model to explore temporal relations in fMRI data. Experiment on real data shows the advantage of CRF, and alludes to functions of anatomical parts of the brain from a statistical perspective.

In the future, we can incorporate high order features defined on subsets of the anatomical ROIs in the model in order to capture the interactions between ROIs. We can also explore model selection methods to automatically balance the model complexity.

References

1. Ogawa, S., Lee, T.M., Kay, A.R., Tank, D.W.: Brain magnetic resonance imaging with contrast dependent on blood oxygenation. Proceedings of the National Academy of Sciences 87, 9868–9872 (1990)
2. Friston, K.J., Holmes, A.P., Poline, J.B., Gransby, P.J., Williams, S.C.R., Frackowiak, R.S.J., Turner, R.: Analysis of fMRI time series revisited. NeuroImage 2(1), 45–43 (1995)
3. Lange, N., Zeger, S.L.: Non-linear Fourier time series analysis for human brain mapping by functional magnetic resonance imaging. Applied Statistics 46(1), 1–29 (1997)
4. Wang, Y., Rajapakse, J.C.: Contextual modeling of functional MR images with conditional random fields. IEEE Transactions on Medical Imaging 25(6), 804–812 (2006)
5. Quiriós, A., Diez, R.M., Wilson, S.P.: Bayesian spatiotemporal model of fMRI data using transfer functions. NeuroImage (2010)
6. Goutte, C., Toft, P., Rostrup, E., Nielsen, F., Hansen, L.K.: On clustering fMRI time series. NeuroImage 9(3), 298–310 (1999)
7. Hansen, L.K., Larsen, J., Nielsen, F.A., Strother, S.C., Rostrup, E., Savoy, R., Lange, N., Sidtis, J., Svarer, C., Paulson, O.B.: Generalizable patterns in neuroimaging: how many principal components? NeuroImage 9(5), 534–544 (1999)
8. Mitchell, T.M., Hutchinson, R., Niculescu, R.S., Pereira, F., Wang, X.: Learning to decode cognitive states from brain images. Machine Learning 57, 145–175 (2004)
9. Sutton, C., Mccallum, A.: An Introduction to Conditional Random Fields for Relational Learning. MIT Press (2006)
10. Vapnik, V.: The Nature of Statistical Learning Theory. Springer, Heidelberg (1995)
11. Murray, L.J., Ranganath, C.: The dorsolateral prefrontal cortex contributes to successful relational memory encoding. Journal of Neuroscience 27, 5515–5522 (2007)
12. Kolb, B., Whishaw, I.: Fundamentals of Human Neuropsychology. Worth Publishers (2003)

Research of EEG from Patients with Temporal Lobe Epilepsy on Causal Analysis of Directional Transfer Functions

Zhi-Jun Qiu[1], Hong-Yan Zhang[2], and Xin Tian[2,*]

[1] Tianjin Neurological Institute,154 An Shan Rd., Heping District, Tianjin 300052, China
[2] Laboratory of Neuroengineering, Research Centre of Basic Medicine,
School of Biomedical Engineering, Tianjin Medical University,
22 Qi Xiang Tai Rd., Heping District, Tianjin 300070, China
`tianx@tijmu.edu.cn`

Abstract. Objective: 16-channel EEG data during intermittent episodes of epilepsy is recoded and analyzed to find lesions source and relationship between brain areas for temporal lobe epilepsy (TLE) patients by causal analysis method.

Methods: There are 8 patients with temporal lobe epilepsy, 5 males and 3 females, aged between 19 to 47 years, the average age of 30.63 years. 16-channel EEG in 8 patients was recorded by Stellate Video EEG. Sample time = 20s (sample points = 4000), Sampling frequency $f_s = 200 Hz$. Directional transfer functions is used to direct the information transduction between each channel of the EEG signals, which can reflect the causal relationship between each channel and determine the location of the lesions source. (In this paper, we used eConnectome software that developed by Biomedical Functional Imaging and Neuroengineering Laboratory at the University of Minnesota, directed by Dr. Bin He).

Results: Causality results of 8 patients during intermittent episodes of EEG 20s are as follows: 6 patients' lesions source are located on channels T5 and F7 in left tempora, One of 5 cases' are located on channel T5 in the left posterior temporal, One of 1 case is located on channel F7 in the left anterior temporal. And 2 patients' lesions source are located on channels T4 and T6 in the right tempora, in the 2 patients, 1 case's lesions source is located on channel T4 in right middle temporal, 1 case's lesions source is located on channel T6 in right posterior temporal. Causality results consistent with the clinical diagnosis.

Conclusions: Research of EEG on causal analysis of directional transfer function can effectively determine the lesions source of seizure, and effectively calculate the transmission direction of the multi-channel information, which is to provide support in the clinic for determine the source of seizure.

Keywords: temporal lobe epilepsy (TLE), Lesions source, EEG, Directed transfer function, causality.

1   Introduction

Temporal lobe epilepsy (TLE) is the common types of epilepsy. Patients with TLE seizure in various forms, such as twitch, absence, paresthesia, mood disorders and so

---

* Corresponding author.

B.-L. Lu, L. Zhang, and J. Kwok (Eds.): ICONIP 2011, Part I, LNCS 7062, pp. 337–344, 2011.
© Springer-Verlag Berlin Heidelberg 2011

on. The Lesions source of TLE is often clear localized and most located in temporal lobe, and a few outside temporal lobes such as in the insular cortex, orbital gyrus and thalamus. TLE can be divided into hippocampal epilepsy, amygdala epilepsy and lateral temporal lobe epilepsy by anatomic site.

Almost 70% of patients with TLE become to refractory epilepsy patients, who were ineffective drug therapy. Surgery is one of the treatments of refractory epilepsy. To determine the lesion source is the key point of successful surgical treatment, and is one of the most hot research issues at home and abroad.

Currently, through the 16-channel dynamic EEG, 128 guided long range video EEG monitoring, CT scan examination, routine MRI examination sequence, SPECT examination, DSA inspection, auxiliary interictal and induced seizures PET / CT examination, and combination with the clinical features of seizures, source of epileptic lesions can be roughly localization.

And through the recording of intracranial EEG, source of epileptic lesions can be precisely localized. Intracranial electrical stimulation can determine the various functional areas of cortex, such as the sports area, language area, sensory area.

Non-invasive EEG reflects the electrical activity of brain or functional state of brain. Abnormal discharge of epileptic lesions can be recorded in EEG. It is the most common neural electrical activity signal. The typical waveforms of epileptic EEG are spike, sharp wave, spike-slow wave integrated, sharp -slow wave integrated, rhythmic changes in seizures.

Nowadays interesting results of EEG synchronization of the nonlinear correlation have been reported in many papers [1]. Numerical analysis and statistical analysis are widely used in EEG analysis. There are too many common analytical tools in EEG analysis, such as correlation dimension; point correlation dimension; mutual dimension; and Ivanov entropy Cole Moge and independent component analysis (ICA) [2]. In recent years, causality analysis of the development is one of the most hot research issues.

The first truly multichannel estimator of direction of propagation--Directed Transfer Function (DTF) based on the multichannel autoregressive model was given by Kamin´ski and Blinowska (1991). DTF is an estimator of the intensity of activity flow between structures, depending on frequency of the signal. DTF function is sensitive to the time delay of signals. If we assume that the propagation of activity is connected with the information transfer we can imply that the high value of DTF indicates the information flow (including direction) between two given structures. Thus, we can detect the influence of one structure onto another [3].

In this paper, we use eConnectome software which developed by Biomedical Functional Imaging and Neuroengineering Laboratory at the University of Minnesota, directed by Dr. Bin He. EConnectome is an open-source software package for mapping and imaging brain functional connectivity from electrophysiological signals. Brain activity is distributed in the three-dimensional space and evolves in time. The spatio-temporal distribution of brain electrical activity and the network behavior provide important information for understanding the brain functions and dysfunctions. As part of the efforts of the Human Connectome project, which shall be aimed at mapping and imaging structural and functional neural circuits and networks, eConnectome provides a tool for investigators to map and image brain functional connectivity from electrophysiological signals, at both the scalp and cortex level. The software can also be used to perform scalp EEG mapping over a generic realistic geometry head model, and cortical current density imaging in a generic realistic geometry boundary element head model constructed from the standard Montreal Neurological Institute brain. The current release allows functional connectivity imaging from EEG. The visualization module is jointly developed with Drs. Fabio Babiloni and Laura Astolfi at the University of Rome "La Sapienza". The development and execution environment of the software requires MATLAB in Windows operating system [4].

2   Methods

2.1   The Clinical 16-Channel EEG Signals from the Patients with TLE

In this paper, 8 cases of TLE by clinically diagnosed were collected (5 males and 3 females; age 19 years to 47 years, the average age of 30.63 years). All the patients' clinical symptoms and EEG consistent with the 1981 and 1989 International League Against Epilepsy's classification of epileptic seizures and epilepsy syndrome standards in TLE.

16-channel video EEG was recorded with Stellate, which is a 16-channel video-EEG recording system, with a bandpass filter of 0.5 - 70Hz. EEG sampling rate is 200Hz. EEG data trails of 20s of duration are acquired. EEG recording electrodes placement is consisted with 10/20 system, as shown in Fig.1.

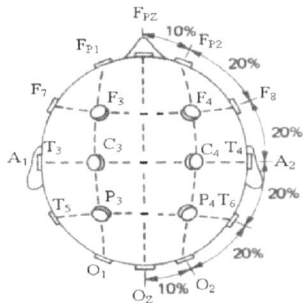

Fig. 1. 16-channel EEG electrodes were placed as shown above

Nasion (between the eyes above the nose concave point) is marked as Nz. Occipital protuberance (head back bulge point) is marked as Lz. Left and right preauricular points are marked as A1 and A2. Then draw a longitude line through that points

(Lz and Nz). The line was divided equally at 10%. Draw parallels through these equal points. Electrodes were placed along the parallel of 10% or 20% of multiple locations.

2.2   Causal Analysis of 16-Channel EEG on Directional Transfer Function

Neural dynamics are often usefully interpreted in the frequency domain. In this paper, a frequency-domain interpretation of Directed Transfer Function (DTF) is used in order to detect and quantify directional relationships between each channel, during the intermittent seizures of epilepsy. DTF was proposed by (Kaminski and Blinowska, 1991) and used to determine the directional influences between any given pair of channels in a multivariate data set. This is an estimator characterizing at the same time direction and spectral properties of the brain signals, and requires only one multivariate autoregressive model to be estimated from all the EEG channel recordings. DTF is an estimator of the intensity of activity flow between structures, depending on frequency of the signal. DTF function is sensitive to the time delay of signals. If we assume that the propagation of activity is connected with the information transfer we can imply that the high value of DTF indicates the information flow (including direction) between two given structures. Thus, we can detect the influence of one structure onto another.

Use $X(t)$ to represents the 16 EEG, and use $x_n(t)$ (n=4000)to represent the n discrete time points of EEG. Specifically, let $X = x_{ij} (i = 16, j = 4000)$ denote the measurement from 16 channels at time 20s.

$$X = \begin{bmatrix} x_{1,1} \cdots x_{1,4000} \\ \vdots \quad \ddots \quad \vdots \\ x_{16,1} \cdots x_{16,4000} \end{bmatrix} \tag{1}$$

Application of autoregressive methods of EEG analysis can be show as follow:

$$X(i) = \sum_{n=1}^{p} A(n)X(n-i) + E(i) \tag{2}$$

Here E is the vector of not uniformly zero mean uncorrelated white noise. The p is 16 in the AR model. In order to make X not equal, we make more convenient way of expressing the same concept is that coefficients A are not uniformly zero under suitable statistical criteria [5].

To examine the causal relations in the spectral domain, we Fourier transform (2) to the frequency domain yields

$$A(f)X(f) = E(f) \tag{3}$$

Causal analysis with directional transfer function calculated for each channel of information between the brain signals the direction of conduction.

The DTF function λ^2_{ij} describing flow from channel i to channel j is defined as the square of the absolute value of the complex transfer function of the AR model, divided by the sum of these values for row i of matrix A [6].

$$\lambda^2_{ij} = \frac{\left|A_{ij}(f)\right|^2}{\sum_1^p \left|A(f)\right|^2} \tag{4}$$

In this paper, we use eConnectome software that developed by Biomedical Functional Imaging and Neuroengineering Laboratory at the University of Minnesota, directed by Dr. Bin He. The DTF method is used in this software. EEG is calculated in the frequency domain and processed by MATLAB program. And the results were obtained. (http://www2.imm.dtu.dk/~pch/Regutools/)

3 Results

3.1 The Results of Patients with Left-TLE

In this paper, 16-channel EEG data which is collected of 6 patients with left-TLE is analyzed accurately. The 20s segments of EEG waveforms of 16 electrodes and the results can be seen in Fig.2. Additionally in Fig.2a (Fig.2c, Fig.2e, Fig.2g, Fig.2i, Fig.2k) we present the EEG registration corresponding to the time span of clinical symptoms. For the first patient, as shown in Fig2.b, the channel T5 in left-temporal lobe is lesions source, indicating the highest flow of signals from this area. The channel P4 in parietal region, the channel C4 in occipital region, and the channel Fp2 in frontal region are influenced by the source electrode. For the second patient, as shown in Fig.2d, the channel T5 in left-temporal lobe is lesions source. The channels O1, O2 in occipital region, the channel Fp1 in frontal region, the channels P3 , p4 in parietal region, and the channels C3,C4 in central region are influenced by the source electrode. For the third patient, as shown in Fig.2f, the channel F7 in left-temporal lobe is lesions source. The channels F3, F4 in frontal region are influenced by the source electrode. For the fourth patient, as shown in Fig.2h, the channel T5 in left-temporal lobe is lesions source. The channels Fp1, Fp2 in prefrontal region, the channels F3, F4 in frontal region, the channels C3, C4 in central region, and the channels P3 , p4 in parietal region ,are influenced by the source electrode. For the fifth patient, as shown in Fig.2j, the channel T5 in left-temporal lobe is lesions source. The channels Fp1, Fp2 in frontal region, the channel F4 in right-frontal region ,and the channel T4 in right-temporal lobe are influenced by the source electrode. For the sixth patient, as shown in Fig.2l, the channel T5 in left-temporal lobe is lesions source. The channels Fp1, Fp2 in prefrontal region, and the channel F4 in right-frontal region are influenced by the source electrode.

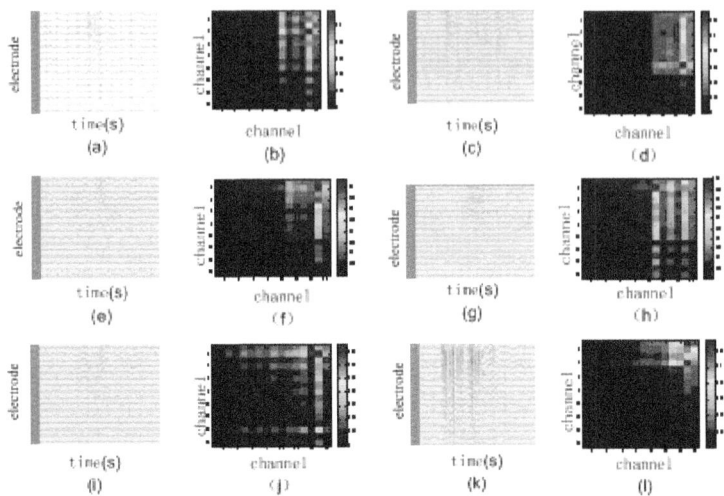

Fig. 2. In the 6 patients with left-TLE group, The 16-channel EEG and causality results are as shown above. (a) (c) (e) (g) (i) (k) are the EEGs of the first patient to sixth patient, (b) (d) (f) (h) (j) (l) are the causality results of the first patient to sixth patient.

3.2   The Results of Patients with Right-TLE

In this paper, 16-channel EEG data which is collected with two patients with right-temporal lobe epilepsy is analyzed exactly. The 20s segments of EEG waveforms of 16 electrodes and the results can be seen in Fig.3. Obviously, in Fig.3a (Fig.3c,) we present the EEG registration corresponding to the time span of clinical symptoms. For the seventh patient, as shown in Fig.3b, the channel T4 in right-temporal lobe is lesions source. The channels O1, O2 in occipital region, and the channel T5 in left-temporal lobe are influenced by the source electrode. For the eighth patient, as shown in Fig.3b, the   channel T6 in right-temporal lobe is lesions source. The channels O1, O2 in occipital region and channels C3, C4 in frontal region, the channels F3, F4 in frontal region and the channels P3, p4 in parietal region are influenced by the source electrode.

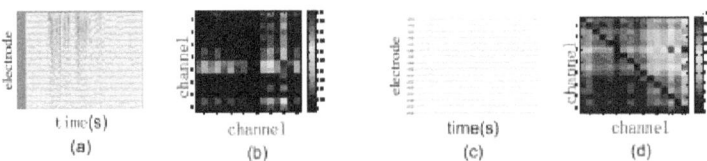

Fig. 3. In the 2 patients with right-TLE group, The 16-channel EEG and causality results are as shown above. (a) (c) are the EEGs of the first patient to sixth patient, (b) (d) are the causality results of the first patient to the sixth patient.

4   Conclusion

Through the analysis of above two groups of patients, DTF algorithm can be used to determine the location of the source of epileptic focus, and effectively calculate transmission direction of information between multi-channel EEG of help to the clinical lesions initially identified the source of seizures.

5   Discussion

In this paper, patients of unilateral epilepsy are selected, whose characters are clear. In addition, we discuss information transmission of patient with bilateral TLE, shown in Fig.4, whose lesions source is the left temporal lobe lesions F7 and right temporal lobe T6.

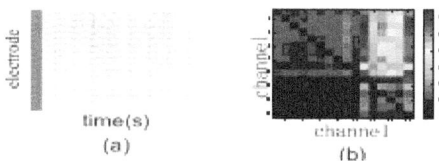

Fig. 4. The 16-channel EEG and causality results are as shown above. (a) is the EEG of the patient , (b) is the causality result of the patient.

Fig.4 shows that patient with bilateral TLE has right medial temporal lesions and limitations of the left temporal lobe perfusion lowered. Therefore, the right medial T6 lesion effects on other brain regions, the limitations of the left temporal parietal F7 reduced blood perfusion, indicating that the lesions are located in these regions. In recent years, cases of brain regions on both sides have been a wide range of researched, bilateral research has important significance.

Acknowledgements. Professor He Bin and all members, university of Minnesota biomedical functional imaging and neural engineering laboratory, made great contributions to this paper. Here, I express my sincere   acknowledge to them. This work was supported by the Natural Science Foundation of China, Grant No. 61074131 and Specialized Research Fund for the Doctoral Program of Higher Education (SRFDP), Grant No. 20101202110007.

References

1. Marinazzo, D., Liao, W., Chen, H., Stramaglia, S.: Nonlinear connectivity by Granger causality. NeuroImage 99, 9–18 (2010)
2. Astolfi, L., Cincotti, F., Mattia, D., Babilonic, C., Carducci, F., Basilisco, A., Rossini, P.M., Salinari, S., Ding, L., Ni, Y., He, B., Babiloni, F.: Assessing cortical functional connectivity by linear inverse estimation and directed transfer function. Simulations and Application to Real Data Clinical Neurophysiology 116, 920–932 (2005)

3. Londei, A., D'Ausilio, A., Basso, D., et al.: Brain network for passive word listening as evaluated with ICA and Granger causality. Brain Research Bulletin 72, 284–292 (2007)
4. Babiloni, F., Cincotti, F., Babiloni, C., Carducci, F., Mattia, D., Astolfi, L., Basilisco, A., Rossini, P.M., Ding, L., Ni, Y., Cheng, J., Christine, K., Sweeney, J., He, B.: Estimation of the cortical functional connectivity with the multimodal integration of high-resolution EEG and fMRI data by directed transfer function. Neuroimage 24(1), 118–131 (2005)
5. Anna, K.: Determination of information flow direction among brain structures by a modified directed transfer function (dDTF) method. Journal of Neuroscience Methods 125, 195–207 (2003)
6. Franaszczuk, P.J., Bergey, G.K.: Application of the Directed Transfer Function Method to Mesial and Lateral Onset Temporal Lobe Seizures. Brain Topography, 11–12 (1998)

Multifractal Analysis of Intracranial EEG in Epilepticus Rats

Tao Zhang and Kunhan Xu

College of Life Sciences, Nankai University, Tianjin, PR China, 300071
zhangtao@nankai.edu.cn

Abstract. Networks of living neurons exhibit diverse patterns of activity, which are always operating far from equilibrium in the mammalian central nervous system. In this study, a blocker of glutamate transporter was employed to detect if nonlinear interactions changed when extra glutamate acted as a potent neurotoxin to neural activity in hippocampus of epileptic Wistar rats. A hypothesis was made that a decrease of complexity of information could be occurred by accumulation of glutamate in hippocampus. An investigation was performed to measure intracranial EEG, which were obtained from two parts of brain in three rat's groups, by using multifractal detrended fluctuation analysis. The results demonstrate that the change of nonlinear interactions in neural network can be clearly detected by multifractal analysis. Moreover, small fluctuation in activity of network exhibited a decrease in multifractal behavior, suggested that the complexity of information transmitting and storing in brain network was weakened by glutamate accumulation.

Keywords: multifractal detrended fluctuation analysis, epileptic rats, glutamate transporter.

1 Introduction

Nonlinear dynamics has been shown to be important in describing complex neural networks and time series, such as electroencephalogram (EEG) [1]. Previous investigations indicated that there were non-trivial long-range correlations within the human EEG signals [2], which was one of fractal features related to self-similar fluctuations [3]. However, neural dynamical systems, driven by multiple-component feedback interactions, actually showed non-equilibrium, variable fractal behaviors, multifractal [4], and their fractal features were hard to be characterized by traditional fractal measurements, such as detrended fluctuation analysis and multiscale entropy [5]. In this case, the method of multifractal analysis becomes a more reasonable approach for measurement of the fractal characteristics of EEG signals. The earliest studies of the nonequilibrium of cardiac interbeat interval time series revealed that the healthy subjects showed more multifractal structure than diseased subjects. Moreover, studying multifractal EEG using wavelet transform modulus maxima (WTMM) method was employed as a new direction for brain dynamics [6]. Recently, Kantelhardt *et al.* developed a new mathematical algorithm named Multifractal detrended

B.-L. Lu, L. Zhang, and J. Kwok (Eds.): ICONIP 2011, Part I, LNCS 7062, pp. 345–351, 2011.
© Springer-Verlag Berlin Heidelberg 2011

fluctuation analysis (MF-DFA), which made analyzing multifractals in nonstationary time series feasible [7]. MF-DFA is based on a generalization of detrended fluctuation analysis (DFA) , and allows more reliable multifractal characterizations for multifractal nonstationary time series than the method of WTMM [8].

Glutamate is the primary excitatory neurotransmitter in the mammalian central nervous system and acts as a potent neurotoxin, implicated as a neurotoxic agent in several neurologic disorders including epilepsy, ischemia, and certain neurodegenerative diseases. The termination of neurotransmission is mediated by sodium-dependent high affinity glutamate transporters, which play an important role in maintaining the extracellular glutamate concentration below neurotoxic levels. A disturbance in glutamate-mediated excitatory neurotransmission has been implicated as a critical factor in the etiology of adult forms of epilepsy. Manipulation of glutamate transporter expression can lead to various neurologic dysfunctions. For example, in epileptic mice deficient in GLT-1, it has shown lethal spontaneous seizures and increased susceptibility to acute injury. Disruption of transporter activity could lead to changes in network activity as a result of enhanced interneuron excitability. Changing glutamate transporter may affect the patterns of complicated EEG and the structure of brain network, particularly when associated with neuronal diseases, such as epilepsy.

In the present study, disruption of glutamate transporter activity could lead to changes in neural network activity as a result of enhanced interneuron excitability, particularly seizures when associated with neuronal diseases such as epilepsy. Therefore, MF-DFA was employed to analyze continuous EEG time series obtained from intracranial depth electrodes placed in the dentate gyrus (DG), which was studied as the focus of temporal lobe epilepsy in hippocampus, and the perforant pathway (PP) investigated as the path for transmitting information from the entorhinal cortex. The changes in nonlinear dynamic were found by multifractal analysis among three groups of animals. And decreasing multifractals caused by hyperexcitability in neural network oscillations were discussed as well.

2   Methods

2.1   Multifractal Detrended Fluctuation Analysis

The approach of multifractal detrended fluctuation analysis (MF-DFA) was employed to calculate the singular spectrum of the intracranial EEG signals, the details of algorithm can be found in the reference [7].

2.2   Surrogate Data Analysis

Whether the MF-DFA analysis truly reflects a multifractal behavior in the EEG or just a broad probability density function for the values is tested by simply shuffling the original intracranial EEG sequences to construct surrogate data. The results are subsequently compared with that of the original sequences for all three groups, respectively. The averaged result was obtained from surrogate data, which were shuffled 20 times for each segment of original intracranial EEG.

2.3 Animal Modeling and Local Field Potential (Intracranial EEG)

The experiments were performed on 18 male Wister rats weighing 304±13 g. The rats were randomly divided into 3 groups, which were control group, epilepsy group and TBOA group. TBOA was administrated intracerebrally in hippocampus for TBOA group. The signals of local field potential (intracranial EEG) were sampled and filtered by high- and low-pass filters set at 0.3Hz and 300 Hz. All the signals were digitized with sampling frequency of 250 Hz (PowerLab/8S, AD Instruments, Australia). All the data are expressed as the Mean ± SEM. Significant differences would be taken when P < 0.05.

3 Decreases of Multifractality from Normal State to Epileptic Model

Fig1a illustrates MF-DFA measurement and multifractal singularity spectrum of intracranial EEG in hippocampal DG, obtained from the control and epilepsy groups. Both $h(q)$ curves show dependence on q, *i.e.* the values of $h(q)$ decrease along with increase of q. It was found that the multifractal scaling exponent sloped faster in the control group compared to that of the epilepsy group (P<0.05 – 0.01, *P<0.05, **P<0.01). Furthermore, to investigate the variation of scaling behavior quantitatively, the values of slope at each scale of $h(q)$ curves was employed to describe the complexity from small fluctuation to large fluctuation. It was found that there were larger slopes at scales from 1 to 9. In addition, the multifractal singularity spectrum (Fig1b), which shows two separated $f(\alpha)$ curves, can be generated via a Legendre transform. There is a wider scaling exponent distribution in the control data. Fig2c shows the group data of $\Delta\alpha$, and it can be seen that the strength of multifractality is greater in the control group than that in the epilepsy group (0.82±0.03 *vs.* 0.68±0.02, *P<0.05).

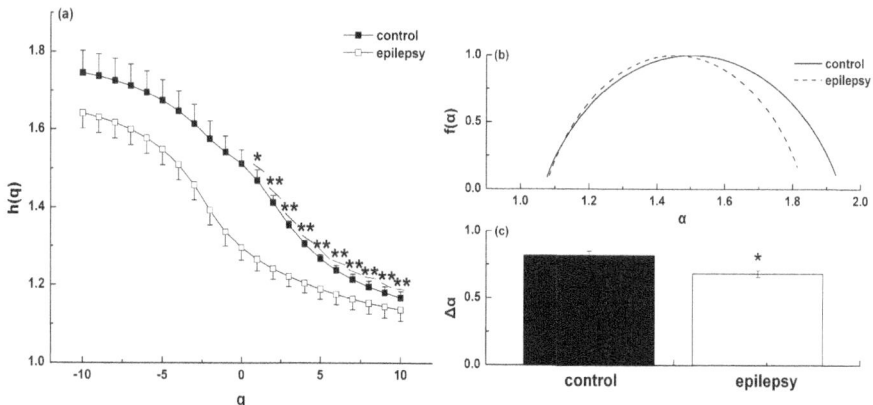

Fig. 1. Comparison of the singularity spectra, obtained from MF-DFA analysis, between the control and epilepsy groups

According to the result from Fig.1c, the $\Delta\alpha$ values indicate significant weakness in the complex structure of dynamic system in epilepsy group over that of the control group. In this case, the complexity of information transmission and storage in DG neural network is weakened. Moreover, the study of segment slope in MF-DFA curves revealed that it was the large fluctuation of oscillation contributing to the weakened complexity in epilepsy group.

4   Changes of Multifractality between Epileptic Rats and TBOA Injection Rats in Hippocampal DG

In the present study, a hypothesis was tested that effects on glutamate transporters were correlated with complexity of neural network oscillations and a loss of multifractality could be associated with brain pathology. Fig2a illustrates the results of MF-DFA measurement of intracranial EEG in hippocampal DG obtained from the epilepsy and TBOA animals, respectively. After TBOA was injected, $h(q)$ curve on most q scales kept relatively constant value, especially for $h(2)$, which was the Hurst exponent (1.24 ± 0.03 $vs.1.20\pm0.03$, P>0.05). However, there are significant differences of the slops at scales -2 and -3 between these two groups (-0.06 ± 0.004 $vs.$ -0.05 ± 0.005, *P < 0.05). Thus, the generalized Hurst exponent in epilepsy group was decreased much greater compared to that in TBOA group. It can be seen that value of $h(q)$ in epilepsy group decreased from 1.64 ± 0.04 to 1.14 ± 0.03 at scales varying from -10 to 10, while it reduced from 1.53 ± 0.02 to 1.11 ± 0.03 in TBOA group. Fig2b shows multifractal singularity spectrums obtained from an epileptic rat and TBOA one, respectively. The group data of $\Delta\alpha$ are represented in Fig2c, however, it was found that there was no significant difference between these two groups (0.68 ± 0.02 $vs.$ 0.64 ± 0.02, P>0.05).

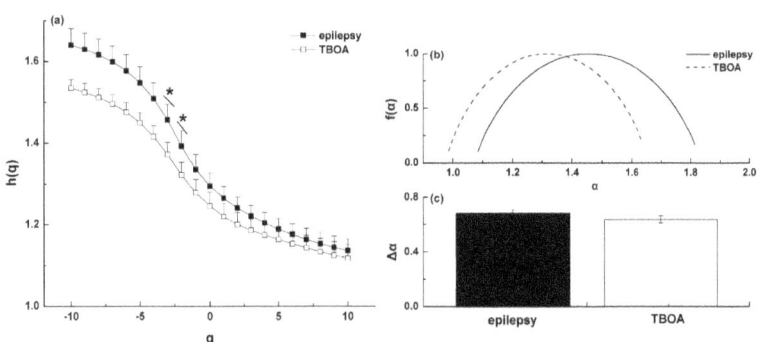

Fig. 2. Comparison of the singularity spectra, obtained from MF-DFA analysis, between the epilepsy and TBOA groups in hippocampal DG

Comparing MF-DFA results from epilepsy group and TBOA group, the constant value of Hurst exponent suggests that the fractal character of dynamic network may not have a change after TBOA injection. However, significant decrease in the slope of

TBOA on relatively small fluctuation MF-DFA curve was observed, indicating that the activities with small amplitude were less complex in TBOA group than that in epilepsy group. Based on the MF-DFA curves through slope of segment in Fig2a, it showed significant difference (P<0.05) between epilepsy group and TBOA group. Thus, according to the decrease of complexity, it was found that accumulation of glutamate brought seizures, and caused the decreasing complexity of information storing and transmitting in the neural network at that time. The glutamate transporters were inhibited by using DL-TBOA, which induced accumulation of glutamate in synaptic space and glial cells dysfunction. The dysfunction, evoked by nerotoxity of glutamate, enhanced neuronal hyperexcitability which led to more regular behavior. This suggests that the reduced complexity of neural network oscillations in hippocampus may be associated with brain dysfunction induced by nerotoxity of glutamate.

5   Decreases of Multifractality from Epileptic Model to TBOA Injection

Fig3a shows MF-DFA measurement and multifractal singularity spectrum of intracranial EEG in hippocampal PP, obtained from the epilepsy and TBOA groups, respectively. It can be seen that the value $h(q)$ in epilepsy group decreased from1.49±0.02 to 0.80±0.02, while the value $h(q)$ in TBOA group reduced from 1.49±0.04 to 0.99±0.05. The generalized Hurst exponent in the epilepsy group declines faster than that in the TBOA group. The slopes of $h(q)$ in the epilepsy group are significantly steeper than that in the TBOA group at scales from -4 to 1 ((P<0.05 – 0.01, *P<0.05, **P<0.01). Fig3b presents multifractal singularity spectrums obtained from an epileptic rat and TBOA one, respectively. It can be seen that the strength of multifractality is greater in the epilepsy group than that in the TBOA group (0.86±0.04 vs.0.69±0.04,* P<0.05, Fig 3c).

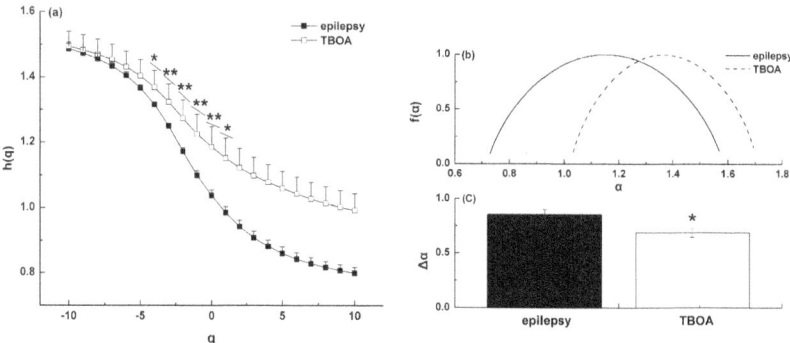

Fig. 3. Comparison of the singularity spectra, obtained from MF-DFA analysis, between the epilepsy and TBOA groups in hippocampal PP

While the use of TBOA inhibited the glutamate transporters, the inhibition caused accumulation of glutamate in neural network of DG, and has the neurotoxicity for dysfunction of neurons and glial cells. The dysfunction evoked by neurotoxicity of glutamate is what we conclude that enhanced excitability leads to more regular behavior, which shows smaller degree of complexity in DG. The seizures were triggered in entorhinal cortex by glutamate accumulation from DG in TBOA group. A previous study indicated that in models of temporal lobe epilepsy, an entorhinal cortex delivers excessive, synchronous, excitatory synaptic input from PP to DG. In the dynamic system of entorhinal cortex, hyperexcitability leads to more synchronized activity, indicating that the input to DG has smaller degree of complexity.

6   Conclusions

In the present study, we discussed the multifractal behavior in the neural network to clarify how the multifractality is related to physiological states in hippocampus of rats. The issue was addresses as to whether multifractal characterizations of intracranial EEG signals, could be detected by the method of MF-DFA. Furthermore, the hypothesis was tested that effects on glutamate transporters were correlated with complexity of neural network oscillations and a loss of multifractality could be associated with brain pathology. Since DL-TBOA can induce glutamate accumulation, which evokes sustained seizures and excitatory neurotoxic effect, the results show that the weakened activity of neuronal network can be detected by MF-DFA. In summary, this study presents that the analysis of MF-DFA can be applied to determine fractal characteristics in EEG signals. Our results demonstrate that (1) Multifractal properties of brain neuron signals obtained from DG and PP in hippocampus were related to the complexity of information transmission and storage in the dynamic neural network in the brain; (2) the complexity of information of neural network in hippocampus could be significantly weakened by epilepsy; (3) the changes of multifractal behavior induced by glutamate accumulation suggests that the neural information could be undermined by disruption of glutamate transporter activity in epileptic animals.

Acknowledgements. This work was supported by grants from the NSFC (30870827, 31070964, 31171053).

References

1. Zhang, T., Turner, D.: A visuomotor reaction time task increases the irregularity and complexity of inspiratory airflow pattern in man. Neurosci. Lett. 297(1), 41–44 (2001)
2. Parish, L., Worrell, G., Cranstoun, S., et al.: Long-range temporal correlations in epileptogenic and non-epileptogenic human hippocampus. Neuroscience 125(4), 1069–1076 (2004)
3. Nikulin, V., Brismar, T.: Long-range temporal correlations in electroencephalographic oscillations: relation to topography, frequency band, age and gender. Neuroscience 130(2), 549–558 (2005)

4. Freeman, W., Vitiello, G.: Nonlinear brain dynamics as macroscopic manifestation of underlying many-body field dynamics. Phys. Life Rev. 3(2), 93–118 (2006)
5. Li, Y., Qiu, J., Yang, Z., et al.: Long-range correlation of renal sympathetic nerve activity in both conscious and anaesthetized rats. J. Neuroscience Method 172(1), 131–136 (2008)
6. Popivanov, D., Jivkova, S., Stomonyakov, V., et al.: Effect of independent component analysis on multifractality of EEG during visual-motor task. Signal Process. 85(11), 2112–2123 (2005)
7. Kantelhardt, J., Zschiegner, S., Koscielny-Bunde, E., et al.: Multifractal detrended fluctuation analysis of nonstationary time series. Physica A 316(1-4), 87–114 (2002)
8. Oswiecimka, P., Kwapien, J., Drozdz, S.: Wavelet versus detrended fluctuation analysis of multifractal structures. Phys. Rev. E 74(1), 16103–16120 (2006)

P300 Response Classification in the Presence of Magnitude and Latency Fluctuations

Wee Lih Lee[1], Yee Hong Leung[1], and Tele Tan[2]

[1] Department of Electrical and Computer Department,
[2] Department of Computing,
Curtin University, WA, Australia

Abstract. The classification of P300 response has been studied extensively in the last decade mainly due to its increasing use in Brain Computer Interfaces (BCIs). Most of the current work involves the classification of P300 response that are produced from the BCI Speller Paradigm. However, this visual stimulation paradigm is ineffective when studying the visual analytics of subjects. Under this situation, studies have shown that the magnitude and latency of the P300 response are affected by the cognitive workload of the task as well as the physiological condition of the subject. In this preliminary study, by using visual oddball paradigm, we investigate and compare the performances of two classifiers, namely the Linear Discriminant Analysis (LDA) and Discrete Hidden Markov Model (DHMM), when the P300 response is affected by variations in magnitude and latency.

Keywords: P300, Oddball Paradigm, Hidden Markov Model, Linear Discriminant Analysis.

1 Introduction

P300 is a characteristic waveform in human EEG, which correlates to the human cognitive process in identifying rare task-relevant stimuli among a series of task-irrelevant stimuli. In recent years, P300 responses have been utilized and widely used in Brain Computer Interface (BCI) to provide disabled people an alternative way to interact and communicate with the outside world [1]. The P300 Speller is one such BCI system that has been widely researched. As described originally in [2], the interface comprises of a 6 x 6 matrix of characters. Each row and each column are sequentially and randomly illuminated which is commonly known as the P300 Speller Paradigm. To spell a character, the user focuses attention on one of the cells in the matrix. The row- and column-wise flashes generate both the rare (target character) and frequent set of EEG data. Although the P300 response in this case is generated from the target illumination, the delayed responses from the preceding illuminations will have a side effect on the shape and form of the P300 and this is affected by the frequency and duration of the flash as well as the spatial separation of the cells. However, other BCI applications like those involving the study of visual analytics rely on

B.-L. Lu, L. Zhang, and J. Kwok (Eds.): ICONIP 2011, Part I, LNCS 7062, pp. 352–359, 2011.

the Oddball Paradigm of stimuli presentation [3]. In this case, the target (rare) and non-target (frequent) stimuli are presented sequentially over time. The presentation rate and duration are usually controlled to minimize spillover signals affecting the potential P300 response. This paper focuses on the P300 response generated by the latter paradigm.

In current P300-based BCI, classical classifiers such as Linear Discriminant Analysis (LDA) [4][5] and Support Vector Machine (SVM) [6] are commonly adopted. They achieve this by making a classification decision based on the value of a linear or non-linear combination of the features subject to respective objective criteria. When building these classifier models for P300 response classifications, critical assumption was made with respect to the P300 response remaining stationary throughout the acquisition period. However, studies have shown that the P300's amplitude, duration and latency fluctuate due to changes with the subject's physiological conditions and the degree of difficulty of the interfacing task of the BCI [7]. Additionally, previous studies revealed that P300 response is usually associated with other types of ERP components such as the N100 and N200 which have different spatial patterns and latency profiles [7][8]. Therefore, we argue that the nonstationary nature of the P300 response may not be adequately addressed by the classifiers mentioned above [4][5]. Alternatively, dynamic Bayesian networks such as the Hidden Markov Model (HMM) has the potential to deal with the dynamic behaviour of ERP components by modelling them as state transitions as described in [7]. However, there have been no reported works in comparing the performances of these classifiers in relation to the classification of P300 response in the presence of magnitude and latency fluctuations. This is also partly attributed to the lack of appropriate datasets that can be used to perform this experimental comparison.

In this paper, we selected the LDA and HMM classifiers for comparison against their ability to maintain desired levels of classification accuracy with varying P300 response characteristics. We also address the implication of training sample size on the performances of these classifiers. We begin by describing the P300 dataset, based on the oddball paradigm visual stimulation concept that was used to facilitate this experiment.

2   Datasets

Existing P300 datasets like those found in the BCI Competition dataset [1] are generated using the BCI Speller Paradigm and are therefore not suitable for our intended study into the classification of Oddball Paradigm generated P300 response. For this reason, we collected an in-house dataset using the Oddball Paradigm of visual stimuli presentations aim originally at assessing the efficacy of 2D and 3D visualisation modalities [9]. This earlier study revealed that the magnitude and latency of the emitted P300 response for the majority of the subjects are affected by the degree of analytic task given. For example, the following observations were made: (a) A lowering of P300 response magnitude was encountered when the analytic task is switched from 2D to 3D visualisation.

(b) A strong increase in the P300 latency when the difficulty of the analytic task increases. Therefore, this dataset is appropriate for our purpose.

2.1 Experiment Setup

To produce the dataset, 11 healthy subjects were recruited. The EEG signals were recorded from 32 channel scalp electrodes following the international 10-20 system. Signals were referenced to the average between both left and right mastoid and acquired at a rate of 1000 Hz.

2.2 Experiment Procedure

In each session, 400 visual stimuli were presented where a cube was used as the non-target (80%) while a sphere was used as the target (20%). Each stimulus lasted for 100 ms, followed by one second of black screen. Subjects were asked to press a key whenever a target stimulus is spotted. Each subject was tested over six sessions on the same day where in each session, the test were carried out at three levels of occlusion (0%, 30% and 70% occlusion) for both 2D and 3D visualization. Fig. 1 shows an example of a subject's grand average P300 signals obtained from channel Cz for different experiment sessions which are depicted as 2d00, 2d30, 2d70, 3d00, 3d30 and 3d70.

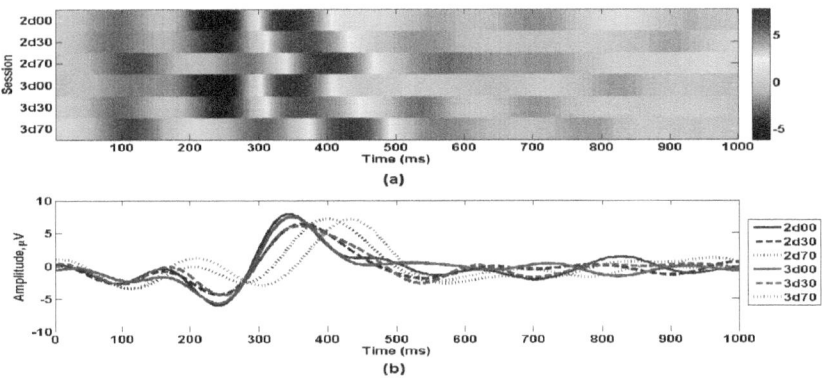

Fig. 1. (a) Amplitude intensity map and (b) the corresponding time signals from a subject's grand average P300 signals at channel Cz for different sessions

2.3 Data Pre-processing and Screening

EEG signals were bandpass filtered between 1-8 Hz and downsampled to 25 Hz. Later, 80 target trials and 320 non-target trials were extracted from each session. Each trial contains 600 ms of EEG signals starting from the stimulus onset. To remove eye blinks, any trial whose signal amplitude in channel Fp1 exceeded the threshold of 50 μV was rejected. Furthermore, after eye blink rejection, to ensure there is sufficient data left for training and building a HMM model, any dataset which has one of its sessions containing less than 50 target trials was again rejected. After screening, the datasets from four different subjects, totaling 24 sessions were used subsequently in our study.

2.4 Single Trial and Average of Multiple Trials

Averaging of multiple trial data is a common technique to deal with noise. To examine the effect of averaging on the classifiers' performance, extra datasets are generated from the recording. In these datasets, all the target and non-target trials are created by averaging across different combination of single trials in the original session (i.e. across three and five trials).

3   Classifiers

3.1 Linear Discriminant Analysis (LDA)

LDA is one of the most popular algorithms used in BCI application due to its low computational requirement and the ease of use [6]. It is a linear classifier which uses hyperplane to separate the data from different classes through the projection of feature vector from high dimensional space into lower dimensional space. To maximize separability, LDA uses the projection w which maximizes the distance between different classes' means and minimizes the interclass variances.

In a two-class problem, by assuming that both classes have an equal covariance matrix, the projection can be easily obtained by using the following equation: $w = S_w^{-1}(m_1 - m_2)$ where S_w is the average covariance matrix from both classes while m_1 and m_2 represent the mean of the feature vectors in each respective class. For classification, a threshold c is calculated where given any input vector x it will be classified to target class if $w^T x > c$ and non-target class if $w^T x < c$, or vice versa. The input vector x is created by concatenating time samples from each selected channel. In this study, LDA was implemented to serve as a benchmark for our HMM classifier.

3.2 Hidden Markov Model (HMM)

In general, a standard HMM is defined and characterized by its model parameters λ. This is usually written in compact notation as $\lambda = \{\pi, A, B\}$ where π represents the initial state probability vector, A represents the state transition probability matrix and B represents the state emission probability matrix.

Since each subject only had to respond to the target stimuli, we assumed that there existed a consistent waveform in the target stimuli while in the non-target stimuli, only random waveform was observed. Based on this assumption, a single HMM model configuration was used. During training, this HMM model was built based on the target training samples. Then, it was tested across the target and non-target training samples. For each training sample, the likelihood of the sample belonging to the HMM model was computed. After that, by finding the mean likelihood scores of the training samples from each respective class, the threshold which separates these two classes was estimated and later used in classification.

To build a HMM model, training usually involves model initialization and re-estimation. Derivations of the formula for likelihood computation and HMM model re-estimation are beyond the scope of this paper. More details can be found in [11]. Our focus is on the issue of the model initialization.

Given a multi-channel EEG signal, in every time instant, amplitude values from different channel electrodes project a particular potential map or spatial pattern onto the scalp. Studies have shown that at single-trial level, P300 signal exhibits semi-stationary temporal structure and these potential maps at different time instant can be described by a few representative spatial patterns [13].

This suggests that a discrete HMM (DHMM) can be built where a shared codebook of spatial patterns can be used to represent these potential maps. To build the shared codebook for DHMM, first, all the EEG signals from both classes were concatenated. Later, L numbers of representative spatial patterns were determined by using the K-mean clustering algorithm. To mimic the sequential time structure of P300 signals, the left-to-right model with a maximum of two state jumps was used. In this HMM model topology, jumping back to previous states is not allowed.

There is no immediate recommendation to justify which is the best HMM model. In this work, DHMM models with different combinations of number of state N and codebook size L were tested. The range of N was between 4 and 10 while the range of L was 16, 32 and 64. Later, the best performing combination was chosen for comparison. In our experiments, 10 states and a codebook size of 64 were used.

4   Performance Metrics

To determine the performance of the P300 response classification, two evaluation measures such as the True Positive Rate (TPR) and False Acceptance Rate (FAR) are used. TPR is defined as the probability of a target stimulus being classified correctly as a target while FAR is associated as the probability of a non-target stimulus being misclassified as a target. However, these two measures are often affected by the classifier's threshold. To depict the trade-off relationship between TPR and FAR for a given classifier, the Receiver's Operating Characteristic (ROC) curve was used in our study. To build the ROC curve for each classifier, the respective TPRs and FARs can be obtained by sweeping through all possible thresholds.

5   Comparison Protocol

We are interested in determining the classifiers' ability to adapt to variations in the magnitude and latency of the P300 response which was observed when the subjects perform different levels of stimulus recognition task [9]. The challenge of BCI systems is to eliminate, if not reduce, the need for user's training and adaptation. We are also interested in studying the impact of the training data on the performance of the classifiers. For this reason, we conducted three tests aim at addressing these issues:

Test 1: Subject specific five-fold cross-validation was performed independently for each session (see Section 2.2) of the dataset.

Test 2: For each subject, the classifier was trained with samples obtained from one selected session and tested across the samples from the remaining sessions. After that, the test was repeated by selecting another training session. The test is completed when all six different sessions have been selected.

Test 3: This test is an extension of Test 1 whereby the five-fold leave-one-out test was performed across all six session datasets with one segment of samples used for training while the remaining four segments were used for testing. In addition, during training, the samples from all other sessions were also included.

For all these tests, only unseen data are used for classification. To present the results of the classification, the ROC curves were averaged across different sessions and subjects. Besides that, the signals from the following eight electrodes: Fz, Cz, Pz, P3, P4, PO7, PO8, Oz were used in our tests because this electrode configuration was shown to provide the best general classification outcome as proposed in [4][5]. In addition, we also performed all the tests on (a) single trial, (b) average of three trials, and (c) average of five trials. These tests are designed to determine the impact of signal averaging on the classification performance.

6   Results

The results of Test 1 reveal the following: (a) Noise or magnitude fluctuation in the EEG signals have a negative impact on all the classifiers (see Fig. 2.a), (b) Classifications involving averaged signals (Fig. 2.b and 2.c) improve significantly suggesting it is a good method for noise suppression. (c) Overall, LDA outperformed the DHMM.

Test 2 attempts to investigate the subject specific generalisation ability of the classifiers as the majority of the test samples are not taken from the same session pool. It can be seen from Fig. 3 that all the classifiers perform relatively poorly compared to Test 1 which is attributed to latency shift in P300 signals. However, there is a relative slight increase in performance across all classifiers when averaged signals are used.

On the contrary, Fig. 4 shows the Test 3 results when we increased the size and representation of P300 training samples. In this case, the LDA classifier improved its robustness for the latency shift, resulting in performance improvement even in single trial samples. The ability to maintain good performances on single trial signal (see Fig. 4.a) also makes LDA a strong candidate for real-time P300-based BCI applications using the Oddball Paradigm.

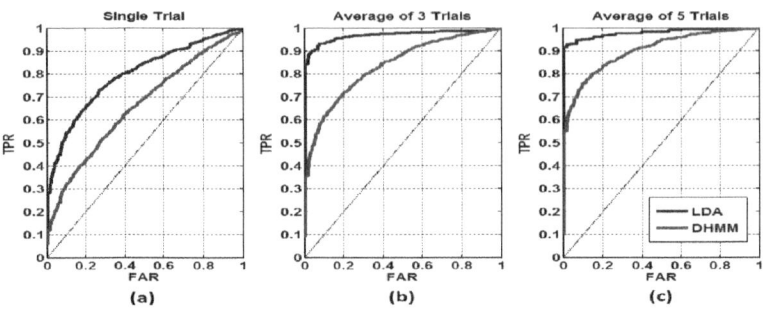

Fig. 2. Averaged ROC curves of LDA and DHMM for Test 1

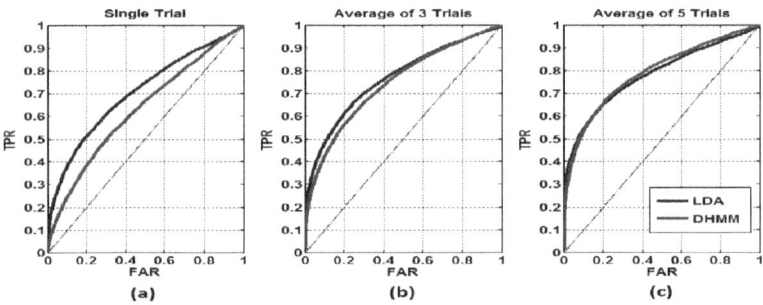

Fig. 3. Averaged ROC curves of LDA and DHMM for Test 2

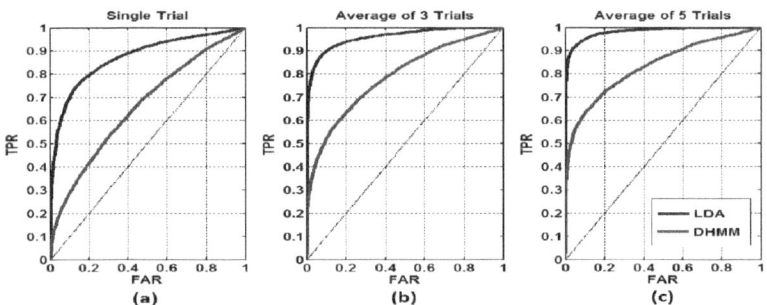

Fig. 4. Averaged ROC curves of LDA and DHMM for Test 3

7   Conclusion

In this study, we investigated the performance of different classifiers in P300 response classification under different scenarios. Based on the observations, it was observed that the fluctuations in P300 magnitude and latency have negative impacts on the classifier's performance. We showed that it is possible to build a robust subject-dependant classifier from a training set which comprises of

P300 signals with different latencies. Although the DHMM generally performs poorer compared to the LDA in most of the tests, it is still a promising classifier to consider especially when limited training data is available. Future work will include adopting these classifiers and implementing them in a visual analytics BCI application.

References

1. Blankertz, B., et al.: The BCI Competition 2003: Progress and Perspectives in Detection and Discrimination of EEG Single Trials. IEEE Transactions on Biomedical Engineering 51, 1044–1051 (2004)
2. Farwell, L.A., Donchin, E.: Talking Off the Top of Your Head: Toward a Mental Prosthesis Utilizing Event-Related Brain Potentials. Electroencephalography and Clinical Neurophysiology 70, 510–523 (1988)
3. MacNamara, A., Foti, D., Hajcak, G.: Tell Me About It: Neural Activity Elicited by Emotional Pictures and Preceding Descriptions. Emotion 9, 531–543 (2009)
4. Krusienski, D.J., et al.: A Comparison of Classification Techniques for the P300 Speller. J. Neural Engineering 3, 299–305 (2006)
5. Hoffmann, U., Vesin, J.M., Ebrahimi, T., Diserens, K.: An Efficient P300-based Brain-Computer Interface for Disabled Subjects. J. Neuroscience Methods 167, 115–125 (2008)
6. Rao, R.P.N., Scherer, R.: Statistical Pattern Recognition and Machine Learning in Brain-Computer Interfaces. In: Karim, G.O. (ed.) Statistical Signal Processing for Neuroscience and Neurotechnology, pp. 335–367. Academic Press, Oxford (2010)
7. Polich, J.: Updating P300: An Integrative Theory of P3a and P3b. Clinical Neurophysiology 118, 2128–2148 (2007)
8. Nelson, C.A., McCleery, J.P.: Use of Event-Related Potentials in the Study of Typical and Atypical Development. Journal of the American Academy of Child & Adolescent Psychiatry 47, 1252–1261 (2008)
9. Ting, S., Tan, T., West, G., Squelch, A.: Quantitative Assessment of 3D and 2D Visualisation Modalities. In: IEEE Conference in Visual Computing and Image Processing, Tainan (2011)
10. Lotte, F., Congedo, M., Lécuyer, A., Lamarche, F., Arnaldi, B.: A Review of Classification Algorithms for EEG-based Brain-Computer Interfaces. J. Neural Eng. 4, R1–R13 (2007)
11. Rabiner, L.R.: A Tutorial on Hidden Markov Models and Selected Applications in Speech Recognition. Proceedings of the IEEE 77, 257–286 (1989)
12. Helmy, S., Al-ani, T., Hamam, Y., El-madbouly, E.: P300 Based Brain-Computer Interface Using Hidden Markov Models. In: 4th International Conference on ISSNIP, Sydney, pp. 127–132 (2008)
13. Lucia, M.D., Michel, C.M., Murray, M.M.: Comparing ICA-based and Single-Trial Topographic ERP Analyses. Brain Topography 23, 119–127 (2010)

Adaptive Classification for Brain-Machine Interface with Reinforcement Learning

Shuichi Matsuzaki, Yusuke Shiina, and Yasuhiro Wada

Nagaoka University of Technology, Nagaoka, Japan
shmatsu@nagaokaut.ac.jp

Abstract. Brain machine interface (BMI) is an interface that uses brain activity to interact with computer-based devices. We introduce a BMI system using electroencephalography (EEG) and the reinforcement learning method, in which event-related potential (ERP) represents a reward reflecting failure or success of BMI operations. In experiments, the P300 speller task was conducted with adding the evaluation process where subjects counted the number of times the speller estimated a wrong character. Results showed that ERPs were evoked in the subjects observing wrong output. Those were estimated by using a support vector machine (SVM) which classified data into two categories. The overall accuracy of classification was approximately 58%. Also, a simulation using the reinforcement learning method was conducted. The result indicated that discriminant accuracy of SVM may improve with the learning process in a way that optimizes the constituent parameters.

Keywords: Brain-machine interface, Event-related potential, P300 speller.

1   Introduction

Rapid development in neuroimaging techniques has followed a considerable number of studies of the brain machine interface (BMI) as well as the development of the practical system. The most recent study has focused on BMI for medical use, such as facilitating communication for people with physical handicaps [1]. However, since present models hardly reflect individual differences of users, further investigation is needed to allow the adaptation to the sensitivity and affectivity of each user. Although a lot of studies have proposed making an adaptable interface through training the user, the accessibility and usability of these interfaces are still poor.

We introduce a BMI system that allows learning and that can be adjusted to users by optimizing the interpretation of measured brain activity. To do this, we used a reinforcement learning method [2]. In our previous study we investigated the BMI system involving a process of reinforcement learning [3], in which the input-output communication of BMI was characterized by whether or not the P300 event related potential (ERP) was elicited which would indicate that it was a reward signal. The result showed that the presence or absence of P300 can be found in the lead learning

B.-L. Lu, L. Zhang, and J. Kwok (Eds.): ICONIP 2011, Part I, LNCS 7062, pp. 360–369, 2011.
© Springer-Verlag Berlin Heidelberg 2011

process in a way similar to other systems using reinforcement learning. To expand those perspectives, we conducted an experiment with the conventional P300 speller system [4] and assessed whether using the reinforcement learning and ERP-based learning process improved its performance.

2   Proposed System

A proposed BMI system estimates reward signals by analyzing the recorded ERP which shows the failure or success of the preceding operation. This section introduces the materials and concepts involved in this system.

2.1   ERPs

Event related potential is a component of brain waves involving a positive/negative deflection in the voltage. It is generally known that ERP is evoked by external stimuli as well as by human cognitive process such as perception, recognition, and attention, those of which has been observed using the oddball paradigm and other similar methods. The most intensively studied component of ERP is P300, in which a positive peak appears at about 300 ms after stimulus onset. We used the oddball paradigm during which frequent stimuli (non-target) and infrequent stimuli (target) were presented in a random sequence, and the user counted the number of times the target was presented.

2.2   P300 Speller

The P300 speller [4][5] is a commonly used system which presents a selection of characters arranged in a 6 x 6 matrix with one of the rows and columns flashing at random while the user focuses attention on one of the 36 character cells on the matrix. The row and column intensifications that intersect at the attended cell represent the target stimuli, which occur with a probability of 1/6. The presentation of characters that is selected by the speller may constitute a P300 ERP evoked in response to a target flash.

2.3   BMI System with Reinforcement Learning

Reinforcement learning is an algorithm involving adjustment of parameters towards maximizing the received reward. As the reward is used to guide the learning process, it plays a critical role in the success of learning. Our previous study has discussed how measured ERP was a reward since it reflected the failure or success of the preceding task [3]. A framework of the conventional BMI system and the proposed BMI is show in Fig. 1. There were two tasks in the experiments: one with a P300 speller and one involving counting errors.

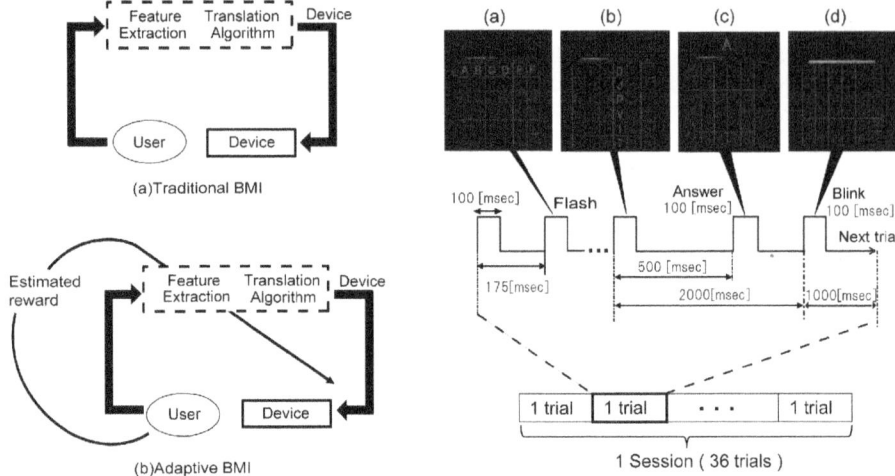

Fig. 1. Typical BMI and proposed system   **Fig. 2.** Experimental design of P300 speller task

3   Method

The ERP response was estimated using a P300 speller task and the applicability of our system to the speller was assessed.

3.1   Experimental Procedure

Nine healthy males (aged 20 - 24) participated in this study, all of whom had no experience using the P300 speller paradigm. Each provided the written informed consent set out in the institutional guidelines of the Nagaoka University of Technology. Subjects sat in front of and watched a computer screen, which presented a 6 x 6 matrix on which cells alphabetic letters were arranged. The EEG was recorded from 64 electrodes placed in accordance with the international 10-20 system.

P300 Speller. The procedures of the P300 speller task is shown in Fig. 2.   Subjects were instructed to determine the target character and when the task started, they focused attention on the target presented on the matrix and counted the number of times it was flashed. Matrices flashed 18 times for each selection. Upon selection, the speller estimated the target character (Fig. 2(c)). The selection and estimate were repeated for 36 sessions. To avoid artifacts due to eye blinking, the blinking period was set between trials (Fig. 2(d)).

Error-counting task. Having finished the P300 speller task, output characters were presented. Those in which the output character was not the target represented target stimuli. The subject counted the number of times that target stimuli were presented. The frequency of the target presentation during the error-counting task was 20%. The experiment consisted of 12 sessions each of which contained 36 trials.

3.2   Experimental Result

The EEG time course during the error-counting task is shown in Fig. 3. The result of the target trial showed a large positive potential (peak latency of 500-ms) and a negative potential (peak latency of 300-ms). This was not the case in the nontarget trials, which suggested that measurement of ERPs during the error-counting task can present failure or success of the preceding task of P300 speller.

3.3   Applicability of Reinforcement Learning

We used support vector machine (SVM) to determine the EEG data of the target and nontarget trials on the basis of whether or not P300 was evoked by the stimulus presentation. The SVM training used data for the first six sessions and testing another six sessions. The measured EEG wave from 0 to 800 ms which was averaged for every 240 ms represented a feature vector; thus, a single measurement site constituted a 10-dimensional space. We computed and used 17 feature vectors with each connected to the following sites, respectively: Fz, Cz, Pz, PO7, Oz, PO8, FCz, C3, C4, CPz, P7, P3, P4, P8, PO3, POz, PO4, O1, and O2.

The number of data used in the analysis. The SVM discrimination rate for target and nontarget trial data are shown in Table 2. The average discrimination accuracy of target trials and non-target trials were 57.2% and 59.7%.

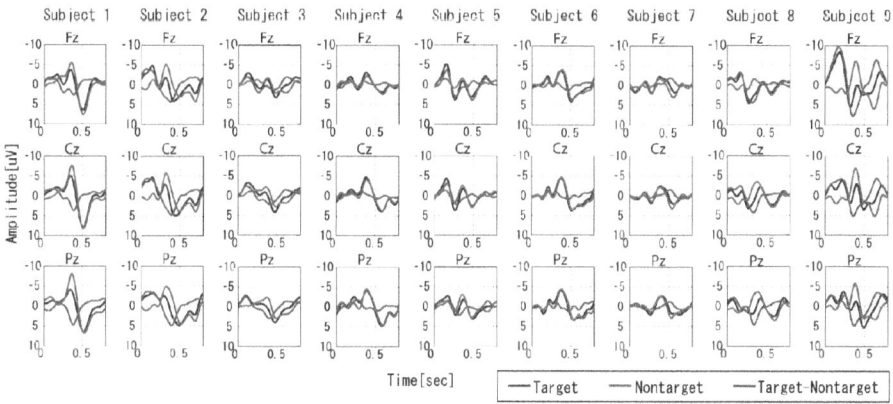

Fig. 3. Average EEG time course during error counting task measured at Fz, Cz and Pz from subject 1 to 10. Blue, red, and green curves represent data from target trial, nontarget trial, and the difference between the two.

3.4   Applicability of Reinforcement Learning

We used support vector machine (SVM) to determine the EEG data of the target and nontarget trials on the basis of whether or not P300 was evoked by the stimulus presentation. The SVM training used data for the first six sessions and testing another six sessions. The measured EEG wave from 0 to 800 ms which was averaged for every 240 ms represented a feature vector; thus, a single measurement site constituted a 10-dimensional space. We computed and used 17 feature vectors with each connected to the following sites, respectively: Fz, Cz, Pz, PO7, Oz, PO8, FCz, C3, C4, CPz, P7, P3, P4, P8, PO3, POz, PO4, O1, and O2.

The number of data used in the analysis. The SVM discrimination rate for target and nontarget trial data are shown in Table 2. The average discrimination accuracy of target trials and non-target trials were 57.2% and 59.7%.

Table 1. Number of training/test data

Subject No.	Trainning data		Test data	
	Target	Nontarget	Target	Nontarget
1	34	165	39	173
2	33	165	40	160
3	39	163	37	173
4	38	178	41	175
5	56	174	40	168
6	35	165	41	173
7	36	162	34	165
8	27	103	38	136
9	37	174	40	174

Table 2. Discrimination accuracy of SVM

Subject No.	C	σ	Target [%]	Nontarget [%]
1	4	38.67	59.5	82.1
2	2	22.55	81.8	69.4
3	2	21.63	73	58.9
4	4	22.89	52.6	61.1
5	4	27.62	2.63	90.9
6	7	29.81	57.1	47.9
7	4	25.81	67.7	47.9
8	2	24.44	44.4	63.1
9	2	31.35	75.7	20.1

4   Reinforcement Learning

We evaluated an actor-critic reinforcement learning method used in the proposed BMI system. The learning algorithm, simulation method, and results are described in the following subsections.

4.1   Learning Model

The proposed model is shown in Fig. 4. An actor-critic method, a kind of Temporal Difference (TD) algorithm, was used which consisted of the two components: one is the actor which selected actions and another was the critic which evaluated the action made by the actor. This evaluation provides the TD error as:

$$E(i) = r(i) + \gamma V(i) - V(i-1) . \tag{1}$$

where r is the reward, V the value function implemented by the critic in a given state, and γ the discount factor. The reward was given as follows:

$$\left\{ \begin{array}{lll} r = & a = 1 & \cdots \quad \text{Target} \\ & b = -1 & \cdots \quad \text{Nontarget} \end{array} \right. \tag{2}$$

The actor-critic method was used to optimize the parameters σ and C of the SVM model which determine the accuracy of the classification. We set candidate values of those parameters as shown in Table 3, and in the estimation process one of those was selected with a given probability. The selected values of σ and C were written as σ_n and C_n, respectively, where n is an index number shown in Table 3. Here, $\Pr_\sigma(n_\sigma)$ and $\Pr_C(n_C/n_\sigma)$ represent the probabilities on which candidates values n_σ and n_C can be selected, respectively. The P300 speller system received those parameters and designed discriminant functions. Calculation of critics used the

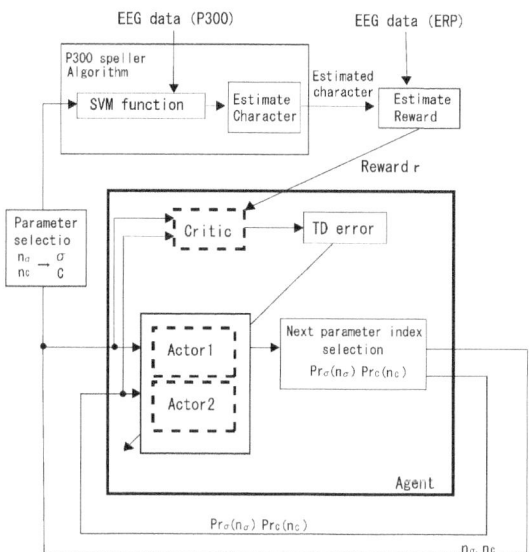

Fig. 4. Proposed model

parameter values σ, C, $\Pr_\sigma(n_\sigma)$, and $\Pr_C(n_C/n_\sigma)$, followed by the result providing the TD-error which was sent to actors. There were two types of actor, Actor-1 and Actor-2, the former updated $\Pr_\sigma(n_\sigma)$ and the latter $\Pr_C(n_C/n_\sigma)$.

4.2   Actor

The computation of the actor used the adaptive Generalized Radial Basis Function (GRBF). The i-th input to actors is expressed as:

$$s_1^A(i) = (\varepsilon \cdot n_\sigma, \Pr_\sigma(n_\sigma))\tag{3}$$

$$s_2^A(i) = (\varepsilon \cdot n_C, \Pr_C(n_C \mid n_\sigma))\tag{4}$$

where $\varepsilon = 0.05$ is a constant. Here, $s_1^A(i)$ and $s_2^A(i)$ represents input to Actor-1 and Actor-2, respectively. The basis function $b_k^A(s)$ is given by the following equation:

$$b_K^A(s^A(i)) = \frac{a_k^A(s^A(i))}{\sum_{l=1}^{L^A} a_l(s^A(i))}\tag{5}$$

$$a_K^A(s^A(i)) = e^{-\frac{1}{2}\left\| M^A(s^A(i) - c_k^A)\right\|^2}\tag{6}$$

Here, c_k^A is the center of the k-th basis function and M^A the matrix that determines a form of the basis function. The initial values $c_1^A = (5.5, 0.1)$ and $M^A = (0.2, 0.08)$

Table 3. Candidate parameter values σ and C

					Parameter (Subject No.3)					
Index number	1	2	3	4	5	6	7	8	9	10
σ	22.72	30.32	33.35	36.69	44.4	48.84	53.72	59.1	65.01	71.52
C	0.1	0.2	0.4	0.7	1	2	4	7	10	20
					Parameter (Subject No.4)					
σ	26.72	39.34	43.29	47.63	52.4	57.67	70.02	77.12	85.08	94.22
C	0.2	0.4	0.7	1	2	4	7	10	20	40

are given. L^A represents the number of units and increases with the learning process; the newborn is added if none of existing basis functions exceed the threshold $b_{th}^A = 0.0075$. A new unit is initialized by $c_k^A = s^A(i)$.

The actor output $\Pr_\sigma(n_\sigma)$ is calculated as follows:

$$P_\sigma(n_\sigma) = \sum_{l=1}^{l_\sigma^A} w_{\sigma,l}^A b_l^A(n_\sigma, \Pr_\sigma(n_\sigma)) \tag{7}$$

$$P_\sigma(n_\sigma) = \frac{P_\sigma(n_\sigma)}{\sum_{m=1}^{N_\sigma} P_\sigma(m)} \tag{8}$$

where N_σ is the numbers of the candidate values. Similarly, $\Pr_C(n_C \mid n_\sigma)$ is calculated by:

$$P_C(n_C \mid n_\sigma) = \sum_{l=1}^{l_C^A} w_{C,l}^A b_l^A(n_C, \Pr_C(n_C \mid n_\sigma)) \tag{9}$$

$$\Pr_C(n_C \mid n_\sigma) = \frac{P_C(n_C)}{\sum_{m=1}^{N_C} P_C(m)} \tag{10}$$

where N_C is the number of the candidate value, and ω^A is the weight of the basis function. A new weight is added if every basis function b_k^A is less than b_{th}^A.

4.3   Critic

The computation of the critic used adaptive GRBF. The i-th input to critics is expressed as:

$$s^C(i) = (\varepsilon \cdot n_\sigma, P_\sigma(n_\sigma), \varepsilon \cdot n_C, P_C(n_C \mid n_\sigma)) \tag{11}$$

The critic output $V(i)$ is calculated by:

$$V(i) = \sum_{m=1}^{l^C} w_{m,}^C b_m^C(s^C(i)) \tag{12}$$

The basis function $b_k^C(s)$ is given by the following equation:

$$b_k^C(s^C(i)) = \frac{a_k^C(s^C(i))}{\sum_{l=1}^{l^C} a_l(s^C(i))} \tag{13}$$

$$`a_k^C(s^C(i)) = e^{-\frac{1}{2}\left\|M_k^C(s^C(i)-c_k^C)\right\|^2} \tag{14}$$

Here, c_k^C represents the center of the k-th basis function, M_k^C the matrix that determines a form of the basis function. The initial values $c_1^C = (5.5, 0.1, 5.5, 0.1)$ and $M^C = (0.2, 0.08, 0.2, 0.08)$ are given. Here, ω^C is the weight of the basis function and L_n^C the number of units. The new unit is added if none of existing basis functions exceed the threshold $b_{th}^C = 0.01$, and it is initialized by $c_k^C = s^C(i)$ and $\omega_k^C = 0$.

4.4   Update

Having calculated the critic output, the actor updates $\omega_{\sigma,k}^A$, $\omega_{C,k}^A$, $\mathrm{Pr}_\sigma(n_\sigma)$ and $\mathrm{Pr}_C(n_C/n_\sigma)$. $\omega_{\sigma,k}^A$ $\omega_{C,k}^A$ are incremented by:

$$\Delta w_{\sigma,k}^A = \eta_a \cdot E(i) \cdot b_{\sigma,k}^A(n_\sigma, \mathrm{Pr}_\sigma(n_\sigma)) \tag{15}$$

$$\Delta w_{C,k}^A = \eta_a \cdot E(i) \cdot b_{C,k}^A(n_C, \mathrm{Pr}_C(n_C \mid n_\sigma)) \tag{16}$$

where $\eta_\sigma = 0.1$ is the learning rate. The estimated value function is updated by:

$$V(i) \leftarrow V(i) + \gamma^C \cdot E(i) \tag{17}$$

where $\gamma^C = 0.01$ is the learning rate. TD error $E(i)$ was minimized using the steepest descent method defined as follows:

$$\frac{\partial(r(i)) + \gamma V(i+1) - V(i))^2}{\partial w_k^C} \propto \eta_c \cdot E(i) \cdot b_k^C = \Delta w_k^C \tag{18}$$

where $\eta_c = 0.001$ is the learning rate.

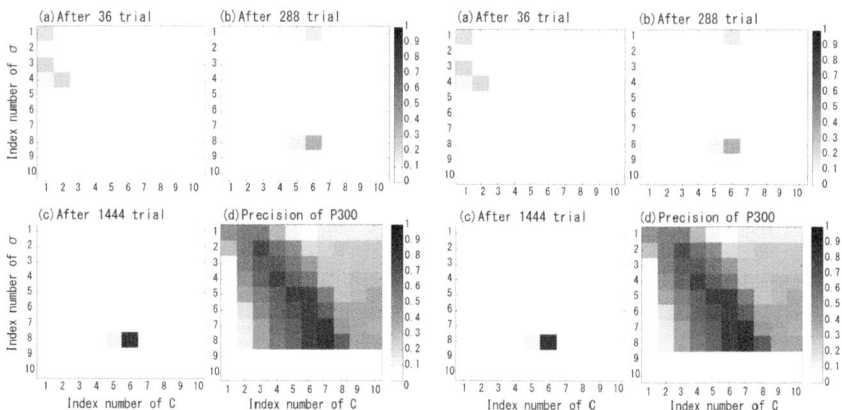

Fig. 5. Selection of SVM parameters (σ and C) by Subject-3 (left) and Subject-4 (right). Candidate values (ID 1 to 10) are listed in Table 3. Colored cells in (a) – (c) represent the frequency when corresponding parameter values are used, and colored cell in (d) represents accuracy rate of P300 speller.

4.5   Simulation of Reinforcement Learning Process

Method. Two healthy males, Subject-3 and -4 in the above-mentioned experiment, participated in this experiment. Subjects undertook the P300 speller task shown in Fig. 2 which repeated two sessions (72 trials); each trial consisted of 180 times the matrix flashed in a random sequence.We used EEG signals for the first session to implement the discriminant function while the second was used for estimation by SVM. The SVM parameters σ and C were selected from the candidate values shown in Table 3. The probability of whether the correct reward value was used constituted the ERP discriminant accuracy rate of the subject shown in Table 2. Thus, for example, learning by Subject-3 takes 73% and 58.9% probabilities that provide reward r = 1 and r = − 1, respectively.

Estimation results. The tendency of the selected parameter values for each subject and the accuracy rate of the speller when applying each parameter value is shown in Fig.5. The results showed that the parameter value that provides a high rate of accuracy is selected in accordance with the increased number of the learning. This correlation suggests the applicability of learning. Changes in the discriminant accuracy of the speller are shown in Fig. 6. The results of Subject-3 showed a marked increase during the early stage of the learning process. This may have resulted from the higher accuracy in P300 detected from Subject-3. On the other hand, the result of Subject-4 showed a slight and overall increase. Since providing wrong rewards may mislead the learning process, further investigation on the ERP-based learning system may involve improving the feature extraction technique for measured ERPs to accurately convey the user's intention to the system.

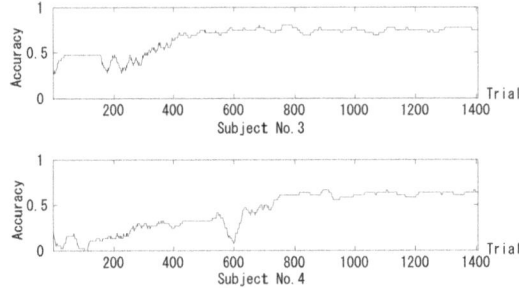

Fig. 6. Changes in level of discriminant accuracy of P300 speller. Each data point indicates the moving average of the last 36 points.

5   Conclusion

We investigated the P300 speller system that involves the process of the reinforcement learning in which algorithm   changes in EEG signals reflected the user's intention and provided the reward. The presence or absence of P300 ERP during a task was used as a measure of the reward and thus indicated whether the output of the speller was correct. The SVM discriminant rate was 57.2% with the target trials and 59.7% with the nontarget trials. The actor-critic learning was used in the proposed system, and altered the SVM parameters and thus reduced the erroneous outputs of the speller. As a result, the output accuracy increased with the learning process. These results suggest that the reinforcement learning system may facilitate adjustment and thus allow the ERP-based BMI system more accessible to users.

Acknowledgments. This research was partially supported by a Grant-in-Aid for Scientific Research (A) from the Ministry of Education, Science, Sports and Culture.

References

[1] Wolpaw, J.R., Birbaumer, N., McFarland, D.J., Pfurtscheller, G., Vaughan, T.M.: Brain computer interfaces for communication and control. Clin. Neurophysiol. 113(6), 767–791 (2002)
[2] Barto, A.G., Sutton, R.S., Anderson, C.W.: Neuronlike adaptive elements that can solve difficult learning control problems. IEEE Trans. Syst. Man Cyber. 13(5), 834–846 (1983)
[3] Yamagishi, Y., Tsubone, T., Wada, Y.: Applicability to Reinforcement Signal of Event-Related Potential. In: Proceedings of 22nd SICE Symposium on Biological and Physiological Engineering, pp. 343–346 (2008)
[4] Farwell, L.A., Donchin, E.: Talking off the top of your head: toward a mental prosthesis utilizing event-related brain potentials. Electroenceph. Clin. Neurophysiol. 70, 510–523 (1988)
[5] Takano, K., Komatsu, T., Hata, N., Nakajima, Y., Kansaku, K.: Visual stimuli for the P300 brain.computer interface: A comparison of white/gray and green/blue flicker matrices. Clin. Neurophysiol. 120(8), 1562–1566 (2009)

Power Laws for Spontaneous Neuronal Activity in Hippocampal CA3 Slice Culture

Toshikazu Samura[1], Yasuomi D. Sato[1,2], Yuji Ikegaya[3], and Hatsuo Hayashi[1]

[1] Graduate School of Life Science and Systems Engineering,
Kyushu Institute of Technology,
2-4 Hibikino, Wakamatsu-ku, Kitakyushu 808-0196, Japan
{samura,sato-y,hayashi}@brain.kyutech.ac.jp
[2] Frankfurt Institute for Advanced Studies (FIAS),
Johann Wolfgang Goethe University,
Ruth-Moufang-Str. 1, Frankfurt am Main, 60438, Germany
sato@fias.uni-frankfurt.de
[3] Laboratory of Chemical Pharmacology,
Graduate School of Pharmaceutical Sciences, The University of Tokyo,
Tokyo 113-0033, Japan
ikegaya@mol.f.u-tokyo.ac.jp

Abstract. We study on computer simulations to infer a network structure of the hippocampal CA3 culture slice, which is not yet found even in physiological experiments. In order to find the network structure, we have to understand dynamical mechanisms of how to establish two power law distributions of spontaneous activities observed in the CA3 cultured slice. The first power law is the probabilistic distribution of firing frequency in a neuron. The second is of synchrony size that means a rate of co-active neurons within a time bin. In this work, we show that the power law observations significantly rely on the two network mechanisms: (1) high-frequency firing of interneurons by feedback from pyramidal cells, (2) log-normal distribution of synaptic weights.

Keywords: Hippocampus, CA3, Spontaneous activity, Power-law distribution, Log-normal distribution, Inhibitory interneurons.

1 Introduction

CA3 in the hippocampus is a significantly important region for information processing such as associative memory [1]. The network structure has been thought as a recurrent network. Pyramidal cells and inhibitory interneurons in the hippocampal CA3 are mutually connected with the other local neurons [2][3].

Takahashi *et al.* have suggested that CA3 network is structured for facilitating synchronization of neuronal activities [4] (The synchronization might be related functionally to associative memory). For this suggestion, high-speed functional multineuron calcium imaging (fMCI) is effectively used to measure spontaneous activities of neurons in a CA3 network on an organotypically cultured slice. The fMCI is allowed to simultaneously record Ca^{2+} activities in hundreds of neurons

B.-L. Lu, L. Zhang, and J. Kwok (Eds.): ICONIP 2011, Part I, LNCS 7062, pp. 370–379, 2011.
© Springer-Verlag Berlin Heidelberg 2011

in an image resolution with the single-neuron level. It can not only detect a temporal sequence of spikes of each neuron, but also identify easily the neuron's position on a CA3 slice. In the experiment using the fMCI, they found that a so-called synchrony size is approximately distributed under the power law. The synchrony size, which is one of important measurements for estimating implicit connectivities between neurons, is defined as a rate of co-active neurons during the spontaneous activities. The CA3 network exhibited a very large synchronous event where around 50% neurons were activated. From the data recordings obtained in the fMCI experiment, connectivities of neurons were also estimated. However, in the experiments, there still exist problems as follows: (1) the probabilistic distribution of firing frequency, (2) the synaptic weight distribution and (3) identification of neuron types (at least, straightforwardly, pyramidal cell or inhibitory interneuron).

In this paper, we do computer simulations on spontaneous activities in CA3 slice culture observed by using the fMCI. In the simulations, an excitatorily and inhibitorily recurrent CA3 network is constructed, using a dynamically plausible spiking model proposed by Izhikevich [5], in order to demonstrate two statistic properties on the simulations: One is the probabilistic distribution of firing frequency while the other is the probabilistic distribution of synchrony size. Interestingly enough, both the statistic properties follow the power law. Essential mechanisms to find the power law of the probabilistic distribution are (1) log-normally distributed synaptic weights and (2) high-frequency firing of inhibitory interneurons driven by pyramidal cell's feedback. More interestingly, in the second we also show that inhibitory interneurons are predominantly in the high-frequency range.

2   Materials and Methods

2.1   Experimental Data

We used experimental data for Ca^{2+} activities in hippocampal CA3 slice culture, which are always available on the following website: http://hippocampus.jp/data/. They are records for the temporal sequence of spikes of all neurons by fMCI. In this work, 14 sampling data of them are employed. The sampling rate is 500 Hz and recording time is 130 sec. The number of neurons for each data is about 53–137. 1,193 is the total number of neurons in 14 slices. Here it is noticed that we do not know neuron type, a pyramidal cell or an interneuron in all of the data, but neuron positions are easily identified on the slice in the data.

We showed statistic properties found in spontaneous activities of 1,193 neurons summed up with 14 slice data (Fig.1). The maximum frequency is 2.331 Hz and the firing frequency decreased smoothly according to the ranking (Fig.1 (a)). 141 neurons did not fire during the recording. Figure 1 (b) shows the probabilistic distribution of firing frequency. The distribution also exhibits the linearity on a log-log scale. The probabilistic distribution of firing frequency is apparently under a power law $P(f) \sim f^{\alpha}$, where f is the firing frequency in neurons.

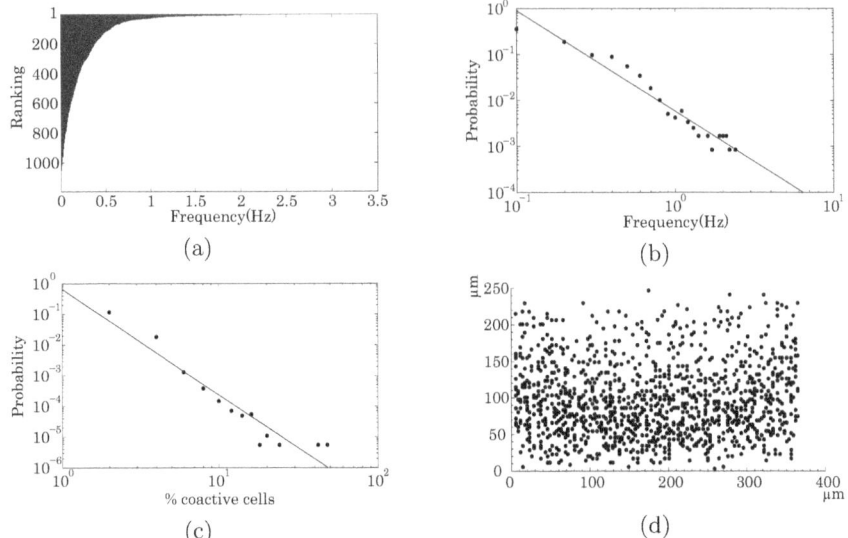

Fig. 1. Stochastic properties of 1, 193 neurons from 14 slices. (a)The distribution of firing frequency. All neurons are ranked according to its firing frequency and sorted by the ranking. (b)The power law distribution of firing frequency ($\alpha = -2.189$). The probability of firing frequency was calculated in each 0.1 Hz bin. (c)The power law distribution of synchrony size ($\beta = -3.451$). Synchrony size is the percentage of co-active neurons within the time bin of 10ms. The probability of synchrony size was calculated in each 2% bin. (d) The location of 1, 193 neurons. We plotted the location of each neuron in an x-y coordinate.

$P(f)$ is the probability function for f (Fig.1 (b) $\alpha = -2.189$). Figure 1 (c) shows the probabilistic distribution of synchrony size. As referred to the previous report [4], the probability obeys a power law distribution $P(n) \sim n^{\beta}$. n is the synchrony size and $P(n)$ is the probability of n. The probability of synchrony size exhibited the linearity on a log-log scale (Fig.1 (c) $\beta = -3.451$). Figure 1 (d) shows the location of all neurons within the local rectangle region in the CA3 slice (250 μm by 400 μm).

2.2   CA3 Slice Model

We construct a CA3 recurrent network consisting of excitatory and inhibitory interneurons, by using Izhikevich's model [5]. The Izhikevich model is given by

$$v'_i = 0.04v_i^2 + 5v_i + 140 - u_i + I_i(t), \tag{1}$$
$$u'_i = a(bv_i - u_i), \tag{2}$$

where i is neuron index. v_i is the membrane potential and u_i is the membrane recovery variable. a is the rate of recovery. b is the sensitivity of the recovery variable. $I_i(t)$ is the inputs from other neurons to ith neuron, and calculated by the following equation:

$$I_i(t) = I_i^{\text{noise}}(t) + \sum_j^{N_i} w_{ij} \sum_k^{N_j^{\text{fired}}} \delta(t - t_j^k - \tau_{ij}), \tag{3}$$

where $I_i^{\text{noise}}(t)$ is the noise input defined in the next section 3. N_i is the total number of presynaptic neurons of ith neuron and w_{ij} means the synaptic weight between ith and jth neurons. N_j^{fired} is the number of firing of jth neuron. $\delta(\cdot)$ is the Dirac delta function and t_j^k is kth firing timing of $j$the neuron. τ_{ij} ($=1$ [ms]) is the synaptic delay between ith and jth neurons.

$$\text{if } v_i \geq 30, \text{ then } \begin{cases} v_i \leftarrow c \\ u_i \leftarrow u_i + d. \end{cases} \tag{4}$$

Let ith neuron fire a spike when v_i arrives at 30. Then v_i and u_i are abruptly reset to c and $u_i + d$, respectively. An excitatory neuron was modeled by an intrinsic bursting neuron, so that we set parameters as follows: $a = 0.02, b = 0.2, c = -55, d = 5$. We also modeled an inhibitory interneuron as a fast spiking neuron, so that we set parameters as follows: $a = 0.1, b = 0.2, c = -65, d = 2$.

Since we assume that the ratio of pyramidal cell to inhibitory interneuron is 10 : 1 in the hippocampus, as referred to [6], our model is thus set up with 100 excitatory neurons and 10 inhibitory interneurons (Fig.2). The locations of neurons are within the region of 250 μm by 400 μm (Fig.1 (d)). Neurons in hippocampal CA3 have over 1,000 μm long axon [2] [3]. Although the long axon enables neurons to connect with other neurons in the region, the axon density of a given neuron is high near the parent neuron [2]. Based on such a neurophysiological experiment result, we have an additional assumption that neurons were locally connected each other, but a part of the connections is the long-range connection. The connectivity density among CA3 pyramidal neurons in a cultured slice is 28.8 % [4]. Excitatory neurons connected to nearest 28 excitatory neurons and all inhibitory interneurons within a range where there are connected excitatory neurons. The connection probability from inhibitory

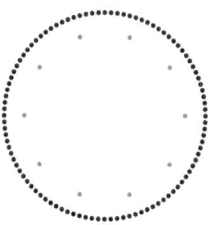

Fig. 2. Structure of CA3 slice model. Black and Gray dots mean an excitatory neuron and an inhibitory interneuron, respectively. As a matter of convenience, 100 excitatory neurons and 10 inhibitory neurons are plotted on outside and inside circle, respectively.

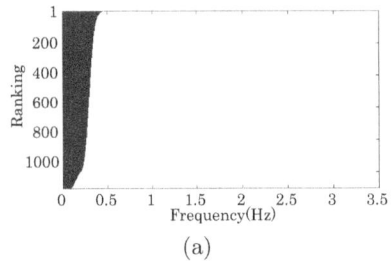
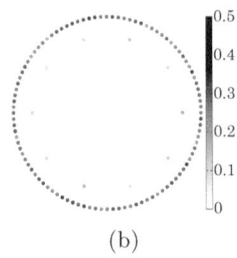

(a)   (b)

Fig. 3. Activities evoked by noise input in CA3 slice model without connections. (a)The distribution of firing frequency (0.069–0.469 Hz). (b)The firing frequency of each neuron in a certain trial. A gray scale means the frequency. Dots on an outside circle mean excitatory neurons and dots on an inside circle mean inhibitory interneurons.

interneurons to excitatory neurons is 60 % [6]. Inhibitory interneurons connected to 61 nearest neighbors of the excitatory neuron. The connections were rewired with probability 25% and attached to randomly selected neurons, but inhibitory interneurons were attached only to excitatory pyramidal cells. Here it is noticed that let there be no inhibitory connections among inhibitory interneurons.

3   Results

Using the CA3 slice model, we obtained simulation results of the activities of 1, 100 neurons from 10 trials in each condition. In each trial, we recorded the activities of 110 neurons for 130 sec. Synaptic weights and noise inputs were defined in each condition.

When all connections are pharmacologically blocked, CA3 pyramidal neurons spontaneously fire at a mean frequency 0.2–0.4 Hz [7]. We defined noise input $I_i^{noise}(t)$ to simulate the spontaneous activity of the slice, in which neurons have no connections with each other. In each time step, we applied the input to each neuron. The strength of input to each neuron comes from normal distribution ($\mu = 0$, $\sigma = 3$).

Figure 3(a) shows the firing frequency of 1, 100 neurons in 10 trials. All synaptic strength was set as 0. All neurons fired at low frequency (< 0.5 Hz) by the noise input applied to excitatory and inhibitory interneurons. The firing frequency of inhibitory neurons is lower than those of excitatory neurons (Fig.3(b)).

We defined synaptic weights of all kinds of connections: excitatory-excitatory (E-E), excitatory-inhibitory (E-I) and inhibitory-excitatory (I-E). The strengths of E-E connections are randomly selected within a range of 0 to 8.1. The strengths of E-I connections are randomly selected within a range of 0 to 16.55. The strengths of I-E connections are randomly selected within a range of -8.1 to 0.

Figure 4 shows the activities of 1, 100 neurons in 10 trials. All synaptic strength was set as above and the noise input is applied to each neuron. The maximum firing frequency (2.377 Hz) was consistent with the experimental data. The firing frequency drastically decreased according with ranking (Fig.4(a)) and

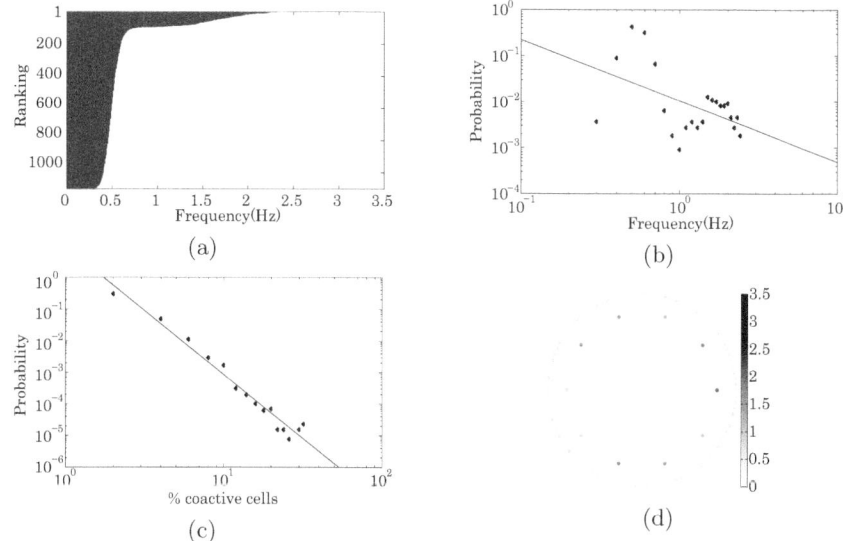

Fig. 4. Activities evoked by noise input in CA3 slice model with connections. (a)The distribution of firing frequency (0.277–2.377 Hz). (b)The probability distribution of firing frequency ($\alpha = -1.331$). (c)The power law distribution of synchrony size ($\beta = -4.0324$). (d)The firing frequency of each neuron in a certain trial.

the distribution was inconsistent with the experimental data. The probabilistic distribution of firing frequency did not exhibit any linearity on a log-log scale (Fig.4(b)). The probabilistic distribution of synchrony size shows the linearity on a log-log scale (Fig.4(c) $\beta = -4.032$). The probabilistic distribution of synchrony size is similar to the experimental data. This indicates that the two distributions do not rely on each other. The firing of inhibitory interneurons evoked by the noise input was at low frequency in CA3 model without connections (Fig.3(a)), but the firing frequency of inhibitory interneurons is relatively higher than those of excitatory neurons (Fig.4(d)). This means that the inhibitory interneurons were driven by excitatory inputs. Inhibitory interneurons ranked in the top of the distribution. The drastic decrease in the distribution comes from the difference between the firing frequencies of excitatory and inhibitory interneurons. Thus it is expected that the high-frequency firing of excitatory neurons may prevent the drastic decrease.

It has been reported that the distribution of synaptic strength in visual cortex layer 5 can be fitted by a log-normal distribution [8]. The distribution has a heavier tail and strong connections probably exist in the network. On the assumption of the log-normal distribution of synaptic weights, it is expected that strong excitatory connections enable excitatory neurons to fire at high-frequency for avoiding the drastic decrease in the distribution. Therefore, we set the synaptic weight of a connection as randomly selected value from a log-normal distribution (E-E connections: $\mu = \log(1.7) + \sigma^2$; E-I connections: $\mu = \log(3.4) + \sigma^2$;

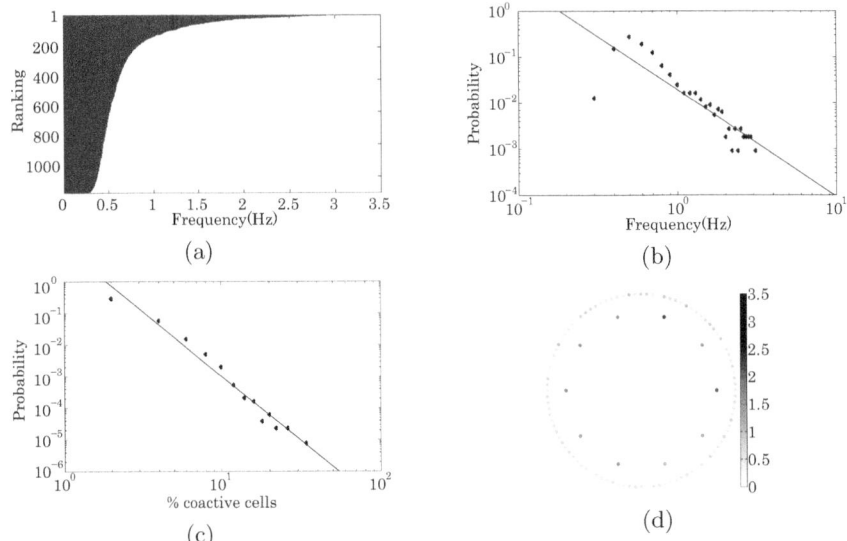

Fig. 5. Activities evoked by noise input in CA3 slice model with log-normally distributed connections . (a)The distribution of firing frequency (0.246–3.008 Hz). (b)The probability distribution of firing frequency($\alpha = -2.318$). (c)The power law distribution of synchrony size ($\beta = -4.082$). (d)The firing frequency of each neuron in a certain trial.

I-E connections: $\mu = \log(1.7) + \sigma^2$, All connections: $\sigma = \log(2.0)$). The synaptic weight of I-E connections was multiplied by -1.

Figure 5 shows the activities of $1,100$ neurons in 10 trials. All synaptic strength obeyed a log-normal distribution and the noise input is applied to each neuron. The maximum firing frequency (3.008 Hz) was higher than the experimental data, but the firing frequency smoothly decreased according with the ranking (Fig.5(a)). Both of distributions exhibit linearity on a log-log scale (Fig.5(b) $\alpha = -2.318$) (Fig.5(c) $\beta = -4.082$). A part of excitatory neurons fired at higher frequency than previous condition (Fig.5(d)). The log-normal distribution of synaptic weights properly caused a gradient of the firing frequency of neurons and the distribution of firing frequency exhibited linearity on a log-log scale. However, the minimum firing frequency was still high.

The minimum firing frequency is similar to the firing frequency evoked by noise input. Thus, we modified noise input to decrease the minimum firing frequency. In each time step, we applied a noise input to each neuron. The strength of the input to each neuron comes from individual normal distribution ($\mu = 0$, $\sigma = 2$–3 randomly selected for each neuron). Figure 6 shows the firing frequency of $1,100$ neurons in 10 trials. All synaptic strength was set as 0.

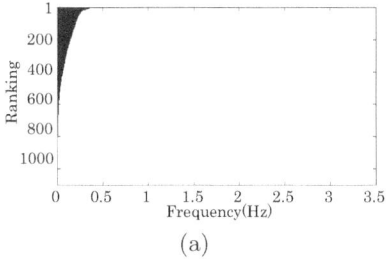 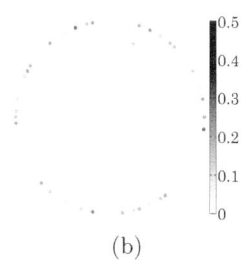

(a)   (b)

Fig. 6. Activities evoked by noise input with gradient in CA3 slice model without connections. (a)The distribution of firing frequency (0–0.385 Hz). (b)The firing frequency of each neuron in a certain trial.

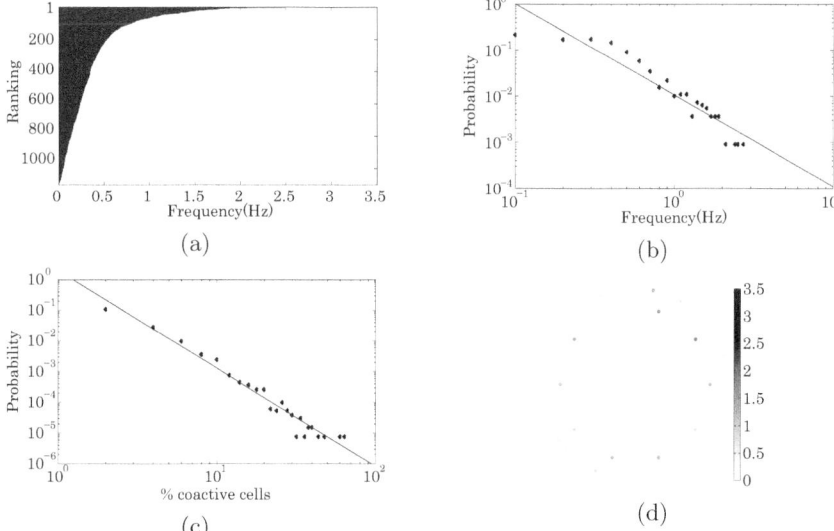

(a)   (b)

(c)   (d)

Fig. 7. Activities evoked by noise input with gradient in CA3 slice model with log-normally distributed synaptic weights. (a)The distribution of firing frequency (0.008–2.6615 Hz). (b)The probability distribution of firing frequency($\alpha = -1.986$). (c)The power law distribution of synchrony size($\beta = -3.194$). (d)The firing frequency of each neuron in a certain trial.

The maximum firing frequency evoked by the noise input was kept, but the minimum firing frequency decreased (Fig.6(a)). Then, most of the inhibitory interneurons and a part of the excitatory neuron did not fire (Fig.6(b)).

We set the synaptic weight of a connection as randomly selected value from a log-normal distribution (E-E connections: $\mu = \log(2.1) + \sigma^2$; E-I connections: $\mu = \log(4.8) + \sigma^2$; I-E connections: $\mu = \log(2.2) + \sigma^2$, All connections: $\sigma = \log(2.0)$). The strength of I-E connections was multiplied by -1.

Figure 7 shows the activities of $1,100$ neurons in 10 trials. The maximum firing frequency (2.662 Hz) was slightly higher than the experimental data (Fig.7 (a)). On the other hand, the minimum firing frequency became low. A few excitatory neurons fired at the same frequency with inhibitory interneurons (Fig.7 (d)). Both of distributions also exhibit linearity on a log-log scale (Fig.7 (b) $\alpha = -1.986$) (Fig.7 (c) $\beta = -3.194$). The distributions of firing frequency and synchrony size were consistent with those of experimental results. The firing of neurons in CA3 cultured slices was reproduced in the CA3 slice model.

4   Discussion and Conclusion

In this work, we have found two important network structures to find power law distributions of firing frequency or of the synchrony size: the high firing frequency of interneurons driven by pyramidal cell's feedback input and the log-normal distribution of the synaptic weights. The two structures were predicted within a framework of the CA3 slice network model simulating on spontaneous activities of neurons in the hippocampal CA3 cultured slices [4]. Power law phenomena are observed in the other brain regions (e.g. [9]). Therefore, these two findings may be common features in the brain, but they are not yet found even in developmental experiments using the fMCI. We will have to wait for experimental confirmation for our results. On the other hand, the topology of CA3 slice is already estimated from the experiments (e.g.[4]). The investigation of the relationship between the topology and the spontaneous activity of CA3 slice is needed in future research.

We have found that the high-frequency level of rankings in the distribution of firing frequency was occupied by inhibitory interneurons. Inhibitory interneurons did not fire at high frequency by noise input, but their high-frequency firings were driven by excitatory neurons. This may indicate the experimental finding that inhibitory interneurons in CA3 fire relatively high-frequency than excitatory neurons and they may be driven by excitatory neurons.

In the slice model, we have assumed that log-normally distributed synaptic weights are observed among neurons in the hippocampal CA3 because such log-normal distribution of synaptic weights is already found among neurons in visual cortex [8]. Under the assumption of the log-normal distributed synaptic weights, unbalance between the firing frequencies of excitatory and inhibitory neurons was reduced. The reduction smoothed the probabilistic distribution curve of firing frequency and reproduced the linearity of the distribution on a log-log scale. This may indicate that the synaptic weights among neurons in the hippocampal CA3 also obey a log-normal distribution.

Acknowledgements. Y.D.S. was supported by the Grant-in-Aid for Young Scientist (B) No. 22700237. Y.I. was supported by the Funding Program for Next Generation World-Leading Researchers (no. LS023).

References

1. McNaughton, B.L., Morris, R.G.M.: Hippocampal Synaptic Enhancement and Information Storage Within a Distributed Memory System. Trends Neurosci. 10, 408–415 (1987)
2. Li, X.G., Somogyi, P., Ylinen, A., Buzsáki, G.: The Hippocampal CA3 Network: an in Vivo Intracellular Labeling Study. J. Comp. Neurol. 339, 181–208 (1994)
3. Gulyás, A.I., Miles, R., Hájos, N., Freund, T.F.: Precision and Variability in Postsynaptic Target Selection of Inhibitory Cells in the Hippocampal CA3 Region. Eur. J. Neurosci. 5, 1729–1751 (1993)
4. Takahashi, N., Sasaki, T., Matsumoto, W., Matsuki, N., Ikegaya, Y.: Circuit Topology for Synchronizing Neurons in Spontaneously Active Networks. Proc. Natl. Acad. Sci. U. S. A. 107, 10244–10249 (2010)
5. Izhikevich, E.M.: Simple Model of Spiking Neurons. IEEE Trans. Neural Netw. 14, 1569–1572 (2003)
6. Traub, R.D., Miles, R.: Neuronal Networks of the Hippocampus. Cambridge Univ. Press (1991)
7. Häusser, M., Raman, I.M., Otis, T., Smith, S.L., Nelson, A., du Lac, S., Loewenstein, Y., Mahon, S., Pennartz, C., Cohen, I., Yarom, Y.: The beat goes on: spontaneous firing in mammalian neuronal microcircuits. J. Neurosci. 24, 9215–9219 (2004)
8. Song, S., Sjöström, P.J., Reigl, M., Nelson, S., Chklovskii, D.B.: Highly nonrandom features of synaptic connectivity in local cortical circuits. PLoS Biol. 3, 68 (2005)
9. Beggs, J.M., Plenz, D.: Neuronal Avalanches in Neocortical Circuits. J. Neurosci. 23, 11167–11177 (2003)

An Integrated Hierarchical Gaussian Mixture Model to Estimate Vigilance Level Based on EEG Recordings

Jing-Nan Gu, Hong-Jun Liu, Hong-Tao Lu, and Bao-Liang Lu

MOE-Microsoft Laboratory for Intelligent Computing and Intelligent Systems,
Department of Computer Science and Engineering, Shanghai Jiao Tong University,
Shanghai 200240, China
{albertgu,scorpioliu,htlu}@sjtu.edu.cn,
bllu@cs.sjtu.edu.cn

Abstract. Effective vigilance level estimation can be used to prevent disastrous accident occurred frequently in high-risk tasks. Brain Computer Interface(BCI) based on ElectroEncephalo-Graph(EEG) is a relatively reliable and convenient mechanism to reflect one's psychological phenomena. In this paper we propose a new integrated approach to predict human vigilance level, which incorporate an automatically artifact removing pre-process, a novel vigilance quantification method and finally a hierarchical Gaussian Mixed Model(hGMM) to discover the underlying pattern of EEG signals. A reasonable high classification performance (88.46% over 12 data sets) is obtained using this integrated approach. The hGMM is proved to be more powerful against Support Vector Machine(SVM) and Linear Discriminant Analysis(LDA) under complicated probability distributions.

Keywords: EEG, Vigilance estimation, Quantification, Mixture model.

1 Introduction

EEG recordings can be interpreted for both medical diagnosis and psychological activity estimation used in BCI systems. An EEG-based BCI system provides an effective way to predict the vigilance level and is proved to be powerful in avoiding serious accidents incurred by losing vigilance [1]. Estimation of human vigilance level is a typical pattern recognition task that involves data preprocessing, feature extraction and finally classifying to a specific category. Unfortunately, several distinguished properties related to EEG signals present great challenges to predict vigilance level. The difficulties involved in processing real world EEG signal include its sensitivity to artifacts, ineffective learning due to inaccurate labels and the great diversity of patterns between different people or even between different time with the same person [2].

Previous studies have shown that disposal of artifacts plays an important role to guarantee the robustness of vigilance prediction [3]. Artifact avoidance, rejection and removal are the 3 mostly used strategies to handle artifacts. An

B.-L. Lu, L. Zhang, and J. Kwok (Eds.): ICONIP 2011, Part I, LNCS 7062, pp. 380–387, 2011.

efficient algorithm aimed at detecting and removing artifacts automatically was developed in [4]. The work will lay a foundation to improve and ensure our classification performance. With a relatively clean EEG recordings, many researchers focused on feature extraction intending to reveal neurophysiological phenomena [5]. Time-series information and power spectral density are the generally-accepted features related to EEG recordings [6]. Both of them will be explored in our study to provide adequate information for subsequent analysis. Also many researchers engaged in specific learning strategies. An infinite Gaussian Mixture Model based on Bayes inference was proposed to avoid overfitting in the training process when dealing with high dimensional data [7]. However, it turns out unpractical for its high computational complexity due to the process of model inference.

In this study, we propose an integrated approach to analyze EEG recordings for the purpose of predicting vigilance levels. To handle the undesired artifacts that made significantly impact on EEG signals, a Blind Source Separation(BSS)-based artifact removal approach was used for preprocessing [4]. Since there's few acceptable rules for labeling vigilance scales, we develope a novel vigilance qualifying method which is described in detail in Section 2. Both the spatial and spectral information were implemented for our feature extractor. Finally the hierarchical Gaussian Mixture Model was investigated to accomplish both classification and regression task [8]. The posterior probability presents the reliability of each classification rather than only a result of category information. To verify the expressive ability provided by hGMM, we compared classification performance with other general methods such as SVM and LDA over 12 data sets. Regression performance based on posterior probability calculated by hGMM was also explored to prove the correctness of mixture model.

The rest of this paper is organized as follows. The experiment environment and data acquisition is presented in detail in Section 2. In Section 3, we discussed the main framework of our integrated approach. Experiment results are discussed for both classification and regression task in Section 4. Finally, conclusions and directions for future work are given in Section 5.

2   Data Acquisition and Quantification

To verify the performance and reliability of the integrated approach we proposed, we conducted extensive experiments to collect adequate data based on our simulated driving system. Each experiment collects one data set once a day at noon. A subject is asked to perform a monotonous task sustained about 1 hour after lunch with inadequate sleep in the previous night [9]. We developed a reasonably new approach to label the vigilance scales associated to EEG signals. Having these EEG recordings and its related labels, we could go deep into our classification study, which is discussed in section 4.

2.1   Data Acquisition

Data acquisition will account for surprisingly large part of the cost of our study. To collect an adequately large and representative set of samples for the subsequent training and testing, we invited 10 healthy volunteers to participate in our experiment. Each of them accomplished 2 experimental trials with interval more than 7 days. It took us nearly one month to complete this group of experiments.

Simulated Driving System. Our simulated driving system mainly consists of a software-based simulator, a signal collection system and other necessary facilities(e.g. a 19 inch LCD, a comfortable chair, etc.) [9]. The whole system is located in an isolated noise-free room which has normal brightness and constant temperature between 24°C~26°C.

During the experiment, the simulator would emit a series of traffic signs randomly on the computer screen. The sign is rectangle or triangle-shaped and the main component will be in one of the 4 colors(red, blue, yellow or green). The sign was shown every 7±0.5 sec for a duration of 1 sec successively, with the interval the screen being filled with pure black. There is also a rectangle panel with 4 colored buttons that the subject should hold during the experiment. The subject is supposed to push the corresponding button in their panel accurately and promptly once the sign appeared on the screen.

The 64-channel NeuroScan system sampling at 500Hz was employed for collecting EEG signals to ensure the integrity of neurophysiological information and being available for research in the future. With data recorded in our previous studies, we have 40 sets of data available.

2.2   Quantification Strategy

There is no gold standard for scoring vigilance scales[10]. But for classification task later, we must find an appropriate strategy to quantify vigilance levels. We developed a reasonable qualifying approach considering both the respond time and the error rate he/she pushes the button within a window along 30sec. Our quantification method presents good agreement with the subject's real vigilance state which is monitored by a camera. The mistake credit is assigned as follows:

- sti− >acc− >sti− >acc− >...   // normal responding sequence
  if (diff< 0.2s) $mistake_i = 1$ // ineffective input.(too fast)
  else if (diff< 2s)   // effective input.
      if (sti != acc) $mistake_i = 1$ // wrong input.
      else $mistake_i = 0$ // right and prompt input.
  else mistake = 2   // ineffective input.(too slow)
- sti− >acc− >acc− >...− >acc− >sti // abnormal responding sequence
  $mistake_i = 0$ // too nervous and is regarded as maintaining
                          // high vigilance level. (rarely happened)
- acc− >sti− >sti− >...− >sti− >acc   // abnormal responding sequence
  $mistake_i = 3$ //asleep or distracted

Here 'sti' stands for a sign emitting on the screen and the 'acc' means the subject's response. 'diff' stands for the time between a 'sti' followed by an 'acc'. $mistake_i$ is assigned to a penalty(0,1,2,3) when it encountered with a sti. The error rate of a particular moment is defined as follows:

$$Err = \frac{\sum_{i=1}^{|sti|}(mistake_i)}{3*|sti|} \tag{1}$$

where $|sti|$ represents the number of 'sti' within a 30sec long window and Err always lies in 0~1. Note that the vigilance level presents negative correlation with Err, with high Err means low vigilance level which is dangerous. The corresponding EEG recordings will be labeled every 5sec. Low vigilance level would be labeled as 3 if Err>0.4 at the moment, middle vigilance level labeled as 2 if 0.2<Err<0.4, otherwise high vigilance level labeled as 1.

3 Methods

Given the raw EEG recordings and associated labels, our vigilance estimation system mainly entails the following 3 steps. Firstly, the raw signal was preprocessed to remove artifacts caused by EOG and EMG and filtered to specific frequency bands of our interest. Secondly, feature extraction was done to find relevant and effective features for subsequent classification and regression tasks. Both the spatial and spectral information are considered to be promising distinguishing features and are implemented in our study. Finally, we employed the hierarchical Gaussian Mixture Model to discover the underlying probabilistic representation of EEG signals. Consequently the approximate model can be used to classify the vigilance levels and to calculate the posterior probability to illustrate the reliability of each classification, which can be converted to regression easily.

3.1 Filtering and Artifact Removal

Table1 listed our prior knowledge about the EEG rhythms. Having known that meaningful information involved in EEG signals were mainly lying in frequency band 1~40(Hz) [11] and the signals were sampled at high frequency 500Hz, it could be firstly downsampled to 100Hz to accelerate the speed of computation in the following process without loss of significant neural activities [12].

Table 1. Frequency band of EEG

δ rhythms	θ rhythms	α_1 rhythms	α_2 rhythms	β rhythms	γ rhythms
0.5-3.5Hz	4-7Hz	8-11Hz	11-13Hz	14-25Hz	26-35Hz

Also we know that artifacts, such as EMG and EOG, would distort the original weak EEG signal seriously. It must be detected reliably and the adverse effect caused by it is supposed to be eliminated before data analysis. Of the 3 mostly used methods[3], artifact removing strategy was applied to tackle our problem for its automaticity. Specifically, a BSS-based technique was used to distinguish the EEG-related signals and EEG-unrelated signals and then we used SVM to identify and remove artifact automatically. For a thorough description of this preprocess, please reference to [4].

3.2   Feature Extraction

Having obtained the artifact-free EEG signals, we could now devoted ourself to selecting and designing features. Note the fact that changes of one's physiology status would produce corresponding changes of power density of specific spectral band[13]. Thus we use Short Time Fourier Transformation(STFT) to calculate the power spectral density and recognize the density as features[14].

Also the coefficients of AutoRegreesion model could supply the spatial information about EEG signals. To incorporated adequate information involved in EEG signals,we combine both temporal and spatial information as features[6]. Note that the order of autoregression is a smoothing parameter that control the dimension of the feature. There's a tradeoff between adequate information and the curse of dimensionality[15]. The order of autoregression model was finally determined by 10 fold cross-validation.

3.3   Hierarchical Gaussian Mixture Model

Although the underlying probability densities of real world EEG data are often difficult to approximate, it would be powerful to both classification and regression problems if we could obtain it. For this reason we introduced latent variables to form a hierarchical Gaussian Mixture Model to express arbitrarily complicated model[8].

$$p(\mathbf{x} \mid \omega_k) = \sum_{i=1}^{|\omega_k|} \pi_i \mathcal{N}(\mathbf{x} \mid \mu_i, \Sigma_i) \tag{2}$$

where ω_k is the kth category and the class conditional distribution $p(\mathbf{x} \mid \omega_k)$ is organized as gaussian mixture model. We applied the well-known EM algorithm to calculate the unknown parameters $(\pi_i, \mu_i, \Sigma_i \mid \omega_k)$ to maximize the log likelihood[16].

$$lnp(\mathbf{X}_k \mid \pi_k, \mu_k, \Sigma_k) = \sum_{n=1}^{N} ln\{\sum_{i=1}^{|\omega_k|} \pi_i \mathcal{N}(\mathbf{x}_n \mid \mu_i, \Sigma_i)\} \tag{3}$$

Given the class conditional probability above, we could calculate the posterior probability quite easily as follows. The posterior probability will be used to indicate the reliability of each classification and to form the regression.

$$p(\omega_k \mid \mathbf{x}) = \frac{p(\mathbf{x} \mid \omega_k)p(\omega_k)}{\sum_{k=1}^{K} p(\mathbf{x} \mid \omega_k)p(\omega_k)} \tag{4}$$

4   Experiment and Results

In this section, we will validate the effectiveness of the integrated hGMM approach we proposed above. Firstly 12 data sets with readily distinguished features were selected for analysis. Both of them either hold their one vigilance level

Table 2. Average Classification Rate over 12 data sets

	hGMM	SVM	LDA
a1	88.46%	80.77%	73.19%
b4	77.81%	72.95%	59.28%
c10	63.95%	58.16%	52.41%

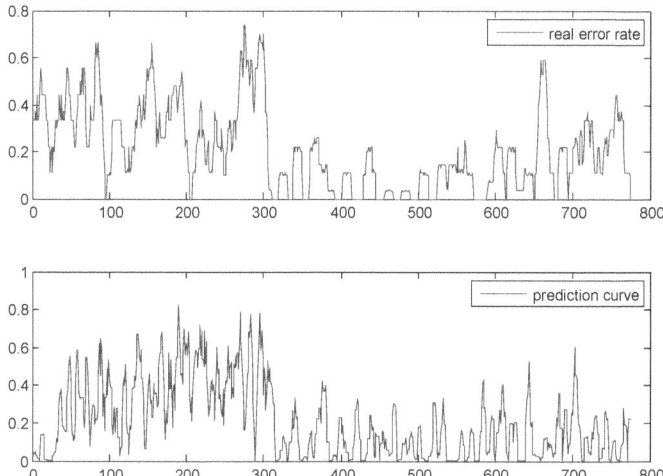

Fig. 1. The vertical axis indicates error rate the subject responded, with low error rate represent high vigilance level. The horizontal axis is the time course of this trial lasting about 5×800sec. This figure shows good agreement between quantified vigilance level and the prediction result.

all the while or have regular changes. The starting and ending 5min recordings were removed to avoid environmental noise. Because too many information is no better than no information at all, which is known as occum's razor rule[15], we decided to use relatively small number of channels located in the central part of scalp(a1 means channel {10}, b4 means channel {10,18:20}, c10 means channel{10,18,20,26,30,46,50,55:57}). This scenario also makes sense in the real-world applications where too many electrodes on scalp is impossible.

The widely used classification approach involving SVM and LDA were used for the purpose of comparison with hGMM. A RBF-kernel SVM was used for classification[17]. To select optimal parameters with each of the 3 models, extensive experiments were conducted based on 10 fold cross-validation, with 2 representative data sets merged for testing data.

We conducted extensive classification experiments on selected 12 data sets to show the performance of the integrated approach. Table2 shows that high classification rate can be obtained when using hGMM with only one channel. With increasing number of channels, which means increase of features, the classification rate was degraded for all of the 3 methods consistently. We believe it is a promising property to utilize as we pointed out above.

Finally, to predict vigilance level continuously and reliably, we investigated the regression ability based on hGMM. The posterior probability of label 3 were supposed to reflect the subject's vigilance level. The posterior probability is averaged over a 30sec long window to avoid extensive oscillation. We obtained a continuous prediction curve which could catch the main trend of vigilance level, as illustrated in Fig.1.

5   Conclusions

In this paper, we proposed a new integrated approach to predict vigilance level automatically and continuously based on EEG signals. With adequate and effective preprocessing, we obtained a reliable and reasonable high classification performance against 2 traditional method SVM and LDA over 12 data sets. The regression performance based on posterior probability of hGMM is proved to be promising and needed further study to improve its generalization ability. Furthermore, to deal with artifact in EEG recordings more naturally, we will investigate the hierarchical Dirichlet Mixture Model to improve the robustness of our method.

Acknowledgments. This study is funded by the National High Technology Research and Development Program of China (No.2008AA02Z310) and Science and Technology Commission of Shanghai Municipality(09511502400). Also special thanks to Prof. Bao-Liang Lu and other researchers in his laboratory for their helpful work on EEG data acquisition.

References

1. Wolpaw, J., Birbaumer, N., McFarland, D., Pfurtscheller, G., Vaughan, T.: Brain-computer Interfaces for Communication and Control. Clinical Neurophysiology 113, 767–791 (2002)
2. Fruhstorfer, H., Bergström, R.: Human vigilance and auditory evoked responses. Electroencephalography and Clinical Neurophysiology 27, 346–355 (1969)
3. Fatourechi, M., Bashashati, A., Ward, R., Birch, G.: EMG and EOG Artifacts in Brain Computer Interface Systems: A Survey. Clinical Neurophysiology 118, 480–494 (2007)
4. Bartels, G., Shi, L.C., Lu, B.L.: Automatic Artifact Removal from EEG-a Mixed Approach Based on Double Blind Source Separation and Support Vector Machine. In: 32nd IEEE Engineering in Medicine and Biology Society, pp. 5383–5386. IEEE Press, Buenos Aires (2010)
5. Grosse-Wentrup, M., Gramann, K., Buss, M.: Adaptive Spatial Filters with Predefined Region of Interest for EEG Based Brain-Computer-Interfaces. In: Advances in Neural Information Processing Systems, vol. 19, pp. 537–544 (2007)
6. Dornhege, G., Blankertz, B., Curio, G., Muller, K.: Combining Features for BCI. In: Advances in Neural Information Processing Systems, pp. 1139–1146 (2003)

7. Rasmussen, C.: The Infinite Gaussian Mixture Model. In: Advances in Neural Information Processing Systems, vol. 12, pp. 554–560 (2000)
8. Rosipal, R., Peters, B., Kecklund, G., Åkerstedt, T., Gruber, G., Woertz, M., Anderer, P., Dorffner, G.: EEG-Based Drivers' Drowsiness Monitoring using a Hierarchical Gaussian Mixture Model. In: Schmorrow, D.D., Reeves, L.M. (eds.) HCII 2007 and FAC 2007. LNCS (LNAI), vol. 4565, pp. 294–303. Springer, Heidelberg (2007)
9. Yu, H., Shi, L.C., Lu, B.L.: Vigilance Estimation Based on EEG Signals (2008)
10. Shi, L.C., Yu, H., Lu, B.L.: Semi-Supervised Clustering for Vigilance Analysis Based on EEG. In: 20th IEEE International Joint Conference on Neural Networks, pp. 1518–1523. IEEE Press, Florida (2007)
11. Omerhodzic, I., Avdakovic, S., Nuhanovic, A., Dizdarevic, K.: Energy Distribution of EEG Signals: EEG Signal Wavelet-Neural Network Classifier
12. Liu, H.J., Ren, Q.S., Lu, H.T.: Estimate Vigilance in Driving Simulation Based on Detection of Light Drowsiness. Bioinformatics, 131–134 (2010)
13. Belyavin, A., Wright, N.: Changes in Electrical Activity of the Brain with Vigilance. Electroencephalography and Clinical Neurophysiology 66, 137–144 (1987)
14. Ramoser, H., Muller-Gerking, J., Pfurtscheller, G.: Optimal Spatial Filtering of Single Trial EEG during Imagined Hand Movement. Rehabilitation Engineering 8, 441–446 (2000)
15. Rasmussen, C., Ghahramani, Z.: Occams Razor. In: 13th Advances in Neural Information Processing Systems: Proceedings of the 2000 Conference, pp. 294–300. The MIT Press, Cambridge (2001)
16. Nabney, I.: NETLAB: Algorithms for Pattern Recognition. Springer, Heidelberg (2002)
17. Chang, C., Lin, C.: Libsvm: a Library for Support Vector Machines (2001)

EEG Analysis of the Navigation Strategies in a 3D Tunnel Task

Michal Vavrečka, Václav Gerla, and Lenka Lhotská

Faculty of Electrical Engineering
Czech Technical University, Prague
vavrecka,gerla,lhotska@fel.cvut.cz

Abstract. This paper focus on the analysis of the navigation task in a 3D virtual tunnel. We localized the neural structures responsible for egocentric and allocentric reference frame processing in horizontal and vertical plane and also analyzed specific segments of the tunnel traverse. We identified intrahemispheric coherences in occipital-parietal and temporal-parietal areas as the most discriminative features for this task. They have 10% lower error rate comparing to single electrode features. The behavioral analysis of navigation reveals that 35% of the participants adopted two types of egocentric reference frames in the vertical plane.

Keywords: frames of reference, spatial navigation, EEG, 3D environment, coherence.

1   Introduction

The human ability to represent space, orient in an environment, create spatial mental images and talk about the environment from various perspectives is related to the adoption of various frames of reference. The basic classification of the reference frames involves spatial coordinate systems based on egocentric and allocentric frames of reference [4]. In the egocentric frame the position of objects is encoded with reference to the body of the observer or to relevant body parts [5]. Spatial positions can also be coded in allocentric coordinates that are independent from the observer's current position.

The fMRI research in humans revealed differences between the utilization of egocentric and allocentric frames of reference. Committeri et al. [2] attribute egocentric coding mainly to the dorsal stream (BA 7) and connected frontal areas, whereas allocentric coding requires both dorsal and ventral regions. The recent fMRI study [6] implicates posterior parietal and frontal associated regions involved in processing the egocentric frame. Allocentric navigation is attributed to the specific parietal subregions and also to the hippocampal and retrosplenial region. These studies are based on a static stimuli experiments. Schönebeck et al. [8] researched reference frames in dynamic environment and conducted a task built in a virtual reality environment consisting of a traverse through a tunnel with straight and turned segments. Gramann et al. [10] replicated this scenario and recorded an EEG signal to identify neural correlates responsible for specific

B.-L. Lu, L. Zhang, and J. Kwok (Eds.): ICONIP 2011, Part I, LNCS 7062, pp. 388–395, 2011.
© Springer-Verlag Berlin Heidelberg 2011

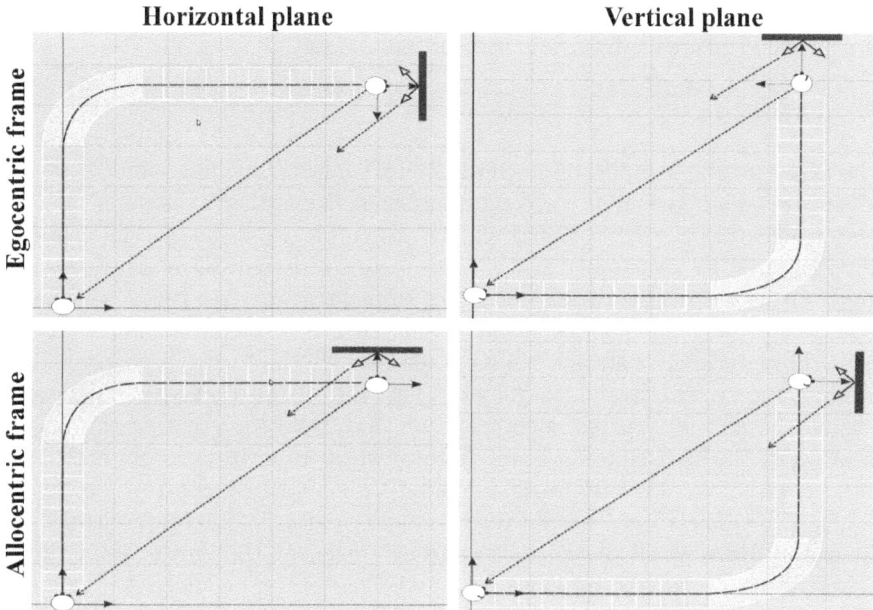

Fig. 1. Visualization of the tunnel task in a 3D environment. The head position is shown at the beginning and the end of the tunnel traverse for specific frames of reference (egocentric vs. allocentric) and specific planes (horizontal vs. vertical). The dark bar represents a computer screen with two arrows standing for the selection period.

frames of reference processing. They localized higher mean source activation in the BA 7 (parietal cortex) for subjects adopting an egocentric frame and BA 32 (anterior cingulate gyrus) for subjects adopting an allocentric frame of reference. In the most recent EEG research [11] based on the tunnel task, stronger alpha blocking was identified in or near the right primary visual cortex (BA 17) for subjects adopting an egocentric reference frame in the turned segment of the tunnel and stronger alpha blocking of the occipito-temporal, bilateral inferior parietal (BA 7), and retrosplenial cortical areas (BA 26,29,30) in the first straight and turned segments for participants adopting an allocentric frame.

Our research is the extension of the mentioned studies. We would like to compare the specific parts of the tunnel traverse in horizontal plane with Gramann et al. results [10] [11], but we want also to administer the tunnel task in vertical plane. Vidal et al. [9] administered a reference frame study in a 3D environment and concluded that the spatial updating process was more accurately performed for a terrestrial strategy (the head turns only in the horizontal plane) and to some extent a subaquatic strategy (the head turns in both the horizontal and vertical plane) than for a weightless (yaw and pitch turns) navigational style. Our goal is the identification of neural correlates responsible for the terrestrial navigation (resembling the allocentric strategy in the tunnel task administered in the vertical plane) and subaquatic navigation (resembling the egocentric strategy in vertical tunnels). We hypothesize different brain areas involved in processing of mentioned reference frames in the vertical plane.

2   Methods

The experimental sample consisted of 38 participants (7 females and 31 males). All subjects had normal or corrected-to-normal vision, they were under no medication affecting the EEG signal and were neurologically intact. The subjects were required to keep the track of their implied virtual 3D position with respect to their starting position within the tunnel traverse (see Fig. 1). A fixation cross was present for 6s prior to each trial. Each tunnel consisted of a 10s traverse through the first straight segment, 6s through the turned segment and 10s through the second straight segment. The bend of the turned passage varied between 30 and 90 degrees at intervals of 15 degrees representing the angular rotation of the participant's head. A total of 20 tunnels were presented to a subject, i.e. 5 tunnels with variable curvature for 4 directions (up, down, left, right). The tunnels were administered pseudo-randomly. There were two three-dimensional arrows displayed on the black screen at the end of each tunnel, representing the correct response within the egocentric or allocentric reference frame. The subject's choice is answers the question: what reference frame did he/she adopt as the navigation system. The answers were evaluated to identify the subject's preferred reference frame. Participants selecting the same frame of reference in above 80 % were considered as representative (native) users of the particular reference frame.

The EEG signal was recorded from 19 electrodes, positioned under the 10-20 system. We performed a visual inspection of each EEG signal prior to data analysis in order to detect obvious technical and biological artifacts. The signal was divided into the segments of constant length (1s), and the following parameters were calculated: statistical parameters, mean and maximum values of the first and second derivation of the samples and absolute/relative power for five EEG frequency bands. The EEG coherences, the correlation between the EEG electrodes, and the mean and maximum correlation values for each segment were computed. The wavelet transform was also applied to the signal segments. Daubechies 4 was used as the mother wavelet, and the signals were decomposed into 4 levels standing for standard EEG frequency bands. We also calculated the mean and maximum values of the wavelet coefficients obtained after applying the wavelet transform to the first and second derivative of the EEG signal. The data was processed in PSGLab Toolbox that is developed in our laboratory [7].

The next step was feature selection. The input matrix for each subject consisted of 1916 features (93 features per electrode + correlations and coherences) for the duration of the experiment. We employed PRTools [12] for this part of the analysis. We removed outliers and normalized the data. The features were individually evaluated using the inter/intra distance to preselect 50 best features. In the next step we applied forward feature selection algorithms to choose 5 best features from the preselected subset. Then we performed 7-fold cross-validation and employed naive Bayes classifier to calculate error rate of the best features.

3   Results

The administration the tunnel task in the vertical plane resulted in new navigation strategies compared to the horizontal plane. The experimental sample consisted of participants natively adopting the egocentric reference frame in both planes (24 %), participants natively adopting the allocentric frame in both frames (22 %) and participants adopting an egocentric frame for horizontal navigation and an allocentric frame for vertical navigation (11 %). Some subjects (30%) reinterpreted the instruction and did the mental u-turn at the end of tunnel and then adopted egocentric strategy. We also excluded 5 subjects (13 %) who answered randomly.

At the first stage we analyzed data from 17 participants adopting egocentric and allocentric strategy in both planes (9 egocentric and 8 allocentric frames of reference). The mean error rate of the best 5 features for both planes was 7.55 %. The dataset for whole tunnel traverse was split to the two subsets and the best features for the horizontal and vertical plane were calculated. The coherences in the theta and gamma band were the most discriminative features distinguishing egocentric and allocentric strategies for the horizontal plane, namely theta and gamma intrahemispheric coherence in the right temporal lobe and gamma coherence in the right frontal lobe. There were also interhemispheric coherences between the orbitofrontal electrodes in the theta band and the temporal electrodes in the delta band. The most discriminating features for the vertical plane were interhemispheric coherence between the temporal lobes in the theta band. Unlike the horizontal plane, there were intrahemispheric coherences between the right parietal and occipital area and the left temporo-occipital area in the beta band. A timeline of best feature (excluding coherences) is shown in Fig. 2. It can be seen that the group mean values differentiate between the navigation strategies.

We did also the analysis for the separate parts of tunnel. The best features for the first segment of the tunnel were concordant with the results for the whole tunnel, namely intrahemispheric theta coherences in left temporal lobe, beta coherence in the right parieto-occipital area and gamma coherence in left frontal lobe. The exception was the gamma activity in the left orbitofrontal lobe as the discriminative feature for the first straight segment. The best feature in the turned segment was the interferometric coherence in theta band between orbitofrontal electrodes. The other features were similar to the first straight segment. The same results were obtained for the second straight segment except the signal changes in the left temporal lobe.

The summarized differences between the allocentric and egocentric groups were visualized on scalp projections. Fig. 3 shows the mean activity in five spectral bands for allocentric and egocentric strategies (columns 1 and 2) and their difference maps (columns 3-8) for the specific planes or parts of the tunnel. There are no visible differences between the horizontal and vertical planes (columns 4 and 5), but there is a change in the beta band activity for the second straight segment of the tunnel (column 8).

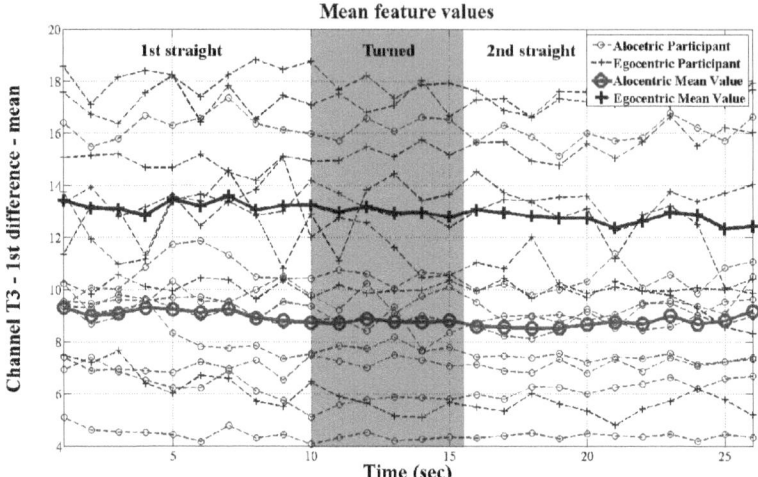

Fig. 2. Timeline of the best feature changes (coherences excluded) within a tunnel traverse. There is a visualization of the mean feature values for specific strategies and averaged feature values for each participant. The x-axis represents time in seconds, and the y-axis represents feature values.

4   Discussion

The experiment revealed new findings regarding the adoption of egocentric and allocentric reference frames in a 3D environment. We identified native adoption of terrestrial navigation [9] (resembling allocentric strategy) in the vertical plane for the group of participants navigating egocetrically in horizontal plane. They represented body in the upright position at the end of the vertically oriented tunnels, so there was no angular rotation of the head direction. On the other hand none of the subjects adopted an allocentric frame in the horizontal plane and an egocentric frame (resembling subaquatic navigation) in the vertical plane.

The neurobehavioral results of our study are partially consistent with previous studies. Gramann et al. [10] localized higher mean source activity in BA 7 for the egocentric frame of reference, but the allocentric strategy was linked to activation in the anterior cingulate cortex (BA 32). A different type of analysis based on event-related spectral perturbations [11] revealed stronger alpha blocking in BA 17 for subjects adopting an egocentric reference frame in the turned segment of the tunnel and stronger alpha blocking in BA 7 and BA 26, 29, 30 for participants adopting an allocentric reference frame. We detected differences in beta band coherence in the left occipital-parietal lobe. The coherences were higher for the allocentric strategy, but the detail analysis of beta activity in the separate electrodes (P4 and O2) revealed higher values for the egocentric group of participant. We can interpret this as coherent low beta activity in this electrodes for the allocentric frame of reference, but the egocentric strategy resulted in non-coherent higher activity in beta band.

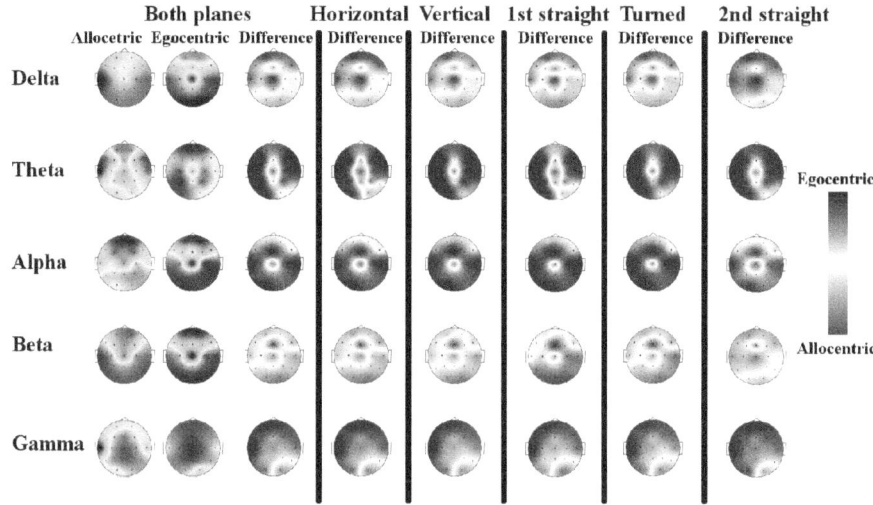

Fig. 3. Difference maps for the horizontal and vertical planes, and for both planes, and specific parts of the tunnel. The rows represent basic spectral bands. Average values for the egocentric and allocentric groups are given in columns 1 and 2. Columns 3-8 stand for difference maps. The values were calculated as the mean of all tunnel traverses.

The data for the horizontal plane indicates as the discriminative features intrahemispheric coherences in the gamma band in the left temporal lobe. Left intrahemispheric coherence in the gamma band stands for higher coherence of the egocentric group, but further analysis revealed higher gamma activity in both electrodes (T3-T5) for the allocentric group. We can interpret this as low coherent gamma activity for the egocentric frame of reference, while the allocentric strategy resulted in non-coherent higher activity in these electrodes. Gamma activity is associated with cognitive functions and multimodal integration. These results differ from the EEG study that adopts the tunnel task [10]. The list of best features also includes interhemispheric coherences but it is not possible to interpret the contralateral scalp locations in terms of brain structures, because they can reflect indirect, common/shared activity of the subcortical brain regions. In contrast to previous studies, we detected changes in the gamma band in the left frontal areas (electrodes Fp1 and F7) in horizontal navigation and also in vertical plane navigation. There was higher coherence for the egocentric strategy but analyses of specific electrodes revealed low coherent activity for the egocentric frame of reference and non-coherent higher activity in gamma band for the allocentric reference frame. The left frontopolar area is involved in memorizing tasks [3]. Thus, the difference in left frontal gamma coherence that we observed may be related to the different memory processing involved in the allocentric strategy.

At the next stage we compared the discriminative features for the horizontal and vertical plane. The differences between specific planes are manifested in the absence of theta coherence in the orbito-frontal cortex (Fp1-Fp2) for the vertical plane. It may be attributed to the lower number of eye movements in the vertically oriented tunnels. There is also a shift in theta coherence from T5 and T6 for the horizontal plane to T3 and T4 for the vertical plane. The most interesting observation is the absence of coherence in the beta band between P4 and O2 in horizontal plane. The results in the vertical plane cannot be compared with other EEG studies, as this is the first attempt to measure the frames of reference in a 3D environment.

We also compared results for the separate segments of tunnel traverse with the previous studies. For the first straight part of the tunnel Gramann et al. [10] research higher mean source activity in bilateral occipital-temporal network, with additional activation in frontal cortex for the egocentric frame and activation within a bilateral temporal-occipital network for the allocentric strategy. The similar activity for both strategies were located in BA 19,29 and 21. Our results revealed differences in the interhemispheric theta band coherences (T3-T4 and T5-T6) in terms of higher coherences for the allocentric strategy. From the spectral point of view we observed lower theta activity in both temporal lobes for allocentric group. There was also coherence in the beta band between P4 and O2 electrode and coherence in frontal left orbito-frontal lobe as a discriminative feature similar to analysis of the whole tunnel.

In the turned segment Gramann et al. [10] localized higher activity in fronto-parietal network (BA 7), with dominance over the left hemisphere for egocentric strategy and left anterior cyngulate gyrus (BA 32) for allocentric group. We localized areas similar to the first segment, but also coherence in theta band between orbito-frontal areas (Fp1, Fp2) in terms of higher activity for egocentric strategy. Theta activity is associated with the heading changes [1] so the differences between allocentric and egocentric group in theta band should stand for different processing in the turned segment. The analysis of the whole tunnel traverse in specific planes uncovered absence of intrahemipheric theta coherence in orbito-frontal areas for vertical plane that confirms our hypothesis about different eye movements for specific planes.

At the second straight segment Gramann et al. [10] attributed egocentric strategy to the activity bilaterally within a fronto-parietal network (BA 7) including regions that were activated both with the onset of the tunnel movement and during the turn in the tunnel passage. Allocentric strategy was manifested in right hemispheric activation pattern comprising the temporal cortex. We found similar set of features for this part of tunnel as for the turned segment. There were not difference in the theta coherence of frontal areas (Fp1-Fp2) that would confirm the eye movement artifact hypothesis for the turned segment.

5   Conclusion

Administering the tunnel task in the vertical plane provided new insights into the area of spatial navigation. We should conclude that there are differences in the EEG activity for the navigation in the horizontal and vertical plane. Future steps in our research will focus on the difference when two types of egocentric reference frame are adopted within navigation in the vertical plane in order to identify the neural correlates of these strategies.

Acknowledgment. This work has been supported by the Ministry of Education, Youth and Sports of the Czech Republic (under project No. MSM 6840770012 "Transdisciplinary Biomedical Engineering Research II") and GAČR grant No. P407/11/P696.

References

1. Bischof, W.F., Boulanger, P.: Spatial Navigation in Virtual Reality Environments: An EEG Analysis. Cyberpsychology and Behaviour 6, 487–496 (2003)
2. Committeri, G., Galati, G., Paradis, A.L., Pizzamiglio, L., Berthoz, A., LeBihan, D.: Reference frames for spatial cognition: different brain areas are involved in viewer-, object-, and andmark-centered judgments about object location. Journal of Cognitive Neuroscience 16(9), 1517–1535 (2004)
3. Fuster, J.M.: The Prefrontal Cortex, 2nd edn. Raven Press, New York (1989)
4. Howard, I.P., Templeton, W.B.: Human spatial orientation. Wiley, London (1966)
5. Galati, G., Lobel, E., Vallar, G., Berthoz, A., Pizzamiglio, L., Le Bihan, D.: The neural basis of egocentric and allocentric coding of space in humans: a functional magnetic resonance study. Experimental Brain Research 133, 156–164 (2000)
6. Galati, G., Pelle, G., Berthoz, A., Committeri, G.: Multiple reference frames used by the human brain for spatial perception and memory. Experimental Brain Research 206(2), 109–120 (2010)
7. Gerla, V., Djordjevic, V., Lhotská, L., Krajča, V.: PSGLab Matlab Toolbox for Polysomnographic Data Processing: Development and Practical Application. In: Proceedings of 10th IEEE International Conference on Information Technology and Applications in Biomedicine (2010)
8. Schönebeck, B., Thanhauser, J., Debus, G.: Die Tunnelaufgabe: eine Methode zur Untersuchung rumlicher Orientierungsleistungen 48, 339–364 (2001)
9. Vidal, M., Amorim, M.A., Berthoz, A.: Navigating in a virtual three dimensional maze: How do egocentric and allocentric reference frames interact? Cognitive Brain Research 19, 244–258 (2004)
10. Gramann, K., Müller, H., Schönebeck, B., Debus, G.: The neural basis of egocentric and allocentric reference frames in spatial navigation: Evidence from spatio-coupled current density reconstruction. Brain Research 1118, 116–129 (2006)
11. Gramann, K., Onton, J., Riccobon, D., Müller, H.J., Bardins, S., Makeig, S.: Human brain dynamics accompanying use of egocentric and allocentric reference frames during navigation. Journal of Cognitive Neuroscience 22(12), 2836–2849 (2010)
12. van der Heijden, F., Duin, R., Ridder D.d., Tax, D.M.J.: Classification, Parameter Estimation and State Estimation: An Engineering Approach Using MATLAB. John Wiley & Sons (2004)

Reading Your Mind: EEG during Reading Task

Tan Vo and Tom Gedeon

School of Computer Science, The Australian National University,
Acton, Canberra, ACT 0200, Australia
{tan.vo,tom.gedeon}@anu.edu.au

Abstract. This paper demonstrates the ability to study the human reading behaviors with the use of Electroencephalography (EEG). This is a relatively new research direction because, obviously, gaze-tracking technologies are used specifically for those types of studies. We suspect, EEG, with the capability of recording brain-wave activities from the human scalp, in theory, could exhibit potential attributes to replace gaze-tracking in such research. To prove the concept, in this paper, we organized a BCI experiment and propose a model for effective classifying EEG data in comparison to the accuracy of gaze-tracking. The results show that by using EEG, we could achieve comparable results against the more established methods while demonstrating a potential live EEG applications. This paper also discusses certain points of consideration for using EEG in this work.

Keywords: BCI, Artificial Neural Network, EEG, Reading tasks, Signal Processing.

1 Introduction

We conducted an experiment where we capture test participants EEG activities while they perform reading tasks. We analyzed a set of EEG features such as the frequency activations of EEG alpha, theta, beta bands etc.... to identify the key factors showing user engagement level in reading. The results are analyzed for each individual participant against the whole set of participants. Our aim is to verify the following hypotheses: "For each participant, can we effectively identify the link between EEG brain activities and his level of engagement in the reading task?" and "Overall, can we achieve a general method to effectively identify the link between EEG brain activities and the level of engagement in the reading task?" For this paper, together with analyzing these hypotheses, we also propose a method of processing EEG signals that is effective enough to be considered for a real-time classification system. We use an Artificial Neural Network (ANN) technique to validate the efficiency of the proposed approach.

2 Backgrounds

2.1 General

Reading is an activity that most human today perform on a very regular basis. We read and process information so much that reading skill becomes an almost

B.-L. Lu, L. Zhang, and J. Kwok (Eds.): ICONIP 2011, Part I, LNCS 7062, pp. 396–403, 2011.

second nature to us. The conjecture we propose here is based on that statement. So comprehensive and comfortable are we in reading words, texts, that we would show, intentionally or not, certain behaviors that could be used to interpret our perception of the contents we read. Understanding the meaning of words in sentences and paragraphs places a certain strain on a person's cognitive process. Depends on various contexts, such strains could go unnoticed by most of us. An example of such process would be, in order to comprehend a text, a person needs to build up linkages of information that he previously obtained with the current text. That cognitive process would be more significant if the text contains more information that he would, deliberately or not, associate back to. That assumption is reasonably correct because a normal person can only keep seven pieces of information (± 2) in their short-term memory [4]. We had some success in [6] in identifying such a process using gaze-tracking technique. For this experiment, however, we would like to validate the results with EEG. Such use of EEG is a brain-computer interaction technique that monitors brain-wave activities from the human scalp. We suspect that the EEG signals would exhibit those aforementioned cognitive activities during reading tasks. A point for consideration in doing research with EEG is to deal with eye movement artifacts in EEG signals. Eye muscles produce considerable EEG signal noises and traditionally, EEG researchers would remove them from the signal analysis[8]. For this paper, however, we would like to propose a different approach to that by not eliminating the effect of eye movements from our analysis of reading tasks. Reading tasks have one unique characteristic that supports our view: a persons eye movements tie quite robustly to their engagement to the contents being read. The increase/decrease in the amount of skipping forward and back-tracking activities found in the gaze correlate with the increase/decrease of the cognitive load in reading [6]. Studying of reading without regard to eye-gaze could limit the potential outcome. As our aim is to identify the same link through the use of EEG instead of gaze-tracking technology, we would like to take advantage of this "good noise.

Another point for consideration is that the EEG signal, by nature, is stochastic. In regards to this experiment, it suggests that the 19 different participants EEG data should be processed and analyzed individually. In this paper, however, we will try both approaches: consider each participant individually vs. all participants as a whole. We then in turn compare the outcomes. We also compare the results with the gaze data we collected from our previous experiment [6].

2.2 Preliminary Analysis

For the initial analysis, we ran time-frequency analysis on the raw EEG data, as suggested by Makeig[7]. We would like to observe the differences in time-frequency distributions of EEG signals captured from a person reading a relevant against irrelevant piece of text (English paragraphs). This analysis is performed on the first raw EEG channel (Fp1) of each participant using Fourier Transform and the time resolution is going to be (time taken to read, in milliseconds, over 512 epochs). The initial observation has revealed that there is a lack of apparent and consistent features that could help distinguished the two classes. Having said that, there is a minor difference that can be spotted by observing those spectrographs - that is, there are more drops in amplitudes found on the spectrographs of EEG signals recorded from reading irrelevant paragraphs compared to the one obtained from reading relevant paragraphs.

Figure 1 demonstrates the above observation. It includes spectrograms generated by analyzing the EEG signal a participant reading two relevant paragraphs (Left-hand Side) and two irrelevant paragraphs(Right-hand side). The circled spots are some of the locations where the drops in amplitude can be identified.

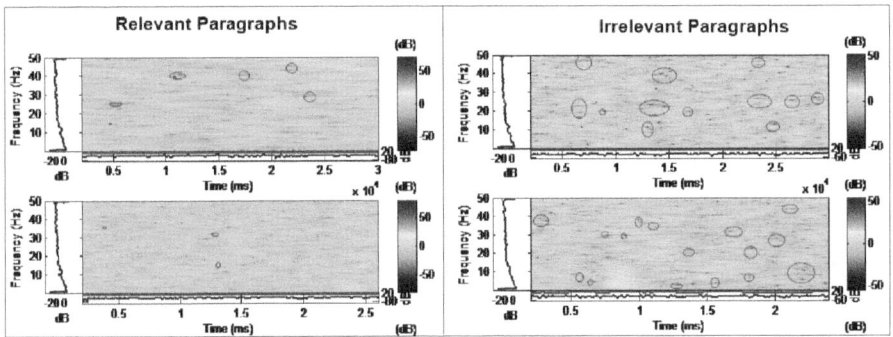

Fig. 1. A distribution of cell movement distances throughout reading activity of a paragraph

The initial analysis has showed to us that there is a possibility that, using statistical machine learning techniques, we could effectively classify the two cases. For that purpose, we are going to use a standard Artificial Neural Network configuration as the foundation. Optimization will be considered if the initial results are promising.

3   Experiment

We have 19 participants for this experiment. The experiment involves the participant reading some paragraphs from a computer screen while the computer captures their brain-wave activities via EEG equipment. In total there were ten paragraphs for the participants to read. Seven of the paragraphs were taken from the paper "Keyboard before Head Tracking Depresses User Success in Remote Camera Control" by Zhu et al.[1]. The remaining three paragraphs were extracts from various sources (miscellaneous paragraphs).

Five of the paragraphs from the paper were chosen for the amount of useful information that was contained within and they are relevant to each other. The other five paragraphs (two from the aforementioned paper and the three miscellaneous ones) were chosen because of their generality and lack of specific technical information - they are irrelevant with the other five and also are irrelevant between themselves. Care was taken to make sure that this fact was not obvious to experiment participants.

3.1 Setups

For each of the 19 volunteer participants, the general instructions are to read as if they were just reading any regular piece of text, and that they would not be questioned about the paragraphs read at the end of the trial. For recording EEG signals, which are very sensitive, we also paid attention to eliminate as much external distraction as possible during each trial. The experiment is designed to help show which participants could look at the bigger picture even when the information is out of sequence and scattered. Hence, the paragraphs were presented to participants in different orders to prevent any specific paragraph ordering from affecting the results. Figure 2(a) shows one of the paragraphs that each participant read. The screen for reading is about 72 cm away from the participant face. The head position of participant is secured with a chin rest. This is to minimize head/face movements (intentionally or not) - which could greatly affect the EEG signals. The EEG equipment we used in this experiment is BioSemi ActiveTwo. We recorded with 16 channels marked and placed according to the 10-20 system .These 16 channels are as followed: Fp1, Fp2, F4, Fz, F3, T7, C3, Cz, C4, T8, P4, Pz, P3, O1, Oz, O2. Figure 2(b) shows one participant setup with the 16 electrodes. The recording was continuously throughout the trial of

In this research, we focus on importing computer vision technology to undertake head tracking in interface design for teleoperation activities. The common remote control situation described above is modelled by using a physical game analogue: playing a table soccer game with two handles. This has the advantage of being more compelling for our student experimental subjects than a more abstract task. We use student experimental subjects as we have limited access to the operators. We then propose a novel design applying natural human head gestures for controlling a Pan-Tilt-Zoom camera as an effective approach to solve the camera control problem.

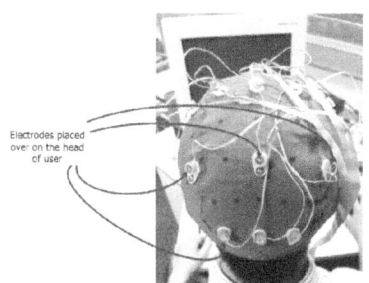

(a) One of the reading paragraphs   (b) EEG electrodes layout

Fig. 2. One example of the paragraphs and The EEG electrodes

each participant. After reading through 10 paragraph (a trial) - the recording is stopped and stored for offline analysis. Timestamps are marked to indicate the start and end of each reading of a paragraph.

3.2 Signal Processing

The EEG signals were originally collected at 1024 Hz, which in turn were downsampled to a rate of 256 samples per second. We ran a low-pass filter of 60Hz to eliminate unwanted EEG frequencies. The time-domain EEG signal is then be broken into epochs (windows) of one second. Fourier-Transformation (FFT) is performed on each window - for each of the 16 channels. The data from the FFT is binned into four frequency ranges: Delta, (0-4Hz), Theta (4-8Hz), Alpha (8-13Hz) and Beta(13-30Hz). Figure   3(b) show the output of FFT with the color-coded ranges. A peak detector (sensitivity of 0.7) is run on the FFT bins

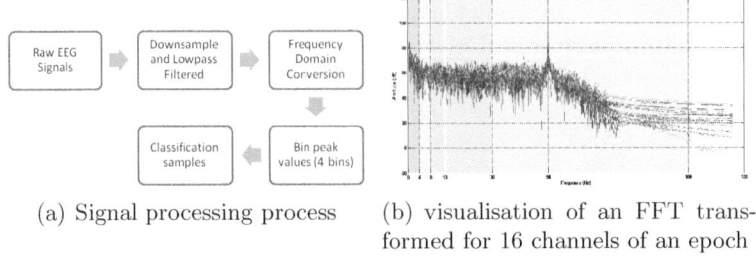

(a) Signal processing process   (b) visualisation of an FFT transformed for 16 channels of an epoch

Fig. 3. Signal Processing Blocks and FFT transformation

to detect peak values for each EEG band. Since we are interested in the activities of the four EEG bands (Alpha, Beta, Theta and Gamma), a window (epoch) will have four features representing it. Since we have 16 channels, our regular dataset will have 64 features (4 x 16). The same process 3(a) , will be applied on the subsequence epochs until the end of EEG recording. The samples are labeled "1" to indicate reading relevant text against "0" for reading irrelevant text. We choose FFT because it is a very efficient transformation algorithm. Using it results in a signal processing capable of perform live (real-time) EEG signal processing. Optimization of window size could further improve the performance of this model.

3.3   Classification

Artifical Neural Network. With the dataset we captured, we think it will be sufficient for us to utilize a standard ANN configuration. The ANN setup we constructed for this experiment is a feed-forward, back-propagation network. This network has one hidden layer containing 20 hidden neurons and one output. As for the neural network optimization algorithm, we took the advantage of the Levenberg-Marquardt optimization (ML) training algorithm. The output value is described as [3],[6],[5]:

$$y_1^T = g_O(b_1 + \sum_j W_{1j} \cdot g_H(b_j + \sum_k w_{jk} \cdot x_k^T)), \tag{1}$$

The back-propagation training algorithm, being Levenberg-Marquardt optimization, will be represented by the formula[3] [6]:

$$\delta w = (J^T J + I \cdot \mu)^{-1} J^T e \tag{2}$$

The training performance is calculated using the Mean Square Error function. The training process is expected to stop once the performance (error) is minimized to the goal.

4   Evaluation and Comparison

4.1   For Individual Participants

We divided our dataset into smaller groups by participant - we called them P1, P2 all the way to P19. The average sample size of reach group is about 57 samples. For each group, we ran our constructed ANN with 10-Fold cross-validation. The results are shown in table 1.

Table 1. ANN classification results for 19 participants (individual)

	Accuracy	Specificity	Sensitivity		Accuracy	Specificity	Sensitivity
P1	0.928	0.966	0.872	**P11**	1.000	1.000	1.000
P2	0.918	0.891	0.963	**P12**	0.927	1.000	0.833
P3	0.986	0.977	1.000	**P13**	0.968	1.000	0.920
P4	0.924	0.921	0.929	**P14**	1.000	1.000	1.000
P5	0.781	0.667	0.929	**P15**	1.000	1.000	1.000
P6	0.931	0.889	1.000	**P16**	0.972	1.000	0.941
P7	0.984	1.000	0.958	**P17**	0.966	0.962	0.972
P8	1.000	1.000	1.000	**P18**	1.000	1.000	1.000
P9	0.900	0.871	0.947	**P19**	0.985	1.000	0.963
P10	0.979	1.000	0.944	**Average**	0.955	0.955	0.956

The results table above demonstrates the effectiveness of our method (data processing and classification). The average accuracy rate is about 95 percent, which is quite encouraging for the task of classifying EEG signals. There is still minor inconsistency in the achieved results - with P5 achiving about high 70% accuracies. It indicates that further studies could be done to investigate the profiles of these participants.

In relation to a real-time system for predicting this kind of scenario, we show that even with a relatively small effort of training i.e. reading tasks of only 10 paragraphs, we can still achieve a quite successful classification result. The potential for a working system based around this experiment is very promising.

4.2   For the Whole Dataset

This section explains our initial attempts in this work i.e. to actively identify an EEG pattern for the particular scenario - regardless of the individuals from whom we collected the EEG signal. For that purpose, we classified the whole dataset with the same ANN setup. This is to confirm our hypothesis that there

Table 2. ANN classification results for 19 participants (together)

Method	Accuracy	Specificity	Sensitivity
EEG	**0.817**	0.889	0.716
Gazetrack	**0.8283**	0.8687	0.7879

is a general" EEG pattern. Similarly, we validated the results with 10-Fold cross validation.

In this section, we also compare this result with EEG with the results we achieved with gaze-tracking data collected from the previous experiment [6]. The comparison is made with the result of Group B dataset where the experiment setup is almost identical to this experiment. The results are as followed

The results we got show that the overall dataset provides lower accuracy than the individual dataset. This is expected because EEG signals are variable or subjective in a number of aspects (time, person to person, mood, etc). We can also see that the accuracy we obtain with EEG is almost on par with the one obtained from gaze-tracking devices. This is very encouraging.

The accuracy we got here also suggests that by increasing the amount of training together with further optimization of ANN configurations, we could improve it to a more desirable level . This result has laid the foundation for further research work in this area.

5   Conclusion

This paper exhibits the capability of using EEG and statistical machine learning algorithms to distinguish different types of human brain activities. The results of this study, from the preliminary analysis stage throughout the final results have shown that there are certain relationships between EEG signals captured from the human brain to the way a person reads or perceives the information while reading. It may not give the definite answer to our hypotheses; and still suggests improvement could be made toward confirming them.

The scenarios we defined in this paper maybe a bit unnatural and forced, but it shows that we are in the right direction on the quest to establish a better model of the thinking brain from the BCI perspective.

References

1. Zhu, D., Gedeon, T., Taylor, K.: Keyboard before Head Tracking Depresses User Success in Remote Camera Control. In: Gross, T., Gulliksen, J., Kotzé, P., Oestreicher, L., Palanque, P., Prates, R.O., Winckler, M. (eds.) INTERACT 2009. LNCS, vol. 5727, pp. 319–331. Springer, Heidelberg (2009)
2. Byvatov, E., Fechner, U., Sadowski, J., Schneider, G.: Comparison of Support Vector Machine and Artificial Neural Network Systems for Drug/Nondrug Classification. Journal of Chemical Information and Computer Sciences, 1882–1889 (2003)

3. Hagan, M.T., Menhaj, M.B.: Training feedforward networks with the Marquardt algorithm. IEEE Transactions on Neural Networks 5(6), 989–993 (1994), doi:10.1109/72.329697
4. Miller, G.A.: The Magical Number Seven, Plus or Minus Two: Some Limits on Our Capacity for Processing Information. In: The Psychology of Communication: Seven Essays. Penguin Books, Inc. (1970)
5. Mendis, B.S.U., Gedeon, T.D., Koczy, L.T.: Learning Generalized Weighted Relevance Aggregation Operators Using Levenberg-Marquardt Method. In: International Conference on Hybrid Intelligent Systems (2006)
6. Vo, T., Mendis, B.S.U., Gedeon, T.: Gaze Pattern and Reading Comprehension. In: Wong, K.W., Mendis, B.S.U., Bouzerdoum, A. (eds.) ICONIP 2010, Part II. LNCS, vol. 6444, pp. 124–131. Springer, Heidelberg (2010)
7. Makeig, S., Debener, S., Onton, J., Delorme, A.: Mining event-related brain dynamics. Trends in Cognitive Science 8, 204–210 (2004); Summarizes benefits and pitfalls of combining ICA, time/frequency analysis, and ERP-image visualization. (2004)
8. Hallez, H., et al.: Muscle and eye movement artifact removal prior to EEG source localization. In: Conference Proceedings of the International Conference of IEEE Engineering in Medicine and Biology Society, vol. 1, pp. 1002–1005 (2006)

Vigilance Estimation Based on Statistic Learning with One ICA Component of EEG Signal

Hongbin Yu and Hong-Tao Lu

Department of Computer Science and Engineering, Shanghai Jiao Tong University,
800 Dongchuan Road Min Hang District, China

Abstract. EEG signal has been regarded as an reliable signal for vigilance estimation for humans who engage in monotonous and attention demanding jobs or tasks, research work in this area have made satisfying progress and most of these methods or algorithms are based on the pattern recognition and clustering principal. Inspired by the HMM(Hiden Markov Model), we proposed a probability method based on the (PSD) Power Spectral Density distribution of the energy changes to estimate the vigilance level of humans using only one ICA(Independent Component Analysis) component of EEG signal. We firstly extract the specific frequency band energy feature using (CWT)Continuous Wavelet Transform, then analyze different vigilance states energy data to get the energy distribution information and vigilance states transformation probability matrix, finally use the energy distribution and vigilance states transformation matrixes to estimate vigilance level. Experiments result show that the proposed method promising and efficiently.

Keywords: Band Frequency, Energy, Histogram, Probability.

1 Introduction

It's a very difficult task for humans keeping their vigilance level at a certain level who take on monotonous and attention demanding jobs or tasks such as driving,guarding, etc., so research on vigilance level estimation is a very important work for reducing the traffic accidents and avoiding the potential accident disaster. Many research work have been done in this area which based on biomedical feature such as head position estimation, eyelid movement, face orientation and gaze movement (pupil movement)[1], etc. and get many satisfying experiment result. EEG signal have been regarded as an efficient methods used for vigilance estimation and many efforts have been made in this field and got some significant progress, however, due to the nature of low signal-to-noise ratio for EEG signals, vigilance estimation based on EEG into application still face many obstacle to remove. Many algorithms and approaches have been proposed or used in the EEG signal analysis, CSP(common spectral pattern), ICA and PCA(Principal Component Analysis)have been used in dimension reduction or feature selection. FFT(Fast Fourier Transform), DWT(Discrete Fourier Transform) and STFT(Short Time Fourier Transform) are also applied into the

B.-L. Lu, L. Zhang, and J. Kwok (Eds.): ICONIP 2011, Part I, LNCS 7062, pp. 404–412, 2011.

feature extraction and some other pre-processing work and many classifier such as SVM(Support Vector Machine), HMM and SR(Sparse Representation) are the mostly used for estimating the vigilance level[5]. some supervise cluster and un-supervise cluster method have also been applied into the research work. However, Most of previous research are based on pattern recognition and clustering prin-cipal, human vigilance is divided into 3 or more degrees such as wake, drowsy and sleepy), so the humans' vigilance estimation work is to classify the EEG signal into different classes which corresponding to different vigilance degrees. There exist two drawbacks in this algorithm framework, first, as we all know that vigilance changes is a continuous process, mechanically segmenting the pro-cess from wake to sleep into different classes can not accurately reflect the true characteristics of these changes. second, it is not an easy job to label the vigi-lance level exactly. Many researchers combine the EEG and EOG or other signal to label the vigilance states and then a variety of biological characteristics are used to predict the degree of alertness and get a relatively high estimation ac-curacy, however, which will result in much computation cost. In this paper, we consider to use both the error rate and response time of the participant's test in simulation environments to determine different vigilance states. As mentioned above, EEG recoding is a non-stationary and extremely sensitive signal and can easily be polluted by the noisy, so noisy remove preprocessing work is impor-tant for the following vigilance estimation stage.In this paper we adopted the method proposed by [2] to remove the artifacts based on the pattern recognition theory. EEG data used in this paper are collected in a simulation environment from one hundred people aged 18-28 and we only use one ICA component and a complex wavelet transformation was used to extract the γ-band frequency, 1-3 order transformation probability matrix and different states generator probabil-ity matrix based on energy value are also got from the above band frequency. Then we apply the state transformation matrix and states generator probability to estimate vigilance degree.

This paper is structured as follows: In section 2 the experiment environment for EEG data acquisition is introduced. In section 3, the EEG preprocessing methods are given, and in section 4 the experiment result is presented. Finally, conclusions and future work are given in section 5.

2   Data Acquisition

More than one hundred young men aged 18-28 participated in our EEG vigilance analysis data collection experiment. This experiment is a monotonous visual task, Subjects were required to sit in a comfortable chair, two feet away from the LCD and wear a special hat with 64 electrodes connected to the amplifier of the NeuoScan system. In front of the subjects is a LCD which four colors of traffic signs were presented randomly and each color has more than 40 different traffic signs. The interval and duration of the traffic signs displayed on the screen is 5.5~7.5 including 5~7 seconds black screen and 500 millisecond respectively. The subjects are asked to recognize the signs' color on the screen and press the corresponding color button laid on the response pad. We do these experiments in a small soundproof room with normal illumination. Each experiment started at 13:00 after lunch and lasted for one hour or more. Every subject participated two experiments, One data was for training and the other is for testing.

For each session, the visual stimulus sequence and response sequence are recorded by the NeuroScan Scan software sampled at 500Hz. Meanwhile, a total of 62 EEG channels are recorded by the NeuroScan system sampled at 500Hz synchronously, and filtered between 0.1 and 40Hz using band-pass filter. The electrodes are arranged based on extended 10/20 system with a reference on the top of the scalp.[3]

3   Method

In this section, we will describe the method we used to estimate humans' vigilance degree based on EEG recording. To begin with, we will give the introduction of the pre-process work. Then we will present our vigilance degree estimation algorithm based on the γ-band frequency energy distribution.

3.1   Data Preprocessing

In the data pre-processing stage, we first remove the EEG signals collected using the damaged channels and then eliminate the artifacts caused by the EMG and EOG signal by the method in [2]. As we all know that the potential generated on the scalp where the electrode placed on is the summation of the potential value generated around it even all of the scalp, so the ICA (Independent Component Analysis) method is applied to the EEG signals to get the mapping matrix which used to find the approximate independent components.

3.2   Feature Extraction

In this stage, we will implement the feature extraction operation. Due to the non-stationary nature of the EEG signal from each electrode of the NeuroScan system, the sampling data of EEG are divided into many overlapping epochs and each epoch contains 200 new sampling data and 200 duplicate data from the previous epoch, so each epoch corresponds to EEG signal of 4 seconds and for each epochs, we use the ICA mapping matrix to get the independent component. We choose only one component from the ICA components and then to extract the energy feature using the Continuous Wavelet Transformation. The wavelet function we used in our experiment is the complex Morlet wavelet function because of its good resolution both in time domain and frequency domain. The function defined [4] by

$$\psi(x) = \frac{1}{\sqrt{\pi f_b}} exp\{2i\pi f_c x\} exp\{-\frac{x^2}{f_b}\} \tag{1}$$

where f_c denote denote the central frequency, and f_b is a bandwidth parameter, f_c and the variance σ_f are related by $f_b = \frac{1}{2\pi^2\sigma_f^2}$. In this way we can accurately obtain the wavelet coefficient of the EEG signals at the specific time and frequency-band by adjusting the parameter values of the Morlet wavelet function. In this paper, we tend to set $f_c = 35$ and $f_b = 5$, which corresponding to the γ-band frequency, because the γ-band energy can give a better estimation result. After continues wavelet transformation operation, we get the energy value of the γ-band frequency of all of vigilance states data including wake, drowsy, sleep states and the transition between states.[5] We don't want to give an exact division about human's vigilance, what we need is just the task of non-wake states detection.

3.3   States Determination

Generally speaking, precision state division is an almost impossible job, it is the same for us. Here we consider both the response time of the recognition behavior and recognition error ration of the subjects' in the task of color recognition as the indicator for vigilance state determination. The following is the response time and the recognition error rates of one subject. From the left two charts we can see that the first experiment data is divided into 1171 overlapping epochs and from the 1st to the 561th epochs, in this session the subject is in the drowsy states, during which the subject cannot response to the color stimuli timely and exactly. Drowsy state in this experiment data can easily be determined, however, not all of experiment data in such case. There are a total of 1935 epochs in the second experiment data and the top is response time graphic and the below is the color judgement error rate graphic. In the response time graphics, the shortest response time is about 300 milliseconds(at the 394th epochs) and the longest is 1000 milliseconds(at the 968th epochs). In the session between 1st

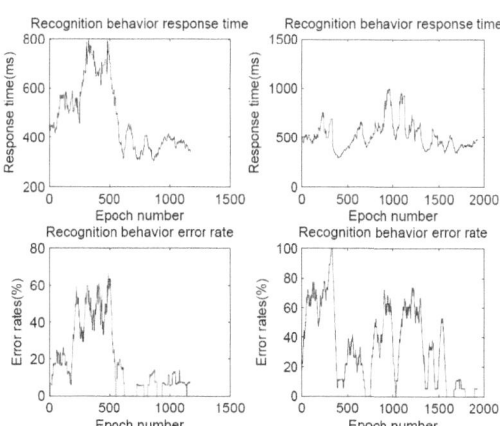

Fig. 1. Response time and recognition error rate. Figure 1 is the four charts about response time and error rate of one subject in two experiments. The left two charts are the response time and error rate charts respectively of one subject in the first experiment and the right two are the second experiment result.

and the 394th epochs the subject's response time is not very long, but in the error rate graphic the judge error rate is above 70%. We regarded this session to be the drowsy states in which states the subject can able to timely respond to external color stimuli but can hardly make the right judgments. From the 395th epochs to the 1586th, the error rates is very hight and only a few session has a relatively low error rate, in this period of time, we can believe that the subjects in a state of extreme fatigue, the session between the 1587 ∼ 1935 epochs, the subject in the state of clear-headed. As mentioned previously, We compute the wavelet coefficient of the γ-band frequency find that human's vigilance states have a close relationship with the energy's distribution. We find that the energy value is inversely-proportional relation to the vigilance degree, generally speaking, energy is low when the humans in the state of clearly-headed and vice versa at the γ-band frequency. In order to give more detail evidence, We also give

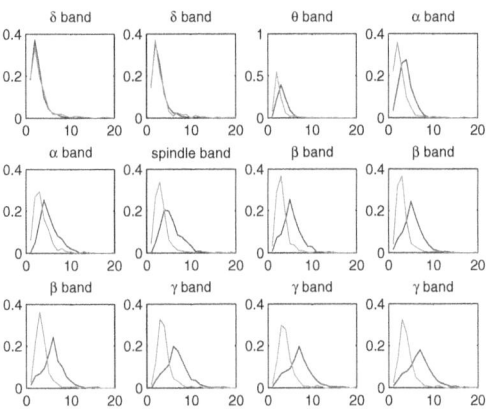

Fig. 2. Energy distribution of wake data and drowsy data. In each graphic, the red line denote the wake state energy distribution, while the green line is the drowsy state energy distribution.From Figure 2, we know that it is nearly impossible to distinguish between the wake and drowsy states using the δ-band frequency, the β or γ-band frequency may be the optimal choice in the vigilance estimation work based on energy distribution.

the energy distribution of EEG recordings in wake and drowsy states, actually, because of the difficulties in states determination work, we can only give approximate estimation about the energy distribution. In Figure 2, we illustrator the energy distribution of wake and drowsy states at different band frequency using histogram with 20 bins.

3.4   Probability Matrix Computation

Inspired by HMM, we try to estimate human's vigilance degree using the probability method. Our algorithm do it by the following step:

1. we calculate the energy distribution's histogram in different vigilance states as well as the whole process histogram get three vectors V_c, V_d, V_w where V_c corresponding to the clearly-headed state, V_d to drowsy state and V_w to the whole process. $V_i(i = c, d, w)$ are matrixes where each element denote the number of the energy in the corresponding to range.
2. Compute the probability in the time sequence by:

$$P_c = V_c/V_w, \quad P_d = V_d/V_w \tag{2}$$

Actually, P_c and P_d are the posterior probability which means that the probability of the subject in the clear-headed states or drowsy state given the specific band frequency.
3. We can also get 1-order, 2-order, 3-order states transformation matrix T_1, T_2, T_3 by the event probability P_c and P_d. Basically, we can just only compute posterior probability one matrix P_d or P_c to detect the drowsy degree or vigilance level, when the human's drowsy degree is high or vigilance level is low the controlling center will send out a warning.
4. Compute the states probability given the vector V by the following:

$$p_d = P_d(V), \text{or} \quad p_c = P_c(V) \tag{3}$$

We can also combine the previous state into the computation of the states probability by:

$$p_d = P_d(V) + \sum_{i=-N}^{-1} T_i(V)W_i \tag{4}$$

where T_i denote the transformation probability matrix, W_i is the weight and N is the order.

4   Experiment Result

In order to test our algorithm, we used two data set collected from the same subject in two experiment, firstly, we compute the wake and drowsy state PSD of the training data by using histogram method and get the state occurrence probability matrix and 1-order, 2-order and 3-order states transformation matrixes; then we used the fatigue state occurrence probability matrixes and 1-order, 2-order and 3-order states transformation matrixes to estimate the wake state degree from the first data set, the experiment result is give by Figure 3. From the Figure 3, we can see that, basically, our algorithm can make a relative accurate estimation to the human's vigilance changes,and just like what we thought, the β, λ-band frequency give better predictions result than the other band frequencies energy because of the the β, λ band frequency have more exact determination bounder which the illustrator displayed in Figure 2.

Then we use the wake state occurrence probability matrixes and 1-order, 2-order and 3-order states transformation matrixes to estimate the wake state degree from the first data set and also give a satisfy estimation result which

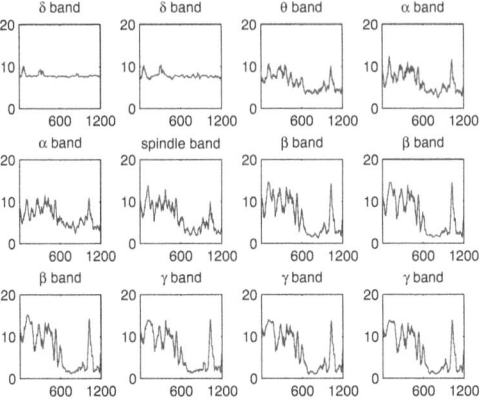

Fig. 3. Drowsy State Estimation Result over Train Data. The horizontal coordinate denote the number of epoch and vertical coordinate is the probability score of drowsy state occurrence.

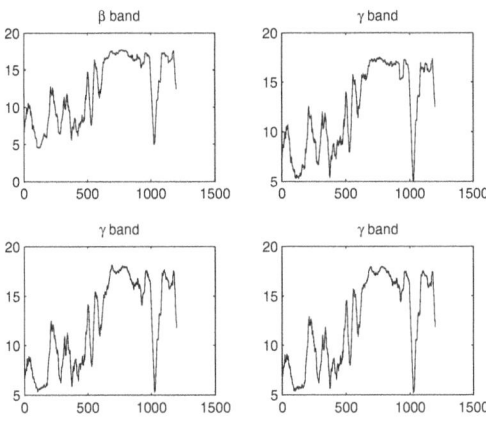

Fig. 4. Wake State Estimation Result over Train Data. Y axis denote the wake degree, while In Figure 3 it denote drowsy degree.

displayed in Figure 5. In Figure 5 we only displayed two estimation result which based on β, λ band frequency energy data. In the above figure, we haven't normalized the Y axis values or given the estimation score instead of the initial calculated data, because we cannot give an appropriate method based on our algorithm framework right now. In our future work, we will committed to solve this problem. In order to further verify the validity of our proposed algorithm, we used the obtained state occurrence and 1-3 order transformation matrixes calculated by the training data to validate the test data, the experiment result are given in Figure 5 and Figure 6. Figure 5 is the fatigue degree and Figure 6 is the wake degree. Comparing Figure 1- the subject's response time and color recognition error rate, we can find that the proposed algorithm can basically estimate the subject's vigilance level. Since only one ICA component is used in the experiment, our algorithm having a good real time capabilities.

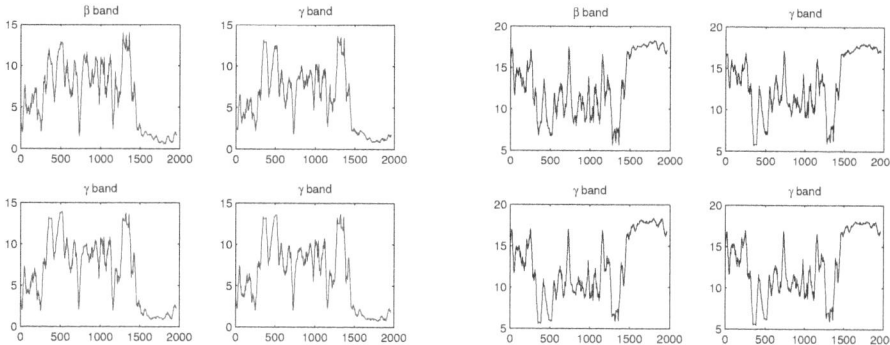

Fig. 5. Estimation Result of Drowsy Degree over Test Data

Fig. 6. Estimation Result of Wake Degree over Test Data

5   Conclusion and Future Work

In this paper, we proposed a statistical algorithm for vigilance estimation based on EEG recordings. We use only one ICA component and then extract energy feature of the λ frequency band with Morlet Wavelet. Inspired by the markov model, histogram method is used to compute the wake and drowsy state probability and 1,2 and order transformation matrix. We used one data for template and another for test, the experiment result show that our proposed algorithm frame reliable and effective. Our job is just at an initial stage, more effort should be paid in solving the following problem:

1. More accurate and reliable method for vigilance states division.More accurate and reliable method for vigilance states division.
2. Design better algorithms based on the work and give a more accurate experiment result.
3. More reliable criterion to assess the estimation results based on our algorithm framework.

References

1. Bergasa, L.M., Nuevo, J., Sotelo, M.A., Barea, R., López, M.E.: Real-time system for monitoring driver vigilance. IEEE Transactions on Intelligent Transportation Systems (March 2006)

2. Bartels, G., Shi, L.-C., Lu, B.-L.: Automatic Artifact Removal from EEG - a Mixed Approach Based on Double Blind Source Separation and Support Vector Machine. In: Proceedings of 32nd Annual International Conference of the IEEE Engineering in Medicine and Biology Society (EMBC 2010), Buenos Aires, Argentina, August 31-September 4, pp. 5383–5386 (2010)
3. Shi, L.-C., Lu, B.-L.: Off-Line and On-Line Vigilance Estimation Based on Linear Dynamical System and Manifold Learning. In: Proc. of 32nd Annual International Conference of the IEEE Engineering in Medicine and Biology Society (EMBC 2010), Buenos Aires, Argentina, August 31- September 4, pp. 6587–6590 (2010)
4. Teolis, A.: Computational signal processing with wavelets, p. 65. Birkhauser (1998)
5. Yu, H., Lu, H., Ouyang, T., Liu, H., Lu, B.-L.: Vigilance Detection Based on Sparse Representation of EEG. In: Proceedings of 32nd Annual International Conference of the IEEE Engineering in Medicine and Biology Society (EMBC 2010), Buenos Aires, Argentina, August 31-September 4, pp. 2439–2442 (2010)

A Recurrent Multimodal Network for Binding Written Words and Sensory-Based Semantics into Concepts

Andrew P. Papliński[1], Lennart Gustafsson[2], and William M. Mount[1]

[1] Monash University, Australia
`Andrew.Paplinski@monash.edu`
[2] Luleå University of Technology, Sweden
`Lennart.Gustafsson@ltu.se`

Abstract. We present a recurrent multimodal model of binding written words to mental objects and investigate the capability of the network in reading misspelt but categorically related words. Our model consists of three mutually interconnected association modules which store mental objects, represent their written names and bind these together to form mental concepts. A feedback gain controlling top-down influence is incorporated into the model architecture and it is shown that correct settings for this during map formation and simulated reading experiments is necessary for correct interpretation and semantic binding of the written words.

Keywords: Words and Concepts, Multimodal binding, Self-Organizing networks, Bigrams.

1 Introduction

We use a network of recurrent self-organizing modules to model aspects of the reading process within the cortex. As perceptions of the world around us are experienced by combining sensory inputs of different modalities with internal world-models learned within the mind, our network consists of models of five cortical areas, two of which process the sensory information and three others represent a two-level hierarchical model of the world. Such an architecture is motivated by the fact that the neural processing first takes place in mainly unimodal (visual, auditory, etc.) hierarchies in the brainstem and sensory cortices of the cerebrum. The unimodal percepts then converge in multimodal association areas such as STS (Superior Temporal Cortex). At this level we have highly abstracted, semantic representations of objects. We attach words to these mental objects and thus build conceptual representations.

The world in our work consists of a set of animals defined in terms of perceptual features and qualities and we bind the written animal names to the learned mental objects representing them. The processing and binding of written names to mental objects follows a similar methodology to [19,9]. However, we have improved working of the network by adding feedback gains to control the level of modulation feedback from the modeled bimodal integration area.

B.-L. Lu, L. Zhang, and J. Kwok (Eds.): ICONIP 2011, Part I, LNCS 7062, pp. 413–422, 2011.

The neuronal circuitry involved in reading is undoubtedly complex and much current research concentrates on an area called VWFA (Visual Word Form Area) in left fusiform gyrus [13] where prelexical, i.e., strings of letters, and lexical processing of word forms [5] takes place. One of the more complete representations of cortical areas involved in the process of reading resulting from intensive fMRI investigation is given in [4]. Fig.2.1 (available electronically) presents 13 interconnected cortical areas, arranged in five groups: visual input, visual word form, access to meaning, access to pronunciation and articulation, and top-down attention and serial reading. In our work we use a much simplified model consisting of just five 'cortical' areas. One of the basic premises of our modelling framework is the concept of a ubiquitous "neuronal code", which implies a unified way of representing information exchanged by modules of the network.

At the beginning of our consideration is the problem of coding words in neocortex, since several methods based on letter combinations and positioning with increasing sophistication have been proposed [3,8,20]. We will employ here a relatively straightforward method called open bigram coding as presented in [20,4] and described further in Section 3.1. The other fundamental consideration is how and where conceptual representations are stored in neocortex. We adopt the unitary system hypothesis as argued in [1].

2   Model Description

As a hierarchical model of reading combining bottom-up sensory integration with top-down processing our model follows principles similar to those presented in [7], which describes a multi-layered model for processing of features, letters and words in cortex. For a related neurocomputational account of map-based processing and its role in language and speech comprehension and production, see also [11,12].

In our approach, which stems from our earlier works on multimodal integration [19,2,18] self-organized modules form low dimensional labels. These labels are used as afferent signals to up-stream modules, and may represent any type of perceptual, conceptual or lexical 'features'. Such universal feature labelling is consistent with the neurocomputational modelling approach of [6].

As noted above, we do not attempt to represent each and every cortical area taking part in perceptual and semantic processing of mental objects or in higher-level language and related cognitive tasks. Rather our aim is to represent a subset of language processing in a smaller number of modules aggregating the processes of several cortical areas. The model consists of 5 processing modules divided into 3 layers, connected as shown in Figure 1. The processing pathway for visual and associated perceptual information for the mental object is depicted on the left hand side of the figure (P and MO) while word processing paths are included on the right (Wrd and UW). These pathways then converge and binding occurs within a hypothetical high-level bimodal association area, M+W.

Each module consists of a number of artificial neuronal units randomly located in a circular area. Relative positions of 'neurons' inside the circle are described by a position matrix $\mathbf{V}$. The total number of neurons is selected in proportion

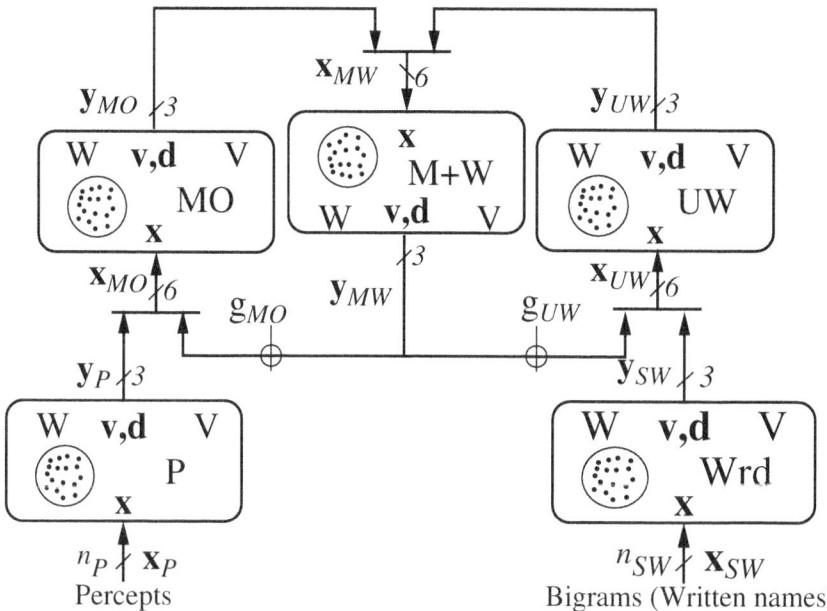

Fig. 1. The network of simulated cortical maps: Wrd – Sensory Word map, UW – Unimodal Association Word map, P – Sensory Percepts map, MO – Mental Objects map, M+W – Bimodal Association map: mental objects and written names

to the number of objects represented by the module, with rows in the weight matrices **W** characterizing the synaptic strengths in each module. The main functions for each map is described below:

- Percepts map (P) encodes basic perceptual features, such as size, colour, form of locomotion and social behaviour for each object within the category: *Animals.*
- Mental Objects map (MO) is a map of perceivable "mental objects" and semantic relationships for the object category, i.e. a topographically organised map of a set of 30 animals arranged according to perceptual features.
- Word map (Wrd) models letter and letter position processing in VWFA. Input to this module take the form of word bigrams based on the possible pairings of 26 English/Latin letters.
- Unimodal Word map (UW) performs sub-lexical processing of words and projects these as written names for lexical binding within the bimodal association map.
- 'Bimodal' association map (M+W) is responsible for binding perceivable mental objects to their written names in order to form a set of labelled mental concepts.

Sensory modules in Figure 1 operate on a relatively large number of afferent signals and produce a low-dimensionality efferent signal labeling the sensory object. Association modules, in turn use these labels as their afferent signals and produce efferent labels encoding positions of activated neuronal patches within these maps. The operation of a cortical module is functionally equivalent to mapping the higher dimensionality input space to a three-dimensional output space. More specifically, the label information consists of a three-dimensional vector encoding the relative location of the most excited neuron within a given cortical patch, supplemented by the post-synaptic activity level of this neuron.

3   Operation of Recurrent Network

The processing modules within the network model are interconnected via feed-forward sensory processing pathways and feedback connections from the bimodal integration module (M+W) (see Figure 1). Feedback gain terms, g_{mo} and g_{uw} are also shown on these recurrent pathways. The effect of these is discussed below and in Section 3.2.

It is assumed that during normal operation association modules perform static non-linear mappings of the form:

$$\mathbf{y_{MW}} = f(\mathbf{W_{MW}} \cdot \mathbf{x_{MW}}) \,, \quad \mathbf{x_{MW}} = [\mathbf{y_{MO}}, \mathbf{y_{UW}}]$$
$$\mathbf{y_{MO}} = f(g_{MO} \cdot \mathbf{W_{MO_{MW}}} \cdot \mathbf{y_{MW}} + \mathbf{W_{MO_P}} \cdot \mathbf{y_P})$$
$$\mathbf{y_{UW}} = f(g_{UW} \cdot \mathbf{W_{UW_{UW}}} \cdot \mathbf{y_{MW}} + \mathbf{W_{UW_{SW}}} \cdot \mathbf{y_{SW}})$$

where $\mathbf{x}$ and $\mathbf{y}$ represent input and output signals, respectively and $f(\cdot)$ describes the Winner-Takes-All function. For the bimodal association module we can write the following dynamic equation:

$$\mathbf{y_{MW}}(t+1) = \mathcal{M}(\mathbf{y_{MW}}(t), \mathbf{y_{MO}}(t), \mathbf{y_{UW}}(t)) \tag{1}$$

which formally describe the recurrent and non-linear nature of the network that may result in complex time behaviour. From the simulation perspective, the trained network is observed to settle immediately if we apply exogenous stimuli that are congruent with the endogenous thoughts or initial conditions. In other words, if the labels in the network are known and congruent, the network quickly converges to a stable state. See [19] for behaviour of a similar network for incongruent inputs.

3.1   Preprocessing of Percepts and Word Bigrams

Prior to training of the maps, a preprocessing step is performed to produce the sensory-based semantic and letter bigram information for the separate perceptual and written word/lexical pathways. Open bigram encoding [20,4] is used in the present model. The purpose of this is to encode attribute lists of the former and relative letter positioning of associated words for the later into a consistent numerical format for the self-organised maps. To ensure computational efficiency all inputs and weight vectors are projected on the unity hypersphere. Working with unity vectors makes it easy to compare them by calculating relevant inner products and allows us to use a simplified dot-product learning law [10].

3.2   Sequential Development of Maps

The map training sequence approximately follows that of the widely accepted model of neural ontogenesis and cortical map formation. In general, maps and their interconnections are trained using the Kohonen learning law[10], normalised Hebbian learning [14,15,16], or combination of both as in [11,12]. In particular, the processing and adaptation to sensory and learned label information is propagated in a feedforward or bottom-up direction and feedback processing subsequently comes into play in a recurrent optimisation of the higher level maps.[1]

Feedforward training of unimodal sensory maps. The first training step involves initial organisation of the unimodal maps, MO and UW. To train the map of mental objects, MO, feature vectors $\mathbf{x_P}$ describing the animals are first encoded by the auxiliary module, P as a dimensionally reduced label, $\mathbf{y_P}$ and used as input to MO. A competitive learning process then encodes these inputs as a map of mental objects organised according to their perceptual semantics. Following training, the object categorizing module P is disconnected and the signal $\mathbf{y_P}$ is interpreted as *thought commands* used for recalling items stored within module MO. The UW map is trained independently using the topographically organised word bigram representations from the Wrd module.

Feedforward training of bimodal integration map. The next step is to train the bimodal map M+W using combined inputs from MO and UW. Through statistical pairing of randomly presented mental object and word label information from each of the unimodal maps, an initial bimodal map of lexically-bound *concepts*, in this case of a named set of animals, is formed.

Feedback training of sensory and bimodal maps. Following completion of the feedforward training steps, the three association modules forming the recurrent part of the network are trained together. This time, after each learning step we perform several relaxation steps running the network as in Eqn (1), until all efferent signals in the network are constant. There is a limit on the number of iterations imposed that is important in the initial learning stages. Once the maps are fully developed, the network stabilizes quickly after a small number of iterations depending on congruence between the perceived mental objects and associated written words or names.

  As a result of this training step, the unimodal maps, MO and UW are optimised and re-organised to reflect contextual information transmitted from the bimodal integration layer. The M+W module in turn is adjusted so as to represent the statistical correlations between the unimodal data encoded in the label information $\mathbf{y_{MO}}$ and $\mathbf{y_{UW}}$. A notable new feature of the model architecture presented in Figure 1 are the feedback gain terms, g_{MO} and g_{UW}. The effect of reducing the feedback gain is to attenuate the significance of the feedback relative to feed-forward signals.

---

[1] Note that while no recurrent feedback from higher level modules is used in the initial feedforward training steps, self-organisation of the maps assumes local recurrent connections across the map output layer in order to implement the required competitive learning process.

In order to ensure that inputs to the MO and UW modules are properly bounded and that weight vectors remain on the unity hypersphere, a novel algorithm is employed in which the inputs and gain values are effectively spatially decomposed and then re-assembled within the neuronal units.

3.3   Testing Response of Network to Word Stimuli

Following training of the maps, the operation of the network is essentially as follows: The written word excites cortical patches in the word bigram map, Wrd, and via the forward path from this low-level sensory area, patches within the unimodal association area UW and bimodal binding area, M+W are also excited. A similar pattern of activation is produced via the feed-forward pathway from the 'perceptual' mental objects area MO to M+W. These direct feed-forward paths assure that the binding process is rapid – at least in the case of congruous thoughts and inputs. In the case of discrepancy or incongruity between the mental object and a written word (for example misspelt word, or mismatch between word and perceived object) a feedback loop is automatically activated from the M+W area back to the MO and UW modules. This initiates a recurrent cycle that typically converges on a globally sensible solution, assuming that no contradictory input is presented and that one actually exists.

As in the feedback training step, the gain terms g_{MO} and g_{UW} can be set separately during the testing or operational phase. In this way the effect of controllable feedback upon the hierarchically organised set of trained maps, or equivalently, top-down influence on the simulated reading process can be explored. An initial set of results showing some of these effects are presented below.

4   Simulation Results

Preliminary results show the development of activities over time in the modules of our simulated network when thought commands about different mental objects, in this case animals and their associated written names are simultaneously presented. This perceptual or mental binding task is considered central to language understanding during the activity of reading. The simulation scenario is also similar to the fMRI-based study of recognition of spoken words representing animals when subjects are cross-modally primed for different animals [17].

As an example of the operation of the network in recognition of words corresponding to known mental objects, consider the situation where the multimodal computational network is initialised to a particular animal, while being presented with the perceptual features of another animal and a misspelt version of the word for that animal. The response of the network to this situation is presented in Figures 2 and 3.

In the maps depicted in these figures, neuronal positions are marked with the yellow dots and the map area is tessellated with respect to the peaks of neuronal activities for each stimulus. We can identify the neuronal patches for various animals, e.g., 'dog', 'frog' and 'goat'. The objects or written names are 'placed' in the relevant cortical area during the learning process. Once the learning is

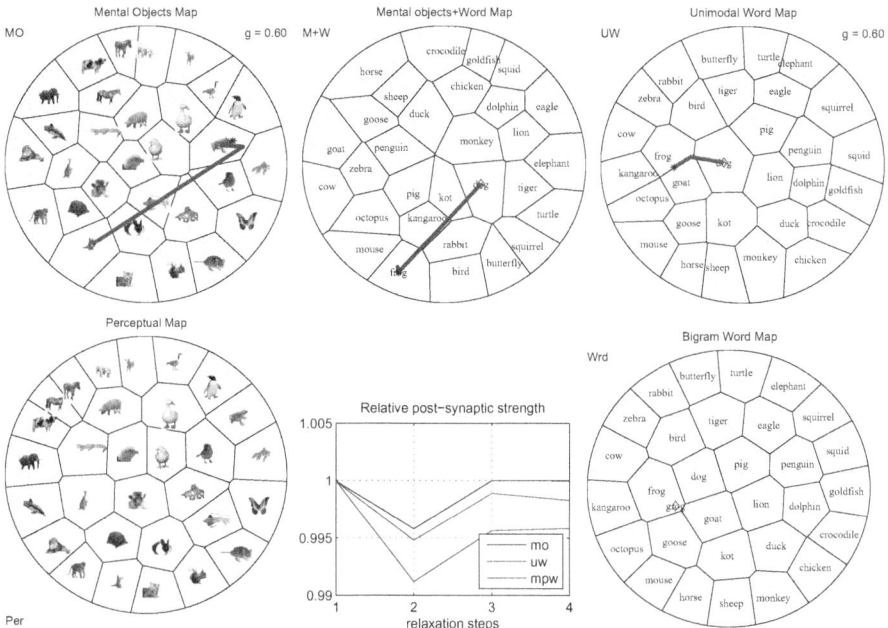

Fig. 2. The relaxation trajectories in the association maps when the word map, Wrd, presents a misspelled word "grog", the perceptual map, Per, presents object "frog", and the initial state was "dog". Both feedback gains are as during training, $(g=0.6)$.

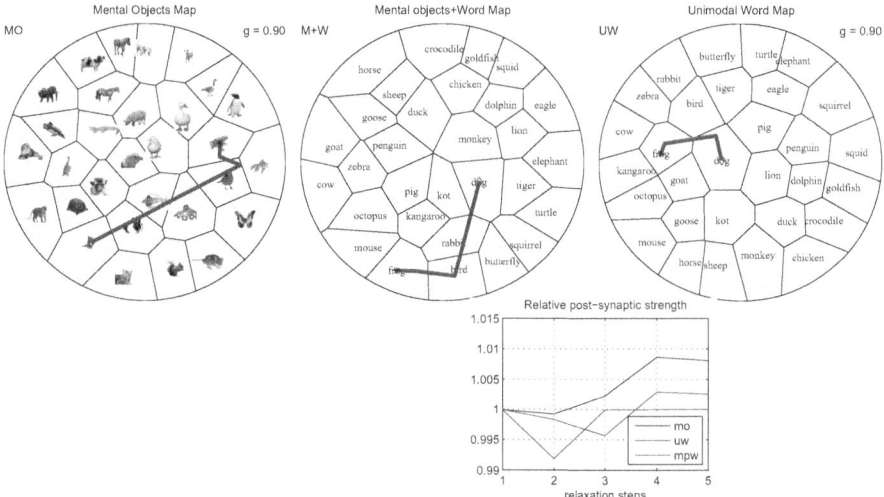

Fig. 3. The same "frog–grog–dog" trajectories for the high feedback gains, $g=0.9$

completed, the cortical area responds with an activity pattern characteristic to each stimulus.

At the starting point, the MO module *thinks* 'dog', whereas the unimodal word association module UW *reads* the orthographically similar word 'grog'. The bimodal association map, M+W is arbitrarily initialized with a mental object 'dog'. This initial state is marked on three maps in Figures 2 and 3 with the '$\diamond$'. Since these positions have been learned during the training procedure, the relative post-synaptic strengths are initially at their maximum value, as seen in the upper right hand side of both figures.

In the first case, Figure 2, relatively low feedback gains, $g_{UW} = g_{MO} = 0.6$ have been used during the reading test phase. In this scenario, although the bimodal map M+W and mental objects map, MO converge to a 'frog' solution, the response of the unimodal word area UW arrives at a point on the class boundary between 'frog' and 'goat'. This indicates uncertainty in recognition of the word for the animal most closely matching the stimulus ('grog'), implying that the network as a whole has not been able to successfully bind the correct name 'frog' to the corresponding mental object. This level of uncertainty is indicated by the final value of relative post-synaptic strengths, in which the response of MO and M+W is stronger than that of UW.

Now compare this with the situation in Figure 3 when a greater level of feedback gain, $g_{UW} = g_{MO} = 0.9$ has been used during testing. In this case the network quickly settles to a solution when the thought prevails; the final response of all maps including UW is consistent and the network as whole converges rapidly on the 'frog' conclusion. The level of confidence in this outcome is further indicated by the high levels of the relative post-synaptic strengths.

Effectively, the network has acted to correct the spelling of the distractor word through application of contextual knowledge contained within the interconnected association modules. This result is a simple demonstration of how *semantic priming* on a known set of mental objects within a category can be effectively represented within the model. In probabilistic (empirical Bayes) terms, it also suggests how new beliefs, hypotheses or perceptions of the world can be inferred when a network conditioned by given prior beliefs or initial conditions is presented with new sensory evidence.

5   Discussion

For the direct comparison above to be possible, it is necessary that exactly the same trained hierarchical network be used in each case. In this example, a feedback gain setting of $g_{UW} = g_{MO} = 0.6$ was used during recurrent phase in the formation process (as described in Section 3.2). More varied simulation results can be obtained if different feedback gain values are used, however due to space limitations examples of the types of aberrant behaviours that can be produced as a result are not considered at this time.

In general, reducing the feedback gain during map formation will result in a overall network that responds well to new and less predictable inputs (such as words and non-words) but which lacks the contextual knowledge required to correctly associate these inputs with cross-modal percepts and mental objects. Conversely, applying too great a level of feedback during recurrent training step results in a network with a tendency to become "locked up" in previously known states or *thoughts* and less able to adapt to new sensory information.

This suggests that an optimal level of feedback gain is required in order to realise a reading network which can effectively employ previously learned knowledge to correctly perceive and learn new words.

One possibility for future research would be to use the feedback gain within an incremental learning regime in which global reinforcement feedback is used to assess the utility of the learned set of maps at a particular setting of feedback versus feedforward bias. The feedback gain g could then be decreased if the network became 'stuck' and unable to adapt to new inputs or sensory evidence and increased if a stronger belief in the prior state or conditions was deemed to result in a better overall performance. Adopting of such a 'self-supervised' approach could be a way to incorporate a process analogous to *selective attention* in a straightforward and integrated way which works to optimise the efficiency of the learning process.

The complete simulation software is available from the first author.

6   Conclusion

We have presented a model for binding written names to perceptually-based semantic objects and provide preliminary results to demonstrate how this can be used in modelling cognitive functions basic to reading. This includes automatic 'correction' of misspelt words when a similar known word that is bound to an active mental object or by extension, object category is attended to. Such cognitive processes are of fundamental importance to the particular human activity of reading. The introduction of controllable feedback gain increases the behavioural repertoire of the model, presenting an opportunity to explore a number of other effects on learning and cognitive behaviour within the outlined computational framework.

In the interest of maintaining structural simplicity, several assumptions have been made in the model. For convenience specific visual and other perceptual modalities, conceptual categories and semantic relationships are combined in a "mental objects map". This simplification is computationally efficient as it allows this information to be encoded as arbitrary lists of attributes. In a more comprehensive and biologically realistic model, the auxiliary 'P' module could be divided into specific sensory modalities or sub-modalities, used to represent visual, tactile, spatial or other features related to mental object categories such as plants, animals or tools.

The representation of the mental objects would then involve a multimodal integration of such features and binding of these to their associated names. From the lessons gained through this modelling exercise and through experiments in lexical binding of spoken names to mental objects, we hope to extend the model to integrate structurally separate processing pathways and perform multimodal and *transmodal* binding across auditory and written language modalities.

References

1. Bright, P., Moss, H.E., Tyler, L.K.: Unitary vs multiple semantics: PET studies of word and picture processing. Cerebral Cortex 89, 417–432 (2004)
2. Chou, S., Papliński, A.P., Gustafsson, L.: Speaker-dependent bimodal integration of Chinese phonemes and letters using multimodal self-organizing networks. In: Proc. Int. Joint Conf. Neural Networks, Orlando, Florida (August 2007)
3. Davis, C.J., Bowers, J.S.: Contrasting five different theories of letter position coding: Evidence from orthographic similarity effects. J. Exp. Psych.: Human Perception and Performance 32(3), 535–557 (2006)
4. Dehaene, S.: Reading in the Brain. Viking (2009), http://pagesperso-orange.fr/readinginthebrain/figures.htm
5. Glezer, L.S., Jiang, X., Riesenhuber, M.: Evidence for highly selective neuronal tuning to whole words in the "Visual Word Form Area". Neuron 62(2), 199–204 (2009)
6. Gliozzi, V., Mayor, J., Hu, J.F., Plunkett, K.: Labels as features (not names) for infant categorisation: A neuro-computational approach. Cog. Sci. 33(3), 709–738 (2009)
7. Graboi, D., Lisman, J.: Recognition by top-down and bottom-up processing in cortex: The control of selective attention. J. Neurophysiol. 90, 798–810 (2003)
8. Grainger, J.: Cracking the orthographic code: An introduction. Language and Cogn. Processes 23(1), 1–35 (2007)
9. Jantvik, T., Gustafsson, L., Papliński, A.P.: A self-organized artificial neural network architecture for sensory integration with applications to letter–phoneme integration. Neural Computation, 1–39 (2011), doi:10.1162/NECO_a_00149
10. Kohonen, T.: Self-Organising Maps, 3rd edn. Springer, Berlin (2001)
11. Li, P., Zhao, X., MacWhinney, B.: Dynamic self-organization and early lexical development in children. Neuron 31, 581–612 (2007)
12. Mayor, J., Plunkett, K.: A neurocomputational account of taxonomic responding and fast mapping in early word learning. Psychol. Rev. 117(1), 1–31 (2010)
13. McCandliss, B.D., Cohen, L., Dehaene, S.: The visual word form area: expertise for reading in the fusiform gyrus. TRENDS Cog. Sci. 7(7), 293–299 (2003)
14. Miikkulainen, R.: Dyslexic and category-specific aphasic impairments in a self-organizing feature map model of the lexicon. Brain and Language, 334–366 (1997), http://nn.cs.utexas.edu/miikkulainen:bl97
15. Miikkulainen, R., Kiran, S.: Modeling the Bilingual Lexicon of an Individual Subject. In: Príncipe, J.C., Miikkulainen, R. (eds.) WSOM 2009. LNCS, vol. 5629, pp. 191–199. Springer, Heidelberg (2009)
16. Monner, D., Reggia, J.A.: An unsupervised learning method for representing simple sentences. In: Proc. Int. Joint Conf. Neural Net., Atlanta, USA, pp. 2133–2140 (June 2009)
17. Noppeney, U., Josephs, O., Hocking, J., Price, C., Friston, K.: The effect of prior visual information on recognition of speech and sounds. Cerebral Cortex 18, 598–609 (2008)
18. Papliński, A.P., Gustafsson, L.: Feedback in Multimodal Self-organizing Networks Enhances Perception of Corrupted Stimuli. In: Sattar, A., Kang, B.-h. (eds.) AI 2006. LNCS (LNAI), vol. 4304, pp. 19–28. Springer, Heidelberg (2006)
19. Papliński, A.P., Gustafsson, L., Mount, W.M.: A model of binding concepts to spoken names. Aust. Journal of Intelligent Information Processing Systems 11(2), 1–5 (2010)
20. Whitney, C.: Comparison of the SERIOL and SOLAR theories of letter-position encoding. Brain and Language 107, 170–178 (2008)

Analysis of Beliefs of Survivors of the 7/7 London Bombings: Application of a Formal Model for Contagion of Mental States

Tibor Bosse[1], Vikas Chandra[2], Eve Mitleton-Kelly[2], and C. Natalie van der Wal[1]

[1] Vrije Universiteit Amsterdam, Department of Artificial Intelligence
de Boelelaan 1081a, 1081 HV Amsterdam, The Netherlands
{tbosse,cn.van.der.wal}@few.vu.nl
[2] London School of Economics and Political Science, LSE Complexity Group
Houghton Street, London, WC2A 2AE, United Kingdom
{E.Mitleton-Kelly,V.Chandra}@lse.ac.uk

Abstract. During emergency scenarios, the large number of possible influences *inter se* between cognitive and affective states of the individuals involved makes it difficult to analyse their (collective) behaviour. To study the behaviour of collectives of individuals during emergencies, this paper proposes a methodology based on formalisation of empirical transcripts and agent-based simulation, and applies this to a case study in the domain of the 7/7 London bombings in 2005. For this domain, first a number of survivor statements have been formalised. Next, an existing agent-based model has been applied to simulate the scenarios described in the statements. Via a formal comparison, the model was found capable of closely reproducing the real world scenarios.

Keywords: London bombings, agent-based simulation, contagion.

1 Introduction

During large-scale emergencies such as terrorist attacks or natural disasters, the involved persons may behave in unexpected ways. For example, some individuals may immediately start panicking and 'lose control over their actions', whereas others may emerge as 'calm leaders' helping other people. Especially in larger crowds, the numerous possible influences of mental states within individuals (e.g., person A has the belief that he will die, and therefore starts panicking) and between individuals (e.g., person B manages to calm down person A) makes it very difficult to predict how a certain crowd will behave in a particular situation. Nevertheless, gaining more insight into the dynamics of these processes is very useful, since it enables policy makers to explore possibilities for developing procedures and interventions that may minimise the number of casualties in such emergency scenarios (e.g., providing emergency exits at appropriate locations, or equipping patrollers with intelligent devices that recommend escape routes). In line with recent developments [2,13], this paper proposes to study such dynamics using agent-based simulations.

More specifically, to be able to analyse the dynamics of mental states and their intra- and interpersonal interaction in emergency scenarios, an agent-based simulation

B.-L. Lu, L. Zhang, and J. Kwok (Eds.): ICONIP 2011, Part I, LNCS 7062, pp. 423–434, 2011.
© Springer-Verlag Berlin Heidelberg 2011

model ASCRIBE (Agent-based Social Contagion Regarding Intention Beliefs and Emotions) has been developed [9]. This model has been inspired by several concepts from Social Neuroscience [6,7], including the concepts of *mirror neuron* (i.e., a type of neuron that fires not only when an individual performs an action, but also when he/she observes this action performed by someone else [10,11]) and *somatic marker* (i.e., a feeling induced by a certain decision option considered by an individual, which helps the individual make decisions by biasing that option [1,7]). Based on these concepts, the ASCRIBE model describes how for different individuals in a crowd, the strength of their beliefs, intentions and emotions may evolve.

The main goal of the current paper is to show how the model can be used to analyse the dynamics of individuals' mental states for a real world incident. To this end, a case study is undertaken which analyses the London bombings of July 7th, 2005. To test the applicability of the model to this case, a research methodology is followed that consists of a number of steps. First, a set of survivor statements which were extracted from the 'Report of the 7 July Review Committee' [12], have been formalised using a dedicated ontology. Next, the ASCRIBE model has been applied to generate a number of simulation runs for fragments of the scenarios described in the survivor statements. And finally, the results of the simulations have been compared with the formalised survivor statements, both in an informal and in a formal manner (using an automated tool).

The remainder of this paper is organised as follows: Section 2 provides a brief description of the London bombings. Section 3 explains how statements of survivors of the attack were obtained and converted to formal notation. Section 4 summarises the main mechanisms of the ASCRIBE model and Section 5 shows how the model was applied to the London bombings scenario. Section 6 discusses the (formal) comparison between the simulation runs and the formalised statements and Section 7 concludes the paper with a discussion.

2 London Bombings

The London bombings of July 7, 2005 (also referred to as 7/7) involved 4 suicide bombers triggering explosions on the London Underground and Bus transport network. Two of these bombings took place on underground trains outside Liverpool Street and Edgeware Road stations and a third one between King's Cross and Russell Square. These bombs went off at around 8:50 in the morning during the 'rush hour' when most commuters travel to their workplaces. The fourth bomb went off on a double-decker bus at Tavistock Square about an hour later. 52 people were killed and more than 770 were injured, see [12].

3 Formalisation of Survivor Statements

Below, Section 3.1 describes how statements of survivors of the attack were obtained, and Section 3.2 explains how these were converted to formal notation.

3.1 Survivor Statements

The July 7 Review Committee was set up to 'identify the successes and failings of the response to the bombings and to help improve things for the future...' [12] and submitted its report to the London Assembly in June 2006. Information from nearly 85 individuals was obtained as part of this report to the London Assembly. These accounts consist of unstructured narratives from individuals involved in the incident and run into 299 pages of text. Of these, 21 are fairly detailed accounts of the experiences of the respective survivors depending on the proximity to the explosion of the concerned survivor, the evacuation process and after-effects on survivors including the psychological. 12 accounts relate to a public hearing held on 23 March 2006 and 9 relate to private meetings with the chairman of the Review Committee. The rest of the accounts consist of information provided by survivors and affected persons through email and letters.

The July 7 Review Committee also obtained information and views from nearly 40 organisations. These accounts consist of unstructured narratives and written submissions of officials from a broad range of organisations including the police, fire brigade, ambulance, hospitals, local authorities, telecommunication companies and business associations and run into 284 pages of text. For the purposes of this paper, only transcripts of individual survivors in their original form have been included in the analysis.

Statements of survivors are publicly accessible and available as a consolidated Volume 3 of the July 7 Review Report, in pdf as well as rich text format. The statements have been anonymised and so the names in the statements do not refer to the actual identity of the survivor. An example of a transcript of a survivor given the name 'John' and who was at the Edgeware Road Station site of the bombings, is shown in Figure 1 below.

Once again, can I thank you all for coming? John from Edgware Road, I believe you are going to start the proceedings.

John (Edgware Road): Thank you. Just after the train left Edgware station, there was a massive bang followed by two smaller bangs and then an orange fireball. I put my hands and arms over my ears and head as the windows and the doors of the carriage shattered from the blast. Splintered and broken glass flew through the air towards me and other passengers. I was pushed sideways as the train came to a sudden halt. I thought I was going to die. Horrific loud cries and screams filled the air, together with

Fig. 1. Extract from John's transcript at the July 7 Review Committee hearing

The transcript was parsed into phrases that as far as possible conveyed a single idea leaving the statement in its original form. These phrases were treated as indications for 'cues' that help explain the behaviour and thoughts of the survivor. References to the location, time and elapsed time were also put alongside the cues. These have been either explicitly stated or inferred from surrounding statements in the transcript for the survivor. Each of the phrases was then formalised according to the scheme explained in the following sub-section. An extract from the parsing table for 'John's transcript' is shown below in Figure 2.

John: Edgeware Road

Time and location	Cue	Belief held by survivor	Belief held by neighbour	External origin	Internal origin	Action	Thought short of action
Just after the train left Edgeware station	Just after the train left Edgeware Station, there was a massive bang, followed by two smaller bangs, then an orange fireball	Blast had occured		Explosion			
	I put my hands and arms over my ears as and head	Risk of injury Blast had occured				Protective measure	
	as the windows and doors of the carriage shattered from the blast			Shattering of windows and doors			
	Splintered and broken glass flew through the air towards me and other passengers	Risk to other passengers		Effect of blast, glass and splinters			
	I was pushed sideways as the train came to a halt			Involuntary push			
	I thought I was going to die	Going to die			Grave risk to life		
	Horrific loud cries and screams filled the air		Risk to life	Screams			

Fig. 2. Parsing table for John's transcript

3.2 Formalisation

As a first step towards formalisation of the survivor statements, a time stamp has been assigned to each cue. Since little information is known about the actual time and duration of the events, we simply used natural numbers to describe the timing of the subsequent events (i.e., we say that they took place at time point 0, 1, 2, and so on). After that, the content of the cues was analysed in more detail, to make an inventory of the classes of concepts they refer to. In general, each cue turned out to refer either to a *belief* or an *action*. Moreover, each belief or action belonged either to the *survivor* himself, or *another individual* at the scene. For example, the statement 'there was a massive bang' refers to a belief of the speaker himself (namely that a blast had occurred), whereas the statement 'I put my hands and arms over my ears and head' refers to an action of the speaker. Similarly, the statement 'a young woman sitting next to me asked me if I was OK' refers to an action of another individual. Furthermore, two types of beliefs could be distinguished, namely, beliefs that are triggered by an *external* stimulus (e.g., 'there was a massive bang') and those triggered by an *internal* stimulus or thought (e.g., 'I thought I was going to die').

Table 1. Domain Ontology

Predicate	informal meaning
has_belief(a:AGENT, b:BELIEF)	agent a has belief b
has_internal_belief(a:AGENT, b:BELIEF)	agent a has (internally triggered) belief b
performed(a:AGENT, ac:ACTION)	agent a performs action ac
Sort	**elements**
AGENT	{john, man_in_front, young_woman, ...}
ACTION	{protect_head, rub_eyes, ...}
BELIEF	{blast_has_occurred, risk_of_injury, ...}

Based on this analysis of the content of the cues, a formal domain ontology (or signature) has been developed. For this purpose, the LEADSTO language has been used, which is an extension of order-sorted predicate logic [4]. In this language, the domain under analysis can be described in terms of sets of sorts and subsorts relations, constants in sorts, functions, and logical predicates over sorts. An overview of the domain ontology developed for the current case study is provided in Table 1.

Note that the predicates have been chosen in such a way that they can be easily mapped onto concepts in the ASCRIBE model. These predicates have generic names. The elements of the sorts are domain-specific, and depend on the particular scenario.

After development of the domain ontology, the actual formalisation of the cues was done. To this end, for each survivor, the following algorithm was executed(?) (described in pseudo-code):

```
start with an empty specification
for t = time step 1 to last-time do
  1.  determine whether the cue at time t refers to a belief (either 'internal' or
      'external') or action
  2.  determine to which agent the cue belongs
  3.  select the appropriate predicate from the domain ontology
  4.  express the cue formally using that predicate, and add the result to the
      specification, annotated with time step t
end
```

As an illustration, Figure 3(a) shows (a visualisation of) the resulting formalisation of the survivor statement that was shown in the earlier Figure 1, in an example trace. In this figure, which contains a fragment of 30 time steps, time is on the horizontal axis; a box on a line indicates that an event is true at that time point.

As a final step, the events included in the formal traces needed to be connected to concepts within the ASCRIBE model, enabling us to apply the model to the scenarios under investigation. The main concepts present in ASCRIBE are *beliefs*, *intentions*, and *emotions*, which may be related either to specific *world states* or to *decision options* (see Section 4 for details). Thus, as an example, the 'external beliefs' were translated into 'beliefs about the positiveness of the situation' and 'belief options' in ASCRIBE, the 'internal beliefs' were translated into 'emotions' (of *fear*) in ASCRIBE, and the 'actions' were translated into 'intention options' in ASCRIBE. For the belief and intention options, two types of actions were distinguished, namely 'protective actions' (e.g., covering one's ears) and 'social actions' (e.g., comforting another passenger). Moreover, during these translations, numerical values (from the set $\{0, 0.1, 0.2, \ldots, 0.9, 1\}$, where 0.5 represents a neutral value) have been assigned to the strength of each state. For example, the belief that a blast has occurred clearly refers to a very negative situation (e.g., value 0.1), whereas the belief that help is underway refers to a positive situation (e.g., value 0.9). To guarantee inter-observer reliability, as a pre-test, part of the survivor statements have been formalised separately by two different observers. When comparing the results, the differences turned out to be small: besides minor interpretation errors, the distance between the numerical scores of the two observers never were greater than 0.2.

An example of the outcome of this final step is shown in Figure 3(b). Note that this figure corresponds to the same scenario as in Figure 3(a), but that a larger fragment (of 70 time steps) has been taken. As shown in the first graph, the positiveness of this agent (named John) fluctuates during the scenario. Initially (i.e. right after the explosion), he has some rather negative beliefs about the situation, but based on the development of the events, he starts to have some more positive beliefs from time point 20. The same pattern is repeated in the period between time point 40 and 70. Similarly, the other graphs show John's level of emotion (fear in this case), and the extent to which his actions are 'protective' or 'social' actions. Note that the graphs only show some values, that is when the information has been available; at the other time points nothing is shown. In Section 5, these kinds of information bits will be used for simulating the scenarios. In particular, the information shown in the first graph (beliefs about the situation) will be used as input for the ASCRIBE model, whereas the information from the other three graphs (emotions, protective and social actions) will be used to compare with the output of the model.

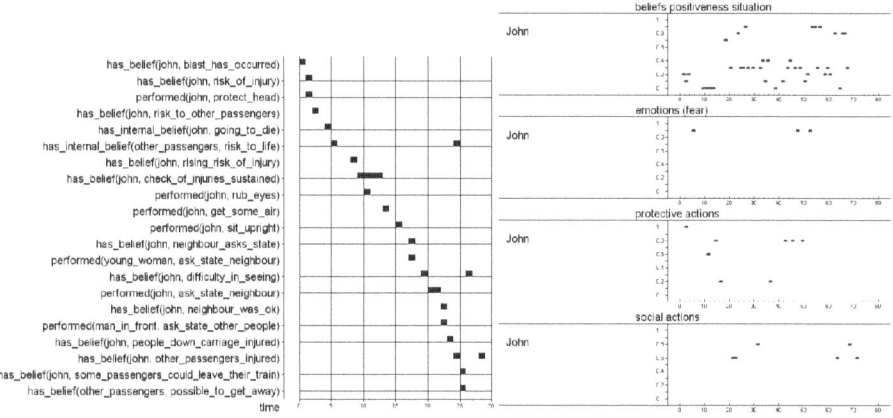

Fig. 3. Example formal trace – qualitative (a) and quantitative (b) information

4   Simulation Model

To simulate the dynamics of beliefs, emotions and intentions of individuals involved in the 7/7 London bombings of 2005, the agent-based model ASCRIBE [9] was used and implemented in Matlab. For a complete overview of ASCRIBE, see [9]. In this section, the model is only briefly summarised and explained in terms of how it was tailored to the 7/7 London bombings case. The main concepts present in the original ASCRIBE model [9] are *beliefs*, *intentions*, and *emotions*. For the current purpose, the following specific states for the agents were taken, namely 1 emotional state per agent (fear), 2 intentional states per agent (either to perform a protective or social action) and 3 beliefs (one about the 'positiveness' of the situation and two about whether an agent should perform a protective or social action):

fear of agent A $\qquad\qquad q_{fearA}(t)$
intention indication for action option O of agent A $\qquad\qquad q_{intention(O)A}(t)$
belief in X (either about situation or action) of agent A $\qquad\qquad q_{belief(X)A}(t)$

In Figure 4, which is adapted from [9], an overview of the interplay of these different states within the model is shown. It is assumed that at the individual level the strength of an intention for a certain action option depends on the person's beliefs (*cognitive responding*) in relation to that option. It is also assumed that beliefs may generate certain emotions (*affective responding*), for example that of fear, that in turn may affect the strength of beliefs (*affective biasing*). Note that it is assumed that these latter emotions are independent of the different action options. The contagion of all the different states between individuals is based on the concept of a *mirror neuron* (e.g., [10,11]) in Neuroscience. When states of other persons are mirrored by some of the person's own states, which at the same time play a role in generating their own behaviour, then this provides an effective basic mechanism for understanding how in a social context, individuals affect each other's mental states and behaviour.

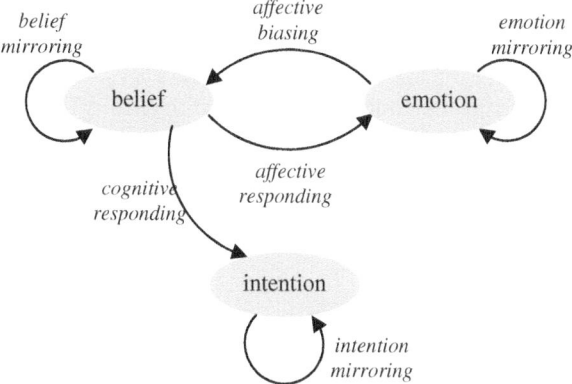

Fig. 4. The interplay of beliefs, emotions and intentions in the 7/7 London bombings context

Note that all mirroring processes take place through interaction between agents, whereas the other processes shown in Figure 4 occur internally, within an individual agent. An overview of the different intra- and interpersonal interaction processes is given in Table 2.

Table 2. The different types of processes in the model

from S	to S'	type	description
belief(X)	*fear*	internal	affective response on information; for example, on threats and possibilities to escape
fear	*fear*	interaction	emotion mirroring by nonverbal and verbal interaction; for example, fear contagion
fear	*belief(X)*	internal	affective biasing; for example, adapting openness or expressiveness
belief(X)	*belief(X)*	interaction	belief mirroring by nonverbal and verbal interaction; for example, of information on threats and action options
belief(X)	*intention(O)*	internal	cognitive response on information; for example, aiming for a protective action based on the danger of the situation
intention(O)	*intention(O)*	interaction	intention mirroring by nonverbal and verbal interaction; for example, of tendency to aim for a social action

The central idea of the model is based upon the notion of contagion strength γ_{SBA} which is the strength with which an agent B influences agent A with respect to a certain mental state S (which, for example, can be an emotion, a belief, or an intention). It depends on the *expressiveness* (ε_{SB}) of the sender B, the strength of the channel (α_{SBA}) from sender B to receiver A and the *openness* (δ_{SA}) of the receiver: γ_{SBA} = $\varepsilon_{SB}\ \alpha_{SBA}\ \delta_{SA}$. The level q_{SA} for mental state S of agent A is updated using the overall contagion strength of all agents B not equal to agent A: $\gamma_{SA} = \Sigma_{B \neq A}\ \gamma_{SBA}$. Then the weighed external impact q_{SA}^*: for the mental state S of all the agents B upon agent A, is determined by: $q_{SA}^* = \Sigma_{B \neq A}\ \gamma_{SBA}\ q_{SB} /\ \gamma_{SA}$. Then, state S for an agent A is updated by:

$$q_{SA}(t+\Delta t) = q_{SA}(t) + \psi_{SA}\ \gamma_{SA}\ [\ f(q_{SA}^*(t), q_{SA}(t)) - q_{SA}(t)]\ \Delta t \tag{1}$$

Here ψ_{SA} is an update speed factor for S, and $f(V_1, V_2)$ a combination function. This expresses that the value for q_{SA} is defined by taking the old value, and adding the change term, which basically is based on the difference between $f(q_{SA}^*(t), q_{SA}(t))$ and $q_{SA}(t)$. The change also depends on two factors: the overall contagion strength γ_{SA} (i.e., the higher this γ_{SA}, the more rapid the change) and the speed factor ψ_{SA}.

Within the definition of the combination function $f(V_1, V_2)$ a number of further personality characteristics determine the precise influence of the contagion. First, a factor η_{SA} is distinguished which expresses the tendency of an agent to absorb or amplify the level of a state S, whereas another personality characteristic β_{SA} represents the bias towards reducing or increasing the value of the state S. Thus, the combination function $f(V_1, V_2)$ is defined as follows:

$$f(V_1, V_2) = \eta_{SA}\ [\ \beta_{SA}\ (1 - (1 - V_1)(1 - V_2)) + (1-\beta_{SA})\ V_1 V_2\ ] + (1 - \eta_{SA})\ V_1 \tag{2}$$

In the ASCRIBE model, the effects of emotions on beliefs are calculated with the formulae in Section 4.1 of [9]. Instead of using these formulae here, the values for beliefs about the situation and action options were taken from the empirical data as explained in Section 2 and 3. Here, we assume the effects of emotions on beliefs are implicitly present in these input values.

The effect of the emotion fear on beliefs is expressed by the following formula:

$$\begin{aligned} q_{fear,A}^*(t) = v_A \cdot (\Sigma_{B \neq A}\ \gamma_{fearBA} \cdot q_{fearB} /\ \gamma_{fearA}) + \\ (1 - v_A) \cdot (\Sigma_X\ \omega_{X,fear,A} \cdot (1 - p_{XA}) \cdot r_{XA} \cdot q_{belief(X)A}) \end{aligned} \tag{3}$$

In formula 3, information has an increasing effect on fear if it is relevant and non positive, through informational state characteristics r_{XA} denoting how relevant, and p_{XA} denoting how positive information X is for person A. The influence depends on the impact from the emotion fear by others (the first factor, with weight v_A) in combination with the influence of the belief present within the person. This $q_{fear,A}^*(t)$ is used in the equation describing the dynamics of fear:

$$q_{fearA}(t+\Delta t) = q_{fearA}(t) + \gamma_{fearA}\ [\ f(q_{fearA}^*(t), q_{fearA}(t)) - q_{fearA}(t)]\ \Delta t$$

with

$$f(q_{fearA}^*(t), q_{fearA}(t)) = \eta_{fearA}\ [\ \beta_{fearA}\ (1 - (1 - q_{fearA}^*(t))(1 - q_{fearA}(t))) + (1-\beta_{fearA})\ q_{SA}^*(t)\ q_{SA}(t)\ ]$$
$$+ (1 - \eta_{fearA})\ q_{fearA}^*(t)$$

Furthermore, the specific state $q_{emotion(O)A}$ was left out of the current model, since this state was not mentioned in the survivor reports and it is not realistic to use in these simulations. Therefore, the effect of emotions on intentions in ASCRIBE is left out in the current model, leaving the effect of beliefs on intentions calculated as follows:

$$q_{beliefsfor(O)A}(t) = \Sigma_X\ \omega_{XOA}\ q_{belief(X)A} /\ \Sigma_X\ \omega_{XOA}$$

where ω_{XOA} indicates how supportive information X is for option O. The combination of the group's aggregated intentions with an agent's own belief for option O is made by a weighted average of the two:

$$q_{intention(O)A}^{**}(t) = (\omega_{OIA}/\omega_{OIBA})q_{intention(O)A}^{*}(t) + (\omega_{OBA}/\omega_{OIBA}) q_{beliefsfor(O)A}(t) \tag{4}$$

$$\gamma_{intention(O)A}^{*} = \omega_{OIBA} \gamma_{intention(O)A} \tag{5}$$

where ω_{OIA} and ω_{OIBA} are the weights for the contributions of the group intention impact (by mirroring) and the own belief impact on the intention of A for O, respectively, and

$$\omega_{OIBA} = \omega_{OIA} + \omega_{OBA}$$

The overall model for the dynamics of intentions for options becomes:

$$q_{intention(O)A}(t + \Delta t) = q_{intention(O)A}(t) + \gamma_{intention(O)A}^{*} [\eta_{intention(O)A} (\beta_{intention(O)A} (1 - (1-$$
$$q_{intention(O)A}^{**}(t))(1-q_{intention(O)A}(t))) + (1-\beta_{intention(O)A}) q_{intention(O)A}^{**}(t) q_{intention(O)A}(t)) \tag{6}$$
$$+ (1 - \eta_{intention(O)A}) q_{intention(O)A}^{**}(t) - q_{intention(O)A}(t)] \cdot \Delta t$$

5 Simulation Results

Multiple survivor reports of the London bombings at 7-7-2005 were formalised, as described in Section 2 and 3. As an illustration, in this section the simulation results of the ASCRIBE model for one particular instance of this data is shown, namely for the scenario described in Section 3, involving the survivor named John. In the survivor report of John, the beliefs, emotions and intentions of 3 other persons were mentioned as well, therefore the simulation in Matlab was made for 4 agents in total. The beliefs of the situation and for the two action options (social action or protective action) were taken as inputs of the model. The fear value of John, and the values for his intentions to act in a protective or social manner, were produced by the ASCRIBE model as outputs. These output values (all between 0 and 1) are shown in Figure 5 and can be compared to the emotion fear and social and protective actions stated in the survivor report, which were formalised and are shown in Figure 3(b). The patterns in Figure 5, outputted by the ASCRIBE model, correspond quite well with the patterns in the formalised empirical data from the survivor report in Figure 3(b). For example, in Figure 3(b) it can be seen that survivor John had a high fear level of 0.9 at three points in his report. In the left graph in Figure 5 it can be seen that through the interactions with the other agents and the internal affective responding, agent John also has a high fear value, fluctuating between 0.7 and 0.9. The right graph in Figure 5 shows that at the beginning, John aims more for protective actions than social actions, which seems the logical thing to do in a dangerous situation. Over all time steps, John shows a decrease in his aiming for protective actions in the first 10 time steps, followed by an increase till time step 15 and than another decrease till time step 30. This pattern can also be seen in Figure 3(b), where John's stated protective actions in his report started high, then decreased, increased and finally decreased. In Figure 3(b) it can also be seen that John stated that he performed social actions, formalised by the value 0.6, around time steps 20-25 and 60-70. Figure 5 shows that the ASCRIBE model as well outputs social actions around the value 0.6, around time steps 20-25 and 60-70. The difference between Figure 3(b) and Figure 5 is, that in Figure 5 all values change dynamically over time, they are continuous, and in Figure 3(b) the values are only available for certain points in time, taken from the survivor report. As a consequence, the total pattern of the real world data is not directly visible in the formalisation, like in Figure 3(b), but is visible when simulated by the ASCRIBE model. To further validate the ASCRIBE model against the real world data, a formal check was performed, where the real world data and the simulation results from the ASCRIBE model were compared automatically. This is explained in the next section.

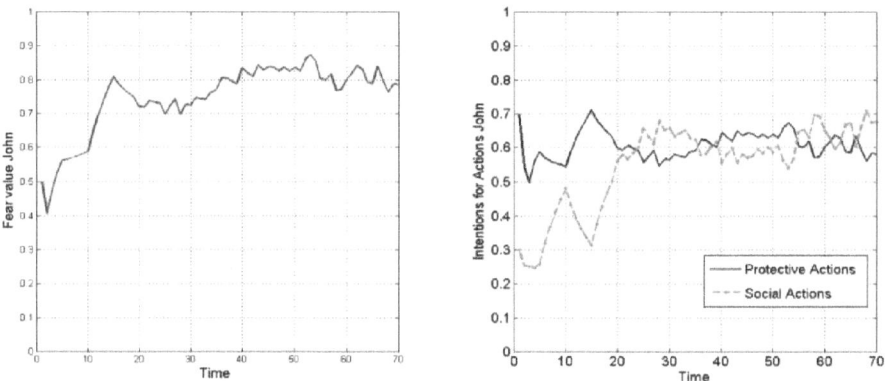

Fig. 5. The values for fear and intentions for actions of survivor John

6   Formal Comparison

To formally compare the simulation results in Section 5 with the formalised transcripts presented in Section 3.2, the TTL Checker Tool [3] has been used. This piece of software enables the researcher to check whether certain expected (dynamic) properties, expressed as statements in the Temporal Trace Language (TTL) [3], hold for a given *trace* (defined as a time-indexed sequence of states). Since the tool can take both simulated and empirical traces as input, it can be used to check (automatically) whether the generated simulation runs show similar patterns to the real world transcripts.

   Using the TTL Checker Tool, a number of dynamic properties have been verified against the traces described in Section 3.2 and 5 (which we will refer to as *empirical traces* and *simulation traces*, respectively). Some of these properties are presented below. To enhance readability, they are represented here in an informal notation, instead of a formal TTL notation. Note that the letters mentioned in the round brackets are parameters, which can be filled in when checking the property using the Checker Tool.

P1(a:agent, i1,i2:interval, m:trace) - 'More positiveness implies more social actions'
For intervals i1 and i2 within trace m, if the average positiveness of agent a's beliefs about the situation is higher in i1 than in i2, then agent a will perform more social actions in i1 than in i2.

P2(a:agent, i1,i2:interval, m:trace) - 'More positiveness implies less protective actions'
For intervals i1 and i2 within trace m, if the average positiveness of agent a's beliefs about the situation is higher in i1 than in i2, then agent a will perform more protective actions in i2 than in i1.

These dynamic properties (among several others, which are not shown due to space limitations) have been checked against the empirical and the simulation traces (where for all agents a, the interviewed persons were filled in). To create the intervals, all traces have been split up into relevant sub-scenarios (e.g., a part in which a person is present within a train carriage, or is present outside the train), and each sub-scenario has been cut into two equal halves, which we call intervals. Thus, by checking property P1 and P2 for all sub-scenarios, we basically checked whether it was the case that people who became more positive during a sub-scenario stopped protecting themselves and started to help others, and vice versa. Surprisingly, this property turned out to hold for almost all sub-scenarios of the empirical traces. This is an interesting finding, which can be potentially explained by the phenomenon that positive people are more open to external stimuli [8]. In addition, the property holds true for the simulated traces for the exact same sub-scenarios as in the empirical traces. Although this is obviously not an exhaustive proof of the correctness of the ASCRIBE model, it illustrates that the model can be used to reproduce similar patterns found in realistic scenarios.

7   Discussion

In this paper, it has been shown how the dynamics of individuals' mental states in a real world emergency can be analysed, through formalising survivor reports of the 7/7 London bombings in 2005 and evaluating them against generated simulations of the same case study with the ASCRIBE model. It is quite rare to work with this type of real world data of survivors of a terroristic attack. Nevertheless, the ASCRIBE simulations in Section 5 showed that it can simulate corresponding patterns in the empirical data of the 7/7 London bombings. The formal check of dynamic properties in Section 6 also shows that the ASCRIBE model can be used to reproduce similar patterns found in emergency scenarios, as in evacuation after a terrorist attack.

So far, the results show that the ASCRIBE model can reproduce patterns in the dynamics of beliefs, intentions and emotions of people involved in a terroristic attack in the real world. The results are promising, and although the transcription work is quite time consuming, the current analysis model has been set up in a generic manner, which means large parts can be re-used for the analysis of other real world incidents or disasters.

The current paper should mainly be seen as a proof-of-concept. The methodology turns out to be applicable to analysis of parts of the 7/7 bombings case study. In future work, the authors intend to analyse reports of a larger number of survivors, and to address more and different case studies, thereby better testing the robustness of the model. In addition, a more extensive evaluation is planned, using a quantitative measure for the correctness of the simulation results.

Acknowledgements. This research has been conducted as part of the FP7 ICT Future Enabling Technologies program of the European Commission under grant agreement No 231288 (SOCIONICAL).

References

1. Bechara, A., Damasio, A.: The Somatic Marker Hypothesis: a neural theory of economic decision. Games and Economic Behavior 52, 336–372 (2004)
2. Bosse, T., Hoogendoorn, M., Klein, M.C.A., Treur, J., van der Wal, C.N.: Agent-Based Analysis of Patterns in Crowd Behaviour Involving Contagion of Mental States. In: Mehrotra, K.G., Mohan, C.K., Oh, J.C., Varshney, P.K., Ali, M. (eds.) IEA/AIE 2011, Part II. LNCS, vol. 6704, pp. 566–577. Springer, Heidelberg (2011)
3. Bosse, T., Jonker, C.M., van der Meij, L., Sharpanskykh, A., Treur, J.: Specification and Verification of Dynamics in Agent Models. International Journal of Cooperative Information Systems 18, 167–193 (2009)
4. Bosse, T., Jonker, C.M., van der Meij, L., Treur, J.: A Language and Environment for Analysis of Dynamics by SimulaTiOn. International Journal of AI Tools 16(3), 435–464 (2007)
5. BBC News website: July 7 Bombings: London Attacks in Depth, http://news.bbc.co.uk/1/shared/spl/hi/uk/05/london_blasts/ what_happened/html/ (Undated, last accessed on May 29, 2011)
6. Cacioppo, J.T., Visser, P.S., Pickett, C.L.: Social neuroscience: People thinking about thinking people. MIT Press, Cambridge (2006)
7. Damasio, A.: The Somatic Marker Hypothesis and the Possible Functions of the Prefrontal Cortex. Philosophical Transactions of the Royal Society: Biological Sciences 351, 1413–1420 (1996)
8. Frederickson, B.L.: The role of positive emotions in positive psychology: The broaden-and-build theory of positive emotions. American Psychologist 56, 218–226 (2001)
9. Hoogendoorn, M., Treur, J., van der Wal, C.N., van Wissen, A.: Modelling the Interplay of Emotions, Beliefs and Intentions within Collective Decision Making Based on Insights from Social Neuroscience. In: Wong, K.W., Mendis, B.S.U., Bouzerdoum, A. (eds.) ICONIP 2010, Part I. LNCS (LNAI), vol. 6443, pp. 196–206. Springer, Heidelberg (2010)
10. Iacoboni, M.: Mirroring People. Farrar, Straus & Giroux, New York (2008)
11. Rizzolatti, G.: The mirror-neuron system and imitation. In: Hurley, S., Chater, N. (eds.) Perspectives on Imitation: from Cognitive Neuroscience to Social Science, vol. 1, pp. 55–76. MIT Press (2005)
12. Report of July 7 Review Committee, http://www.london.gov.uk/ who-runs-london/the-london-assembly/publications/ safety-policing/report-7-july-review-committee (June 2006, last accessed on May 29, 2011)
13. Tsai, J., Fridman, N., Bowring, E., Brown, M., Epstein, S., Kaminka, G., Marsella, S., Ogden, A., Rika, I., Sheel, A., Taylor, M.E., Wang, X., Zilka, A., Tambe, M.: ESCAPES - Evacuation Simulation with Children, Authorities, Parents, Emotions, and Social comparison. In: Tumer, K., Yolum, P., Sonenberg, L., Stone, P. (eds.) Proceedings of the 10th International Conference on Autonomous Agents and Multiagent Systems (AAMAS 2011), Innovative Applications Track, pp. 457–464. ACM Press (2011)

Modular Scale-Free Function Subnetworks in Auditory Areas

Sanming Song and Hongxun Yao[*]

School of Computer Science and Technology, Harbin Institute of Technology,
150001, Harbin, China
{ssoong,h.yao}@hit.edu.cn

Abstract. Function connectivity analysis is set to probe the whole-brain network architecture. Only several specific areas have to be focused when a specific modal is considered. To explore the microscopic subnetworks in auditory modality, the mean shift algorithm is proposed to cluster the fMRI time courses in the corresponding activation areas and several heuristic conclusions are obtained. 1) The voxel degree distribution supports scale-free hypothesis, but the exponential is relatively small. 2) More global subnetworks appear in the more abstract cognition process. 3) At least half of the subnetworks are local networks and they seldom cross with each other, acting as independent modules.

Keywords: Function subnetwork, fMRI, scale-free, modular.

1 Introduction

Our brain experiences continuous and complex cognition processes every day. The cortical network dynamics could be peeked from fMRI, which digitize the blood oxygen level dependent (BOLD) signal. Exploring the network architecture, the mechanisms underlying neuronal signal integration and modularization not only do help to the understanding of the human cognition, but also are heuristic for artificial intelligence.

The whole-brain functional connectivity supports complex network hypothesis [1]. Eguíluz et al.[2] firstly reported the scale-free topological structure of a large-scale brain network by calculating the temporal correlation between voxels. By calculating the degree distribution of voxels, Buckner et al. [3] proposed there may be hierarchical function hubs in the brain networks. And interesting, latter, Tomasi et al. [4] obtained similar results by calculating the function connection density, enhancing the scale-free hypothesis.

Besides, the brain has hierarchical modular network architecture in macro-scopic scale. Adams et al. [5] used visual and auditory object matching tasks to identify the brain areas underlying basic and subordinate cognition process. Davis et al. [6-7] used multiple speech conditions to explore the brain regions that are involved in spoken language comprehension, found that the hierarchical auditory system could be fractionated into sound based and more abstract

---

[*] This work is supported by National Natural Science Foundation of China (61071180).

B.-L. Lu, L. Zhang, and J. Kwok (Eds.): ICONIP 2011, Part I, LNCS 7062, pp. 435–442, 2011.

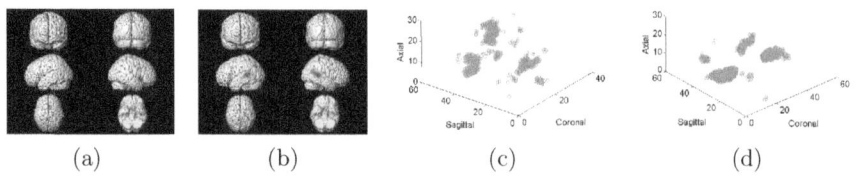

(a)	(b)	(c)	(d)

Fig. 1. Activated areas for S-B and A-S contrasts in auditory modality is shown in (a) and (b) respectively. The ROIs are shown in (c) and (d).

higher-level process. Later, Okada's experiments [8] also supported the experiment results.

But what's the network likes in each area and in each modality? Only several specific areas have to be focused when a specific modal is considered. For example, if we are mainly interested in semantic dynamics, it's better to analyze the network structure in the corresponding areas. In this paper, we try to explore the function subnetwork architecture in auditory task study. Our conclusions on scale-free network and local subnetwork modularization are heuristic for auditory function subnetwork modeling.

To explore the function subnetwork architecture, we detect the activation area by SPM8 [9] and then the function subnetworks is analyzed by mean shift [10] clustering. The experiment material is introduced in Section 2. Mean shift algorithm is introduced and applied to functional space clustering in Section 3. We discuss in detail the subnetwork architecture in Section 4. Finally, we conclude in Section 5.

2   Material

Analysis were conducted based on an open dataset available at the fMRIDC site[11], which was used by Adams and Janata in 2002 to compare the neural circuits of auditory and visual object categorization[5]. Briefly, 12 undergraduate volunteers were presented with 3 runs, each consisting of 4 blocks of stimulus trials. Each bock represented a different task (auditory, visual, or semantic) and was divided into 2 epochs of 15 trials each. During one epoch, participants matched objects to subordinate-level words and in the other they matched objects to basic-level words. 270 acquisitions were made in each run from each subject. Each acquisition consisted of 64 contiguous slices using a gradient EPI pulse sequence with the following parameters: TR = 2s; TE =35 ms; matrix size =64 × 64; resolution =3.75 × 3.75 × 5.0 mm^3; interslice spacing = 0 mm. Two sets of high-resolution T1-weighted anatomical images were also obtained. And we only adopted the high resolution set with 124 sagittal slices.

SPM8 was the main tool for image pre-processing [9]. All runs were realigned to the first volume of the first run. The structural image was co-registered to this mean (T2) image. Finally all the images were spatially normalized to a standard

Tailarach template [12] and smoothed using a 6mm full width at half maximum (FWHM) isotropic Gaussian kernel. And also, we use the default $p > 0.05$(FWE) to detect the activated voxels. SUBORDINATE-BASIC (S-B)contrasts for auditory modality and AUDITORY-SEMANTIC (A-S) contrasts were computed. And the contrasts were then entered into a random-effects analysis and SPMs for the group data were created. SPMs were thresholded at $p < 0.01$ (uncorrected).

The statistical significant cortex areas are shown in Fig.1. With S-B contrast, the areas mainly located in IFS (inferior frontal sulcus) and IFG (inferior frontal gyrus) , and with the A-S contrast, the activated areas mainly located in STG(superior temporal gyrus), STS (superior temporal sulcus) and FG (fusiform gyrus). The two contrasts seldom overlap with each other, A-S contrast mainly reflects the basic-level auditory cognition whereas the S-B contrast reflects the subordinate-level auditory cognition.

3   Mean Shift Clustering

To explore the fMRI time courses, we have to learn their distribution at first. Though many clustering techniques, like Fuzzy C-Means (FCM) and Gussian Mixture model (GMM), have been applied to fMRI time courses analysis, the similarity measure and cluster number are hard to choose [13]. The mean shift algorithm estimates the density adaptively [11], and we use it to analyze the fMRI time courses. We firstly briefly review it as follows.

3.1   Basic Mean Shift

Assume $x_i, i = 1, \cdots, N$ are in d dimensional space $\mathcal{R}^d$. The multivariate kernel density estimator

$$\hat{f}_{\mathcal{K}}(x) = \frac{1}{nh^d} \sum_{i=1}^{N} k\left(\left\|\frac{x - x_i}{h}\right\|^2\right) \tag{1}$$

based on a spherically symmetric kernel $\mathcal{K}$ with bounded support satisfying $\mathcal{K}(x) = c_{k,d} k\left(\|x\|^2\right) > 0 \quad \|x\| < 1$ is an adaptive nonparametric estimator of the density at location x in the feature space. The function $k(x)$, only for $x \geq 0$,is called the profile of the kernel, and the normalization constant $c_{k,d}$ makes $\mathcal{K}(x)$ integrates to one. The function $g(x) = -\dot{k}(x)$ can be defined when the derivative of $k(x)$ exists. Using $g(x)$ for profile, the kernel $\mathcal{G}(x)$ is defined as $\mathcal{G}(x) = c_{g,d} g\left(\|x\|^2\right)$.

By taking the gradient of (1),

$$m_{\mathcal{G}}(x) = \frac{\sum_{i=1}^{N} g\left(\left\|\frac{x_i - x}{h_i}\right\|^2\right) x_i}{\sum_{i=1}^{N} g\left(\left\|\frac{x_i - x}{h_i}\right\|^2\right)} - x \tag{2}$$

Input: Normalized $x_i, i = 1, \cdots, N$

1 **for** $i = 1; i \leq N; i + +$ **do**

2      $x \leftarrow x_i$;

3   Initialize $m(x)$ $w.r.t$ $\|m(x)\| > \varepsilon$;

4   **while** $\|m(x)\| > \varepsilon$ **do**

5          $m(x) = \dfrac{\sum\limits_{i=1}^{N} f_x(x_i) g\left(\frac{d^2(x, x_i)}{h_x}\right)}{\sum\limits_{i=1}^{N} g\left(\frac{d^2(x, x_i)}{h_x}\right)} - x$;

6          $x = f^{-1}{}_x (x + m(x))$;

7   Save x;

8   **end**

9 **end**

Output: All x

Algorithm 1. Basic Mean Shift Algorithm

is called the mean shift vector. Since mean shift always points to the higher density regions, it can be used as a hill climbing method in optimization, for example, clustering. In some condition, mapping the samples into another space would facilitate the analysis. If $f_x(x_i)$ is the mapping function, and $f_x^{-1}(x_i)$ is the inverse mapping, then formulae (2) could be written as shift calculation formulae (line 5) in Algorithm 1. And $d(x, x_i)$ is the distance measure in mapping space.

3.2 Functional Mean Shift

The fMRI signal at location i is denoted $s_i = \left(s_i^1, \cdots, s_i^L\right), i = 1, \cdots, N$. L is the length of scans. Since the functional networks are widely considered as the synchronization between remote issues, we try to probe the similarities between fMRI courses.

Similar to [14], it's better to discuss the time courses in functional space, which is a unit sphere $\Im^{N-2}$ with N normalized points. In fMRI study, each normalized time course could be considered as a point scattered in the functional space. In Riemannian geometry, the sphere is a simple manifold, and geodesic distance (formulae (4)) could be used to describe how close two points are [13]

$$d(x, y) = \theta, \cos(\theta) = \langle x, y \rangle. \tag{3}$$

With exponential mapping function

$$y = \exp_x(v) = x \cos\theta + v \sin\theta / \|v\|, \tag{4}$$

where $\theta = \|v\|$, and logarithm inverse mapping

$$v = \log_x y = \theta (y - x \cos\theta) / \|y - x \cos\theta\|, \tag{5}$$

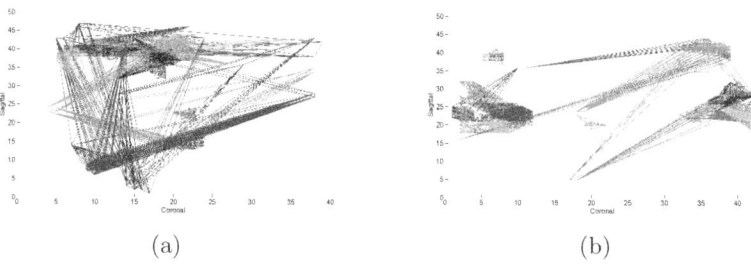

(a)   (b)

Fig. 2. All subnetworks in the activation areas for run 1 and subject 1. No matter in S-B (left), or in A-S (right), some subnetworks are limitted in a single region but other subnetworks extends over multiple regions.

the mean shift vector (2) could be written as

$$m_{\mathcal{G}}(\mathbf{x}) = \frac{\sum_{i=1}^{N} \log_{\mathbf{x}}(\mathbf{x}_i)\, g\left(\frac{d^2(\mathbf{x},\mathbf{x}_i)}{h_{\mathbf{x}}}\right)}{\sum_{i=1}^{N} g\left(\frac{d^2(\mathbf{x},\mathbf{x}_i)}{h_{\mathbf{x}}}\right)}. \tag{6}$$

When the mean shift iteration stops, the voxels points to the same target voxel having very similar time courses and could be considered to belong to the same subnetwork. Then the activation areas could be fractionated to many subnetworks. The clustered subnetworks for run 1 and subject 1 are shown in Fig.2. According to the extent, the subnetworks could be divided into local networks and global networks. The local networks are limited into a continuous activation region but the global extends to several regions. And the local networks seldom overlap with each other. We will analyze these statistical features of the subnetworks.

4 Characterize the Function Subnetwork

According to the above analysis, the activation areas could be fragmented into many subnetworks. And we would like to quantify these issues in this part.

4.1 Scale-Free Subnetwork

If one voxel points to another voxel in one mean-shift step, we draw a direction edge between them[1]. Then the voxel degrees are calculated by summing the income edges. It is curious that the voxels with larger degree seldom have stable

---

[1] The edge could be understood as the correlation between voxels. When many voxels converge to the same voxel, they could be seen as the deformations of the voxel. The deformations may be induced by physical movement, biochemical process, or channel noises. And note that only one-step mean-shift-vector is used.

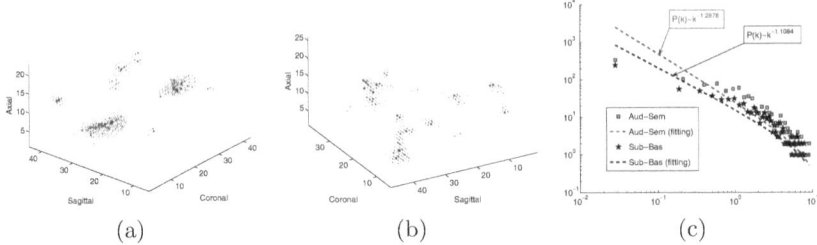

Fig. 3. The average voxel degrees for all subjects for A-S and S-B are shown in (a) and (b). The size of degree is represented by the size of red sphere. And the degree distributions are plotted in (c).

income edges. That is, for example, voxel $\mathcal{A}$ coverges to voxel $\mathcal{B}$ in one subject, but it converges to voxel $\mathcal{C}$ in another subject. This may be explained by the local function area differences between brains. The same cortical sub-area may play different roles between subjects.

But, the phenomenon is not random. Because in statistics, it can be found easily from Fig.3(a) and (b) that some voxels have larger income degrees, while some other voxels have smaller income degrees. So we speculate that these "hub" voxels may have very similar roles for all subjects.

The voxel distribution is plotted in Fig.3(c). It's interesting that S-B and A-S are scale-free networks with very similar fitting curves. Furthermore, the exponentials, -1.2876 for A-S and -1.1084 for S-B, are smaller when compared to [2], which demonstrates that much less voxels have very large degree and much more voxels have very small degree in auditory areas. Whether these statistical features are universe in all modalities is waiting for future work.

4.2   Global Subnetworks

One of the aims of original name verification experiments is to identify the neural circuitry involved in the process of auditory identification and categorization. Generally speaking, the subordinate categorization needs more widespread participation of regions, because subordinate cognition may involve the feedback from high-level cortical areas, like IFS, IFG, and so on. To integrate neural information from related cortical regions, more remote interactions are necessary. So, more global subnetworks would appear in S-B case. We calculate the proportion of global network for both S-B and A-S cases. And the bar map is shown in Fig.4 (a). As it easily can be seen, the proportion of global networks in S-B is far more than that in A-S case. The average rate is 14.28% for A-S, and 38.84% for S-B. Our result could be seen as neurobiological evidence for the global participation in more abstract cognition process.

4.3   Network Modularization

Though many global works exists in the activated areas, we should note that at least half subnetworks are local networks (Fig.4 (a)). What's more, the most

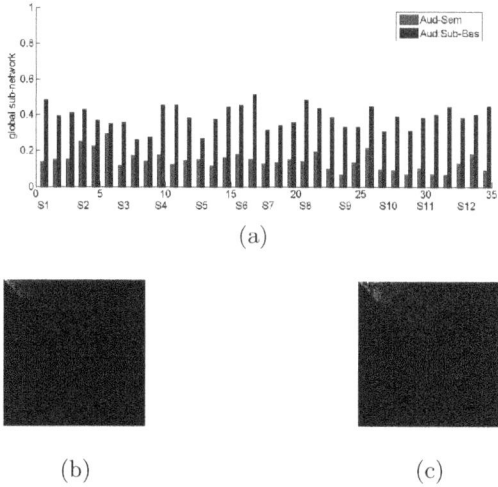

(a)

(b)   (c)

Fig. 4. Features of subnetworks. The proportion of global subnetworks for all runs and all subjects are plotted in (a). And the overlaps between the 100 largest subnetworks are shown in (b) (S-B case) and (c) (A-S case).

interesting is that these local subnetworks nearly don't cross with each other. To explore the relationships between subnetworks, we calculated overlaps between the 100 largest subnetworks for all runs and all subjects. When calculating the overlap matrix, the isolated voxels are excluded. Now that the overlap matrix is symmetric, only the above triangle part is shown in Fig.4 (b) and Fig.4(c) for convenience.

Two conclusions can be drawn easily. Firstly, comparison of the overlap matrix between S-B case and A-S case shows that more network intersections appears in more abstract cognition process, which support the conclusion of section 4.2. Secondly, larger networks tend to overlap with several other larger networks, while the smaller networks are nearly isolated. So the ROI network may be constructed by a few larger intersected sub-networks and many modularized smaller sub-networks.

Modularization means that these local subnetworks play more independent roles in cognition process, which is a necessary base for hierarchical architecture hypothesis and is typical for scale-free networks [1].

5   Conclusions

In this paper, we try to explore the function subnetwork architecture in auditory modality. We re-analyze the name verification experiments and many heuristic conclusions are obtained. It should be noted that there must be some suppression mechanism in the thalamencephalon that activate cortical areas selectively. But it is beyond our topic here and only the network architecture in auditory activated areas is focused. The conclusions on scale-free network and modularization would be heuristic for artificial networks. Our future works will focus on the function subnetworks modeling and the more microscopic level architecture analysis.

Acknowledgments. We thank *fMRIDC* group for approval use of the 54th dataset, and we also thank *Reginald B.Adams* and *Petr Janata* for critical discussions on processing the data. Special thanks go to *Yakov Kazanovich* and *Cheng-Yuan Liou* for helpful comments and suggestions for the manuscript.

References

1. Bullmore, E., Sporns, O.: Complex brain networks: graph theoretical analysis of structural and functional systems. Nat. Rev. Neurosci. 10, 186–198 (2009)
2. Eguíuz, V.M., Chialvo, D.R., Cecchi, G.A., Baliki, M., Apkarian, A.V.: Scale-free brain functional networks. Phys. Rev. Lett. 94, 018102 (2005)
3. Buckner, R.L., Sepulcre, J., Talukdar, T., Krienen, F.M., Liu, H., Hedden, T., Andrews-Hanna, J.R., Sperling, R.A., Johnson, K.A.: Cortical hubs revealed by intrinsic functional connectivity: mapping, assessment of stability, and relation to Alzheimer's disease. J. Neurosci. 29, 1860–1873 (2009)
4. Tomasi, D., Volkow, N.D.: Association between Functional Connectivity Hubs and Brain Networks. Cereb. Cortex (in press, 2011)
5. Adams, R.B., Janata, P.: A comparison of neural circuits underlying auditory and visual object categorization. Neuroimage 16, 361–377 (2002)
6. Peelle, J.E., Johnsrude, I.S., Davis, M.H.: Hierarchical processing for speech in human auditory cortex and beyond. Front. Hum. Neurosci. 4, 51–53 (2010)
7. Davis, M.H., Johnsrude, I.S.: Hearing speech sounds: Top-down influences on the interface between audition and speech perception. Hearing Research 229, 132–147 (2007)
8. Okada, K., Rong, F., Venezia, J., Matchin, W., Hsich, I.-H., Saberi, K., Serrences, J.T., Hickok, G.: Hierarchical organization of human auditory cortex: Evidence from acoustic invariance in the response to intelligible speech. Cereb. Cortex (2010), doi:10.1093/cercor/bhp318
9. Friston, K.J., Holmes, A.P., Ashburner, J., Poline, J.B.: SPM8, http://www.fil.ion.ucl.ac.uk/spm/
10. Comaniciu, D., Meer, P.: Mean shift: a robust approach toward feature space analysis. IEEE Transactions on Pattern Analysis and Machine Intelligence 24(5) (2002)
11. fMRIDC site, http://www.fmridc.org/f/fmridc/54.html
12. Talairach, P., Tournoux, J.: A stereotactic coplanaratlas of the human brain. Thieme Verlag (1988)
13. Cheng, J., Shi, F., Wang, K., Song, M., Jiang, J., Xu, L., Jiang, T.: Nonparametric Mean Shift Functional Detection on Functional Space for Task and Resting-state fMRI. In: MICCAI 2009 Workshop on fMRI Data Analysis (2009)
14. Friston, K.J.: Statistical parametrical mapping: the analysis of functional brain images. Academic Press (2007)

Bio-inspired Model of Spatial Cognition

Michal Vavrečka[1], Igor Farkaš[2], and Lenka Lhotská[1]

[1] Faculty of Electrical Engineering
Czech Technical University, Prague
{vavrecka,lhotska}@fel.cvut.cz
[2] Faculty of Mathematics, Physics and Informatics
Comenius University, Bratislava
farkas@fmph.uniba.sk

Abstract. We present the results of an ongoing research in the area of symbol grounding. We develop a biologically inspired model for grounding the spatial terms that employs separate visual *what* and *where* subsystems that are integrated with the symbolic linguistic subsystem in the simplified neural model. The model grounds color, shape and spatial relations of two objects in 2D space. The images with two objects are presented to an artificial retina and five-word sentences describing them (e.g. "Red box above green circle") with phonological encoding serve as auditory inputs. The integrating multimodal module is implemented by Self-Organizing Map or Neural Gas algorithms in the second layer. We found out that using NG leads to better performance especially in case of the scenes with higher complexity, and current simulations also reveal that splitting the visual information and simplifying the objects to rectangular monochromatic boxes facilitates the performance of the *where* system and hence the overall functionality of the model.

Keywords: self-organization, categorization, symbol grounding, spatial relations, linguistic description.

1 Introduction

The core problem of embodied cognitive science is how to ground symbols to the external world. We are looking for a system interacting with the environment that is able to understand its internal representations which should preserve constant attributes of the environment, store them as concepts, and connect these to the symbolic level. This approach to the meaning representation is different from the classical symbolic theory based on formal semantics of truth values, which cannot guarantee correspondence of the symbolic level with the external world.

In this article we propose an extended version of the classical grounding architecture [1] that implements the multimodal representations in the framework of the perceptual symbol system proposed by Barsalou [2]. The main innovation is the processing of symbolic input by a separate auditory subsystem and the integration of auditory and visual information in a multimodal layer that

B.-L. Lu, L. Zhang, and J. Kwok (Eds.): ICONIP 2011, Part I, LNCS 7062, pp. 443–450, 2011.

incorporates the process of identification of symbols with concepts. Our theory is similar to grounding transfer approach [3] but unlike it, our model works in a fully unsupervised manner.

Our approach was tested in the area of spatial cognition. In our models, we consider the evidence that the information about the location and identification of an object in space are processed separately. Studies with humans [4] revealed two separate pathways involved in processing of visual and spatial information: The dorsal *where* pathway is assumed to be responsible for spatial representation of the object location, while the ventral *what* stream is involved in object recognition and form representation.

In our previous experiment [5] we compared two versions of the visual subsystem, analyzing the distinction between *what* and *where* pathways, by proposing different ways how to represent object features (shape and color) and object position (in a spatial quadrant). Model I contained a single self-organizing map (SOM; [6]) that learned to capture both *what* and *where* information. Model II consisted of two SOMs for processing *what* information (foveal input) and *where* information (retinal input). Comparison of both models confirmed the effectiveness of separate visual processing of shape and spatial properties that led to a significant decrease of errors in the multimodal layer.

Both models assume the existence of the higher layer that integrates the information from two primary modalities. This assumption makes the units in the higher layer bimodal (i.e. they can be stimulated by any of the primary layers) and their activation can be forwarded for further processing. Bimodal (and multimodal) neurons are known to be ubiquitous in the association areas of the brain [7]. See also discussion in [5] for the relation of our model to several other connectionist models.

2   Motivation

The first goal of experiments presented here is the more detailed analysis of the information processing in the *where* system. We tested two types of inputs for this subsystem, namely full retinal images projected to *where* system (the same as previously) and simplified version of retinal projections (the color information was omitted and the object shapes were simplified to rectangular monochromatic boxes. The results should help us decide, whether this simplification is important for enhancing the overall model performance.

In [5] we also identified the difference in the effectiveness of the SOM comparing to Neural Gas (NG) algorithm [8] in the multimodal layer in favour of lower NG error rates. The higher error rate in SOM should be attributed to its fixed neighborhood function (while NG uses flexible neighborhood) that imposes constraints to the learning process in multimodal layer. The second goal of the current experiment is hence the analysis of the neighborhood function in SOM. We presented stimuli with increasing fuzziness in the spatial location (see Fig. 2) and compared error rates of SOM and NG algorithms.

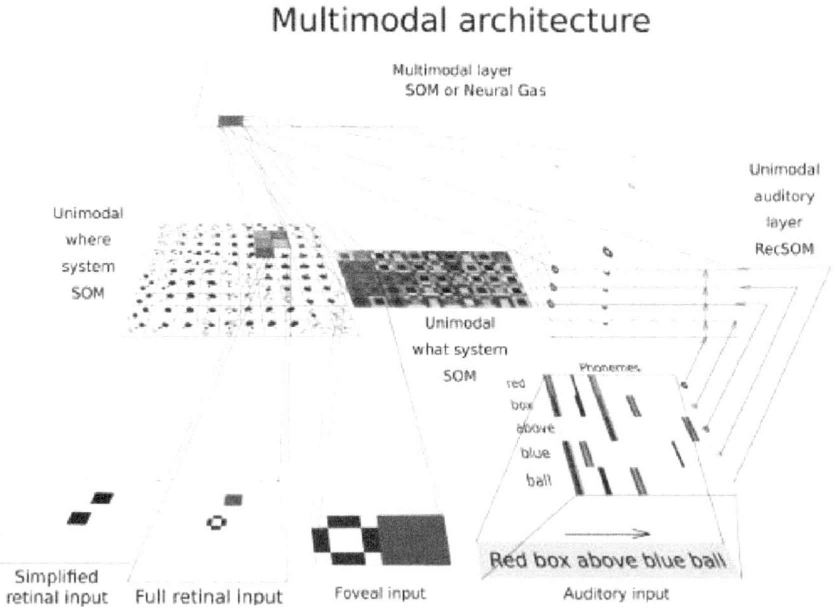

Multimodal architecture

Fig. 1. The two-layer multimodal architecture used in our experiments

3   The Model

We adopted the model from [5] to test the architecture with a simplified type of inputs and variable level of fuzziness. The inputs (similar to previous models) consisted of two objects in 2D environment and their linguistic descriptions. The scenes contained the trajector and the base object in different spatial configurations. The position of the base is fixed in the center of the scene and the trajector position is fuzzy with variable level of fuzziness (Fig. 2). We trained the model using scenes with 3 colors (red, green, blue), 5 object types (box, ball, table, cup, bed) and 4 spatial relations (above, bellow, left, right) that means 840 combinations of two different objects in the scene. There were 42000 examples (50 instances per spatial configuration) in the training set.

3.1   Visual Subsystem

The visual subsystem is formed by an artificial retina ($28{\times}28$ neurons) and an artificial fovea (two visual fields consisting of $4{\times}4$ neurons) that project visual and spatial information about the trajector and the base to the the primary unimodal visual layers. These layers are both made of SOMs that differentiate various positions of two objects (resembling *where* pathway) from retinal projection and color and shape of objects (resembling *what* pathway) from foveal projection. The color of each pixel was encoded by the activity level. Both maps were trained for 100 epochs with decreasing parameter values (unit neighborhood radius, learning rate).

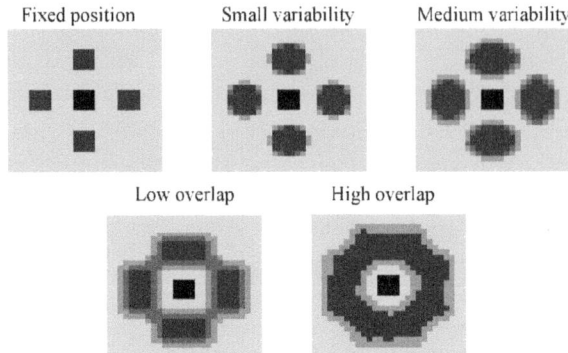

Fig. 2. Simplified visual inputs with varying levels of spatial fuzziness

3.2 Auditory Subsytem

Auditory inputs (English sentences) were encoded as phonological patterns representing word forms using PatPho, a generic phonological pattern generator that fits every word (up to trisyllables) onto a template according to its vowel-consonant structure [9]. It uses the concept of syllabic template: a word representation is formed by combinations of syllables in a metrical grid, and the slots in each grid are made up by bundles of features that correspond to consonants and vowels. In our case, each sentence consists of five 54-dimensional vectors with component values in the interval (0,1). These inputs are sequentially fed to RecSOM [10] that learns to represent inputs (words) in the temporal context (hence capturing sequential information). RecSOM output, in terms of map activation, feeds to the multimodal layer, to be integrated with the visual pathway. RecSOM units become sequence detectors after training, topographically organized according to the suffix (the last words).

Since RecSOM, unlike SOM, is not common, we provide its mathematical description here. Each neuron $i \in \{1, 2, ..., N\}$ in RecSOM has two associated weight vectors: $\mathbf{w}_i \in \mathbb{R}^n$ – linked with an n-dimensional input $\mathbf{s}(t)$ (in our case, the current word, $n = 54$) feeding the network at time t and $\mathbf{c}_i \in \mathbb{R}^N$ – linked with the context $\mathbf{y}(t-1) = [y_1(t-1), y_2(t-1), ..., y_N(t-1)]$ containing map activations $y_i(t-1)$ from the previous time step.

The output of a unit i at time t is $y_i(t) = \exp(-d_i(t))$, where

$$d_i(t) = \alpha \cdot \|\mathbf{s}(t) - \mathbf{w}_i\|^2 + \beta \cdot \|\mathbf{y}(t-1) - \mathbf{c}_i\|^2.$$

Here, $\|\cdot\|$ denotes the Euclidean norm, $\alpha > 0$ and $\beta > 0$ are model parameters that respectively influence the effect of the input and the context upon neuron's profile. Their suitable values are usually found heuristically (in our model, we used $\alpha = \beta = 0.1$). Both weight vectors are updated using the same form of SOM learning rule:

$$\Delta\mathbf{w}_i = \gamma \cdot h_{ik} \cdot (\mathbf{s}(t) - \mathbf{w}_i),$$

$$\Delta\mathbf{c}_i = \gamma \cdot h_{ik} \cdot (\mathbf{y}(t-1) - \mathbf{c}_i),$$

where b is an index of the best matching unit at time t, $b = \text{argmin}_i\{d_i(t)\}$, and $0 < \gamma < 1$ is the learning rate. Neighborhood function h_{ib} is a Gaussian (of width σ) on the distance $d(i,b)$ of units i and b in the map: $h_{ib} = \exp(-d(i,b)^2/\sigma^2)$. The 'neighborhood width', σ, linearly decreases in time to allow for forming topographic representation of input sequences.

3.3 Multimodal Integration

Outputs from both visual SOMs and auditory RecSOM are projected to the multimodal layer (SOM or NG). The main task for the multimodal layer is to find and learn the categories by merging different sources of information. We compared SOM and NG algorithms that are both unsupervised and based on the competition among units, but NG uses a flexible neighborhood function, as opposed to the fixed neighborhood in SOM.

For clarity, we explain NG algorithm briefly here. NG shares with SOM a number of fetaures. In each iteration t, an input vector $\mathbf{m}(t)$ is randomly chosen from the training dataset. Subsequently, for all units in the multimodal layer we compute $d_i(t) = \|\mathbf{m}(t) - \mathbf{z}_i\|$ and sort the NG units according to their increasing distances d_i, using indices $l = 0, 1, ...$ (where $l(0)$ corresponds to unit b, the current winner). Then we update all weight vectors $\mathbf{z}_i$ according to

$$\Delta\mathbf{z}_i = \epsilon \cdot \exp(-l(i)/\lambda) \cdot (\mathbf{m}(t) - \mathbf{z}_i) \qquad (1)$$

with ϵ as the adaptation step size and λ as the so-called neighborhood range. We used $\epsilon = 0.5$ and $\lambda = n/2$ where n is number of neurons. Both parameters are reduced with increasing t. It is known that after sufficiently many adaptation steps the feature vectors cover the data space with minimum representation error [8]. The adaptation step of the NG can be interpreted as gradient descent on a cost function.

Inputs for the multimodal layer are taken as unimodal activations (from both modalities) using the k-WTA (i.e. winner-takes-all) mechanism, where k most active units are proportionally turned on, and all other units are reset to zero (in the models, we used $k = 6$ for visual layers). The motivation for this type of output representation consists in introducing overlaps between similar patterns to facilitate generalization. On the other hand, the output representation in the multimodal layer is chosen to be localist for better interpretation of results and the calculation of error rate.

4 Results

We trained the system with the fixed sizes of unimodal layers (30×30 units) and the multimodal layer (29×29). After the training phase, the system was tested by a novel set of inputs. All inputs were indexed for the error calculation in the second layer. Then we measured the effectiveness of this system, based on the percentage of correctly classified test inputs. To calculate the accuracy of neuron responses, we applied a voting algorithm after training to label each neuron in the layer based on its most frequent response. Then we measured the accuracy of this system, based on the percentage of correctly classified test inputs. At first, we compared the error rate in unimodal *where* layer trained with full retinal images or simplified monochromatic rectangles standing for objects in the scene. Results are shown in Fig. 3. It can be seen that simplified input significantly reduce the error for all levels of spatial fuzziness which could be explained by reduced variability of inputs that are topographically mapped in the SOM.

The analysis of the model behavior revealed that the trajector shape and the spatial term representations are the most difficult task components for visual unimodal systems which is caused by their variability and fuzziness. The model analysis also confirmed that simplified projection of retinal images to the *where* system resulted in lower error rates compared to full retinal images (Fig. 3). This leads us to the conclusion that it is possible to simplify the information projected to the *where* system to optimize the speed and effectiveness of our architecture.

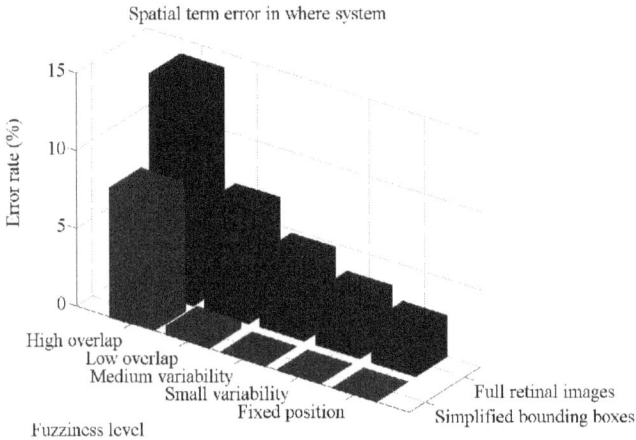

Fig. 3. The error rates in the *where* system as a function of input types and the levels of spatial fuzziness

Next we compared SOM and NG algorithms in the multimodal layer using the simplified *where* system. The calculation of the error rates was the same as for the unimodal layers. Fig. 4 shows a lower error rate for NG in all levels of fuzziness and the high error rates for SOM regardless of the fuzziness level.

The poorer result of multimodal SOM compared to NG could most probably be attributed to the fixed neighborhood function which imposes constraints the learned nonlinear mapping. There was a 70% error rate for all levels of fuzzy inputs, so the multimodal SOM is able to represent neither fuzzy inputs nor distinct inputs. We observed a different type of clustering in unimodal layers

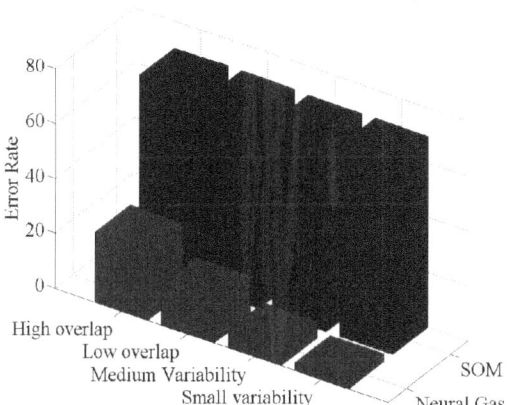

Comparison of algorithms in multimodal layer

Fig. 4. Comparison of two models for different type of inputs. The error rates in the *where* system as a function of levels of spatial fuzziness.

that are transferred to the multimodal layer at which the SOM is not able to adapt to the joint outputs from unimodal layers. The results of NG algorithm for the same input data confirm this hypothesis. There was a 25% error rate only for highly overlaping inputs (compared to 70% error rate for all type of inputs for SOM. The effectiveness of NG for less fuzzy inputs was even better.

5   Conclusion

Previous models of symbol grounding (see Discussion in [5]) deal with the lexical level but our model goes beyond words because it is able to represent sentences with fixed grammar via RecSOM. It finds the mapping of the particular words to the concepts in the multimodal layer without any prior knowledge, so the system proposes the solution to the binding problem. The system design allows us in principle to append other modalities and still represent discrete multimodal categories. The hierarchical representation of the sign components is the important advantage of our model. It guarantees better processing and storing of representations because the sign (multimodal level) is modifiable from both modalities (the sequential "symbolic" auditory level and the parallel "conceptual" visual level). The separate multimodal level provides a platform for the development of subsequent stages of this system (e.g. inference mechanisms). Further tests of this approach should focus on scaling up this architecture to more complex mappings.

Acknowledgment. This research was supported by the research program MSM 6840770012 of the Czech Technical University in Prague, and by SAIA scholarship (M.V.) and by Slovak Grant Agency for Science, no. 1/0439/11 (I.F.).

References

1. Harnad, S.: The Symbol Grounding Problem. Physica D 42, 335–346 (1990)
2. Barsalou, L.: Perceptual symbol systems. Behavioral and Brain Sciences 22(4), 577–660 (1999)
3. Riga, T., Cangelosi, A., Greco, A.: Symbol grounding transfer with hybrid self-organizing/supervised neural networks. In: Int. Joint Conf. on Neural Networks (2004)
4. Millner, A., Goodale, M.: The Visual Brain in Action. Oxford University Press (1995)
5. Vavrečka, M., Farkaš, I.: Unsupervised grounding of spatial relations. In: Proceedings of European Conference on Cognitive Science, Sofia, Bulgaria (2011)
6. Kohonen, T.: Self-Organizing Map (3rd, extended edition). Springer, New York (2011)
7. Stein, B., Meredith, M.: Merging of the Senses. MIT Press, Cambridge
8. Martinetz, T., Schulten, K.: A neural-gas network learns topologies. In: Kohonen, T., et al. (eds.) Int. Conf. on Artificial Neural Networks, pp. 397–402. North-Holland, Amsterdam (1991)
9. Li, P., McWhinney, B.: PatPho: A phonological pattern generator for neural networks. Behavior Research Methods, Instruments and Computers 34(3), 408–415 (2002)
10. Voegtlin, T.: Recursive self-organizing maps. Neural Networks 15(8-9), 979–991 (2002)

EEG Classification with BSA Spike Encoding Algorithm and Evolving Probabilistic Spiking Neural Network

Nuttapod Nuntalid[1], Kshitij Dhoble[1], and Nikola Kasabov[1,2]

[1] Knowledge Engineering and Discovery Research Institute,
Auckland University of Technology,
Private Bag 92006, Auckland 1010, New Zealand
{nnuntali,kdhoble,nkasabov}@aut.ac.nz
[2] Institute for Neuroinformatics, University of Zurich and ETH Zurich

Abstract. This study investigates the feasibility of Bens Spike Algorithm (BSA) to encode continuous EEG spatio-temporal data into input spike streams for a classification in a spiking neural network classifier. A novel evolving probabilistic spiking neural network reservoir (epSNNr) architecture is used for the purpose of learning and classifying the EEG signals after the BSA transformation. Experiments are conducted with EEG data measuring a cognitive state of a single individual under 4 different stimuli. A comparison is drawn between using traditional machine learning algorithms and using BSA plus epSNNr, when different probabilistic models of neurons are utilised. The comparison demonstrates that: (1) The BSA is a suitable transformation for EEG data into spike trains; (2) The performance of the epSNNr improves when a probabilistic model of a neuron is used, compared to the use of a deterministic LIF model of a neuron; (3) The classification accuracy of the EEG data in an epSNNr depends on the type of the probabilistic neuronal model used. The results suggest that an epSNNr can be optimised in terms of neuronal models used and parameters that would better match the noise and the dynamics of EEG data. Potential applications of the proposed method for BCI and medical studies are briefly discussed.

Keywords: Spatio-Temporal Patterns, Electroencephalograms (EEG), Stochastic neuron models, evolving probabilistic spiking neural networks.

1 Introduction

EEG refers to the recording of electrical brain signal activity that is acquired along the head scalp as a result of neuronal activity in brain. In clinical aspect, EEG have been used in clinical recording of the brain electrical activity over a period of time and usually employs 19 electrodes / channels that are placed over various locations on the scalp. In neurology, the main diagnostic application of EEG is in the case of epilepsy, as epileptic seizure creates clear spike activities that can be measured in standard EEG equipment [1], [2].

B.-L. Lu, L. Zhang, and J. Kwok (Eds.): ICONIP 2011, Part I, LNCS 7062, pp. 451–460, 2011.

A clinical application of EEG is to show the type and location of the activity in the brain during a seizure. It is also commonly used to evaluate people who are having problems related to brain functionality such as coma, tumours, long-term memory loss, or stroke. In computer science, there have been many studies that focus on EEG applications for Brain Computer Interfaces (BCI). EEG data analysis has been explored for a better understanding of the information processing capability of the mammalian brain. EEG can also be potentially used in biometric systems. Given that the brain-wave pattern of every individual is unique, EEG can be used for developing person identification or authentication systems [3], [4].

Having in mind the importance of the accurate analysis and study of EEG signals, we are aiming in this paper to propose a method for EEG data transformation into spike trains and their accurate classification.

The rest of the paper is structured as follows: Section 2 describes related researches and motivations. Section 3 presents the proposed methodology that is been used in this experimentation. Section 4 contains comparative experimental results along with discussions. Finally, Section 5 summarizes the conclusion and future directions of our study.

2   Related Works

Spiking neuron networks (SNN) have been used for EEG analysis in some researches, and have shown remarkable performance in comparison to other traditional methods for classification task. In [5], the authors have proposed a method for the creation of spiking neural networks for EEG classification of epilepsy data for the purpose of epileptic seizure detection. Their experiment used a simple 3-layer feed-forward architecture (having input layer, hidden layer, output layer) which resulted in an average classification accuracy of approximately 90.7%.

In a recent study [6], the researchers analysed rat's EEG data using reservoir computing approach (echo state network) for epileptic seizure detection in real-time, based on a data from 4 EEG channels. It is a two class problem, where they had to classify the EEG signals for detection of seizure and tonic-seizure. The reservoir was made of 200 Leaky Integrate-and Fire (LIF) neurons, where 20% and 80% of the EEG data was used for training and testing respectively. The results of this study claimed that the performance was higher than the other four traditional methods in terms of detection time, which was around 85% accuracy in 0.5 seconds for seizure and 85% accuracy in 3 seconds for tonic-seizure. However, this study was done by using EEG data from a rat, acquired from only 4 channels and the frequency for detecting seizure/s was known in advance (8, 16 and 24 Hz).

Hence, recent studies on SNN and reservoir computing for EEG application shows that many of them produced comparatively good results while utilizing a deterministic neuronal model. However, in one of our recent work [7], we have shown that replacing the deterministic with the probabilistic spiking neuron models yields better results.

In this study we aim to analyze the feasibility of BSA spike encoding scheme along with a SNN reservoir such as LSM using probabilistic spiking neuron models for complex spatio-temporal human EEG data, acquired from 64 channels. In the following we have described the design of our experiment and its setup.

3   Methodology

The framework for classification of spatio-temporal data based on evolving probabilistic spiking neural network reservoir (epSNNr) paradigm is presented in Fig.1. At first, each channel of spatial-temporal data (EEG) is transformed into trains of spikes by the encoder module. Then the trains of spikes are distributed into spatio-temporal filter which employs the latest reservoir paradigm (i.e. LSM) that utilizes several stochastic neuron models as liquid generators [7]. Further, the filter generates liquid state for each time step. These states are fed into a readout function for training and testing the classification performance using a pre-defined type of a classifier.

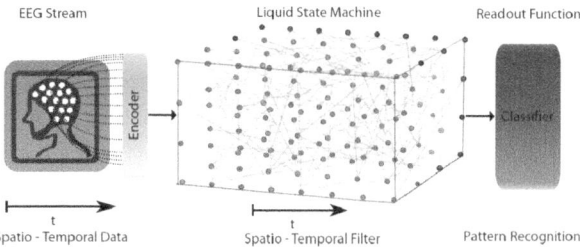

Fig. 1. Framework for EEG spatio-temporal pattern learning and classification based on epSNNr

3.1   Spike Encoder

In our methodology we have incorporated BSA spike encoding scheme. So far, this encoding scheme has only been used for encoding sound data. However, since EEG signals also fall under the frequency domain, we hypothesised that BSA encoding will be suitable to transform EEG signals into spike representation. The key benefit of using BSA is that the frequency and amplitude features are smoother in comparison to the HSA (Hough Spiker Algorithm) spike encoding scheme [8]. Moreover, due to the smoother threshold optimization curve, it is also less susceptible to changes in the filter and the threshold [8]. Studies have shown that this method offers an improvement of 10dB-15dB over the HSA spike encoding scheme. According to [8], the stimulus is estimated from the spike train by

$$s_{est} = (h \times x)(t) = \int_{+\infty}^{+\infty} x(t - \tau)h(\tau)d\tau = \sum_{k=1}^{N} h(t - t_k) \qquad (1)$$

where, t_k represents the neurons firing time, $h(t)$ denotes the linear filters impulse response and $x(t)$ is the spike of the neuron that can be calculated as

$$x(t) = \sum_{k=1}^{N} \delta(t - t_k) \tag{2}$$

For this particular dataset, we have set the Finite Impulse Response (FIR) filter size to 20, and the BSA threshold to 0.955.

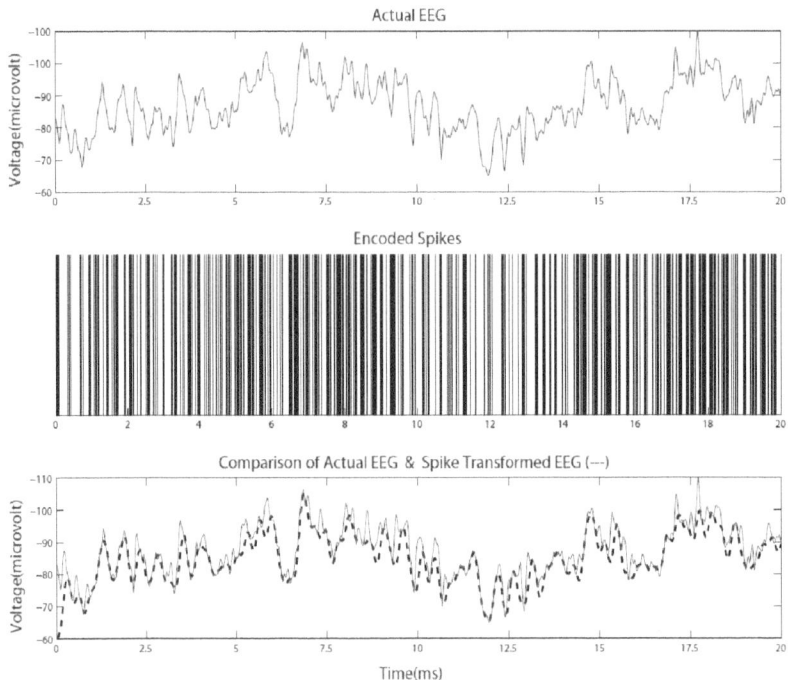

Fig. 2. The top figure shows the Actual one channel EEG signal for the duration of 20ms. The middle figure is the spike representation of the above figure obtained using BSA. The bottom figure shows the actual one channel EEG signal that has been superimposed with another signal (dashed lines) which represents the reconstructed EEG signal from the BSA encoded spikes. The similarity between the two signals is obvious that illustrates the applicability of the BSA transformation.

However, when the spike train $x(t)$ is applied with a discrete FIR filter, the Eq.2 can be represented as

$$o(t) = (h \times x)(t) = \sum_{k=0}^{M} x(t - k)h(k) \tag{3}$$

where, M refers to the number of filter taps. A more detailed explanation is given in [8].

3.2   Evolving Probabilistic Spiking Neuron Network Reservoir (epSNNr)

In epSNNr, we have replaced the deterministic LIF neurons of a traditional LSM with probabilistic neural models that have been comprehensively described in [7]. The probabilistic approach has been inspired by biological neurons that exhibit substantial stochastic characteristics. Therefore, incorporation of non-deterministic elements into the neural model may provide us with advantage due to the brain-like information processing system. In our reservoir we have used the standard Leaky Integrate and Fire (LIF) neuron model as well as probabilistic models such as Noisy Reset (NR), Step-wise Noisy Threshold (ST) and Continuous Noisy Threshold (CT) (see fig.3. for an illustration of the difference between the three stochastic neuronal models). The advantage of stochastic neural models has been demonstrated in a previous study [7].

Table 1. The following table provides the parameter setting that has been used in our experimental setup for the epSNNr

Parameters	Value/s
For Neuron	
Time Constance	10 ms
Reset Potential	0 mV
Initial Firing Threshold	10 mV
Standard Deviation of reset fluctuation	3 mV
Standard Deviation of Step-wise Firing Threshold	2 mV
Standard Deviation of Continuous Firing Threshold	1 mV
For LSM	
Simulation Time	500 ms
Number of Neurons	135
Excitatory to Inhibitory Neuron Ratio	4:1
Input Neurons Connection Probability	0.02
Input Neurons Connection Weight	1.62 mV
Time-bin for Liquid Responses	25 ms

3.3   Dataset

RIKEN EEG Dataset was collected in the RIKEN Brain Science Institute in Japan. It includes 4 stimulus conditions (4 classes): Class1 - Auditory stimulus; Class2 -Visual stimulus; Class3 - Mixed auditory and visual stimuli; Class4 -No stimulus. The EEG data were acquired using a 64 electrode EEG system that was filtered using a 0.05Hz to 500 Hz band- pass filter and sampled at 2KHz. According to the sample rate, the dataset is instable. In this preliminary proof of concept investigation we collected a small number of data points: 11 epochs from 50 epochs (1988-2153 samples/epoch/50ms) of each class (4 classes are 44 epochs) which have closer rate as possible. We used 80% (9 epochs) and 20% (2 epochs) for training and testing respectively.

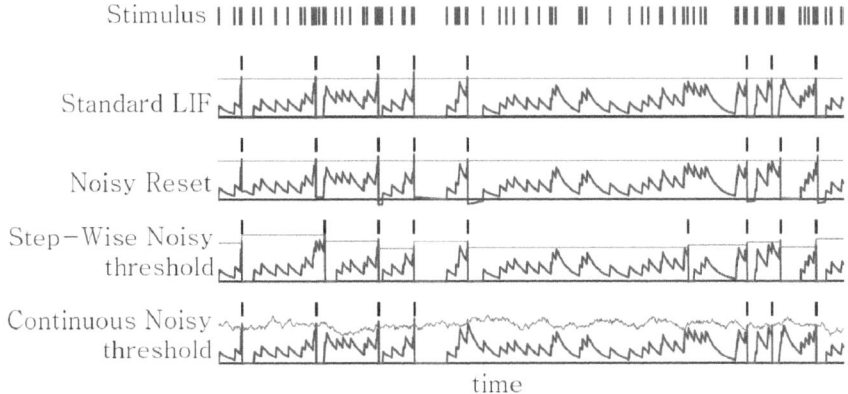

Fig. 3. Evolution of the post-synaptic potential *u(t)* and the firing threshold over time (blue (dark) and yellow (light) curves respectively recorded from a single neuron of each of the three stochastic neural models used in this paper vs the standard LIF model. The input stimulus for each neuron is shown at the top of the diagram. The output spikes of each neuron are shown as thick vertical lines above the corresponding threshold curve (from [7])

4   Experiments and Discussions

The experimental setup of this study is presented in Fig.1. All networks have the same network topology and the same connection weight matrix. A detailed description of the network generation and parameterisation is given in Table 1. We construct a reservoir having a small-world interconnectivity pattern as described in [9]. In order to make a standard comparison in our further investigation, the recurrent SNN is generated by using Brain[10] whose grid alignment is similar to the CSIM's (A neural Circuit Simulation) default LSM setting having 135 neurons in a three-dimensional grid of size 9 × 5 × 3. In this grid, two neurons A and B are connected with a connection probability

$$P(A, B) = C \times e^{\frac{-d(A, B)}{\lambda^2}} \tag{4}$$

The sample rate of the EEG dataset is extremely higher than usual EEG datasets, where each epoch belonging to a class (having 1988-2153 samples/epoch) was encoded into 50ms spike trains which are then transformed to 500ms, in order to normalise the parameters and simulation time steps.

The liquid responses from the network, which are shown in fig.4, were mapped into 25ms time-bins (20 time–bins/epoch). This particular setting resulted in an optimal accuracy for this experimental setup. There are two readout functions in this investigation. The first is none-adaptive Naivebayes whose Numeric estimator precision values are chosen based on analysis of the training data. The Second is Multi-Layered Perceptron(MLP) which utilizes 139 sigmoid nodes for hidden layer (number of input attributes plus 4 stimuli), 0.3 for learning rate, 0.2 for momentum, 500 training iterations, and validate threshold was set to 20.

In conventional method, parameters of Naivebayes and MLP were setup in the same way as the proposed method but MLP included only 68 sigmoid nodes for hidden layer (64 input plus 4 stimuli).

The state time-bins from 1st to 9th epoch were used for training set (equivalent to 80%) and 10th to 11th epoch were used as test set. From table 2, it can be seen that the traditional classifiers do not perform optimally on the raw EEG data. However, when the raw EEG is applied with BSA spike encoder and is passed through epSNNr with various stochastic models and classifiers such as Naivebayes and MLP, they perform especially well. For our experiment we had considered various other classifiers but they were found to be inappropriate due to their inability to handle complex spatio-temporal EEG data.

The accuracy obtained from epSNNr that utilizes Naivebayes have the same result for all the neuronal models, however the root mean squared error (RMSE) values (see Fig.5.) shows significant difference particularly for the ST model with Naivebayes, which is found to be the lowest, signifying the highest performance and stability in comparison with deterministic and other probabilistic models for this experiment.

Our main results prove that transforming EEG signals into spike trains using the BSA spike encoding scheme results is significantly higher classification accuracy. A second result is that using a stochastic neuronal model in the epSNNr (e.g. the ST model) may lead to an improved accuracy (see the classification results for the MLP classifier from Table 2 and the root mean square error results from Fig.5).

5   Conclusion and Future Works

In this study, we have shown that BSA spike encoding scheme is suitable for encoding EEG data stream. Moreover, we have also addressed the question whether probabilistic neural models are principally suitable liquids in the context of LSM. We have experimentally shown that, the performance of the epSNNr improves when a probabilistic model of a neuron is used, compared to the use of a deterministic LIF model of a neuron, and the classification performance of the EEG data in an epSNNr depends on the type of the probabilistic neuronal model used. The results suggest that an epSNNr can be optimised in terms of neuronal models used and parameters that would better match the noise and the dynamics of EEG data. Moreover, previous researches have had never incorporated 64 EEG channels. Our results have indicated potential advantages of using epSNNr along with the viability of BSA encoding scheme for EEG data streams that are spatio-temporal in nature which may contribute to BCI and medical studies. However, noise reduction and/or feature extraction methods, and optimization algorithm could also be employed possibly for both local and global optimization in our future study, since various parameters in the framework need to be adjusted.

However, further study on the behavior of the epSNNr architecture under different conditions is needed and more experiments are required to be carried out on EEG datasets.

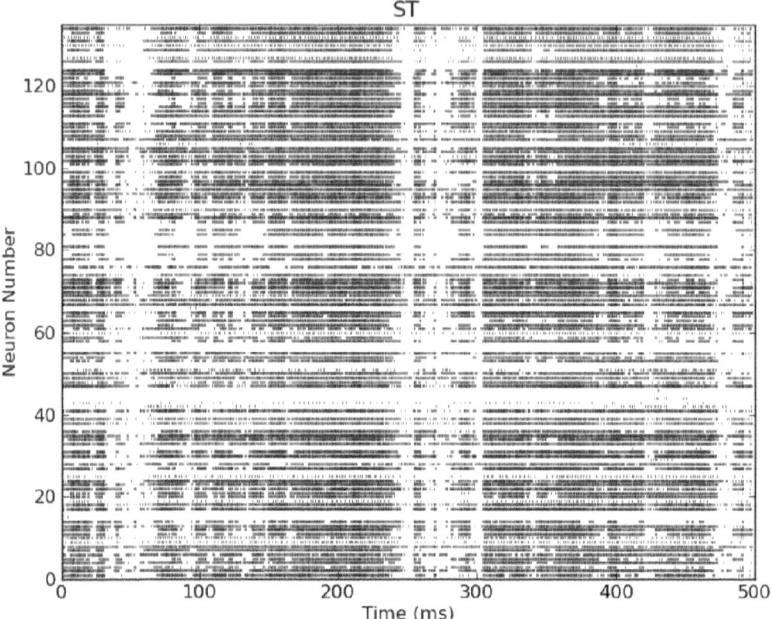

Fig. 4. The reservoir using ST Model response of one epoch of auditory stimulus is shown, where x axis represents time in 500ms and y axis is neurons

Table 2. The following table provides Classification Accuracy (%), for various methods

Methods	Without Reservoir	With Reservoir			
(Classifiers)	Accuracy	LIF Model	NR Model	ST Model	CT Model
Naivebayes	66.90 %	75.00 %	75.00 %	75.00 %	75.00 %
MLP	64.87 %	50.00 %	50.00 %	75.00 %	50.00 %

Several methods will be investigated for the improvement of the epSNNr: Using dynamic selection of the 'chunk' of input data entered into the epSNNr; A new algorithm for an evolving (adaptive) learning in the epSNNr will be developed so that the reservoir learns to discriminate better states that represent different class data. Using more complex probabilistic spiking neuron models, such as [11], would require dynamic optimization of its probabilistic parameters. We intend to use a gene regulatory network (GRN) model to represent the dynamics of these parameters in relation to the dynamics of the spiking activity of the epSNNr as suggested in [12]. Each of the probability parameter, the decay parameter, the threshold and other parameters of the neurons, will be represented as a function of particular genes for a set of genes related to the epSNN model, all genes being linked together in a dynamic GRN model. Furthermore, various parameters such as the connection probability, size and shape of the network topology shall also be tested. In this respect the soft winner-take-all topology will be investigated [13]. For applications that require on line training we intend to use evolving SNN classifier [14], [15]. Finally, implementation of the developed models on existing SNN hardware [16], [17] will be studied especially for on-line learning and object recognition applications such as intelligent mobile robots [18].

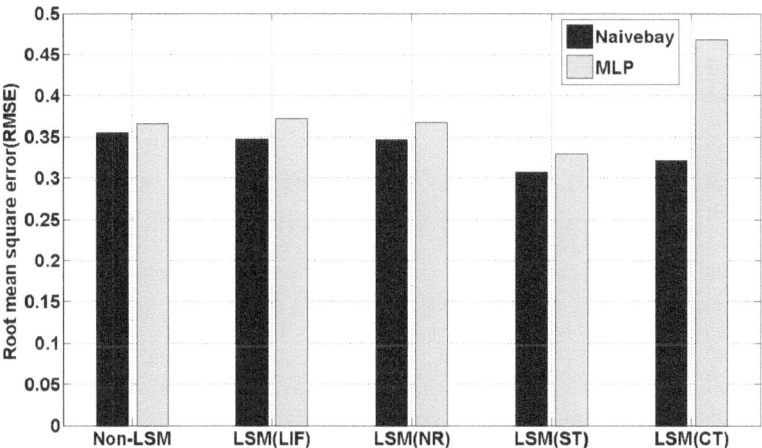

Fig. 5. This figure shows the Root Mean Squared Error (RMSE) for various stochastic neuron models when applied to Naivebayes and MLP classifiers

Acknowledgment. The EEG data used in the experiments were collected in the RIKEN Brain Science Institute, Tokyo by a team lead by Case van Leuven and Andrjei Chihotsky. We developed a software simulator using the software environment Brian [18]. The work on this paper has been supported by the Knowledge Engineering and Discovery Research Institute (KEDRI, www.kedri.info), Auckland University of Technology. N.K. has been supported by a one year Marie Curie International Incoming Fellowship within the 7th European Framework Programme under the project 'EvoSpike', hosted by the Institute for Neuroinformatics at the University of Zurich and ETH Zurich.

References

1. Tatum, W.: Handbook of EEG interpretation. Demos Medical Publishing (2007)
2. Niedermeyer, E., Da Silva, F.: Electroencephalography: basic principles, clinical applications, and related fields. Lippincott Williams & Wilkins (2005)
3. Marcel, S., Millán, J.: Person authentication using brainwaves (EEG) and maximum a posteriori model adaptation. IEEE Transactions on Pattern Analysis and Machine Intelligence, 743–752 (2007)
4. Palaniappan, R., Mandic, D.: EEG based biometric framework for automatic identity verification. The Journal of VLSI Signal Processing 49(2), 243–250 (2007)
5. Ghosh-Dastidar, S., Adeli, H.: A new supervised learning algorithm for multiple spiking neural networks with application in epilepsy and seizure detection. Neural Networks 22(10), 1419–1431 (2009)

6. Buteneers, P., Schrauwen, B., Verstraeten, D., Stroobandt, D.: Real-time epileptic seizure detection on intra-cranial rat data using reservoir computing. In: Advances in Neuro-Information Processing, pp. 56–63 (2009)

7. Schliebs, S., Nuntalid, N., Kasabov, N.: Towards Spatio-Temporal Pattern Recognition Using Evolving Spiking Neural Networks. In: Wong, K.W., Mendis, B.S.U., Bouzerdoum, A. (eds.) ICONIP 2010, Part I. LNCS, vol. 6443, pp. 163–170. Springer, Heidelberg (2010)

8. Schrauwen, B., Van Campenhout, J.: BSA, a fast and accurate spike train encoding scheme. In: Proceedings of the International Joint Conference on Neural Networks, vol. 4, pp. 2825–2830. IEEE (2003)

9. Maass, W., Natschläger, T., Markram, H.: Real-time computing without stable states: A new framework for neural computation based on perturbations. Neural Computation 14(11), 2531–2560 (2002)

10. Goodman, D., Brette, R.: The Brian simulator. Frontiers in Neuroscience 3(2), 192 (2009)

11. Kasabov, N.: To spike or not to spike: A probabilistic spiking neuron model. Neural Networks 23(1), 16–19 (2010)

12. Kasabov, N., Schliebs, R., Kojima, H.: Probabilistic Computational Neurogenetic Framework: From Modelling Cognitive Systems to Alzheimer's Disease. IEEE Transactions on Autonomous Mental Development (2011)

13. Rutishauser, U., Douglas, R., Slotine, J.: Collective Stability of Networks of Winner-Take-All Circuits. Neural Computation, 1–39 (2011)

14. Kasabov, N.: Evolving connectionist systems: the knowledge engineering approach. Springer-Verlag New York Inc. (2007)

15. Schliebs, S., Kasabov, N., Defoin-Platel, M.: On the Probabilistic Optimization of Spiking Neural Networks. International Journal of Neural Systems 20(6), 481–500 (2010)

16. Indiveri, G., Stefanini, F., Chicca, E.: Spike-based learning with a generalized integrate and fire silicon neuron. In: Proceedings of 2010 IEEE International Symposium on Circuits and Systems (ISCAS), pp. 1951–1954. IEEE (2010)

17. Indiveri, G., Chicca, E., Douglas, R.: Artificial cognitive systems: From vlsi networks of spiking neurons to neuromorphic cognition. Cognitive Computation 1(2), 119–127 (2009)

18. Bellas, F., Duro, R., Faina, A., Souto, D.: Multilevel Darwinist Brain (MDB): Artificial Evolution in a Cognitive Architecture for Real Robots. IEEE Transactions on Autonomous Mental Development 2(4), 340–354 (2010)

A New Learning Algorithm for Adaptive Spiking Neural Networks

J. Wang, A. Belatreche, L.P. Maguire, and T.M. McGinnity

Intelligent Systems Research Centre (ISRC)
Faculty of Computing and Engineering
University of Ulster, Magee Campus
Northland Road BT48 7JL Derry, United Kingdom, UK
Wang-J1@email.ulster.ac.uk,
{A.Belatreche,LP.Maguire,TM.McGinnity}@ulster.ac.uk

Abstract. This paper presents a new learning algorithm with an adaptive structure for Spiking Neural Networks (SNNs). STDP and anti-STDP learning windows were combined with a 'virtual' supervisory neuron which remotely controls whether the STDP or anti-STDP window is used to adjust the synaptic efficacies of the connections between the hidden and the output layer. A simple new technique for updating the centres of hidden neurons is embedded in the hidden layer. The structure is dynamically adapted based on how close are the centres of hidden neurons to the incoming sample. Lateral inhibitory connections are used between neurons of the output layer to achieve competitive learning and make the network converge quickly. The proposed learning algorithm was demonstrated on the IRIS and the Wisconsin Breast Cancer benchmark datasets. Preliminary results show that the proposed algorithm can learn incoming data samples in one epoch only and with comparable accuracy to other existing training algorithms.

Keywords: spiking neurons; supervised learning; spike response model; online learning; offline learning, adaptive structure, classification.

1 Introduction

Inspired by the human brain, artificial neural networks (ANNs) acquire information from the surrounding environment and through learning store the knowledge in synaptic weights [1]. The main interest for studying SNNs lies in their close resemblance with biological neural networks. While significant progress has already been made in understanding neuronal dynamics, considerably less has been achieved in developing efficient spiking neural learning mechanisms. To date, a number of supervised and unsupervised learning methods have been developed; unfortunately, most of them do not scale up and require retraining in a continuously changing environment. The development of efficient learning approaches for SNNs is crucially important in order to increase their applicability to solve real world problems and create intelligent system that are capable of

B.-L. Lu, L. Zhang, and J. Kwok (Eds.): ICONIP 2011, Part I, LNCS 7062, pp. 461–468, 2011.

handling continuous streams of information, scaling up and adapting to continuously changing environments.

SpikeProp was an adaptation of the classical backpropagation algorithm. Performance on several benchmark datasets demonstrated that SNNs with fast temporal coding can achieve comparable results to rate-coded networks [2]. But it is slow if used in an online setting. Belatreche et al. [3] proposed a derivative-free supervised learning algorithm. An evolutionary strategy (ES) was used to minimise the error between the output firing times and the corresponding desired firing times. This algorithm achieved a better performance than the SpikeProp algorithm, however, since the algorithm was an ES-based iterative process, the training procedure was extremely time-consuming and is not suitable for online learning.

Legenstein et al. [4] presented a Supervised-Hebbian learning method (SHL) that enabled different transformation from input spike trains to output spike trains quite well although parameters continued to be changed even if the neuron had already fired at the particular points in time. ReSuMe developed by Ponulak [5] integrated the idea of learning-windows with the novel concept of remote supervision. It was shown that the desired temporal sequences of spikes can be learned efficiently by this method with a LSM (Liquid State Machine) network. The authors claim that since the synaptic weights are updated in an incremental manner the method is suitable for online processing. Desired precise output spike timing is crucial to ReSuMe learning. Glackin [6] presented a three layer feedforward FSNN (Fuzzy SNN) topology for classification and comparable results with other existing approaches were produced. The authors claim ReSuMe training was employed between the hidden neurons and output neurons, however it is not the precise spike times that are relevant instead the number of spikes of output neurons are important.

Fixed network architectures were employed in the above approaches, with the number of neurons in the hidden layer is set in advance. However for maximum adaptability, as is needed in a changing environment, it is desirable that the structure as well as the weights adapts to new data.

This paper provides a new learning procedure with an adaptive structure for classification using a SNN. STDP and anti-STDP learning windows were integrated with a 'virtual' supervisory neuron that controls remotely whether the STDP or anti-STDP window is used and the synaptic efficacies are adjusted between the hidden layer and output layer. No actual supervisory spike train existed. A simple new technique for updating the centres of hidden neurons is embedded in the hidden layer. The structure of the hidden layer is dynamically adapted based on how close to the centers of hidden neurons are the incoming sample. Lateral inhibitory connections are used between neurons of the output layer to achieve competitive learning and make the network converge quickly.

The remainder of this paper is structured as follows: Section 2 describes the structure design and the learning procedure. Section 3 presents preliminary test results for training of SNNs using the IRIS and Wisconsin Breast Cancer Benchmark datasets. Finally section 4 concludes the paper.

2   Learning and Structural Adaptation

2.1   Network Topology and Simulation Setup

Figure 1 presents the proposed network topology that consists of a three layer feedforward SNN. There are no neurons in the hidden layer initially, as they are added dynamically as the incoming data is received. Every input neuron represents a feature in each instance, every output neuron represents a class. There are lateral inhibitory connections in the output layer neurons.

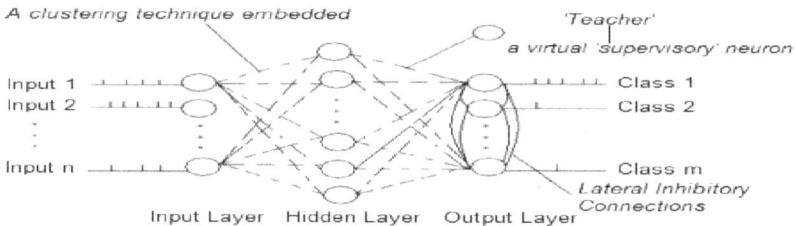

Fig. 1. A three layer feedforward SNN with adaptive structure

There is a vector representing the centre of every added hidden neuron. A new simple technique for updating the centres of the hidden neurons is embedded in the hidden layer, and the structure of the hidden layer is dynamically adapted based on how close the incoming information is to the centres of the hidden neurons. The weights between the input layer and hidden layer remains fixed once initialised. The centres of the hidden neurons are continuously adjusted (i.e. learned) as the samples are propagated into the network.

The weights between the hidden layer and output layer are learned in a supervised mode using equation (1) . The two opposite learning processes STDP and anti-STDP are combined together to update the relevant synapses. Figure 2 shows the learning windows of STDP and anti-STDP. The STDP window is used to increase the weights when a spike is produced at the desired output neuron and the anti-STDP window is used to decrease the weights when a spike is produced at the undesired neuron. The 'teacher' controls just whether STDP or anti-STDP window is used, no actual supervisory spike train exists. Rate decoding are employed and the sample is treated as correctly classified when the maximum number of spikes are produced at the correct output class neuron.

$$
\begin{aligned}
W^d(s^d) &= \begin{cases} +A^d * \exp(\frac{-s^d}{\tau^d}) & \text{if } s^d > 0 \\ 0 & \text{if } s^d \leq 0 \end{cases} \\
W^l(s^l) &= \begin{cases} -A^l * \exp(\frac{-s^l}{\tau^l}) & \text{if } s^l > 0 \\ 0 & \text{if } s^l \leq 0 \end{cases}
\end{aligned}
\tag{1}
$$

In Equation 1 above s^d denotes the relative timing of the post-synaptic spike produced on desired output neuron and pre-synaptic spikes, s^l denotes the relative timing of the post-synaptic spike produced on undesired output neuron and

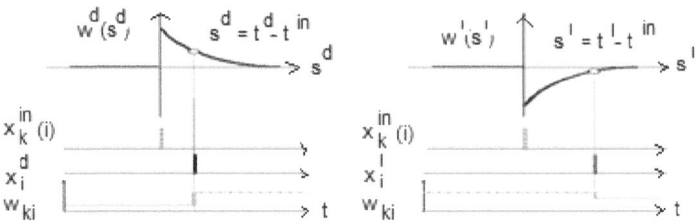

Fig. 2. Learning windows for STDP and anti-STDP [5]

pre-synaptic spikes. A^d and A^l determine the maximum amounts of synaptic modification. τ^d and τ^l determine the range of pre- to postsynaptic inter spike intervals over which synaptic strengthening and weakening occur.

Adding lateral inhibitory connections is the simplest way to ensure competitive learning of neurons. When lateral inhibitory connections are added between output layer neurons, the spikes produced by one output neuron inhibit the other output neurons. So the speed of convergence is greatly improved.

2.2   Learning Procedure

The following sequential steps describe this new learning procedure.

1. The first data sample $X_1 = [x_{11}, x_{12}, , x_{1m}]$ is normalized to the range [10 40]. This range is chosen arbitrarily, and m is the dimension of the dataset; Poissonian encoding is employed and is the encoded data is passed for training into the hidden layer. The first hidden neuron ($Num\_neuron\_H = 1$) is instantiated and its weights initialised. Then the weights relevant to this neuron, and the weights between this neuron and the neurons in the output layer are trained using the described method. The weights between the input neurons and hidden neuron will be kept the same after their initialisation. The centre of the first hidden neuron is given by the vector $C_i = \langle x_{1i}, x_{2i}, \cdots x_{mi} \rangle$, and $x_{1i} = x_{11}, \cdots, x_{mi} = x_{1m}$. Set $N_{sample}(i) = 1$, i=1 represents the first hidden neuron.

2. A training sample (n^{th} observation, $n > 1$)$X_n = [x_{n1}, x_{n2}, \cdots, x_{nm}]$ is normalised to the range [10 40]. Poissonian encoding is employed and is the encoded data is passed for training into the hidden layer. The distance between this sample and the centre C_i (i is in the range of $[1, Num\_neuron\_H]$) of every hidden neuron using the Matlab Euclidean distance weight function (dist) is calculated and we find the hidden neuron (i) that has minimum distance (D_min) with this sample. The minimum distance (D_min) with a chosen distance value of threshold (Threshold) is compared.

3. If D_min is less than the Threshold, this sample will be routed to the hidden neuron that has minimum distance with this sample. The weights between this hidden neuron and the neurons in the output layer will be trained using

the described method. The centre of this hidden neuron will be updated based on equation 2; $N_{sample}(i)$ will be updated by increasing one.

$$X_{1i} = \frac{X_{1i} * N_{sample}(i) + X_{n1}}{N_{sample}(i) + 1}, \cdots, X_{mi} = \frac{X_{mi} * N_{sample}(i) + X_{nm}}{N_{sample}(i) + 1} \quad (2)$$

4. But if D_min is greater than the Threshold, a new hidden neuron (the jth hidden neuron) in the hidden layer will be added $(Num\_neuron\_H = Num\_neuron\_H + 1)$, the weights relevant to this new added neuron will be initialised, the network structure will be updated, and the weights between the added hidden neuron and output neurons will be trained using the described method. The centre of the added hidden neuron is given by the vector $C_j = \langle x_{1j}, x_{2j}, \cdots x_{mj} \rangle$, and $x_{1j} = x_{n1}, \cdots, x_{mj} = x_{nm}$. Set $N_{sample}(j) = 1$.

3   Experiments and Results

The proposed learning procedure has been applied to a spiking network of SRM (Spike Response Model) neurons as shown in figure 1. The IRIS and Wisconsin Breast Cancer datasets were used to evaluate the performance of the proposed procedure and the results obtained have been compared with previous work.

3.1   IRIS Dataset Classification

Data Preparation. The IRIS dataset consists of 150 instances with four features, there are 50 instances for each class. Four input neurons represent four features in each instance, three output neurons represent three classes. The dataset was partitioned for cross-validation to test for generalisation. Five folds of data were formed in such a way that there were 10 data samples for class 1, class 2 and class 3 respectively in each fold. Four folds of data samples were employed for training, and one fold for testing.

Simulation Results. Figure 3 shows the change of number of hidden neurons during training with threshold set at 0.4 (left) and 0.8 (right). Table 1 presents the IRIS results for all 5 folds with the distance value of threshold set at 0.8 and 0.4. Table 2 compares the proposed method with other algorithms.

Table 1. IRIS Results (*iteration* = 1)

Threshold	Fold	1	2	3	4	5	AVG	SD
	Number of hidden neurons	13	14	14	13	15		
0.8	Accuracy for Trainingset(%)	95.8	97.5	98.3	95.8	95.0	96.5	1.4
	Accuracy for Testingset (%)	100	93.3	96.7	93.3	96.7	96.0	2.8
	Number of hidden neurons	49	48	50	45	48		
0.4	Accuracy for Trainingset(%)	96.7	98.3	99.2	97.5	97.5	97.8	0.9
	Accuracy for Testingset(%)	100	96.7	93.3	100	96.7	97.3	2.8

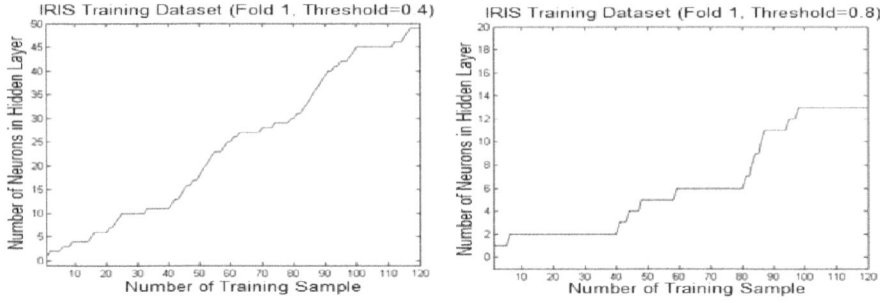

Fig. 3. Evolution of the number of RBF hidden neurons during training

Table 2. Compared the proposed method with other algorithms

Algorithm	Trainingset	Testingset	Properties
SpikeProp	97.4%+/- 0.1	96.1%+/- 0.1	Supervised,batch,fixed structure
FSNN	97.8%	95.0%	Supervised,batch,fixed structure
Proposed method	97.8%+/- 0.9	97.3%+/- 2.8	Hybrid,one epoch,adaptive(Th=0.4)
	96.5%+/- 1.4	96.0%+/- 2.8	Hybrid,one epoch,adaptive(Th=0.8)

3.2   Wisconsin Breast Cancer Dataset Classification

Data Preparation. The Wisconsin Breast Cancer dataset consists of 699 instances with 9 features with values in the range $[1, 10]$. Nine input neurons represent nine features in each instance, two output neurons represent two classes. The dataset was partitioned for cross-validation to test for generalisation. Five folds of data were formed in such a way that there were 88 class 1 and 47 class 2 data samples in each fold. Four folds of data samples were employed for training, and one fold of data samples were employed for testing.

Simulation Results. Figure 4 shows the change of number of hidden neurons during training with threshold set at 10.77 (left) and 10 (right). Table 3 presents the Wisconsin results for all 5 folds with the distance value of threshold set at 10.77 and 10. Table 4 compares the proposed method with other algorithms.

Table 3. Wisconsin Results ($iteration = 1$)

Threshold	Fold	1	2	3	4	5	AVG	SD
	Number of hidden neurons	6	6	6	6	7		
10.77	Accuracy for Trainingset(%)	97.4	97.2	97.2	97.0	97.2	97.2	0.1
	Accuracy for Testingset(%)	98.6	97.9	97.9	95.8	94.4	96.9	1.8
	Number of hidden neurons	10	11	10	8	11		
10	Accuracy for Trainingset(%)	97.2	97.0	97.4	97.4	97.6	97.3	0.2
	Accuracy for Testingset(%)	98.6	98.6	97.2	95.8	95.8	97.2	1.4

Table 4. Compared the proposed method with other algorithms

Algorithm	Trainingset	Testingset	Properties
SpikeProp	97.6%+/- 0.2	97.0%+/- 0.6	Supervised,batch,fixed structure
FSNN	97.6%	95.3%	Supervised,batch,fixed structure
Proposed method	97.3%+/- 0.2	97.2%+/- 1.4	hybrid,one epoch adaptive(Th=10)
	97.2%+/- 0.1	96.9%+/- 1.8	hybrid,one epoch adaptive(Th=10.77)

3.3   Analysis of Results

Preliminary experiments have shown that the network employing this learning procedure, with only one pass training, can reach similar levels of performance as compared with previously presented work that employed batch training. This new algorithm is a one pass training, but previously acquired knowledge is able to be retained, it would not be required to access previously used data if subsequent training sessions were conducted. As a cost, one additional parameter (e.g. Threshold) needs to be tuned. The learning for hidden neurons is performed independently of the class of the presented sample, the smaller the value of threshold, the more the number of hidden neuron grow and a better performance of the network for training datasets is attained. How to choose a value for the threshold will depend on the demanded size of the obtained network and the acceptable level of performance required. A smaller threshold can correctly classify two input patterns that become the closest to the centre of one hidden neuron, however more hidden neurons would be added. Although a bigger threshold can be selected to limit the number of hidden neurons that will be added to the network; however it is very hard to obtain the desired performance by just varying the threshold. For large datasets, many added hidden neurons may affect the learning speed, the resultant network may be over-trained as a number of the hidden neurons may be redundant. Embedded pruning process would solve this problem. In the learning processes, it is possible that some centres of hidden neurons are very close to each other, the hidden neuron is pruned to reduce the complexity of the resulting network. The removal of unnecessary hidden neurons may enable knowledge to be extracted more easily, and reduce the number of free parameters which in turn, improve learning speed. It's especially important for online system.

For both the IRIS and Wisconsin Breast Cancer datasets, we use five-fold cross-validation, the results also show very good stability of the learning algorithm.

4   Conclusion

This paper presents a simple training algorithm that is suitable for online learning for spiking neural networks which use spike trains to encode input data. It was shown that the proposed algorithm can benchmark datasets in one pass using a simple three layer feedforward spiking neural network with a dynamically adaptive hidden layer. A new simple clustering technique was used for dynamically adapting the hidden layer structure. In addition, lateral inhibitory connections were used between the output neurons for competitive learning.

The same procedure works on the IRIS and Wisconsin Breast Cancer datasets. The algorithm determines the number of hidden neurons during training stage dynamically. The primary goal of this research is to develop an efficient and scalable online learning strategies for biologically plausible spiking neural networks based on synaptic plasticity. In that direction, this work presents a significant contribution. In this paper, because the benchmark datasets were used for simulation, the value of Threshold is chosen assuming that the whole training dataset is known. But in a continuously changing environment, the data is presented temporally. In order to meet this requirement and set this value efficiently, we intend to employ a small value of Threshold at the beginning and adapt the value of Threshold during training based on the performance and also a pruning process will be embedded during training. Future work will demonstrate these online capabilities of the algorithm using time varying and dynamic datasets.

References

1. Haykin, S.: Neural Networks: A Comprehensive Foundation. Macmillan College Publishing, New York (1999)
2. Bohte, S.M., Kok, J.N., Poutre, H.L.: Error-backprogation in Temporally Encoded Networks of spiking neurons. In: Neurocomputing 48, 17–37 (2002)
3. Belatreche, A., Maguire, L.P., McGinnity, M., Wu, Q.: A Method for Supervised Training of Spiking Neural Networks. In: Proceedings of IEEE Cybernetics Intelligence - Challenges and Advances (CICA), Reading, UK, pp. 39–44 (2003)
4. Legenstein, R., Naeger, C., Maass, W.: What can a Neuron Learn with Spike-Timing-Dependent Plasticity? Neural Computation 17, 2337–2382 (2005)
5. Ponulak, F., Kasinski, A.: A novel approach towards movement control with Spiking Neural Networks. In: 3rd International Symposium on Adaptive Motion in Animals and MachinesIlmenau (2005)
6. Glackin, C., McDaid, L., Maguire, L., Sayers, H.: Implementing Fuzzy Reasoning on a Spiking Neural Network. In: Kůrková, V., Neruda, R., Koutník, J. (eds.) ICANN 2008, Part II. LNCS, vol. 5164, pp. 258–267. Springer, Heidelberg (2008)

Axonal Slow Integration Induced Persistent Firing Neuron Model

Ning Ning, Kaijun Yi, Kejie Huang, and Luping Shi

Data Storage Institute,
5 Engineering Drive 1, Singapore 117608
`ning_ning@dsi.a-star.edu.sg`

Abstract. We present a minimal neuron model that captures the essence of the persistent firing behavior of interneurons as discovered recently in the field of Neuroscience. The mathematical model reproduces the phenomenon that slow integration in distal axon of interneurons on a timescale of tens of seconds to minutes, leads to persistent firing of axonal action potentials lasted for similar duration. In this model, we consider the axon as a slow leaky integrator, which is capable of dynamically switching the neuronal firing states between normal firing and persistent firing, through axonal computation. This model is based on the Izhikevich neuron model and includes additional equations and parameters to represent the persistent firing dynamics, making it computationally efficient yet bio-plausible, and thus well suitable for large scale spiking network simulations.

Keywords: neuron model, slow integration, persistent firing, spiking network, axonal computation.

1 Introduction

In the classic viewpoint about the information flow in the nervous system, synaptic inputs are received and integrated in the dendrites on a timescale of milliseconds to seconds, and when the depolarized somatic membrane potential exceeds the threshold, action potentials are triggered at the axon hillock and propagate along the axon. However, this convention view of neuron was challenged by the recent findings [21] of slow integration induced persistent firing in distal axons of rodent hippocampal and neocortical interneurons. It was found that the slow integration from tens of seconds to minutes in distal axon, leads to persistent firing of action potentials lasted for similar duration [21]. To trigger such persistent firing, axonal action potential firing was required but somatic depolarization was not, implying that axon may perform its own neural computations without any involvement from soma or dendrites. Interestingly, in paired recording with one neuron being stimulated, persistent firings were observed in the other unstimulated cell, suggesting that axons could communicate with each other, although such communication mechanism remains unknown yet.

The possible functions of the persistent firing were suggested to be related to working memory [21]. The ability to maintain the persistent firing of action

B.-L. Lu, L. Zhang, and J. Kwok (Eds.): ICONIP 2011, Part I, LNCS 7062, pp. 469–476, 2011.

potentials without on-going stimulation provides a mechanism of storing the information for a short period of time. This mechanism is similar to our working memory [1,5,6], which actively holds a limited amount of information [16] in the absence of stimuli. For example, when someone tells you a telephone number and you try to remember it, the working memory keeps the number series on-line in your head before you dial it, for a period of several seconds to tens of seconds without rehearsal [18]. Working memory has been extensively explored from perspectives of highly abstract top levels in the domains of psychology, neuroscience and anatomy, but there are much less works from perspectives of bottom level of biological neurons [4,9,15,23]. It is still unknown how working memory is represented within a population of cortical neurons.

To understand the neural correlative of working memory, we need to combine the experimental studies of animal and human nervous system with numerical simulation of the animal or human brain based on bio-plausible spiking neurons. Over the past decades, there are many biologically inspired neural model proposed and utilized in large scale brain simulations [2,14,13,8]. Depending on the level of abstraction, there are roughly two kinds of neuron models. One of them focuses on detailed and complex description of channel dynamics and bio-physics basis, for example, the Hodgkin-Huxley model [10] and compartment models [3]. The second kind of model focuses on capture the spiking nature and essential elements of the behavior with simplified complexity, for example, leaky integrated and fire model and spiking neuron models such as Izhikevich [12], Wilson [25], Hindmarsh-Rose [19], Morris-Lecar [17], FitzHugh-Nagumo [7], and Resonate-and-Fire [11] models. Among these models, Izhikevich neuron model is of particular interest for large-scale neural simulation as the model is capable to reproduce rich spiking and bursting behaviors and computationally efficient.

In this paper, we present a new neuron model that captures the essence of the persistent firing behavior of neurons, and remains to be computationally efficient. This persistent firing neuron model is based on the Izhikevich model and includes additional equations and parameters to represent the persistent firing dynamics. Section 2 describes the mathematical model of persistent firing neurons. Section 3 describes the simulation results of the neuron model with two stimulation protocols. In the final section we conclude and discuss the implications of this work.

2   Persistent Firing Neuron Model

This persistent firing neuron model is based on the Izhikevich model, which can be described by the equations (1)-(3) [12]:

$$v' = 0.04v^2 + 5v + 140 - u + I \tag{1}$$
$$u' = a(bv - u) \tag{2}$$

with the after-spike resetting:

$$\text{if } v \geq 30mV, \text{then } \begin{cases} v \leftarrow c, \\ u \leftarrow u + d \end{cases} \tag{3}$$

where v and u are the membrane potential and membrane recovery variable, respectively. $' = d/dt$, where t is the time, a describes the time scale of u, b represent the sensitivity of u to the subthreshold fluctuations of v, c is the after-spike reset value of membrane potential, and d describes the after-spike reset of u.

By modifying the a, b, c, d variables in equations (1) - (3), different firing patterns can be generated. Thus, it is possible to define different firing states of a neuron, for example, normal firing or persistent firing. One important concept of our model is to dynamically switch the neuronal firing states by selecting different variable set of a, b, c, and d on the fly.

The concept of the neuron in the model is illustrated in Fig. 1. We consider the axon as a slow leaky integrator, which is capable to alter the neuronal firing states through its own computation. We can model the persistent firing behavior as the axonal leaky integrator controlled switching of normal firing and persistent firing states of a neuron. In the normal firing state, the neuron dynamics are described by the equations (1) - (3). In most cases, stimulus is required to trigger and maintain the firing activities. In the persistent firing state, the parameters of the model are chosen such that the neuron is in an oscillatory state, in which the neuron can keep firing action potentials without the presence of any stimulus. This may explain the in the persistent firing phenomenon as observed in Sheffield's experiment [21], the firings of axonal action potentials can be sustained without any dendritic inputs.

Mathematically, the model is described by equations (1), (2), (4) - (7).

$$w' = -fw \tag{4}$$

$$\text{if } v \geq 30mV, \quad \text{then} \begin{cases} v \leftarrow c, \\ u \leftarrow u + d, \\ w \leftarrow w + e \end{cases} \tag{5}$$

$$\text{if } w \geq w_p, \quad \text{then } (a, b, c, d, e) \leftarrow (a, b, c, d, e)_p \tag{6}$$

$$\text{if } w \leq w_n, \quad \text{then} \begin{cases} (a, b, c, d, e) \leftarrow (a, b, c, d, e)_n, \\ w \leftarrow w_n \end{cases} \tag{7}$$

where w is dimensionless variable that describes the potential of the axonal leaky integrator, e describes the after-somatic-spike axonal accumulation in the normal firing state, f describes the rate of the leak, w_p is the upper threshold value of w to trigger the persistent firing, w_n is the lower threshold value of w to stop the persistent firing, $(a, b, c, d, e)_p$ and $(a, b, c, d, e)_n$ describe the parameter sets of a, b, c, d, e variables in persistent firing state and in normal firing state, respectively.

In contrast to the somatic leaky integrator which accounts for the integration of dendritic inputs, the axonal leaky integrator has a larger time constant for its integration and leakage, as described by e and f. Therefore, the axon integrates the incoming spikes generated in the axon hillock on a larger timescale, from

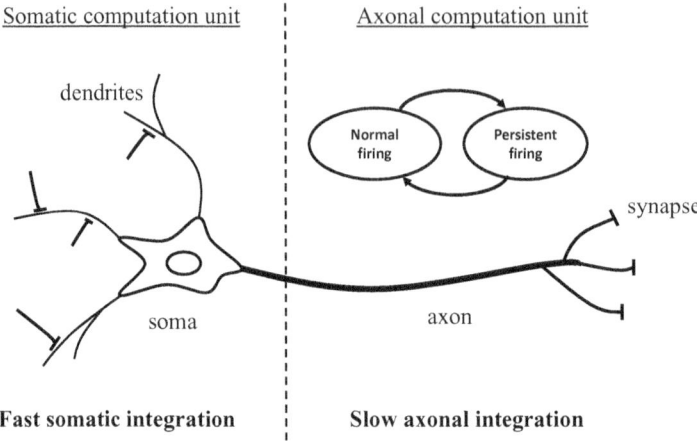

Fig. 1. The concept of neuron in the persistent firing neuron model. There are two computational units, *somatic computation unit* and *axonal computation unit*, accounting for fast and slow integrations, respectively.

tens of seconds to minutes. When the potential in the axonal leaky integrator exceeds the axonal threshold of persistent firing, w_p, persistent firings will be triggered, and the neuron will be in an oscillatory state as determined by the $(a, b, c, d, e)_p$ variables.

When the neuron is in the persistent firing state, if there is no somatic spikes accumulated in the axon, the potential w decreases at the rate of f due to the leaky nature of the axonal integrator. When w reaches the lower threshold, w_n, the neuron will return to the normal firing state, and the a, b, c, d, e variables will be reset to $(a, b, c, d, e)_n$.

3   Simulation Results

Using our model, the persistent firing neuronal behaviors can be simulated with various stimulation protocols. We have implemented a MATLAB program based on equations (1), (2), (4) - (7). The parameters used in the simulation is summarized in Table 1. As the persistent firing behaviors were observed in the inhibitory interneurons [21], which exhibit the fast spiking firing pattern [12], we choose the parameters in the model to reproduce the fast spiking behavior for interneurons in the firing state. When the neuron is persistently firing, the chosen parameters set the neuron in oscillatory state.

The simulation results with step/pause stimulation protocol [21] are shown in Fig. 2. We applied 1-second current step of 15 mA during each 5-second sweep to the simulation model. During the sweeps of step/pause input stimulus, it is observed that the fast spiking pattern appears when the synaptic input stimulus is presented and disappears when the stimulus is removed. When w accumulates to a level higher than the upper threshold, w_p, persistent firing occurs, after

Table 1. Simulation parameters

Step/pause protocol (Fig. 2)	Long pulse protocol (Fig. 3)
$(a,b,c,d,e)_p = (0.1, 0.3, -85, 0, 0)$	$(a,b,c,d,e)_p = (0.1, 0.3, -85, 0, 0)$
$(a,b,c,d,e)_n = (0.1, 0.2, -65, 2, 0.02)$	$(a,b,c,d,e)_n = (0.1, 0.2, -65, 2, 0.012)$
$f = 5 \times 10^{-4}$	$f = 8 \times 10^{-4}$
$w_n = 0.2, w_p = 1.9$	$w_n = 0.2, w_p = 2.1$

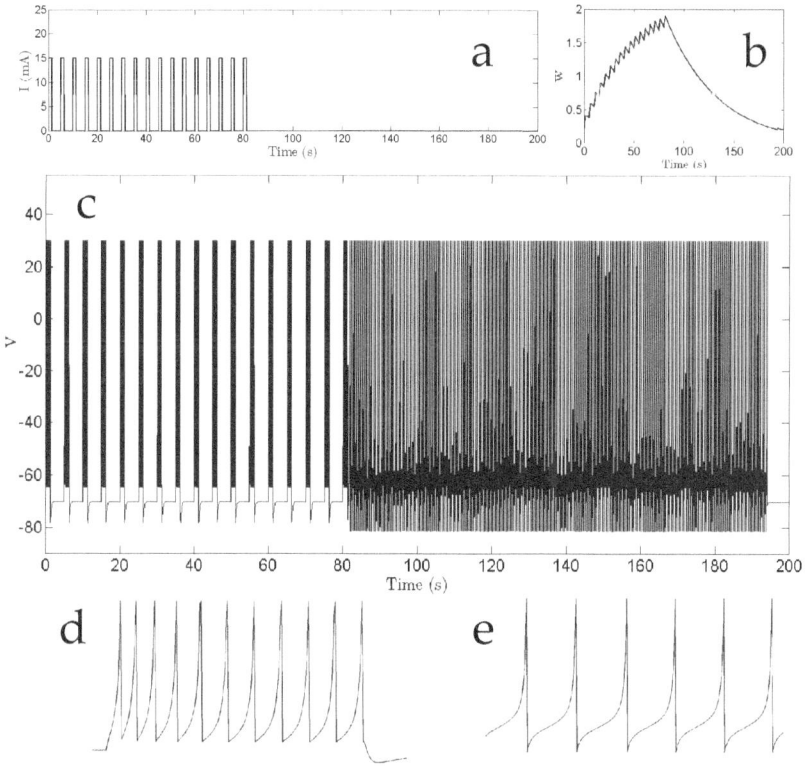

Fig. 2. Simulation of persistent firing neuronal behaviors with step/pause protocol. (a) The waveform of the input stimulation of *synaptic current I*, consisting of a pulse train with 1-second current step of 15 mA during each 5-second sweep. There are 17 sweeps in this run. (b) The time response of w, which represents the potential of the axonal leaky integrator. (c) The time response of v, which describes the membrane or axonal potential. Persistent firing is observed after the 17th sweep. (d) The fast spiking waveform of v as observed in (c) when the 15mA current step is present. (e) The oscillatory waveform of v when the neuron is in the persistent firing state.

which w decreases monotonically due to axonal leakage and no stimulus, as described in equation (4). The persistent firing ends when w is lower than the lower threshold of persistent firing, w_n. It should be noted that in this model, when the interneuron is in the persistent firing state, the resting potential is about 20 mV below the one for the fast spiking mode, which is intended to match the data from patch-clamp recordings in the persistent firing neurons [21]. We can emulate this phenomenon through switching the parameter c in equation (5) between c_p and c_n.

We also applied the stimulation with a 40-second long pulse of 15 mA current to the persistent firing neuron model with a different parameter set (see Table 1). The simulation results are shown in Fig. 3. The persistent firing is triggered shortly after input stimulus is removed, and lasts for more than a minute.

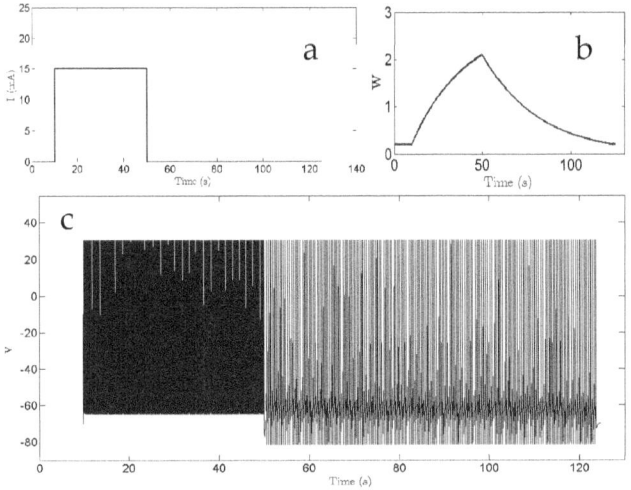

Fig. 3. Simulation of persistent firing neuronal behaviors with long pulse stimulation protocol. (a) The input stimulus. A pulse of current with 15mA amplitude lasts 40 seconds as the synaptic input to the neuron. (b) Time response of w, the variable describing the potential of the axonal leaky integrator. (c) Time response of v, which describes the membrane or axon potential. The waveforms of fast spiking and persistent spiking are the same as Fig. 2 (d) and (e), respectively.

By tuning the parameters of e, f, w_p and w_n, we can set different time scales for the axonal slow integration, allowing the model to accommodate different types of neurons with persistent firing behavior.

4   Discussion

In this paper, we present a minimal model that captures the essence of the persistent firing behavior of neurons. The mathematical model presented in this paper reproduces rich spiking behavior of biological persistent firing neurons through only three equations, two of which describe the Izhikevich neuron model. The model is computationally efficient yet bio-plausible, and thus well suitable for large scale spiking network simulation, enabling further investigation of possible functions of persistent firing and their roles in animal and human brain.

For example, we can study the relationship between working memory and persistent firing and neural correlative of working memory with computer simulation. By incorporating the persistent firing neuron model in spiking network simulations, we can simulate a more bio-plausible and realistic memory system in animal or human cortex. Recently, we proposed an initial framework of artificial cognitive memory with the objective of developing a novel function-driven memory technology in comparison to conventional density-driven storage technology [22]. The model presented in this paper can be used in the simulation of cognitive memory architectures under the framework of artificial cognitive memory.

Due to the computational simplicity of our model, it is rather straightforward to implement the model in hardware, either in digital circuits or analog circuits. There have been several circuit implementations of Izhikevich neuron model [20,24]. It is intuitive to add a leaky integrator emulating the axon in the circuits, for controlling the a, b, c, d, e parameters which may be stored in memory devices, e.g. non-volatile memory, or current sources in analog circuits. The silicon implementation of this neuron model and spiking network will enable a considerably faster emulation of the neural systems in a highly parallel manner.

Acknowledgments. The work was supported by Agency for Science, Technology, and Research (A*STAR), Singapore (Grant No: 0921570130).

References

1. Baddeley, A.: Working memory. Science 255(5044), 556 (1992)
2. Boahen, K.: Neuromorphic microchips. Scientific American 292(5), 56–63 (2005)
3. Brette, R., Rudolph, M., Carnevale, T., Hines, M., Beeman, D., Bower, J.M., Diesmann, M., Morrison, A., Goodman, P.H., Harris, F.C., et al.: Simulation of networks of spiking neurons: A review of tools and strategies. Journal of Computational Neuroscience 23(3), 349–398 (2007)
4. D'Esposito, M., Detre, J.A., Alsop, D.C., Shin, R.K., Atlas, S., Grossman, M.: The neural basis of the central executive system of working memory. Nature 378(6554), 279–281 (1995)
5. Durstewitz, D., Seamans, J.K., Sejnowski, T.J.: Neurocomputational models of working memory. Nature Neuroscience 3, 1184–1191 (2000)
6. Egorov, A.V., Hamam, B.N., Fransén, E., Hasselmo, M.E., Alonso, A.A.: Graded persistent activity in entorhinal cortex neurons. Nature 420(6912), 173–178 (2002)
7. Fitzhugh, R.: Impulses and physiological states in theoretical models of nerve membrane. Biophysical Journal 1(6), 445–466 (1961)
8. de Garis, H., Shuo, C., Goertzel, B., Ruiting, L.: A world survey of artificial brain projects part I: Large-scale brain simulations. Neurocomputing 74(1-3), 3–29 (2010)
9. Goldman-Rakic, P.S.: Cellular basis of working memory. Neuron 14(3), 477 (1995)

10. Hodgkin, A.L., Huxley, A.F.: A quantitative description of membrane current and its application to conduction and excitation in nerve. The Journal of Physiology 117(4), 500 (1952)
11. Izhikevich, E.M.: Resonate-and-fire neurons. Neural Networks 14(6-7), 883–894 (2001)
12. Izhikevich, E.M.: Simple model of spiking neurons. IEEE Transactions on Neural Networks 14(6), 1569–1572 (2003)
13. Izhikevich, E.M., Edelman, G.M.: Large-scale model of mammalian thalamocortical systems. Proceedings of the National Academy of Sciences 105(9), 3593 (2008)
14. Markram, H.: The blue brain project. Nature Reviews Neuroscience 7(2), 153–160 (2006)
15. Miller, E.K., Erickson, C.A., Desimone, R.: Neural mechanisms of visual working memory in prefrontal cortex of the macaque. The Journal of Neuroscience 16(16), 5154 (1996)
16. Miller, G.A.: The magical number seven, plus or minus two: some limits on our capacity for processing information. Psychological Review 63(2), 81 (1956)
17. Morris, C., Lecar, H.: Voltage oscillations in the barnacle giant muscle fiber. Biophysical Journal 35(1), 193–213 (1981)
18. Peterson, L., Peterson, M.J.: Short-term retention of individual verbal items. Journal of Experimental Psychology 58(3), 193 (1959)
19. Rose, R.M., Hindmarsh, J.L.: The assembly of ionic currents in a thalamic neuron i. the three-dimensional model. Proceedings of the Royal Society of London B Biological Sciences 237(1288), 267 (1989)
20. Van Schaik, A., Jin, C., Hamilton, T.J.: A log-domain implementation of the Izhikevich neuron model. In: IEEE International Symposium on Circuits and Systems, pp. 4253–4256 (2010)
21. Sheffield, M.E., Best, T.K., Mensh, B.D., Kath, W.L., Spruston, N.: Slow integration leads to persistent action potential firing in distal axons of coupled interneurons. Nature Neuroscience 14(2), 200–207 (2010)
22. Shi, L.P., Yi, K.J., Ramanathan, K., Zhao, R., Ning, N., Ding, D., Chong, T.C.: Artificial cognitive memory–changing from density driven to functionality driven. Applied Physics A: Materials Science & Processing 102(4), 865–875 (2011)
23. Szatmáry, B., Izhikevich, E.M.: Spike-timing theory of working memory. PLoS Computational Biology 6(8) (2010)
24. Wijekoon, J.H., Dudek, P.: Compact silicon neuron circuit with spiking and bursting behaviour. Neural Networks 21(2-3), 524–534 (2008)
25. Wilson, H.R.: Simplified dynamics of human and mammalian neocortical neurons. Journal of Theoretical Biology 200(4), 375–388 (1999)

Brain Inspired Cognitive System for Learning and Memory

Huajin Tang and Weiwei Huang

Cognitive Computing Group
Institute for Infocomm Research,
Agency for Science Technology and Research,
Singapore 138632

Abstract. Hippocampus, a major component of brain, plays an important role in learning and memory. In this paper, we present our brain-inspired cognitive model which combines a hippocampal circuitry together with hierarchical vision architecture. The structure of the simulated hippocampus is designed based on an approximate mammalian neuroanatomy. The connectivity between neural areas is based on known anatomical measurements. The proposed model could be used to explore the memory property and its corresponding neuron activities in hippocampus. In our simulation test, the model shows the ability of recalling character images that it had been learned before.

Keywords: Hippocampus, Brain inspired model, Hierarchical vision architecture, memory, HMAX, Object recognition, Learning.

1 Introduction

There are many things humans find easy to do while they are tough for computers, such as visual pattern recognition, understanding spoken language and navigating in a complex world, due to difficulty of the preprogrammed intelligent systems in learning, adaptation and cognition. This challenge has intrigued a lot of research approaches in artificial intelligence (e.g., see [1], [2]). Despite decades of research, there are few algorithms that achieves human-like performance on a computer. In human-like intelligence research, memory is considered to be a fundamental part of the intelligent systems [3]. Hippocampus, which is a major component of the brain, plays an important role in the formation of memories and spatial navigation [4]. A hippocampal-like structure can greatly contribute to the development of human-like intelligence system.

The research of hippocampus has been carried out for hundreds of years. Many subregions of hippocampus have been identified. At a macroscopic level, highly processed neocortical information from all sensory inputs converges onto the media temporal lobe where the hippocampus resides [5]. These processed signals enter the hippocampus via the entorhinal cortex (EC). Within the hippocampus [6][7][8], there are connections from the EC to all fields of the hippocampal formation, including the dentate gyrus (DG), CA3 and CA1 through perforant

B.-L. Lu, L. Zhang, and J. Kwok (Eds.): ICONIP 2011, Part I, LNCS 7062, pp. 477–484, 2011.

pathway, from the dentate gyrus to CA3 through mossy fibers, from CA3 to CA1 through schaffer collaterals, and then from CA1 back to entorhinal cortex. There are also strong recurrent connections within the CA3 region. Based on the research work of hippocampus, many cognitive tasks have been tested with the hippocampal-like structure, especially in spatial navigation tasks [9][10][11].

On the other side, recognition of visual objects is a fundamental cognitive task for intelligence behavior. Computer vision systems are far from viewing as well as human beings doing. In neuroscience point of view, the problem of object recognition is also very challenging. It requires several levels of understanding like: biological mechanisms, level of circuits and information processing. Generally, vision processing in brain is modeled as a hierarchical process with information flowing from the retina to the lateral geniculate nucleus, occipital and temporal regions of the cortex [12]. Based this idea, many interesting hierarchical model of object recognition in cortex have been proposed such as HMAX model [13] and GWM model [14].

In this paper, we describe our brain inspired cognitive system which combines a hippocampal circuitry together with hierarchical vision architecture. The proposed cognitive structure is designed for the vision based intelligence task. In this design, the structure of the simulated hippocampus is inspired from Darwin series' [15] hippocampus model which is based on an approximate mammalian neuroanatomy. The HMAX [13] hierarchical vision architecture is adopted as a computational model of object recognition in cortex. This brain inspired model not only enables us to study the intelligence behavior such as learning and memory in the brain, but also enables us to record and analyzed all the neuron activities in each sub-area of hippocampus. In the simulation test, the model shows the ability of recalling character images that it had been learned before.

2   Brain Inspired Cognitive System

The brain inspired cognitive model includes two parts: hierarchical vision architecture and hippocampual circuitry. The schematic of the neural structure is shown in Fig. 2. The hierarchical vision architecture is often used to model the information process in vision system biologically. The hippocampal circuitry is inspired from the Darwin series' [9] hippocampus model which includes DG CA3 and CA1. EC connects the vision cortex and hippocampal circuitry which composes of two layers; one is for input and the other is for output.

2.1   Vision Input

Visual information in cortex is considered to be processed through ventral visual pathway [16] running from primary visual cortex V1 over extrastriate visual area V2 and V4 to inferotemporal cortex IT. It is classically modeled as a hierarchical-layer structure. Here, we adopt the HMAX [13] hierarchical vision architecture which is a computational model of object recognition in cortex. A description of information process in HMAX model is given in Fig. 1. It consists of four layers with linear and non-linear operations. The first layer, namely S1, performs a linear oriented filter and normalization to the input image. In the next layer (C1), outputs of S1 with same orientation and near location are selected by a maximum operation. In the next stage (S2), outputs from C1 with near location are combined to form more complex features. The C2 layer is similar to C1 layer: by pooling output S2 with same type and near location by a maximum operation together. The information in C2 becomes more invariant to position and scales but preserve feature selectivity which may correspond roughly to V4 in visual cortex [13]. The output of the C2 layer is mapped to the inferotemporal cortex (IT) as in a ventral cortical pathway (HMAX→IT). Then the neuronal states of IT are projected to the hippocampus via EC.

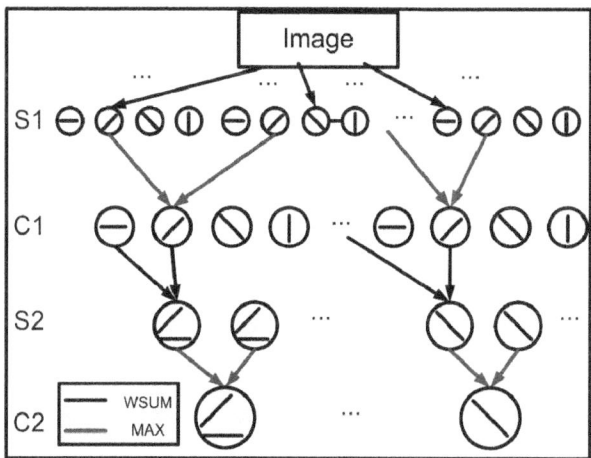

Fig. 1. HMAX model for the object recognition [13]. The circuitry consists of a hierarchy of layers of visual process by using two different type of pooling method: weighted sum (WSUM) and maximum (MAX). The first layer S1 performs a linear oriented filter and normalization to the input image. In the next layer C1, outputs of S1 with same orientation and near location are selected by a maximum operation. In the stage S2, outputs from C1 with near location are combined to form more complex features. The C2 layer is similar to C1 layer: by pooling together output of S2 with same type and near location by a maximum operation. The output of the C2 layer is mapped to the IT as in a ventral cortical pathway.

2.2 Hippocampus Circuitry

Connection Structure. Inside the structure, inputs from all sensory regions converge on the input layerd of EC (ECin). The ECin connects to all fields of the hippocampal formation, including the DG, CA3 and CA1 through perforant pathway (ECin→DG, ECin→CA3, ECin→CA1). The DG connects with CA3 through mossy fibers (DG→CA3). The CA3 connects with CA1 through schaffer collaterals (CA3→CA1), and CA1 connects back to entorhinal cortex output

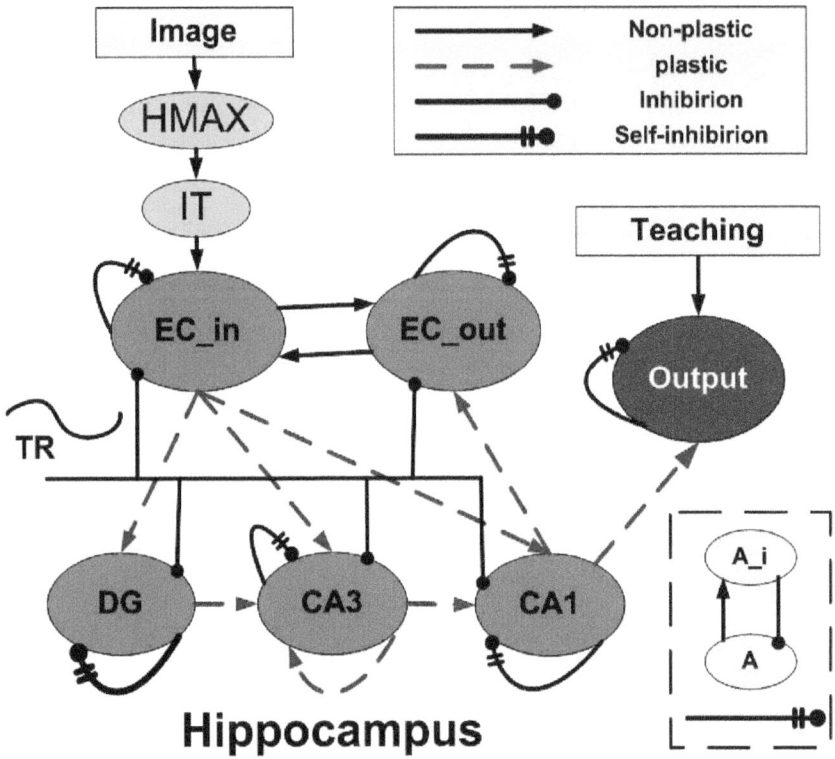

Fig. 2. Schematic of the regional and functional neuroanatomy of brain inspired structure. (A) Ellipses denote different neural area; boxes denote different device; arrows demote projections from one area to another. Input to the neural system is an image. The neural structure contains neural areas such as simulated visual cortex (HMAX); inferotemporal cortex (IT). Inside the hippocampus, there are neural areas including dentate gyrus (DG), CA3 and CA1 subfields which receive inputs via entorhinal cortex;. In this system, a theta rhythm (TR) signal is used to inhibit neural areas to keep activity level stable.

(CA1→ECout). There are also strong recurrent connections with in the CA3 region (CA3→CA3). In the system structure, a theta rhythm signal (TR) is used to inhibit neural areas to keep activity level stable [17] (TR→ECin, Ecout, DG, CA3, CA1). The TR activity follows a half cycle of sinusoidal wave:

$$TR(n) = sin(mod(\frac{n\pi}{N}, \pi))$$ (1)

where n is the time step; N is the number of steps that are required for the hippocampus to reach a stable state for a new input.

Neuronal Dynamic. Neuronal units in this system are simulated by a mean firing rate model. The mean firing rate range of each neuronal unit is from 0 (no firing) to 1 (maximal firing). The state of a neuron unit is calculated based on its current state and contributions from other neuron units. The postsynaptic influence on unit i is calculated based on equation:

$$Post_i(t) = \sum_{j=1}^{M}[w_{ij}s_j(t)]$$ (2)

where $s_j(t)$ is the activity of unit j; w_{ij} is the connection strength from unit j to unit i; $Post_i(t)$ is the postsynaptic influence on unit i; M is the number of connection to unit i.

The new activity is determined by the following activation function:

$$s_i(t+1) = \Phi(\tanh(Post_i(t) + \varepsilon s_i(t)))$$ (3)

$$\Phi(x) = \begin{cases} 0 & x < \delta_i^{fire} \\ x & otherwise \end{cases}$$ (4)

where ε controls the persistence of unit activity from previous state; δ_i^{fire} is firing threshold.

Synaptic Plastic. In learning and memory, synaptic plasticity is one of the key issue for the neural network to learn and store the memory. Experimental data from the visual cortex led to a synaptic modification rule, namely BCM rule. The model has two main features [18]: (1), synapses between neuron units with strong correlated firing rates are potentiated; (2),it assumes that a neuron processes a synaptic modification threshold which control the direction of weight modification. In this paper, the synaptic plastic is based on a modified BCM learning rule [9].

$$\Delta c_{ij}(t+1) = \eta s_i(t)s_j(t)BCM(s_i(t))$$ (5)

$$BCM(s) = \begin{cases} -s & s \leq \frac{\Theta}{2} \\ s - \Theta & \frac{\Theta}{2} > s \leq \Theta \\ \tanh(s - \Theta) & otherwise \end{cases}$$ (6)

where $s_i(t)$ and $s_j(t)$ are activities of postsynaptic and presynaptic units, respectively; η is the learning rate. The threshold is adjusted based on the postsynaptic activity.

$$\Delta\Theta = 0.25(s_i^2(t) - \Theta)$$ (7)

3   Learning of an Input Image

In this cognitive model design, the HMAX returns the key information of the input image. EC processes this vision information and passes it to the hippocampus region including DG, CA3 and CA1. DG is often considered to have the function of capturing the abstract information from EC area. In our design, the self-inhibition of DG is very strong. It inhibits more surrounding neurons comparing to the self-inhibition in CA3 and CA1 area. Due to the competitive learning process, the key information from EC will be remained in DG area. The CA3 is considered to store the memory information in hippocampus. In the design, a strong recurrent connection is included in CA3 model. The stabilized response of CA3 will then activate the corresponding pattern of CA1.

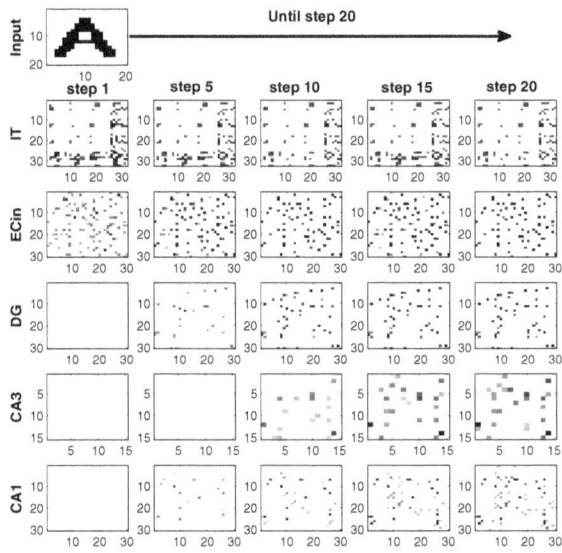

Fig. 3. The response of neurons in hippocampus to input image 'A' in a simulation cycle with different steps; top to bottom: input image, neuron activity of IT, ECin, DG, CA3 and CA1.

In the simulation test, a 20x20 pixel black and white character image 'A' is input to the system. In the hippocampus, the neuron areas require several steps to stabilize to a new input. As shown in Fig. 3, after about 20 steps, the whole system becomes stable. Here, the theta cycle is chosen to be 20 in the simulation. In CA1 area, the neuron shows a stable pattern to the input character. This pattern can be recalled if the input image is 'A' character.

To associate the pattern in CA1 to the input character, we create an output neuron area as shown in Fig. 2. It has the same size as the input image. The connection weight between CA1 neuron and output neuron is updated according to the BCM learning rule by Equation (5) and (6). As shown in Fig. 4, with two round of training, the hippocampus can recall the input character and show it in output neuron area.

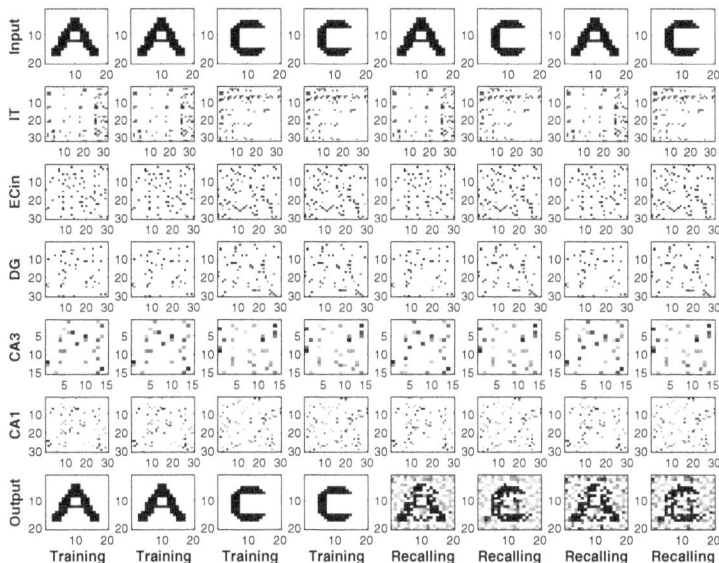

Fig. 4. Learning and recalling the input image by hippocampus; First two cycle is training for input 'A', the next two cycle is training for input 'C', in the testing stage image 'A' and 'C' are input to the system alternatively; top to bottom: input image, neuron activity of IT, ECin, DG, CA3 and CA1.

4   Conclusion

Hippocampus is a major component of the brain which plays an important role in learning and memory. In this study, we develop a brain inspired neural model to analyze these properties. The model includes a hippocampal circuitry and hierarchical vision architecture. This system allows us to track all the neural activities during different scenario of experiments. The system is found to be able to recall the input image that it has been trained previously. The analysis of this study may contribute to a better understanding of learning and memory functions in the hippocampus region.

References

1. Arel, I., Rose, D.C., Karnowski, T.P.: Deep Machine Learning–A New Frontier in Artificial Intelligence Research. IEEE Computational Intelligence Magazine 5(4), 13–18 (2010)
2. Coyle, D.: Neural Network Based Auto Association and Time Series Prediction for Biosignal Processing in Brain Computer Interfaces. IEEE Computational Intelligence Magazine 4(4), 47–59 (2009)
3. Tang, H., Li, H., Yan, R.: Memory Dynamics in Attractor Networks with Saliency Weight. Neural Computation 22, 1899–1926 (2010)

4. Frank, L.M., Brown, E.N., Wilson, M.: Trajectory encoding in the hippocampus and entorhinal cortex. Neuron 27, 169–178 (2000)
5. Lavenex, P., Amaral, D.G.: Hippocampal neocortical interaction: a hierarchy of associativity. Hippocampus 10, 420–430 (2000)
6. Amaral, D.G., Ishizuka, N., Claiborne, B.: Neurons, numbers and the hippocampal network. In: Progress in Brain Research, pp. 1–11. Elsevier Science, Amsterdam (1990)
7. Bernard, C., Wheal, H.V.: Model of local connectivity patterns in CA3 and CA1 areas of the hippocampus. Hippocampus 4, 497–529 (1994)
8. Treves, A., Rolls, E.T.: Computational analysis of the role of the hippocampus in memory. Hippocampus 4, 374–391 (1994)
9. Fleischer, J.G., Krichmar, J.L.: Sensory Intergation and Remapping in a Model of the Medial Temporal Lobe during Maze Navigation by a Brain-based Device. Journal of Integrative Neuroscience 6(3), 403–431 (2007)
10. Barrera, A., Weitzenfeld, A.: Biological Inspired Neural Controller for Robot Learning and Mapping. In: Proc. IEEE Int. Joint Conf. on Neural Networks, pp. 3664–3671. IEEE Press, New York (2006)
11. Wyeth, B.G., Milford, M.: Spatial Cognition for Robots. IEEE Robotics & Automation Magazine 16(3), 24–32 (2009)
12. Kandel, E.R., Schwartz, J.H., Jessel, T.M.: Principles of Neural Science. MIT Press, Cambridge (2000)
13. Riesenhuber, M., Poggio, T.: Hierarchical Models of Object Recognition in Cortex. Nature Neuroscience 2(11), 1019–1025 (1999)
14. Murray, J.F., Kreutz-Delgado, K.: Visual Recognition and Inference Using Dynamic Overcomplete Sparse Learning. Neural Computation 19, 2301–2352 (2007)
15. Krichmar, J.L., Seth, A.K., Nitz, D.A., Fleischer, J.G., Edelman, G.M.: Spatial Navigation and Causal Analysis in a Brain-Based Device Modeling Cortical-Hippocampal Interactions. Neuroinformatics 3(3), 197–221 (2005)
16. Ungerleider, L.G., Haxby, J.V.: "What" and "Where" in the human brain. Curr. Opin. Neurobiol. 4, 157–165 (1994)
17. Skaggs, W.E., McNaughton, B.L., Wilson, M.A., Barnes, C.A.: Theta phase precession in hippocampal neuronal populations and the compression of temporal sequences. Hippocampus 6(2), 149–156 (1996)
18. Jedlicka, P.: Synaptic plasticity, metaplasticity and BCM theory. Bratisl Lek Listy 103, 137–143 (2002)

A Neuro-cognitive Robot for Spatial Navigation

Weiwei Huang, Huajin Tang, Jiali Yu, and Chin Hiong Tan

Cognitive Computing Group
Institute for Infocomm Research,
Agency for Science Technology and Research,
Singapore 138632

Abstract. This paper presents a brain-inspired neural architecture with spatial cognition and navigation capability. It captures some navigation properties of rat brain in hidden goal hunting. The brain-inspired system consists of two main parts. One part is hippocampal circuitry and the other part is hierarchical vision architecture. The hippocampus is mainly responsible for the memory and spatial navigation in the brain. The vision system provides the key information about the environment. In the experiment, the cognitive model is implemented in a mobile robot which is placed in a spatial memory task. During the navigation, the neurons in CA1 area show a place dependent response. This place-dependent pattern of CA1 guides the motor neuronal area which then dictates the robot move to the goal location. The results of current study could contribute to the development of brain-inspired cognitive map which enables the mobile robot to perform a rodent-like behavior in the navigation task.

Keywords: Hippocampus, Brain-inspired model, Place-dependent response, Spatial memory, HMAX, Object recognition, Neurobotics.

1   Introduction

The autonomous capabilites of robots and learning systems, for example, autonomous navigation, learning and autonomous self-reconfiguration, have attracted increasing interests [1,2]. Navigating in a complex world is easy for humans, yet it is a tough problem for the robots. A cognitive mechanism that converts the odometric and perceptual information into a guild map is still a challenging problem in robotics area. To address this problem, simultaneous localization and mapping (SLAM) is widely studied. In this approach, a mobile robot develops a map of its environment while simultaneously localizes itself within this map [3]. In the implementation, SLAM algorithm often requires a precise reference location. Many multi-hypothesis techniques such as Kalman filter [4] or Particle filter [5] are used to have a practical and robust SLAM algorithm.

Compared with SLAM in robotics area, animals such as rat and primates have inborn ability to navigate in a complex environment. Neurophysiological studies suggest that hippocampus, a major component of brain, plays an important role

B.-L. Lu, L. Zhang, and J. Kwok (Eds.): ICONIP 2011, Part I, LNCS 7062, pp. 485–492, 2011.

in memory and spatial navigation [6][7]. The discovery of hippocampus neural activities in rat navigation experiment led to a theory that the hippocampus might act as a cognitive map: a neural representation of the layout of the environment [8]. This spatial navigation ability arouses a great interest to incorporate rat's adaptive navigation model in mobiles robots. Barrera et. al [9] have proposed a neural structure to mimic "place field" property of hippocampus and guild a robot to search for the goal. Wyeth and Milford [10] proposed a more biological based spatial cognition model which includes two parts: pose cells which function as grid cells in entorhinal cortex (EC) and experience map which functions as place cells in CA1 and CA3. The robot controlled by the hippocampal-like system achieved a great performance of simultaneous navigating and localization in the lab.

Inspired from the rat's navigation and prior art on brain-inspired navigation method, we target to develop a brain based cognitive model for the robot spatial navigation. The proposed model should have a brain-inspired neural architecture which can build, update, integrate and use the map simultaneously throughout its traveling period. In our architecture, the brain model is designed based on an approximate mammalian neuroanatomy [11][12][13]. The key part of the brain system is hippocampus circuitry which includes dentate gyrus (DG), CA3 and CA1. The hippocampus receives inputs from virtually all association areas in the neocortex via entorhinal cortex (EC). The connections between all these neural areas are based on known anatomical measurements. In this design, vision and orientation are implemented as sensory inputs. Vision input is an important information for the robot navigation. In our design, a computational model of object recognition in cortex is adopted to work as hierarchical vision system.

In this paper, we will describe the implementation of our brain-inspired cognitive model. To explore the spatial navigation ability of the robot, we implement the model in a mobile robot which is placed in a spatial memory task. The experiment environment is a plus maze where the robot is commanded to search for a hidden goal. This maze experiment has been used in the rodent studies of spatial memory [14]. In the test, the region of CA1 shows a place-dependent response during the maze navigation. This place-dependent pattern of CA1 guides the motor neuronal area which dictates the robot move to the goal location. In the goal searching test, the robot is found to be able to remember the direction of the goal that it found previously. Because of the simulated model of hippocampus, we can trace the population activities of the neurons and functional connections inside the hippocampus. This analysis reveals the contributions of the place-dependent response in hippocampus. The study results could be greatly helpful in developing the cognitive map for the navigation.

2   Brain-inspired Cognitive System

The Brain-inspired neural structure incorporates a hippocampal circuitry, Entorhinal cortex, sensory cortical regions and motor cortical regions. The schematic of the neural structure is shown in Fig. 1. The hippocampal circuitry is inspired from the Darwin series' [15] hippocampus model which includes DG CA3 and

CA1. The model for sensory inputs contains two regions; visual pathway for camera input and orientation pathway for compass input.

Visual information in cortex is considered to be processed through ventral visual pathway [16] running from primary visual cortex V1 over extrastriate visual area V2 and V4 to inferotemporal cortex IT. It is classically modeled as a hierarchical-layer structure. Here, we adopt the HMAX [17] hierarchical vision architecture which is a computational model of object recognition in cortex. It consists of four layers (S1, C1, S2, C2) with linear and non-linear operations (shown in Fig. 1 B). The output of the C2 layer is mapped to the inferotemporal cortex (IT) as in a ventral cortical pathway. Another visual input is the color information where four colors, red, green, blue, yellow, are filtered from camera input. The filtered images are mapped to IT in a ventral cortical pathway. Then the neuronal states of IT are projected to the hippocampus via EC.

Head direction is activated by the compass input and mapped to the anterior thalamic nucleus (ANT) and to motor area (Mhdg). The neural states of ANT are also projected to the hippocampus via EC.

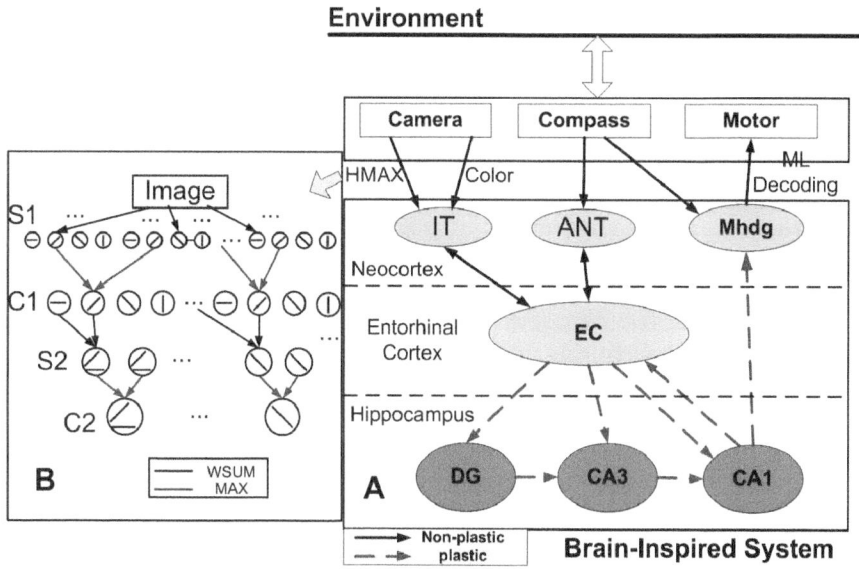

Fig. 1. Schematic of the regional and functional neuroanatomy of brain-inspired structure. (A) Ellipses denote different neural area; boxes denote different device; arrows demote projections from one area to another. The neural structure contains neocortex, entorhinal cortex and hippocampus circuitry. (B) HMAX model for the object recognition [17]. The circuitry consists of a hierarchy of layers of visual process by using two different type of pooling method: weighted sum (WSUM) and maximum (MAX).

Inside the hippocampus, inputs from all these sensory regions converge on the input layerd of EC. The EC connects to all fields of the hippocampal formation, including the dentate gyrus (DG), CA3 and CA1 through perforant pathway.

The DG connects with CA3 through mossy fibers. The CA3 connects with CA1 through schaffer collaterals, and CA1 connects back to entorhinal cortex output. There are also strong recurrent connections with in CA3 regions. Synaptic strengths of the plastic connection inside the hippocampus are modified according to a BCM learning rule [15]. The synaptic strengths between CA1 to motor cortical area (Mhdg) are modified by a temporal difference reinforcement learning rule [15] based on the reward system.

The motor cortical area dictates the moving direction of the robot. To estimate the optimal moving direction from the motor cortical area, we adopt a Maximum Likelihood (ML) estimator method. Deneve et al. [18] have shown that ML estimation can be done with a biologically plausible neural network. In this paper, we use a recurrent neural network to read the neuronal activities in motor cortical area. The recurrent neural network is composed of a population of neurons with bell-shaped tuning curves. Neurons in the network communicated through symmetric lateral connections. The input of the network is the neuronal activities of Mhdg which are multi-modal noisy hill. By tuning the parameters of the network, the activity of the network should converge over time to a smooth hill which is unimodal. Once the smooth hill is obtained, its peak position is an estimate of the direction of motion.

3   Device, Task and Environment in the Simulation

Analyzing the hippocampus function in brain is challenging due to the difficulty of recording neuronal activities simultaneously from many neuronal areas and multiple neuronal layers. A possible solution is to build an artificial hippocampus model and implement it in real environment [15]. Many neuro-robotic systems [19][15][20][21][22] have been developed whose behaviors are guided by the simulated neural systems. In this paper, instead of carrying out the experiment in real environment, we test the system in a simulated environment. An advantage of doing the experiments in the simulated environment is that the experiment can run as long as we need without worrying about the damage of hardware. Also, the cost is much lower comparing to the real hardware implementation. In this implementation, the task is to navigate the plus-maze as shown in Fig. 2. The simulated environment is developed in Webots which is a dynamic simulation software based on open dynamic engine (ODE). All the real world sensors such as vision, compass, distance sensor and DC motors can be modeled in this environment.

The maze is similar to the experiment of Darwin XI [15] which was imitating the environment used in the studies of rodent hippocampal place activity [14]. The robot explored the maze autonomously and made a choice of direction at the intersection. On each trial, the robot starts from any arm as indicated by the "★" in the Fig. 2 and enters an arm among the rest of the arms. A hidden goal platform is placed randomly at end of a maze arm. To provide the cue for the robot, visible walls are placed outside the maze arm which have different shapes and colors. As for the experimental robot, a camera is mounted on the head to provide the visual input; a compass is also mounted on the head to provide the orientation input; IR sensors are mounted on the body to detect the intersection and 1 additional IR sensor is mounded at the bottom to detect the hidden goal platform. In the experiment, the robot moves based on the visual and orientation input. A reward system is given to the neural system depending on the goal hunting results.

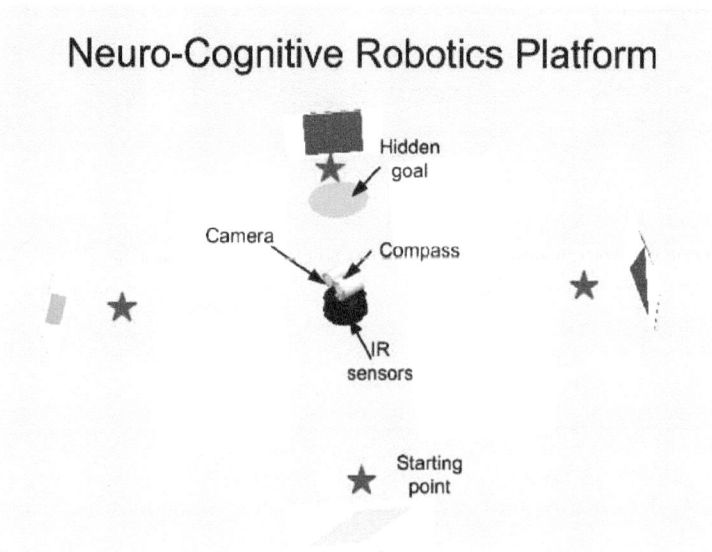

Fig. 2. Neuro-Cognitive Robotics Platform. The maze has four cue walls with different shapes and colors; a hidden platform is placed randomly at the end of a maze arm; the camera and compass mounted on the robot head provide the vision and orientation inputs for the robot; the IR sensors are only used to detect the interception; in the experiment, the robot can start journey at any starting points marked by the red "★" in the maze.

4   Simulation Results

We implement the brain-inspired model in a mobile robot which is placed inside a simulated plus maze environment and commanded to search for a hidden goal (shown in Fig. 2). The scenario of the simulation is designed as following: the robot starts from a starting point as marked by the red "★" in Fig. 2; checks the three possible directions at the intersection; makes the choice based on the neuron activities; checks the reward when the robot reaches the end of the maze arm. This scenario is repeated to see how the robot searches the hidden goal in an unknown environment.

In the experiment, the hippocampus shows a place dependent response in CA1 area. When the robot moves to the end of maze arm and looks at the image shown

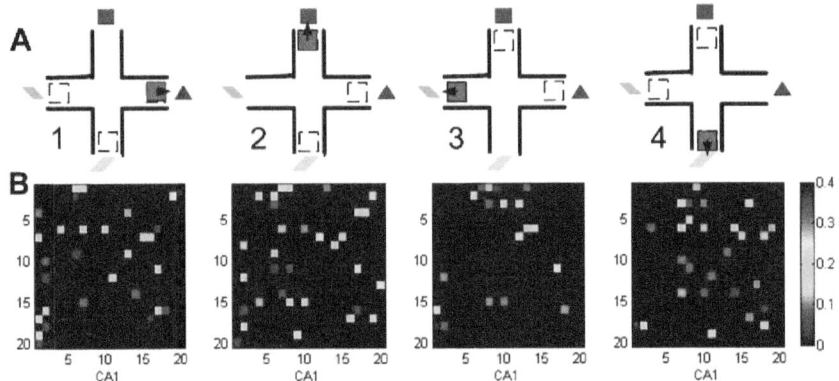

Fig. 3. Place dependent pattern of CA1 in different location. (A) Robot reaches the end of each arm. (B) The response of CA1 at each arm. The arrow indicates the facing direction.

on the wall. The neuron activity in CA1 area shows different pattern for each maze arm. This place related pattern is shown in Fig. 3 Since the hippocampus shows different pattern when it stands at different location, this property helps the robot to remember the location it has seen before. In this place field response, about 57 neurons in total in CA1 area out of 400 neurons are fired when pass all four different locations. The ratio is similar to the experiment of Darwin XI [15].

When the robot reaches the intersection, it looks to the left, front and right sequentially. Fig. 4 B shows the neuron activity in CA1 area when the robot stays at intersection point. For the same heading direction, the response of CA1 is similar with the case when the robot reaches at end of the wall as shown in Fig. 3. It shows that although the robot see the image in different distance, it still can recognize the image. Then, based on the neuron response in Mhdg area, the robot chooses a direction with the highest neuron activity. Mhdg area consists 60 neurons which corresponding to 0-359 degree in direction. If the environment is new to the robot, the four directions have similar neuron activity level (as shown in Fig. 4 C upper sub-figure). After the robot has found a hidden goal in the maze arm, a positive reward will be given to this direction. The robot will get the knowledge of this goal location according to the reinforcement learning rule. In the case when the robot has not found the hidden goal in a maze, a negative reward will be given to the system. The information of no goal in this arm will be learned. Fig. 4 C (bottom sub-figure)shows the neuron activity of Mhdg area after given the reward. After the robot has found the hidden goal in west arm, a positive reward is given. The neuron activity around 270 degree becomes stronger. When the robot reaches at the end of north arm, no hidden goal is found. In this case, a negative reward is given. The neuron activity around 180 degree becomes weaker.

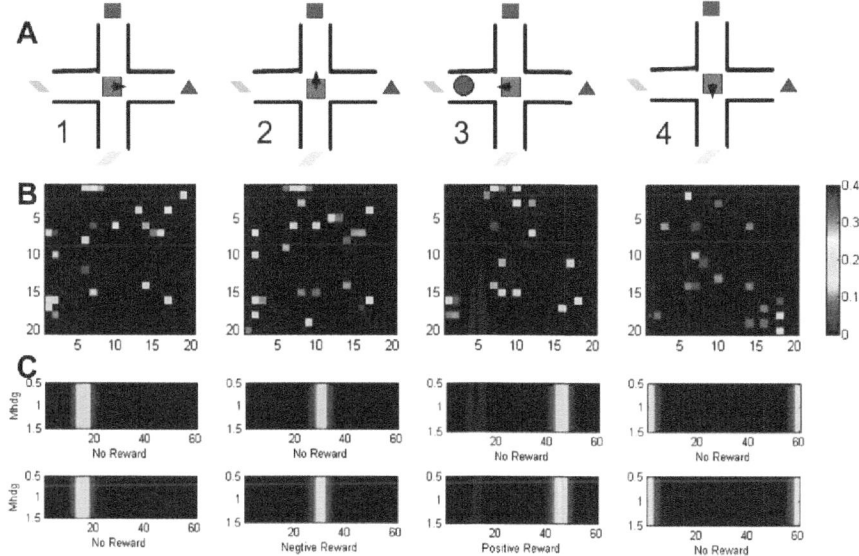

Fig. 4. A comparison of positive reward and negative reward to the neuron states in Mhdg area. (A) robot stays at interception point and make the choice of direction (B) The response of CA1 with different heading (C) The neuron state of Mhdg area before knowing the environ (upper sub-figure) and after reward training (lower sub-figure).

5   Conclusion

Hippocampus is a major component of the brain which plays an important role in spatial navigation. In this paper, we have presented our brain-inspired neural architecture for the spatial navigation. The model includes a hippocampal circuitry, hierarchical vision architecture and other sensory input cortical regions and motor output cortical regions. This system allows us to track all the neural activities during different scenario of experiments. In the experiments, the place-dependent response is observed in CA1 area in hippocampus model. The robot is found to be able to remember the direction of the goal that it found previously. The analysis of this study may contribute to a better understanding of place-dependent response and spatial memory in the hippocampus region.

References

1. Arel, I., Rose, D.C., Karnowski, T.P.: Deep Machine Learning–A New Frontier in Artificial Intelligence Research. IEEE Computational Intelligence Magazine 5(4), 13–18 (2010)
2. Meng, Y., Zhang, Y., Jin, Y.: Autonomous Self-Reconfiguration of Modular Robots by Evolving a Hierarchical Mechanochemical Model. IEEE Computational Intelligence Magazine 6(1), 43–54 (2011)

3. Dissanayake, M.W.M.G., Newman, P., Clark, S., Durrant-Whyte, H.F., Csorba, M.: A Solution to the Simultaneous Localization and Map Building (SLAM) Problem. IEEE Transactions on Robotics and Automation 17(3), 229–241 (2001)
4. Bar-Shalom, Y., Fortmann, T.E.: Tracking and Data Association. Academic Press, San Diego (1988)
5. Montemerlo, M., Thrun, S.: Simultaneous localization and mapping with unknown data association using FastSLAM. In: Proc. IEEE Int. Conference on Robotics and Automation (ICRA), pp. 1985–1991. IEEE Press, New York (2003)
6. Tang, H., Li, H., Yan, R.: Memory Dynamics in Attractor Networks with Saliency Weight. Neural Computation 22, 1899–1926 (2010)
7. Frank, L.M., Brown, E.N., Wilson, M.: Trajectory encoding in the hippocampus and entorhinal cortex. Neuron 27, 169–178 (2000)
8. O'Keefe, J., Nadel, L.: The Hippocampus as a Cognitive Map. Oxford University Press, Oxford (1978)
9. Barrera, A., Weitzenfeld, A.: Biological Inspired Neural Controller for Robot Learning and Mapping. In: Proc. IEEE Int. Joint Conf. on Neural Networks, pp. 3664–3671. IEEE Press, New York (2006)
10. Wyeth, B.G., Milford, M.: Spatial Cognition for Robots. IEEE Robotics & Automation Magazine 16(3), 24–32 (2009)
11. Amaral, D.G., Ishizuka, N., Claiborne, B.: Neurons, numbers and the hippocampal network. Progress in Brain Research 83, 1–11 (1990)
12. Bernard, C., Wheal, H.V.: Model of Local Connectivity Patterns in CA3 and CA1 Areas of The Hippocampus. Hippocampus 4, 497–529 (1994)
13. Treves, A., Rolls, E.T.: Computational Analysis of The Role of The Hippocampus in Memory. Hippocampus 4, 374–391 (1994)
14. Ferbinteanu, J., Shapiro, M.L.: Prospective and Retrospective Memory Coding in The Hippocampus. Neuron 40(6), 1227–1239 (2003)
15. Fleischer, J.G., Krichmar, J.L.: Sensory Intergation and Remapping in a Model of the Medial Temporal Lobe during Maze Navigation by a Brain-based Device. Journal of Integrative Neuroscience 6(3), 403–431 (2007)
16. Ungerleider, L.G., Haxby, J.V.: "What" and "Where" in The Human Brain. Curr. Opin. Neurobiol. 4, 157–165 (1994)
17. Riesenhuber, M., Poggio, T.: Hierarchical Models of Object Recognition in Cortex. Nature Neuroscience 2(11), 1019–1025 (1999)
18. Deneve, S., Latham, P.E., Pouget, A.: Reading Population Codes: A Neural Implementation of Ideal Observers. Nature Neuroscience 8(2), 740–755 (1999)
19. Krichmar, J.L., Seth, A.K., Nitz, D.A., Fleischer, J.G., Edelman, G.M.: Spatial Navigation and Causal Analysis in a Brain-Based Device Modeling Cortical-Hippocampal Interactions. Neuroinformatics 3(3), 197–221 (2005)
20. Pfeifer, R., Scheier, C.: Sensory-motor coordination: The metaphor and beyond. Robot. Auton. Syst. 20, 157–178 (1997)
21. Floreano, D., Mondada, F.: Evolutionary neurocontrollers for autonomous mobile robots. Neural Netw. 11, 1461–1478 (1998)
22. Gaussier, P., Revel, A., Banquet, J.P., Babeau, V.: From view cells and place cells to cognitive map learning: processing stages of the hippocampal system. Biol. Cybern. 86, 15–28 (2002)

Associative Memory Model of Hippocampus CA3 Using Spike Response Neurons

Chin Hiong Tan[1], Eng Yeow Cheu[1], Jun Hu[2], Qiang Yu[2], and Huajin Tang[1]

[1] Cognitive Computing Group, Institute for Infocomm Research, A*STAR, Singapore
[2] National University of Singapore, Singapore

Abstract. One of the functional roles of the hippocampus is the storage and recall of associative memories. The hippocampus CA3 region has been hypothesized to function as an associative network. Dual oscillations have been recorded in brain regions involved in memory function in which a low frequency theta oscillation is subdivided into about seven subcycles of high frequency gamma oscillation. In this paper, the computational model of hippocampus CA3 proposed by Jensen et al. is realized using the Spike Response Model (SRM). The SRM-based network is able to demonstrate the same memory storage capability with added simplicity and flexibility. Different short term memory items are encoded by different subset of principal neurons and long term associative memory is maintained in the synaptic modifications of recurrent collaterals. The formation of associative memory is demonstrated in simulations.

Keywords: Hippocampus, CA3, associative memory, spike response model, theta cycle, gamma cycle.

1 Introduction

Mimicking brain-style intelligence has intrigued enormous efforts in the past decades (see, e.g. [1,2]). One of the most widely studied regions of the brain, the hippocampus, is believed to be the storage location of declarative memories. Computational modelling of these brain regions using biologically realistic neurons provides insights to how the brain handles such memory functions.

Pantic et al. examined the use of dynamic synapses in a stochastic Hopfield-like network to store and retrieve memories [19]. Sato & Yamaguchi demonstrated a hierarchical architecture for object-place association by modelling the EC layer II and CA3 regions [22]. Sommer & Wennekers demonstrated a biologically realistic CA3 network which uses gamma oscillations as a means to retrieve excitatory associative memories [23]. Cutsuridis et al. explored the biophysical mechanisms to achieve the storage and recall of spatial-temporal patterns using a microcircuit model of the CA1 region [6]. Fleischer & Krichmar studied how sensory information is mapped using a model of the medial temporal lobe and the hippocampus [8].

In this paper, we propose a SRM-based [14] computational model of the hippocampus CA3 region for the storage and retrieval of a sub-category of declarative memory known as associative memory. The architecture of the hippocampus

B.-L. Lu, L. Zhang, and J. Kwok (Eds.): ICONIP 2011, Part I, LNCS 7062, pp. 493–500, 2011.
© Springer-Verlag Berlin Heidelberg 2011

model is inspired by Jensen *et al.* [10]. The SRM neuron is chosen for its simplicity in implementation as well as flexibility in defining varied neuron characteristics. The SRM is simple to implement because it is fully defined by only two continuous kernel functions. In addition, both synaptic plasticity and intrinsic neuronal plasticity can easily be realized using the SRM neuron.

2 Hippocampus

The hippocampus resides within the medial temporal lobe of the brain. At the macroscopic level, highly processed neocortical information from all sensory inputs converges onto the medial temporal lobe [12]. These processed signals enter the hippocampus via the entorhinal cortex (EC). Within the hippocampus [21], there are connections from the EC to all parts of the hippocampus, including the dentate gyrus (DG), CA3 and CA1 through perforant pathway, from the DG to CA3 through mossy fibres, from CA3 to CA1 through schaffer collaterals, and then from CA1 back to EC. There are also strong recurrent connections within the DG and CA3 regions. Figure 1 depicts the broad overview of hippocampus.

Dual oscillations have been recorded in hippocampus in which a low frequency theta oscillation is subdivided into about seven subcycles of high frequency gamma oscillation [4]. The theta rhythm in the hippocampus refers to the regular oscillations of the local field potential at frequencies of 4-12 Hz which has been observed in rodents [25]. It is thought that different information can be stored at different phases of a theta cycle [18]. This type of neural information representation is commonly known as phase encoding. It is also proposed that the theta rhythm could work in combination with another brain rhythm known as the gamma rhythm, of frequencies 40-100 Hz [4], to actively maintain auto-associative memories [13,10].

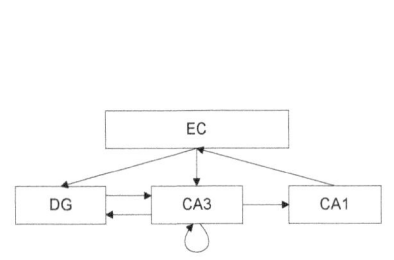

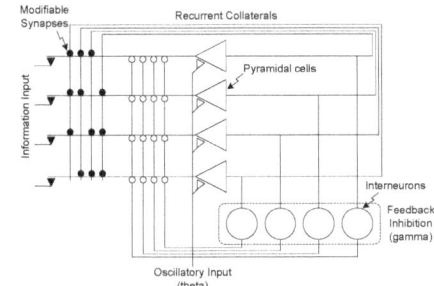

Fig. 1. Block diagram of hippocampus   **Fig. 2.** Overview of CA3 model using SRM

3   Model of CA3

The proposed computational model is a simplified network of the subcortical area and hippocampus that incorporates two major network components; namely: the synaptic input from EC to CA3, and CA3 itself. The network architecture and dynamics is inspired by the works of Jensen et al. [10]. An overview of the hippocampal CA3 architecture used in this paper is shown in Fig. 2. The EC and DG is modelled as the input layer while the hippocampal CA3 region is modelled as a recurrent network. All pyramidal cells in the CA3 also accept an oscillatory input that is used to model the theta rhythm. Feedback inhibition from interneurons is applied to all pyramidal cells.

3.1   Spike Response Model

This paper presents the CA3 model using the Spiking Response Model (SRM) neurons [14]. The SRM neuron is preferred to the integrate and fire (I&F) neuron model due to its flexibility in defining the neuron response and the synaptic response. While the characteristics of an I&F neuron is defined by several parameters, the characteristics of a SRM neuron is defined by a continuous kernel function. The kernel function may be designed to model neurons with different characteristics like those found in a biological nervous system. Neurons in the hippocampus has been observed to change in intrinsic neuronal excitability [26,15]. It was postulated that changes in the after-hyperpolarization (AHP) in the cells caused its firing frequency to change [11]. Persistent concomitant changes in the intrinsic neuronal excitability and long term synaptic modifications have been shown [7]. Similar to synaptic plasticity, intrinsic plasticity is bidirectional and input- or cell- specific. Intrinsic plasticity increases the reliability of the input-output function and improves the temporal precision of the neuronal discharge [5]. While intrinsic plasticity is modelled as a static function in this paper, the SRM neuron model has the potential to be extended to incorporate such forms of intrinsic plasticity dependent learning in future works.

Mathematically, the membrane potential of a neuron i under the SRM model is described by a state variable u_i. A spike is modelled as an instantaneous event that occurs when the membrane potential u_i exceeds a threshold V_{thres}. The time at which u_i crosses V_{thres} is said to be the firing time $t_i^{(f)}$. The set of all firing times of neuron i is denoted by

$$\mathcal{F}_i = \left\{ t_i^{(f)}; 1 \leq f \leq n \right\} = \{ t | u_i(t) = V_{\text{thres}} \,\wedge\, u_i'(t) > 0 \} \ , \tag{1}$$

where n is the length of the simulation. After a spike has occurred at $t_i^{(f)}$, the state variable u_i will be reset by adding a negative contribution $\eta_i(t - t_i^{(f)})$ to u_i. The kernel $\eta_i(s)$, known as the refractory function, vanishes for $s \leq 0$ and decays to zero for $s \to \infty$. The refractory kernel defines a refractory period immediately following a spike during which the neuron will be incapable of firing another spike. The neuron may also receive input from presynaptic neurons $j \in \Gamma_i$ where

$$\Gamma_i = \{ j | j \text{ presynaptic to } i \} \ . \tag{2}$$

A presynaptic spike increases or decreases the state variable u_i of neuron i by an amount $w_{ij}\epsilon_{ij}(t - t_j^{(f)})$. The weight w_{ij} is known as the synaptic weight and it characterises the strength of the connection from neuron j to neuron i. The kernel $\epsilon_{ij}(s)$ models the response of neuron i to presynaptic spikes from neurons $j \in \Gamma_i$ and vanishes for $s \leq 0$. In addition to spike input from other neurons, a neuron may receive external input h^{ext}, for example from non-spiking sensory neuron. Under the SRM model [14], the state $u_i(t)$ of neuron i at time t is hence given by (3).

$$u_i(t) = \sum_{t_i^{(f)} \in \mathcal{F}_i} \eta_i \left(t - t_i^{(f)} \right) + \sum_{j \in \Gamma_i} \sum_{t_j^{(f)} \in \mathcal{F}_j} w_{ij} \epsilon_{ij} \left(t - t_j^{(f)} \right) + h^{\text{ext}}(t) \ . \quad (3)$$

The kernels $\eta_i(s)$ and $\epsilon_{ij}(s)$ fully define neuron i under the SRM neuron model described in (1) to (3). The response kernels can be adapted to give rise to different neuronal characteristics. The kernels can be configured to adapt SRM to function like the I&F model. With appropriate selection of the response kernels, the SRM neuron can even approximate the Hodgkin-Huxley conductance-based neuron model [14]. Hence, the SRM offers flexibility in defining neurons with different characteristics.

3.2   Pyramidal Cells

A pyramidal cell is fired when its membrane potential u_i exceeds threshold V_{thres} = 10 mV (see [9] for approximate value). In hippocampal pyramidal cells, action potentials are followed by after-hyperpolarizations (AHPs) [20]. In addition, pyramidal cells exhibit an after-depolarization (ADP) after spike during cholinergic [24] or serotonergic modulation [3] or as a result of metabotropic glutamate receptors involved in the conversion of AHP to ADP [20]. In the SRM neuron, both AHP and ADP are modelled by the refractory kernel $\eta_i(t - t_i^{(f)})$ of a pyramidal cell i. (see (4)).

$$\eta_i(s) = V_{\text{refr}}(s) = A_{\text{AHP}} \exp \left(-\frac{s}{\tau_{\text{AHP}}} \right) + A_{\text{ADP}} \frac{s}{\tau_{\text{ADP}}} \exp \left(1 - \frac{s}{\tau_{\text{ADP}}} \right) \ , \quad (4)$$

where $A_{\text{AHP}} = -3.96$ mV, $\tau_{\text{AHP}} = 5$ ms (see [24] for approximate value), A_{ADP} = 9.9 mV, and $\tau_{\text{ADP}} = 200$ ms (see [20] for approximate value).

The theta oscillatory signal in Fig. 2 is modelled by (5).

$$h^{\text{ext}}(t) = V_{\text{theta}}(t) = A_{\text{theta}} \sin \left(2\pi f t + \phi \right) \ , \quad (5)$$

where $A_{\text{theta}} = 4.95$ mV, $f = 6$ Hz, and $\phi = -\pi/2$.

For synaptic transmission between pyramidal cells in the recurrent collaterals, the response kernel ϵ_{ij} denotes an excitatory postsynaptic potential (EPSP) V_{EPSP}. The synaptic transmission is mediated by synaptically released glutamate binding to AMPA (α-amino-3-hydroxy-5-methyl-4-isoazole-proprionic acid). AMPA receptors activate and deactivate within a few milliseconds of presynaptic glutamate release. This synaptic input is modelled by (6).

$$\epsilon_{ij}(t) = V_{\mathrm{EPSP}}(t) = \frac{A_{\mathrm{AMPA}}}{aN} \left(\frac{t - t_j^{(f)} - t_{\mathrm{delay}}}{\tau_{\mathrm{AMPA}}} \right) \exp \left(1 - \frac{t - t_j^{(f)} - t_{\mathrm{delay}}}{\tau_{\mathrm{AMPA}}} \right) .$$

$$(6)$$

where $A_{\mathrm{AMPA}} = 23.1$ mV, $\tau_{\mathrm{AMPA}} = 1.5$ ms (see [10] for approximate value), N denotes the number of pyramidal cells/interneurons in the network, and the delay in the recurrent feedback is $t_{\mathrm{delay}} = 0.5$ ms. The constant $a = M/N$ denotes the sparseness of pyramidal cells for information coding with M denoting the number of cells representing a memory pattern.

For synaptic transmission from an interneuron to pyramidal cells, the response kernel ϵ_{ij} denotes an inhibitory postsynaptic potential (IPSP) V_{IPSP}. V_{IPSP} represents the net GABAergic inhibitory feedback to the pyramidal cells from one interneuron. This inhibitory effect is modelled in (7) by assuming the interneurons fire in synchrony by recurrent excitation of the pyramidal cells.

$$\epsilon_{ij}(t) = V_{\mathrm{IPSP}}(t) = \frac{A_{\mathrm{GABA}}}{aN} \left(\frac{t - t_j^{(f)}}{\tau_{\mathrm{GABA}}} \right) \exp \left(1 - \frac{t - t_j^{(f)}}{\tau_{\mathrm{GABA}}} \right) , \qquad (7)$$

where $A_{\mathrm{GABA}} = -5.94$ mV, $t_j^{(f)}$ refers to the spike time of interneuron j, $\tau_{\mathrm{GABA}} = 4$ ms (see [16] for approximate value).

3.3   Interneurons

The response kernel ϵ_{ij} of interneuron i to presynaptic spikes from pyramidal cell j is modelled by (8). The synaptic transmission w_{ij} from pyramidal cell j to interneuron i is assumed with unit weight. The interneuron is also assumed with no refractory ($\eta(s) = 0$). This set-up simply serves as a signal from a pyramidal cell to initiate IPSP to other pyramidal cells via the inhibitory feedback from the interneurons.

$$\epsilon_{ij}(t - t_j^{(f)}) = \begin{cases} 4 \text{ mV (see [16] for approximate value)} & \text{if } (t - t_j^{(f)}) = 0 \ , \\ 0 \text{ mV} & \text{if } (t - t_j^{(f)}) \neq 0 \ . \end{cases}$$

$$(8)$$

3.4   Synaptic Modification

Associative LTM memory storage is achieved by synaptic modifications that follow the Hebb-rule; simultaneous presynaptic and postsynaptic activity enhances synaptic efficacies. The following equations ((9)-(11)) are defined for synaptic modifications [10]. The synaptic strength from pyramidal neuron i to neuron j is determined by (11).

$$i_{\mathrm{post}}(s) = \frac{s}{\tau_{\mathrm{post}}} \exp \left(1 - \frac{s}{\tau_{\mathrm{post}}} \right) , \qquad (9)$$

$$b_{\text{glu}}(s) = \exp\left(-\frac{s}{\tau_{\text{NMDA,f}}}\right)\left(1 - \exp\left(-\frac{s}{\tau_{\text{NMDA,r}}}\right)\right) , \qquad (10)$$

$$\frac{\partial w_{ij}}{\partial t} = \frac{i_{\text{post}}(t - t_j^{(f)}).b_{\text{glu}}(t - t_i^{(f)} - t_{\text{delay}})(1 - w_{ij})}{\tau_{\text{pp}}} +$$

$$\left(\frac{i_{\text{post}}(t - t_j^{(f)})}{\tau_{\text{npp}}} + \frac{b_{\text{glu}}(t - t_i^{(f)} - t_{\text{delay}})}{\tau_{\text{pnp}}}\right)(0 - w_{ij}) . \qquad (11)$$

where $\tau_{\text{post}} = 2.0$ ms, $\tau_{\text{pp}} = 50$ ms, $\tau_{\text{pnp}} = \tau_{\text{npp}} = 250$ ms, and τ_{delay} is the time taken for an action potential to travel from the soma to the synapses of the recurrent collaterals, i_{post} is the postsynaptic depolarization dynamics, and b_{glu} is the kinetics of the NMDA channels. Using the kinetics of the NMDA receptors [17], $\tau_{\text{NMDA,f}} = 7.0$ ms, and $\tau_{\text{NMDA,r}} = 1.0$ ms.

4   Results and Discussion

Figure 3 shows the repetitive firings of neurons in the network. A memory item is represented by the coincident firing of a subset of neurons. In Fig. 3, external input V_{in} (simulation of spikes from input layer) representing seven different memories are introduced into the network. Each of the input patterns is encoded by five pyramidal neurons ($M = 5$).

The first memory is inserted into the memory buffer at 160 ms by an external input that synchronously fires five of the pyramidal neurons. This firing triggers a short AHP and then a slowly rising ADP in the neurons. The ADP subsequently causes the neurons to fire on next theta cycle. The membrane potential of each group of five neurons is represented by a line in the membrane potential subplot of Fig. 3. When a neuron fires, the ADP is reset, making it possible for the same processes to occur on the next theta cycle. The second memory is inserted at 500 ms which causes the synchronous firing of the another group of five neurons. This memory is repeated in this sequence in the next theta cycle. The ADP causes persistent firing and controls the timing of the firing of each group of neurons such that the neurons that encode the first memory fires before neurons that encode the second memory and so forth for the rest of the memory items. The feedback inhibition V_{IPSP} follows each action potential from the pyramidal neurons. This inhibition serves to restrict the firing of each group of neurons to discrete phases of the theta oscillation. Each memory item is repeated in the order of its introduction to the network.

Figure 4 illustrates the synaptic matrix; the size of the square at location i, j denotes the synaptic strength of the connection from neuron j (y-axis) to neuron i (x-axis). The repetition of memories in the short term memory network gradually leads to the build up of synaptic strength in LTM. Here, hetero-associative memory is also gradually formed due to the repeated presentation of the same sequence of memory items. Figure 4 also illustrates neuron group (neuron 1 to 5) of the first memory item are hetero-associated with neuron group of second

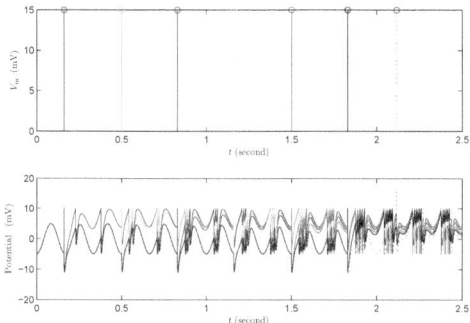

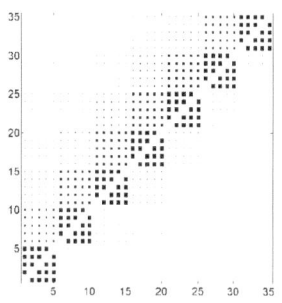

Fig. 3. Mechanism of STM network

Fig. 4. Synaptic matrix representation of LTM

memory item. Each neuron group in the CA3 network is hetero-associated with the next neuron group. This results in the formation of associative memory in the form sequential memory in LTM.

5 Conclusion

Simulation results showed that the proposed CA3 model using SRM neurons can capture multiple memory items in real time and incorporate them into LTM. The interaction between AHP, ADP and feedback inhibition led to the sustained firing of memory items in their order of introduction once every theta cycle. Consequently, the repeated firing of memory items led to the gradual incorporation of each item in LTM via modification of the synaptic strengths between groups of neurons. The simplicity of the SRM neuron model is demonstrated without any compromise in network functionality.

References

1. Arel, I., Rose, D.C., Karnowski, T.P.: Deep Machine Learning–A New Frontier in Artificial Intelligence Research. IEEE Comput. Intell. Mag. 5(4), 13–18 (2010)
2. Coyle, D.: Neural Network Based Auto Association and Time Series Prediction for Biosignal Processing in Brain Computer Interfaces. IEEE Computat. Intell. Mag. 4(4), 47–59 (2009)
3. Araneda, R., Andrade, R.: 5-hydroxytryptamine2 and 5-hydroxytryptamine1a receptors mediate opposing responses on membrane excitability in rat association cortex. Neuroscience 40(2), 399–412 (1991)
4. Bragin, A., Jando, G., Nadasdy, Z., Hetke, J., Wise, K., Buzsaki, G.: Gamma (40-100 hz) oscillation in the hippocampus of the behaving rat. The Journal of Neuroscience 15(1), 47–60 (1995)
5. Campanac, E., Debanne, D.: Plasticity of neuronal excitability: Hebbian rules beyond the synapse. Archives Italiennes de Biologie 145, 277–287 (2007)
6. Cutsuridis, V., Cobb, S., Graham, B.P.: Encoding and retrieval in a model of the hippocampal CA1 microcircuit. Hippocampus 20(3), 423–446 (2010)

7. Debanne, D., Poo, M.M.: Spike-timing dependent plasticity beyond synapse pre- and post-synaptic plasticity of intrinsic neuronal excitability. Frontiers in Synaptic Neurosci. 2(21), 1–5 (2010)

8. Fleischer, J.G., Krichmar, J.L.: Sensory integration and remapping in a model of the medial temporal lobe during maze navigation by a brain-based device. Journal of Integrative Neurosci. 6(3), 403–431 (2007)

9. Jensen, M.S., Azouz, R., Yaari, Y.: Spike after-depolarization and burst generation in adult rat hippocampal CA1 pyramidal cells. Journal of Physiology 492, 199–210 (1996)

10. Jensen, O., Idiart, M.A., Lisman, J.E.: Physiologically realistic formation of autoas- sociative memory in networks with theta/gamma oscillations: role of fast NMDA channels. Learning & Memory 3(2-3), 243–256 (1996)

11. Kandel, E.R., Spencer, W.A.: Electrophysiology of hippocampal neurons. ii. after- potentials and repetitive firing. Journal of Neurophysio. 24(3), 243–259 (1961)

12. Lavenex, P., Amaral, D.G.: Hippocampal-neocortical interaction: A hierarchy of associativity. Hippocampus 10(4), 420–430 (2000)

13. Lisman, J., Idiart, M.: Storage of 7 +/- 2 short-term memories in oscillatory sub- cycles. Science 267(5203), 1512–1515 (1995)

14. Maass, W., Bishop, C.M. (eds.): Pulsed Neural Networks. MIT Press, Cambridge (1998)

15. McKay, B.M., Matthews, E.A., Oliveira, F.A., Disterhoft, J.F.: Intrinsic neuronal excitability is reversibly altered by a single experience in fear conditioning. Journal of Neurophysio. 102(5), 2763–2770 (2009)

16. Miles, R.: Synaptic excitation of inhibitory cells by single CA3 hippocampal pyra- midal cells of the guinea-pig in vitro. Journal of Physiology 428(1), 61–77 (1990)

17. Monyer, H., Burnashev, N., Laurie, D.J., Sakmann, B., Seeburg, P.H.: Develop- mental and regional expression in the rat brain and functional properties of four NMDA receptors. Neuron 12(3), 529–540 (1994)

18. O'Keefe, J., Recce, M.L.: Phase relationship between hippocampal place units and the eeg theta rhythm. Hippocampus 3(3), 317–330 (1993)

19. Pantic, L., Torres, J., Kappen, H.J., Gielen, S.C.: Associative memory with dy- namic synapses. Neural Computation 14(12), 2903–2923 (2002)

20. Park, J.Y., Remy, S., Varela, J., Cooper, D.C., Chung, S., Kang, H.W., Lee, J.H., Spruston, N.: A post-burst afterdepolarization is mediated by group I metabotropic glutamate receptor-dependent upregulation of $Ca_v 2.3$ R-type calcium channels in CA1 pyramidal neurons. PLoS Biol. 8(11), e1000534 (2010)

21. Rolls, E.T.: A computational theory of episodic memory formation in the hip- pocampus. Behavioural Brain Research 215(2), 180–196 (2010)

22. Sato, N., Yamaguchi, Y.: Spatial-area selective retrieval of multiple object-place as- sociations in a hierarchical cognitive map formed by theta phase coding. Cognitive Neurodynamics 3(2), 131–140 (2009)

23. Sommer, F.T., Wennekers, T.: Associative memory in networks of spiking neurons. Neural Netw. 14(6-7), 825–834 (2001)

24. Storm, J.F.: An after-hyperpolarization of medium duration in rat hippocampal pyramidal cells. Journal of Physiology 409(1), 171–190 (1989)

25. Vanderwolf, C.: Hippocampal electrical activity and voluntary movement in the rat. Electroencephalography and Clinical Neurophysiology 26(4), 407–418 (1969)

26. Xu, J., Kang, N., Jiang, L., Nedergaard, M., Kang, J.: Activity-dependent long- term potentiation of intrinsic excitability in hippocampal CA1 pyramidal neurons. Journal of Neuros. 25(7), 1750–1760 (2005)

Goal-Oriented Behavior Generation for Visually-Guided Manipulation Task

Sungmoon Jeong[1], Yunjung Park[2], Hiroaki Arie[3], Jun Tani[3], and Minho Lee[1,2]

[1] School of Electronics Engineering
[2] School of Robotics Engineering, Kyungpook National University,
1370 Sankyuk-Dong, Puk-Gu, Taegu 702-701 Korea
[3] Brain Science Institute, RIKEN,
2-1 Hirosawa, Wako-shi, Saitama, 351-0198 Japan
{jeongsm,yj-park}@ee.knu.ac.kr, arie@bdc.brain.riken.jp,
tani@brain.riken.jp, mholee@knu.ac.kr

Abstract. We propose a new neuro-robotics network architecture that can generate goal-oriented behavior for visually-guided multiple object manipulation task by a humanoid robot. For examples, given a "sequential hit" multiple objects task, the proposed network is able to modulate a humanoid robot's behavior by taking advantage of suitable timing for gazing, approaching and hitting the object and again for the other object. To solve a multiple object manipulation task via learning by examples, the current study considers two important mechanisms: (1) stereo visual attention with depth estimation for movement generation, dynamic neural networks for behavior generation and (2) their adaptive coordination. Stereo visual attention provides a goal-directed shift sequence in a visual scan path, and it can guide the generation of a behavior plan considering depth information for robot movement. The proposed model can simultaneously generate the corresponding sequences of goal-directed visual attention shifts and robot movement timing with regards to the current sensory states including visual stimuli and body postures. The experiments show that the proposed network can solve a multiple object manipulation task through learning, by which some novel behaviors without prior learning can be successfully generated.

Keywords: Multiple object manipulation task, visual attention shifts, behavior generation, stereo visual attention, multiple time-scale recurrent neural networks.

1   Introduction

To achieve the visual-guided manipulation tasks for multiple objects by a humanoid robot, for example, approaching and hitting one object then again another object, the visual attention and robot movement need to switch to a specific object corresponding to a specified task in time. Humans control gaze shifts and fixations (visual attention) proactively to gather visual information for guiding movements, which is highly related to a specified task [1]. In addition, visual attention can effortlessly detect

B.-L. Lu, L. Zhang, and J. Kwok (Eds.): ICONIP 2011, Part I, LNCS 7062, pp. 501–508, 2011.

(location) and help to recognize (identification) an interesting area or object within natural or cluttered scenes through the selective attention mechanism with various visual features [2] – [4].

The current study, based on the previous study [5], examines how a humanoid robot can learn to manipulate multiple objects in a sequence by acquiring adequate visual attention shifts and movement timing with depth information. The stereo visual attention system selects a specific object region with depth information among several candidate areas with multiple objects from input images [5], [6]. The behavior system sequentially plays pre-defined specific behavior such as walking and approaching an object. Another essential idea is to utilize a functional hierarchy and to integrate the stereo visual attention and behavior sequence generation by employing a new dynamic neural network model called the multiple timescale recurrent neural network (MTRNN) [7], [8]. In this work, Yamashita et al.'s work for single object manipulation is extended to multiple objects by localizing each object's characteristics, through time, by visually selective attention. And previous Jeong et al.'s work for multiple object manipulation in the same location is extended to different object locations with depth information. It means that a current task needs to correctly choose the walking in a distance and approaching behavior commands to the specific object through time.

Previous research has shown that a set of primitive behaviors are acquired in the fast dynamic network for lower-level activities, while the sequencing of these primitive behaviors in the slow dynamic network develop into higher-level activities [5]. In the current model, the function of a sequence generation by a visual attention shift and behavior shift are considered to be attained in the slow dynamics network along with the one for sequencing of the primitive behaviors such as walking to and hitting the object. The output command sequences of the visual shift and behavior shift from this slow dynamics network are sent to a hard-wired gaze system and a behavior system to attend target objects to localize the target objects and to achieve behavior control through time, respectively. In our experiment, the robot learns to perform visual attention shifts with depth information and behavior shifts followed by acquired behavior patterns through supervisor; the robot sequentially attends and approaches to an object before hitting it and then to the other object.

This paper is organized as follows: In Section 2, we present the proposed new neuro-robotics architecture that is combined with a stereo visual attention system, behavior system and a new dynamic neural network model known as MTRNN. In Section 3, we present the experiment and its results. The discussion and conclusions follow in Section 4.

2   Visually-Guided Behavior Planning Generation

2.1   Overall Architecture

Fig. 1 describes the overall information flow in the proposed model. When receiving a desired task as the inputs in the slow dynamics of MTRNN, the MTRNN simultaneously predicts how the robot behavior command (b_{t+1}) and visual attention command (v_{t+1}) change through time. The inputs for the MTRNN consist of visual attention command (v_t), robot behavior command (b_t) and a visual stimuli (s_t). Here, the visual attention command represents 3 object categories (red, green, blue colors) to be attended and behavior command represents 3 different behavior categories (walk, turn, hit).

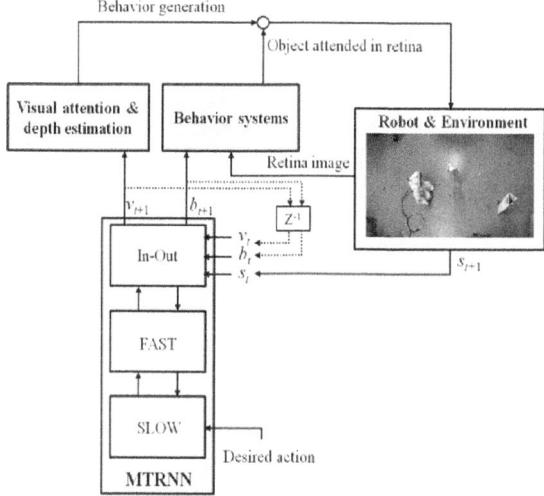

Fig. 1. Proposed neuro-robotics architecture. Inputs for the MTRNN: v_t (visual attention command at time t: 3 dimensions vector (red, green, and blue)), s_t (visual stimuli at time t: 2 dimensions vector (encoding object location information such as angle and depth)), b_t (behavior command at time t: 3 dimensions vector (walk, turn, and hit)), Outputs of the MTRNN: v_{t+1}, b_{t+1}. Z^{-1} (time delay unit), In-Out (input and output unit), FAST (fast context unit: time constant τ=2), SLOW (slow context unit: time constant τ=30).

Each vector with 2 and 3 dimensions is transformed by a topology preserving map (TPM) to cluster as 64 TPM units and 16 TPM units in our experiments. The stereo visual attention system receives a visual attention command from the MTRNN and the retina image from the robot's vision. The spatial location of a target object is encoded by using an angle between robot body and objects together with depth information obtained from the stereo-type visual attention system [4], [6]. The spatial location of a target object is used for helping the robot movement shifts such as turning or walking to the object. The spatial location of a target object, the visual attention command and behavior command are fed back to the inputs of the MTRNN.

2.2   Stereo Saliency Map for Visual Attention and Depth Estimation

In the current study, the dynamic neural network generates not only a goal-directed arm behavior signal, but also a visual attention command to a vision system with a stereo visual attention to find a specified object and complete the task. In the course of detecting an object to achieve the object manipulation task, stereo bottom-up processing work together for selective attention on a specified object region in an input scene. The MTRNN generates a goal-directed visual attention sequence in time at a test mode. Then, a localized area selected by a bottom-up SM model is tested for matching how much the selected area meets the visual characteristics of an object for a specific manipulation task generated by the MTRNN. For example, if a visual attention command from the MTRNN is to find a blue object, then only the blue characteristic is intensified in a bottom-up SM (wanted feature information), the other color is inhibited (unwanted feature information). After successfully localizing corresponding landmarks on both the left image and right image, we are able to get depth information by means of the stereo visual attention model [6].

2.3   MTRNN for Coordination of Attention and Behavior Sequences

The MTRNN is a type of the continuous time recurrent neural network (CTRNN) model in which neurons have different time scales; therefore, the MTRNN has the functional hierarchy characteristic [7], [9], [10]. From this characteristic, neurons with a fast time constant encode a set of primitive behaviors, and neurons with a slow time constant prepare for the compositional sequences of these primitives behavior. The MTRNN has three groups of neural units in present study, namely input-output units (68), fast context units (70) and slow context units (30). Among the input units, the first 36 units (i = 1-36) correspond to the visual input (angle and depth), the next 16 units (i = 37-52) correspond to the visual attention command and the last 16 units (i = 53-68) correspond to the behavior command, respectively. The 8 dimensional inputs were thus transformed into 68 dimensional sparsely encoded vectors by a topology preserving map (TPM) with 3×10^6 training epochs [11], [12]. This transformation reduces the redundancy of the input trajectories for units. The size of the TPMs is 36 (6×6) for vision sense information, 16 (4×4) for the visual attention command and 16 (4×4) for the behavior command. The fast context units are connected with the input-output units of which synaptic weights are determined through learning by examples. The activation of these neurons is calculated by Eq. (1)

$$\tau_i (du_{i,t} / dt) = -u_{i,t} + \sum_j w_{ij} x_{j,t} \qquad (1)$$

where $u_{i,t}$ is the membrane potential of each i-th neural unit at time step t and $x_{j,t}$ is the neural state of the j-th unit, and w_{ij} is synaptic weight from the j-th unit to the i-th unit. The time constant τ is defined as the decay rate of a unit's membrane potential. If the τ value is large, the activation of the unit changes slowly, because the internal state potential is strongly affected by history of the unit's potential. On the other hand, if the τ value is small, the effect of history of the unit's potential is also small, and thus it is possible for activation of the unit to change quickly. The fast context units with small time constant (τ = 2) whose activity changed quickly, whereas the slow context unit with a large time constant (τ = 30) whose activity, in contrast, changed much more slowly.

3   Experiments and Results

3.1   Experiment Setting; Humanoid Robot and Workbench

A humanoid robot NAO was used in the role of a physical body interacting with the environment. To calculate the disparity information for depth, we installed stereo camera modules on the NAO's head as shown in Fig. 2.

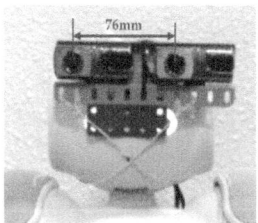

Fig. 2. Stereo camera module on NAO's head

The stereo cameras have 640x480 resolutions, 76mm distance between two cameras, 5.1mm focal length and 4.8mm CCD width. A workbench was set as shown in Fig. 3.

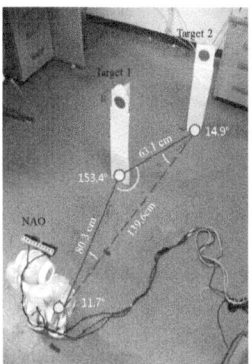

Fig. 3. Workbench for robot experiments

Two cardboard towers with different color marks are located in similar height (52cm from floor) at the same level as the robot's eye. The positions of the color marks were fixed and the NAO's initial position is shown in Fig. 3. Two target color marks are located about 80cm and 139cm in front of NAO, respectively.

3.2   Visually-Guided Multiple Objects Manipulation Task

In the current experiment, the robot was trained for multiple objects manipulation task as shown in Fig. 4. Our multiple object manipulation task considering visually guided behavior information is; first, NAO attends and approaches to the first target of a red mark, then hitting the object as shown in Figs. 4 (a) and (b). Second, NAO detects and approaches the other target 2 of a blue mark, then again hitting the second object as shown in Figs. 4 (c) and (d). After training the behavior sequences, the tests were conducted by a regeneration of them. Then, the experiment was further conducted by a generalization test with untrained situation. Four teaching sequences were used to train for each NAO position that was learned by the error backpropagation through time (BPTT) algorithm, with 5×10^3 training epochs [7]. It was considered that a trial was successful if the two objects were sequentially and successfully knocked down by the robot during the course of the experiment.

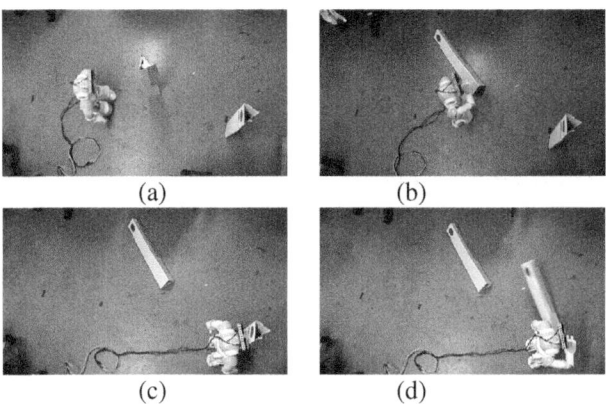

(a)   (b)

(c)   (d)

Fig. 4. Multiple objects manipulation task: (a) attending and approaching first target, (b) hitting the first target, (c) attending and approaching the second target, (d) hitting the second target

3.3   Task generation Results

Fig. 5 shows examples of behavioral sequences of the given task. In vision sensation as shown in Fig. 5 (a), solid line represents the angle between robot body and a target object and dashed line represents depth information of selected object. In the case of vision sensation, the angle and depth information are fed back to robot according to a robot interaction with environment. In the visual attention command as shown in Fig. 5 (b), the solid, dashed and gray lines are the sequences of attention command signals to focus on the red color, the green color and the blue color marks, respectively. In the behavior command as shown in Fig. 5 (c), the solid, dashed and gray lines are the sequences of behavior commands for walking, turning and hitting of the objects, respectively. The detail robot behavior is as follows: first, the robot was initially set by home position. The MTRNN simultaneously generates a sequence both for the visual attention command and for the behavior command. Using the output values of MTRNN, the robot can autonomously shift the visual attention and behavior generation.

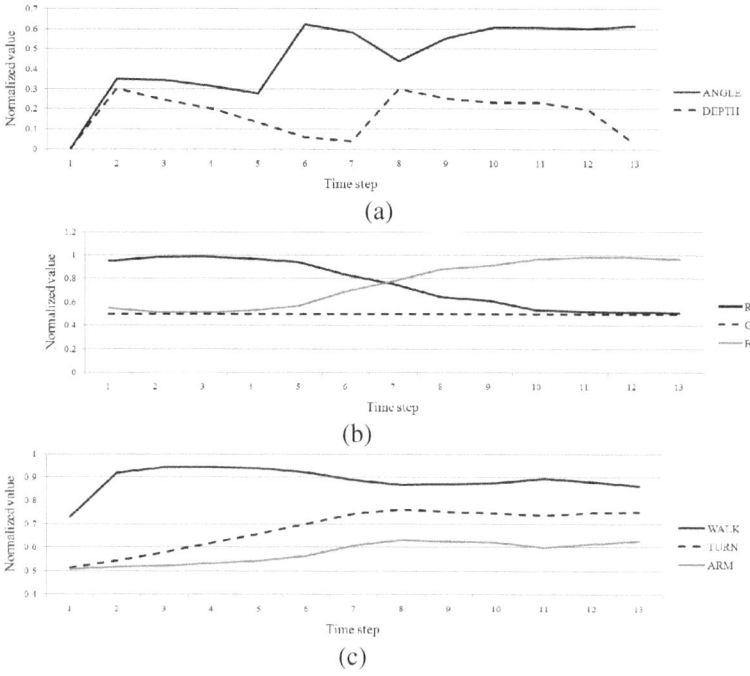

Fig. 5. The results of behavioral sequences of the given task

(1) From the initial step to 6th step: according to the visual attention (red attention command is dominant) and behavior command (walk behavior command is dominant), the robot detects, attends, and walks to the red target object. (2) At the 7th step: after encoding the location of the target object considering, humanoid robot hits the red target object by robot hand, in which the robot is finding to the red target object (behavior command for hitting is dominant). (3) From the 8th step to 12th step: the robot detects, attends and walks to the blue target object by suitable behavior command (blue attention command and walk behavior command are dominant). (4) At the final 13th step: the robot hits the target blue object (behavior command for hitting is dominant).

4   Conclusion and Further Works

For achieving the visually-guided multiple object manipulation tasks for neuro-robotics via learning by example, we proposed a new dynamic behavior sequence generation model in which the visual attention shift and motor behaviors shifts are associated with task specific manners. The proposed model can generate goal-directed behavior concerning the current sensory states including visual stimuli through interaction with a real environment. We will consider an active perception of an object which utilizes texture and appearance of objects to generate the complex top-down attention using high-level visual cognition for achieving the high-level object manipulation task. Also, we will introduce a reinforcement learning paradigm for acquiring an improved perception with depth estimation and behavior shift skills, which has been inspired by McCallum's model.

Acknowledgments. This research was supported by the Converging Research Center Program funded by the Ministry of Education, Science and Technology (2011K000659) (50%) and also the Original Technology Research Program for Brain Science through the National Research Foundation of Korea (NRF) funded by the Ministry of Education, Science and Technology (2011-0018292) (50%). Also, the authors would like to thank to Mr. Timothy A. Mann for his help in revision.

References

1. Johansson, R.S., Westling, G., Bäckström, A.: Eye-hand coordination in object manipulation. The Journal of Neuroscience 21, 6917–6932 (2001)
2. Ballard, D.H., Hayhoe, M.M., Li, F., Whitehead, S.D., Frisby, J.P., Taylor, J.G., Fisher, R.B.: Hand-eye coordination during sequential tasks [and Discussion]. Philosophical Transactions: Biological Sciences Royal Society 337, 331–339 (1992)
3. Berkinblit, M.B., Fookson, O.I., Smetanin, B., Adamovich, S.V., Poizner, H.: The interaction of visual and proprioceptive inputs in pointing to actual and remembered targets. Exp. Brain Res. 107, 326–330 (1995)
4. Jeong, S., Ban, S.-W., Lee, M.: Stereo saliency map considering affective factors and selective motion analysis in a dynamic environment. Neural Networks 21, 1420–1430 (2008)
5. Jeong, S., Lee, M., Arie, H., Tani, J.: Developmental learning of integrating visual attention shifts and bimanual object grasping and manipulation tasks. In: Proc. IEEE Int. Conf. Develop. Learn., Ann Arbor, MI, pp. 165–170 (2010)
6. Choi, S.B., Jung, B.S., Ban, S.W., Niitsuma, H., Lee, M.: Biologically motivated vergence control system using human-like selective attention model. Neurocomputing 69, 537–558 (2006)
7. Yamashita, Y., Tani, J.: Emergence of functional hierarchy in a multiple timescale neural network model: A humanoid robot experiment. PLoS Comput. Biol. 4(11), 1–18 (2008)
8. Nishimoto, R., Tani, J.: Development of hierarchical structures for actions and motor imagery: a constructivist view from synthetic neurorobotics study. Psychological Research 73, 545–558 (2009)
9. Williams, R.J., Zipser, D.: A learning algorithm for continually running fully recurrent neural networks. Neural Computation 1(2), 270–280 (1989)
10. Doya, K., Yoshizawa, S.: Memorizing oscillatory patterns in the analog neuron network. In: Proceedings of 1989 International Joint Conference on Neural Networks, Washington, D.C, vol. I, pp. 27–32 (1989)
11. Saarinen, J., Kohonen, T.: Self-organized formation of colour maps in a model cortex. Perception 14(6), 711–719 (1985)
12. Kohonen, T.: Self-Organizing Maps. Springer Series in Information Sciences, vol. 30. Springer, Heidelberg (2001)

Dynamic Complex-Valued Associative Memory with Strong Bias Terms

Yozo Suzuki, Michimasa Kitahara, and Masaki Kobayashi

University of Yamanashi, Interdisciplinary Graduate School of Medicine and
Engineering,
4-3-11 Takeda, Kofu, Yamanashi 400-8511, Japan
{t08kf018,g11dm001,k-masaki}@yamanashi.ac.jp

Abstract. Complex-valued associative memory (CAM) can store multi-level patterns. Dynamic complex-valued associative memory (DCAM) can recall all stored patterns. The CAM stores the rotated patterns, which are typical spurious states, in addition to given training patterns. So DCAM recalls all the rotated patterns in the recall process. We introduce strong bias terms to avoid recalling the rotated patterns. By computer simulations, we can see that strong bias terms can avoid recalling the rotated patterns unlike simple bias terms.

Keywords: complex-valued associative memory, dynamic neuron, rotated patterns, bias terms.

1 Introduction

Several advanced associative memory models have been proposed since Hopfield proposed Hopfield associative memory (HAM) (Hopfield[1,2]). Complex-valued associative memory (CAM) is one of them (Jankowski et al.[3]). Chaotic associative memories can recall all the stored patterns. Nakada and Osana[8] proposed Chaotic complex-valued associative memory (CCAM). Kitahara et al.[4] quantified and simplified CCAM after Nagumo and Sato[7]. In this paper, the quantified version is referred to as Dynamic complex-valued associative memory (DCAM).

CAM has an inherent property of rotation invariance. For a pattern vector $\mathbf{z}$, patterns $\exp(\theta\sqrt{-1})\mathbf{z}$ are called the rotated patterns. CAM stores not only training patterns but also their rotated patterns. Rotated patterns correspond to inversed patterns of the HAM. When the complex-valued neurons are K-level neurons, there are $K-1$ rotated patterns for each training pattern. In recall process, DCAM recalls all the rotated patterns. In general, we cannot tell training patterns from rotated patterns because of rotation invariance.

We introduce bias terms to DCAM in order to avoid recalling the rotated patterns. Lee[5] and Muezzinoglu et al.[6] have simply introduced bias terms. It is equivalent to appending a neuron which constantly outputs one. Simple use of bias terms, however, is almost in vain for DCAM. So we introduce strong bias terms. We examined the effect of strong bias term by computer simulations.

B.-L. Lu, L. Zhang, and J. Kwok (Eds.): ICONIP 2011, Part I, LNCS 7062, pp. 509–518, 2011.

The appearance of rotated patterns gradually reduced as the bias terms became stronger. For too strong bias terms, however, DCAM recalled some spurious patterns which are not rotated patterns.

2   Complex-Valued Associative Memory

First, we construct complex-valued neurons. A complex-valued neuron receives and produces a complex number. Let K be a positive integer greater than two. And let a real number θ_K and complex numbers s_k be as follows:

$$\theta_K = \frac{\pi}{K}, \tag{1}$$

$$s_k = \exp(2k\theta_K). \tag{2}$$

The integer K is a resolution factor to divides the complex unit circle into K quantization levels. The set $\{s_k\}$ is the set of states of complex-valued neurons. The states of complex-valued neurons are determined by a complex-signum function $f(\cdot)$. The complex-signum function $f(\cdot)$ is defined as follows:

$$f(z) = \begin{cases} s_0 & 0 \le z < \theta_K \\ s_1 & \theta_K \le z < 3\theta_K \\ s_2 & 3\theta_K \le z < 5\theta_K \\ \vdots & \\ s_{K-1} & (2K-3)\theta_K \le z < (2K-1)\theta_K \\ s_0 & (2K-1)\theta_K \le z < 2\pi \end{cases} \tag{3}$$

Next, we construct complex-valued associative memory (CAM). Let a complex number w_{ji} be the connection weight from the neuron i to the neuron j. And let z_i be the state of the neuron i. Then the input sum S_j to the neuron j is defined as follows:

$$S_j = \sum_{i \neq j} w_{ji} z_i. \tag{4}$$

The neurons are updated in sequence. We restrict the connection weights as follows:

$$w_{ji} = \overline{w}_{ij}, \tag{5}$$

where $\overline{w}$ stands for the complex conjugate of w. This restriction makes CAM reach a stable state.

Suppose the state $\mathbf{z} = (z_1, z_2, \cdots, z_N)$, where N is the number of neurons, is stable. Then the following equation holds for all j:

$$z_j = f(S_j). \tag{6}$$

Consider a rotated pattern of $\mathbf{z}$. There exists k such that the rotated pattern is expressed by $s_k \mathbf{z} = (s_k z_1, s_k z_2, \cdots, s_k z_N)$. The following equation holds true for the rotated pattern:

$$f\left(\sum_{i \neq j} w_{ji} s_k z_i\right) = s_k f\left(\sum_{i \neq j} w_{ji} z_i\right) = s_k z_j. \tag{7}$$

Therefore the rotated patterns are also stable.

Finally, we describe the learning method. Hebbian rule has often been used. Let $\mathbf{z}_p = (z_{p1}, z_{p2}, \cdots, z_{pN})$ be the training pattern p. Then the connection weights w_{ji} are given as follows:

$$w_{ji} = \sum_p z_{pj} \bar{z}_{pi}. \tag{8}$$

Given a training pattern $\mathbf{z}_q$ to CAM, we can calculate the input sum S_j to the neuron j as follows:

$$S_j = \sum_p z_{pj} \bar{z}_{pi} z_{qi} \tag{9}$$

$$= (N-1) z_{pj} + \sum_{p \neq q} z_{pj} \bar{z}_{pi} z_{qi}. \tag{10}$$

The first term help CAM recall the training pattern q and the second term is the cross-talk term.

3   Dynamic Complex-Valued Associative Memory

CAM surely reaches a stable state. we introduce dynamic complex-valued neuron to get out of stable states and search all stable states. Let $z(t)$ and $S(t)$ be the neuron output and input in time t, respectively. A dynamic complex-valued neuron behaves as follows:

$$z(t+1) = f\left(S(t+1) - \alpha \sum_{d=0}^{t} k^d z(t-d)\right), \tag{11}$$

where α and k are the scaling factor and the damping factor, respectively. The second term help CAM move to the opposite state. When CAM keep a stable state, the effect of the second term increases and CAM can get out of the stable state.

DCAM can recall all the training patterns. However it also recalls all the rotated patterns of training patterns. We need the system which can recall only training patterns.

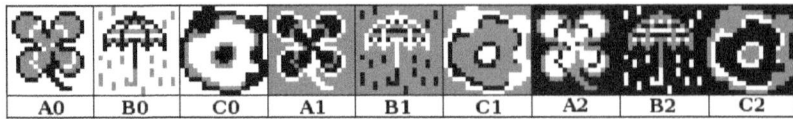

Fig. 1. Patterns A0, B0 and C0 are the training patterns. Patterns A1, B1 and C1 are the rotated patterns by $2\pi/3$. Patterns A2, B2 and C2 are the rotated patterns by $4\pi/3$.

4   Bias Terms

To destroy rotation invariance, bias terms have often been used. Bias terms are considered as neurons which constantly produce one. Let w_{j0} be the bias term of the neuron j. Then the neuron input S_j is defined as follows:

$$S_j = \sum_{i \neq j} w_{ji} z_i + w_{j0}. \tag{12}$$

By hebbian rule, w_{j0} are often determined as follows:

$$w_{j0} = \sum_p z_{pj}. \tag{13}$$

The effect of the bias terms is few, because of quantification. We introduce strong bias terms as follows:

$$w_{j0} = C \sum_p z_{pj}, \tag{14}$$

where C is a parameter to control the strength of bias terms. In case of $C = 0$, there exist no bias terms. In case of $C = 1$, conventional bias terms are used.

5   Computer Simulations

We carried out computer simulations to examine the effect of bias terms. The conditions were as follows:

1. $K = 3$
2. $N = 400$
3. $\alpha = 17$
4. $k = 0.98$
5. The number of training patterns was three.

To visualize the recall process, we used the training patterns A0, B0 and C0 in Fig. 1. Each dot stands for the state of the neuron. White, gray and black correspond to s_0, s_1 and s_2, respectively. The patterns A1, B1 and C1 are the rotated patterns of A0, B0 and C0 by $2\pi/3$, respectively. The ones A2, B2 and

Table 1. Appeared pattterns in case of C = 0

patterns	periods				
A0	0 - 17 125 - 130				
B0	19 - 31 141 - 154				
C0	158 - 171				
A1	65 - 79				
B1	46 - 59				
A2	94 - 112				
B2	114 - 117				
C2	180 - 192				
others	35 - 43	62 - 63	82 - 91	119 - 123	133 - 138
	174 - 175	198 - 199			

Table 2. Appeared patterns in case of C = 100

patterns	periods				
A0	0 - 9 59 - 68 160 - 170				
B0	12 - 26 173 - 191				
C0	30 - 56 121 - 143 194 - 199				
C1	75 - 89				
C2	99 - 118				
others	70 - 71	93 - 97	146 - 147	149 - 150	154 - 155

C2 are the rotated patterns of A0, B0 and C0 by $4\pi/3$, respectively. In recall process, we regarded the continuously appeared patterns as recalled patterns.

We carried out computer simulations in case of $C = 0, 1, 100, 200, 300$ and 400. In case of $C = 0$, DCAM has no bias terms. In case of $C = 1$, it has conventional bias terms. In case of $C = 100, 200, 300$ and 400, it has strong bias terms. Figures 4-7 show the simulation results between $t = 0$ and $t = 199$.

In case of $C = 0$ (Fig. 2), the simulation result is Table 1. Others stand for spurious patterns except rotated patterns. The typical patterns of them are mixture patterns. All the training patterns appeared. However, many rotated patterns also appeared. Moreover, other patterns including mixture patterns also appeared. The simulation result in case of $C = 1$ (Fig. 3) is almost same as that in case of $C = 0$. Therefore simple bias term does not help avoiding spurious patterns.

In case of $C = 100$ (Fig. 4), the simulation result is Table 2. All the training patterns appeared. Rotated patterns of A0 and B0 disappeared. The other spurious patterns less frequently appeared. However, the appearances of spurious patterns remained.

In case of $C = 200$ (Fig. 5), the simulation result is Table 3. All the training patterns appeared and any rotated patterns disappeared. The other spurious patterns appeared only in initial periods and disappeared after $t = 14$. The stable states recalled in initial periods are considered to be attracted by the bias terms, because the bias terms prompt DCAM to recall the mixture pattern of all

Table 3. Appeared patterns in case of C = 200

patterns	periods
A0	0 - 3 51 - 73 129 - 146
B0	15 - 25 76 - 98 149 - 172
C0	29 - 47 101 - 126 178 - 199
others	5 - 7 8 - 13

Table 4. Appeared patterns in case of C = 400

patterns	periods					
A0	70 - 88	130 - 145	191 - 199			
B0	35 - 43	91 - 105	148 - 161			
C0	52 - 67	111 - 126	171 - 186			
others	2 - 7	8 - 13	14 - 16	17 - 21	23 - 30	31 - 32
	46 - 50	165 - 168				

Table 5. The frequency of appeared patterns

C	A0	B0	C0	A1	B1	C1	A2	B2	C2	others
0	88	88	4	87	84	3	85	99	4	232
1	86	92	3	86	84	3	86	83	3	234
100	82	103	95	19	22	39	30	36	45	245
200	134	134	134	0	0	0	0	0	0	2
300	150	150	149	0	0	0	0	0	0	5
400	184	184	125	0	0	0	0	0	0	246

the training patterns. After the mixture pattern was taken into the second term of expression (11), the mixture pattern disappeared. Therefore, the strong bias terms helped DCAM avoid appearances of spurious patterns. The simulation result in case of $C = 300$ (Fig. 6) is almost same as that in case of $C = 200$. So we can see that the behavior is not sensitive to the parameter C.

In case of $C = 400$ (Fig. 7), the simulation result is Table 4. All the training patterns appeared and any rotated patterns disappeared. However the other spurious patterns more frequently appeared. Therefore the parameter C is considered to be too large.

Moreover we summarize the simulation results until $t < 10000$. Table 5 shows that frequencies of training patterns, rotated patterns and the other patterns for each C. In case of $C = 0$ or 1, the pattern C0 hardly appeared. In case of large C, all training patterns almost equally appeared. In case of $C = 100$, the appearances of rotated patterns reduced. In case of $C \geq 200$, all the rotated patterns disappeared. In case of $C = 200$ or 300, we can see that the other patterns disappeared after initial periods. from Table 5 and Fig. 5 and 6. These are considered as ideal cases. In case of $C = 400$, many other patterns appeared.

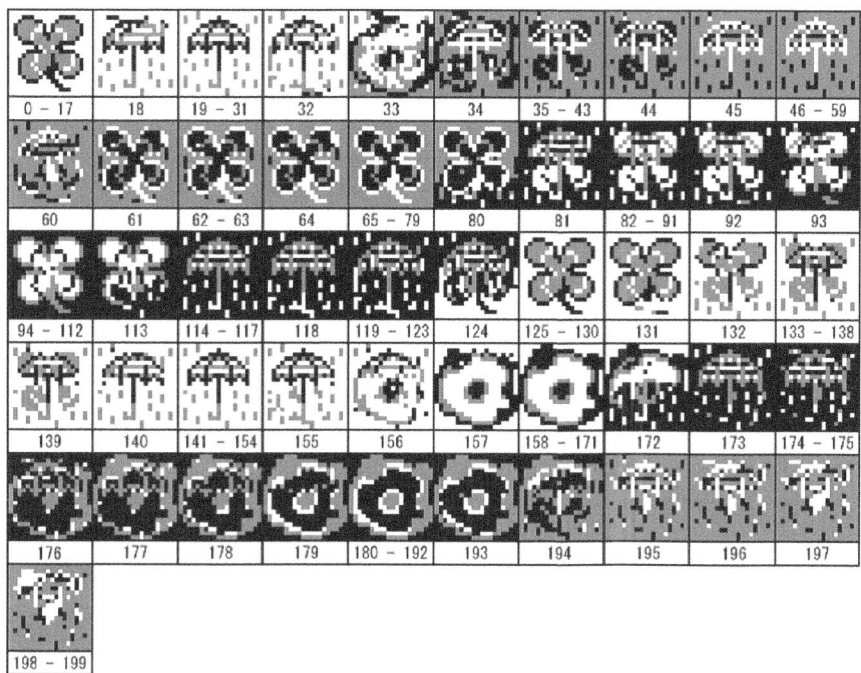

Fig. 2. Recall process in case of $C = 0$. In this case, there were no bias terms.

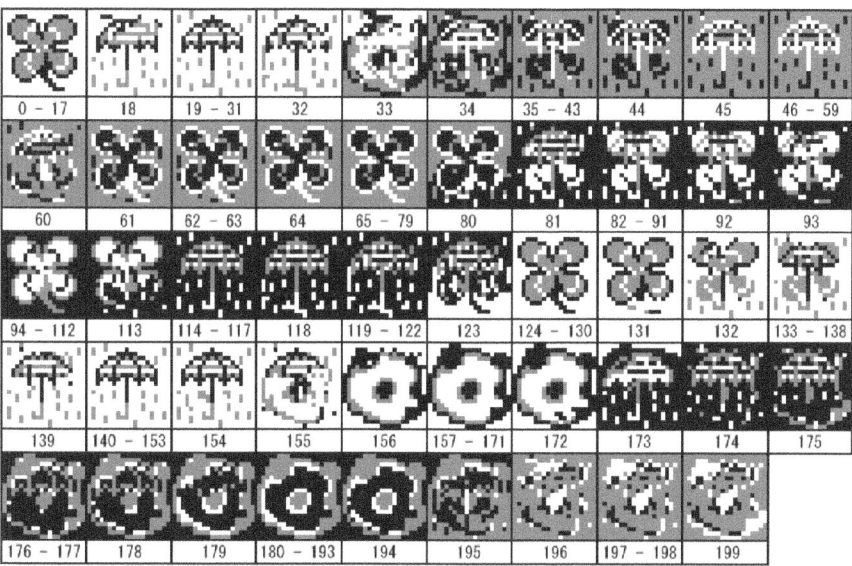

Fig. 3. Recall process in case of $C = 1$. In this case, the conventional bias terms were used.

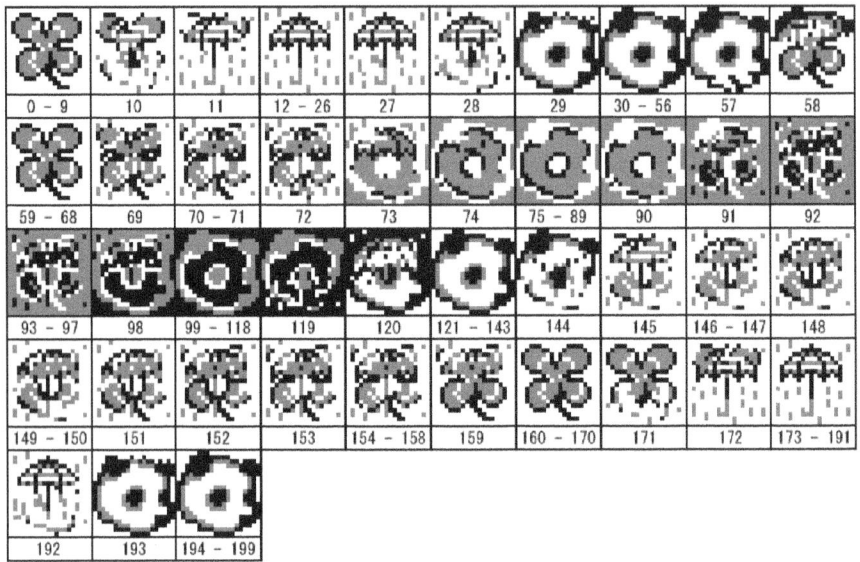

Fig. 4. Recall process in case of $C = 100$

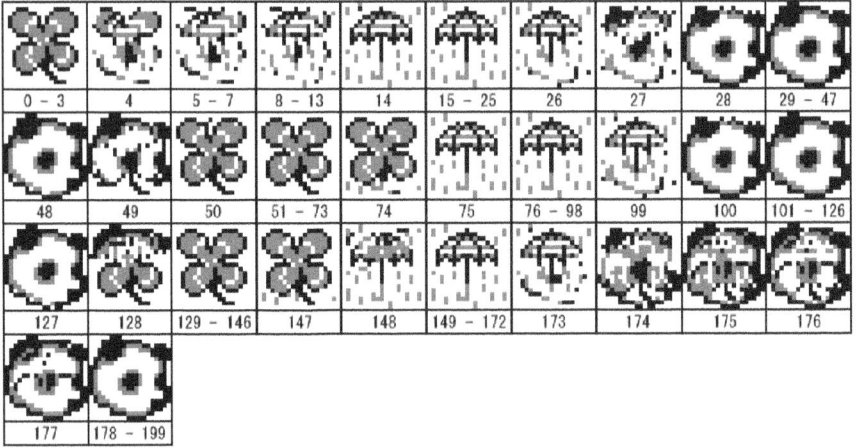

Fig. 5. Recall process in case of $C = 200$

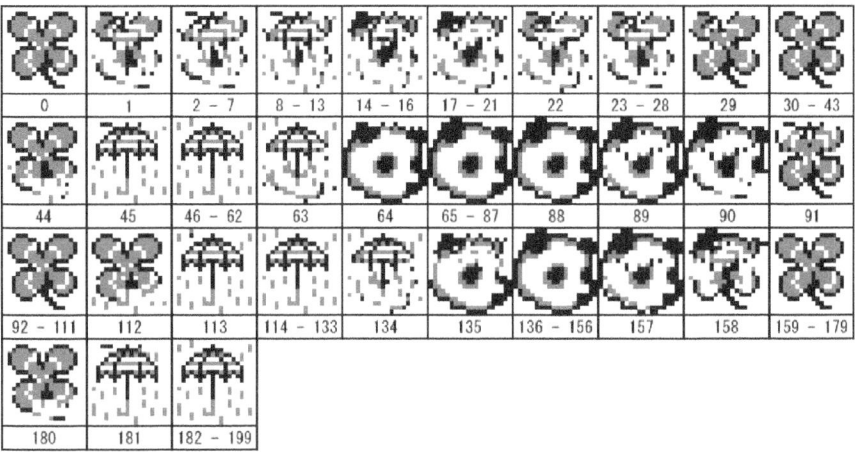

Fig. 6. Recall process in case of $C = 300$

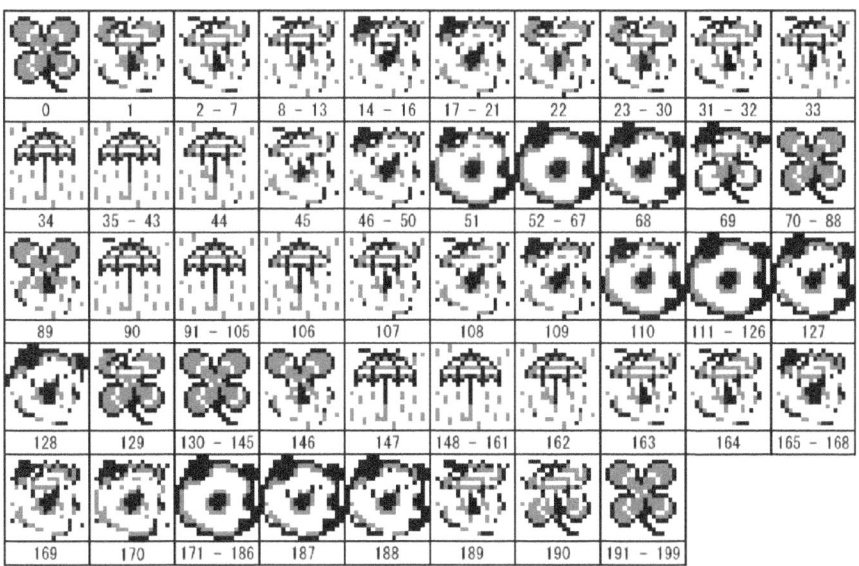

Fig. 7. Recall process in case of $C = 400$

6   Conclusion

The conventional DCAM recalls not only training patterns but also their rotated patterns. We introduced strong bias terms to avoid recalling the rotated patterns. When the proper bias terms were used, DCAM behaved as follows.

1. All the training patterns were recalled.
2. Any the rotated patterns of training patterns disappeared.
3. Only in initial periods, mixture patterns appeared. After that, no spurious patterns appeared.

Moreover, we can see that the behavior is not sensitive to the parameter C.

Our proposed method can be regarded as appending C constant neurons producing one. Then the substantive number of neurons is $N+C$ and the correlations between training patterns become higher.

References

1. Hopfield, J.J.: Neural networks and physical systems with emergent collective computational abilities. Proceedings of the National Academy of Sciences of the United States of America 79(8), 2554–2558 (1982)
2. Hopfield, J.J.: Neurons with graded response have collective computational properties like those of two-state neurons. National Academy of Sciences of the United States of America 81(10), 3088–3092 (1984)
3. Jankowski, S., Lozowski, A., Zurada, J.M.: Complex-valued multistate neural sssociative memory. IEEE Transaction on Neural Networks 7(6), 1491–1496 (1996)
4. Kitahara, M., Kobayashi, M., Hattori, M.: Chaotic rotor associative memory. In: Proceedings of International Symposium on Nonlinear Theory and its Applications, pp. 399–402 (2009)
5. Lee, D.L.: Improvements of complex-valued hopfield associative memory by using generalized projection rules. IEEE Transaction on Neural Networks 17(5), 1341–1347 (2006)
6. Muezzinoglu, M.K., Guzelis, C., Zurada, J.M.: A new design method for the complex-valued multistate hopfield associative memory. IEEE Transaction on Neural Networks 14(4), 891–899 (2003)
7. Nagumo, J., Sato, S.: On a response characteristic of a mathmatical neural model. Kybernetik 10(3), 155–164 (1972)
8. Nakada, M., Osana, Y.: Chaotic complex-valued associative memory. In: Proc. International Symposium on Nonlinear Theory and its Applications, pp. 493–496 (2007)
9. Zemel, R.S., Williams, C.K.I., Mozer, M.C.: Lending direction to neural networks. Neural Networks 8(4), 503–512 (1995)

Widely Linear Processing
of Hypercomplex Signals

Tohru Nitta

National Institute of Advanced Industrial Science and Technology (AIST),
Tsukuba Central 2, 1-1-1 Umezono, Tsukuba-shi, Ibaraki, 305-8568 Japan
tohru-nitta@aist.go.jp

Abstract. In this paper, we formulate a Clifford-valued widely linear estimation framework. Clifford number is a hypercomplex number that generalizes real, complex numbers, quaternions, and higher dimensional numbers. And also, as a first step, we will give a theoretical foundation for a quaternion-valued widely linear estimation framework. The estimation error obtained with the quaternion-valued widely linear estimation method is proven to be smaller than that obtained using the usual quaternion-valued linear estimation method.

1   Introduction

The widely linear (WL) estimation method has been proven mathematically to be effective for estimation problems using complex-valued data. Estimation using WL uses complex conjugate parameters in addition to complex-valued parameters [1]. It has been applied to communications [2,3] and adaptive filters [4], together with so-called *augmented complex statistics*, a concept introduced by Picinbono et al. Moreover, an extension of the WL method to quaternion-valued case has been presented [5], which fully exploits available statistical information. A quaternion, a four-dimensional number invented by W. R. Hamilton in 1843, is an extension of a complex number. Quaternion algebra has been used in fields such as robotics, computer vision, neural networks, signal processing, and communications (e.g. [6,7]).

In this paper, we formulate a Clifford-valued widely linear estimation framework. And also, as a first step, we present a theoretical foundation for quaternion-valued WL estimation: it is proved that the estimation error obtained using the quaternion-valued WL estimation method is smaller than that obtained using the usual quaternion-valued estimation method.

2   The Clifford Algebra

This section briefly describes the Clifford algebra or geometric algebra [8].

Clifford algebra $Cl_{p,q}$ is an extension of real, complex numbers, and quaternions to higher dimensions, and has 2^n basis elements. The subscripts p, q such that $p + q = n$ determine the characteristics of the Clifford algebra. Note that

B.-L. Lu, L. Zhang, and J. Kwok (Eds.): ICONIP 2011, Part I, LNCS 7062, pp. 519–525, 2011.
© Springer-Verlag Berlin Heidelberg 2011

commutativity does not hold generally. For example, in the case of $n = 2, p = 0, q = 2$, the number of basis is $2^2 = 4$, which corresponds to the basis of quaternions.

The quaternion is defined over $\boldsymbol{R}$ with three imaginary units: i, j, k such that

$$ij = -ji = k, \ jk = -kj = i, \ ki = -ik = j, \ i^2 = j^2 = k^2 = ijk = -1 \quad (1)$$

where $\boldsymbol{R}$ denotes the set of real numbers. Every quaternion can be written explicitly as

$$q = a + bi + cj + dk \in \boldsymbol{Q}, \qquad\qquad a, b, c, d \in \boldsymbol{R} \qquad (2)$$

and has a quaternion conjugate

$$q^* = a - bi - cj - dk \qquad (3)$$

where $\boldsymbol{Q}$ denotes the set of quaternions. This leads to the norm of $q \in \boldsymbol{Q}$

$$|q| = \sqrt{qq^*}$$
$$= \sqrt{a^2 + b^2 + c^2 + d^2}. \qquad (4)$$

The commutativity does not hold: $pq \neq qp$ for any $p, q \in \boldsymbol{Q}$. The above description on quaternions may help with the understanding of the Clifford algebra.

We now define the Clifford algebra. Let the space $\boldsymbol{R}^{n+1}$ be given with the basis $\{e_0, \cdots, e_n\}$. And also, let $p \in \{0, \cdots, n\}, q \overset{\text{def}}{=} n - p$ be given. Assume that the following rules on the multiplication hold:

$$e_0 e_i = e_i e_0 = e_i \ \ (i = 1, \cdots, n), \qquad (5)$$
$$e_i e_j = -e_j e_i \ \ (i \neq j; \ i, j = 1, \cdots, n), \qquad (6)$$
$$e_0^2 = e_1^2 = \cdots = e_p^2 = 1, \qquad (7)$$
$$e_{p+1}^2 = \cdots = e_{p+q}^2 = -1. \qquad (8)$$

Then, the 2^n basis elements of the Clifford algebra $Cl_{p,q}$ are obtained:

$$e_0; \ e_1, \cdots, e_n; \ e_1 e_2, \cdots, e_{n-1} e_n; \ e_1 e_2 e_3, \cdots; \ \cdots; \ e_1 e_2 \cdots e_n, \qquad (9)$$

where e_0 is a unit element. The addition and the multiplication with a real number are defined coordinatewise. For example, for $x = a_3 e_3 + a_9 e_9 \in Cl_{9,3}$ and $y = b_9 e_9 + b_{32} e_3 e_2 \in Cl_{9,3}$,

$$x + y = (a_3 e_3 + a_9 e_9) + (b_9 e_9 + b_{32} e_3 e_2)$$
$$= a_3 e_3 + (a_9 + b_9) e_9 + b_{32} e_3 e_2. \qquad (10)$$

And, for any $\beta \in \boldsymbol{R}$ and $x = a_{14} e_1 e_4 + a_{361} e_3 e_6 e_1 \in Cl_{9,3}$,

$$\beta x = \beta(a_{14} e_1 e_4 + a_{361} e_3 e_6 e_1)$$
$$= \beta a_{14} e_1 e_4 + \beta a_{361} e_3 e_6 e_1. \qquad (11)$$

Furthermore, we assume that the following condition holds:

$$e_1 e_2 \cdots e_n \neq \pm 1 \quad \text{if} \quad p - q \equiv 1 \ (\text{mod } 4). \tag{12}$$

The algebra thus obtained is called Clifford algebra $Cl_{p,q}$.

The conjugation in $Cl_{p,q}$ is defined in the following way. For any $x \in Cl_{p,q}$, we describe

$$x = [x]_0 + [x]_1 + \cdots + [x]_n, \tag{13}$$

where

$$[x]_k \overset{\text{def}}{=} \sum_{\substack{A \in P_n \\ |A| = k}} x_A e_A, \tag{14}$$

P_n is the set containing the subsets of $\{1, \cdots, n\}$,

$$e_{i_1 i_2 \cdots i_p} \overset{\text{def}}{=} e_{i_1} e_{i_2} \cdots e_{i_p}, \tag{15}$$

and $i_1 i_2 \cdots i_p$ of $e_{i_1 i_2 \cdots i_p}$ in the left hand side of (15) means a set $\{i_1, i_2, \cdots, i_p\}$. Then, for any $x \in Cl_{p,q}$, the Clifford conjugation $x^* \in Cl_{p,q}$ is given as follows:

$$x^* = [x]_0 - [x]_1 - [x]_2 + [x]_3 + [x]_4 - \cdots \tag{16}$$

i.e.,

$$([x]_k)^* = [x]_k \quad \text{for} \quad k \equiv 0, 3 \ (\text{mod } 4), \tag{17}$$
$$([x]_k)^* = -[x]_k \quad \text{for} \quad k \equiv 1, 2 \ (\text{mod } 4). \tag{18}$$

3   The WL Model

In this section, the complex-valued WL model and the quaternion-valued WL model are reviewed, and the Clifford-valued WL model is formulated.

3.1   The Complex-Valued WL Model

Let $y \in C$ be a scalar complex-valued random variable to be estimated in terms of an observation that is a complex-valued random vector $x \in C^N$ where C is a set of complex numbers, and N is a natural number. That is, y is a true value and x is an observed value. In complex-valued linear mean square estimation (LMSE), the problem is to find an estimate written as

$$\hat{y}_L = h^H x, \tag{19}$$

where $h \in C^N$, and H represents the complex conjugation and transposition. Then, the objective of the problem is to find the parameter $h \in C^N$ that minimizes the estimation error $E|y - \hat{y}_L|^2$.

In the meantime, the complex-valued widely linear mean square estimation (WLMSE) problem can be stated as follows: Consider the scalar $\hat{y}$ defined as

$$\hat{y} = \boldsymbol{h}^H \boldsymbol{x} + \boldsymbol{g}^H \boldsymbol{x}^*, \tag{20}$$

where $\boldsymbol{g}, \boldsymbol{h} \in \boldsymbol{C}^N$, and $v^* \overset{\text{def}}{=} a - bi$ is the complex conjugate of $v = a + bi \in \boldsymbol{C}$. Then, the objective of the problem is to find parameters $\boldsymbol{g}, \boldsymbol{h} \in \boldsymbol{C}^N$ that minimize the estimation error $E|y - \hat{y}|^2$.

Picinbono et al. gave a theoretical foundation for the complex-valued WLMSE described above: it was proved that the estimation error obtained using the complex-valued WLMSE method is smaller than that obtained using the usual complex-valued LMSE method: $E|y - \hat{y}_L|^2 \geq E|y - \hat{y}|^2$ where the equality holds only in exceptional cases [1].

3.2   The Quaternion-Valued WL Model

The quaternion-valued WL model is a natural extension of the complex-valued WL model described in Sect. 3.1. Let $y \in \boldsymbol{Q}$ be a scalar quaternion-valued random variable to be estimated in terms of an observation that is a quaternion-valued random vector $\boldsymbol{x} \in \boldsymbol{Q}^N$. That is, y is a true value and $\boldsymbol{x}$ is an observed value.

In quaternion-valued linear mean square estimation (LMSE), the problem is to find an estimate written as

$$\hat{y}_L = \boldsymbol{h}^H \boldsymbol{x}, \tag{21}$$

where $\boldsymbol{h} \in \boldsymbol{Q}^N$, N is a natural number, and H represents the quaternionic conjugation and transposition.

The quaternion-valued WLMSE problem can be stated as follows: Consider the scalar $\hat{y}$ defined as

$$\hat{y} = \boldsymbol{h}^H \boldsymbol{x} + \boldsymbol{g}^H \boldsymbol{x}^*, \tag{22}$$

where $\boldsymbol{g}, \boldsymbol{h} \in \boldsymbol{Q}^N$, N is a natural number, H represents the quaternionic conjugation and transposition, and $v^* \overset{\text{def}}{=} a - bi - cj - dk$ is the quaternionic conjugate of $v = a + bi + cj + dk \in \boldsymbol{Q}$. Then, the objective of the problem is to find parameters $\boldsymbol{g}, \boldsymbol{h} \in \boldsymbol{Q}^N$ that minimize $E|y - \hat{y}|^2$.

Took and Mandic derived an augmented quaternion least mean squares (AQLMS) algorithm for quaternion-valued adaptive filters based on the quaternion-valued WL model, and confirmed its effectiveness via computer simulations [5]. Actually, the experimental results on the Lorenz attractor, real-world wind forecasting, and data fusion via quaternion spaces support the approach. Consequently, computer simulations proved that the quaternion-valued WL estimation method is superior to the usual quaternion-valued linear estimation method. However, no theoretical proof for the superiority of the quaternion-valued WL estimation method on estimation errors has been given to date, as it has been for the complex-valued WL estimation method. In the complex-valued setting, Picinbono et al. proved that the estimation error obtained with the complex-valued WLMSE is smaller than the error obtained using the complex-valued LMSE [1].

3.3   The Clifford-Valued WL Model

In this section, the Clifford-valued WL model is formulated, which is a generalization of the complex-valued and quaternion-valued WL models.

Let $y \in Cl_{p,q}$ be a scalar Clifford-valued random variable to be estimated in terms of an observation that is a Clifford-valued random vector $\boldsymbol{x} \in Cl_{p,q}^N$ where N is a natural number. That is, y is a true value and $\boldsymbol{x}$ is an observed value. In Clifford-valued linear mean square estimation (LMSE), the problem is to find an estimate written as

$$\hat{y}_L = \boldsymbol{h}^H \boldsymbol{x}, \tag{23}$$

where $\boldsymbol{h} \in Cl_{p,q}^N$, and H represents the Clifford conjugation and transposition. In the meantime, the Clifford-valued *widely linear* mean square estimation (WLMSE) problem can be stated as follows: Consider the scalar $\hat{y}$ defined as

$$\hat{y} - \boldsymbol{h}^H \boldsymbol{x} + \boldsymbol{g}^H \boldsymbol{x}^*, \tag{24}$$

where $\boldsymbol{g}, \boldsymbol{h} \in Cl_{p,q}^N$, and v^* is the Clifford conjugate of $v \in Cl_{p,q}$. Then, the objective of the problem is to find parameters $\boldsymbol{g}, \boldsymbol{h} \in Cl_{p,q}^N$ that minimize $E|y - \hat{y}|^2$.

4   A Theoretical Foundation of the Quaternion-Valued WL Model

In this section, we show the superiority of the quaternion-valued WLMSE method as a first step to investigate the property of the Clifford-valued WLMSE method formulated in Sect. 3.3. The main result is as follows: the estimation error obtained using the quaternion-valued WL estimation method is smaller than that obtained using the usual quaternion-valued linear estimation method, except in rare cases. The result is obtainable in the same manner described by [1]. However, the noncommutativity of the quaternion products must be considered during the analysis ($xy \neq yx$ for any $x, y \in \boldsymbol{Q}$).
Define

$$X \stackrel{\text{def}}{=} \left\{ Z(\omega) = \boldsymbol{h}^H \boldsymbol{x}(\omega) + \boldsymbol{g}^H \boldsymbol{x}^*(\omega) \,\middle|\, \boldsymbol{g}, \boldsymbol{h} \in \boldsymbol{Q}^N \right\}. \tag{25}$$

Therein, X is a set of scalar quaternion-valued random variables that constitutes a linear space, and which becomes a Hilbert subspace of the one-dimensional quaternion-valued Hilbert space $Y = \{z(\omega) \in \boldsymbol{Q}\}$ with the scalar product $< z, w > \stackrel{\text{def}}{=} E[zw^*]$ $(z, w \in X)$. Then, for a true value $y \in Y$, an observed value $\boldsymbol{x} \in \boldsymbol{Q}^N$, and the estimate $\hat{y} \in X$, the following equations hold:

$$(y - \hat{y}) \amalg \boldsymbol{x}, \tag{26}$$

$$(y - \hat{y}) \amalg \boldsymbol{x}^*, \tag{27}$$

where $u \amalg v$ means that all the components of v are orthogonal to u with the scalar product $< \cdot , \cdot > (u \in Q, v \in Q^N)$. From (26) and (27), we obtain

$$E[xy^*] = E[x\hat{y}^*], \tag{28}$$
$$E[x^*y^*] = E[x^*\hat{y}^*]. \tag{29}$$

Consequently, from (22), (28), (29), the following equations hold:

$$\Gamma_1 h + Cg = r, \tag{30}$$
$$C^H h + \Gamma_2 g = s^*, \tag{31}$$

where $\Gamma_1 \overset{\text{def}}{=} E[xx^H], \Gamma_2 \overset{\text{def}}{=} E[x^*x^T], C \overset{\text{def}}{=} E[xx^T], r \overset{\text{def}}{=} E[xy^*], s \overset{\text{def}}{=} E[yx]$. Eqs. (30) and (31) yield

$$g = (\Gamma_2 - C^H \Gamma_1^{-1} C)^{-1} \cdot (s^* - C^H \Gamma_1^{-1} r), \tag{32}$$
$$h = (\Gamma_1 - C\Gamma_2^{-1} C^H)^{-1} \cdot (r - C\Gamma_2^{-1} s^*). \tag{33}$$

Then, the estimation error ε_{WL} is calculable from (22), (30), and (31) as follows:

$$\varepsilon_{WL}^2 \overset{\text{def}}{=} E|y - \hat{y}|^2$$
$$= E|y|^2 - (r^H h + s^T g). \tag{34}$$

The estimation error ε_L in the quaternion-valued LMSE can be obtained using (21) as shown below.

$$\varepsilon_L^2 \overset{\text{def}}{=} E|y - \hat{y}_L|^2$$
$$= E|y|^2 - r^H \Gamma_1^{-1} r. \tag{35}$$

Then, from (30) and (32), (34), (35), the quantity $\delta\varepsilon^2$ can be expressed as

$$\delta\varepsilon^2 \overset{\text{def}}{=} \varepsilon_L^2 - \varepsilon_{WL}^2$$
$$= (s^* - C^H \Gamma_1^{-1} r)^H \cdot (\Gamma_2 - C^H \Gamma_1^{-1} C)^{-1} \cdot (s^* - C^H \Gamma_1^{-1} r), \tag{36}$$

which is the difference between the estimation error of the quaternion-valued LMSE and that of the quaternion-valued WLMSE. Eq. (36) is nonnegative because the matrix $\Gamma_2 - C^H \Gamma_1^{-1} C$ is positive-semidefinite. Furthermore, (36) is equal to zero only when one of the following conditions holds:

$$s^* - C^H \Gamma_1^{-1} r = 0, \tag{37}$$
$$\hat{y} = y. \tag{38}$$

Eq. (37) is an exceptional case, and (38) means that the true value y can be estimated with probability of one, which is a rare case.

5   Conclusions

We have formulated a Clifford-valued WL model, and have presented a theoretical foundation for the quaternion-valued WL estimation method. It was proved that the estimation error obtained using the quaternion-valued WL estimation method is smaller than that obtained using the usual quaternion-valued linear estimation method, except in rare cases. In our future studies, we will proceed with analyses of the Clifford-valued WL estimation.

Acknowledgements. The author extends special thanks to Prof. B. Picinbono, the Laboratoire des Signaux et Systèmes, Supélec, Gifsur Yvette, France for help in resolving several questions.

References

1. Picinbono, B., Chevalier, P.: Widely Linear Estimation with Complex Data. IEEE Trans. Signal Processing 43(8), 2030–2033 (1995)
2. Gerstacker, H., Schober, R., Lampe, A.: Receivers with Widely Linear Processing for Frequency-Selective Channels. IEEE Trans. Communication 51(9), 1512–1523 (2003)
3. Schober, R., Gerstacker, W.H., Lampe, L.H.-J.: Data-Aided and Blind Stochastic Gradient Algorithms for Widely Linear MMSE MAI Suppression for DS-CDMA. IEEE Trans. Signal Processing 52(3), 746–756 (2004)
4. Mandic, D.P., Goh, V.S.L.: Complex Valued Nonlinear Adaptive Filters: Noncircularity, Widely Linear and Neural Models. John Wiley and Sons Ltd. (2009)
5. Took, C.C., Mandic, D.P.: The Quaternion LMS Algorithm for Adaptive Filtering of Hypercomplex Processes. IEEE Trans. Signal Processing 57(4), 1316–1327 (2009)
6. Nitta, T.: A Quaternary Version of the Back-Propagation Algorithm. In: In Proc. IEEE Int. Conf. on Neural Networks, ICNN 1995-Perth, November 27-December 1, vol. 5, pp. 2753–2756 (1995)
7. Arena, P., Fortuna, L., Muscato, G., Xibilia, M.G.: Neural Networks in Multidimensional Domains. LNCIS, vol. 234. Springer, Heidelberg (1998)
8. Gürlebeck, K., Habetha, K., Sprößig, W.: Holomorphic Functions in the Plane and N-Dimensional Space. Birkhäuser (2008)

Comparison of Complex- and Real-Valued Feedforward Neural Networks in Their Generalization Ability

Akira Hirose and Shotaro Yoshida

Department of Electrical Engineering and Information Systems, The University of Tokyo, Japan
http://www.eis.t.u-tokyo.ac.jp/

Abstract. We compare the generalization characteristics of complex-valued and real-valued feedforward neural networks when they deal with wave-related signals. We assume a task of function approximation. Experiments demonstrate that complex-valued neural networks show smaller generalization error than real-valued ones in particular when the signals have high degree of wave nature.

Keywords: Complex-valued neural network, function approximation, generalization.

1 Introduction

Researches on general complex-valued networks have revealed various aspects of their dynamics. However, at the same time, it is true that a complex number is represented by a pair of real numbers, namely, real and imaginary parts, or amplitude and phase. Actually a variety of useful neural dynamics are obtained by paying attention to the real and imaginary parts [7] [1] [8] or amplitude and phase [2] [3] . This fact sometimes leads to an assumption that a complex-valued neural network is almost equivalent to a double-dimensional real-valued neural network. However, complex-valued networks has only smaller degree of freedom at the synaptic weighting.

In this paper, we compare complex- and real-valued neural networks by focusing on their generalization characteristics. Generalization is one of the features most useful and specific to neural networks widely. We investigate the generalization ability of feedforward complex-valued and double-dimensional real-valued neural networks, in particular when they learn and process wave-related data for function approximation or filtering. We observe the characteristics by feeding signals that have various degrees of wave nature by mixing a sinusoidal wave and white noise. Computer experiments demonstrate that the generalization characteristics of complex-valued neural networks are different from those of double-dimensional real-valued neural networks dependently on the degree of wave nature of the signals.

2 Construction of Experiments and Learning Dynamics

We organize our experiment as follows.

- Preparation of input signals: Variously weighted summation of (A) sinusoidal wave and (B) non-wave data, i.e., white noise having random amplitude and phase (or real and imaginary parts).

B.-L. Lu, L. Zhang, and J. Kwok (Eds.): ICONIP 2011, Part I, LNCS 7062, pp. 526–531, 2011.

- Definition of task to learn: Identity mapping, which is expected to show the learning characteristics clearly, for the above signals with various degrees of wave nature.
- Evaluation of generalization: Observation of the generalization error when the input signals shifts in time, or when the amplitude is changed.

2.1   Forward Processing and Learning Dynamics

Complex-valued neural network. We consider a layered feedforward network having input terminals, hidden neurons, and output neurons. In the case of a CVNN, we employ a phase-amplitude-type sigmoid activation function and the teacher-signal-backpropagation learning process [3] [6] with notations of

$$z^{\mathrm{I}} = [z_1, ..., z_i, ..., z_I, z_{I+1}]^T \quad \text{(Input signal vector)} \tag{1}$$

$$z^{\mathrm{H}} = [z_1, ..., z_h, ..., z_H, z_{H+1}]^T \quad \text{(Hidden-layer output signal vector)} \tag{2}$$

$$z^{\mathrm{O}} = [z_1, ..., z_o, ..., z_O]^T \quad \text{(Output-layer signal vector)} \tag{3}$$

$$\mathbf{W}^{\mathrm{H}} = [w_{hi}] \quad \text{(Hidden neuron weight matrix)} \tag{4}$$

$$\mathbf{W}^{\mathrm{O}} = [w_{oh}] \quad \text{(Output neuron weight matrix)} \tag{5}$$

where $[\cdot]^T$ means transpose. In (4) and (5), the weight matrices include additional weights $w_{h\ I+1}$ and $w_{o\ H+1}$, equivalent to neural thresholds, where we add formal constant inputs $z_{I+1} = 1$ and $z_{H+1} = 1$ in (1) and (2), respectively. Respective signal vectors and synaptic weights are connected with one another through an activation function $f(z)$ as

$$z^{\mathrm{H}} = f\left(\mathbf{W}^{\mathrm{H}} z^{\mathrm{I}}\right), \quad z^{\mathrm{O}} = f\left(\mathbf{W}^{\mathrm{O}} z^{\mathrm{H}}\right) \tag{6}$$

where $f(z)$ is a function of each vector element $z\ (\in C)$ defined as

$$f(z) = \tanh\left(|z|\right) \exp\left(\sqrt{-1}\ \arg z\right) \tag{7}$$

Figure 1 is a diagram to explain the supervised learning process. We prepare a set of teacher signals at the input $\hat{z}_s^{\mathrm{I}} = [\hat{z}_{1s}, ..., \hat{z}_{is}, ..., \hat{z}_{Is}, \hat{z}_{I+1\ s}]^T$ and the output $\hat{z}_s^{\mathrm{O}} = [\hat{z}_{1s}, ..., \hat{z}_{os}, ..., \hat{z}_{Os}]^T$ $(s = 1, ..., s, ...S)$ for which we employ the teacher-signal backpropagation learning. We define an error function E to obtain the dynamics [6] [3] as

$$E \equiv \frac{1}{2} \sum_{s=1}^{S} \sum_{o=1}^{O} \left| z_o(\hat{z}_s^{\mathrm{I}}) - \hat{z}_{os} \right|^2 \tag{8}$$

$$\left| w_{oh}^{\mathrm{new}} \right| = \left| w_{oh}^{\mathrm{old}} \right| - K \frac{\partial E}{\partial |w_{oh}|} \tag{9}$$

$$\arg w_{oh}^{\mathrm{new}} = \arg w_{oh}^{\mathrm{old}} - K \frac{1}{|w_{oh}|} \frac{\partial E}{\partial(\arg w_{oh})} \tag{10}$$

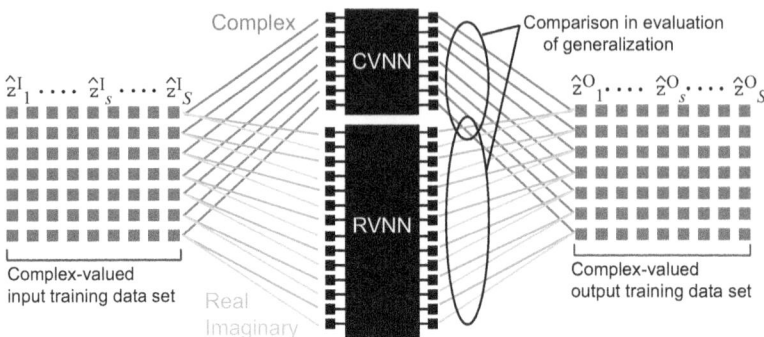

Fig. 1. Schematic diagram of the learning process of complex- and double-dimensional real-valued feedforward neural networks for pairs of input-output teachers

$$\frac{\partial E}{\partial |w_{oh}|} =$$

$$\left(1 - |z_o|^2\right) \left(|z_o| - |\hat{z}_o| \cos\left(\arg z_o - \arg \hat{z}_o\right)\right) |z_h| \cos\left(\arg z_o - \arg \hat{z}_o - \arg w_{oh}\right)$$

$$- |z_o| |\hat{z}_o| \sin\left(\arg z_o - \arg \hat{z}_o\right) \frac{|z_h|}{\tanh^{-1}|z_o|} \sin\left(\arg z_o - \arg \hat{z}_o - \arg w_{oh}\right) \quad (11)$$

$$\frac{1}{|w_{oh}|} \frac{\partial E}{\partial(\arg w_{oh})} =$$

$$\left(1 - |z_o|^2\right) \left(|z_o| - |\hat{z}_o| \cos\left(\arg z_o - \arg \hat{z}_o\right)\right) |z_h| \sin\left(\arg z_o - \arg \hat{z}_o - \arg w_{oh}\right)$$

$$+ |z_o| |\hat{z}_o| \sin\left(\arg z_o - \arg \hat{z}_o\right) \frac{|z_h|}{\tanh^{-1}|z_o|} \cos\left(\arg z_o - \arg \hat{z}_o - \arg w_{oh}\right) \quad (12)$$

where $(\cdot)^{\text{new}}$ and $(\cdot)^{\text{old}}$ indicates the update of the weights from $(\cdot)^{\text{old}}$ to $(\cdot)^{\text{new}}$, and K is a learning constant. The teacher signals at the hidden layer $\hat{z}^{\text{H}} = [\hat{z}_1, ..., \hat{z}_h, ..., \hat{z}_H]^T$ is obtained by making the output teacher vector itself $\hat{z}^{\text{O}}$ propagate backward as

$$\hat{z}^{\text{H}} = \left(f\left(\left(\hat{z}^{\text{O}}\right)^* \hat{\mathbf{w}}^{\text{O}}\right)\right)^* \quad (13)$$

where $(\cdot)^*$ denotes Hermite conjugate. Using $\hat{z}^{\text{H}}$, the hidden layer neurons change their weights by following (9)–(12) with replacement of the suffixes o, h with h, i [4] [5].

Double-dimensional real-valued neural network. Similarly, the forward processing and learning of a double-dimensional real-valued neural network are explained as follows. Figure 1 includes also this case. We represent a complex number as a pair of real numbers as $z_i = x_{2i-1} + \sqrt{-1}\, x_{2i}$. That is, we have a double-dimensional real input vector z_R^I, a double-dimensional hidden signal vector z_R^H, and a double-dimensional output signal vector z_R^O. A forward signal processing connects the signal vectors as well as hidden neuron weights $\mathbf{W}_R^H$ and output neuron weights $\mathbf{W}_R^O$ through a real-valued activation function f_R as

$$z_R^I = [\ \overbrace{x_1,\quad x_2}^{\text{real \& imaginary}}\ , ..., x_{2i-1}, x_{2i}, ..., x_{2I-1}, x_{2I}, x_{2I+1}, x_{2I+2}]^T$$
$$\left(= z^I\right) \quad \text{(Input signal vector)} \tag{14}$$

$$z_R^H = [x_1, x_2, ..., x_{2h-1}, x_{2h}, ..., x_{2H-1}, x_{2H}, x_{2H+1}, x_{2H+2}]^T$$
$$\text{(Hidden-layer output signal vector)} \tag{15}$$

$$z_R^O = [x_1, x_2, ..., x_{2o-1}, x_{2o}, ..., x_{2O-1}, x_{2O}]^T$$
$$\text{(Output-layer signal vector)} \tag{16}$$

$$\mathbf{W}_R^H = [w_{Rhi}] \quad \text{(Hidden neuron weight matrix)} \tag{17}$$

$$\mathbf{W}_R^O = [w_{Roh}] \quad \text{(Output neuron weight mateix)} \tag{18}$$

$$z_R^H = f_R\left(\mathbf{W}_R^H z_R^I\right), \quad z_R^O = f_R\left(\mathbf{W}_R^O z_R^H\right) \tag{19}$$

$$f_R(x) = \tanh(x) \tag{20}$$

where the thresholds are $w_{R\,h\,2I+1}$, $w_{R\,h\,2I+2}$, $w_{R\,h\,2H+1}$, and $w_{R\,h\,2H+2}$ with formal additional inputs $x_{2H+1} = 1$, $x_{2H+2} = 1$, $x_{2H+1} = 1$, and $x_{2H+2} = 1$. We employ the conventional error backpropagation learning.

3 Computer Experiments

3.1 Experimental Setup

We add white Gaussian noise to a sinusoidal wave with various weighting. Then the degree of wave nature is expressed as the signal-to-noise ratio S/N where $S/N = \infty$ means complete wave, while $S/N = 0$ corresponds to complete non-wave. The network parameters are as follows: Number of input neurons $I = 16$, hidden neurons $H = 25$, output neurons $O = 16$, learning constant $K = 0.01$, and the learning iteration is 3,000.

3.2 Result

Figure 2(a) shows an example of the learning curve when $S/N = \infty$, i.e., the signal is sinusoidal. We find that the learning is successfully completed for both of the CVNN and RVNN. The learning errors converge almost at zero, which means that there is only slight residual error at the learning teacher points.

After the learning, we input other input signals to investigate the generalization. As mentioned above, the wavelength is adjusted to span over the 16 neural input terminals. For example, we gradually move the input signal forward in time while keeping the

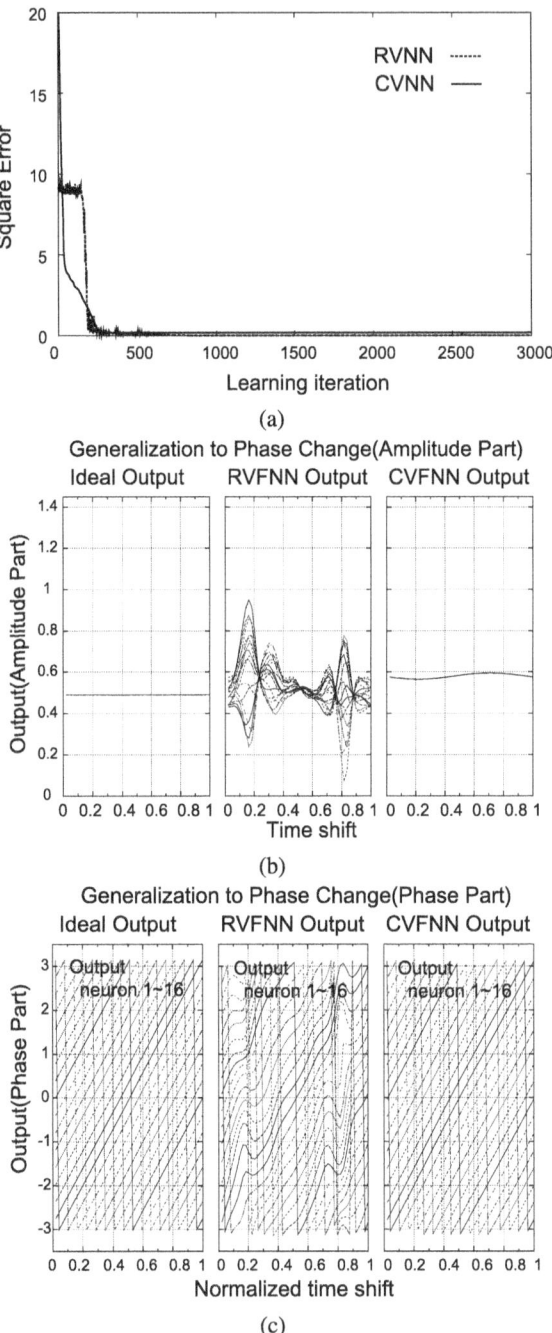

Fig. 2. An example of (a)learning curve, and (b)amplitude and (c)phase when the input signal gradually sifts in time in the real-valued and complex-valued neural networks (RVNN and CVNN) when no noise is added to sinusoidal signals ($S/N = \infty$)

amplitude unchanged at $a = 0.5$. Figures 2(b) and (c) present examples of the output amplitude and phase, respectively, showing from left-hand side to the right-hand side the ideal output of the identity mapping, the RVNN outputs, and CVNN outputs of the 16 output neurons. The horizontal axes present the time shift t normalized by the unit-wave duration.

In Fig.2(b), we find that the output signals of the RVNN deviate greatly from the ideal ones. The learning points are plotted at $t = 0.5$, where the output amplitude is almost 0.5 for all the neurons. However, with the time course, the amplitude values fluctuate largely. Contrarily, the CVNN amplitude stays almost constant. At the learning point $t = 0.5$, the value is slightly larger than 0.5, corresponding to the slight non-zero value of the residual error in the learning curve. In Fig.2(c), the ideal output phase values on the left-hand side exhibit linear increase in time. In the RVNN case, though the phase values at $t = 0.5$ are the same as those of ideal outputs, the values sometimes swing strongly. In contrast, the CVNN output phase values increase orderly, which is almost identical with the ideal values. In summary, the CVNN presents much better generalization characteristics than the RVNN when the degree of wave nature is high, i.e., $S/N = \infty$.

We also investigated the results for various S/N, which represents the degree of wave nature. We find that in the parameter region of high degree of wave nature, the generalization of the CVNN is much better than that of the RVNN.

4   Conclusion

This paper investigated numerically the generalization characteristics in the feedforward complex-valued and real-valued neural networks (CVNN and RVNN). We compared a CVNN and a double-dimensional RVNN in a simple case where the network works for function approximation or as a filter. Computer experiments demonstrated that the CVNN exhibits better generalization characteristics in particular for signals having high degree of wave nature.

References

1. Benvenuto, N., Piazza, F.: On the complex backpropagation algorithm. IEEE Transactions on Signal Processing 40, 967–969 (1992)
2. Georgiou, G.M., Koutsougeras, C.: Complex domain backpropagation. IEEE Transactions on Circuits and Systems II 39(5), 330–334 (1992)
3. Hirose, A.: Continuous complex-valued back-propagation learning. Electronics Letters 28(20), 1854–1855 (1992)
4. Hirose, A.: Applications of complex-valued neural networks to coherent optical computing using phase-sensitive detection scheme. Information Sciences –Applications 2, 103–117 (1994)
5. Hirose, A.: Complex-Valued Neural Networks. Springer, Heidelberg (2006)
6. Hirose, A., Eckmiller, R.: Behavior control of coherent-type neural networks by carrier-frequency modulation. IEEE Transactions on Neural Networks 7(4), 1032–1034 (1996)
7. Leung, H., Haykin, S.: The complex backpropagation algorithm. IEEE Transactions on Signal Processing 39, 2101–2104 (1991)
8. Nitta, T.: An extension of the back-propagation algorithm to complex numbers. Neural Networks 10, 1391–1415 (1997)

An Automatic Music Transcription Based on Translation of Spectrum and Sound Path Estimation

Ryota Ikeuchi and Kazushi Ikeda

Nara Institute of Science and Technology
Ikoma, 630-0192 Nara, Japan
{ryota-i,kazushi}@is.naist.jp
http://hawaii.naist.jp/

Abstract. An automatic music transcription method is proposed. The method is based on a generative model that takes into account the translation of spectrum for an instrument and the sound path from the instrument to a microphone. The fundamental frequency (note), the spectrum of the instrument (basis pattern) and the sound path are estimated simultaneously using an extended complex nonnegative matrix factorization. The effectiveness of the proposed method is confirmed by synthetic data.

Keywords: Music transcription, Non-negative matrix factorization, Translation of spectrum, Sound path estimation.

1 Introduction

One of the important functions of music information retrieval (MIR) systems to be equipped is automatic music transcription [1–7].

Although an instrument's sounds have less variety in spectrum compared to speech signals, music sounds have more harmonics and consist of several tones of several instruments, that makes the problem more difficult than it seems.

Another difference of MIR problems is the existence of sound paths from source signals to microphones, which can be neglected in speech recognition since a speaker often stands in front of a microphone. In our cases, however, the condition under which a music was recorded strongly affects the transcription performance. For example, the performance of a transcription system for a music is much lower when the system is trained with musics in different CD than when with musics in the same CD [8, 9]. This fact has little been considered in constructing transcription systems so far.

We tackle the problem of automatic music transcription by making a generative model that takes the facts mentioned above into account. We set a sound path to a microphone after the conventional model, that is, the product of sound pattern activations and basis patterns Another difference from the conventional model is to introduce a prior knowledge to basis patterns. We assume that the basis pattern of a tone is similar to (or almost the same as) those of other tones for the same instrument (Fig. 1).

B.-L. Lu, L. Zhang, and J. Kwok (Eds.): ICONIP 2011, Part I, LNCS 7062, pp. 532–540, 2011.
© Springer-Verlag Berlin Heidelberg 2011

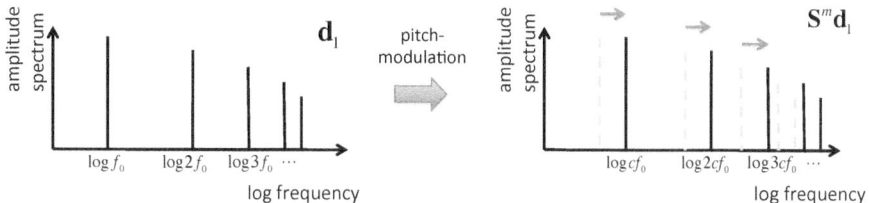

Fig. 1. Spectra of different tones are almost translation of each other

The estimation of the model parameters from a given spectrogram is formulated as a matrix factorization. Non-negative matrix factorization (NMF) methods [10] would seem suitable for this problem since the power spectrum of a sound is non-negative but sound signals and their spectra are essentially complex values and additivity does not hold in our cases as shown later. A complex NMF proposed in [11] is a method to cope with this difficulty. We extend the complex NMF so that it is applicable to our model.

The rest of the paper is organized as follows: Section 2 shows a generative model of instrument sounds that combines activations, basis patterns and sound paths. In section 3, a method for estimating each component of the model simultaneously is described, which is based on complex non-negative matrix factorization. Some results of experiments are given in section 4 and section 5 concludes the paper.

2   A Generative Model for Instrument Sounds

Our model consists of a set of activations that correspond to notes, a set of basis patterns each of which expresses the spectrum of a tone for an instrument, and a set of sound paths that are represented by finite impulse response (FIR) filters. The complex signal $Y_{f,t}$ of frequency $f = 1, \ldots, F$ at time $t = 1, \ldots, T$ is modeled as

$$Y_{f,t} \approx X_{f,t} = r_f \sum_k B_{f,k} A_{k,t} \exp\left(j\phi_{k,f,t}\right) \tag{1}$$

where $A_{k,t}$, $B_{f,k}$ and r_f denote the activations, basis patterns and frequency responses of a sound path, respectively. Their details are shown in the following subsections.

2.1   Activations

In this work, we consider only Western polyphonic music and assume that music consists of note symbols. Activations are expressed as a matrix $A = \{A_{k,t}\}$ where $A_{k,t}$ is the loudness of a specific note k at time t that corresponds to an instrument and a pitch. Note that A has a sparse representation, that is, $A_{k,t} = 0$ for almost all k and t.

2.2   Basis Patterns

A row vector B_k of a basis-pattern matrix $B = \{B_{f,k}\}$ represents the spectrum of a note of an instrument. For a specific instrument, the spectra for different notes have similar forms, that is, a spectrum vector is approximately expressed as a translation of another. In the following, we consider only one instrument but the discussion can easily be applied to the cases with more instruments.

The similarity of spectra has been utilized in the literature [12–14]. Eggert et al. restricted the basis-patterns B_k to translation-invariant ones within an instrument. This means that B_k is expressed as

$$B_k = S^{k-1}b \tag{2}$$

where S is a shift matrix

$$S = \begin{pmatrix} 0 & 1 & 0 & \cdots & 0 \\ 0 & 0 & 1 & \ddots & 0 \\ \vdots & \ddots & \ddots & \ddots & \vdots \\ 0 & 0 & \ddots & \ddots & 1 \\ 0 & 0 & \cdots & \cdots & 0 \end{pmatrix} \tag{3}$$

and b is a constant vector to be estimated [12]. Note that the lowest and left-most element should be one for making theoretical treatment easier [9]. Raczynski et al. initialized the basis matrix with zeros everywhere but at positions corresponding to harmonic frequencies of consequent notes of the equal temperament scale [13]. Ochiai et al. introduced a penalty function that evaluates a similarity of two spectra for adjacent pitches, taking into account the pitch shift [14]. Our model employs the simplest way used in [12].

2.3   Sound Paths

Our model settles a sound path between an instrument and a microphone (Fig. 2). The path depends on the condition under which each music was recorded and has a frequency response denoted by $r = \{r_f\}$.

3   Complex Non-negative Matrix Factorization

Such a model as (1) is solved by non-negative matrix factorization (NMF) methods [10]. There are a lot of work on MIR systems based on NMF [12–14], however, conventional NMF methods are not suitable in treating sound signals. This is because the facts that sound signals are essentially complex and the absolute value of the sum of two complex numbers is not equal to the sum of the absolute values of the two, $|a + b| \neq |a| + |b|$, in general.

Kameoka et al. proposed a complex NMF to cope with this problem, that introduces a time-varying phase spectrum $\phi_{k,f,t}$ to each factor $B_{f,k}A_{k,t}$ [11].

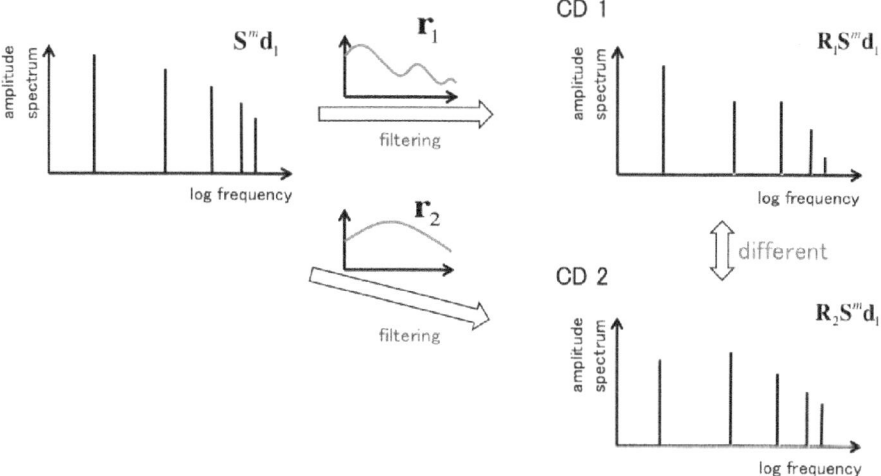

Fig. 2. Schematic view of the proposed model. A sound path is settled from an instrument to a microphone

The factor is replaced with $r_f B_{f,k} A_{k,t}$ due to a sound path r in our case as shown in (1). Hence, we modify the complex NMF so as to match our model in the following.

Considering the maximum likelihood estimation and additive white Gaussian noise, we minimize the squared errors,

$$\Psi(\theta) = \sum_{f,t} |Y_{f,t} - X_{f,t}|^2 + 2\lambda U \tag{4}$$

$$= \sum_{f,t} \left| Y_{f,t} - r_f \sum_{k} B_{f,k} A_{k,t} \exp\left(j\phi_{k,f,t}\right) \right|^2 + 2\lambda U, \tag{5}$$

where λ controls the balance of two terms and $\theta = \{r, b, A, \phi\}$, under the constraints

$$\sum_{f} r_f^2 = 1, \qquad\qquad \sum_{f} b_f^2 = 1, \tag{6}$$

that remove the ambiguity of scales. Here,

$$U = \sum_{k,t} |A_{k,t}|^p \qquad\qquad (0 < p < 2) \tag{7}$$

is a regularization term to introduce sparseness to the matrix A.

Thanks to the property of translation-invariance on B, r and b do not have ambiguity under mild conditions. Intuitively speaking, r and b have only $2F$ degrees of freedom (DoG) in total while the matrix $\mathrm{diag}(r)B$ has FT DoG.

An efficient iterative algorithm for minimizing $\Psi(\theta)$ can be derived using an auxiliary function $\Psi^+(\theta, \bar{\theta})$ in a similar way to [3, 11]. When $\Psi^+(\theta, \bar{\theta})$ satisfies

$$\Psi(\theta) = \min_{\bar{\theta}} \Psi^+(\theta, \bar{\theta}), \tag{8}$$

it is known that $\Psi(\theta)$ is non-increasing under the updates,

$$\bar{\theta} \leftarrow \arg\min_{\bar{\theta}} \Psi^+(\theta, \bar{\theta}), \qquad \theta \leftarrow \arg\min_{\theta} \Psi^+(\theta, \bar{\theta}). \tag{9}$$

In this study, we employed an auxiliary function derived using Jensen's inequality,

$$\Psi^+(\theta, \bar{\theta}) \equiv \sum_{k,f,t} \frac{1}{\beta_{k,f,t}} \left| \bar{Y}_{k,f,t} - r_f \sum_{f'} (S^{k-1})_{f,f'} b_{f'} A_{k,t} \exp\left(j\phi_{k,f,t}\right) \right|^2 + U^+ \tag{10}$$

$$\geq \sum_{f,t} \left| Y_{f,t} - r_f \sum_{k} \sum_{f'} (S^{k-1})_{f,f'} b_{f'} A_{k,t} \exp\left(j\phi_{k,f,t}\right) \right|^2 + U \tag{11}$$

$$= \Psi(\theta), \tag{12}$$

where

$$U^+ = \sum_{k,t} \left(\frac{p|\bar{A}_{k,t}|^{p-2}}{2} A_{k,t}^2 + |\bar{A}_{k,t}|^p - \frac{p|\bar{A}_{k,t}|^p}{2} \right) \geq U \tag{13}$$

and

$$\sum_{k} \bar{Y}_{k,f,t} = Y_{f,t}, \qquad \sum_{k} \beta_{k,f,t} = 1 \qquad \beta_{k,f,t} > 0. \tag{14}$$

Note that equality in (11) holds only when

$$\bar{Y}_{k,f,t} = r_f \sum_{f'} (S^{k-1})_{f,f'} b_{f'} A_{k,t} \exp\left(j\phi_{k,f,t}\right)$$

$$+ \beta_{f,t,k} \left(Y_{f,t} - \sum_{k} r_f \sum_{f'} (S^{k-1})_{f,f'} b_{f'} A_{k,t} \exp\left(j\phi_{k,f,t}\right) \right) \tag{15}$$

and $\bar{A}_{k,t} = A_{k,t}$.

The explicit update rules of r, b, A and ϕ are given below. Note that we relax the constraints in (6) at first and then normalize r and b. Solving the equations from necessary conditions,

$$\frac{\partial \Psi^+}{\partial r_f} = 0, \qquad \frac{\partial \Psi^+}{\partial b_l} = 0, \qquad \frac{\partial \Psi^+}{\partial A_{k,t}} = 0, \tag{16}$$

we get

$$r_f = \frac{\sum_{k,t} \frac{A_{k,t}}{\beta_{k,f,t}} \sum_{f'} (S^{k-1})_{f,f'} b_{f'} \Re\left[\bar{Y}^*_{k,ft} \exp(j\phi_{k,f,t})\right]}{\sum_{k,t} \frac{A^2_{k,t}}{\beta_{k,f,t}} \left(\sum_{f'} (S^{k-1})_{f,f'} b_{f'}\right)^2}, \tag{17}$$

$$b_l = \frac{\sum_{k,f,t} \frac{r_f A_{k,t} (S^{k-1})_{f,l}}{\beta_{k,f,t}} \Re\left[\bar{Y}^*_{k,f,t} \exp(j\phi_{k,f,t})\right]}{\sum_{k,f,t} \frac{r_f^2 A^2_{k,t} (S^{k-1})^2_{f,l}}{\beta_{k,f,t}}}$$

$$- \frac{\sum_{k,f,t} \frac{r_f A^2_{k,t} (S^{k-1})_{f,l}}{\beta_{k,f,t}} \left(\sum_{f'} (S^{k-1})_{f,f'} b_{f'} - (S^{k-1})_{f,l} b_l\right)}{\sum_{k,f,t} \frac{r_f^2 A^2_{k,t} (S^{k-1})^2_{f,l}}{\beta_{k,f,t}}}, \tag{18}$$

$$A_{k,t} = \frac{\sum_f \frac{r_f}{\beta_{k,f,t}} \sum_{f'} (S^{k-1})_{f,f'} b_{f'} \Re\left[\bar{Y}^*_{k,ft} \exp(j\phi_{k,f,t})\right]}{\sum_f \frac{r_f^2}{\beta_{k,f,t}} \left(\sum_{f'} (S^{k-1})_{f,f'} b_{f'}\right)^2 + \lambda p |\bar{A}_{k,t}|^{p-2}}. \tag{19}$$

The fact of

$$(S^{k-1})_{f,l} = \begin{cases} 1 & \text{for } k = l - f + 1 \\ 0 & \text{otherwise} \end{cases} \tag{20}$$

simplifies (18) to

$$b_l = \frac{\sum_{f,t} \frac{r_f A_{k',t}}{\beta_{k',f,t}} \Re\left[\bar{Y}^*_{f,t} \exp(j\phi_{k',f,t})\right] - \sum_{f,t} \frac{r_f A^2_{k',t}}{\beta_{k',f,t}} \left(\sum_{f'} S_{f,f',k'} b_{f'} - S_{f,l,k'} b_l\right)}{\sum_{f,t} \frac{r_f^2 A^2_{k',t}}{\beta_{k',f,t}}}, \tag{21}$$

where

$$k' = l - f + 1, \quad 1 \le k' \le K. \tag{22}$$

As for $\phi_{k,f,t}$, we rewrite Ψ^+ by separating terms dependent of ϕ and the rest, that is,

$$\Psi^+ = c - 2 \sum_{k,f,t} \frac{r_f A_{k,t}}{\beta_{k,f,t}} \sum_{f'} (S^{k-1})_{f,f'} b_{f'} |\bar{Y}_{k,f,t}| \cos\left(C_{k,f,t} - \phi_{k,f,t}\right) \tag{23}$$

where $C_{k,f,t}$ is determined by

$$\cos C_{k,f,t} = \frac{\Re\left[\bar{Y}_{k,f,t}\right]}{|\bar{Y}_{k,f,t}|}, \qquad \sin C_{k,f,t} = \frac{\Im\left[\bar{Y}_{k,f,t}\right]}{|\bar{Y}_{k,f,t}|} \tag{24}$$

and c is a constant independent of $\phi_{k,f,t}$. (23) means that Ψ^+ is minimized when $\phi_{k,f,t} = C_{k,f,t}$.

The free parameter β in (14) is adaptively set as

$$\beta_{k,f,t} = \frac{A_{k,t} \sum_{f'} (S^k)_{f'} b_{f'}}{\sum_n A_{n,t} \sum_{f'} (S^k)_{f'} b_{f'}} \tag{25}$$

according to [11].

4   Experiments and Results

To confirm the effectiveness of our algorithm, we carried out some experiments as below.

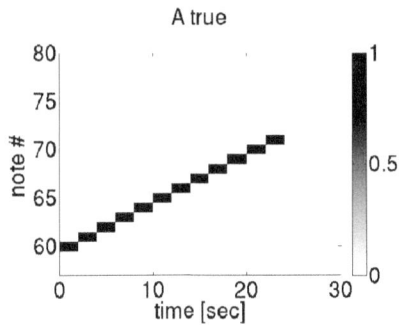

(a) Note activations. Each column shows the note id at a time.

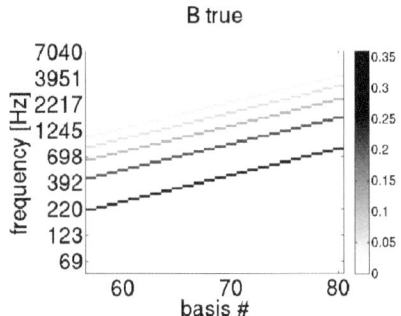

(b) Basis patterns. Each column shows the spectrum of a note.

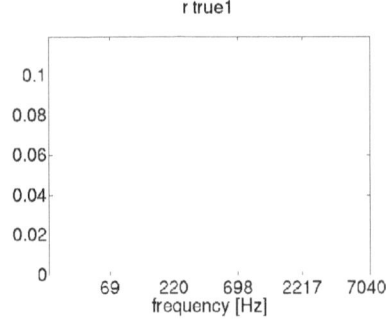

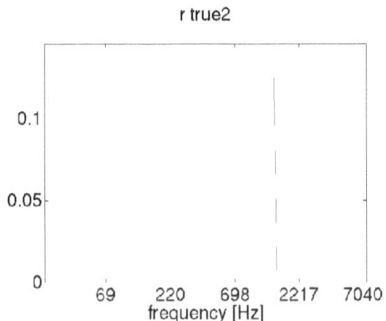

(c) Frequency responses of sound paths for cases 1 and 2.

Fig. 3. True note activations, basis patterns and frequency responses (two cases) of sound paths used in the experiments

Since the problem of music transcription is an ill-posed problem, we synthesized monaural sound samples to clarify true note activations, basis patterns and frequency responses of sound paths (Fig. 3).

Figure 4 shows the results of our algorithm for $\lambda = 0.1$. The algorithm can estimate the true note activations and frequency responses almost correctly. Note that the frequency response takes zero at frequencies where no sound signals exist.

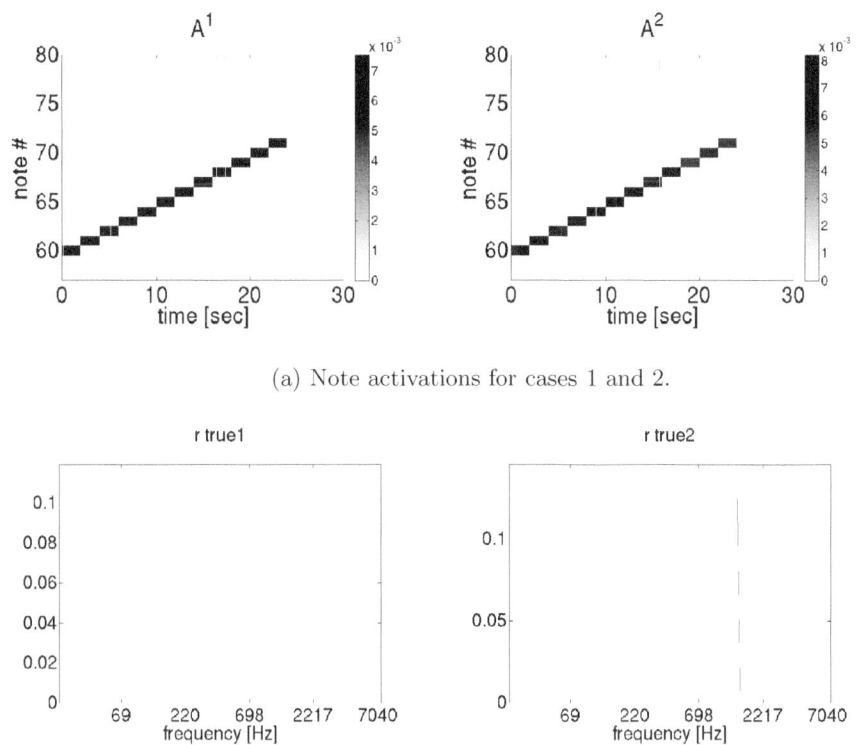

(a) Note activations for cases 1 and 2.

(b) Frequency responses of sound paths for cases 1 and 2.

Fig. 4. Estimated note activations and frequency responses of sound paths for sparseness parameter $\lambda = 0.1$

5   Conclusions

We proposed an automatic music transcription method based on a generative model that takes into account the translation of spectrum for an instrument and the sound path from the instrument to a microphone. The activations and the basis patterns of the instrument are estimated simultaneously as well as the response frequency of the sound path, by extending a complex nonnegative matrix factorization. The effectiveness of the proposed method was confirmed by experiments with synthetic data.

References

1. Klapuri, A., Davy, M. (eds.): Signal Processing Methods for Music Transcription. Springer, Heidelberg (2006)
2. Lee, D.D., Seung, H.S.: Learning the Parts of Objects by Non-Negative Matrix Factorization. Nature 401, 788–791 (1999)
3. Lee, D.D., Seung, H.S.: Algorithms for Nonnegative Matrix Factorization. In: NIPS, pp. 556–562 (2000)
4. Plumbley, M., Abdallah, S., Bello, J., Davies, M., Monti, G., Sandler, M.: Automatic Music Transcription and Audio Source Separation. Cybernetics and Systems 33(6), 603–627 (2002)
5. Blumensath, T., Davies, M.: Sparse and Shift-Invariant Representations of Music. IEEE Trans. ASLP 14(1), 50–57 (2006)
6. Ihara, M., Maeda, S., Ishii, S.: Estimation of the Source-Filter Model Using Temporal Dynamics. In: Proc. International Joint Conference on Neural Networks, pp. 3098–3103 (2007)
7. Ihara, M., Maeda, S., Ishii, S.: Instrument Identification in Monophonic Music Using Spectral Information. In: International Symposium on Signal Processing and Information Technology, pp. 607–611 (2007)
8. Ihara, M.: Statistical Approach to the Single-Channel Sound Source Extraction. Ph.D. Thesis. Nara Institute of Science and Technology (2010)
9. Ihara, M., Maeda, S., Ikeda, K., Ishii, S.: Low-Dimensional Feature Representation for Monophonic Music Instrument Identification (to appear)
10. Cichocki, A., Zdunek, R., Phan, A., Amari, S.: Nonnegative Matrix and Tensor Factorizations. John Wiley & Sons Ltd. (2009)
11. Kameoka, H., Ono, N., Kashino, K., Sagayama, S.: Complex NMF: A New Sparse Representation for Acoustic Signals. In: International Conference on Acoustics, Speech and Signal Processing, pp. 3437–3440 (2009)
12. Eggert, J., Wersing, H., Koerner, E.: Transformation-Invariant Representation and NMF. In: International Joint Conference on Neural Networks (2004)
13. Raczynski, S.A., Ono, N., Sagayama, S.: Multipitch Analysis with Harmonic Nonnegative Matrix Approximation. In: International Symposium on Music Information Retrieval, p. 381 (2007)
14. Ochiai, K., Nakano, M., Ono, N., Sagayama, S.: Parallel Nonnegative Matrix Factorization of High-Time-Resolution and High-Frequency-Resolution Spectrograms for Multipitch Analysis of Music Signals. In: Acoustic Society of Japan Spring Conference, pp. 705–723 (2011)

Real-Time Hand Gesture Recognition Using Complex-Valued Neural Network (CVNN)

Abdul Rahman Hafiz, Md. Faijul Amin, and Kazuyuki Murase

Department of Human and Artificial Intelligence System,
Graduate School of Engineering, University of Fukui, Japan
{abdul,amin,murase}@synapse.his.u-fukui.ac.jp
http://www.synapse.his.u-fukui.ac.jp

Abstract. Computer vision system is one of the newest approaches for human computer interaction. Recently, the direct use of our hands as natural input devices has shown promising progress. Toward this progress, we introduce a hand gesture recognition system in this study to recognize real time gesture in unconstrained environments. The system consists of three components: real time hand tracking, hand-tree construction, and hand gesture recognition. Our main contribution includes: (1) a simple way to represent the hand gesture after applying thinning algorithm to the image, and (2) using a model of complex-valued neural network (CVNN) for real-valued classification. We have tested our system to 26 different gestures to evaluate the effectiveness of our approach. The results show that the classification ability of single-layered CVNN on unseen data is comparable to the conventional real-valued neural network (RVNN) having one hidden layer. Moreover, convergence of the CVNN is much faster than that of the RVNN in most cases.

Keywords: Hand gesture recognition, Human computer interaction, Complex-valued neural network.

1 Introduction

The advent of new technologies of human computer interaction has proved that conventional approaches like the keyboard, mouse and pen are not the most efficient and natural ways to interact with the computer. Even though the mouse and the keyboard have made computers more accessible for many decades, but the growing interest in new computer usage has brought to light the need for natural way of human computer interaction. The human hand is the main tool for natural communication, therefore a large variety of techniques have been used for modeling the hand. This model has to be simple so that the computation can be done in real time and as much descriptive that it can give varieties of gestures.

Starner and Pentland [1] studied Human-computer interaction using hand gestures, their algorithm could show promising results in recognizing American Sign Language(ASL), but the algorithm is computationally expensive and limits the resolution of the input frames when applied in real time.

B.-L. Lu, L. Zhang, and J. Kwok (Eds.): ICONIP 2011, Part I, LNCS 7062, pp. 541–549, 2011.
© Springer-Verlag Berlin Heidelberg 2011

The use of inductive learning for hand gesture recognition has been explored in [2], which can learn the temporal spatial information of the hand gesture in real time, Nevertheless, it has the weakness of limited gesture that can be modeled.

While in [3], a DataGlove with 18 sensors is used to recognize the hand patterns, which exist in the raw sensor data of the DataGlove. A pattern of 300 hand gestures is used to train and test different gestures, using back-propagation learning algorithm. The recognition system achieves good performance in real time, the weakness of this approach is the requirement of special hardware like DataGloves.

In our approach, we try to achieve an accurate result for large numbers of gestures (for example, English characters) in real time by using Kinect camera [4]. The system can be applied to various background, changeable lighting of the environment and different kinds of human colors. To achieve that, we construct simple representation of human hand (Hand-Tree) after applying edge detection with thinning algorithm (Fig.1) to the input image, we then define gestures for each English characters that is distinguishable by our algorithm. At the same time, these gestures are designed to make the user change between each representation with minimum effort so that the user can type using his hand instead of the keyboard.

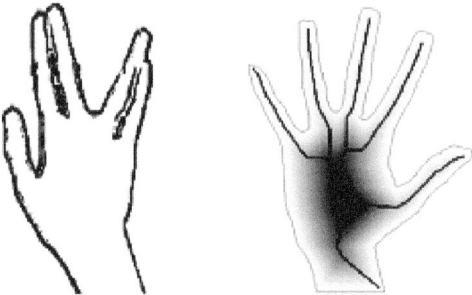

Fig. 1. The left image is a human hand after applying edge detection algorithm, while the image on the right is showing the branches that were produced after applying thinning algorithm to the image

We used two layer complex-valued neural network CVNN [5], due to the nature of the data that we can collect from the generated Hand-Tree. Where each real-valued input is encoded into a phase between 0 and π of a complex number of unity magnitude, and multiplied by a complex-valued weight. The weighted sum of inputs is then fed to an activation function. The activation functions map complex values into real values, and their role is to divide the net-input space into multiple regions representing the classes of input patterns.

2   Procedures

In the project presented, the hardware we used is a Kinect's camera [4], while the software platform is OpenCV [6], for real time image frame capturing and processing.

Fig. 2, shows the modules that represent the complete system. First, the image from the Video Input module projects to Hand Location module as well as the Image Processing module, the Image Processing module processes the area of the image where the hand have been located. After that, the image will be feed to Hand Tree Construction module that feed backs to Locate Hand module about the accurate location of the hand. The final step is the recognition part that uses complex-valued neural network (CVNN) to recognize the gesture. The following subsections describe in detail each stage of the system.

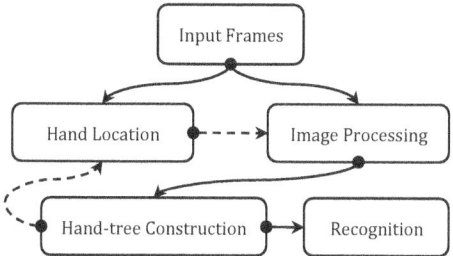

Fig. 2. Human hand gesture recognition system, showing its module and the connections between them

2.1   Detecting and Tracing the Human Hand

The first step is to detect where the hand is among the other things in the image. To do that, we should first delete the background by using the depth map that Kinect's camera provides. The result is shown in Fig. 3. We see only the completed human body in the result.

Fig. 3. Simple illustration of background deletion, the image in the left is the source image, the image in the middle is the depth map that Kinect's camera provides, and the image in the right is the result after deleting the background

Next step, is to delete the areas which don't have the same color as human skin. The result of this process shows the location of the human hand and face. we use HSL representation of color to identify the color of human skin, since HSL representation identifies the color of human skin more accurately than RGB representation. Lastly, we identify the hand from face (Fig. 4).

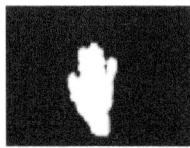 

Fig. 4. The result after skin detection, showing human hand with two states, in the left is the human hand with the fingers are together, and in the right when the fingers are apart

The hand area has a different hue comparing with that of the face, but the difference is small. So we have to support it by other method, i.e. , at the beginning the system depends on the motion of the hand to distinguish it from the face, after that the system knows the location of the hand by the feed back of the Hand Tree Construction part of the system.

From Fig. 4, we can notice that when the fingers are gathered we lose some information about each finger state, to solve this problem we use a sequence of image processing algorithms to gather the information about the human hand fingers, as described in next section.

2.2   Image Processing

After locating the human hand in the image, the system filters the location as shown in Fig. 5.

First, the system applies an edge detection filter to the location of the hand (Sobel Edge Detection [7]). This filter scan the image for any pixels that have different value than its neighbor pixels and assigns a high value for that pixel.

Next, by thresholding the result of the edge detection algorithm the system deletes the noisy edges. but it causes some disconnected parts in the edges of the hand which will effect the result of thinning algorithm. To avoid that the program makes a Dilation on the image, this results in the connection between the disconnected parts.

After that, thinning algorithm is applied [8], this algorithm generates one line of pixels representing each segment of the image.

Finally, we got a hand representation as lines that connects with each other with nodes, now this step aims to read that representation. The system creates pairs of data by detecting the nodes and traces the line that connects these nodes to calculate the length and the angle of these lines, the length of the line which connects two nodes and its relative angle can be calculated as shown in Fig. 6.

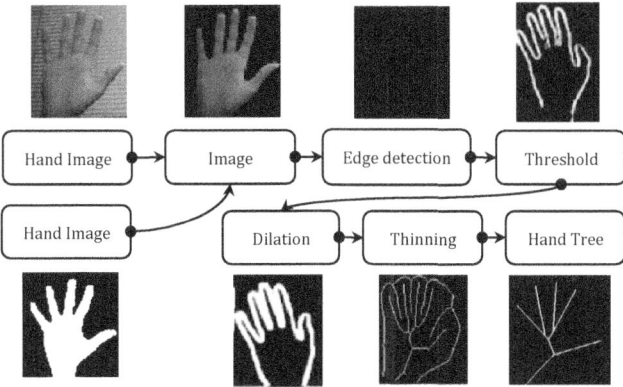

Fig. 5. The image filters that applied in real-time for the hand location, that produce connected branches of line which represent the fingers of human hand

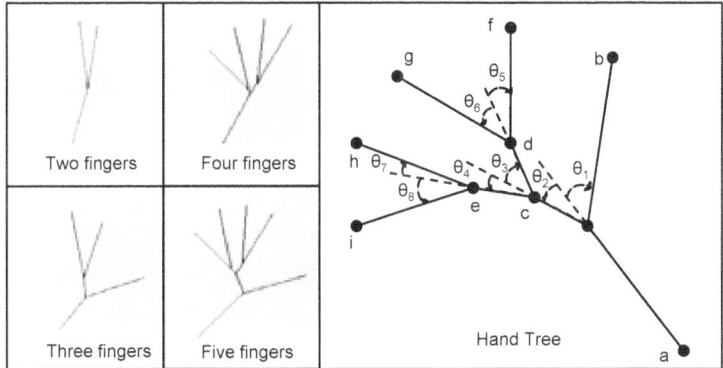

Fig. 6. The left side of the figure is the Hand-Tree branches and nodes that have been generated from the result of the thinning algorithm, in the right side is the pairs of data that can be abstracted from these branches

2.3   Hand Gesture Representation

Since the CVNN processes complex-valued information, it is necessary to map real input values to complex values in order to solve real-valued classification problems. After such mapping, the neuron processes information in a way similar to the conventional neuron model except that all the parameters and variables are complex-valued, and computations are performed according to complex algebra. As illustrated in Fig. 7, the neuron, therefore, first sums up the weighted complex-valued inputs and the threshold value to represent its internal state for the given input pattern, and then the weighted sum is fed to an activation function which maps the internal state (complex-valued weighted sum) to a real value. Here, the activation function combines the real and imaginary parts of the weighted sum.

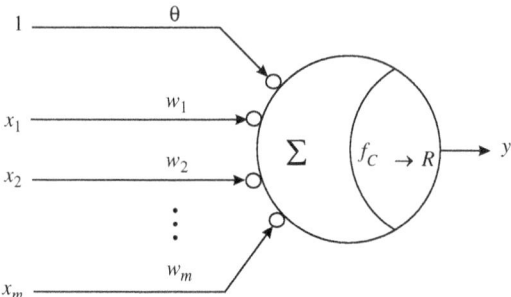

Fig. 7. Model of a complex neuron. The sign $\sum$ sums up the weighted inputs $w_j x_j (1 \leq j \leq m)$ and the bias θ. $f_{C \to R}$ is an activation function that maps the complex- valued internal state to a real-valued output y [5].

By Eulers formula, as given by Eq.1, a complex value z is obtained.

$$z = e^{i\varphi} = sin(\varphi) + i * cos(\varphi). \tag{1}$$

Let the net-input of a complex neuron be $z = u + iv$, then we define the activation functions, Eq.2:

$$f_{C \to R}(z) = (f_R(u) - f_R(V))^2. \tag{2}$$

where $f_R(x)$ is defined as Eq.3:

$$f_R(x) = 1/(1 + exp(-x)). \tag{3}$$

where x, u, $v \in R$, the activation functions combine the real and imaginary parts. The real and imaginary parts are first passed through the same sigmoid function individually. Thus each part becomes bounded within the interval (0,1).

The learning rule is derived from gradient-descent learning rule for a CVNN similar to the work in [6]. We deal with the CVNN without any hidden units.

3   Results

A pattern set with 26 hand gestures are collected, each gesture with 10 samples from one minute video, The set is used to train the neural network, and the testing stage is done online.

We test the system robustness in simple data collection and noise deletion tasks, the system could work in 10 frames per second (fps) and could read the hand state for 90% of the time.

For recognizing the English character, we defined distinguishable gestures of the hand to represent each character, these gestures have been chosen so that it will be easier for the system to recognize them while taking in consideration a human's natural skills to move from one gesture to other, and since our algorithm detects the edges between the fingers even though the fingers are stuck together. This allows us to design a simple representation for each character as shown in Fig. 8.

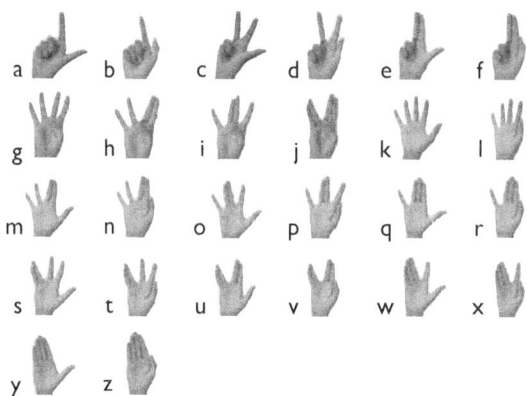

Fig. 8. Hand gesture for each English character, we can notice that the gestures are differing in the number of fingers and the angles they make with each other

First, we test the system with real-valued neural network RVNN to recognize the patterns that exist in these data. The selected design approach for the neural network topology uses a feed-forward network with a single hidden layer. We use supervised learning algorithm for modifying the weight of the neural network. The training method selected is back-propagation using a variable learning rate. These choices are selected because of their simple structure and low computational cost.

The number of input neurons are determined by the maximum number of Hand Tree branches, which is 9, and because each branch has both length and angle that means the input layer of our network should have at least 18 neurons. Accordingly, neural network model has 18 input and 26 output nodes.

The determination of the optimal number of neurons in the hidden layer is achieved by starting at a very low value and increasing until the network can be trained to an acceptable error level. The optimal number of neurons in the hidden layer is determined to be 30 in our study.

Table 1 show us the result of testing 26 english characters for one minute in 10 fps, and that after trained the neural network with the training set of these character with samples taken for one second (10 frames) for each gesture, as we can see each gesture have different recognition rate, and that due to the edges that have created from each finger. In the table, the correct percentage refer to percentage that the output neuron fires to the correct input pattern, incorrect is when the wrong neuron fires, and undetected is when no neuron reach its threshold to produce output.

Table 1. The result of testing 26 english character in real time

Gesture	Correct	Incorrect	Undetected
a	64%	9%	27%
b	59%	11%	30%
c	84%	5%	11%
d	81%	12%	7%
e	71%	16%	13%
f	68%	6%	26%
g	89%	8%	3%
h	82%	13%	5%
i	79%	15%	6%
j	76%	16%	8%
k	96%	3%	1%
l	93%	7%	0%
m	94%	4%	2%
n	89%	7%	4%
o	92%	6%	2%
p	87%	10%	3%
q	89%	8%	3%
r	90%	7%	3%
s	92%	7%	1%
t	85%	3%	12%
u	90%	7%	3%
v	84%	4%	12%
w	82%	8%	10%
x	80%	6%	14%
y	75%	5%	20%
z	64%	9%	27%
Total	82%	8%	10%

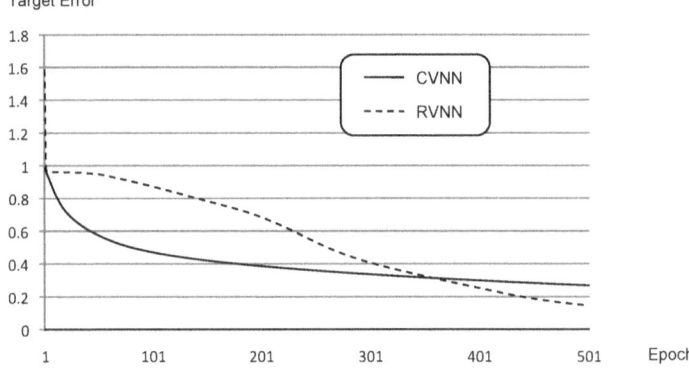

Fig. 9. Learning processes of the two-layered RVN, and single-layered CVN

Second, we test the system with two layer complex-valued neural network CVNN to compare it with RVNN. The input of the CVNN is in complex domain, which allows as to represent each fingers length and angle as one complex number, and because we have 5 fingers and 3 structural branches (Fig. 6) the input layer consist of eight complex-valued neurons, while the output layer consist of 26 neurons that represent english alphabets.

Regarding the learning convergence, single-layered CVNNs required far less training cycles (epochs), in almost all the cases, to reach the minimum validation error than the RVNN counterpart as can be seen in Fig. 9. In other words, learning convergence of CVNN is faster than that of the RVNN.

4 Conclusion

We have developed a hand gesture recognition system, which is shown to be robust for detecting various gestures, and we tested our system to distinguish among 26 differed gestures (English Alphabet). The system is fully automatic and it works in real-time. It is fairly robust to ignore the noise. The advantage of the system lies in the ease of its usage. The Users do not need to wear a glove, neither is there a need for an uniform background.

By using Kinect depth map and the human skin caller we could isolate human hand from the rest of the image, then we used a sequence of image filters to generate a descriptive representation of human hand we call it "Hand-Tree", this representation allows as to use a CVNN for learning and recognition stage, the results shows that a single-layered CVNN is much faster in learning than the RVNN in terms of number of epochs for reaching the minimum validation error. Generalization ability of the single-layered CVNN is comparable to two-layered RVNN in such problem.

Furthermore we can improve the result by adding hidden neurons to CVNN for better Generalization.

Acknowledgments. This study was supported by grants to K.M. from Japanese Society for promotion of Sciences and Technology, and the University of Fukui.

References

1. Starner, T., Pentland: Real-Time American Sign Language Recognition from Video Using Hidden Markov Models, Technical Report 375, MIT Media Lab, Perceptual Computing Group (1995)
2. Zhao, M., Quek, F.K.H., Wu, X.: Recursive induction learning in hand gesture recognition. IEEE Trans. Pattern Anal. Mach. Intell. 20, 1174–1185 (1998)
3. Xu, D.: A Neural Network Approach for Hand Gesture Recognition in Virtual Reality Driving Training System of SPG. In: 8th International Conference on Pattern Recognition, vol. 3 (2006)
4. Shotton, J., Sharp, T.: Real-Time Human Pose Recognition in Parts from Single Depth Images. Statosuedu 2 (2011), retrieved from
   http://www.stat.osu.edu/dmsl/BodyPartRecognition.pdf
5. Amin, M.F., Murase, K.: Single-layered complex-valued neural network for real-valued classification problems. Neurocomputing 72(46), 945–955 (2009)
6. Intel Corporation, Open Source Computer Vision Library, Reference Manual, Copyright 1999-2001, http://www.developer.intel.com
7. Sobel, I., Feldman, G.: A 3x3 Isotropic Gradient Operator for Image Processing. Presented at a Talk at the Stanford Artificial Project (1968); unpublished but often cited
8. Holt, C.M., Stewart, A., Clint, M., Perrott, R.H.: An improved parallel thinning algorithm. Communications of the ACM 30(2), 156–160 (1987)

Wirtinger Calculus Based Gradient Descent and Levenberg-Marquardt Learning Algorithms in Complex-Valued Neural Networks

Md. Faijul Amin, Muhammad Ilias Amin,
A.Y.H. Al-Nuaimi, and Kazuyuki Murase

Department of System Design Engineering,
University of Fukui, Japan
{amin,ahmedyarub,murase}@synapse.his.u-fukui.ac.jp

Abstract. Complex-valued neural networks (CVNNs) bring in nonholomorphic functions in two ways: (i) through their loss functions and (ii) the widely used activation functions. The derivatives of such functions are defined in Wirtinger calculus. In this paper, we derive two popular algorithms—the gradient descent and the Levenberg-Marquardt (LM) algorithm—for parameter optimization in the feedforward CVNNs using the Wirtinger calculus, which is simpler than the conventional derivation that considers the problem in real domain. While deriving the LM algorithm, we solve and use the result of a least squares problem in the complex domain, $\|\mathbf{b} - (\mathbf{Az} + \mathbf{Bz}^*)\|_{\min_{\mathbf{z}}}$, which is more general than the $\|\mathbf{b} - \mathbf{Az}\|_{\min_{\mathbf{z}}}$. Computer simulation results exhibit that as with the real-valued case, the complex-LM algorithm provides much faster learning with higher accuracy than the complex gradient descent algorithm.

Keywords: Complex-valued neural networks (CVNNs), Wirtinger calculus, gradient descent, Levenberg-Marquardt, least squares.

1 Introduction

With the advancement in technology, we see that complex-valued data arise in various practical contexts, such as array signal processing [1], radar and magnetic resonance data processing [2,3], communication systems [4], signal representation in complex baseband [5], and processing data in the frequency domain [2]. Since neural networks are very efficient models in adaptive and nonlinear signal processing, the extension of real-valued neural networks (RVNNs) to complex-valued neural networks (CVNNs) has gained a considerable research interest in the recent years. The key features of CVNNs are that their parameters are complex numbers and they use complex algebraic computations. Although complex-valued data can be processed with RVNNs by considering the data as double dimensional real-valued data, several studies have shown that CVNNs are much more preferable in terms of nonlinear mapping ability, learning convergence, number of parameters, and generalization ability [6].

B.-L. Lu, L. Zhang, and J. Kwok (Eds.): ICONIP 2011, Part I, LNCS 7062, pp. 550–559, 2011.

An important fact in the CVNNs is that they bring in *nonholomorphic* functions in two ways: (i) with the loss function to be minimized over the complex parameters and (ii) the most widely used activation functions. The former is completely unavoidable as the loss function is necessarily real-valued. The second source of nonholomorphism arises because boundedness and analiticity cannot be achieved at the same time in the complex domain, and it is the boundedness that is often preferred over analyticity for the activation functions [6]. Although some researchers have proposed some holomorphic activation functions having singularities [7], a general consideration is that the activation functions can be nonholomorphic. In such a scenario, optimization algorithms are unable to use standard complex derivatives since the derivatives do not exist (i.e., the *Cauchy-Riemann equations* do not hold). As an alternative, conventional approaches cast the optimization problem in the real domain and use the real derivatives, which often requires a tedious computational labor. Here computational labor is meant for human efforts associated with the calculation of derivatives in analytic form.

An elegant approach that can save computational labor in dealing with nonholomorphic functions is to use Wirtinger calculus [8], which uses conjugate coordinate system. A pioneering work that utilizes the concept of conjugate coordinates is by Brandwood [9]. The author formally defines complex gradient and the condition for stationary point. The work is further extended by van dan Bos showing that complex gradient and Hessian are related to their real counterparts by a simple linear transformation [10]. However, neither of the authors has cited the contribution of Wilhelm Wirtinger, a German mathematician, who originally developed the the central idea. Today the Witinger calculus is well appreciated and has been fruitfully exploited by several recent works [11,12].

Although the Wirtinger calculus can be a useful tool in adapting well known first- and second-order optimization algorithms used in the RVNN to the CVNN framework, only few studies can be found in the literature [13]. In [13], the Wirtinger calculus has been utilized to derive a gradient descent algorithm for a feedforward CVNN. The authors employ holomorphic activation functions and show that the derivation is simplified only because of the holomorphic functions. It is further stated that the evaluation of gradient in nonholomorphic case has to be performed in the real domain as it is done traditionally.

In this paper, we argue that the Wirtinger calculus can simplify the gradient evaluation in nonholomorphic activation functions too, which is the original motivation of the Wirtinger calculus. Our gradient evaluation is more general and the CVNN with holomorphic activation function becomes a special case. A major contribution of this paper is that we derive a popular second-order learning method, Levenberg-Marquardt (LM) algorithm [14], for CVNN parameter optimization. We find that a key step of LM algorithm is a solution to the least squares problem $\|\mathbf{b} - (\mathbf{Az} + \mathbf{Bz}^*)\|_{\min_{\mathbf{z}}}$ in the complex domain, which is more general than the $\|\mathbf{b} - \mathbf{Az}\|_{\min_{\mathbf{z}}}$. Here $\mathbf{z}^*$ denotes the complex conjugate of a column vector $\mathbf{z}$. A solution to the least squares problem has been carried out with a proof in this paper. All computations regarding gradient descent and LM algorithm are carried out in complex matrix-vector form that can be easily implemented in any computing environment where computations are optimized for matrix operations.

An important aspect of our derivations is that we use functional dependency graph for a visual evaluation method of derivatives, which is particularly useful in multilayer CVNNs. Because the Wirtinger calculus essentially employs conjugate coordinates, a coordinate transformation matrix between the real and conjugate coordinates system plays an important role in adapting optimization algorithms in the RVNNs to the CVNNs. It turns out that the Wirtinger calculus, the coordinate transformation matrix, and the functional dependency graph are three useful tools for deriving algorithms in the CVNN framework.

The remainder of the paper is organized as follows. Section 2 presents complex domain derivations of two popular algorithms—the gradient descent and the LM algorithm —for CVNN parameter optimization, along with a brief discussion of the Wirtinger calculus. Computer simulation results are discussed in Section 3. Finally, concluding remarks are given in Section 4.

2   Complex Gradient Descent and Complex-LM Algorithm Using Wirtinger Calculus

2.1   Wirtinger Calculus

Any function of a complex variable z can be defined as $f(z) = u(x, y) + jv(x, y)$, where $z = x + jy$ and $j = \sqrt{-1}$. The function is said to be holomorphic (complex derivative exists) if the Cauchy-Riemann equations holds, i.e., $u_x = v_y$, $v_x = -u_y$; otherwise, it is nonholomorphic (complex derivative does not exist). Nonholomorphic functions, however, can be dealt with conjugate coordinates, which are related to the real coordinates by

$$\begin{pmatrix} z \\ z^* \end{pmatrix} = \begin{pmatrix} 1 & j \\ 1 & -j \end{pmatrix} \begin{pmatrix} x \\ y \end{pmatrix} . \tag{1}$$

From the inverse relations, $x = (z + z^*)/2$ and $y = -j(z - z^*)/2$, Wirtinger defines the following pair of derivatives for a function $f(z) = f(z, z^*)$:

$$\frac{\partial f}{\partial z} = \frac{1}{2}\left(\frac{\partial f}{\partial x} - j\frac{\partial f}{\partial y}\right), \quad \frac{\partial f}{\partial z^*} = \frac{1}{2}\left(\frac{\partial f}{\partial x} + j\frac{\partial f}{\partial y}\right) . \tag{2}$$

The derivatives are called $\mathbb{R}$-derivative and conjugate $\mathbb{R}$-derivative, respectively.

Wirtinger calculus generalizes the concept of derivatives in the complex domain. It is easy to see that the Cauchy-Riemann equations are equivalent to $\frac{\partial f}{\partial z^*} = 0$. Rigorous description of the Wirtinger calculus with applications can be found in [16,17]. The attractiveness of Wirtinger calculus is that it enables us to perform all computations directly in the complex domain, and the derivatives obey all rules of conventional calculus, including the chain rule, differentiation of products and quotients. In the evaluation of $\frac{\partial f}{\partial z}$, z^* is considered as a constant and vice versa. Here are some useful identities that we use extensively in the derivation of learning algorithms presented in the next subsections.

$$\left(\frac{\partial f}{\partial z}\right)^* = \frac{\partial f^*}{\partial z^*}; \quad \text{for } f \text{ is real,} \quad \left(\frac{\partial f}{\partial z}\right)^* = \frac{\partial f}{\partial z^*} \tag{3}$$

$$df = \frac{\partial f}{\partial z}dz + \frac{\partial f}{\partial z^*}dz^* \quad \text{Differential rule} \tag{4}$$

$$\frac{\partial h(g)}{\partial z} = \frac{\partial h}{\partial g}\frac{\partial g}{\partial z} + \frac{\partial h}{\partial g^*}\frac{\partial g^*}{\partial z} \quad \text{Chain rule} \tag{5}$$

$$\frac{\partial h(g)}{\partial z^*} = \frac{\partial h}{\partial g}\frac{\partial g}{\partial z^*} + \frac{\partial h}{\partial g^*}\frac{\partial g^*}{\partial z^*} \quad \text{Chain rule} \tag{6}$$

2.2 The Complex Gradient Descent Algorithm

The gradient of a real-valued scalar function of several complex variables can evaluated in both real and conjugate coordinates systems. In fact, there is a one to one correspondence between the coordinate systems. Let $\mathbf{z}$ be a n-dimensional column vector, i.e., $\mathbf{z} = (z_1, z_2, \ldots, z_n)^T \in \mathbb{C}^n$, where $z_i = x_i + jy_i$, $i = 1, \ldots, n$ and $j = \sqrt{-1}$. Then

$$\mathbf{c} \triangleq \begin{pmatrix} \mathbf{z} \\ \mathbf{z}^* \end{pmatrix} \Leftrightarrow \mathbf{r} \triangleq \begin{pmatrix} \mathbf{x} \\ \mathbf{y} \end{pmatrix} = \begin{pmatrix} \Re(\mathbf{z}) \\ \Im(\mathbf{z}) \end{pmatrix}$$

and they are related as follows

$$\mathbf{c} = \begin{pmatrix} \mathbf{z} \\ \mathbf{z}^* \end{pmatrix} = \begin{pmatrix} \mathbf{I} & j\mathbf{I} \\ \mathbf{I} & -j\mathbf{I} \end{pmatrix} \begin{pmatrix} \mathbf{x} \\ \mathbf{y} \end{pmatrix} = \mathbf{J}\begin{pmatrix} \mathbf{x} \\ \mathbf{y} \end{pmatrix} = \mathbf{Jr} \tag{7}$$

Note that $\mathbf{J}^{-1} = \frac{1}{2}\mathbf{J}^H$, where H denotes the Hermitian transpose. Consequently, the real gradient and the complex gradient are related by a linear transformation. The relation guides the algorithm derivation directly in the complex domain. It is established that the complex-gradient of a real-valued function is evaluated as $\nabla_{\mathbf{z}^*} f = 2\frac{\partial f}{\partial \mathbf{z}^*}$ [13].

In the following derivation of complex gradient descent algorithm, we will consider a single hidden layer CVNN for the sake of notational convenience only. The forward equations for signal passing through the network are as follows

$$\mathbf{y} = \mathbf{Vx} + \mathbf{a}; \quad \mathbf{h} = \phi(\mathbf{y}); \quad \mathbf{v} = \mathbf{Wh} + \mathbf{b}; \quad \mathbf{g} = \phi(\mathbf{v}) . \tag{8}$$

Here $\mathbf{x}$ is the input signal and $\mathbf{h}$ and $\mathbf{g}$ are the outputs at hidden and output layer, respectively; the weight matrix $\mathbf{V}$ connects the input units to the hidden units, while the matrix $\mathbf{W}$ connects the hidden units to the output units; the column vectors $\mathbf{a}$ and $\mathbf{b}$ are the biases to the hidden and output units, respectively; and ϕ is any activation function (*holomorphic* or *nonholomorphic*) having real partial derivatives. When the function takes a vector as its argument, each component is mapped individually yielding another vector. The gradient descent algorithm minimizes a real-valued loss function

$$l(\mathbf{z}, \mathbf{z}^*) = \frac{1}{2} \sum_k e_k^* e_k = \frac{1}{2} \mathbf{e}^H \mathbf{e} , \tag{9}$$

where $\mathbf{e} = \mathbf{d} - \mathbf{g}$ denotes the complex error and $\mathbf{d}$ the desired signal. The negative of complex gradient can be written as

$$-2 \frac{\partial l}{\partial \mathbf{z}^*} = -2 \left(\frac{\partial l}{\partial \mathbf{z}} \right)^* = -2 \left(\frac{\partial l}{\partial \mathbf{z}^T} \right)^H . \tag{10}$$

We find that it is convenient to take derivative of a scalar or a column vector with respect to a row vector as it gives the Jacobian naturally. Now

$$-2 \frac{\partial l}{\partial \mathbf{z}^T} = -\frac{\partial \left(\mathbf{e}^H \mathbf{e} \right)}{\partial \mathbf{e}^T} \frac{\partial \mathbf{e}}{\partial \mathbf{z}^T} - \frac{\partial \left(\mathbf{e}^H \mathbf{e} \right)}{\partial \left(\mathbf{e}^T \right)^*} \frac{\partial \mathbf{e}^*}{\partial \mathbf{z}^T} \quad \text{[Eq. (5) in vector form]} \tag{11}$$

$$= \mathbf{e}^H \mathbf{J}_{\mathbf{z}} + \mathbf{e}^T \left(\mathbf{J}_{\mathbf{z}^*} \right)^* \quad \text{[Eq. (3)]} \tag{12}$$

Here, we define two Jacobian matrices, $\mathbf{J}_{\mathbf{z}} = \dfrac{\partial \mathbf{g}}{\partial \mathbf{z}^T}$ and $\mathbf{J}_{\mathbf{z}^*} = \dfrac{\partial \mathbf{g}}{\partial \left(\mathbf{z}^* \right)^T}$. Taking Hermitian transpose to both side of Eq. (12) yields the negative of complex-gradient

$$-\nabla_{\mathbf{z}^*} l = \mathbf{J}_{\mathbf{z}}^H \mathbf{e} + \left(\mathbf{J}_{\mathbf{z}^*}^H \mathbf{e} \right)^* \tag{13}$$

It is clear from Eq. (13) that in order to evaluate complex-gradient all we need is to compute a pair of Jacobians, $\mathbf{J}_{\mathbf{z}}$ and $\mathbf{J}_{\mathbf{z}^*}$. It should be noted that the Jacobians have the form of $\mathbf{P} + j\mathbf{Q}$ and $\mathbf{P} - j\mathbf{Q}$, respectively, because of the definition of derivatives in the Wirtinger calculus. Here $\mathbf{P}$ and $\mathbf{Q}$ are complex matrices since $\mathbf{g}$ is complex-valued. Consequently, we can compute the other one while computing one of the Jacobians. In the feedforward CVNN, parameters are structured in layers. We find that it is very much efficient to compute the Jacobians visually if we draw a functional dependency graph, where each node denotes a vector-valued functional and each edge connecting two nodes is labeled with the Jacobian of one node w.r.t. the other. Figure 1 depicts the functional dependency graph for a single hidden layer CVNN. To evaluate the Jacobian of the right most node (i.e., network's output) w.r.t. any other node to its left, we just need to follow the possible paths from the right most node to that node. Then the desired Jacobian is the sum of all possible paths, where for each path the labeled Jacobians from right to left are multiplied successively. For example, in Fig. 1,

$$\mathbf{J}_{\mathbf{y}} = \frac{\partial \mathbf{g}}{\partial \mathbf{y}^T} = \Lambda_{\mathbf{v}} \mathbf{W} \Lambda_{\mathbf{y}} + \Lambda_{\mathbf{v}^*} \mathbf{W}^* \left(\Lambda_{\mathbf{y}^*} \right)^* \tag{14}$$

where $\Lambda_{\mathbf{v}}$ denotes the Jacobian of right node $\mathbf{g}$ w.r.t. its left node $\mathbf{v}$. Fortunately, we can reuse the computation from right to left. We only need to look for Jacobian to the immediate rightmost nodes, presumably the Jacobian are already computed there. Thus Eq. (14) can be alternatively computed as

$$\mathbf{J}_{\mathbf{y}} = \mathbf{J}_{\mathbf{h}} \Lambda_{\mathbf{y}} + \mathbf{J}_{\mathbf{h}^*} \left(\Lambda_{\mathbf{y}^*} \right)^*$$

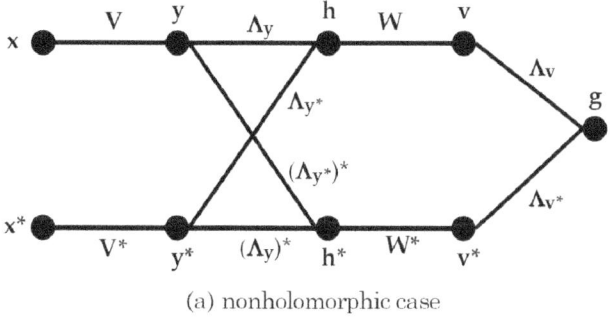

(a) nonholomorphic case

(b) holomorphic case

Fig. 1. Functional dependency graph of a single hidden layer CVNN

It is now a simple task to find the update rule for the CVNN parameters. We note from Eq. (8) that $\mathbf{J_b} = \mathbf{J_v}$ and $\mathbf{J_a} = \mathbf{J_y}$. Thus update rules for the biases at hidden and output layer are

$$\Delta\mathbf{a} = \alpha \left(\mathbf{J}_\mathbf{a}^H \mathbf{e} + \left(\mathbf{J}_{\mathbf{a}^*}^H \mathbf{e} \right)^* \right); \quad \Delta\mathbf{b} = \alpha \left(\mathbf{J}_\mathbf{b}^H \mathbf{e} + \left(\mathbf{J}_{\mathbf{b}^*}^H \mathbf{e} \right)^* \right) \tag{15}$$

Here α is the learning rate. Extending the notation for vector gradient to matrix gradient of a real-valued scalar function [13] and using Eq. (8), the update rules for hidden and output layer weight matrices are given by

$$\Delta\mathbf{V} = (\Delta\mathbf{a})\mathbf{x}^H; \quad \Delta\mathbf{W} = (\Delta\mathbf{b})\mathbf{h}^H . \tag{16}$$

2.3   The Complex LM Algorithm

The LM algorithm is a widely used batch-mode fast learning algorithm in the neural networks that also yields higher accuracy in function approximation than the gradient descent algorithm [15]. Basically, it is an extension of Gauss-Newton algorithm. Therefore, the Gauss-Newton algorithm will be first derived in the complex-domain.

The Gauss-Newton method iteratively *re-linearizes* the nonlinear model and updates the current parameter set according to a least squares solution to the linearized model. In the CVNN, the linearized model of network output $\mathbf{g}(\mathbf{z}, \mathbf{z}^*)$ around $(\hat{\mathbf{z}}, \hat{\mathbf{z}}^*)$ is given by

$$\mathbf{g}(\hat{\mathbf{z}} + \Delta\mathbf{z}, \hat{\mathbf{z}}^* + \Delta\mathbf{z}^*) \approx \hat{\mathbf{g}} + \mathbf{J_z}\Delta\mathbf{z} + \mathbf{J_{z^*}}\Delta\mathbf{z}^* \quad [\text{Eq. (4) in vector form}] \tag{17}$$

The error associated with the linearized model is $\mathbf{e} = \hat{\mathbf{e}} - (\mathbf{J_z}\Delta\mathbf{z} + \mathbf{J_{z^*}}\Delta\mathbf{z}^*)$ Then the Gauss-Newton update rule is given by the least squares solution to

$\|\hat{\mathbf{e}} - (\mathbf{J}_{\mathbf{z}}\Delta\mathbf{z} + \mathbf{J}_{\mathbf{z}^*}\Delta\mathbf{z}^*)\|$. So we encounter a more general least squares problem, $\|\mathbf{b} - (\mathbf{A}\mathbf{z} + \mathbf{B}\mathbf{z}^*)\|_{\min}$, than the well known problem, $\|\mathbf{b} - \mathbf{A}\mathbf{z}\|_{\min}$.

Proposition 1. *Let* $\mathbf{A}$ *and* $\mathbf{B}$ *are arbitrary complex matrices of same dimension. Then a solution to the least squares problem,* $\|\mathbf{b} - (\mathbf{A}\mathbf{z} + \mathbf{B}\mathbf{z}^*)\|_{\min}$, *is given by the following normal equation*

$$\mathbf{C}^H \begin{pmatrix} \mathbf{b} \\ \mathbf{b}^* \end{pmatrix} = \mathbf{C}^H \mathbf{C} \begin{pmatrix} \mathbf{z} \\ \mathbf{z}^* \end{pmatrix}; \quad where \ \mathbf{C} = \begin{pmatrix} \mathbf{A} & \mathbf{B} \\ \mathbf{B}^* & \mathbf{A}^* \end{pmatrix}. \tag{18}$$

Proof. From Eq. (7), we know that the conjugate coordinates are related to the real coordinates by the transformation matrix $\mathbf{J}$, while $\mathbf{J}^{-1} = \frac{1}{2}\mathbf{J}^H$. The error equation and its complex conjugate associated to the least squares problem are

$$\mathbf{e} = \mathbf{b} - (\mathbf{A}\mathbf{z} + \mathbf{B}\mathbf{z}^*) \tag{19}$$
$$\mathbf{e}^* = \mathbf{b}^* - (\mathbf{A}^*\mathbf{z}^* + \mathbf{B}^*\mathbf{z}) \tag{20}$$

Combining Eqs. (19) and (20) to form a single matrix equation and applying the coordinate transformation, the problem can be transformed to real coordinate system, for which the normal equation for least squares problem is well known. This gives the following equation

$$\mathbf{J}^H \begin{pmatrix} \mathbf{e} \\ \mathbf{e}^* \end{pmatrix} = \mathbf{J}^H \begin{pmatrix} \mathbf{b} \\ \mathbf{b}^* \end{pmatrix} - \mathbf{J}^H \mathbf{C} \left(\frac{1}{2}\mathbf{J}\mathbf{J}^H\right) \begin{pmatrix} \mathbf{z} \\ \mathbf{z}^* \end{pmatrix}; \quad \left[\frac{1}{2}\mathbf{J}\mathbf{J}^H = I\right] \tag{21}$$

It can be shown that $\mathbf{J}^H \mathbf{C}\frac{1}{2}\mathbf{J}$ is a real-valued matrix. Because it is now completely in the real coordinate system, we can readily apply the normal equation of the form $\mathbf{P}^T\mathbf{q} = \mathbf{P}^T\mathbf{P}\mathbf{x}$, for real-valued matrices. Noting that the ordinary transpose and Hermitian transpose is the same in the real-valued matrices, the normal equation for the least squares problem of Eq. (21) is

$$\frac{1}{2}\mathbf{J}^H \mathbf{C}^H \mathbf{J}\mathbf{J}^H \begin{pmatrix} \mathbf{b} \\ \mathbf{b}^* \end{pmatrix} = \frac{1}{2}\mathbf{J}^H \mathbf{C}^H \mathbf{J}\mathbf{J}^H \mathbf{C} \left(\frac{1}{2}\mathbf{J}\mathbf{J}^H\right) \begin{pmatrix} \mathbf{z} \\ \mathbf{z}^* \end{pmatrix} \tag{22}$$

Equation (22) immediately yields the following complex normal equation

$$\mathbf{C}^H \begin{pmatrix} \mathbf{b} \\ \mathbf{b}^* \end{pmatrix} = \mathbf{C}^H \mathbf{C} \begin{pmatrix} \mathbf{z} \\ \mathbf{z}^* \end{pmatrix} \qquad \square$$

According to Proposition 1, the least squares solution to $\|\hat{\mathbf{e}} - (\mathbf{J}_{\mathbf{z}}\Delta\mathbf{z} + \mathbf{J}_{\mathbf{z}^*}\Delta\mathbf{z}^*)\|$ gives the the following Gauss-Newton update rule

$$\begin{pmatrix} \Delta\mathbf{z} \\ \Delta\mathbf{z}^* \end{pmatrix} = \mathbf{H}^{-1}\mathbf{G}^H \begin{pmatrix} \hat{\mathbf{e}} \\ \hat{\mathbf{e}}^* \end{pmatrix} \tag{23}$$

where $\mathbf{G} = \begin{pmatrix} \mathbf{J}_{\mathbf{z}} & \mathbf{J}_{\mathbf{z}^*} \\ (\mathbf{J}_{\mathbf{z}^*})^* & (\mathbf{J}_{\mathbf{z}})^* \end{pmatrix}$ and $\mathbf{H} = \mathbf{G}^H\mathbf{G} = \begin{pmatrix} \mathbf{H}_{\mathbf{z}\mathbf{z}} & \mathbf{H}_{\mathbf{z}^*\mathbf{z}} \\ \mathbf{H}_{\mathbf{z}\mathbf{z}^*} & \mathbf{H}_{\mathbf{z}^*\mathbf{z}^*} \end{pmatrix}$

The matrix $\mathbf{H}$ can be considered as an approximation to the Hessian matrix that would result from the Newton method. Note that when $\mathbf{H} = \mathbf{I}$, the Gauss-Newton update rule reduces to the gradient descent algorithm. There is also a pseudo-Gauss-Newton algorithm [17], where the off-diagonal block matrices of $\mathbf{H}$ are $\mathbf{0}$. The pseudo-Gauss-Newton update rule then takes a simpler form

$$\Delta \mathbf{z}^{pseudo-Gauss-Newton} = \mathbf{H}_{\mathbf{zz}}^{-1} \left(\mathbf{J}_{\mathbf{z}}^{H} \hat{\mathbf{e}} + \left(\mathbf{J}_{\mathbf{z}^{*}}^{H} \hat{\mathbf{e}} \right)^{*} \right) . \qquad (24)$$

When the activation functions in the CVNN are holomorphic, the output function $\mathbf{g}(\mathbf{z})$ is also holomorphic. Then the Gauss-Newton update rule resembles the real-valued case

$$\Delta \mathbf{z}^{holomorphic} = \left(\mathbf{J}_{\mathbf{z}}^{H} \mathbf{J}_{\mathbf{z}} \right)^{-1} \mathbf{J}_{\mathbf{z}}^{H} \hat{\mathbf{e}} \qquad (25)$$

It can be observed that all the computations use the Jacobian matrices extensively, which can be evaluated visually and efficiently from the functional dependency graph of Fig. 1. It thus shows the simplicity of our derivation using the Wirtinger calculus.

The LM algorithm makes a simple modification to the Gauss-Newton algorithm of Eq. (23) in the following way

$$\begin{pmatrix} \Delta \mathbf{z} \\ \Delta \mathbf{z}^{*} \end{pmatrix} = \left(\mathbf{G}^{H} \mathbf{G} + \mu \mathbf{I} \right)^{-1} \mathbf{G}^{H} \begin{pmatrix} \hat{\mathbf{e}} \\ \hat{\mathbf{e}}^{*} \end{pmatrix} \qquad (26)$$

The parameter μ is varied over the iterations. Whenever a step increases the error rather than decreasing, μ is multiplied by a factor β. Otherwise, μ is divided by β. A popular choice for the initial value of μ is 0.01 and $\beta = 10$. The algorithmic steps of complex-LM is same as the steps of real-LM [15], except for the parameter update rule derived above.

3   Computer Simulation Results

Computer experiments were performed in order to verify the algorithms presented in the previous section. The algorithms were implemented in Matlab to utilize its matrix computing environment. We took a problem for experiment from [6], where the task is to learn a set of given patterns (see Table 3 of [6]). We trained a 2-4-1 CVNN using the complex gradient descent learning algorithm and two variants of complex-LM algorithm. Note that the LM algorithm uses Gauss-Newton update rule as its basic constituent, which has a variant called pseudo-Gauss-Newton method. Accordingly, we call the variants as complex-LM and pseudo-complex-LM. For all learning algorithms, the training was stopped when the error (mean squared error (MSE)) goal was met, or a maximum number of iteration has been passed. We used the same activation function of [6] in the hidden and output layer, which is nonholomorphic.

The number of iterations required to meet error goal (MSE = 0.001) were 32 and 82 for the complex-LM and the pseudo-complex-LM, respectively.

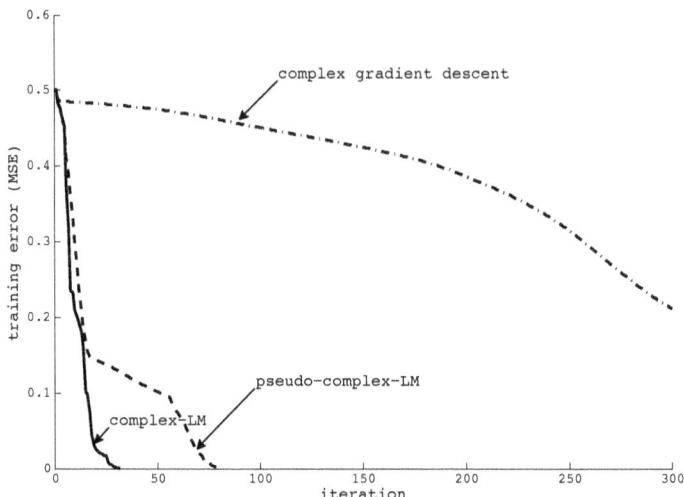

Fig. 2. Learning curves of complex gradient descent, complex-LM, and pseudo-complex-LM algorithm

The complex gradient descent, however, failed to reach the error goal. Figure 2 shows the learning curves for first 300 iterations. Note that the LM algorithms stopped long before. The experimental result suggests that the complex-LM along with its variant has very fast learning convergence with much lower MSE than the complex gradient descent algorithm, which resembles the learning characteristics in the RVNNs. Furthermore, the complex-LM is faster than the pseudo-complex-LM since the latter is an approximation to the former. But the pseudo-complex-LM involves less computation.

The application of complex-LM algorithms, however, is limited by the number of parameters of CVNN, because computation of a matrix inversion is involved in each iteration, while the matrix dimension is in the order of total number of learning parameters. Therefore, the complex-LM algorithms are very useful when the number of parameters are reasonable for matrix inversion and/or a high accuracy in the network mapping is required, such as system identification and time-series prediction problems in the complex domain.

4   Conclusions

In this paper, the complex gradient descent and the complex-LM algorithm have been derived using the Wirtinger calculus, which enables performing all computations directly in the complex domain. The use of the Wirtinger calculus also simplifies the derivation. In course of complex-LM derivation, we have encountered a general least squares problem in the complex domain. We have presented a solution with proof and used the result for derivation in this paper. We identify that the Wirtinger calculus, the coordinate transformation between conjugate and real coordinates, and the functional dependency graph of Jacobians greatly simplify adaptation of the algorithms known for the RVNN to the CVNN framework. Computer simulation results are provided to verify the derivations, and it is shown that the complex-LM as well as its variant, the pseudo-complex-LM, have much faster learning convergence with higher accuracy in nonlinear mapping by the CVNN.

References

1. van Trees, H.L.: Optimum Array Processing. Wiley Interscience, New York (2002)
2. Hirose, A.: Complex-Valued Neural Networks. Springer, Heidelberg (2006)
3. Calhoun, V.D., Adali, T., Pearlson, G.D., van Zijl, P.C.M., Pekar, J.J.: Independent component analysis of fMRI data in the complex domain. Magnetic Resonance in Medicine 48(1), 180–192 (2002)
4. Stuber, G.L.: Principles of Mobile Communication. Kluwer, Boston (2001)
5. Helstrom, C.W.: Elements of Signal Detection and Estimation. Prentice Hall, New Jersey (1995)
6. Nitta, T.: An Extension of the Back-Propagation Algorithm to Complex Numbers. Neural Networks 10(8), 1391–1415 (1997)
7. Kim, T., Adali, T.: Approximation by Fully-Complex Multilayer Perceptrons. Neural Computation 15(7), 1641–1666 (2003)
8. Wirtinger, W.: Zur Formalen Theorie der Funktionen von mehr Complexen Veränderlichen. Mathematische Annalen 97, 357–375 (1927) (in German)
9. Brandwood, D.H.: Complex Gradient Operator and its Application in Adaptive Array Theory. IEE Proceedings F (Communications, Radar and Signal Processing) 130(1), 11–16 (1983)
10. van den Bos, A.: Complex Gradient and Hessian. IEEE Proceedings (Vision, Image and Signal Processing) 141(6), 380–382 (1994)
11. Bouboulis, P., Theodoridis, S.: Extension of Wirtinger's Calculus to Reproducing Kernel Hilbert Spaces and the Complex Kernel LMS. IEEE Trans. on Signal Processing 59(3), 964–978 (2011)
12. Li, H., Adali, T.: Algorithms for Complex ML ICA and Their Stability Analysis Using Wirtinger Calculus. IEEE Trans. on Signal Processing 58(12), 6156–6167 (2010)
13. Li, H., Adali, T.: Complex-Valued Adaptive Signal Processing Using Nonlinear Functions. EURASIP Journal on Advances in Signal Processing 2008, 1–9 (2008)
14. Marquardt, D.: An Algorithm for Least-Squares Estimation of Nonlinear Parameters. SIAM Journal on Applied Mathematics 11(2), 431–441 (1963)
15. Hagan, M., Menhaj, M.: Training Feedforward Networks with the Marquardt Algorithm. IEEE Trans. on Neural Networks 5(6), 989–993 (1994)
16. Remmert, R.: Theory of Complex Functions. Springer, New York (1991)
17. Kreutz-Delgado, K.: The Complex Gradient Operator and the CR-Calculus, http://dsp.ucsd.edu/~kreutz/PEI05.htm

Models of Hopfield-Type Clifford Neural Networks and Their Energy Functions - Hyperbolic and Dual Valued Networks -

Yasuaki Kuroe, Shinpei Tanigawa, and Hitoshi Iima

Department of Information Science, Kyoto Institute of Technology
Matsugasaki, Sakyo-ku, Kyoto 606-8585, Japan
kuroe@kit.ac.jp

Abstract. Recently, models of neural networks in the real domain have been extended into the high dimensional domain such as the complex and quaternion domain, and several high-dimensional models have been proposed. These extensions are generalized by introducing Clifford algebra (geometric algebra). In this paper we extend conventional real-valued models of recurrent neural networks into the domain defined by Clifford algebra and discuss their dynamics. We present models of fully connected recurrent neural networks, which are extensions of the real-valued Hopfield type neural networks to the domain defined by Clifford algebra. We study dynamics of the models from the point view of existence conditions of an energy function. We derive existence conditions of an energy function for two classes of the Hopfield type Clifford neural networks.

Keywords: Clifford algebra, Hopfield neural network, Clifford neural network, Energy function.

1 Introduction

In recent years, there have been increasing research interests of artificial neural networks and many efforts have been made on applications of neural networks to various fields. As applications of the neural networks spread more widely, developing neural network models which can directly deal with complex numbers is desired in various fields. Several models of complex-valued neural networks have been proposed and their abilities of information processing have been investigated[1,2]. Moreover those studies are extended into the quaternion numbers domain, and models of quaternion neural networks are proposed and actively studied[2,11]. These extensions are generalized by introducing Clifford algebra (geometric algebra). Recently Clifford algebra has been recognized to be powerful and practical framework for the representation and solutions of geometrical problems. It has been applied to various problems in science and engineering [13,14]. Neural computation with Clifford algebra is, therefore, expected to possess superior ability of information processing and to realize superior computational intelligence.

B.-L. Lu, L. Zhang, and J. Kwok (Eds.): ICONIP 2011, Part I, LNCS 7062, pp. 560–569, 2011.
© Springer-Verlag Berlin Heidelberg 2011

In this paper we extend conventional real-valued models of recurrent neural networks into the domain defined by Clifford algebra and discuss their dynamics. We present models of fully connected recurrent neural networks, which are extensions of the real-valued Hopfield type neural networks to the domain defined by Clifford algebra. We also discuss dynamics of those models from the point view of existence of an energy function. We have already derived existence conditions and proposed energy functions for Hopfield type complex and quaternion valued neural networks [9,10,11]. Those results can be revisited from the point of view of Clifford algebra models [12]. In this paper we discuss existence conditions of an energy function for two classes of the Hopfield type Clifford neural networks: hyperbolic and dual valued neural networks. Comparisons are also made among the results of those neural networks and that of complex valued neural networks.

2   Clifford Algebra

In this paper we consider the finite dimensional Clifford algebra defined over the real field $\mathbb{R}$. Its outline is given in this section. See [4,5] in detail.

2.1   Definition

Let $\mathbb{R}^{p,q,r}$ denote a $(p+q+r)$-dimensional vector space over the real field $\mathbb{R}$. Let a commutative scalar product be defined as $* : \mathbb{R}^{p,q,r} \times \mathbb{R}^{p,q,r} \to \mathbb{R}$. That is,

$$\boldsymbol{a} * \boldsymbol{b} = \boldsymbol{b} * \boldsymbol{a} \in \mathbb{R} \quad \text{for} \quad \boldsymbol{a}, \boldsymbol{b} \in \mathbb{R}^{p,q,r}.$$

For $\mathbb{R}^{p,q,r}$, the canonical basis, denoted by $\overline{\mathbb{R}}^{p,q,r}$, is defined as totally ordered set

$$\overline{\mathbb{R}}^{p,q,r} := \{\boldsymbol{e}_1, \cdots, \boldsymbol{e}_p, \boldsymbol{e}_{p+1}, \cdots, \boldsymbol{e}_{p+q}, \boldsymbol{e}_{p+q+1}, \cdots, \boldsymbol{e}_{p+q+r}\} \subset \mathbb{R}^{p,q,r} \quad (1)$$

where the $\{\boldsymbol{e}_i\}$ have the property

$$\boldsymbol{e}_i * \boldsymbol{e}_j = \begin{cases} +1, 1 \leq i = j \leq p, \\ -1, p < i = j \leq p + q, \\ 0, p + q < i = j \leq p + q + r, \\ 0, i \neq j \end{cases} \quad (2)$$

The combination of a vector space with a scalar product is called a quadratic space denoted by $(\mathbb{R}^{p,q,r}, *)$. Clifford algebra is defined over the quadratic space $(\mathbb{R}^{p,q,r}, *)$ by introducing so called Clifford product (geometric product) denoted by $\circ$. The Clifford algebra over $(\mathbb{R}^{p,q,r}, *)$ is denoted by $\mathbb{G}(\mathbb{R}^{p,q,r})$ or simply $\mathbb{G}_{p,q,r}$.

[Definition of Clifford Algebra $\mathbb{G}_{p,q,r}$]

Let $\mathbb{G}_{p,q,r}$ denote the associative algebra over the quadratic space $(\mathbb{R}^{p,q,r}, *)$ and let $\circ$ denote the product. Note that the field $\mathbb{R}$ and the vector space $\mathbb{R}^{p,q,r}$ can both be regarded as subspaces of $\mathbb{G}_{p,q,r}$. The $\mathbb{G}_{p,q,r}$ is said to be Clifford algebra if the followings are satisfied.

- $\mathbb{G}_{p,q,r}$ is a vector space equipped with vector addition $+$ and multiplication with scalar ($\alpha \in \mathbb{R}$).
- There exists the product $\circ$ which satisfies the following properties.
  1. The algebra is closed under the product $\circ$, that is, $a \circ b \in \mathbb{G}_{p,q,r}$ for all $a, b \in \mathbb{G}_{p,q,r}$.
  2. Associativity: $(a \circ b) \circ c = a \circ (b \circ c)$ for all $a, b, c \in \mathbb{G}_{p,q,r}$.
  3. Distributivity: $a \circ (b + c) = a \circ b + a \circ c$ for all $a, b, c \in \mathbb{G}_{p,q,r}$.
  4. Scalar multiplication: $\alpha \circ a = a \circ \alpha = \alpha a$, for all $a \in \mathbb{G}_{p,q,r}, \alpha \in \mathbb{R}$.
  5. Let $a \in \mathbb{R}^{p,q,r} \subset \mathbb{G}_{p,q,r}$; then $a \circ a = a * a \in \mathbb{R}$

Note that the commutativity is not imposed on the Clifford product $\circ$, that is, it is non-commutative.

2.2   Basic Properties and Algebraic Basis

The elements of Clifford algebra $\mathbb{G}_{p,q,r}$ are called multivectors whereas the elements of $\mathbb{R}^{p,q,r}$ are called vectors. For multivectors $a, b \in \mathbb{G}_{p,q,r}$, the Clifford product $a \circ b$ is expressed as a sum of its symmetric and antisymmetric parts:

$$a \circ b = \frac{1}{2}(a \circ b + b \circ a) + \frac{1}{2}(a \circ b - b \circ a).$$

If a and b are vectors, that is, $a, b \in \mathbb{R}^{p,q,r}$, the following relation holds.

$$(a + b) \circ (a + b) = (a + b) * (a + b)$$

$$\Leftrightarrow a \circ a + a \circ b + b \circ a + b \circ b = a * a + 2a * b + b * b$$

$$\Leftrightarrow \frac{1}{2}(a \circ b + b \circ a) = a * b$$

Let express the antisymmetric part as $a \wedge b := \frac{1}{2}(a \circ b - b \circ a)$, then $a \circ b = a * b + a \wedge b$. The product $\wedge$ is called the outer or wedge product. In particular, for basis vectors e_i, e_j in $\mathbb{R}^{p,q,r}$, $e_i \circ e_j = e_i \wedge e_j$ since $e_i * e_j = 0$ ($i \neq j$) from (2), which implies

$$e_i \circ e_j = -e_j \circ e_i. \tag{3}$$

We are now in the position to construct a basis of the Clifford algebra $\mathbb{G}_{p,q,r}$, which is called an algebraic basis of $\mathbb{G}_{p,q,r}$. From now on, the Clifford product will be denoted by juxtaposition of symbols. For example, $a \circ b$ is now written as ab. Since the Clifford product is associative, $(a \circ b) \circ c$ or $a \circ (b \circ c)$ is written as abc. Also, the product operator $\prod$ refers to the Clifford product of the operands, for example, $\prod_{i=1}^{3} a_i = a_1 a_2 a_3$.

Consider the Clifford product of a number of different elements of the canonical basis $\overline{\mathbb{R}}^{p,q,r}$ of $\mathbb{R}^{p,q,r}$, called a basis blade, which plays an important role to construct a basis of the Clifford algebra $\mathbb{G}_{p,q,r}$.

Let $\mathbb{A}$ be an ordered set and let $\mathbb{A}[i]$ denote the ith elements of $\mathbb{A}$. That is, if $\mathbb{A} = \{2, 3, 1\}$, then $\mathbb{A}[2] = 3$. A basis blade in $\mathbb{G}_{p,q,r}$ denoted by $e_{\mathbb{A}}$ is defined; let $\mathbb{A} \subset \{1, 2, \cdots, p + q + r\}$, then

$$e_{\mathbb{A}} = \prod_{i=1}^{|\mathbb{A}|} \overline{\mathbb{R}}^{p,q,r}[\mathbb{A}[i]]. \tag{4}$$

where $|\mathbb{A}|$ denotes the number of elements of the set $\mathbb{A}$. The number of the factors under the Clifford products in each basis blade $e_{\mathbb{A}}$, that is, $|\mathbb{A}|$ is called grade. For example, if $\mathbb{A} = \{2,3,1\}$, then $e_{\mathbb{A}} = e_2 e_3 e_1$ and its grade is 3.

Given a vector space $\mathbb{R}^{p,q,r}$ with a canonical basis $\overline{\mathbb{R}}^{p,q,r}$ in (1), there are 2^{p+q+r} ways to combine the $\{e_i\}$ with the Clifford product such that no two of these products are linearly independent, that is, there exist 2^{p+q+r} linearly independent basis blades. The collection of 2^{p+q+r} linearly independent basis blades forms an algebraic basis of $\mathbb{G}_{p,q,r}$. The choice of basis is arbitrary, however, it is useful to choose an ordered basis, called the canonical algebraic basis, defined as follows. Let $\mathbb{I} = \{1, 2, \cdots, p+q+r\}$, its power set is denoted by $\mathcal{P}[\mathbb{I}]$ and its ordered power set is denoted by $\mathcal{P}_{\mathcal{O}}[\mathbb{I}]$. For example, if $\mathbb{I} = \{1,2,3\}$, then $\mathcal{P}_{\mathcal{O}}[\mathbb{I}] = \{\{\emptyset\}, \{1\}, \{2\}, \{3\}, \{1,2\}, \{1,3\}, \{2,3\}, \{1,2,3\}\}$ where $\emptyset$ is the empty set. The canonical algebraic basis of $\mathbb{G}_{p,q,r}$, denoted by $\overline{\mathbb{G}}_{p,q,r}$ is defined:

$$\overline{\mathbb{G}}_{p,q,r} := \{e_{\mathbb{A}} : \mathbb{A} \in \mathcal{P}_{\mathcal{O}}[\mathbb{I}]\}$$

where we let $e_{\emptyset} = 1 \in \mathbb{R}$. For example, let $p + q + r = 3$ and consider $\mathbb{R}^3 :=$ $\mathbb{R}^{p,q,r}$ with a canonical basis $\overline{\mathbb{R}}^3 = \{e_1, e_2, e_3\}$. The canonical algebraic basis $\overline{\mathbb{G}}_3$ of $\mathbb{G}_3$ is then given by $\overline{\mathbb{G}}_3 = \{1, e_1, e_2, e_3, e_1 e_2, e_1 e_3, e_2 e_3, e_1 e_2 e_3\}$. A general multivector of $\mathbb{G}_{p,q,r}$ is written as a linear combination of the elements of the canonical algebraic basis $\overline{\mathbb{G}}_{p,q,r}$ thus defined, that is, $a \in \mathbb{G}_{p,q,r}$ is written as:

$$a = \sum_{i=1}^{2^{p+q+r}} a^{(i)} \overline{\mathbb{G}}_{p,q,r}[i] \tag{5}$$

where $a^{(i)} \in \mathbb{R}$. Recall that $\overline{\mathbb{G}}_{p,q,r}[i]$ denotes the ith element of $\overline{\mathbb{G}}_{p,q,r}$.

The absolute value (modulus) of $a \in \mathbb{G}_{p,q,r}$, denoted by $|a|$, is defined as

$$|a| = \left(\sum_{i=1}^{2^{p+q+r}} a^{(i)^2} \right)^{1/2}.$$

3   Models of Hopfield Type Clifford Neural Networks

In this section we present models of fully connected recurrent neural networks, which are extensions of real valued Hopfield neural networks into the domain of the Clifford algebra. In the previous section we write the multivectors, that is, elements of the Clifford algebra, in boldface like $a \in \mathbb{G}_{p,q,r}$. From now on, we write them in normal face like $a \in \mathbb{G}_{p,q,r}$ for simplicity.

We consider a class of Clifford neural networks described by differential equations of the form:

$$\begin{cases} \tau_i \dfrac{du_i}{dt} = -u_i + \sum_{j=1}^{n} w_{ij} v_j + b_i \\ v_i = f(u_i) \qquad (i = 1, 2, \cdots, n) \end{cases} \tag{6}$$

where n is the number of neurons, τ_i is the time constant of the ith neuron, u_i and v_i are the state and the output of the ith neuron at time t, respectively, b_i is the threshold value, w_{ij} is the connection weight coefficient from the jth neuron to the ith one and $f(\cdot)$ is the activation function of the neurons. In the model u_i, v_i, b_i and w_{ij} are multivectors, that is, the elements of the Clifford algebra $\mathbb{G}_{p,q,r}$: $u_i \in \mathbb{G}_{p,q,r}$, $v_i \in \mathbb{G}_{p,q,r}$, $b_i \in \mathbb{G}_{p,q,r}$, $w_{ij} \in \mathbb{G}_{p,q,r}$. The time constant τ_i is a positive real number: $\tau_i \in \mathbb{R}, \tau_i > 0$. $w_{ij}v_j$ is the Clifford product of w_{ij} and v_j in $\mathbb{G}_{p,q,r}$: $w_{ij} \circ v_j$. The activation function $f(\cdot)$ is a nonlinear function which maps from a multivector to a multivector: $f : \mathbb{G}_{p,q,r} \to \mathbb{G}_{p,q,r}$. For a multivector $u(t) = \sum_{i=1}^{2^{p+q+r}} u^{(i)}(t)\overline{\mathbb{G}}_{p,q,r}[i]$, its time derivative is defined by

$$\frac{d}{dt}u(t) := \sum_{i=1}^{2^{p+q+r}} \frac{d}{dt}u^{(i)}(t)\overline{\mathbb{G}}_{p,q,r}[i].$$

Note that the neural network described by (6) is a direct Clifford domain extension of the real-valued neural network of Hopfield type.

Since in $\mathbb{G}_{p,q,r}$, the Clifford product is non-commutative, we can consider other two models: one is the model in which the second term $\sum_{j=1}^{n} w_{ij}v_j$ of the right side of (6) is replaced by $\sum_{j=1}^{n} v_j w_{ij}$ and the other is the model in which the second term is replaced by $\sum_{j=1}^{n} w_{ij}^* v_j w_{ij}$ where w_{ij}^* could generally be any multivecter in $\mathbb{G}_{p,q,r}$ different from w_{ij}, but it is useful in the Clifford algebra to let $*$ be an involution operator. The involution is an operation that maps an operand to itself when applied twice: $(w^*)^* = w$.

In the followings we investigate dynamics of Hopfield-type Clifford neural networks described by (6). In particular we discuss existence conditions of an energy function for three basic classes of the networks: Clifford neural network of class $\mathbb{G}_{1,0,0}$, $\mathbb{G}_{0,1,0}$ and $\mathbb{G}_{0,0,1}$ which are isomorphic to the hyperbolic, complex and dual numbers, respectively. Note that the Clifford product of $\mathbb{G}_{1,0,0}$, $\mathbb{G}_{0,1,0}$ and $\mathbb{G}_{0,0,1}$ is commutative.

4   Existence Conditions of Energy Functions

4.1   Definition of Energy Functions

It is well known that one of the pioneering works that triggered the research interests of neural networks in the last two decades is a proposal of models for neural networks by Hopfield et. al. [6,7,8]. He introduced the idea of an energy function to formulate a way of understanding the computation performed by fully connected recurrent neural networks and showed that a combinatorial optimization problem can be solved by them. The energy functions have been applied to various problems such as qualitative analysis of neural networks, synthesis of associative memories, optimization problems etc. ever since. It is, therefore, of great interest to investigate existence conditions of energy functions and to obtain energy functions for the neural networks (6).

One of the important factors to characterize dynamics of recurrent neural networks is their activation functions which are nonlinear functions. It is therefore, important to discuss which type of nonlinear functions is chosen as activation functions for Clifford neural networks (6). In the real-valued neural networks, the activation is usually chosen to be a smooth and bounded function such as a sigmoidal function. Recall that, in the complex domain, the Liouvill's theorem says that 'if $f(\cdot)$ is analytic at all points of the complex plane and bounded, then $f(\cdot)$ is constant'. Since a suitable $f(\cdot)$ should be bounded, it follows from the theorem that if we choose an analytic function for $f(\cdot)$, it is constant over the entire complex plain, which is clearly not suitable. In the complex-valued neural networks in [9], in place of analytic function, a function whose real and imaginary parts are continuously differentiable with respect to the real and imaginary variables of its argument, respectively, is chosen for the activation function and the existence conditions of an energy function are derived[9].

Although the Liouvill's theorem is in general not valid for Clifford algebras, in this paper as a first step we choose the following functions as activation functions for the Clifford neural networks (6). Letting $f^{(i)}$, $i = 1, 2, \cdots, 2^{p+q+r}$ be real value functions: $f^{(i)} : \mathbb{R}^{2^{p+q+r}} \to \mathbb{R}$, the nonlinear function on the Clifford algebra $f(u) : \mathbb{G}_{p,q,r} \to \mathbb{G}_{p,q,r}$ is described as follows:

$$f(u) = \sum_{i=1}^{2^{p+q+r}} f^{(i)}(u^{(1)}, u^{(2)}, \cdots, u^{(2^{p+q+r})}) \overline{\mathbb{G}}_{p,q,r}[i] \tag{7}$$

where

$$u = \sum_{i=1}^{2^{p+q+r}} u^{(i)} \overline{\mathbb{G}}_{p,q,r}[i]. \tag{8}$$

We assume the following conditions on the activation function $f(u) : \mathbb{G}_{p,q,r} \to \mathbb{G}_{p,q,r}$ of the neural networks (6).

(i) $f^{(l)}(\cdot), (l = 1, 2, \cdots, 2^{p+q+r})$ are continuously differentiable with respect to $u^{(m)}, (m = 1, 2, \cdots, 2^{p+q+r})$.

(ii) $f(\cdot)$ is a bounded function, that is, there exists some $M > 0$ such that $|f(\cdot)| \leq M$.

We are now in the position to give the definition of energy functions for the Clifford neural networks (6). If the neural network (6) is real valued, that is, u_i, v_i, b_i and w_{ij} are all real, $u_i \in \mathbb{R}$, $v_i \in \mathbb{R}$, $b_i \in \mathbb{R}$, $w_{ij} \in \mathbb{R}$ and the activation function is a real nonlinear function $f : \mathbb{R} \to \mathbb{R}$, the existence condition of an energy function which Hopfield et. al. obtained is that the weight matrix $W = \{w_{ij}\}$ is a symmetric matrix $(w_{ij} = w_{ji})$ and the activation function is continuously differentiable, bounded and monotonically increasing. The following function $E : \mathbb{R}^n \to \mathbb{R}$ was proposed as an energy function for the network.

$$E(\boldsymbol{v}) = -\frac{1}{2} \sum_{i=1}^{n} \sum_{j=1}^{n} w_{ij} v_i v_j - \sum_{i=1}^{n} b_i v_i + \sum_{i=1}^{n} \int_0^{v_i} f^{-1}(\rho) d\rho \tag{9}$$

where $\boldsymbol{v} = [v_1, v_2, \cdots, v_n] \in \mathbb{R}^n$ and f^{-1} is the inverse function of f. Hopfield et. al. showed that, if the existence conditions hold, the network (6) has the function $E(\boldsymbol{v})$ and it has the following property; the time derivative of E along the trajectories of (6), denoted by $\frac{dE}{dt}\big|_{(6)^R}$ is less or equal to 0, $\frac{dE}{dt}\big|_{(6)^R} \leq 0$, and furthermore $\frac{dE}{dt}\big|_{(6)^R} = 0$ if and only if $\frac{dv_i}{dt} = 0$ ($i = 1, 2, \cdots, n$).

We define an energy function for the Clifford neural networks (6) by the analogy to that for Hopfield type real-valued neural networks as follows.

Definition 1. *Consider the Clifford neural network (6). E is an energy function of the Clifford neural network (6), if the following conditions are satisfied.*

(i) *$E(\cdot)$ is a mapping, $E : \mathbb{G}_{p,q,r} \to \mathbb{R}$, and bounded from below.*
(ii) *The derivative of E along the trajectories of the network (6), denoted by $\frac{dE}{dt}\big|_{(6)}$, satisfies $\frac{dE}{dt}\big|_{(6)} \leq 0$. Furthermore, $\frac{dE}{dt}\big|_{(6)} = 0$ if and only if $\frac{dv_i}{dt} = 0$ ($i = 1, 2, \cdots, n$).*

4.2 Existence Conditions for Clifford Neural Networks of Classes $\mathbb{G}_{1,0,0}$, $\mathbb{G}_{0,1,0}$ and $\mathbb{G}_{0,0,1}$

The canonical basis of the Clifford algebra $\mathbb{G}_{1,0,0}$, $\mathbb{G}_{0,1,0}$ or $\mathbb{G}_{0,0,1}$ is given by

$$\overline{\mathbb{G}}_{p,q,r} = \{1, \boldsymbol{e}_1\}.$$

where $\boldsymbol{e}_1\boldsymbol{e}_1 = 1$ for $\mathbb{G}_{1,0,0}$, $\boldsymbol{e}_1\boldsymbol{e}_1 = -1$ for $\mathbb{G}_{0,1,0}$ and $\boldsymbol{e}_1\boldsymbol{e}_1 = 0$ for $\mathbb{G}_{0,0,1}$. An element of $\mathbb{G}_{1,0,0}$, $\mathbb{G}_{0,1,0}$ or $\mathbb{G}_{0,0,1}$ is described as follows.

$$x = x^{(0)} + x^{(1)}\boldsymbol{e}_1. \tag{10}$$

We need the following assumptions on the weight coefficients and the activation functions of (6).

Assumption 1. *The weight coefficients of the Clifford neural networks (6) of the class $\mathbb{G}_{1,0,0}$, $\mathbb{G}_{0,1,0}$ or $\mathbb{G}_{0,0,1}$ satisfy the following conditions.*
For the class $\mathbb{G}_{1,0,0}$,

$$w_{ji} = w_{ij} \quad (i, j = 1, 2, \cdots, n). \tag{11}$$

For the class $\mathbb{G}_{0,1,0}$,

$$w_{ji} = w_{ij}^* \quad (i, j = 1, 2, \cdots, n) \tag{12}$$

where $*$ is defined: for $w = x^{(0)} + x^{(1)}\boldsymbol{e}_1 \in \mathbb{G}_{0,1,0}$, $w^* = x^{(0)} - x^{(1)}\boldsymbol{e}_1$.
For the class $\mathbb{G}_{0,0,1}$,

$$w_{ji} = w_{ij} \quad and \quad w_{ij}^{(1)} = 0 \quad (i, j = 1, 2, \cdots, n) \tag{13}$$

where $w_{ij} = w_{ij}^{(0)} + w_{ij}^{(1)}\boldsymbol{e}_1$.

The activation function (6) in the Clifford Algebra $\mathbb{G}_{1,0,0}$, $\mathbb{G}_{0,1,0}$ and $\mathbb{G}_{0,0,1}$ is described by

$$f(u) = f^{(0)}(u^{(0)}, u^{(1)}) + f^{(1)}(u^{(0)}, u^{(1)})e_1,$$

where $u = u^{(0)} + u^{(1)}e_1$.

Assumption 2. *The activation function f of the Clifford neural networks (6) of the class $\mathbb{G}_{1,0,0}$, $\mathbb{G}_{0,1,0}$ or $\mathbb{G}_{0,0,1}$ is an injective function and satisfies*

$$\text{(i)} \quad \frac{\partial f^{(0)}}{\partial u^{(0)}} > 0, \quad \text{(ii)} \quad \frac{\partial f^{(0)}}{\partial u^{(1)}} = \frac{\partial f^{(1)}}{\partial u^{(0)}}, \quad \text{(iii)} \quad \frac{\partial f^{(0)}}{\partial u^{(0)}}\frac{\partial f^{(1)}}{\partial u^{(1)}} - \frac{\partial f^{(0)}}{\partial u^{(1)}}\frac{\partial f^{(1)}}{\partial u^{(0)}} > 0 \quad (14)$$

for all $u \in \mathbb{G}_{1,0,0}$, $u \in \mathbb{G}_{0,1,0}$ or $u \in \mathbb{G}_{0,0,1}$.

Because of the injectivity and boundedness of f, there exists the inverse function of f, denoted by $g = f^{-1} : \mathbb{G}_{1,0,0} \to \mathbb{G}_{1,0,0}$, $\mathbb{G}_{0,1,0} \to \mathbb{G}_{0,1,0}$ or $\mathbb{G}_{0,0,1} \to \mathbb{G}_{0,0,1}$. We express g as $u = g(v)$:

$$g(v) = g^{(0)}(v^{(0)}, v^{(1)}) + g^{(1)}(v^{(0)}, v^{(1)})e_1 \quad (15)$$

The following lemma holds.

Lemma 1. *If f satisfies Assumption 2, there exists a scalar function $G(\cdot)$: $\mathbb{G}_{1,0,0} \to \mathbb{R}$, $\mathbb{G}_{0,1,0} \to \mathbb{R}$ or $\mathbb{G}_{0,0,1} \to \mathbb{R}$ such that*

$$\frac{\partial G}{\partial v^{(0)}} = g^{(0)}(v^{(0)}, v^{(1)}), \qquad \frac{\partial G}{\partial v^{(1)}} = g^{(1)}(v^{(0)}, v^{(1)}) \quad (16)$$

This lemma can be proved by defining the function $G(v)$ as

$$G(v) := \int_0^{v^{(0)}} g^{(0)}(\rho, 0)d\rho + \int_0^{v^{(1)}} g^{(0)}(v^{(0)}, \rho)d\rho \quad (17)$$

We now propose candidates of the energy functions for the Clifford neural networks (6) of the classes $\mathbb{G}_{1,0,0}$, $\mathbb{G}_{0,1,0}$ and $\mathbb{G}_{0,0,1}$ as follows. For the network of the classes $\mathbb{G}_{1,0,0}$ and $\mathbb{G}_{0,0,1}$,

$$E(v) = -\sum_{i=1}^{n}\sum_{j=1}^{n}\left\{\frac{1}{2}Sc\left(v_i w_{ij} v_j + 2b_i v_i\right) - G(v_i)\right\} \quad (18)$$

where $v = [v_1, v_2, \cdots, v_n]^T \in \mathbb{G}_{1,0,0}^n$ or $v = [v_1, v_2, \cdots, v_n]^T \in \mathbb{G}_{0,0,1}^n$ and $Sc(\cdot)$ is defined; for $x \in \mathbb{G}_{p,q,r}$, $Sc(x) = x^{(0)}$. For the network of the class $\mathbb{G}_{0,1,0}$,

$$E(v) = -\sum_{i=1}^{n}\sum_{j=1}^{n}\left\{\frac{1}{2}Sc\left(v_i^* w_{ij} v_j + 2b_i^* v_i\right) - G(v_i)\right\} \quad (19)$$

where $v = [v_1, v_2, \cdots, v_n]^T \in \mathbb{G}_{0,1,0}^n$.
    The following theorem is obtained.

Theorem 1. *If the Clifford neural networks (6) of the classes* $\mathbb{G}_{1,0,0}$, $\mathbb{G}_{0,1,0}$ *and* $\mathbb{G}_{0,0,1}$ *satisfy Assumptions 1 and 2, then there exists an energy function which satisfies Definition 1.*

This theorem can be proved as follows. Calculating the time derivatives of the function (18) or (19) for the networks (6) and using Lemma 1, we can show that the conditions of Definition 1 of energy functions hold.

The existence conditions of energy functions thus obtained are ones on the connection weight coefficients w_{ij} and the activation function $f(\cdot)$. Note that the existence conditions of energy functions for the Clifford neural networks of the class $\mathbb{G}_{0,1,0}$ are equivalent to those of the complex valued neural networks in [9]. As examples of the functions which satisfy Assumption 2,

$$f(u) = \frac{u}{1 + |u|} \tag{20}$$

$$f(u) = \tanh(u^{(0)}) + \tanh(u^{(1)})e_1 \tag{21}$$

can be considered. Equation (20) has the same form as that of the complex-valued function which is often used in the complex-valued neural networks[9]. The function (21) is a split activation function, that is, each component of its argument is transformed separately.

It is expected that the energy functions (18) and (19) can be applied to various problems. In the real valued neural networks energy functions have been applied to various problems such as qualitative analysis of neural networks, synthesis of associative memories and optimization problems. In [9] and [11], qualitative analysis of the complex valued and quaternion valued networks is performed by utilizing energy functions and some results are obtained. The similar results can be obtained for the Clifford neural networks of classes $\mathbb{G}_{1,0,0}$, $\mathbb{G}_{0,1,0}$ and $\mathbb{G}_{0,0,1}$ by utilizing the energy functions (18) and (19).

5   Conclusions

Recently, models of neural networks in the real domain have been extended into the high dimensional domain such as the complex and quaternion domain. These extensions are generalized by introducing Clifford algebra (geometric algebra). In this paper we extend conventional real-valued models of recurrent neural networks into the domain defined by Clifford algebra and discuss their dynamics. We presented models of fully connected recurrent Clifford neural networks, which are extensions of the real-valued Hopfield type neural networks to the domain defined by Clifford algebra. We also studied dynamics of the proposed models from the point view of existence conditions of an energy function. The existence conditions were discussed for three classes of Hopfield type Clifford neural networks defined in $\mathbb{G}_{1,0,0}$, $\mathbb{G}_{0,1,0}$ and $\mathbb{G}_{0,0,1}$. Further work is underway in deriving the existence conditions for more general classes of Hopfield type Clifford neural networks.

References

1. Hirose, A. (ed.): Complex-Valued Neural Networks Theories and Applications. World Scientific (2003)
2. Nitta, T. (ed.): Complex-Valued Neural Networks Utilizing High-Dimensional Parameters. IGI Global (2009)
3. Buchholz, S.: A Theory of Neural Computation with Clifford Algebra. Ph.D. Thesis, University. of Kiel (2005)
4. Lounesto, P.: Clifford Algebras and Spinors, 2nd edn. Cambridge Univ. Press (2001)
5. Perwass, C.: Geometric Algebra with Applications in Engineering. Springer, Heidelberg (2009)
6. Hopfield, J.J.: Neurons with graded response have collective computational properties like those of two-state neurons. Proc. Natl. Acad. Sci. USA 81, 3088–3092 (1984)
7. Hopfield, J.J., Tank, D.W.: Neural' computation of decisions in optimization problems. Biol. Cybern. 52, 141–152 (1985)
8. Tank, D.W., Hopfield, J.J.: Simple 'neural' optimization networks: an A/D converter, signal decision circuit, and a linear programming circuit. IEEE Trans. Circuits Syst. 33, 533–541 (1986)
9. Hashimoto, N., Kuroe, Y., Mori, T.: On Energy Function for Complex-Valued Neural Networks. Trans. of the Institute of System, Control and Information Engineers 15(10), 559–565 (2002)
10. Kuroe, Y., Yoshida, M., Mori, T.: On Activation Functions for Complex-Valued Neural Networks - Existence of Energy Functions. In: Kaynak, O., Alpaydın, E., Oja, E., Xu, L. (eds.) ICANN 2003 and ICONIP 2003. LNCS, vol. 2714, pp. 985–992. Springer, Heidelberg (2003)
11. Yoshida, M., Kuroe, Y., Mori, T.: Models of Hopfield-Type Quaternion Neural Networks and Their Energy Functions. International Journal of Neural Systems 15(1&2), 129–135 (2005)
12. Kuroe, Y.: Models of Clifford Recurrent Neural Networks and Their Dynamics. In: Proceedings of 2011 International Joint Conference on Neural Networks, pp. 1035–1041 (2011)
13. Dorst, L., Fontijne, D., Mann, S.: Geometric Algebra for Computer Science An object-oriented Approach to Geometry. Morgan Kaufmann (2007)
14. Bayro-Corrochano, E., Scheuermann, G. (eds.): Geometric Algebra Computing in Engineering and Computer Science. Springer, Heidelberg (2010)

Simultaneous Pattern and Variable Weighting during Topological Clustering

Nistor Grozavu[1,2] and Younès Bennani[1,2]

[1] Université Paris 13,
99, av. J-B Clément, 93430 Villetaneuse
[2] LIPN-UMR 7030, Université Paris 13,
99, av. J-B Clément, 93430 Villetaneuse, France
{firstname.secondname}@lipn.univ-paris13.fr

Abstract. This paper addresses the problem of detecting a subset of the most relevant features and observations from a dataset through a local weighted learning paradigm. We introduce a new learning approach, which provides simultaneously Self-Organizing Map (SOM) and double local weighting. The proposed approach is computationally simple, and learns a different features vector weights for each cell (relevance vector) and an observation weighting matrix. Based on the lwo-SOM and lwd-SOM [7], we present a new weighting approach allowing to take into account the importance of the observations and of the variables simultaneously called dlw-SOM. After the learning phase, a selection method is used with weight vectors to prune the irrelevant variables and thus we can characterize the clusters. A number of synthetic and real data are experimented on to show the benefits of the proposed double local weighting using self-organizing models.

Keywords: self-organizing maps, weighting, feature selection.

1 Introduction

The data size can be measured in two dimensions, the size of features and the size of observations. Both dimensions can take very high values, which can cause problems during the exploration and analysis of the dataset. Models and tools are therefore required to process data for an improved understanding.

Feature selection is commonly used in machine learning, wherein a subset of the features available from the data are selected for application of a learning algorithm. The best subset contains the features that give the highest accuracy score.

In order to find out relevant features, we combine feature weighting with variable selection techniques. In variable selection, the task is reduced to simply eliminating variables which are completely irrelevant. Variable weighting is an extension of the selection process where the variables are associated to continuous weights which can be regarded as degrees of relevance. Continuous weighting provides a richer feature relevance representation. Consequently, it is necessary to develop a simultaneous algorithm of clustering and variables weighting/selection.

The models that interest us in this paper are those that could make at the same time the dimensionality reduction and clustering using Self-Organizing Maps (SOM) [12] in

B.-L. Lu, L. Zhang, and J. Kwok (Eds.): ICONIP 2011, Part I, LNCS 7062, pp. 570–579, 2011.

order to characterize clusters. SOM models are often used for visualization and unsupervised topological clustering. Its allow projection in small spaces that are generally two dimensional. Some extensions and reformulations of the SOM model have been described in the literature [2], [13], [15].

We find several important research topics in cluster analysis and variable weighting [6], [14], [9], [4], [8]. In [4], the authors propose a probabilistic formalism for variable selection in unsupervised learning using Expectation-Maximization (EM). Grozavu et al. [7] proposed two local weighting unsupervised clustering algorithms (lwo-SOM and lwd-SOM) to categorize the unlabelled data and determine the best feature weights within each cluster. Similar techniques, based on k-means and weighting have been developed by other researchers [14], [10].

The both adaptive local weighting approaches, lwo-SOM and lwd-SOM depend on the initial data, if the confidence is given to the observations we will weight the observations using the lwo-SOM, contrarily we will look on the data distribution when weighting the distance using the lwd-SOM. It is difficult to extract this information from a dataset, and sometimes weighting the observations can give better results than weighting the distances and vice versa, relatives to the dataset. Therefore, we introduce another weighting approach, which integrates both adaptive local weighting methods called dlw-SOM (double local weighting Self-Organizing Map).

Hence, these weight vectors are used for local variable selection that allows us to characterize clusters with the best subset of variables. For variable selection task we use the statistical approach Scree Test of Cattell which is initially proposed to select the principal components [3].

The rest of this paper is organized as follows: we present the proposed approach dlw-SOM (double local weighting) in section 3, after introducing the classical Self-Organizing Maps (SOM) in section 2. In the section 4, we show the experimental results on several data sets. These data sets allow us to illustrate the use of this algorithm for topological clustering. Finally we offer some concluding comments of proposed method and further research.

2   Classical Self-Organizing Map (SOM)

Self-organizing maps are increasingly used as tools for the visualization of data, as they allow projection in low, typically bi-dimensional spaces. The basic model proposed by Kohonen consists of a discrete set $\mathcal{C}$ of cells called "map". This map has a specific topology defined by an undirected graph, which is usually a regular, two-dimensional grid. For each pair of cells (j,k) on the map, the distance $\delta(j,k)$ is defined as the length of the shortest chain linking cells j and k on the grid. For each cell j this distance defines a neighbouring cell; a kernel positive function $\mathcal{K}$ ($\mathcal{K} \geq 0$ and $\lim_{|y|\to\infty} \mathcal{K}(y) = 0$) is introduced to determine the neighbouring area. We define the mutual influence of two cells j and k by $\mathcal{K}_{j,k}$. In practice, as for classical topological maps, we use a smooth function to determine the size of the neighbouring area: $\mathcal{K}_{j,k} = \exp(\frac{-\delta(j,k)}{T})$. Using this kernel function, T becomes a parameter of the model. As in the Kohonen algorithm, we decrease T from an initial value T_{max} to a final value T_{min}.

Let $\Re^m$ be the Euclidean data space and $E = \{\mathbf{x}_i; i = 1, \ldots, N\}$ a set of observations, where each observation $\mathbf{x}_i = (x_i^1, x_i^2, \ldots, x_i^m)$ is a vector in $\Re^m$. For each cell j of the grid (map), we associate a referent vector (prototype) $\mathbf{w}_i = (w_i^1, w_i^2, \ldots, w_i^m)$ which characterizes one cluster associated to cell i. The set of referent vectors is denoted by $\mathcal{W} = \{\mathbf{w}_j, \mathbf{w}_j \in \Re^m\}_{j=1}^{|\mathcal{W}|}$. Unlike k-means, the SOM is not optimizing any well-defined cost function [5]. However, SOM can be seen as a constrained k-means, in which the distance is weighted using a neighbouring function $\mathcal{K}$. In this case, we determine the set of parameters $\mathcal{W}$ by minimizing the objective function:

$$R(\chi, \mathcal{W}) = \sum_{i=1}^{N} \sum_{j=1}^{|\mathcal{W}|} \mathcal{K}_{j,\chi(\mathbf{x}_i)} \|\mathbf{x}_i - \mathbf{w}_j\|^2 \qquad (1)$$

where χ assigns each observation $\mathbf{x}_i$ to a single cell in the map $\mathcal{C}$. This cost function can be minimized using both stochastic and batch techniques [12].

3   Double Local Weighting SOM : dlw-SOM

One of the significant limitations of the classical SOM algorithms is that they treat all features equally. This is not desirable for many applications of clustering, in which observations are defined by a large number of features. A cluster provided by SOM is often characterized by only a subset of features rather than by the entire features set. The presence of other features may therefore prevent the discovery of the specific cluster structure associated to each cell. The relevance of each observation and feature changes from one cluster to another.

dlw-SOM provides a principal alternative to classical SOM and overcomes some limitations mentioned previously. Indeed, the proposed clustering algorithm and feature weighting aims to select the optimal prototypes, observations and feature weights at the same time. Each prototype $\mathbf{w}_j = (w_j^1, w_j^2, \ldots, w_j^m)$ corresponding to cell j is allowed to have its own set of local features weights $\pi_j^{(o)} = (\pi_j^{(o)1}, \pi_j^{(o)2}, \ldots, \pi_j^{(o)m})$ and its own set of local distance weights $\pi_j^{(d)} = (\pi_j^{(d)1}, \pi_j^{(d)2}, \ldots, \pi_j^{(d)m})$ respectively. We denote the set of weight vectors ($|\Pi| = |W|$) by $\Pi = \{\pi_j, \pi_j \in \Re^m\}_{j=1}^{|\Pi|}$ for both observation and distance weighting.

For the double local weighting process, we introduce the both weights in the SOM objective function, and we obtain:

$$R_{dlw-SOM}(\chi, \mathcal{W}, \Pi^{(d)}, \Pi^{(o)}) = \sum_{i=1}^{|N|} \sum_{j=1}^{|\mathcal{W}|} \mathcal{K}_{j,\chi(\mathbf{x}_i)} (\pi_j^{(d)})^\beta \|\pi_j^{(o)} \mathbf{x}_i - \mathbf{w}_j\|^2 \qquad (2)$$

where $\Pi^{(d)}$ are the distance weights, $\Pi^{(o)}$ the observations weights and β is the discrimination coefficient.

As we combined two types of the weighting techniques, contrarily to precedent weighting approaches, the minimization of the $R_{dlw-SOM}$ objective function is made in four steps:

1. Minimize $R_{dlw}(\chi, \hat{\mathcal{W}}, \hat{\Pi}^{(d)}, \hat{\Pi}^{(o)})$ with respect to χ by fixing $\hat{\mathcal{W}}$, $\hat{\Pi}^{(d)}$ and $\hat{\Pi}^{(o)}$. The expression is defined as follows:

$$\chi(\mathbf{x}_i) = \arg\min_j \left((\pi_j^{(d)})^\beta \| \pi_j^{(o)} \mathbf{x}_i - \mathbf{w}_j \|^2 \right) \tag{3}$$

2. Minimize $R_{dlw-SOM}(\hat{\chi}, W, \hat{\Pi}^d, \hat{\Pi}^o)$ with respect to W by fixing $\hat{\chi}$, $\hat{\Pi}^{(d)}$ and $\hat{\Pi}^{(o)}$. The prototype's vectors are updated using the following expression:

$$\mathbf{w}_j(t+1) = \mathbf{w}_j(t) + \epsilon(t)\mathcal{K}_{j,\chi(\mathbf{x}_i)}(\pi_j^{(d)})^\beta \left(\pi_j^{(o)} \mathbf{x}_i - \mathbf{w}_j(t) \right) \tag{4}$$

3. Minimize $R_{dlw-SOM}(\hat{\chi}, \hat{\mathcal{W}}, \hat{\Pi}^{(d)}, \Pi^{(o)})$ with respect to $\Pi^{(o)}$ by fixing $\hat{\chi}$, $\hat{\mathcal{W}}$ and $\hat{\Pi}^{(d)}$. The update of the observation weights vectors $\pi^{(o)}{}_j(t+1)$ are made using the following expression:

$$\pi_j^{(o)}(t+1) = \pi_j^{(o)}(t) + \epsilon(t)\mathcal{K}_{j,\chi(\mathbf{x}_i)}(\pi_j^{(d)}(t))^\beta \mathbf{x}_i \left(\pi_j^{(o)}(t)\mathbf{x}_i - \mathbf{w}_j(t) \right) \tag{5}$$

4. Minimize $R_{dlw-SOM}(\hat{\chi}, \hat{\mathcal{W}}, \Pi^{(d)}, \hat{\Pi}^{(o)})$ with respect to $\Pi^{(d)}$ by fixing $\hat{\chi}$, $\hat{\mathcal{W}}$ and $\hat{\Pi}^{(o)}$. The update of the distance weights vectors $\pi^{(d)}{}_j(t+1)$ are made using the following expression:

$$\pi^{(d)}{}_j(t+1) = \pi_j^{(d)}(t) + \epsilon(t)\mathcal{K}_{j,\chi(\mathbf{x}_i)}\beta(\pi_j^{(d)}(t))^{\beta-1} \left(\pi_j^{(o)}(t)\mathbf{x}_i - \mathbf{w}_j(t) \right) \tag{6}$$

The proposed method is presented in the Algorithm 1.

3.1 Complexity of the dlw-SOM

As the dlw-SOM has four optimization steps it is clearly that we increase the computational complexity compared to the both lwo-SOM and lwd-SOM weighting approaches. Analyzing the computational complexity for all phases we can extract:

- Finding the best matching unit will take into account the weights (double weighting distance) and it will be : $O(\pi^{(d)} + (\pi^{(o)} + N \times m \times |\mathcal{W}|))$
- The affectation phase will use the both matrix weights, and the complexity cost is: $O(\pi^{(d)} + (\pi^{(o)} + N \times m \times |\mathcal{W}|))$;
- For the observations weights computing, the cost is: $O(\pi^{(d)} + (\pi^{(o)} + m \times N \times |\mathcal{W}|))$;
- Finally, the computational cost for the distance weights is: $O(\pi^{(d)} + \pi^{(o)} + N \times m \times |\mathcal{W}|))$

As the size of the weights matrix are the same for the map size, the total computational cost for the dlw-SOM is $O(2|W| + |W| \times N \times m)$. Even dlw-SOM increases with two phases the learning of the SOM, the algorithm is still efficient due to its linear complexity and improves a good scalability. As we can see, the complexity depends in a bigger part on the map size (number of $|\mathcal{W}|$) because for each cell, the algorithm will compute the prototype and the both weights, but, thanks to the usually small number of cells, the complexity time does not grow significantly.

Algorithme 1. The dlw-SOM learning algorithm

Input: Data set X; Iter - number of iterations
Initialization Phase:
Randomly initialize the prototype matrix W;
Randomly initialize the observation weight matrix $\Pi^{(o)}$;
Randomly initialize the distance weight matrix $\Pi^{(d)}$;

for $t = 1$ to $Iter$ **do**
    Learning Phase:
    Present a learning example x and find the BMU (Best Matching Unit) computing the double
    weighted Euclidean distance (expression 3);
    Updating Phase:
    Compute the new prototypes w using the expression 4;
    Compute the observations weights $\pi^{(o)}$ using the expression 5
    Compute the distance weights $\pi^{(d)}$ using the expression 6
end for

3.2   Cluster characterization

In our case, we use the weight set Π^o and prototype set $\mathcal{W}$ provided by dlw-SOM. We apply the process of characterization using selection in clusters associated to cells and group of cells after clustering the map. For map clustering we use traditional hierarchical clustering combined with Davies-Bouldin index to choose optimal partition [16], and the prototype matrix weighted by the Π^d.

We then used an established statistical method, *scree method*, to select the most important features. The subjective scree test is a graphical method first proposed by [3].

The basic idea of scree test is to generate, for a principal components analysis (PCA), a curve associated with eigenvalues, allowing random behaviour to be identified (a simple line plot). Cattell suggests to find the place where the smooth decrease of eigenvalues appears to level off to the right of the plot. To the right of this point, presumably, one finds only "factorial scree". The acceleration factor indicates where the elbow of the scree plot appears. It corresponds to the acceleration of the curve, i.e. the second derivative. Frequently this scree is appearing where the slope of the hill change drastically to generate the scree. It is why many researches choose the criterion eigenvalue where the slope change quickly to determine the number of components for a PCA. It is what Cattell named the elbow. So, they look for the place where the positive acceleration of the curve is at his maximum. Hence, in our case, we use this method to choose the variables represented by their relevance vector Π. The purpose is to detect, the 'scree' where the slope of the relevance graph changes radically which corresponds to the position of the variable from which the pertinence π becomes not significant. The number of variables retained is equal to the number of values preceding this 'scree'. We therefore needed to identify the point of maximum deceleration in the curve.

4   Experimental Results

We have performed several experiments on five known problems from the UCI Repository of machine learning databases [1].

- Waveform data set: This data set consists of 5000 instances divided into 3 classes. The original base included 40 variables, 19 are all noise attributes with mean 0 and variance 1. Each class is generated from a combination of 2 of 3 "base" waves.
- Wisconsin Diagnostic Breast Cancer (WDBC): This data has 569 instances with 32 variables (ID, diagnosis, 30 real-valued input variables). Each data observation is labelled as benign (357) or malignant (212). Variables are computed from a digitized image of a fine needle aspirate (FNA) of a breast mass. They describe characteristics of the cell nuclei present in the image.
- Isolet data set: This data set was generated as follows. 150 subjects spoke the name of each letter of the alphabet twice. Hence, we have 52 training examples from each speaker. The speakers are grouped into sets of 30 speakers each, and are referred to as isolet1, isolet2, isolet3, isolet4, and isolet5. The data consists of 1559 instances and 617 variables. All variables are continuous, real-valued variables scaled into the range -1.0 to 1.0.
- Madelon data set: MADELON is an artificial dataset, which was part of the NIPS 2003 feature selection challenge. This is a two-class classification problem with continuous input variables. MADELON is an artificial dataset containing data points grouped in 32 clusters placed on the vertices of a five dimensional hypercube and randomly labelled +1 or -1. The five dimensions constitute 5 informative features. 15 linear combinations of those features were added to form a set of 20 (redundant) informative features. Based on those 20 features one must separate the examples into the 2 classes (corresponding to the +/-1 labels).
- SpamBase data set: The SpamBase data set is composed from 4601 observations described by 57 variables. Every variable described an e-mail and its category: spam or not-spam. Most of the attributes indicate whether a particular word or character was frequently occurring in the e-mail. The run-length attributes (55-57) measure the length of sequences of consecutive capital letters.

To evaluate the quality of clustering, we adopt the approach of comparing the results to a "ground truth". We use the clustering accuracy for measuring the clustering results. In general, the result of clustering is usually assessed on the basis of some external knowledge about how clusters should be structured. The only way to assess the usefulness of a clustering result is indirect validation, whereby clusters are applied to the solution of a problem and the correctness is evaluated against objective external knowledge. This procedure is defined by [11] as "validating clustering by extrinsic classification", and has been followed in many other studies. Thus, to adopt this approach we need labelled data sets, where the external (extrinsic) knowledge is the class information provided by labels. Hence, if dlw-SOM finds significant clusters in the data, these will be reflected by the distribution of classes. Thus a purity score can be expressed as the percentage of elements of the assigned class in a cluster.

In this paper, we also validate our approaches in supervised case. We used the K-fold cross validation technique, repeated s times for $s = 5$ and $K = 3$, to estimate the performance of dlw-SOM. For each run, the dataset was split into three disjoint subsets of equal size (we do 15 runs for each data set). We used two subsets for training and then tested the model on the remaining subset using all variables and selected variables (selected on the cells or on the clusters). The labels generated were compared to the real labels of the test set for each run. As, the result of the proposed method is a topological map it is difficult to compare to well known clustering methods in literature, and we chose to compare the results with the classical SOM, lwo-SOM and lwo-SOM by fixing same parameters for all approaches.

4.1   Results on Waveform

We used this data set to show the good progress of the dlw-SOM. All observations are used for learning a map with 26×14 cells dimension. The learning algorithm provides three vectors for each cell: the referent vector $\mathbf{w}_j = (w_j^1, w_j^2, ..., w_j^m)$, the observations weight vector $\pi_j^{(o)} = (\pi_j^{(o)1}, \pi_j^{(o)2}, ..., \pi_j^{(o)m})$, and the distance weight vector $\pi_j^{(d)} = (\pi_j^{(d)1}, \pi_j^{(d)2}, ..., \pi_j^{(d)m})$ where $m = 40$.

Before analyzing the dlw-SOM result, we display on 3D visualization the referent vector and weight vectors provided by the traditional SOM and our approach (Figure 1). The axes X and Y indicate respectively the variables and the referent indexes. The amplitude indicates the mean value of each component. Observing the graph (1(b)) we find that the noise represented by variables from 22 to 40 is clearly detected with low amplitudes. This visual analysis of results are clearly with the new proposed dlw-SOM algorithm. The graph of prototypes $\mathcal{W}$ show visually that variables associated to noise is irrelevant with low amplitude. The dlw-SOM algorithm provides well results because the weight vectors work as filter for observation and estimates the referents which reflect this filtering. In order to verify that it is possible to select automatically the variables using our algorithms, we apply the selection task on all parameters of the map before and after map clustering. This task consists of detecting the abrupt changes for each input vector represented as a signal graph. The prototypes of the map changes and it can be seen that the variables $[1 - 21]$ has a bigger importance compared to variables $[22 - 40]$ that means that dlw-SOM detect easy the relevant variables. Analyzing the distance weights (Figure 1(c)) we can detect the influence of the observations weights (Figure 1(d)) on the distance: the distance weights is bigger where observations weights are more important. The distance weights depend on the observations weights because the observations computation phase is done before the assignment of the distance weights during the learning process.

In order to characterize clusters obtained with the dlw-SOM map we apply the feature selection ScreeTest on the prototypes matrix. The test shows that the relevant features are $[2 - 21]$ and irrelevant are 1 and $[22 - 40]$ which are representatives for waveform dataset. The figure 2 shows the relevance of each corresponding variable coupled with a relevant prototype. The red colour (darker) corresponds to the most relevant features compared to green colour (less dark) - for irrelevant features. Compared to SOM, lwo-SOM and lwd-SOM, we can detect easily the form of waves and the variable's relevance for each prototype. This is due to the combination of the both local weights.

In order to evaluate the relevance of selected variables, we compute purity score by running a 3-fold cross-validation five times, and we obtain a purity score equals to 82.5% for the SOM, 83.4% for lwd-SOM, 83.9% for lwo-SOM and finally, 84.5 for the dlw-SOM.

After dlw-SOM map clustering with the referents $\mathcal{W}$, which are already double weighted, we obtain 3 clusters which is significant for our example. This means that when here is no cluster (labels) knowledge, the observation and variable weighting helps to find the pureness clusters, and to characterize them.

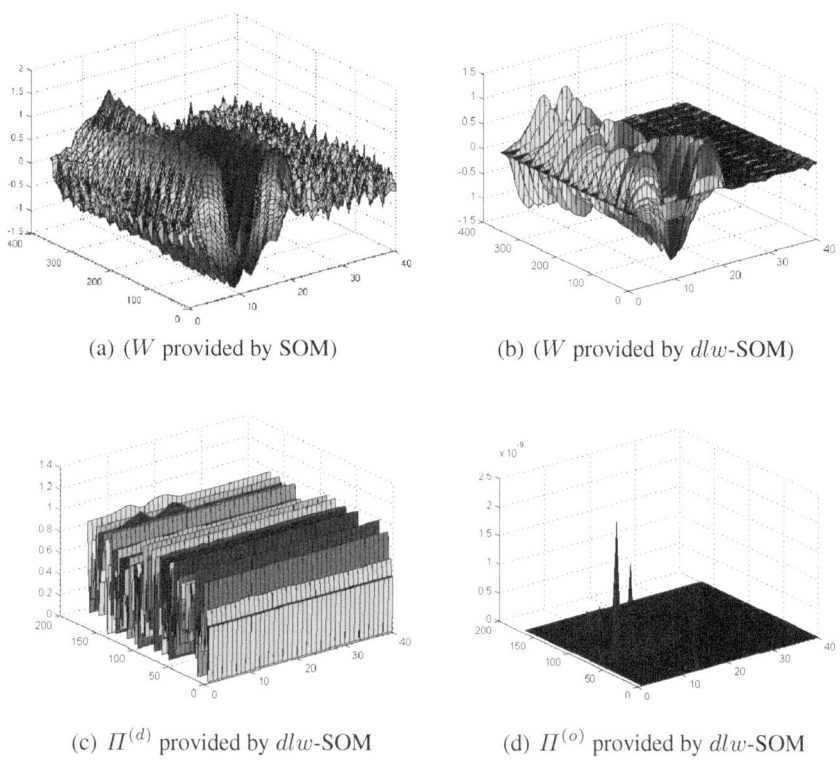

(a) (W provided by SOM)   (b) (W provided by dlw-SOM)

(c) $\Pi^{(d)}$ provided by dlw-SOM   (d) $\Pi^{(o)}$ provided by dlw-SOM

Fig. 1. 3D visualization of the referent matrix and weight matrices. The axes X and Y indicate respectively the variables and the referent indexes. The amplitude indicates the mean value of each component of map 26×14 (364 cells).

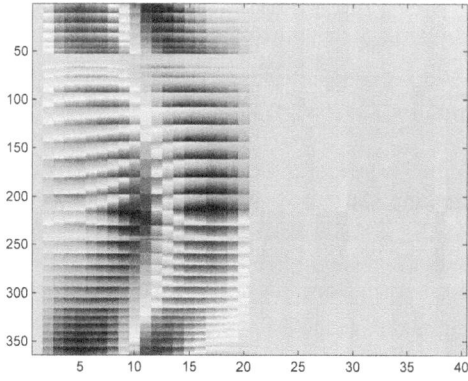

Fig. 2. Cluster characterization using double weighting SOM

5   Results on Others Data Sets

We tested our algorithm on additional data sets with different characteristics. To demonstrate the interest of the simultaneous clustering and double weighting, we use the referent and weight vector to cluster the map. Using the proposed algorithm, we show the results obtained after feature selection on the Isolet, Madelon, WDBC and SpamBase data sets.

Table 1 provides a comparative result of classification accuracy corresponding to different data set after running a 3-fold cross-validation five times. We compare different case of variables selected with the proposed approach(dlw-SOM) and the traditional SOM, lwo-SOM and lwd-SOM. We observe the performance of dlw-SOM is superior compared to other three methods in different case (using all variables, selected variables by cell and selected variables by cluster). We observed that the proposed method dlw-SOM has significantly better performance in different cases and they are more stable.

Table 1. Comparison of purity score with $\pm$SD after running a 3-fold cross-validation five times (15 runs for each). b/a - before and after segmentation; Sel f. - selected variables by the cell; Sel cl. - selected variables by cluster

Db.	b/a sel/cl	method			
		SOM	lwo-SOM	lwd-SOM	dlw-SOM
Isolet	b.	0.7786$\pm$0.05	0.7975$\pm$0.04	0.7792$\pm$0.047	0.7991$\pm$0.037
	Sel f.	0.7409$\pm$0.052	0.7863$\pm$0.043	0.7608$\pm$0.041	0.7906$\pm$0.036
	Sel cl.	0.6786$\pm$0.061	0.7821$\pm$0.047	0.7796$\pm$0.048	0.7926$\pm$0.043
wdbc	b.	0.8941$\pm$0.042	0.9203$\pm$0.037	0.9052$\pm$0.041	0.9274$\pm$0.034
	Sel f.	0.8923$\pm$0.047	0.9152$\pm$0.04	0.9023$\pm$0.043	0.9207$\pm$0.041
	Sel cl.	0.891$\pm$0.046	0.9145$\pm$0.041	0.9014$\pm$0.042	0.9185$\pm$0.041
Spam	b.	0.8958$\pm$0.041	0.8669$\pm$0.041	0.8568$\pm$0.043	0.8727$\pm$0.038
	Sel f.	0.8579$\pm$0.039	0.8754$\pm$0.04	0.8727$\pm$0.043	0.8761$\pm$0.04
	Sel cl.	0.6184$\pm$0.044	0.8564$\pm$0.041	0.8534$\pm$0.042	0.869$\pm$0.039
made-lon	b.	0.6541$\pm$0.041	0.6803$\pm$0.04	0.6752$\pm$0.039	0.6917$\pm$0.042
	Sel f.	0.6608$\pm$0.038	0.7017$\pm$0.041	0.6914$\pm$0.04	0.6984$\pm$0.04
	Sel cl.	0.6524$\pm$0.052	0.7163$\pm$0.042	0.7089$\pm$0.047	0.7025$\pm$0.045

6   Conclusions and Future Work

In this paper, we presented a new approach to perform SOM clustering and observations and variable weighting simultaneously. Compared to lwo-SOM and lwd-SOM, the proposed dlw-SOM algorithm can characterize clusters by determining the variable weights and observations weights within each cluster.

Our approach demonstrates its efficiency and effectiveness in dealing with high dimensional data for simultaneous clustering and weighting. We also show that through different means of visualization, that the weighted learning approaches provide various information that could be used in practical applications. The superiority of dlw-SOM is established experimentally. The distance used is valid for continuous feature; its extension to accommodate other kinds of variables (categorical, mixed) may be investigated.

References

1. Asuncion, A., Newman, D.: Uci machine learning repository (2007),
   `http://www.ics.uci.edu/~mlearn/MLRepository.html`
2. Bishop, C.M., Svensén, M., Williams, C.K.: Gtm: The generative topographic mapping. Neural Comput. 10(1), 215–234 (1998)
3. Cattell, R.: The scree test for the number of factors. Multivariate Behavioral Research 1, 245–276 (1966)
4. Dy, J.G., Brodley, C.E.: Feature selection for unsupervised learning. JMLR 5, 845–889 (2004)
5. Erwin, E., Obermayer, K., Schulten, K.: Self-organizing maps: Ordering, convergence properties and energy functions. Biological Cybernetics 67, 47–55 (1992)
6. Frigui, H., Nasraoui, O.: Unsupervised learning of prototypes and attribute weights. Pattern Recognition 37(3), 567–581 (2004)
7. Grozavu, N., Bennani, Y., Lebbah, M.: From variable weighting to cluster characterization in topographic unsupervised learning. In: Proc. of IJCNN 2009, International Joint Conference on Neural Network (2009)
8. Guérif, S., Bennani, Y., Janvier, E.: μ-SOM: Weighting features during clustering. In: Proceedings of the 5th Workshop On Self-Organizing Maps (WSOM 2005), pp. 397–404. Paris 1 Panthéon-Sorbonne University, France (2005)
9. Huang, J.Z., Ng, M.K., Rong, H., Li, Z.: Automated variable weighting in k-means type clustering. IEEE Transactions on Pattern Analysis and Machine Intelligence 27(5), 657–668 (2005)
10. Huh, M.H., Lim, Y.B.: Weighting variables in k-means clustering. Journal of Applied Statistics 36(1), 67–78 (2009)
11. Jain, A.K., Dubes, R.C.: Algorithms for clustering data. Prentice-Hall, Inc., Upper Saddle River (1988)
12. Kohonen, T.: Self-organizing Maps. Springer, Berlin (2001)
13. Lebbah, M., Rogovschi, N., Bennani, Y.: Besom: Bernoulli on self organizing map. In: IJCNN 2007, Orlando, Florida (2007)
14. Tsai, C.Y., Chiu, C.C.: Developing a feature weight self-adjustment mechanism for a k-means clustering algorithm. Comput. Stat. Data Anal. 52(10), 4658–4672 (2008)
15. Verbeek, J., Vlassis, N., Krose, B.: Self-organizing mixture models. Neurocomputing 63, 99–123 (2005)
16. Vesanto, J., Alhoniemi, E.: Clustering of the self-organizing map. IEEE Transactions on Neural Networks 11(3), 586–600 (2000)

Predicting Concept Changes Using a Committee of Experts

Ghazal Jaber[1,2], Antoine Cornuéjols[2], and Philippe Tarroux[1]

[1] Université de Paris-Sud, LIMSI, Bâtiment 508, F-91405 Orsay Cedex, France
[2] AgroParisTech, Dept. MMIP, 16 rue Claude Bernard, 75005 Paris Cedex, France

Abstract. In on-line machine learning, predicting changes is not a trivial task. In this paper, a novel prediction approach is presented, that relies on a committee of experts. Each expert is trained on a specific history of changes and tries to predict future changes. The experts are constantly modified based on their performance and the committee as a whole is thus dynamic and can adapt to a large variety of changes. Experimental results based on synthetic data show three advantages: (a) it can adapt to different types of changes, (b) it can use different types of prediction models and (c) the committee outperforms predictors trained on a priori fixed size history of changes.

Keywords: Committee of experts, on-line learning, concept changes, prediction.

1   Introduction

Different types of concept changes exist in the litterature. *Concept drifts* [9] refer to the change of the statistical properties of the concept's target value. For example, the behavior of customers in shopping might evolve with time and thus the concept capturing this behavior evolves as well. The speed of the change can be gradual or sudden. A sudden drift is sometimes referred to as a *concept shift* [7,8]. Another type of change, known as *virtual concept drift* [6], *pseudo-concept drift* [4], *covariate shift* [1] or *sample selection bias* [3] occurs when the distribution of the training data changes with time. In this context, several problems should be adressed:

- What is the optimal size of the memory in order to predict well the future trends? A long history may allow for more precise predictions, but it can also be misleading in case of sudden change of regime. This is the basis of the well-known *stability-plasticity* dilemma.
- What is the nature of the change ruling the evolving environment? Do we consider change as a stable function or can the change itself vary with time?

Most current approaches assume that the change is a temporal function that can be learnt using standard temporal series methods, for instance using linear regression [2] or a hidden markov model [8]. Therefore, they consider the change as a stable function that can be predicted using time information only.

B.-L. Lu, L. Zhang, and J. Kwok (Eds.): ICONIP 2011, Part I, LNCS 7062, pp. 580–588, 2011.
© Springer-Verlag Berlin Heidelberg 2011

In this paper, we suggest a general approach to anticipate how a concept changes with time. We solve the stability-plasticity dilemma by using a committee of experts where each expert is a predictor trained on its own history size. Our method is inspired by [5] where a dynamic committee of experts is used to learn under *concept drift*. By analogy, we use a committee of experts to learn under *concept change drift*. Therefore, we look at the bigger picture by considering *the change* itself as a *dynamic function* that can vary with time.

The rest of this paper is organized as follows. Section 2 discusses related works. We present our approach in Section 3. In Section 4, we test our approach on two scenarios: the first is designed to show how our committee allows a fast adaptation to sudden changes while preserving a good prediction accuracy on gradual changes, the second mixes different types of predictors: neural networks and polynomial regression models. Finally, Section 5 summarizes our results.

2   Related Work

In learning under concept drift, some approaches aim at removing the effect of change [4] while others suggest techniques to detect change and adapt their model accordingly [5]. The PreDet [2] algorithm is one of the few works directly related to ours. It anticipates future decision trees by predicting for each node the evaluation measure of each attribute, this value being used to determine which attribute will split the node. The system uses linear regression models trained on a fixed size history to predict future changes. The size of the history requires a priori knowledge on the speed of change. Another prediction system, RePro [8], stores the observed concepts as a markov chain. Once a change is detected, it uses the markov chain to predict the future concept. It assumes that concepts repeat over time.

3   Prediction Algorithm

The prediction scenario works as follows. The training examples are received in data sets or *batches* of fixed size. A *training example* is represented by a pair $(\mathbf{x}, y)$, where $\mathbf{x} \in R^p$ is a vector in a p-dimensional feature space and y is the desired output or target.

For each batch S_i a concept C_i is learnt. The concept C_i can be a classification rule, a decision tree or any other model that learns the model of the training data in S_i. By analyzing the sequence of concepts $(C_1, ..., C_i)$, the future concept C_{i+1} is predicted. The main parts of the prediction algorithm are presented next. The pseudo-code is shown in Algorithm 1.

To simplify the discussion, we represent a concept C as a vector of parameters of dimension n: $C = [c_1, c_2, ..., c_n]$. If the concept is a neural network for example, it can be represented as a vector of the network weight values.

After each batch S_i is received, a concept C_i is learnt. A change sample δ_t corresponds to a change between two consecutive concepts: $\delta_t = (C_t, C_{t+1})$. At timestep $t + 1$, the total history of changes is the sequence $(\delta_1, ..., \delta_t)$.

3.1   Building the Committee

In our approach, a committee of predictors is learned, each of which trained on a different history size.

The algorithm starts by defining a set of predictors with possibly different structures. For instance, the predictors can be neural networks with different numbers of hidden neurons and activation functions, or they can be decision trees, linear regression models, Bayes rules, SVMs etc. The set of predictors will be referred to as the "base of predictors" and has size n_base.

The committee P is initially empty. When the first change sample δ_1 is observed, a base of predictors b_1 trained on δ_1 is added to the committee. When the second change sample δ_2 is observed, each predictor in the base b_1 adds δ_2 to its history and is retrained. In addition, a new base of predictors b_2, trained on δ_2 only, is added to the committee. This process continues until a maximum number of base of predictors max_base in the committee is reached. At this point, the base of predictors $\{b1, b2, ..., b_{max\_base}\}$ have history sizes: $\{max\_base, max\_base - 1, ..., 1\}$ respectively. In subsequent steps, the committee P is updated. At each timestep t, a small but significant number of predictors (e.g. 25%) is selected and their history size is increased; the history size of the remaining predictors remains unchanged. The worst n_base predictors are removed from the committee and replaced with a new base of predictors trained on the last seen change δ_{t-1} only.

3.2   Prediction

At timestep t, each committee member $p \in P$ give its prediction of the next concept $\tilde{C}_{t+1}^p$.

$$\tilde{C}_{t+1}^p = [\tilde{c}_{(t+1,1)}^p, ..., \tilde{c}_{(t+1,n)}^p] \tag{1}$$

where $\tilde{c}_{(t+1,i)}^p$ is the i-th parameter of the concept C_{t+1}, predicted by the p-th predictor for timestep $t + 1$.

Each predictor in the committee predicts all the parameters of the next concept. However, the final prediction of the committee is formed by selecting the best prediction for each parameter $c_{(t+1,i)}$ of the concept independently. This is motivated by the fact that selecting the whole predictions $\tilde{C}_{t+1}^p = [\tilde{c}_{(t+1,1)}^p, ..., \tilde{c}_{(t+1,n)}^p]$ of a predictor p assumes that all the parameters $c_{(t+1,i)}$ evolve at the same speed, which may not be the case. The final prediction of the committee is: $\tilde{C}_{t+1} = [\tilde{c}_{(t+1,1)}, ..., \tilde{c}_{(t+1,n)}]$, where $\forall i \in \{1, .., n\}$

$$\tilde{c}_{(t+1,i)} = \arg\min_{p \in P} \| c_{(t+1,i)} - \tilde{c}_{(t+1,i)}^p \| \tag{2}$$

We choose the best predictions by defining an evaluation function $eval(p, i)$ that measures the performance of a predictor p in predicting the parameter i of the concept C. Equation 2 then becomes: $\tilde{c}_{(t+1,i)} = \tilde{c}_{(t+1,i)}^{p^*}$, where $p^* = \arg\max_{p \in P} (eval(p, i))$.

Algorithm 1. The Concept Change Prediction Algorithm

$P \leftarrow \phi$, $t \leftarrow 2$, $maxsize\_P = n\_base * max\_base$
$C_1 \leftarrow \text{train}(C_1, S_1)$

while batches are received **do**
    $C_t \leftarrow \text{train}(C_t, S_t)$
    $\delta_{t-1} = (C_{t-1}, C_t)$ {the last sample change}

    {Remove the lowest performing predictors}
    if size(P) $>= maxsize\_P$ **then**
        for $k = 1 \rightarrow n\_base$ **do**
            {$p \in P$ has the lowest prediction performance}
            $P \leftarrow P \setminus p$
        end for
    end if

    {Update remaining predictors}
    for $k = 1 \rightarrow$ size(P) **do**
        $p_k \in P$ is the k_{th} predictor
        $hist_k$ is the history size of p_k
        $r \leftarrow rand[0, 1]$
        if $r >= 0.75$ **then**
            $hist_k \leftarrow hist_k + 1$
        end if
        Retrain p_k on the last $hist_k$ sample changes
    end for

    {Add new predictors}
    b_t is a base of predictors trained on δ_t
    $P \leftarrow P \cup b_t$

    {Predict next concept change}
    $H \leftarrow \phi$ is the set of predictions
    for $k = 1 \rightarrow$ size(P) **do**
        $p_k \in P$ is the k_{th} predictor
        $H \leftarrow H \cup \tilde{C}_{t+1}^{p_k}$
    end for

    $\tilde{C}_t \leftarrow \phi$ is the final prediction
    if $H \neq \phi$ **then**
        for $i = 1 \rightarrow$ size(n) **do**
            $p^* = \arg\max_{p \in P} \left(eval(p, i) \right)$
            $\tilde{c}_{(t,i)} = \tilde{c}_{(t,i)}^{p^*}$
        end for
    end if

end while

---

4   Experiments

The first experiment in Section 4.1 is designed to show how the prediction committee adapts to different types of change. In the second experiment, in Section
4.2, we show that by mixing different types of predictors in our committee (neural networks, polynomial regression models), we take advantage of each type of
predictor and get better prediction results. Finally, we show that our predictors,
whose history size change dynamically with time, outperform predictors trained
on a fixed size window.

4.1   Experiment 1

We simulate a *concept drift* by continuously moving the hyperplane corresponding to a decision function (the target concept) in a d-dimensional space. A hyperplane is described by the equation $\sum_{i=1}^{d-1} w_i x_i = w_0$.

Sequence of concepts. For each slowly modified hyperplane, we generate a
batch of 1000 training examples $(\mathbf{x}, y)$, where $\mathbf{x} \in [0, 1]^d$ is a randomly generated
vector of dimention d and $y = sign(\sum_{i=1}^{d} w_i x_i - w_0)$. The value of w_0 is set to
$1/2 \sum_{i=1}^{d} w_i$ so that nearly half of the y's are positive and the other are negative.
In this experiment, $d = 6$ and the hyperplane weights w_i are initially set to
random values, that are gradually incremented, not necessarily with the same
increment for each weight, until time step 101, then decreased until they reach
their initial values. We learn a sequence of perceptrons each of which trained on
the corresponding batch of training examples.

Prediction. In order to predict the perceptron changes, we use feed-forward
neural networks that anticipate the new values of the perceptron weights given
the current ones. The maximum size of the committee is set to 10 predictors.
    At timestep t, each predictor in the committee gives its prediction of the
next hyperplane weights. For each weight w_i, we define the best predictor as
the committee member that predicted w_i with the least mean error on previous
timesteps $t-1$, $t-2$ and $t-3$[1]. At timestep t, we also compute for each predictor
its mean square error on all the weights to predict. The worst predictor on
previous timesteps $t-3$, $t-2$ and $t-1$ is removed.

Results and Discussion. The prediction results for the perceptron weight w_1
is presented in Figure 4.1. The same behavior is observed for the other weights.
The committee predicts well when the weight value increases. When the values
start suddenly to decrease at time step 101, it corrects its prediction error rapidly
and regains its former prediction performance afterwards. This ability to adapt
to rapid changes is due to the dynamic committee of predictors whose members

---

[1] In all our experiments, we evaluate the predictors' performance on a window size
of 3. The window size is set to a small value to adapt to the recent predictors'
performance

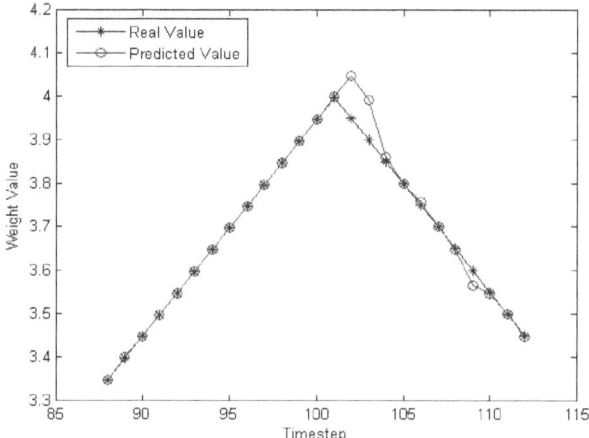

Fig. 1. The prediction results of the perceptron weight w_1 in the time interval $[85, 115]$, during which the weight suddenly starts decreasing. The line with asterisks represents the real value of w_1 whereas the line with circles represents the predicted value.

are trained on different history sizes. We show in Figure 2 the evolution in the history size of three of the committee members.

The predictor A is added to the committee at timestep 2. Its history keeps growing until timestep 105 where it is replaced by a new predictor, trained on a history of size 1. Indeed, the sudden change in the weight value deteriorates the predictor's performance, causing its elimination from the committee. Predictor B is added at timestep 8, and is also removed soon after the sudden change. Predictor C, added to the committee at timestep 5, is replaced before the sudden change because it is the lowest performing committee member. It is also common for a newly added predictor to be replaced soon after it is added to the committee, as we see for predictor C during time interval $[90, 130]$. This occurs when the change is gradual and thus the performance of a newly added predictor will be bad compared to the other committee members.

4.2   Experiment 2

In this section, the hyperplane in a six-dimensional space undergoes more complex changes in the weight values than in experiment 4.1 (see Figure 3). A sequence of 250 different hyperplanes $(\boldsymbol{H}_1, \boldsymbol{H}_2, ..., \boldsymbol{H}_{250})$ is generated, where each hyperplane $\boldsymbol{H_i} = [w_{i,1}, w_{i,2}, ..., w_{i,6}]$ is represented by its weight vector.

Neural Networks vs Polynomial Regression. In this *first set of experiments*, we tested our prediction approach using different types of predictors. We were specially interested in comparing neural networks and polynomial regression models as predictors.

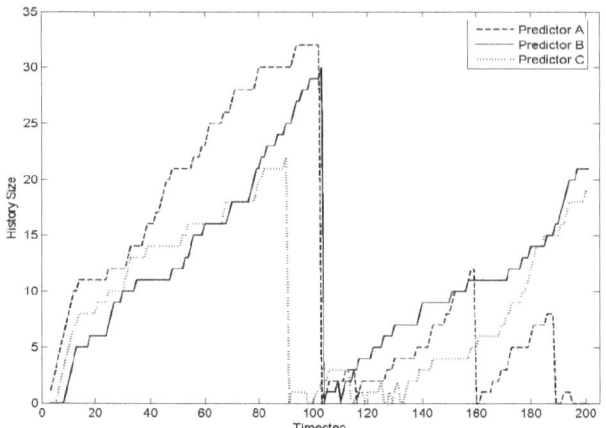

Fig. 2. The evolution in the history size of three of the committee predictors

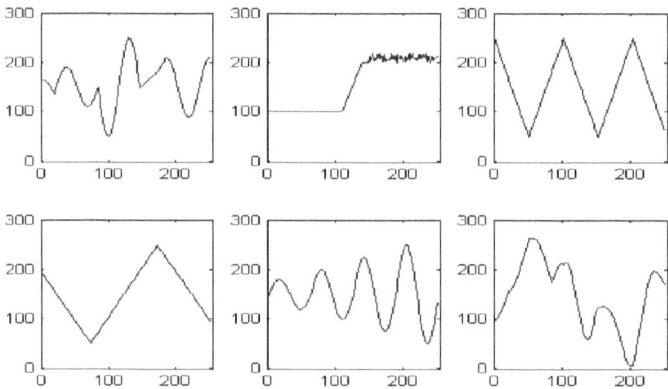

Fig. 3. The evolution of the six weights of the hyperplane described in experiment 4.2

In each experiment, we compared feed-forward neural predictors to simple predictors which consider the next hyperplane equals to the current one. The prediction results are reported in exp. 1,2 and 3 of Table 1. Our prediction approach beats the simple prediction approach in nearly 85.6% of the time.

In the *second set of experiments*, we tested our prediction approach using polynomial regression models instead of neural networks as predictors. We repeated the previous tests using a base of predictors that consists of 3 polynomial predictors with degree 1,2 and 3 respectively. The results are reported in exp. 4, 5 and 6 of Table 1. Globally, neural networks beat polynomial regression models by having a smaller prediction error when they are better than the simple prediction scenario. On the other hand, polynomial regression models beat the neural networks by having a smaller prediction error on average.

Table 1. The prediction results with different predictor types and committee sizes, using our prediction approach. *Exp* is the index of the experience. *Base of Pred.* is the base of predictors; 1 FF stands for one feed forward neural network and 3 PR stands for three polynomial regression models with degree 1,2 and 3 respectively. *Max. Base of Pred.* is the maximum number of base of predictors in the committee. During the experiments, we predict the weights of 250 hyperplanes. For each prediction, we compute the prediction MSE: the mean square error between the predicted values and the real values. The *S_b_O MSE* is the percentage of time our prediction MSE is smaller than the simple prediction MSE. The *S_b_O MSE ratio* is the ratio between the simple prediction MSE and our prediction MSE, when our prediction MSE is smaller than the simple prediction MSE. The *S_O MSE ratio* is the ratio between the simple prediction MSE and our prediction MSE.

Exp.	Base of Pred.	Max. Base of Pred.	S_b_O Per. (%)	S_b_O MSE ratio	S_O MSE ratio
1	1 FF	8	85.94	3.61	2.05
2	1 FF	13	85.94	4.03	0.26
3	1 FF	20	85.14	5.02	0.77
4	3 PR	8	89.95	2.11	1.68
5	3 PR	13	91.16	2.19	1.76
6	3 PR	20	89.55	2.27	1.77
7	1 FF, 3 PR	8	87.95	3.18	1.30
8	1 FF, 3 PR	13	89.95	3.38	1.77
9	1 FF, 3 PR	20	85.94	3.64	1.88

Table 2. The prediction results using predictors with a fixed size history. *Exp* is the index of the experience. *Predictor* is the type of predictor used in the experience; 1 FF stands for one feed forward neural network. *History Size* is the fixed history size of the predictor. The last three columns are explained in Table 1.

Exp	Predictor	History Size	S_b_O Per. (%)	S_b_O MSE ratio	S_O MSE ratio
1	1 FF	3	82.32%	4.4	0.21
2	1 FF	5	87.95%	3.9	1.7
3	1 FF	9	70.28%	2.44	1.07
4	1 FF	15	17.67%	1.4	0.29
5	1 FF	*growing*	4.47%	1.2	0.0108

In the *third set of experiments*, we mixed both type of predictors: the base of predictors contains a feed forward neural network and 3 polynomial regression models with degree 1, 2 and 3 respectively. The prediction results are reported in exp. 7, 8 and 9 of Table 1. By mixing neural networks with polynomial regression models, we take advantage of both types of predictors: the *S_O MSE ratio* and the *S_b_O MSE Per.* increase compared to when we only used neural networks while the *S_b_O MSE ratio* increases compared to when we only used polynomial regression models.

Dynamic History Size vs Fixed History Size. Prediction performances are compared with our committee and with predictors using a fixed history size. Five experiments were conducted. In the first four experiments, the history size was set to 2, 4, 8 and 15 respectively. In the fifth experiment, the history size of the predictor grows with time. The results are reported in Table 2. It appears that using fixed window size predictors requires a priori knowledge of the suitable window size for the prediction task. Choosing the wrong window size might give catastrophic results.

5 Conclusion

We have presented an approach to predict future concept changes using a dynamic and diverse committee of experts. Each expert in the committee is a predictor that anticipates the future changes of an evolving concept, taking into account the observed history of changes. The committee can be comprised of different types of experts (neural networks, polynomial regression models, SVMs etc...) with different history sizes. It is also *dynamic* by constantly updating its members. The experiments show that the diversity in the history size allows us to adapt to different types of changes while using multiple types of experts improves the prediction results.

References

1. Bickel, S., Bruckner, M., Scheffer, T.: Discriminative learning under covariate shift. The Journal of Machine Learning Research 10, 2137–2155 (2009)
2. Bottcher, M., Spott, M., Kruse, R.: Predicting future decision trees from evolving data. In: The Eighth IEEE International Conference on Data Mining, pp. 33–42 (2008)
3. Fan, W., Davidson, I., Zadrozny, B., Yu, P.S.: An improved categorization of classifier's sensitivity on sample selection bias. In: Fifth IEEE International Conference on Data Mining (2005)
4. Ruping, S.: Incremental learning with support vector machines. In: The 2001 IEEE International Conference on Data Mining, pp. 641–642 (2001)
5. Stanley, K.O.: Learning concept drift with a committee of decision trees. Technical report, Computer Science Department, University of Texas at Austin, USA (2003)
6. Syed, N.A., Liu, H., Sung, K.K.: Handling concept drifts in incremental learning with support vector machines. In: The Fifth ACM SIGKDD International Conference on Knowledge Discovery and Data Mining, pp. 317–321. ACM Press (1999)
7. Widmer, G., Kubat, M.: Learning in the presence of concept drift and hidden contexts. Machine Learning 23, 69–101 (1996)
8. Yang, Y., Wu, X., Zhu, X.: Mining in anticipation for concept change: Proactive-reactive prediction in data streams. Data Mining and Knowledge Discovery 13, 261–289 (2006)
9. Zliobaite, I.: Learning under concept drift: an overview. Technical report, Vilnius University, Faculty of Mathematics and Informatics (2009)

Feature Relationships Hypergraph for Multimodal Recognition

Luming Zhang[1], Mingli Song[1], Wei Bian[2], Dacheng Tao[2], Xiao Liu[1],
Jiajun Bu[1], and Chun Chen[1]

[1] Zhejiang Provincial Key Laboratory of Service Robot,
Computer Science College, Zhejiang University
{zglumg,brooksong,ender_liux,bjj,chenc}@cs.zju.edu.cn
[2] Centre for Quantum Computation and Information Systems,
University of Technology, Sydney
wei.bian@student.uts.edu.au, dacheng.tao@uts.edu.au

Abstract. Utilizing multimodal features to describe multimedia data is a natural way for accurate pattern recognition. However, how to deal with the complex relationships caused by the tremendous multimodal features and the curse of dimensionality are still two crucial challenges. To solve the two problems, a new multimodal features integration method is proposed. Firstly, a so-called Feature Relationships Hypergraph (FRH) is proposed to model the high-order correlations among the multimodal features. Then, based on FRH, the multimodal features are clustered into a set of low-dimensional partitions. And two types of matrices, the inter-partition matrix and intra-partition matrix, are computed to quantify the inter- and intra- partition relationships. Finally, a multi-class boosting strategy is developed to obtain a strong classifier by combining the weak classifiers learned from the intra- partition matrices. The experimental results on different datasets validate the effectiveness of our approach.

Keywords: hypergraph, multimodal features, boosting.

1 Introduction

To recognize objects better, human's cognitive system usually combines different types of features. For example, it is difficult to separate pear from banana by using color feature alone because both are of yellow. Similarly, it is difficult to separate apple from pear by using the shape feature alone. However, if we use both the color and shape features, these fruits can be easily classified. Motivated by such instance, researchers have been working to improve the recognition accuracy by integrating multiple features in the recogntion process. In contrast with the conventional single modal based approaches, the multimodal based approaches exploit richer cues from different types of features. If such different types of features are integrated optimally, a great improvement can be made in the process of pattern recognition.

The existing multimodal feature integration can be grouped into two categories based on the way these features are represented, i.e., multi-cue integration and modality identification integration.

B.-L. Lu, L. Zhang, and J. Kwok (Eds.): ICONIP 2011, Part I, LNCS 7062, pp. 589–598, 2011.

Multi-cue integration treats each type of feature as one modality. In the literature, a series of multi-cue integration [1,2,3] have been proposed. Multiple Kernel Learning(MKL) [1] linearly combines kernels computed from different types of features into a more expressive one. As a generalized version of MKL, LPBoost [2] adopts a boosting strategy to integrate multimodal features. By exploring the complementary property of different types of features, Multi-view Spectral Embedding [3] obtains a physical meaningful embedding of the multimodal features. It is noticeable that, multi-cue integration treats each type of feature as one modality, which is heuristic-based, because the interdependencies between different types of features are left unexplored.

To overcome the limitation of multi-cue integration, modality identification integration [17,4] is proposed. Zhou et al. [17] concatenates different types of features into one modality. Aiming at an optimal combination of multimodal features, Wu et al. [4] rearranges features in different modalities by a modality identification step. And different modalities of features are further integrated together in the kernel space by Support Vector Machine(SVM). Although better recognition accuracy is observed in [4], the binary correlation between different multimodal features is not consistent with the real condition, e.g., an object tracked by multiple cameras.

In this paper, by constructing a Feature Relationships Hypergraph(FRH) to model the high-order correlations among the multimodal features, a new feature integration method is proposed. To alleviate the curse of dimensionality, we intend to obtain a set of low-dimensional partitions from the input multimodal features. To deal with the complex relationships among multimodal features, we intend to obtain a classifier which describes the inter- and intra- partition relationships. Targeting this aim, firstly, a new measure called shared entropy, is proposed to describe the high-order correlations among multimodal features. And an efficient FRH construction algorithm is presented accordingly. Next, based on the FRH, we cluster the multimodal features into a set of low-dimensional partitions and obtain the inter- and intra- partition matrices. To model the inter- and intra- partition's structure, a multi-class boosting strategy is developed to combine the weak classifiers learned by the intra-partition matrices into a strong one. And the combining process is constrained by the inter-partition matrix.

2   Feature Relationships Hypergraph

As mentioned in Section 1, the relationships among the multimodal features are very helpful to find their complementation for recognition. But the relationships among the multimodal features are too complicated to be described by using the binary correlation only. For instance, when recognizing an object by using the multi-view images, the multimodal features describe the object from different aspects/views, thus the binary correlation is unable to capture the complementary relationships among these features. Therefore, it is necessary to find a new measure to describe the high-order correlation among the multimodal features.

2.1 High-Order Correlations among Multimodal Features

Given a set of multimodal features $X_1, X_2, \cdots, X_m$ $(m \geq 1)$, the joint entropy [8] measures the amount of information contained in $X_1, X_2, \cdots, X_m$, i.e.,

$$J(X_1, X_2, \cdots, X_m) = \sum_{x_1, x_2, \cdots, x_m} P(x_1, x_2, \cdots, x_m) \log_2 P(x_1, x_2, \cdots, x_m) \quad (1)$$

where $P(x_1, x_2, \cdots, x_m)$ is the joint probability of $X_1, X_2, \cdots, X_m$. Based on joint entropy, a new measure called shared entropy, is defined to discribe the high-order correlation among $X_1, X_2, \cdots, X_m$, i.e.,

$$S(X_1, X_2, \cdots, X_m) = (-1)^0 \sum_{i=1}^{m} J_i + (-1)^1 \sum_{1 \leq i < j \leq m} J_{ij} + \cdots + (-1)^{m-1} J_{1,2,\cdots,m}$$
$$(2)$$

Shared entropy measures the amount of information shared in $X_1, X_2, \cdots, X_m$, and a larger $S(X_1, X_2, \cdots, X_m)$ implies the higher $X_1, X_2, \cdots, X_m$ are correlated. An visual illustration of the shared entropy is given in Fig. 1. When $m = 1$, shared entropy reduces to the entropy of a single feature, and when $m = 2$, shared entropy reduces to information gain [8]. Note that, as an entropy-based measure, shared entropy only measures the correlation among nominal features directly. To measure continuous features, it is necessary to discretize them into nominal features beforehand, as in [6].

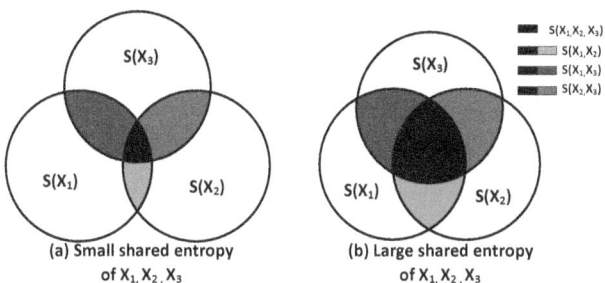

Fig. 1. A graphical illustration of shared entropy(The three circles represent $S(X_1), S(X_2)$ and $S(X_3)$ respectively. $S(X_1, X_2), S(X_1, X_3)$ and $S(X_2, X_3)$ are represented by the overlaps colored with gray+green, gray+red and gray+blue respectively. $S(X_1, X_2, X_3)$ is represented by the overlap colored with gray. The larger the gray overlap is, the higher X_1, X_2 and X_3 are correlated.)

2.2 Hypergraph

Hypergraph [5] is a generalization of classic graph wherein the edges, called hyperedges, are arbitrary non-empty subsets of the vertex set. In a hypergraph $\mathcal{G} = (\mathcal{V}, \mathcal{E})$, $\mathcal{V}$ is the vertex set; $\mathcal{E}$ is the hyperedge set wherein each hyperedge $e \in \mathcal{E}$ is a subset of $\mathcal{V}$. The degree of each vertex $v \in \mathcal{V}$, is defined as: $\sigma(v) = \sum_{v \in e, e \in \mathcal{E}} w(e)$, where $w(e)$ is the weight associated with hyperedge $e \in \mathcal{E}$. For each hyperedge $e \in \mathcal{E}$, its degree $\delta(e)$ is the number of vertices connected by this hyperedge. In our approach, the largest degree of a hyperedge within hypergraph $\mathcal{G}$ is defined as the depth of $\mathcal{G}$. Given a δ-degree hyperedge e and a $(\delta-1)$-degree hyperedge e', if $e' \subset e$, then e is called the child hyperedge of e' and conversely, e' is called the parent hyperedge of e.

2.3   Efficient FRH Construction

As the shared entropy measures the high-order correlation among the multimodal features, it is natural to analogize this correlation to a weighted hyperedge which connects multiple vertices. Motivated by this analogy, we propose Feature Relationship Hypergraph(FRH) to model multimodal features and their correlations. In detail, each vertex of an FRH represents a multimodal feature; each hyperedge of an FRH connects multiple multimodal features and represents the correlation among them; the correlation, measured by shared entropy, is assigned as the weight of this hyperedge.

Based on the concept of FRH, intuitively, given a set of input multimodal features $X_1, X_2, \cdots, X_m$, the construction of an FRH can be deemed as the construction of a set of hyperedges, i.e., identifying whether multimodal features in a candidate hyperedge are highly correlated. Unfortunately, if we construct an i-degree hyperedge ($1 \leq i \leq \xi$) in a straightforward way, we have to evaluate C_m^i candidate hyperedges, where ξ is the degree of an FRH. That is to say, to construct an ξ-depth FRH, we have to evaluate $\sum_{i=1}^{\xi} C_m^i$ candidate hyperedges, which is computational intractable when ξ is large.

To accelerate the construction of FRH, we make use of the observation that the shared entropy among a set of multimodal features $X_1, X_2, \cdots, X_\xi$ is lower and upper bounded, i.e.,

$$\eta \leq S(X_1, X_2, \cdots, X_\xi) \leq S(\mathcal{X}) \tag{3}$$

where $\mathcal{X} \subseteq \{X_1, X_2, \cdots, X_\xi\}$; η is a threshold representing the minimum shared entropy of hyperedges in FRH, that is to say, hyperedge whose shared entropy is less than η fails to be constructed. (3) means that if we want to construct a hyperedge connects $X_1, X_2, \cdots, X_\xi$, we must ensure that all subsets of $X_1, X_2, \cdots, X_\xi$ are connected beforehand. Thus a dynamic programming [7] based algorithm is proposed to accelerate the construction of FRH.Based on the concept of dynamic programming, we decompose the construction of a hyperedge into a set of separate sub-procedures, each sub-procedure identifies whether the candidate hyperedges' one parent hyperedge can be constructed. If one parent hyperedge fails to be constructed, then the construction of the current candidate hyperedge is terminated. Therefore, in contrast with the aforementioned straightforward way, the dynamic programming based hyperedge construction is accelerated since fewer candidate hyperedges are evaluated. Based on the theoretical analysis above, we present the details of efficient FRH construction in Algorithm 1.

Table 1. Efficient FRH construction(Algorithm 1)

---

input: m multimodal features: $X_1, X_2, \cdots, X_m$;The depth of FRH: ξ;
  The minimum value of shared entropy of FRH:η
output: A set of hyperedges $\mathcal{L}_1, \mathcal{L}_2, \cdots, \mathcal{L}_\xi$

---

begin:
1. $\mathcal{L}_1 = \{X_1, X_2, \cdots, X_m\}$
2. **for** $i = 2$ to ξ
   Set e as the first hyperedge in $\mathcal{L}_{i-1}$;
   do begin
      Set e' as the hyperedge next to e in $\mathcal{L}_{i-1}$;
      do begin
3.   **if** $(|e \cap e'| == 1)$
         Construct a candidate hyperedge: $e^c \leftarrow e \cup e'$;
         else
            Break;
         Reset $count$ to 0 and search the parent hyperedges of e^c in $\mathcal{L}_{i-1}$;
         If one parent hyperedge is found, then $count \leftarrow count + 1$;
4.   **if**$(count == i - 2$ && $\eta \le S(e^c))$
         Insert e^c into $\mathcal{L}_i$; Assign $S(e^c)$ as the weight of e^c;
         Mark all parent hyperedges of e^c as removing hyperedges;
         else
            Set e' as the hyperedge next to e' in $\mathcal{L}_{i-1}$;
      end until$(e' == null)$
      Set e as the hyperedge next to e in $\mathcal{L}_{i-1}$;
   end until$(e == null)$
5. Remove all marked hyperedges;
end for;
end

---

3   Boosting Compositional Partitions

To alleviate the curse of dimensionality, based on FRH, we cluster the multi-modal features into a set of partitions, each containing limited number of multimodal features. In our approach, we employ Community Learning by Graph Approximate (CLGA) [10] to cluster on FRH. In detail, given the adjacent matrix A of an FRH as computed in [5], CLGA outputs a set of partitions $\{P_i\}_{i=1}^{\gamma}$ as well as an inter-partition matrix $B \in \mathbb{R}^{\gamma \times \gamma}$, and a set of intra-partition matrix $\{D_i\}_{i=1}^{\gamma}$ can be obtained accordingly.

3.1   Intra-Partition Kernel Matrices

Since each intra-partition matrix D_i ($i \in [1, \gamma]$) corresponds to a graph $\mathcal{G}_i$, we learn the walk kernels [9] corresponding to the γ graphs $\{\mathcal{G}_i\}_{i=1}^{\gamma}$, each representing the structure partition P_i. Specifically, given a pair of intra-partition graphs $\mathcal{G}_i$ and $\mathcal{G}_i'$ corresponding to a pair of samples, their walk kernel is learned by comparing all possible walks, a finite sequence of neighboring vertices, between $\mathcal{G}_i$ and $\mathcal{G}_i'$. Thus walk kernel [9] encodes the sub-structure of $\mathcal{G}_i$. The computation of the p-length($p \ge 1$) walk kernel between $\mathcal{G}_i$ and $\mathcal{G}_i'$ is formulated from (4) to (6):

The 1-length walk kernel between $\mathcal{G}_i$ and $\mathcal{G}_i'$ starting from vertex v in $\mathcal{G}_i$ and vertex v' in $\mathcal{G}_i'$ is computed as:

$$k^1(\mathcal{G}_i, \mathcal{G}_i', v, v') = k(v, v') \tag{4}$$

where $k(v, v')$ is the basis kernel between a pair of vertices v and v'. In this paper, RBF kernel is used as the basis kernel. Based on Eq. 4, the p-length walk kernel between $\mathcal{G}_i$ and $\mathcal{G}_i'$ starting from vertex v in $\mathcal{G}_i$ and vertex v' in $\mathcal{G}_i'$ is computed recursively as:

$$k^p(\mathcal{G}_i, \mathcal{G}_i', v, v') = k(v, v') * \sum_{r \in \mathcal{N}_{\mathcal{G}_i}^v, r' \in \mathcal{N}_{\mathcal{G}_i'}^{v'}} k^{p-1}(\mathcal{G}_i, \mathcal{G}_i', v, v') \tag{5}$$

where $\mathcal{N}_{\mathcal{G}_i}^v$ is the set of neighboring vertices of v in graph $\mathcal{G}_i$. The final walk kernel between $\mathcal{G}_i$ and $\mathcal{G}_i'$ is:

$$k^p(\mathcal{G}_i, \mathcal{G}_i') = \sum_{v \in \mathcal{V}_{\mathcal{G}_i}, v' \in \mathcal{V}_{\mathcal{G}_i'}} k^p(\mathcal{G}_i, \mathcal{G}_i', v, v') \tag{6}$$

where $\mathcal{V}_{\mathcal{G}_i}$ is the set of vertices in $\mathcal{G}_i$.

3.2 Boosting Compositional Partitions

The intra-partition kernel matrices encode the structure within each partition. To encode the structure between partitions, we make use of the inter-partition matrix B, a symmetric matrix whose diagonal elements are 1(representing the relevance of each partition with itself) and the off-diagonal elements are between 0 and 1 (representing the relevance of two different partitions). Specifically, given γ partitions $\{P_i\}_{i=1}^{\gamma}$, we construct a compositional partition CP with τ partitions that all pairs of partitions in CP are highly related, i.e.,

$$\prod_{P \in \{P_i\}, P' \in \{P_i\}} B(P, P') > \delta_p \tag{7}$$

where δ_p is a tuning parameter. We collect all CPs into the set $\mathcal{CP}$:

$$\mathcal{CP} = \{CP | \prod_{P \in \{P_i\}, P' \in \{P_i\}} B(P, P') > \delta_p \wedge |CP| = \tau\} \tag{8}$$

For each $CP \in \mathcal{CP}$, we train a SVM classifier $\mathcal{C}$. Based on $\{\mathcal{C}_i\}_{i=1}^{\lambda} (\lambda = |\mathcal{CP}|$ is the number of compositional partitions in $\mathcal{CP}$), we develop a multi-class boosting algorithm to integrate the λ weak classifiers$\{\mathcal{C}_i\}_{i=1}^{\lambda}$ into a strong one $\mathcal{C}$. The details of the boosting algorithm are presented in Algorithm 2.

4   Experimental Results and Analysis

To validate the performance of our approach, we experiment on two datasets, i.e., Corel [11] and our emotional speech dataset. The experiment runs on a system equipped with Intel E8500 CPU and 4GB RAM. The algorithm of our approach is implemented on Matlab 2008b platform.

Table 2. Boosting compositional partitions(Algorithm 2)

---

input: λ classifiers $\{\mathcal{C}_i\}_{i=1}^{\lambda}$; Training data $\{\mathbf{x}_j, l_j\}_{j=1}^{N_{tr}}$
Iteration number of boosting: T
output: A strong classifier $\mathcal{C}(x)$

---

begin:
1.Set the training sample weights $w_j = \frac{1}{N_{tr}}$, $j = 1, 2, \cdots, N_{tr}$;
2.**for** $t = 1$ to T
    a).Select a classifier $\mathcal{C}^{(t)}$ from $\{\mathcal{C}_i\}_{i=1}^{\lambda}$ by: $\arg\min_{\mathcal{C}^{(t)} \in \{\mathcal{C}_i\}} \sum_{j=1}^{N_{tr}} w_j \prod (l_j \neq \mathcal{C}^{(t)}(x_j))$;
    b).Compute weighted training error: $err^t = \frac{\sum_{j=1}^{N_{tr}} w_j \prod (l_j \neq \mathcal{C}^{(t)}(x_j))}{\sum_{j=1}^{N_{tr}} w_j}$;
    c).$a^t \leftarrow \log \frac{(1-err^t)}{err^t} + \log(K-1)$;
    d).Update the sample weight: $w_j \leftarrow w_j \cdot \exp[a^t \prod (l_j \neq \mathcal{C}^{(t)}(\mathbf{x}_j)]$;
    e). Re-normalize w_j;
3.Return $\mathcal{C}(x) = \arg\max_k \sum_{t=1}^{T} a^t \cdot \prod (l_j = \mathcal{C}^{(t)}(x))$;
  end for;
end

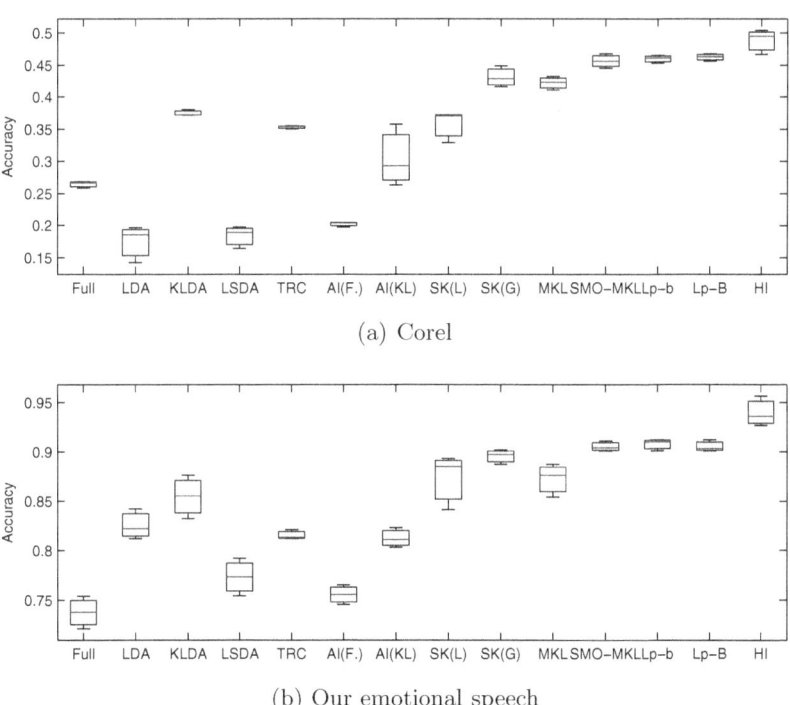

(a) Corel

(b) Our emotional speech

Fig. 2. Average recognition accuracy of the compared methods(Full means no integration is applied; AI(F.) and AI(KL) mean adaption feature integration with Frobenius norm and KL-divergence respectively; SK(L) and SK(G) mean super kernel integration with linear and Gaussian kernel respectively; LP-β and LP-B are two versions of Multiclass LPboost [2]; HI means our FRH based integration.)

4.1 Multimodal Object Recognition on Corel

Corel [11] contains 7700 images from 77 categories, which makes it a large and heterogeneous image dataset. For each category, we employ 50 images for training and leave the rest for testing. Five types of features, i.e., 1).64-dim color histogram; 2).64-dim color texture moment; 3).300-dim SIFT histogram on grayscale images; 4).1000-dim SIFT histogram on H,S,V images; 5).3400-dim PHOG [19] histogram on 4 levels of a pyramid, are extracted to describe each image.

In Fig. 2(a), we compare our approach with 4 well-known dimensionality reduction methods, i.e., linear LDA [14], kernel LDA [14] (with RBF kernel), Locality Sensitive Discriminative Analysis (LSDA) [12] and Trace Ratio Criterion(TRC) [20]. Besides, to demonstrate the advantage of our approach over other feature integration methods, we further compare our approach with 5 well-known feature integration methods, i.e., adaptation cue integration [16], super kernel integration [4], MKL [1], SMO-MKL [21] and LPboost [2]. Towards a fair comparison, recognition on multimodal features with no integration is evaluated also.

The experimental settings are as follows: linear SVM [14] is chosen as the classifier; the number of nearest neighbours in LSDA is tuned from 0 to 10; for MKL [1], SMO-MKL [21] and LPBoost [2], we follow the experimental settings in [2] and [21]; for our approach, the depth of FRH, ξ, is set to 4; the number of partition, γ, is tuned from 1 to 10 and the best recognition accuracy is achieved when $\gamma = 5$; the length of walk kenrel, p, is tuned from 1 to 10 and the best recognition accuracy is achieved when $p = 6$; the iteration number of compositional partition boosting, T, is set to 200.

4.2 Multimodal Speech Emotion Recognition

Our emotional speech dataset contains 450 sentences emotional speech labeled with five different emotions: happiness, surprise, neutral, sad and anger. We select 150 sentences for training and leave the rest for testing. six types of acoustic features [13], i.e., pitch, log energy, the first three formant frequencies, 13-dim MFCCs, 13-dim PLCCs and 10-dim LSFs are extracted.

Table 3. Number of hyperedges, recognition accuracy and time consumption of FRH construction v.s. the value of η($\eta < 0.2$ is not evaluated because over 36h is consumed)

η	0.2	0.3	0.4	0.5	0.6	0.7	0.8	0.9
# of hyperedges	98432	56433	23432	5615	1452	532	110	17
Accuracy(%)	92.6	92.2	91.3	91.0	90.6	88.2	85.2	78.3
Time consumption(s)	73244	43223	20011	5000	976	332	100	14

First of all, using the same experimental settings as in Experiment 1 (except $\gamma = 3$), we report the recognition accuracy of the 10 methods in Fig. 2(b).

In Table 3, we report the number of hyperedges in FRH, recognition accuracy and the time consumption in FRH construction,under different value of η, the threshold of shared entropy in FRH. As seen, a smaller value of η implies more number of hyperedges, more time consumption of FRH construction and a higher recognition accuracy.

In Fig. 3(a), we report the recognition accuracy with increasing value of p, the length of walk kernel($\eta = 0.6$). As seen, recognition accuracy increases when $p > 1$ and peaked in $p = 5$; however, when $p > 5$ recognition accuracy decreases. This observation is consistent with the theoretical analysis in [9].

In Fig. 3(b), we report the error rate with increasing value of T, the iteration number of compositional partition boosting. For comparison, we also present the error rate of boosting the multimodal features directly(with no integration). As seen, the error rate(both training and testing) of boosting multiple compositional partition is significantly lower than boosting multimodal features directly, which demonstrates a better generalization ability of the former.

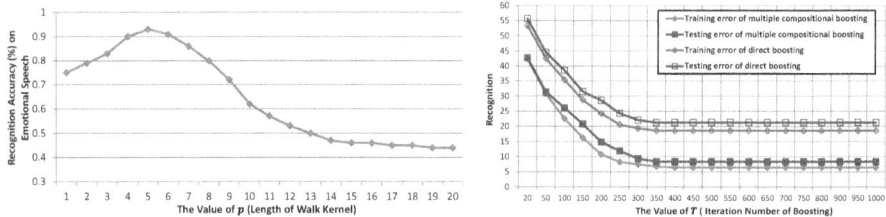

Fig. 3. Recognition accuracy v.s. the value of p(left) and Error rate v.s. the value of T(right)

5   Conclusions

This paper introduces a new multimodal feature integration method by exploiting the high-order correlations among the multimodal features. A new measure, called shared entropy, is proposed to discribe high-order correlations among multimodal features. And an efficient FRH construction algorithm is proposed based on dynamic programming. Experimental results demonstrate that multimodal features relationships are better represented and the curse of dimensionality is efficiently alleviated.

Acknowledgments. This work is supported by National Natural Science Foundation of China (60873124), Program for New Century Excellent Talents in University (NCET-09-0685), and the Natural Science Foundation of Zhejiang Province (Y1090516).

References

1. Bach, F.R., Lanckriet, G.R.G., Jordan, M.I.: Multiple kernel learning, conic duality, and the SMO algorithm. In: Proc. of ICML (2004)
2. Gehler, P., Nowozin, S.: On Feature Combination for Multiclass Object Classification. In: Proc. of ICCV (2009)

3. Xia, T., Tao, D., Mei, T., Zhang, Y.: On Combining Classifier. IEEE TPAMI 17(10), 226–239 (1998)
4. Wu, Y., Chang, E.Y., Chang, K.C., Smith, J.R.: Optimal multimodal fusion for multimedia data analysis. In: Proc. of ACM Mulitmedia (2004)
5. Zhou, D., Huang, J., Scholkopf, B.: Learning with hypergraphs: Clustering, Classification, and embedding. In: Proc. of NIPS (2006)
6. Liu, H., Hussain, F., Tan, C.L., Dash, M.: Discretization: An enabling technique. In: Data Mining and Knowledge Discovery, pp. 393–423 (2002)
7. Wolsey, L.A., Nemhauser, G.L.: Integer and Combinatorial Optimization. John Wiley (1998)
8. MacKay, D.: Information Theory, Inference and Learning Algorithms. Cambridge University Press (2003)
9. Gärtner, T., Flach, P.A., Wrobel, S.: On Graph Kernels: Hardness Results and Efficient Alternatives. In: Schölkopf, B., Warmuth, M.K. (eds.) COLT/Kernel 2003. LNCS (LNAI), vol. 2777, pp. 129–143. Springer, Heidelberg (2003)
10. Long, B., Xu, X., Zhang, Z., Yu, P.S.: Community learning by graph approximation. In: Proc. of ICDM, pp. 232–241 (2007)
11. http://corel.digitalriver.com
12. Cai, D., He, X., Zhou, K., Han, J., Bao, H.: Locality Sensitive Discriminant Analysis. In: Proc. of IJCAI, pp. 1713–1726 (2007)
13. Zhang, L., Song, M., Li, N., Bu, J., Chen, C.: Feature Selection for Fast Speech Emotion Recognition. In: Proc. of ACM Multimedia, pp. 753–756 (2009)
14. Duda, R.O., Hart, P.E., Stork, D.G.: Pattern Classification. Wiley-Interscience (2000)
15. Zhou, X., Bhanu, B.: Integrating Face and Gait for Human Recognition. In: Proc. of CVPRW (2006)
16. Sun, Z.: Adaptation For Multiple Cue Integration. In: Proc. of CVPR (2003)
17. Zhou, X., Bhanu, B.: Feature fusion of side face and gait for video-based human identification. Pattern Recognition 41(3), 778–795 (2008)
18. Porway, J., Wang, K., Yao, B., Zhu, S.C.: Scale-invariant shape features for recognition of object categories. In: Proc. of ICCV (2004)
19. Bosch, A., Zisserman, A., Mun, X.: Representing shape with a spatial pyramid kernel. In: Proc. of CIVR (2007)
20. Nie, F., Xiang, S., Jia, Y., Zhang, C., Yan, S.: Trace Ratio Criterion for Feature Selection. In: Proc. of AAAI (2008)
21. Vishwanathan, S.V.N., Sun, Z., Ampornpunt, N., Varma, M.: Multiple Kernel Learning and the SMO Algorithm. In: Proc. of NIPS (2010)

Weighted Topological Clustering for Categorical Data

Nicoleta Rogovschi and Mohamed Nadif

LIPADE, Paris Descartes University
45, rue des Saints Pères, 75006 Paris, France
firstname.secondname@parisdescartes.fr

Abstract. This paper introduces a probabilistic self-organizing map for topographic clustering, analysis of categorical data. By considering a parsimonious mixture model, we present a new probabilistic Self-Organizing Map (SOM). The estimation of parameters is performed by the EM algorithm. Contrary to SOM, our proposed learning algorithm optimizes an objective function. Its performance is evaluated on real datasets.

Keywords: unsupervised learning, mixture models, Self-Organizing Maps, categorical data.

1 Introduction

Data visualization is an important step in the exploratory phase of data analysis. This step is more difficult when it involves binary data and categorical variables [1, 27]. Self-organizing maps are being increasingly used as tools for visualization, as they allow projection over small areas that are generally two dimensional. The basic model proposed by Kohonen [17], was only designed for numerical data, but it has been successfully applied to treating textual data, [20]. This algorithm has also been applied to binary data following transformation of the original data [13, 23]. Developing generative models of the Kohonen map has long been an important goal. These models vary in the form of the interactions, and they assume the hidden generators may follow in generating the observations. Some extensions and reformulations of the Kohonen model have been described in the literature. They include probabilistic self-organizing maps [2] which define a map as a gaussian mixture and use the maximum likelihood approach to define an iterative algorithm.

In [28], the authors propose a probabilistic generalization of Kohonen's SOM which maximizes the variational free-energy that sums data log-likelihood and Kullback-Leibler divergence between a normalized neighbourhood function and the posterior distribution on the given data for the components. We have also Soft topographic vector quantization (STVQ), which uses some measure of divergence between data items and cells to minimize a new error function [9, 10]. Another model, often presented as the probabilistic version of the self-organizing map, is the Generative Topographic Map (GTM) [4, 19]. However, the manner in which GTM achieves the topographic organization is quite different from those used in the SOM models. In GTM mixture components are parameterized by a linear combination of nonlinear functions of the locations of the

B.-L. Lu, L. Zhang, and J. Kwok (Eds.): ICONIP 2011, Part I, LNCS 7062, pp. 599–607, 2011.
© Springer-Verlag Berlin Heidelberg 2011

components in the latent space. The GTM was developed for continuous data. A specific GTM model was subsequently developed for binary data by adopting a variational approximation to the binomial likelihood [9]. Also, in [18], the authors concentrate on modelling binary coded data where only the presence or absence of a variable is of interest. In contrast to other approaches, the model is linear. The model is seen as a Bernoulli analogue of the multinomial decomposition model. In [15], the main of the proposed method is to speed-up convergence of EM, and second to yield same results (or not so far) than traditional EM using categorical data. Others similar techniques have been developed to cluster large data sets [11, 21].

Here, we concentrate on modelling qualitative data using binary coding. This model involves use of the probabilistic formalism of the topological map used in [2]; therefore, it consists of estimating the parameters of the model by maximizing the likelihood of the data set. The learning algorithm that we propose is an application of the EM standard algorithm, [25]. Some variants are proposed to speed-up EM in reducing the time spent in the E-step in the case of categorical data, [15]. In this paper we proposed a new method called WeCSOM (Weighted Categorical Self-Organizing Map) which combine the benefits of SOMs, K-mode [12] algorithm and mixture models to design a new mixture for categorical data.

The rest of this paper is organized as follows: we present the principle of probabilistic map and categorical data in section 2. Our proposed approach is presented in sections 2.1 and 2.2. In sections 3, we present different results and, finally the paper ends with a conclusion and some future works for the proposed methods.

2   Categorical Data and Probabilistic Self-organizing Map

As with a traditional self-organizing map, we assume that the lattice $\mathcal{C}$ has a discrete topology (discrete output space) defined by an undirect graph. Usually, this graph is a regular grid in one or two dimensions. We denote the number of cells in $\mathcal{C}$ as N_{cell}. For each pair of cells (c,r) on the map, the distance $\delta(c, r)$ is defined as the length of the shortest chain linking cells r and c.

2.1   General Probabilistic Formalism

To define the model of topological maps based on mixture models we associate to each cell c of the map $\mathcal{C}$ a density function $\mathbf{f}_c(\mathbf{x}) = p(\mathbf{x}|\theta_c)$ whose parameters are denoted by θ. Following the bayesian formalism, presented in [22, 2], we assume that each observation $\mathbf{x}$ is generated by the following process: We start by associating to each cell $c \in \mathcal{C}$ a probability $p(\mathbf{x}|c)$ where $\mathbf{x}$ is a vector in the data space. Next, we pick a cell c^* from $\mathcal{C}$ according to the prior probability $p(c^*)$. For each cell c^*, we select an associated cell $c \in \mathcal{C}$ following the conditional probability $p(c|c^*)$. All cells $c \in \mathcal{C}$ contribute to the generation of $\mathbf{x}$ with $p(\mathbf{x}|c)$ according to the proximity to c^* described by the probability $p(c|c^*)$. Thus, a high proximity to c^* implies a high probability $p(c|c^*)$, and therefore the contribution of c to the generation of $\mathbf{x}$ is high.

Due to the "Markov" property, $p(\mathbf{x}|c, c^*) = p(\mathbf{x}|c)$, the probability distribution of the observations generated by a cell c^* of $\mathcal{C}$ is a mixture $p_{c^*}(\mathbf{x}|c^*)$ of probabilities completely defined from the map as:

$$p_{c^*}(\mathbf{x}|c^*) = \sum_{c \in \mathcal{C}} p(c|c^*)p(\mathbf{x}|c).$$

The generative model considers the mixture of probabilities, given by :

$$p(\mathbf{x}) = \sum_{c,c^* \in \mathcal{C}} p(c, c^*, \mathbf{x}) = \sum_{c,c^* \in \mathcal{C}} p(\mathbf{x}|c)p(c|c^*)p(c^*) = \sum_{c^* \in \mathcal{C}} p(c^*)p_{c^*}(\mathbf{x}), \quad (1)$$

with

$$p_{c^*}(\mathbf{x}) = p(\mathbf{x}|c^*) = \sum_{c \in \mathcal{C}} p(c|c^*)p(\mathbf{x}|c), \quad (2)$$

where the conditional probability $p(c|c^*)$ is assumed to be known. To introduce the self-organizing process in the mixture model learning, we assume that $p(c|c^*)$ can be defined as:

$$p(c|c^*) = \frac{K^T(\delta(c, c^*))}{\sum_{r \in \mathcal{C}} K^T(\delta(r, c^*))},$$

where K^T is a neighbourhood function depending on the parameter T (called temperature): $K^T(\delta) = K(\delta/T)$, where K is a particular kernel function which is positive and symmetric ($\lim_{|x| \to \infty} K(x) = 0$). Thus K defines for each cell c^* a neighbourhood region in $\mathcal{C}$. The parameter T allows control of the size of the neighbourhood influencing a given cell on the map. As with the Kohonen algorithm, we decrease the value of T between two values T_{max} and T_{min}.

2.2   The Proposed Model

In the following, let we focus on categorical data. Let be a set of N instances $\mathbf{x}_1, \ldots, \mathbf{x}_N$ described by n categorical attributes $\mathbf{x}^1, \ldots, \mathbf{x}^n$. The data matrix is noted x and defined by $\mathbf{x} = \{(x_i^j); i = 1, \ldots, N; k = 1, \ldots, n\}$. Each instance $\mathbf{x}_i$ is represented as $[x_i^1, \ldots, x_i^n]$ and for each attribute $\mathbf{x}^j$, we note c^j the number of categories. We consider a restricted latent class model [16], then the conditional distribution in $p(\mathbf{x}_i|c)$ is now given as the product of univariate single distributions

$$p(\mathbf{x}_i|c) = f_c(\mathbf{x}_i|\mathbf{w}_c, \boldsymbol{\epsilon}_c) = \prod_{k=1}^{n} f_c(x_i^k|w_c^k, \epsilon_c^k),$$

where $\mathbf{w}_c = (w_c^1, \ldots, w_c^n)$ represents the vector of categories and $\boldsymbol{\epsilon}_c = (\varepsilon_c^1, \ldots, \varepsilon_c^n)$ is a vector of probabilities. Taking

$$f_c(x_i^k|w_c^k, \varepsilon_c^k) = (1 - \varepsilon_c^k)^{1-d(x_i^k, w_c^k)} \left(\frac{\varepsilon_c^k}{c^k - 1}\right)^{d(x_i^k, w_c^k)},$$

where $d(a, b) = 1$ if $a = b$ and 0 otherwise, we define a parsimonious model where the parameter c consists of $(\mathbf{w}_c, \boldsymbol{\epsilon}_c)$ with $\mathbf{w}_c$ is the mode of the the component and $\boldsymbol{\epsilon}_c$ is a k-dimensional vector of probabilities indicating the degree of heterogeneity. The density $f_c(\mathbf{x}_i|\mathbf{w}_c, \boldsymbol{\epsilon}_c)$ expresses that, for c, the attribute $\mathbf{x}^k$ takes category w_c^k with the greatest probability $(1 - \epsilon_c^k)$ and takes each other category with the same probability $\frac{\epsilon_c^k}{(c^k-1)}$. Note that, setting the clustering problem under the classification maximum likelihood approach, the authors in [16] have defined a generalization of the $k$modes criterion and proposed better fit criteria. In our situation, we can assume that the parameter ε_c^k depends only on a cell $c \in \mathcal{C}$. Then, the model mixture generator becomes:

$$p(\mathbf{x}) = \sum_{c^* \in \mathcal{C}} p(c^*) \sum_{c \in \mathcal{C}} p(c|c^*) f_c(\mathbf{x}, \mathbf{w}_c, \boldsymbol{\epsilon}_c). \tag{3}$$

Therefore, the parameters $\theta = \theta^{\mathcal{C}} \cup \theta^{\mathcal{C}^*}$ which define the model mixture generator (3) are constituted of the parameters ($\theta^{\mathcal{C}} = \{\theta^c, c = 1..N_{cell}\}$, where $\theta^c = (\mathbf{w}_c, \boldsymbol{\epsilon}_c)$), and all the prior probabilities, also called mixing coefficients ($\theta^{\mathcal{C}^*} = \{\theta^{c^*}, c^* = 1..N_{cell}\}$ where $\theta^{c^*} = p(c^*)$). The difficulty now is to define the cost function and the learning algorithm for estimating all these parameters dedicated to categorical data.

2.3   Cost Function and Optimization Algorithm

The learning algorithm is based on maximizing the likelihood of the observations by applying the EM algorithm [6]. Learning is facilitated by introducing N hidden variables $\Xi = (\xi_1, \ldots, \xi_N)$; each hidden variable $\xi = (c, c^*)$ indicates which of the cell pairs c and c^*, generate the corresponding data observation $\mathbf{x}$. We introduce the hidden variable $\xi = (c, c^*)$ in expression (3):

$$p(\mathbf{x}) = \sum_{\xi \in \mathcal{C} \times \mathcal{C}} p(\mathbf{x}, \xi) = \sum_{c, c^* \in \mathcal{C}} p(c^*) p(c|c^*) f_c(\mathbf{x}, \mathbf{w}_c, \varepsilon_c). \tag{4}$$

We define a binary indicator variable $\alpha_i^{(c,c^*)}$ which indicates the hidden generator that may follow in generating the observation $\mathbf{x}_i$ as: $\alpha_i^{(c,c^*)} = \left\{ \begin{array}{l} 1 \text{ for } \xi_i = (c, c^*) \\ 0 \text{ otherwise} \end{array} \right\}$ Using expression (4), and the binary indicator $\alpha_i^{(c,c^*)}$, we can define the classification likelihood of the observations using the hidden variables as follows:

$$L^T(\mathbf{x}, \Xi; \theta) = \prod_{i=1}^{N} \prod_{c^* \in \mathcal{C}} \prod_{c \in \mathcal{C}} \left[\theta^{c^*} p(c|c^*) f_c(\mathbf{x}, \mathbf{w}_c, \boldsymbol{\epsilon}_c) \right]^{\alpha_i^{(c,c^*)}}.$$

The log-likelihood becomes:

$$\ln L^T(\mathbf{x}, \Xi; \theta) = \sum_{i=1}^{N} \sum_{c, c^* \in \mathcal{C}} \alpha_i^{(c,c^*)} \left[\ln(\theta^{c^*}) + \ln \left(\frac{K^T(\delta(c^*, c))}{T_{c^*}} \right) + \ln(f_c(\mathbf{x}, \mathbf{w}_c, \boldsymbol{\epsilon}_c)) \right],$$

where $T_{c^*} = \sum_{r \in \mathcal{C}} K^T(\delta(r, c^*))$. The application of the EM algorithm [7] for the maximization of log-likelihood requires $Q^T(\theta^t, \theta^{t-1})$ to be maximised for a fixed temperature T defined as:

$$Q^T(\theta^t, \theta^{t-1}) = E\left[\ln L^T(\mathbf{x}, \Xi; \theta^t) | \mathbf{x}, \theta^{t-1}\right],$$

where θ^t is the set of the parameters estimated at the tth step of the learning algorithm. However, the E-step calculates the expectation of log-likelihood with respect to the hidden variable while maintaining the established parameter θ^{t-1}. During the M-step, after updating $Q^T(\theta^t, \theta^{t-1})$ from the previous step, we maximize the $Q^T(\theta^t, \theta^{t-1})$ with respect to θ^t, $(\theta^t = \arg\max_\theta(Q^T(\theta, \theta^{t-1})))$. The two-steps increase the function likelihood. The function $Q^T(\theta^t, \theta^{t-1})$ is defined as:

$$Q^T(\theta^t, \theta^{t-1}) = \sum_{i=1}^{N}\sum_{c^*\in\mathcal{C}}\sum_{c\in\mathcal{C}} E(\alpha_i^{(c,c^*)}|\mathbf{x}_i, \theta^{t-1})$$
$$\times \left[\ln(\theta^{c^*}) + \ln\left(\frac{K^T(\delta(c^*, c))}{T_{c^*}}\right) + \ln(f_c(\mathbf{x}, \mathbf{w}_c, \epsilon_c))\right]$$

where $E(\alpha_i^{(c,c^*)}|\mathbf{x}_i, \theta^{t-1}) = p(\alpha_i^{(c,c^*)} = 1|\mathbf{x}_i, \theta^{t-1}) = p(c, c^*|\mathbf{x}_i, \theta^{t-1})$, with

$$p(c, c^*|\mathbf{x}_i, \theta^{t-1}) = \frac{p(c^*)p(c|c^*)p(\mathbf{x}|c)}{p(\mathbf{x})}.$$

The function $Q^T(\theta^t, \theta^{t-1})$ breaks into three terms

$$Q^T(\theta^t, \theta^{t-1}) = Q_1^T(\theta^{\mathcal{C}}, \theta^{t-1}) + Q_2^T(\theta^{\mathcal{C}^*}, \theta^{t-1}) + Q_3^T(\theta^{t-1}) \qquad (5)$$

where

$$Q_1^T(\theta^{\mathcal{C}}, \theta^{t-1}) = \sum_{k=1}^{n}\sum_{i=1}^{N}\sum_{c\in\mathcal{C}}\sum_{c^*\in\mathcal{C}} p(c, c^*|\mathbf{x}_i, \theta^{t-1})\ln(f_c(x^k, w_c^k, \varepsilon_c^k)),$$

$$Q_2^T(\theta^{\mathcal{C}^*}, \theta^{t-1}) = \sum_{i=1}^{N}\sum_{c^*\in\mathcal{C}}\sum_{c\in\mathcal{C}} p(c, c^*|\mathbf{x}_i, \theta^{t-1})\ln(\theta^{c^*}),$$

$$Q_3^T(\theta^{t-1}) = \sum_{i=1}^{N}\sum_{c^*\in\mathcal{C}}\sum_{c\in\mathcal{C}} p(c, c^*|\mathbf{x}_i, \theta^{t-1})\ln\left(\frac{K^T(\delta(c^*, c))}{T_{c^*}}\right).$$

The parameters $\theta^{\mathcal{C}}$ and $\theta^{\mathcal{C}^*}$ indicate the parameters estimated at the tth step. The first term $Q_1^T(\theta^{\mathcal{C}}, \theta^{t-1})$ depends on $\theta^{c,k} = (w_c^k, \varepsilon_c^k)$; the second term $Q_2^T(\theta^{\mathcal{C}^*}, \theta^{t-1})$ depends on θ^{c^*}, and the third term is constant. Maximizing $Q^T(\theta^t, \theta^{t-1})$ with respect to θ^{c^*} and θ^{c} can be performed separately including the parameter $\mathbf{w}_c$ and ϵ_c. The maximization of $Q^T(\theta^t, \theta^{t-1})$ leads to the updates that are calculated using the parameters estimated at the $t - 1^{th}$ step. The expressions are defined as follows:

$$\theta^{c^*} = p(c^*) = \frac{\sum_{\mathbf{x}_i\in\mathcal{A}} p(c^*|\mathbf{x}_i, \theta^{t-1})}{N} \qquad (6)$$

where
$p(c^*|\mathbf{x}_i, \theta^{t-1}) = \sum_{c\in\mathcal{C}} p(c, c^*|\mathbf{x}_i, \theta^{t-1})$ and $p(c|\mathbf{x}_i, \theta^{t-1}) = \sum_{c^*\in\mathcal{C}} p(c, c^*|\mathbf{x}_i, \theta^{t-1})$.

Each component of $\mathbf{w}_c = (w_c^1, \ldots, w_c^k, \ldots, w_c^n)$ and $\epsilon_c = (\varepsilon_c^1, \varepsilon_c^2, \ldots, \varepsilon_c^k, \ldots, \varepsilon_c^n)$ is then computed as follows:

$$w_c^k = \underset{e=1,\ldots,c^k}{\arg\min} \sum_{i=1}^{N} p(c|\mathbf{x}_i, \theta^{t-1}) d(x_i^k, w_c^k) \tag{7}$$

and

$$\varepsilon_c^k = \frac{\sum_{i=1}^{N} p(c|\mathbf{x}_i, \theta^{t-1}) d(x_i^k, w_c^k)}{\sum_{i=1}^{N} p(c|\mathbf{x}_i, \theta^{t-1})}, \tag{8}$$

The application of EM for the maximization gives rise to the iterative algorihtm of WeCSOM. The version of the WeCSOM algorithm for a fixed T parameter is presented in the following way:

Algorithm 1. Principal stages of the learning algorithm WeCSOM

---

1. **Initialization** (iteration t = 0) Choose the initial parameters (θ^0) and the number of iterations N_{iter}.
2. **Basic Iteration at a constant** T (iteration $t \geq 1$) Calculate all the parameters $\theta^t = \{\theta^{c^*}, \mathbf{w}_c, \epsilon_c\}$ from the previous parameters θ^{t-1} associated with each cell c and c^* by applying the formulas: (6), (7) and (8).
3. **Repeat** the basic iteration until $t > N_{iter}$.

The WeCSOM learning algorithm allows us to estimate the parameters maximizing the log-likelihood function for a fixed T. As in the SOM algorithm, we decrease the value of T between two values T_{max} and T_{min}, to control the size of the neighbourhood influencing a given cell on the map. For each T value, we get a likelihood function L^T, and therefore the expression varies with T. When decreasing T, the learning algorithm of WeCSOM is defined in the Algorithm 2.

Algorithm 2. Algorithm WeCSOM varying T

---

1. **Initialization Phase** (iteration $t = 0$): Choose T_{max}, T_{min} and N_{iter}. Apply the principal stages of WeCSOM algorithm described above for the value of T fixed to T_{max}.
2. **Iterative step:** We assume that the previous parameter θ^t are known. Compute the new value of T by applying the following formula: $T = T_{max} \left(\frac{T_{min}}{T_{max}} \right)^{\frac{t}{N_{iter}-1}}$.
   For fixed value of the parameter T, apply the basic iteration described in the principal stages, which estimates the new parameter θ^{t+1} using the formulas (6), (7) and (8).
3. **Repeat** the Iterative step while $t \leq N_{iter}$.

We can define two steps in the operating of the algorithm:

- The first step corresponds to high T values. In this case, the influencing neighbourhood of each cell c on the map is important and corresponds to higher values of $K^T(\delta(c,r))$. Formulas (6), (7) and (8) use a high number of observations to estimate model parameters. This step provides the topological order.
- The second step corresponds to small T values. The number of observations in formulas (6), (7) and (8) is limited. Therefore, the adaptation is very local. The parameters are accurately computed from the local density of the data.

3   Experimentations and Validations

To evaluate the quality of clustering, we adopt the approach of comparing the results to a "ground truth". We use the clustering accuracy for measuring the clustering results. This is a common approach in the general area of data clustering. This procedure is defined by [14] as "validating clustering by extrinsic classification", and has been followed in many other studies [1,29]. Thus, to adopt this approach we need labeled data sets, where the external (extrinsic) knowledge is the class information provided by labels. Hence, if the WeCSOM finds significant clusters in the data, these will be reflected by the distribution of classes. Therefore we operate a vote step for clusters and compare them to the behavior methods from the literature. The so-called vote step consists in the following. For each cluster $c \in \mathcal{C}$:

- Count the number of observation of each class l (call it N_{cl}).
- Count the total number of observation assigned to the cell c (call it N_c).
- Compute the proportion of observations of each class (call it $S_{cl} = N_{cl}|N_c$).
- Assign to the cluster the label of the most represented class ($l^* = \arg\max_l(S_{cl})$.

A cluster c for which $S_{cl} = 1$ for some class labeled l is usually termed a "pure" cluster, and a purity measure can be expressed as the percentage of elements of the assigned class in a cluster. The experimental results are then expressed as the fraction of observations falling in clusters which are labeled with a class different from that of the observation. This quantity is expressed as a percentage and termed "purity percentage" (indicated as $Purity\%$ in the results).

To test the performance of our approach we used many publics data sets extracted from the UCI repository [3]. The table 1 summarizes a short description of these data sets.

Table 1. Description of the used datasets for the validations

Data set	Size	nb. of classes
Zoo	101×16	7
Congressional vote	435×16	2
Wisconsis-B-C	699×9	2
Nursery	12960×8	2
Car	1728×6	4
Post-Operative	90×8	3

To conduct experimental comparison and to verify the efficacy of our proposed model, we compare our method with the RTC (Relational Topological Clustering), [24]. We choose this method because it is based on the same principle of the Kohonens model (conservation of the data topological order) and uses the Relational Analysis formalism by optimizing a cost function defined by analogy with Condorcet criterion. One disadvantage of the RTC method is that this approach treats all the features equally. We use the same categorical data sets obtained from UCI repository [3] and used in [24].

For each dataset we learned a map of different sizes (from 5x5 to 10x10) and we indicate in the table 2 the purity of clustering for RTC technique and WBSOM. The results illustrate that the proposed technique increase the purity index compared to the RTC and also presents the advantage to treat directly the categorical data without using the binary coding.

We compared also the performance of our method with the result provided in [29] that used a version of K-modes clustering method dedicated to categorical data. Table 3 lists the classification error obtained with different methods. We compute the fraction of observations falling in clusters which are labeled with a class different from that of the observation. We can observe that our results are much better then the results provided by K-modes [29]. Also we improve the error rate compared to BinBatch algorithm which represents the classical SOM approach dedicated to binary data using Hamming distance.

Table 2. Comparison between RTC et WeCSOM using purity index. RTC : Relational Topological Clustering dedicated to categorical data using the Relational Analysis formalism.

Purity: %	Size map	RTC	WeCSOM
Zoo	(5×5)	97.84	98.13
Nursery	(6×6)	78.69	81.52
Car	(10×10)	80.17	82.19
Post-Operative	(5×5)	78.21	81.34

Table 3. Comparison of the classification performances reached by K-modes, BinBatch and WeCSOM clustering algorithms

Error rate: %	K-modes	BinBatch	WeCSOM
Wisconsis-B-C	13.2	3.87	2.34
Zoo	16.6	2.97	1.87
Congressional vote	13.2	5.91	5.77

4   Conclusion

This study reports the development of a computationally efficient EM approach to maximize the likelihood of the data set to estimate the parameters of a probabilistic self-organizing map model dedicated to categorical variables. This algorithm has the advantage of providing a prototype with the same coding as the input data. The extention of the proposed method to the co-clustering will be an interesting future work for dealing with large-scale problems.

References

1. Andreopoulos, B., An, A., Wang, X.: Bi-level clustering of mixed categorical and numerical biomedical data. International Journal of Data Mining and Bioinformatics 1(1), 19–56 (2006)
2. Anouar, F., Badran, F., Thiria, S.: Self-organizing map, a probabilistic approach. In: WSOM 1997-Workshop on Self-Organizing Maps, pp. 339–344 (1997)
3. Blake, C.L., Merz, C.L.: Uci repository of machine learning databases (1998), ftp://ftp.ics.uci.edu/pub/machine-learningdatabases

4. Bishop, C.M., Svensen, M., Williams, C.K.I.: GTM: The generative topographic mapping. Neural Comput. 10(1), 215–234 (1998)
5. Celeux, G., Govaert, G.: A classification em algorithm for clustering and two stochastic versions. Comput. Stat. Data Anal. 14(3), 315–332 (1992)
6. Dempster, A.P., Laird, N.M., Rubin, D.B.: Maximum likelihood from incomplete data via the em algorithm. Roy. Statist. Soc. 39(1), 1–38 (1977)
7. Girolami, M.: The topographic organisation and visualisation of binary data using mutivariate-bernoulli latent variable models. I.E.E.E Transactions on Neural Networks 12(6), 1367–1374 (2001)
8. Govaert, G.: Classification binaire et modeles. Revue de Statistique Applique 38(1), 67–81 (1990)
9. Graepel, T., Burger, M., Obermayer, K.: Self-organizing maps: generalizations and new optimization techniques. Neurocomputing 21, 173–190 (1998)
10. Heskes, T.: Self-organizing maps, vector quantization, and mixture modeling. IEEE Trans. Neural Networks 12, 1299–1305 (2001)
11. Hofmann, T.: Unsupervised learning by probabilistic latent semantic analysis. Machine Learning 42, 177–196 (2001)
12. Huang, Z.: Extensions to the k-means algorithm for clustering large data sets with categorical values. Data Mining and Knowledge Discovery 2, 283–304 (1998)
13. Ibbou, S., Cottrell, M.: Multiple correspondance analysis crosstabulation matrix using the kohonen algorithm. In: Verlaeysen, M. (ed.) Proc. of ESANN 1995, pp. 27–32 (1995)
14. Jain Anil, K., Dubes Richard, C.: Algorithms for clustering data. Prentice-Hall, Inc., Upper Saddle River (1988)
15. Jollois, F.X., Nadif, M.: Speed-up for the expectation-maximization algorithm for clustering categorical data. Journal of Global Optimization 37(4), 513–525 (2007)
16. Jollois, F.X., Nadif, M.: Clustering Large Categorical Data. In: Chen, M.-S., Yu, P.S., Liu, B. (eds.) PAKDD 2002. LNCS (LNAI), vol. 2336, pp. 257–263. Springer, Heidelberg (2002)
17. Kohonen, T.: Self-organizing Maps. Springer, Berlin (2001)
18. Kaban, A., Bingham, E., Hirsimaki, T.: Learning to read between the lines: The aspect bernoulli model. In: Proceedings of the Fourth SIAM International Conference on Data Mining (2004)
19. Kaban, A., Girolami, M.: A combined latent class and trait model for the analysis and visualization of discrete data. IEEE Trans. Pattern Anal. Mach. Intell. 23, 859–872 (2001)
20. Kaski, S., Honkela, T., Lagus, K., Kohonen, T.: Websomself-organizing maps of document collections. Neurocomputing 21, 101–117 (1998)
21. Kostiainen, T., Lampinen, J.: On the generative probability density model in the self-organizing map. Neurocomputing 48, 217–228 (2002)
22. Luttrel, S.P.: A bayesian ananlysis of self-organizing maps. N. Comp. 6, 767–794 (1994)
23. Lebbah, M., Thiria, S., Badran, F.: Topological map for binary data. In: Proc. European Symposium on Artificial Neural Networks-ESANN 2000, pp. 267–272 (2000)
24. Labiod, L., Grozavu, N., Bennani, Y.: Relational Topographic clustering (RTC). In: International Joint Conference on Neural Networks, IJCNN 2010 (2010)
25. McLachlan, G., Krishman, T.: The EM algorithm and Extensions. Wiley, N.Y (1997)
26. Nadif, M., Govaert, G.: Clustering for binary data and mixture models: Choice of the model. Applied Stochastic Models and Data Analysis 13, 269–278 (1998)
27. Saund, E.: A multiple cause mixture model for unsupervisedlearning. Neural Comput. 7(1), 51–71 (1995)
28. Verbeek, J.J., Vlassis, N., Krose, B.J.A.: Self-organizingmixture models. Neurocomputing 63, 99–123 (2005)
29. Khan Shehroz, S., Kant, S.: Computation of initial modes for k-modes clustering algorithm using evidence accumulation. In: IJCAI, pp. 2784–2789 (2007)

Unsupervised Object Ranking
Using Not Even Weak Experts

Antoine Cornuéjols and Christine Martin

AgroParisTech, UMR-MIA-518 AgroParisTech - INRA,
16, rue Claude Bernard, F-75231 Paris Cedex 05, France
{antoine.cornuejols,christine.martin}@agroparistech.fr

Abstract. Many problems, like feature selection, involve evaluating objects while ignoring the relevant underlying properties that determine their true value. Generally, an heuristic evaluating device (e.g. filter, wrapper, etc) is then used with no guarantee on the result. We show in this paper how a set of experts (evaluation function of the objects), not even necessarily weakly positively correlated with the unknown ideal expert, can be used to dramatically improve the accuracy of the selection of positive objects, or of the resulting ranking. Experimental results obtained on both synthetic and real data confirm the validity of the approach. General lessons and possible extensions are discussed.

Keywords: Ensemble methods, unsupervised learning, feature selection, ranking.

1 Introduction

Imagine you are asked to identify students who have taken a course in some subject. All you have is a set of handouts from a collection of students, some of whom have followed the course of interest. Because you do not know the specifics of the target course, you do not know how to evaluate the handouts in order to single out the students you are looking for. However, you can use the services of a set of colleagues who do not know either, *a priori*, how to recognize the "positive" students, but have the ability to grade the copies using their own set of evaluation criteria. These can be quite diverse. For instance, a grader might be sensitive to the color of the paper, the margins, the thickness of the paper and the number of pages. Another could count the number of underlinings and use the length of the name and the color of the ink to evaluate the student's handouts. And a third one could measure the number of dates cited in the copy, the average length of the sentences, and the average space between lines.

Would that help you in identifying the students who have taken a course, unknown to you? And by combining in some way the various evaluations and rankings of your "experts", could you somehow increase your confidence in detecting the right students?

While grading students using a bunch of unknowledgeable and weak experts might certainly interest many of us, the problem outlined above is just one illustration of a much larger set of applications.

B.-L. Lu, L. Zhang, and J. Kwok (Eds.): ICONIP 2011, Part I, LNCS 7062, pp. 608–616, 2011.

Actually our research on this subject started when, some years ago, a biologist came to us for asking our help in discovering the genes that were involved in the response of cells to weak radioactivity level. The organism studied was the yeast *Saccharomyces cerevisiae* and its 6,135 genes. Thanks to microarray technology, their activity level in conditions where low radiation doses were present and conditions where no such radioactivity existed was measured. All together, 23 microarrays were obtained. In a way, our students here were the genes and their handouts were their profile of responses to the 23 conditions. We did not know what kind of profile was relevant to identify the genes that interested the biologist, but we had some techniques, like SAM [1], ANOVA [2,3], RELIEF [4,5] and others, that could be used to rank the genes according to some statistical patterns or regularities found in the profiles of the genes.

This paper shows how, in the absence of supervised information, can we still attain a high precision level in finding the "positive" instances in a large collection, given that we can make use of a varied set of "weak" experts and generate samples according to some null hypothesis.

Our paper is organized as follows. Section 2 provides a more formal definition of the problem and discusses why existing approaches using ensemble techniques, like Rankboost [6] and other similar methods, are inadequate. The bases of our new approach and the algorithm for a single iteration are presented in Section 3. Experimental results, on synthetic data as well as real data, are then reported in Section 4, while Section 5 describes the extension of the method to multiple iterations. Finally, lessons and perspectives are discussed in Section 6.

2   The Problem and Related Works

2.1   Definition of the Problem

We can define the problem we are interested in as follows.

- A sample S of *objects* (e.g. students, genes, features) of which it is strongly suggested that some are "positive" objects, while the others are to be considered as "negative" ones (e.g. students who did not take the target course, genes not sensitive to radioactivity, irrelevant attributes).
- A set $\mathcal{E}$ of *"experts"*, also called graders, who, given an object, return a value according to their own set of criteria.

Note that nothing is known beforehand about the alignment of our "experts" with the ideal grader. As far as we know, some experts may tend to rank objects similarly as the target expert, but other may well rank them in a somewhat reverse order, while still others may be completely orthogonal to the target regularities (e.g. it might be expected that the number of letters of the name of the students do not provide any information about the courses they have taken).

In this truly unsupervised setting, is it then possible to use such a set of experts, or a subset thereof, in order to rank the objects in the sample S with some guarantee about the proximity with the target ranking (one that would put all the positive objects before the negative ones)?

2.2   Related Works

In the ranking problem, the goal is to learn an ordering or ranking over a set of objects. One typical application arises in Information Retrieval where one is to prioritize answers or documents to a given query. This can be approached in several ways. The first one is to learn a *utility* or *scoring* function that evaluates objects relative to the query, and can therefore be used to induce an ordering. This can be seen as a kind of ordinal regression task (see [9]). A second approach consists in learning a *preference* function defined over pairs of objects. Based on this preference function, or partial ordering, one can then try to infer a complete ordering that verify all the known local constraints between pairs of objects.

Both of these approaches need training data which, generally, takes the form of pairs of instances labeled by the relation $\preceq$ (*must precede or be at the same rank*) or $\succ$ (*must follow*). In the case of the bipartite ranking problem, there are only two classes of objects, labeled as positive or negative, and the goal is to learn a scoring function that ranks positive instances higher than negative ones [8]. Ensemble methods developed for supervised classification have thus been adapted to this learning problem using a training sample of ordered pairs of instances. Rankboost [6] is a prominent example of these methods.

Another perspective on the ranking problem assumes that orderings on the set of objects, albeit imperfect and/or incomplete, are available. The goal is then to complete, merge or reconcile these rankings in the hope of getting a better combined one. This is in particular at the core of Collaborative Filtering where training data is composed of partial orderings on the set of objects provided by users. Using similarity measures between objects and between users, the learning task amounts then to completing the matrix under some optimization criterion.

In [7], another approach was taken where, starting from rankings supposedly independent and identically corrupted from an ideal ranking, the latter could be approximated with increasing precision by taking the average rank of each object in an increasing number of rankings. However, the underlying assumptions of independence and of corruption centered on the true target ranking were disputable and the experimental results somewhat disappointing.

Above all, all these methods rely on supervised training data, either in the form of labelled pairs of instances, or in the form of partial rankings. In the latter case, these rankings are supposed to be mild alterations of the target ranking.

In this paper, the problem we tackle does not presuppose any learning data. Furthermore, the evaluation functions or "experts" that are considered are not supposed to be positively correlated with the target evaluation function.

3   A New Method

3.1   The Principle

Let us return to the situation outlined in the introduction where the task is to distinguish the students in a university who have taken a course in some discipline of which you do not know the characteristics. Given our ignorance on both the expert's expertise for the question at hand, and on the target concept, the situation might seem hopeless. However, the key observation is that the sample at hand is not a random sample: the sample of students very likely includes positive instances (students) to some significant extent (e.g. more than could be due to accidental fluctuations in the background of students). This provides us with a lever.

We consider now *pairs* of experts, and measure their correlation both *a priori*, by averaging on all possible ranking problems, and on the target problem. If the two experts are sensitive to the "signal" in the sample, then their correlation on this sample will differ from their *a priori* one.

This observation is at the basis of the ensemble method we propose. Instead of relying on a combination of separate experts, it uses pairs of experts both to detect experts that are relevant and to assign them a weight. The utterly unsupervised, and seemingly hopeless, task is tamed thanks to the measurements of correlations of higher orders between experts.

3.2   Formal Development

Let d be the number of objects in the target sample $\mathcal{S}$ (e.g. the students) and suppose two graders or experts rank the elements of $\mathcal{S}$.

In case the experts were random rankers, the top n elements of each ranking would be equivalent to a random drawing of n elements. Therefore the size k of their intersection would obey the *hypergeometric law*:

$$H(d, n, k) = \frac{\binom{n}{k} \cdot \binom{d-n}{n-k}}{\binom{d}{n}} \tag{1}$$

where $H(d, n, k)$ denotes the probability of observing an intersection of size k when drawing at random two subsets of n elements from a set of size d.

For instance, in the case of the intersection of two subsets of 500 elements randomly drawn from a set of 6,135 elements, the most likely intersection size is 41. It can be noticed that $k/n = n/d$ (e.g. $41/500 \approx 500/6{,}135$).

In other words, if two totally uncorrelated graders were asked to grade 6,135 copies, and if one looked at the intersection of their 500 top ranked students, one would likely find an intersection of 41. However, two graders using exactly the same evaluation criteria would completely agree on their ranking no matter what. Then the intersection size k would be equal to n for all values of n. The opposite case of two anti-correlated graders would yield two opposite rankings for any sample. The intersection size would therefore be zero up to $n = d/2$ and then grow up as $2(n - d/2)/n$.

There is therefore a whole spectrum of possible correlation degrees between pairs of experts, from 'totally correlated', to 'maximally anti-correlated', going through 'uncorrelated' (case of the hypergeometric law) as shown on Figure 1.

As an illustration, Figure 2 shows the curve obtained for the pair of "experts" ANOVA and RELIEF when they ranked samples of possible genes. It appears that the two methods are positively correlated. The curve stands above

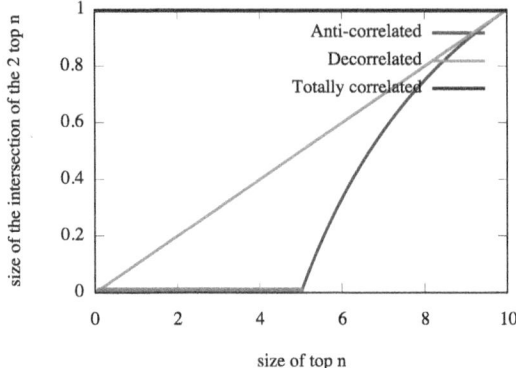

Fig. 1. Illustration of the classes of possible correlation curves between pairs of experts. The x-axis stands for n, the size of the top n ranked elements considered in each ranking. The y-axis represents the size of the intersection $k(n)$. Any curve between the "totally correlated" one and the "anti-correlated" one are possible.

the diagonal) at approximatively two standard deviations above, except for the tail of the curves. This "over-correlation" starts sharply when the intersection is computed on the top few hundreds genes in both rankings, and then it levels off, running approximatively parallel to the *a priori* correlation curve. This suggests that this is in the top of both rankings that the genes present patterns that significantly differ from the patterns that can be observed in the general population of instances. Actually, the fact that the relative difference between the curves is maximal for $n \approx 180$ and $n \approx 540$ would imply that it is best to consider the top$_{180}$ or the top$_{540}$ ranked genes by ANOVA on one hand and by RELIEF on the other hand because they should contain the largest number of genes corresponding to information that is specific to the data.

3.3 Estimating the Number of Positive Instances

Following the computation of the hypergeometric law, one can compute the number of combinations to obtain an intersection of size k when one compares the top n elements of both rankings.

Let us call k_{corr} the number of elements that the two experts tend to take in common in their top n ranked elements, and k_{corr}^{+} the number of positive elements within this set. Then the probability of observing a combination as described on Figure 3 is:

$$H"(k, k_{corr}^{-}, k_{corr}^{+}, p_1, p_2, n, d) =$$

$$\frac{\overbrace{\binom{n - p_1}{k_{corr}^{-}}}^{a} \overbrace{\binom{d - p - (n - p_1)}{n - p_2 - k_{corr}^{-}}}^{b} \overbrace{\binom{p_1}{k^{+} - k_{corr}^{+}}}^{c} \overbrace{\binom{p - p_1}{p_2 - k^{+} - k_{corr}^{+}}}^{d}}{\binom{d}{n}} \tag{2}$$

where the overbraces refer to the sets indicated in Figure 3.

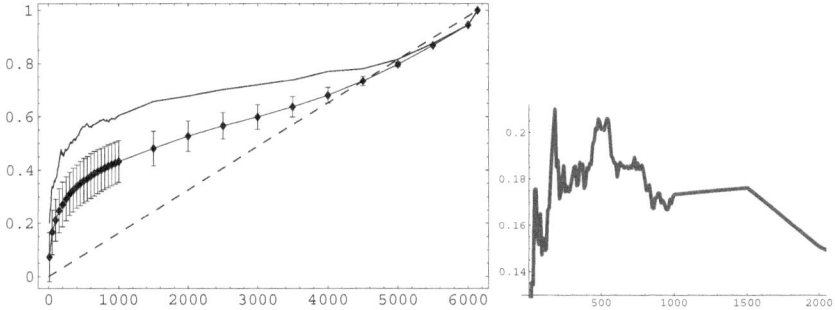

Fig. 2. The x-axis stands for the number n of top-ranked features (e.g. genes) considered. The y-axis stands for the ratio of the intersection size to n. **Left**: (Top curve) the intersection size for the true data. (Center curve): the mean intersection size due to the a priori correlation between ANOVA and RELIEF (with some standard deviation bars). (Lower curve): the intersection size explainable by randomness alone. **Right**: Curve of the relative difference, with respect to n, of the observed intersection size k and the intersection size k_{corr} due to a priori correlation between ANOVA and RELIEF. The curve focuses on the beginning of the curve, for $n < 2,000$, since this is the more interesting part.

In this equation, k, k_{corr}, n and d are known. And since k_{corr} is independent on the classes of the objects, one can estimate that $k_{corr}^{+}/k_{corr} = p/d$. The unknown are therefore: k^{+}, p_1 and p_2. When there is no *a priori* reason to believe otherwise, one can suppose that the two experts are equally correlated to the ideal expert, which translates into $p_1 = p_2$, and there remains two unknowns only, p_1 and k^{+}. Using any optimization software, one can then look for the values of p_1 and k^{+} which yield the maximum of Equation 2.

In the case of the low radiation doses data, the maximum likelihood principle, applied with $d = 6,135, n = 500, k_{corr} = 180$ and $k = 280$ yields $p = 420 \pm 20$ and $p_1 = 340 \pm 20$ as the most likely numbers of total relevant genes and of the relevant genes among the top$_{500}$ ranked by both methods. From these estimations, a biological interpretation of the tissues of the cell affected by low radioactivity was proposed [10].

3.4   Experimental Results on Synthetic Data

In these experiments, $d = 1,000$ genes, or features, were considered, whose value were probabilistically function of the condition. For each feature, 10 values were measured in condition 1 and 10 values in condition 2. The relevant features were such that condition 1 and condition 2 were associated with two distinct gaussian distributions. The difference δ between the means μ_1 and μ_2 could be varied, as well as the variance. The values of the irrelevant features were drawn from a unique gaussian distribution with a given variance. In the experiments reported

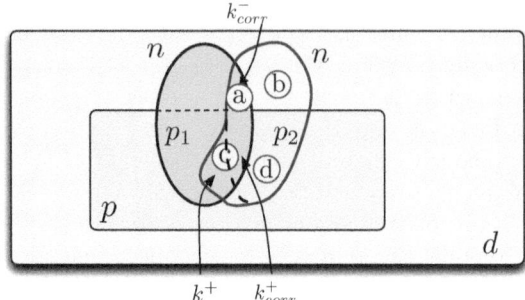

Fig. 3. The sets involved in the generative model of the intersection size k when the correlation *a priori* is taken into account (see Equation 2)

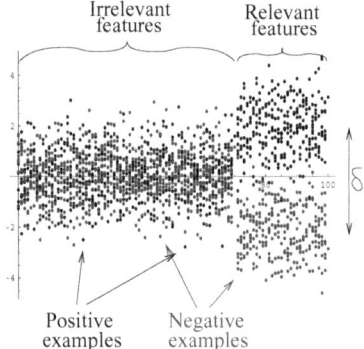

Fig. 4. The values of artificial genes. For each relevant genes, 10 values are drawn from distribution 1, and 10 from distribution 2, while the 20 values for the irrelevant genes are drawn from a unique distribution.

here, the number p of relevant features was varied in the range $[50, 400]$, the difference $\delta \in [0.1, 5]$ and $\sigma \in [1, 5]$. The smaller δ and the larger σ, the more difficult it is to distinguish the relevant features from the irrelevant ones.

Figure 5 shows the results for $p \in \{50, 200, 400\}$, $\delta = 1$ and $\sigma = 1$ (left), and for $\delta = 2$ (right). Predictively, the curves are much more peaked in the case of a larger signal/noise ratio. Indeed, on the right curves, the value of p can be guessed directly. However, even in the less favorable case (Figure 1, left), the value of p can be retrieved quite accurately using equation 2 and the maximum likelihood principle. After computation, the values $p = 50, 200, 400$ emerge rightly as the more likely ones.

These results obtained in this controlled experiments and others not reported here show the value of considering pairs of experts in this strongly unsupervised context. But is it possible to use more than one pair of experts?

Ratio : $\dfrac{\text{intersection size}}{n}$

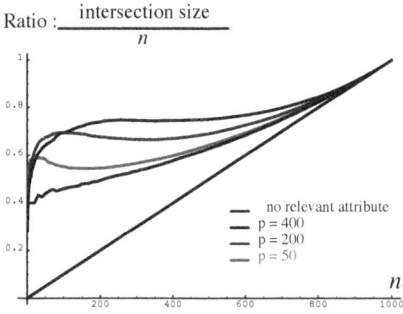

Ratio : $\dfrac{\text{intersection size}}{n}$

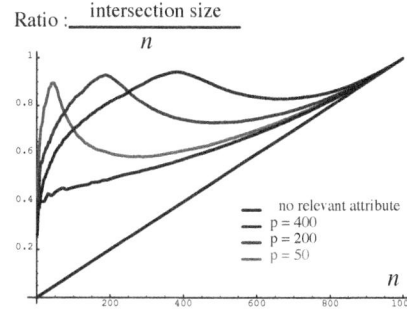

Fig. 5. The correlation curves obtained for various values of p, in a difficult setting (left), and a milder one (right)

4 Towards Higher Order Combinations of Experts and Rankings

A combination of experts must be weighted according to the value of each expert. But how to proceed when no information is *a priori* available on the value of the various experts for the task at hand?

Measuring the difference in the correlation of pairs of experts on average and on the sample $\mathcal{S}$ as described in Section 3 allows one to make such a distinction. On one hand, experts that are blind to the target regularities won't show any over or under correlation with any other expert on the sample $\mathcal{S}$ since they do not distinguish in any way the positive instances. On the other hand, experts that are not blind and tend either to promote positive instances towards the top (resp. the bottom) of the ranking will show over-correlation on $\mathcal{S}$ with experts that do the same, and under-correlation with experts that do the opposite. Two classes of experts will thus appear, the ones positively correlated with the target ranking and the ones negatively correlated. If, in addition, we assume that the positive instances are a minority in $\mathcal{S}$, it is easy to discriminate the "positive" class form the "negative" one. Because we measure correlation by comparing the top n ranked elements of the experts, the correlation curve rises much more sharply for the positive experts, that put the positive instances towards the top, than for the negative ones that put the rest of the instances towards the top.

Knowing the sign of the weight of each expert is however too crude an information in order to obtain a good combined ranking. We are currently working on a scheme to compute the value of the weight that each expert should be given in order to reflect its alignment with the unknown target evaluation function.

5 Lessons and Perspectives

This paper has presented a new totally unsupervised approach for ranking data from experts diversely correlated with the target regularities. One key idea is to measure and exploit the possible difference between a priori correlation existing in pairs of experts and their correlation on the actual data to be ranked. We have shown how to measure these correlations and how to estimate from them relevant parameters through a maximum a posteriori principle.

The use of pairs of experts in order to overcome the lack of supervised information is, to our knowledge, new. The experimental results obtained so far confirm the practical and theoretical interest of the method. We have also suggested ways to use multiple pairs of experts in a boosting like process. Future work will be devoted to the precise design of such an algorithm and to extensive experimentations. They will include comparisons with other co-learning methods specially designed for unsupervised learning (see for instance [11]).

Acknowledgments. We thank Gilles Bisson for many enlightening discussions, and Romaric Gaudel for helping in finding relevant references to related works and for his participation in the early stages of this study. This work has been supported in part by the ANR-Content Holyrisk project.

References

1. Tusher, V.G., Tibshirani, R., Chu, G.: Signficance analysis of microarrays applied to the ionizing radiation response. Proc. Natl Acad. Sci. USA 98, 5116–5121 (2001)
2. Sahai, H.: The Analysis of Variance. Birkhauser, Boston (2000)
3. Pavlidis, P., Noble, W.: Analysis of strain and regional variation in gene expresion in mouse brain. Genome Biol. 2, 1–14 (2001)
4. Kira, K., Rendell, L.: A practical approach to feature selection. In: Sleeman, D., Edwards, P. (eds.) Proceedings of the International Conference on Machine Learning, pp. 249–256. Morgan Kaufmann, Aberdeen (1992)
5. Kononenko, I.: Estimating attributes: analysis and extensions of RELIEF. In: Bergadano, F., De Raedt, L. (eds.) ECML 1994. LNCS, vol. 784, pp. 172–182. Springer, Heidelberg (1994)
6. Freund, Y., Iyer, R., Schapire, R., Singer, Y.: An efficient boosting algorithm for combining preferences. Journal of Machine Learning Research 4, 933–969 (2003)
7. Jong, K., Mary, J., Cornuéjols, A., Marchiori, E., Sebag, M.: Ensemble Feature Ranking. In: Boulicaut, J.-F., Esposito, F., Giannotti, F., Pedreschi, D. (eds.) PKDD 2004. LNCS (LNAI), vol. 3202, pp. 267–278. Springer, Heidelberg (2004)
8. Agarwal, S., Cortes, C., Herbrich, R. (eds.): Proceedings of the NIPS 2005 Workshop on "Learning to Rank" (2005)
9. Herbrich, R., Graepel, T., Obermayer, K.: Large margin rank boundaries for ordinal regression. In: Smola, A., Bartlett, P., Schökopf, B., Schuurmans, D. (eds.) Advances in Large Margin Classifiers, pp. 115–132. MIT Press (2000)
10. Mercier, G., Berthault, N., Mary, J., Peyre, J., Antoniadis, A., Comet, J.-P., Cornuéjols, A., Froidevaux, C., Dutreix, M.: Biological detection of low radiation by combining results of two analysis methods. Nucleic Acids Research (NAR) 32(1), e12 (8 pages) (2004)
11. Grozavu, N., Ghassany, M., Bennani, Y.: Learning Confidence Exchange in Collaborative Clustering. In: Proc. IJCNN, IEEE International Joint Conference on Neural Network, San Jose, California, July 31-August 5 (2011)

Research on Classification Methods of Glycoside Hydrolases Mechanism

Fan Yang[1,*] and Lin Wang[2]

[1] State Key Laboratory of Microbiological Technology, Shandong University,
No.27 Shanda'nan Road,
Jinan, 250100 Shandong, P.R. China
[2] Shandong Provincial Key Laboratory of Network Based Intelligent Computing,
University of Jinan,
Jinan 250022, Shandong, P.R. China
sunnyFan.Y@gmail.com

Abstract. Data mining methods are helpful in analyzing large amount of sequence and structure information of proteins. Classifiers can do a good job in achieving accurate mechanism classification of glycoside hydrolases which have different physicochemical properties. This classification method is not limited by reaction conditions. In this paper, a new method is proposed to classify the catalytic mechanism of a certain glycoside hydrolase according to their sequence and structure features by using several classifiers. Through making a comparison of the classification results achieved by the k-nearest neighbor (kNN) classifier and the Naive Bayes (NB) classifier and Multilayer Perceptron (MLP)classifier, the kNN classifier is approved to be an ideal choice in classifying and predicting the catalytic mechanisms of glycoside hydrolases with various physicochemical properties.

Keywords: Mechanism classification, K-Nearest Neighbor, Glycoside hydrolase.

1 Introduction

Glycoside hydrolases are playing key roles in the development of biofuels as well as in the food and pulp industries. The number of studies on sequences and structures of Glycoside hydrolases is increasing at an amazing speed. It has been proven that the functions and catalytic mechanisms of enzymes are closely related with both the sequences and three dimensional structure information[1]. A reasonable data mining and analysis of those sequence and structure data can help to solve many mysteries about the functions and catalytic process of most glycoside hydrolases. Classification is an important research area in data mining. Compared with traditional experiential methods, a classification by using data mining techniques is not limited by reaction conditions and is suitable for most glycoside hydrolases with various physicochemical properties.

---

* Corresponding author.

B.-L. Lu, L. Zhang, and J. Kwok (Eds.): ICONIP 2011, Part I, LNCS 7062, pp. 617–624, 2011.

In this paper, a classification of catalytic mechanisms of glycoside hydrolases based on their sequence and three-dimensional structure features by using three data mining methods was introduced.It is the first time to make a classification of the mechanisms of glycoside hydrolases according to both their sequences and structure features. This classification will assist the study on figuring out the catalytic process of each glycoside hydrolases and enhance our understanding of the mechanisms of glycoside hydrolases. More than 200 pieces of glycoside hydrolases spreading over 46 GH families were selected. The kNN classifier was chosen to be our primary classifier. The classification results achieved by the kNN classifier were compared with that achieved from the other two classifiers, including the Multilayer Perceptron classifier(MLP) and the Naive Bayes(NB) classifier.

2   Material and Method

Normally there are two mechanisms[2] for glycoside hydrolases to perform their functions. Inverting enzymes perform a single-displacement mechanism with a net inversion of an anomeric carbon configuration, while retaining enzymes adopt a double-displacement mechanism with a net retention of a substrate configuration [3]. Moreover, in most GH families, all the members within the same family adopt same catalytic mechanism[4].Inspired by a new approach proposed by L.C. Borro, S.R.M. Oliveira, etc[5], which is used to predict enzyme classes from protein structure with the help of classifies[6], an experiment to classify and predict the catalytic mechanisms of glycoside hydrolases using data mining techniques was established. The process of this method is shown in Fig. 1. At first, glycoside hydrolase data was collected from the PDB protein database, (http://www.rcsb.org/pdb/home/home) and the PDBsum database (http://www.ebi.ac.uk/pdbsum). Then, the features were selected carefully and the class sizes were balanced to gain a reliable classification result. At last, k-Nearest Neighbor classification method was used to categorize samples into one of catalytic mechanisms. Two other classifiers were also adopted as comparison bases.

2.1   Data Collection

In this research, each glycoside hydrolase was assigned into one of the two different catalytic mechanisms according to their sequence and structure features. We selected 136 pieces of retaining enzymes within 29 GH families and 67 pieces inverting ones spread over another 17 families to make up a data set. We defined that the Class A consists of retaining enzymes, and Class B contains all inverting ones. It is a data set large enough to ensure our classification reliable and accurate.

2.2   Representative Feature Selection and Data Processing

Some attributes that contain the same information or well correlated in the whole data set, will be identified as redundancy features. The accuracy of classification

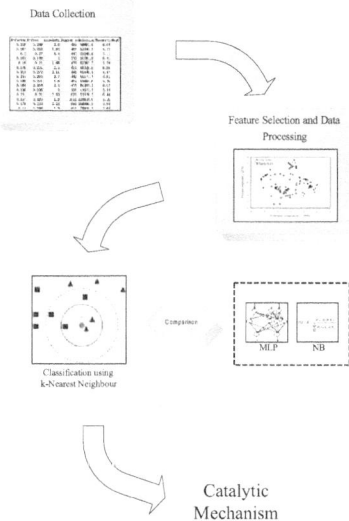

Fig. 1. Process of glycoside hydrolases classification

algorithm would be severely degraded with noisy and irrelevant features. Also if features and their scales are not consistent with their importance, the classification accuracy would be influenced[7]. In our study, many efforts had been done to exclude the factors which would reduce the credibility.

Firstly, the transit sequences, NP bind sequences, region sequences, etc are excluded, since the number of amino acids of a sequence is too small to extract representative features [6]. Also, if there are two or more chains within one sequence, entire chains were included in our data set. Even though some features overlap among these chains, we regarded each chain as a unique one because they may distinguish one kind of mechanism from other one. Many redundancy features were removed after a sound trade off and a carefully selection. 19 features of each glycoside hydrolase were extracted. They are unique and playing important roles in distinguishing retaining glycoside hydrolases from inverting ones. Take the charged residues such as the Asp, Glu for example, they are conserved residues and usually exist in the active sites for enzymes. Retaining enzymes and inverting glycoside hydrolases usually have different kinds of those conserved residues, and the number of charged resides are not the same. Since different glycoside hydrolases have various numbers of charged residues, some retaining glycoside hydrolases may have more Glu, while inverting ones usually have both Asp and Glu in their active sites. The theoretical PI of retaining and inverting enzymes are not same. The features chosen in this research can be divided into three categories:

(1) We extracted some sequence information, including the number of charged residues or the number of amino acids, etc, since the sequence of a protein is

unique to that protein, and defines the structure and function of the protein. These features are listed in Table 1.

Table 1. Sequence features after selection

Attributes	Description
NumOfAAs	Number of amino acids
Carbon	Number of carbon (C)
Hydrogen	Number of hydrogen (H)
Nitrogen	Number of nitrogen (N)
Oxygen	Number of oxygen (O)
Sulfur	Number of sulfur (S)
NumOfAtom	Number of atoms

(2) Structure features are selected because function of a protein is directly dependent on its three dimensional structure. They are searched from the PDB-sum structure database (http://www.ebi.ac.uk/pdbsum/)[8] and displayed in Table 2.

Table 2. Structure features after selection

Attributes	Description
R-factor	Residual factor or reliability factor
R-free	Free R value
Resolution	Resolution

(3) Several physico-chemical parameters of a protein sequence were chosen to construct this data set. They are regarded as important factors in predicting protein functions and the catalytic mechanisms of enzymes. Thses patameters are listed in Table 3.

Table 3. Physico-chemical features after selection

Attributes	Description
MolWeight	Molecular weight
Theoreticalpl	Theoretical pl
NegCharResidue	Number of negatively charged residues
PosCharResidue	Number of positively charged residues
ExtCoefALLCys	Extinction coefficients based on the assumption that all cysteine residues appear as half cystines
ExtCoefNoCys	Extinction coefficients based on the assumption that no cysteine appears as half cystine
InstabilityIndex	Instability index
AliphaticIndex	Aliphatic index
Gravy	Grand average of hydropathicity

Secondly, the range of those selected classification properties is wide, the class sizes should be balanced at first to gain a reliable classification result. All numeric values in the given data were normalized into a range from 0 to 1. Then, uniform distribution of class sizes was made[5].

After data processing, half of the sample population was selected to make up a training data set randomly. The training data set includes both retaining enzymes and inverting ones. This step is known as supervised learning. The remaining half was regarded as the testing data set used for evaluating the classification performance of the learned model.

2.3   Classification Methods

K-Nearest Neighbor (kNN)

The k-Nearest Neighbor algorithm (kNN) is a method for classifying objects based on closest training examples in the feature space. It is a type of instance-based learning, and the nearer neighbors contribute more to the average than the farther ones. This classifier is simple and suitable for a binary classification. In our research, it was adopted as a chief classifier for its predominant features. Given a query vector X_0 and a set of N labeled instances $\{x_i, y_i\}$, the task of the classifier is to predict the class label of x_0 on the predefined P classes. New examples were classified by choosing the majority class among the k closest examples in the training data. It has been demonstrated that through feature selection and data processing effectively, the kNN classifier can improve its performance significantly [9]. In our research, a 2-fold cross-validation was used and the value of "k" was 1.With this classifier, the catalytic mechanisms of glycoside hydrolases are divided into two categories: inverting enzymes and retaining enzymes in our research.

The implementation of the kNN classifier available at Weka version3.6.4 (http://www.cs.waikato.ac.nz/ml/weka) was chosen in this research. Weka is a software environment for knowledge analysis. This software is composed of a collection of machine learning algorithms used for data-mining tasks, clustering, association rules, and visualization.

Comparison between Different Classifiers

Further studies have been carried out to approve that the kNN classifier is sensitive and proper for analyzing our data set[10]. The other two classifiers: the Naive Bayes[11] and the Multilayer Perception classifier[12] are selected as the control. Detailed discussion and analysis on the classification accuracy achieved from those three classifiers were made.

3   Results and Discussion

In this research, the result achieved by the kNN classifier was demonstrated in Figure 2, showing that 69.9507% of 203 enzymes were correctly classified. It was a satisfying result compared with the results from the other two classifiers.

Table 4 displayed a more detailed explanation related to the results achieved by the kNN classifier. The two classes considered in our research were presented in the first column. The number of instances classified correctly in a given class divided by the number of instances in that class is called the TP rate, which is an abbreviation of true-positives rate. Consequently, the false-positive rate is equal to 1 minus the recall of the test, where recall corresponds to the number of true-positives divided by the sum of true-positives and false-negatives. Precision refers to the proportion of true-positives in a given class divided by the total number of enzymes classified in that class. F-measure means a harmonic measure that gets the most of both precision and recall, and it is defined as:

$$F - measure = 2\frac{precision \times recall}{precision + recall} \tag{1}$$

Besides that, the weighted average (Weighted Avg.) of A and B classes were provided in the last row. According to the results, it is known that the kNN classifier is proper and helpful to make an accurate classification. However, we also have to point out that, the kNN classifier seems to do a better job when recognizing retaining enzymes, which owns a larger quantity of more than 130 pieces of enzymes. The TP rate could be as high as 0.831. By contrast, when it was used to classify those inverting enzymes with only 80 pieces available for our study, the TP rate was just 0.433, which was relatively lower. Shown by Figure 2, the classification accuracy is 65.5172 % and 66.9951% using Naive Bayes classifier and Multilayer Perceptron classifier, respectively. The detailed description of classification accuracies by using the two classifiers are given in Table 4.

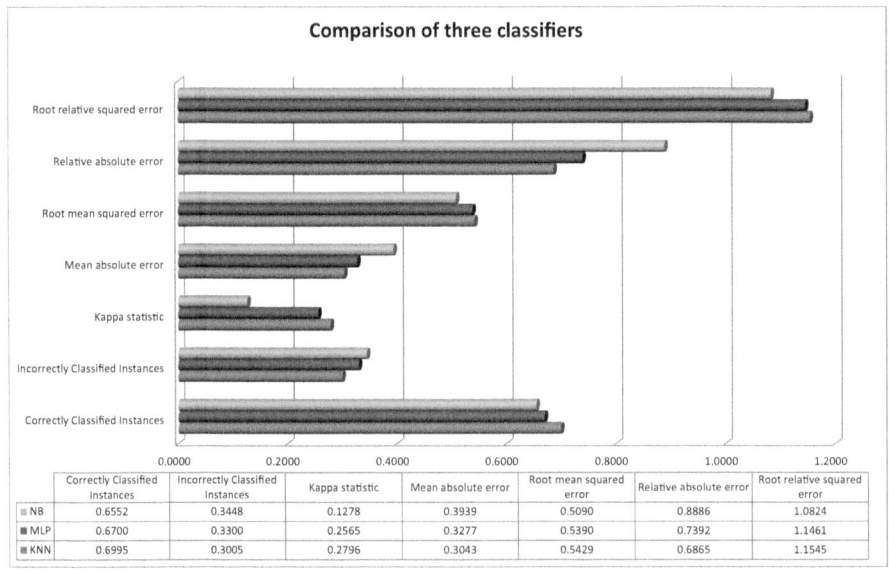

	Correctly Classified Instances	Incorrectly Classified Instances	Kappa statistic	Mean absolute error	Root mean squared error	Relative absolute error	Root relative squared error
NB	0.6552	0.3448	0.1278	0.3939	0.5090	0.8886	1.0824
MLP	0.6700	0.3300	0.2565	0.3277	0.5390	0.7392	1.1461
KNN	0.6995	0.3005	0.2796	0.3043	0.5429	0.6865	1.1545

Fig. 2. Comparison of the three classifiers

Table 4. Detailed accuracy of the three classifiers

Classiffier	class	TP Rate	FP Rate	Precision	Recall	F-Measure	ROC Area
kNN	A	0.831	0.567	0.748	0.831	0.787	0.629
	B	0.433	0.169	0.558	0.433	0.487	0.629
	Weighted Avg.	0.7	0.436	0.685	0.7	0.688	0.629
Naive Bayes	A	0.846	0.731	0.701	0.846	0.767	0.56
	B	0.269	0.154	0.462	0.269	0.34	0.56
	Weighted Avg.	0.655	0.541	0.622	0.655	0.626	0.56
Multilayer Perception	A	0.75	0.493	0.756	0.75	0.753	0.675
	B	0.507	0.25	0.5	0.507	0.504	0.675
	Weighted Avg.	0.63	0.412	0.671	0.63	0.671	0.675

After a careful analysis and comparison of the results achieved from the three classifiers adopted in our research, the kNN classifier demonstrated the highest accuracy and provided a satisfying and reasonable classification of catalytic mechanisms among glycoside hydrolases families.

4 Conclusion

Through selecting 19 attributes from the known features, a data set was constructed and processed. After applying this data set in the kNN classifier, Naive Bayes classifier and the multilayer perception classifier, we approved that the kNN classifier would be the most reliable one in making classification of glycoside hydrolases with an accuracy as high as 69.9507%. This result indicated that once one gets a new piece of glycoside hydrolase, one could be capable to predict its catalytic mechanism by extracting several sequence or three dimensional structural features. The main contributions of this study can be summarized as following:

(1) In contrast with some traditional methods, the method adopted in our study is more convenient and trouble-free. It would not be limited by reaction conditions and is available for most glycoside hydrolases with different physico-chemistry properties.
(2) This method is demonstrated to have a strong mathematical foundation and is easy to implement. Just by selecting proper representative features to consist a data set, and then the kNN classifier could help to make a relatively satisfying classification.
(3) In order to increase the classification accuracy, the data used in this study was preprocessed to balance the class sizes. Both the sequence parameters and the three dimensional features were extracted.

The successful classification in this study donates that if one acquires a new sequence or a sequence whose catalytic mechanism is unknown, one can easily assign them into one of the two classes based on sequence and structure information by using data mining techniques. To move forward a single step, once one knows what kind of catalytic mechanism those enzymes use, one would be able to change the reaction process of several enzymes.

Acknowledgments. This work was partially supported by National Natural Science Foundation of China under grant No.60873089. We greatly acknowledge Prof. LuShan Wang from the State Key Laboratory of Microbial Technology for his discussion, and we also acknowledge Prof. Merschel Sylvia from UCLA for her valuable comments.

References

1. Henrissat, B., Davies, G.: Structural and sequence-based classification of glycoside hydrolases. Current Opinion in Structural Biology 7, 637–644 (1997)
2. Yang, J.K., Yoon, H.J., Ahn, H.J., Il Lee, B., Pedelacq, J., Liong, E.C., Berendzen, J., Laivenieks, M., Vieille, C., Zeikus, G.J.: Crystal Structure of [beta]-d-Xylosidase from Thermoanaerobacterium saccharolyticum, a Family 39 Glycoside Hydrolase. Journal of Molecular Biology 335, 155–165 (2004)
3. Uversky, V.N., Wohlkönig, A., Huet, J., Looze, Y., Wintjens, R.: Structural Relationships in the Lysozyme Superfamily: Significant Evidence for Glycoside Hydrolase Signature Motifs. PLoS ONE 5, e15388 (2010)
4. Honda, Y., Fushinobu, S., Hidaka, M., Wakagi, T., Shoun, H., Taniguchi, H., Kitaoka, M.: Alternative strategy for converting an inverting glycoside hydrolase into a glycosynthase. Glycobiology 18, 325 (2008)
5. Borro, L.C., Oliveira, S.R.M., Yamagishi, M.E.B., Mancini, A.L., Jardine, J.G., Mazoni, I., dos Santos, E.H., Higa, R.H., Kuser, P.R., Neshich, G.: Predicting enzyme class from protein structure using Bayesian classification. Genet. Mol. Res. 5, 193–202 (2006)
6. Lee, B.J., Lee, H.G., Lee, J.Y., Ryu, K.H.: Classification of Enzyme Function from protein sequence based on feature representation. In: IEEE International Conference on Bioinformatics and Bioengineering - BIBE, pp. 741–747. IEEE Press, Boston (2007)
7. Nigsch, F., Bender, A., van Buuren, B., Tissen, J., Nigsch, E., Mitchell, J.B.O.: Melting point prediction employing k-nearest neighbor algorithms and genetic parameter optimization. Journal of Chemical Information and Modeling 46, 2412–2422 (2006)
8. Gasteiger, E., Hoogland, C., Gattiker, A., Duvaud, S., Wilkins, M., Appel, R.D., Bairoch, A.: Protein identification and analysis tools on the ExPASy server. In: The Proteomics Protocols Handbook, pp. 571–607 (2005)
9. Nasibov, E.N., Kandemir-Cavas, C.: Efficiency analysis of KNN and minimum distance-based classifiers in enzyme family prediction. Computational Biology and Chemistry 33, 461–464 (2009)
10. Valavanis, I.K., Spyrou, G.M., Nikita, K.S.: A comparative study of multi-classification methods for protein fold recognition. International Journal of Computational Intelligence in Bioinformatics and Systems Biology 1, 332–346 (2010)
11. Towfic, F., Caragea, C., Dobbs, D., Honavar, V.: Struct-NB: predicting protein-RNA binding sites using structural features. International Journal of Data Mining and Bioinformatics 4, 21–43 (2010)
12. Nanni, L., Lumini, A.: A further step toward an optimal ensemble of classifiers for peptide classification, a case study: HIV protease. Protein and Peptide Letters 16, 163–167 (2009)

A Memetic Approach to Protein Structure Prediction in Triangular Lattices

Md. Kamrul Islam[1], Madhu Chetty[1], A. Dayem Ullah[2], and K. Steinhöfel[2]

[1] GSIT, Monash University, Churchill Vic 3842, Australia
[2] King's College London, Department of Informatics, London WC2R 2LS, UK

Abstract. Protein structure prediction (PSP) remains one of the most challenging open problems in structural bioinformatics. Simplified models in terms of lattice structure and energy function have been proposed to ease the computational hardness of this combinatorial optimization problem. In this paper, we describe a clustered meme-based evolutionary approach for PSP using triangular lattice model. Under the framework of memetic algorithm, the proposed method extracts a pool of cultural information from different regions of the search space using data clustering technique. These highly observed local substructures, termed as *meme*, are then aggregated centrally for further refinements as second stage of evolution. The optimal utilization of 'explore-and-exploit' feature of evolutionary algorithms is ensured by the inherent parallel architecture of the algorithm and subsequent use of cultural information.

Keywords: PSP, evolutionary approach, meme, memetic algorithm, data clustering.

1 Introduction

Proteins are composed of linear chains of amino acids (residues) and regulate almost all cellular functions in an organism. The three dimensional folded structure of proteins (tertiary structures) play key roles in their functionality. According to Anfinsen's thermodynamic hypothesis, proteins fold into tertiary structures of minimum free energy (native state) which can be predicted from the corresponding amino acid sequences [2]. However, incomplete knowledge of folding mechanism, absence of an established perfect energy function as well as apparently complex and irregular structures in three-dimensional space make the PSP problem ever so difficult, which encourages researchers adopting simplified lattice and energy models to ease the computational hardness of the problem so as to explain essential functional properties of proteins. The prevailing strategy to determine protein structures has been to determine a self-avoiding walk (SAW) embedding of amino acids in 3D lattice space (conformation) that results in overall minimum energy. Energy function has been modelled that captures the idea of assigning energy between amino acids pairs placed within neighbouring positions in the lattice. The problem is shown to be NP-Complete [5] even for the simplest of Hydrophobic-Polar (HP) energy function [8] on discrete rectangular lattice model. Being a hard combinatorial optimization problem, PSP has been approached by approximation algorithms [1,9], constraint

B.-L. Lu, L. Zhang, and J. Kwok (Eds.): ICONIP 2011, Part I, LNCS 7062, pp. 625–635, 2011.
© Springer-Verlag Berlin Heidelberg 2011

programming [6,20,28], stochastic local search algorithms (e.g., simulated annealing [4,15], tabu search [4,17], REMC [26]), evolutionary algorithms (e.g., genetic algorithm [19,27], hybrid tabu search [14], ant colony optimization [25], EMC [18]) etc.

Evolutionary algorithms (EAs) in combination with local optimization procedures have been shown to improve precision where EA alone leads to suboptimal solution [21]. "Memetic algorithm" (MA) stems from the term *meme* [7] to denote a smaller part of gene that remains unchanged over the evolution. Hence memes are regarded as units of cultural inheritance [22]. The concept of memetic algorithm has evolved as a family of population-based metaheuristics that employ a local search (LS) step as a distinctive part within its main evolutionary cycle of standard recombination and mutation operators. The emphasis of MAs for PSP thus lies in the use of LS. Krasnogor et al. [16] first proposed a multimeme algorithm for PSP which employs six different LS methods (later extended to fuzzy logic-based local searchers [24]), and self-adaptively select from this set which heuristic to use for different stages of the search or for different individuals in the population. Bazzoli and Tettamanzi [3] presented an MA where LS can self-adaptively act towards either exploitation or diversification of fitness, according to its degree of convergence within the population. More recently, Islam *et al.* [12,13] proposed an MA that incorporates a guided local search based on the exploitation of domain knowledge that can be explained by well-known schema preservation technique. These methods have been tested on the HP model proteins using small-to-medium sized instances on 2D and 3D rectangular lattices [3,12,13,16] and small instances on 2D triangular lattice [16].

In this paper, we intend to reintroduce the concept of original meme definition in the standard MA approach to PSP. We restate "meme" to be highly observed local substructures from diverse areas of landscape which can be utilized to improve individuals under the traditional MA framework. The overall process ideally takes advantage of two-stage data clustering technique that explores the underlying landscape in Stage-1 and exploit the knowledge in Stage-2. In section 2, we describe the outline of our proposed memetic framework to approach PSP on triangular lattices. In the process, we discuss population initialization and meme generation technique, which are unique to our approach. Apart from the reason that research is lacking on triangular lattice PSP with population based EAs, our choice of the triangular lattices is further motivated by the fact that 3D triangular lattice (also known as FCC lattice) has been shown to yield very good approximations of real protein structures [23]. Also triangular lattices do not suffer from the bipartiteness of the rectangular lattice, which allows interactions only between amino acids of opposite parity in the sequence. In section 3, we discuss how memes can be utilized in clustered framework. The simulation results are shown at the end for selected benchmarks. As meme is being suggested as a unit of cultural information; therefore, to avoid any confusion, we will call our proposed approach as "cultural memetic algorithm".

2   Cultural Memetic Algorithm

For the PSP problem, memes are sub-structures that become fossilised when the population converges towards a particular local minima. Memes can preserve domain knowledge of local minima efficiently and can help to explore and exploit the entire search space effectively. However, since genetic drift drives all individuals to one local minima, a single set of population will fail to identify memes corresponding to different regions of the search space.

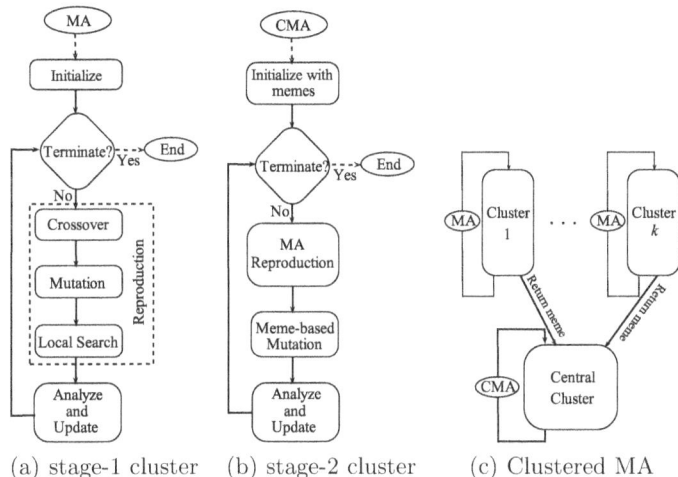

(a) stage-1 cluster   (b) stage-2 cluster   (c) Clustered MA

Fig. 1. Cultural memetic algorithm framework

Our proposed algorithm aims to explore different areas of the landscape in parallel and exploit the acquired knowledge in the form of memes to guide the search process towards global minima. Assuming a powerful enough method is available for combining promising solutions covering important interactions of decision variables in a problem, clustering can be used very efficiently to search the solution space thoroughly. We take advantage of an initial data set clustering technique to help in both exploring multiple local minima and covering as many individuals as possible efficiently. Each cluster is populated capitalizing natural way of protein generation and then undergoes evolution under the memetic framework of standard crossover, mutation and local search operators. Once the population has sufficiently converged following a convergence criteria, memes are extracted from the best set of individuals. Memes from all Stage-1 clusters are then transferred to a Stage-2 cluster where the knowledge, containing memes from different regions, is exploited to generate new individuals. The evolutionary process then continues under the memetic framework with an additional meme-based mutation technique. The overall architecture is shown in Fig. 1.

2.1 Population Initialization

A protein conformation can be specified as a sequence of moves taken on the lattice from one residue to the next. Absolute encoding of a move depends on entire orientation whereas relative encoding depends on the last move taken (see fig. 2(a)). Similar substructures in different conformations are represented by same relative move sequence which may have different absolute move representations depending on its orientation on lattice geometry. Traditionally an individual for PSP is generated using random relative move sequence which can not guarantee a SAW individual, necessitating a SAW-checking procedure to validate and re-generate if required. This process is repeated until the necessary number of individuals is generated.

In general, consider $\mathbb{C} = (\mathbf{A,B,C,E,F})$ to be possible relative move set in 2D triangular lattice model (move $\mathbf{D}$ is prohibited to avoid immediate backward move). An individual is represented by a relative move sequence emanating from $\mathbb{C}$ (see fig. 2(b)). Since a conformation can always start with $\mathbf{A}$, a protein of length n can be defined by $(n-2)$ subsequent moves. There can be $N_p = 5^{n-2}$ possible individuals, some of which may be non-SAW. A sequence of minimum

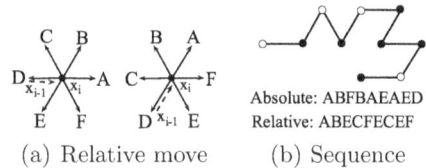

C B   B A
D$\xrightarrow{}$$\overset{}{\underset{x_i}{\times}}$A   C$\xleftarrow{}$$\overset{}{\underset{x_i}{\times}}$F
x$_{i-1}$ x$_i$
E F   D$^{x_{i+1}}$ E

Absolute: ABFBAEAED
Relative: ABECFECEF

(a) Relative move   (b) Sequence

Fig. 2. (a) relative move-set and (b) a conformation for triangular lattice. *Black beads present 'H', white beads present 'P' residues. 3 H-H contacts contributes to total energy of -3.*

4 residues can have two non-SAW conformations by forming loop like structures **ACC** or **AEE**. For $n > 4$, such a loop may form at the beginning in worst case with 5^{n-4} possible moves for remaining residues. Irrespective of the appended moves after the loop, the conformations remain non-SAW. Hence, the possible number of non-SAW conformations is $N_w = 2 \times 5^{n-4}$. The worst case probability of generating a non-SAW individual is $P_w(tri) = \frac{N_w}{N_p} = \frac{2 \times 5^{n-4}}{5^{n-2}} = \frac{2}{5^2} = 0.08$ for 2D triangular lattice and $P_w(fcc) = \frac{4 \times 11^{n-4}}{11^{n-2}} = \frac{4}{11^2} = 0.033$ for FCC lattice.

In our proposed *Dynamic Individual generation* (DIG) method [13] which follows Nature, i.e., amino acids joining one by one to form the sequence, an individual is generated by using relative moves together with coordinates of the lattice points. For each residue r_i at position i, the process stores a possible set of relative moves, $S_{P_i} = (\mathbf{A,B,C,E,F})$, and the set is updated according to the move m_i taken at that position. Before finalizing any random move out of set S_{P_i} for r_i, we verify whether the lattice point is already occupied by another residue $r_{j<i}$ or not. If not occupied, then the move is implemented for r_i and the point is updated accordingly. Otherwise, the move is discarded from S_{P_i} and another random move is chosen from the truncated S_{P_i}. If S_{P_i} becomes empty,

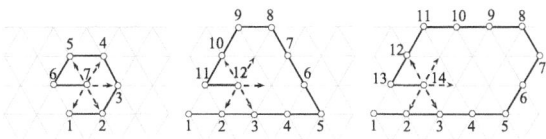

Fig. 3. Examples of close-ended local substructures

the process recognizes that the previous move from residue r_{i-1} is possibly a wrong choice and passes the control back to r_{i-1}. $S_{P_{i-1}}$ is updated by removing that wrong move and a new move is taken from the truncated move set. This process is continued until all moves are generated for the individual. The worst case scenario causing maximum back tracking occurs only with a close-ended local substructure (see fig. 3). If the number of moves forming such structure is L_s then the probability of such formation is $(\frac{1}{5})^{L_s}$. Therefore the probability of being trapped into those structures reduces exponentially with L_s. The proposed method guarantees that it will come out of any close-ended substructures after visiting all the lattice points inside and it will never revisit that path again.

2.2   Meme Generation

The meme generation process requires a systematic identification of substructure that best suits a particular segment of the conformation. Memes identified from different basins of attraction can be treated as a database for transferring knowledge to new or existing individuals. Once the population is perceived as converged, meme identification technique comes into play. In our case, the sufficient population convergence is decided based on two main criteria: (i) Best fitness value in the population remains fixed for a pre-specified n_c generations, and (ii) Difference in fitness values of the top 75% (i.e. $\frac{3}{4}$-th of total individuals) of the individuals are within a pre-specified close range Δ.

Meme Identification. A meme can be identified by finding a highly probable move sequence in particular positions, having at least two consecutive moves each of which have an independent probability greater than a specified threshold to occupy those positions. The structure of a meme is therefore described by a triplet containing the relative move sequence, its start position and end position. In general, consider a conformation C_i in any lattice model and $\mathbb{C}$ to be a set of all possible relative moves with size $|\mathbb{C}|$. The proposed meme technique is generic and applicable to any lattice model.

We first construct a two dimensional matrix $M_{C_i} : (n-2) \times \mathbb{C}$ with rows defining the $n-2$ positions of moves and the columns describing the actual moves $\mathbb{C}_q$. The matrix M_{C_i} is populated as $[a_{rq}]_{r=1,\cdots,n-2;q=1,\cdots,|\mathbb{C}|}$. Now, if the r-th position of a conformation C_i is $\mathbb{C}_q$, then $a_{rq} = \epsilon \times F(C_i)$, otherwise $a_{rq} = 0$. ϵ is fixed to -1 and $F(C_i)$ is the fitness of C_i. To determine a highly probable meme, which is likely to occur in subsequent generation, we obtain matrix $\Gamma = \sum_{i=1}^{N} M_{C_i}$. Multiplying Γ with a column vector $[1 \cdots 1]^T$ results in another vector $\bar{\Gamma} = [\rho_1 \cdots \rho_{n-2}]^T$. The r-th row of $\bar{\Gamma}$ represents the cumulative weight ρ_r of r-th position for all conformations. The probability of occurrence of each move at this r-th position is obtained from multiplying Γ_r by $1/\rho_r$. By doing so for each position, we obtain another matrix Γ' that contains significant information about the probability of occurrence of moves at a given position. To classify whether a move is highly probable in a given position, we define a cut-off value, $\chi = 0.4 + \frac{1}{|\mathbb{C}|}$. If at least two consecutive moves in the matrix Γ' have value greater than χ, then this highly probable set of moves is identified as a meme.

To treat a substructure and its reflection same, a simple modification is applied on Γ'. We add up pairs of columns of matrix Γ' to generate Γ^r where the pairs represent relative move directions that are reflection of each other:

$$\forall i \{\Gamma^r[i, x] = \Gamma^r[i, y] = \Gamma'[i, x] + \Gamma'[i, y]\}, \text{ if } \mathbb{C}_x = \text{reflection}(\mathbb{C}_y)$$

For example, in 2D triangular lattice, relative moves **B** and **F** are reflections of each other; the same applies to moves **C** and **E**. Move **A** has no corresponding reflection (or reflection of **A** is **A** itself). Therefore memes are identified from columns of Γ^r representing moves **A**, **B**, **C** only. Note that, since move **D** is absent in any valid conformations, there is no such corresponding column in Γ^r.

Meme Validation. The memes generated as above need to be validated against individuals with best fitness values. It is also possible that few memes identified from Γ^r are non-SAW. A non-SAW meme can be removed from the meme set or part of it can be redefined as new meme. Let $S(\varepsilon)$ be the identified meme-set and $S(I)$ be the best individual set. The meme validation method checks whether a meme $\varepsilon \in S(\varepsilon)$ exists in any of the best individual or not. If a full match is found, then it will be added to a new meme-set $S_n(\varepsilon)$. Otherwise the portion of the meme that matches will be considered as a new meme and added to $S_n(\varepsilon)$. Note that memes in $S(\varepsilon)$ ignores the reflective part of the moves, therefore during the search, moves in the best individuals are matched against each move in the meme and its reflection. The matched meme are corrected accordingly when added to $S_n(\varepsilon)$. In this process, common features that represent best individuals are extracted from the best individual set and passed to other individuals as improvement features.

The meme validation technique is illustrated here with an arbitrary example sequence HPHPHHHPPPHHHHHPPHH. Five individuals with fitness (E) from the best individual set $S(I)$ at some stage of the evolutionary search are listed below with the corresponding weighted matrix Γ^r. Applying meme identification technique with reflection property into consideration, rows in Γ^r represent moves {**A,B,C,E,F**} and columns represent positions of the moves in conformations. By applying cut-off value, $\chi = 0.6$ on Γ^r, three candidate memes are identified namely (**ACACBB**,1,6), (**BA**,9,10) and (**AACBA**,13,17). However, these memes are generic and might not be present in any of the best individuals or might be present in different reflected form, therefore need to be validated. Applying meme validation technique to these three memes gives rise to the actual memes; (**AEA**,1,3), (**CBF**,4,6), (**BA**,9,10), (**AACBA**,13,17) (see Fig. 4).

	Conformation	E
C1	AEAFECBFEAFBFACBA	-13
C2	ABACBFECACFBAACBA	-13
C3	AECEFBAAFEAACAEFA	-13
C4	AEAFECECBACEAACBA	-12
C5	ABACBFECBACEAACBA	-12

$$\Gamma^r = \begin{bmatrix} A:1\ 0\ .8\ 0\ 0\ 0\ .2\ .2\ .2\ .6\ .2\ .2\ .6\ 1\ 0\ 0\ 1 \\ B:0\ .4\ 0\ .4\ .6\ .6\ .2\ .2\ .6\ 0\ .4\ .4\ .2\ 0\ 0\ 1\ 0 \\ C:0\ .6\ .2\ .6\ .4\ .4\ .6\ .6\ .2\ .4\ .4\ .4\ .2\ 0\ 1\ 0\ 0 \\ E:0\ .6\ .2\ .6\ .4\ .4\ .6\ .6\ .2\ .4\ .4\ .4\ .2\ 0\ 1\ 0\ 0 \\ F:0\ .4\ 0\ .4\ .6\ .6\ .2\ .2\ .6\ 0\ .4\ .4\ .2\ 0\ 0\ 1\ 0 \end{bmatrix}^T$$

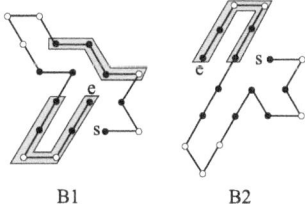

Fig. 4. Examples of individuals with validated memes (gray)

Fig. 5. Examples of optimum structures with memes

The potential gain from incorporating memes are observable from two optimum structures **ACFCFBFBCABEAACBA** and **AEBEFCAAEFAAAACBA** (see Fig. 5) with fitness -15 where meme (**AACBA**,13,17) is present. Also note that (**CFB**,4,6), a reflection of meme (**CBF**,4,6), is present in the first optimum structure. We may conclude that although memes are generated from random basins of attraction, on many occasions the global optimum are led from these basins especially which shows that *memes* remain common to the search regions of interest.

3   Clustering MA

The proposed method for clustering works in two stages. At Stage-1, κ number of clusters are populated and evolved independently for I_κ number of generations. Memes are extracted from best individuals of each cluster and used to populate the single Stage-2 cluster. Here, we generate $N_\kappa = N/\kappa$ individuals incorporating the memes from each cluster, N being the population size at Stage-2. The new individuals will then evolve inside the cluster under the same memetic framework with an additional mutation operation based on meme replacement.

3.1   Population Initialization with Memes

Let the meme-pool returned by the i-th stage-1 cluster be $S_i(\varepsilon)$. An individual is first generated using the DIG method (see Sec. 2.1). Then memes $\varepsilon \in S_i(\varepsilon)$ are sequentially selected to be incorporated into the individual in the appropriate positions. If the individual becomes non-SAW for a ε, then the reflection of ε is tried. If the individual remains non-SAW, next meme from $S_i(\varepsilon)$ is tried. The process is repeatedly applied until a valid individual is generated.

3.2   Meme-Based Mutation

All memes generated from different stage-1 clusters and new memes generated at stage-2 cluster during evolution form the stage-2 meme pool $S_{s2}(\varepsilon)$. An iterative procedure similar to individual generation is employed here that tries to improve an individual by replacing part of the individual by appropriate meme $\varepsilon \in S_{s2}(\varepsilon)$. Each meme (and its reflection) is tried separately to be incorporated into the selected individual and from the new set of mutated individuals, the best one is kept and replaces the original individual. A reflection of each meme is used to ensure no information loss.

4   Experimental Evaluation

Experiments are performed on the selected benchmark sequences from Table 1 with lengths ranging from 36–64. The benchmark is selected to perform a direct comparison with an existing hybrid evolutionary method, namely hybrid genetic algorithm (HGA) with twin removal approach [10,11]. It is noteworthy to mention that, for shorter sequences, Stage-1 clusters are efficient enough to identify global minima without using meme information, therefore we consider only moderate sized sequences for our experiments. Two sets of experiments are done. The first set deals with comparative performance analysis between the proposed individual generation method DIG and the traditional method to generate an initial population of 100 individuals.

Table 1. Benchmark sequences for HP model protein. *Here, $E*$ gives the optimum fitness for 2D triangular and 3D FCC lattices.*

Id	Len.	Sequence	$E*$(Tri)	$E*$(Fcc)
s1	36	3P2H2P2P5P7H2P2H4P2H2PH2P	-24	-38
s2	48	2PH2(P3H)5P10H6P2(2H2P)H2P5H	-43	-74
s3	54	H4(HP)4HP4(H3P)P2(H3P)HP4H4(PH)H	-41	-77
s4	60	2P3HP8H3P10HPH3P12H4P6HPH2(HP)	-70	-130
s5	64	12H2(PH)2P2(2H2P)3(H2PH)P2(P2HP)3(PH)11H	-75	-132

Besides initialization, more than 50% new random individuals are required across generation to generation to replace existing individuals to ensure diversity and combat premature convergence problem. Hence, individual generation process has an overall effect on the performance of MA. Table 2 shows that DIG performs better than traditional method in terms of both runtime and the diversity of population.

Table 2. Performance analysis for DIG in triangular lattices

Id	2D Triangular				FCC			
	Time(hh:mm:ss)		Divergence		Time(hh:mm:ss)		Divergence	
	DIG	Traditional	DIG	Traditional	DIG	Traditional	DIG	Traditional
s1	0.0376	0.398	0.9560	0.9448	0.177	0:25.6	0.9165	0.8779
s2	0.0602	4.872	0.9630	0.9513	0.521	5:22.9	0.9341	0.8823
s3	0.0904	15.966	0.9663	0.9541	0.564	15:30.7	0.936	0.8865
s4	0.1284	49.060	0.9674	0.9560	0.708	1:02:01.7	0.9377	0.8885
s5	0.3576	01:44.800	0.9686	0.9575	0.746	2:09:48.8	0.9392	0.8899

The next set of experiments deals with the performance analysis of our proposed clustered cultural memetic approach. For each sequence, the implemented program is run 5 times with the following parameters:

- Size of the population in each generation, $N = 100$
- Number of Stage-1 clusters, $\kappa = 5$
- Maximum number of generations per cluster $= 20$
- Mutation rate $= 0.05$
- Crossover rate $= 1.0$
- Fraction of population on which local search will be applied $= 0.4$
- Fraction of population removed when premature convergence occurs $= 0.9$
- Number of best individuals used for meme generation $= 10$

We have used one-point mutation, one-point crossover with *roulette-wheel selection strategy* and pull-move based local search introduced in [4]. In Table 3, we report the minimum energy (best fitness), their average and standard deviation observed from Stage-1 and Stage-2 clusters. The results from Stage-1 and Stage-2 clusters are indicators of the improvement that can be obtained from incorporating meme information.

Table 3. Performance analysis for CMA in triangular lattices

Id	2D Triangular				FCC			
	E_{CMA}	$E_{s-1}(\mu \pm \sigma)$	$E_{s-2}(\mu \pm \sigma)$	E_{HGA}	E_{CMA}	$E_{s-1}(\mu \pm \sigma)$	$E_{s-2}(\mu \pm \sigma)$	E_{HGA}
s1	-24	-22.43 $\pm$ 0.77	-23.4 $\pm$ 0.55	-19	-38	-36.92 $\pm$ 0.90	-37.8 $\pm$ 0.45	-51
s2	-43	-40.15 $\pm$ 1.81	-42.6 $\pm$ 0.55	-32	-74	-68.80 $\pm$ 2.17	-73.2 $\pm$ 0.96	-69
s3	-41	-37.84 $\pm$ 1.28	-40.8 $\pm$ 0.45	-23	-73	-66.68 $\pm$ 2.38	-71.4 $\pm$ 1.67	-59
s4	-69	-65.44 $\pm$ 1.87	-68.0 $\pm$ 0.71	-46	-126	-122.75 $\pm$ 2.31	-123.2 $\pm$ 0.45	-117
s5	-71	-65.36 $\pm$ 3.36	-70.6 $\pm$ 0.55	-46	-128	-120.75 $\pm$ 3.03	-124.2 $\pm$ 3.03	-103

5 Conclusion

One of the most important aspects of applying population-based metaheuristics to multimodal functions is identifying as much near-optimal solutions as possible to analyze the underlying fitness landscape and understand the parameters of the problem at hand. Thus, our method is significantly better than related evolutionary algorithms and sets standard for future researchers working on EA based PSP on triangular lattice models. As our proposed clustered algorithm is easily scalable in a grid or distributed architecture environment, it provides another great advantage related to computational cost. The ability to continue evolution process in parallel helps faster exploration and then exploiting the cultural knowledge guide the search process towards newer regions of the search space. Comparison with actual global minimum energy, though, shows that our method has limitations in finding the global minima for longer sequences. This can be overcome partly by fine tuning the various memetic parameters and also by using specialized improvement operators which is being pursued actively within the group. These studies are focused to see whether extracted memes can reflect the secondary structures of proteins when considering elaborate energy models.

References

1. Agarwala, R., et al.: Local rules for protein folding on a triangular lattice and generalized hydrophobicity in the HP model. In: Proc. SODA 1997, pp. 390–399 (1997)
2. Anfinsen, C.B.: Principles that govern the folding of protein chains. Science 181, 223–230 (1973)
3. Bazzoli, A., Tettamanzi, A.G.B.: A Memetic Algorithm for Protein Structure Prediction in a 3D-Lattice HP Model. In: Raidl, G.R., Cagnoni, S., Branke, J., Corne, D.W., Drechsler, R., Jin, Y., Johnson, C.G., Machado, P., Marchiori, E., Rothlauf, F., Smith, G.D., Squillero, G. (eds.) EvoWorkshops 2004. LNCS, vol. 3005, pp. 1–10. Springer, Heidelberg (2004)
4. Böckenhauer, H.-J., Dayem Ullah, A.Z.M., Kapsokalivas, L., Steinhöfel, K.: A Local Move Set for Protein Folding in Triangular Lattice Models. In: Crandall, K.A., Lagergren, J. (eds.) WABI 2008. LNCS (LNBI), vol. 5251, pp. 369–381. Springer, Heidelberg (2008)
5. Crescenzi, P., et al.: On the complexity of protein folding. Journal of Computational Biology 5, 423–465 (1998)
6. Dal Palù, A., et al.: A constraint solver for discrete lattices, its parallelization, and application to protein structure prediction. Software-Practice and Experience 37, 1405–1449 (2007)
7. Dawkins, R.: The Selfish Gene. Oxford University Press, New York (1976)
8. Dill, K.A., et al.: Principles of protein folding - A perspective from simple exact models. Protein Science 4, 561–602 (1995)
9. Hart, W.E., et al.: Fast protein folding in the hydrophobic-hydrophilic model within three-eights of optimal. In: ACM Symposium on Theory of Computing, pp. 157–168 (1995)
10. Hoque, M.T., Chetty, M., Dooley, L.S.: A Hybrid Genetic Algorithm for 2D FCC Hydrophobic-Hydrophilic Lattice Model to Predict Protein Folding. In: Sattar, A., Kang, B.-h. (eds.) AI 2006. LNCS (LNAI), vol. 4304, pp. 867–876. Springer, Heidelberg (2006)
11. Hoque, M.T., et al.: Protein folding prediction in 3d fcc hp lattice model using genetic algorithm. In: Proc. CEC 2007, pp. 4138–4145 (2007)
12. Islam, M. K., Chetty, M.: Novel Memetic Algorithm for Protein Structure Prediction. In: Nicholson, A., Li, X. (eds.) AI 2009. LNCS, vol. 5866, pp. 412–421. Springer, Heidelberg (2009)
13. Islam, M.K., et al.: Clustered Memetic Algorithm for Protein Structure Prediction. In: Proc. CEC 2010, pp. 1–8 (2010)

14. Jiang, T., et al.: Protein folding simulations of the hydrophobic–hydrophilic model by combining tabu search with genetic algorithms. The Journal of Chemical Physics 119(8), 4592–4596 (2003)
15. Kapsokalivas, L., et al.: Two Local Search Methods for Protein Folding Simulation in the HP and the MJ Lattice Models. In: Proc. BIRD 2008, pp. 167–179 (2008)
16. Krasnogor, N., Blackburne, B.P., Burke, E.K., Hirst, J.D.: Multimeme Algorithms for Protein Structure Prediction. In: Guervós, J.J.M., Adamidis, P.A., Beyer, H.-G., Fernández-Villacañas, J.-L., Schwefel, H.-P. (eds.) PPSN 2002. LNCS, vol. 2439, pp. 769–778. Springer, Heidelberg (2002)
17. Lesh, N., et al.: A complete and effective move set for simplified protein folding. In: Proc. ICCB 2003, pp. 188–195 (2003)
18. Liang, F., et al.: Evolutionary Monte Carlo for protein folding simulations. Journal of Chemical Physics 115(7), 3374–3380 (2001)
19. Lopes, H.S., et al.: An enhanced genetic algorithm for protein structure prediction using the 2D hydrophobic-polar mode. In: Proc. AE 2005, pp. 238–246 (2005)
20. Mann, M., et al.: CPSP-tools - Exact and Complete Algorithms for High-throughput 3D Lattice Protein Studies. BMC Bioinformatics 9, 230 (2008)
21. Martínez-estudillo, A., et al.: Hybridization of evolutionary algorithms and local search by means of a clustering method. IEEE Transactions on Systems, Man and Cybernetics, Part B 36, 534–545 (2006)
22. Ong, Y., et al.: Memetic Computation — Past, Present & Future [Research Frontier]. IEEE Computational Intelligence Magazine 5(2), 24–31 (2010)
23. Park, B.H., et al.: The complexity and accuracy of discrete state models of protein structure. Journal of Molecular Biology 249(2), 493–507 (1995)
24. Pelta, D.A., et al.: Multimeme algorithms using fuzzy logic based memes for protein structure prediction. In: Krasnogor, N., Smith, J.E. (eds.) Recent Advances in Memetic Algorithms. Springer, Heidelberg (2004)
25. Shmygelska, A., et al.: An ant colony optimisation algorithm for the 2D and 3D hydrophobic polar protein folding problem. BMC Bioinformatics 6(30) (2005)
26. Thachuk, C., et al.: A replica exchange Monte Carlo algorithm for protein folding in the HP model. BMC Bioinformatics 8, 342 (2007)
27. Unger, R., et al.: Genetic algorithms for protein folding simulations. Journal of Molecular Biology 231(1), 75–81 (1993)
28. Yue, K., et al.: Forces of tertiary structural organization in globular proteins. Proc. Natural Academy of Sciences USA 92, 146–150 (1995)

Conflict Resolution Based Global Search Operators for Long Protein Structures Prediction

Md. Kamrul Islam, Madhu Chetty, and Manzur Murshed

GSIT, Monash University, Churchill, VIC 3842, Australia

Abstract. Most population based evolutionary algorithms (EAs) have struggled to accurately predict structure for long protein sequences. This is because conventional operators, i.e., crossover and mutation, cannot satisfy constraints (e.g., connected chain and self-avoiding-walk) of the complex combinatorial multi-modal problem, protein structure prediction (PSP). In this paper, we present novel crossover and mutation operators based on conflict resolution for handling long protein sequences in PSP using lattice models. To our knowledge, this is a pioneering work to address the PSP limitations for long sequences. Experiments carried out with long PDB sequences show the effectiveness of the proposed method.

Keywords: crossover by conflict resolution, mutation by conflict resolution, clustered memetic algorithm.

1 Introduction

Protein structure prediction (PSP) is one of the oldest and most challenging problems in structural bioinformatics. Since native states of proteins can be determined with the minimum free energy conformation [1], this implies that it is essentially a computational problem. Despite the research over past 30 years, no truly accurate *ab initio* method exists to predict protein structures from amino acid sequence [5]. This is because our knowledge and computation power is simply insufficient to search the entire search space of such a high complexity [6]. Since protein structures have an enormous impact in medicine and pharmaceutical industry, researchers are interested in finding near optimal in-silico solutions [5]. Evolutionary algorithms (EAs) are prominent for finding such near-optimal solutions but they fail to succeed if the protein sequences are long. Different EAs have been used to solve PSP problem including genetic algorithm (GA) [15], ant colony optimization (ACO) [18], immune algorithm (IM) [2], estimation of distribution algorithm (EDA) [17], and memetic algorithm [10,11,12]. A comprehensive review on the advances on ab initio PSP on lattices with natural computing can be found in [21]. So far, none of these EAs attempted to determine long protein structures (greater than 150 amino acids) using their proposed methods.

Multi-modal PSP problem is also a constrained combinatorial problem as it needs to find a sequence of moves so that the structure remains as a connected chain and *self-avoiding-walk* (SAW) on a lattice [21]. This means that conventional crossover or mutation operator may not necessarily be successful in all the cases as they do not guarantee SAW. Moreover, the situation worsens when dealing with long protein sequences [11]. Again, success of crossover depends on the parents' conformations and it

B.-L. Lu, L. Zhang, and J. Kwok (Eds.): ICONIP 2011, Part I, LNCS 7062, pp. 636–645, 2011.

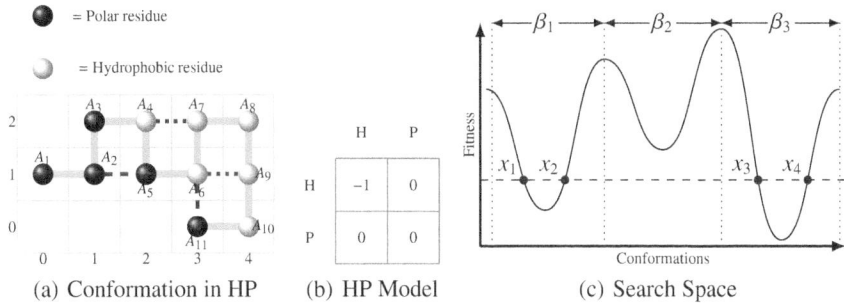

(a) Conformation in HP   (b) HP Model   (c) Search Space

Fig. 1. (a) One possible conformation in HP model protein for the sequence *PPPHPHHHHHP*. Topological contact for HH is shown in dotted line. Total energy for the conformation is -2 in HP model because there are two HH contacts. (b) Table showing contact energy for HP model. (c) a typical search space with three BOAs, β_1, β_2, and β_3, and four different individuals are pointed as x_1, x_2, x_3, and x_4. Individuals x_1 and x_2 are in basin β_1 whereas x_3 and x_4 are in basin β_3.

often fails due to *reproduction error* [16] (when parents come from two different basins of attraction (BOA)) and *twins* [8] (same child as parent). Conventional mutation will also have problem to form valid conformations due to the SAW constraint.

To overcome the problems of conventional crossover and mutation, in this paper we propose a novel generic crossover technique (*crossover-by-conflict-resolution*) and a mutation technique (*mutation-by-conflict-resolution*) for the constrained combinatorial problem. The proposed crossover operator can produce valid child conformation irrespective of the parents' structures with equivalent computational cost as needed for a conventional operation. To the best of our knowledge, no previous attempts to solve longer sequences on lattice models using population based EAs are reported. We have especially focused on longer protein sequences (more than 200 amino acids studied in [20]). The rest of the paper is organized as follows: Section 2 gives a little background of related topics; Section 3 explains proposed method; empirical results are presented in Section 4 and finally the paper concludes with a short summary.

2   Backgrounds

This section provides a brief background on the on-lattice PSP problem and global search operators.

2.1   PSP Using Lattice Models

A lattice-based PSP represents conformations of proteins as non-overlapping en-grafting of the amino-acid sequence on a lattice. One of the most studied lattice models is the HP model [4] where a protein's primary structure is simplified as a linear chain of amino acids identified as either H (hydrophobic) or P (polar); $\sigma = \{H, P\}^+$. In HP model, when two H residues are topological neighbour (TN) — not neighbour in the chain but neighbour in the lattice — lowers the total energy by one (see Fig. 1(b)). Let $\xi(A_i, A_j)$ be the contact energy in the HP model defined as

$$\xi(A_i, A_j) = \begin{cases} -1, & \text{if } A_i \text{ and } A_j \text{ are TNs and both are H residues;} \\ 0, & \text{otherwise.} \end{cases} \tag{1}$$

PSP using lattice model is an optimization problem, which searches for a conformation in the given lattice (Fig. 1(a)) such that fitness of the conformation, in terms of total contact energy, is minimized [17]. This optimization problem is now finally defined in the following equation:

$$\left. \begin{array}{l} \textbf{Minimize } \sum_{i=1}^{L-1} \sum_{j=i+1}^{L} \xi(A_i, A_j) \\ \text{given} \quad (A_1, A_2, \dots, A_L) \text{ is a conformation on a lattice.} \end{array} \right\} \tag{2}$$

1: **procedure** GENERATECHILDREN $(P(t) = \{x_1^t, \cdots, x_{\mathbb{P}}^t\})$ ▷ $P(t)$ is the parent population set at iteration t

---
2:     $C(t) \leftarrow \emptyset$
3:     **while** Termination criteria $\mathfrak{T}$ is NOT satisfied **do**
4:         **while** True **do**
5:             Select x_i & x_j randomly from $P(t)$ where $1 \le i, j \le \mathbb{P}$ and $i \ne j$
6:             Initialize c ▷ c will be the child generated from parent x_i & x_j
7:             $crossPoint \leftarrow$ RAND$(1,\mathbb{L})$
8:             $ret \leftarrow$ DOCROSSOVER$(x_i, x_j$, ref c, $crossPoint)$
9:             **if** ret =True AND $c \notin P(t)$ **then**
10:                 $C(t) \leftarrow C(t) \cup c$
11:                 break

Fig. 2. A generic framework for children generation using crossover operator

2.2   Global Search Operators

Crossover and *mutation* are mainly used to modify individuals where *selection* plays vital role of controlling diversity in the population [14]. Here we mainly focus on the effective modification of individuals in terms of information sharing; besides both operators have to maintain the constraint of SAW and generate output in reasonably short time. EAs e.g., genetic algorithm (GA) and memetic algorithm (MA), maintain a population to increase performance by sharing *information* directly between individuals. This communication is achieved mostly by crossover operator [13]. Design purpose of crossover operator is to communicate and construct individuals with building blocks of the particular BOA where the solution converges; whereas the idea of mutation is to disrupt an individual [7]. As the solution converges to local minima of a BOA, mutation disrupts building blocks to divert the solution to another BOA. So, adaptive control of the mutation rate is the key to control convergence [19].

3   Proposed Method

PSP being a constraint combinatorial problem a conventional crossover operator (e.g., 1-point crossover, n-point crossover) or mutation operator cannot guarantee satisfying combinational constraint namely SAW. We propose new crossover and mutation techniques which will not only serve routine purpose of crossover and mutation but will also satisfy the combinatorial constraint. The proposed crossover and mutation techniques are termed as *crossover-by-conflict-resolution* (CCR) and *mutation-by-conflict-resolution* (MCR) respectively.

1: **procedure** CROSSOVERBYCONFLICTRESOLUTION $(P_1 = (a_1, a_2, \cdots, a_{\mathbb{L}}), P_2 = (b_1, b_2, \cdots, b_{\mathbb{L}}),$ ref C, p,
   $maxAttempts)$ ▷ P_1 & P_2 are parents, C = child, p = crossover point, $maxAttempts$ = maximum
   allowable attempts

2:    **if** $attempts = 0$ **then** ▷ $attempts$ is a global variable initialized to 0
3:       **for** $i = 1$ to p **do**
4:          $C_i \leftarrow a_i$ ▷ ith element of child C is C_i
5:    **else if** $attempts = maxAttempts$ **then**
6:       **return** False
7:    **if** $\mathsf{p} = \mathbb{L} + 1$ **then** ▷ $\mathbb{L}$ is the length of the and individual
8:       **return** True
9:    $\mathsf{e} \leftarrow \emptyset$ ▷ Here e is the set of not allowed moves
10:   $\mathsf{f} \leftarrow$ GETPOSSIBLE$(C_{\mathsf{p}-1})$ ▷ GETPOSSIBLE$(C_{\mathsf{p}-1})$ returns a set of all allowable
    move set $\mathbb{C}$ based on $C_{\mathsf{p}-1}$
11:   **repeat**
12:      $attempts \leftarrow attempts + 1$
13:
$$C_{\mathsf{p}} \leftarrow \begin{cases} select(\{a_{\mathsf{p}}, b_{\mathsf{p}}\} - \mathsf{e}) & \text{if } (\{a_{\mathsf{p}}, b_{\mathsf{p}}\} - \mathsf{e}) = \emptyset \text{ and } rand \leq \alpha \\ select(\mathsf{f} - \mathsf{e}) & \text{otherwise} \end{cases}$$
      ▷ $select(A)$ returns any element of set A with equal likelihood. $rand$ returns a uniformly distributed real
      value from the range $[0, 1)$.
14:      **if** SATISFYCONSTRAINT$((C_1, \cdots, C_{\mathsf{p}})) = $ True **then**
15:         $C \leftarrow (C_1, \cdots, C_{\mathsf{p}})$
16:         **if** CROSSOVERBYCONFLICTRESOLUTION$(P_1, P_2,$ ref C, $\mathsf{p} + 1) = $ True **then**
17:            **return** True
18:      $\mathsf{e} \leftarrow \mathsf{e} \cup C_{\mathsf{p}}$
19:      **if** $\mathsf{f} = \mathsf{e}$ **then**
20:         **return** False
21:   **until** True

Fig. 3. Proposed crossover technique, *crossover-by-conflict-resolution*, for PSP

3.1 Conventional Crossover

Conventional crossover (CC) generates a set of children, $C(t) = \{c_1^t, \cdots, c_{|C|}^t\}$, from the parent population set, $P(t)$, then merges the two sets (*selection*) into one population of size $\mathbb{P}$ based on the best fitness value on a fast convergence architecture [14]. CC operator works by selecting two parents $P_1 = (a_1, a_2, \cdots, a_{\mathbb{L}})$ and $P_2 = (b_1, b_2, \cdots, b_{\mathbb{L}})$ using any random process (e.g., pure random selection, roulette-wheel selection [14]) from $P(t)$. Parents are randomly split at a crossover point p and first part of the first parent is joined with the second part of the second parent to make first child $C_1 = (a_1, a_2, \cdots, a_{\mathsf{p}}, b_{\mathsf{p}+1}, \cdots, b_{\mathbb{L}})$ and the second child is build from the first part of second parent and the second part of the first parent, $C_2 = (b_1, b_2, \cdots, b_{\mathsf{p}}, a_{\mathsf{p}+1}, \cdots, a_{\mathbb{L}})$. Then, based on the fitness comparison of these two children, best one is considered as the new child. It is necessary that the child generated in this way is verified for SAW. So it might happen that, none of the children is a SAW. In algorithm DOCROSSOVER shown in Fig. 2 performs this verification by returning true or false (line 8).

Analysis As the search begins, the individuals are not in their compact structures but as the solution converges, they become compact because the conformation tries to form a *hydrophobic core* [8]. So, combining two individuals to produce a new child often fails for the following two reasons:

1. *Parents from different BOA*: If we chose parents, say $x_1 \in \beta_1$ and $x_3 \in \beta_3$ of Fig. 1(c), where subsections of those individuals have less or no similarity, crossover will fail to produce either a SAW or a fit individual when parts of them join with

each other. This type of error is already identified in literature as *reproduction error* [16].

2. *Parents from same BOA*: If we chose individuals e.g., $x_1 \in \beta_1$ and $x_2 \in \beta_1$ of Fig. 1(c), they, by definition of convergence in genetic like process, will have some common parts in the individuals' chromosome and they will remain unchanged for rest of the convergence [7,3]. An individual x_i can be considered as a sequence of moves $x_i = (a_1, \cdots, a_{\mathbb{L}_c})$. If we consider there are n_m number of common parts, $\varepsilon_1, \cdots \varepsilon_{n_m}$, and each has a start and end position, $\varepsilon_i : (a_{s_{\varepsilon_i}}, \cdots, a_{e_{\varepsilon_i}})$, to identify their location in a chromosome, where $\forall_i (e_{\varepsilon_{i-1}} < s_{\varepsilon_i} < e_{\varepsilon_i} < s_{\varepsilon_{i+1}})$. CC operator will not change these segments, but will increase the number of these stable parts when the two selected parents will have complementary section of part ε_i at the crossover point. This implies that the number of different individuals possible from the outcome of crossover will reduce subsequently, thereby increasing the number of same child generation as the solution converges.

3.2   Conflict Resolution Based Crossover

The purpose of crossover is to share information of the two selected parents in the offspring. There are a number of crossover techniques available in literature including $1-$point crossover [7] to $n-$point crossover and a parametrized (P_0) uniform crossover [3, 19] which end up with either *reproduction error* or no way to explore new individuals of the same BOA. To overcome these limitations, we propose a generic crossover technique (see Fig. 3)) by resolving conflict occurring due to a constraint.

Here, the procedure CROSSOVERBYCONFLICTRESOLUTION (see Fig. 3) is a recursive procedure which checks all possible allowable moves at a particular point with a higher priority (α) to the moves that belong to either of the parents (line 13). If any move violates the constraint, it is added to the "not-allowed" list $\mathfrak{e}$ (line 18). As long as the constraint is satisfied the algorithm moves forward to select the next move recursively (line 16) and if $\mathfrak{e} = \mathfrak{f}$, obviously it means that the previous move was an incorrect move so it goes back to previous call by returning False (line 20).

Analysis. The proposed CCR algorithm will always return in a finite time with either True or False. When a successful crossover happens ($\mathfrak{p} = \mathbb{L} + 1$), it generates a child satisfying the constraint. Child generated by proposed algorithm sometimes do contains genes (moves) which do not belong to any of its parents (*conflict points*) but the child will still carry most of the information from its parents which is the main goal of crossover [13]. This actually helps to overcome the most common problems of conventional crossover:

1. *Parents from different BOAs*: If the parents are from different BOAs (e.g., x_1 & x_4 of Fig. 1(c)), this process will reduce reproduction error for the conflict resolution process.

2. *Parents from same BOA*: If the parents are from the same BOA (e.g., x_1 & x_2 of Fig. 1(c)), it will increase the possibility to generate new individual of that BOA. As explained earlier, in CC, child of same BOA will have the common parts, $\{\varepsilon_1, \cdots \varepsilon_{n_m}\}$, and the rest of the genes will be chosen from either of the parents in conventional crossover. If the total length of these fixed parts is $\mathfrak{m} = \sum_i^m |\varepsilon_i|$ and

the total length of the chromosome (conformation) is $\mathbb{L}$, then number of possible different children of that BOA is $\leq 2 \times (\mathbb{L} - m)$ in CC. On the other hand the proposed CCR technique opens up the scope to choose children from $|f|^{\mathbb{L}}$ where f is the set of all possible moves at a particular position (though parents' moves have higher priority). So, the proposed technique will have longer exploitation range in the search space.

Given a protein of length $\mathbb{L}$, 1-point CC needs on an average $O(\mathbb{L}/4)$ time to verify a SAW (only needs to verify SAW for shorter segment) whereas the proposed CCR method has a complexity of at least $O(\mathbb{L}/2)$ (considering all first moves are valid) and at max $(2\mathbb{L} + |\mathbb{C}|)$ as maximum attempts are set to $2\mathbb{L}$. If we consider an exhaustive CC where all possible crossover points $(\mathbb{L}-1)$ will be taken serially, the total complexity will be $(\mathbb{L} - 1) \times O(\mathbb{L}/4) = O(\mathbb{L}^2/4)$. To make it computationally equal, proposed technique can have at least approximately $\mathbb{L}/8$ runs ($\frac{\mathbb{L}^2/4}{2\mathbb{L}+|\mathbb{C}|} \approx \mathbb{L}/8$). Thus, a single run of CCR is equivalent to 8 runs of 1-point crossover. In general, it can be shown that a single run of CCR is equivalent to $4(n + 1)/n$ runs of n-point crossover. In other words, this means that less number of runs can be applied to CC with the increase of crossover points to make it equivalent to CCR.

3.3   Conflict Resolution Based Mutation

In PSP problem when the solution converges, the structures become more compact as hydrophobic amino acids attempt to form a hydrophobic core. But as the individuals become compact, it gets difficult to mutate them due to SAW constraint. Hence, instead of regular mutation, we propose a *mutation-by-conflict-resolution* (MCR). The algorithm is similar to the crossover one (see Fig. 3) where only one parent is present and priority will be given to parent genes except at the point of mutation where a gene will be chosen other than the parent's gene. In this way, the generation of individual will not be equivalent to a conventional single point mutation; rather it will be similar to macromutation where a number of genes will be mutated to satisfy the constraint (e.g., SAW).

Based on type of encoding, mutation plays different role in PSP. Two main stream of encodings used for PSP problem in lattice models are absolute and relative encoding [11, 12] though there exists a better encoding in terms of reduction of unwanted variation, non-isomorphic encoding [9]. For absolute encoding mutation just change direction at the point of mutation whereas in relative encoding mutation rotates the structure from the point of mutation (shown by example in [8]). This means that mutation in relative encoding will have higher mobility than mutation in absolute encoding. For this, mutation in relative encoding will be more difficult than mutation in absolute encoding.

4   Results

To test the performance of proposed techniques, a simple memetic algorithm (SMA) architecture ($P(t) \rightarrow$ LS (pull move, mutation based) $\rightarrow P'(t) \rightarrow$ GS (crossover($P'(t)$)) $\rightarrow C(t) \rightarrow$ selection($P'(t), C(t)) \rightarrow P(t + 1)$) has been used. Based on the experimental

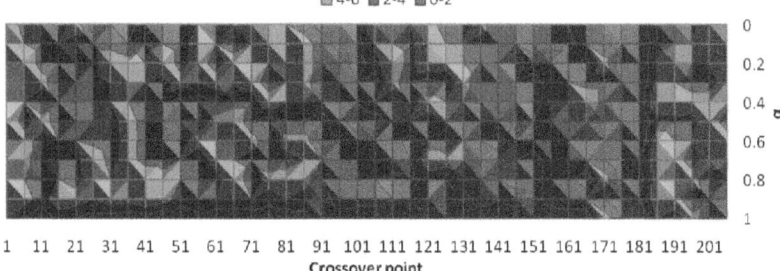

Fig. 4. Sensitivity test for α

result, a reasonable combination of the parameter values are applied on our previously proposed more robust EA framework, *clustered memetic algorithm* (CMA) [11, 12].

4.1 Crossover

To choose a suitable value for α which determines the amount of priority that will be given to parents' moves, we have run a sensitivity test on α — run crossover for each of the possible values of α, on each of the possible crossover points; and record how many new children can be successfully produced from a given number of parent sets (6 in our case). A population size of 4 is maintained in SMA. In each iteration t, we run CCR on 6 possible parent pairs with $\alpha = (0, 0.1, \cdots, 1)$ and for each value of α CCR has been done at $(1, 6, 11, \cdots, \lfloor L/5 \rfloor 5)$ points and number of *successes* are recorded. A *success* is defined as a new valid conformation that does not exist in the existing population. A surface plot (see Fig. 4) has been taken for the number of average *success* for iterations 6 to 15 found by applying CCR on *PDB 1BQS* [20] on a 2D square lattice. Fig. 4 shows that if α is too high (100% – 90%) crossover fails most of the time and for rest of the values of α, no such influence on success. Another important finding came out from the experiment — *success* rate is higher when crossover point lies in the first half of the sequence or at the end of the sequence.

Complexity of CC and CCR$_{\alpha=0.5}$ are equalled by running 1-point CC sequentially for all point n and CCR$_{\alpha=0.5}$ for $n/8$ times. Whenever a valid child is found using any of the methods, it is locally optimized using steepest ascent local search with pull moves. As a performance measure of the CCR$_{\alpha=0.5}$ and 1-point CC, the number of successful child generated from same set of parents is calculated. For experiment, following parameters are considered: population size = 4, number of crossover per iteration = 6, maximum iteration = 100, $\alpha = 0.5$ in a SMA. Simulation terminates when either both of the crossovers fails to produce a *success* on a particular iteration, or iteration count reaches to maximum iteration. In the experiment we have used two long sequences (*PDB 1B0F* (218 length) and *PDB 1BQS* (209 length)) from [20] on a 2D square lattice. The lattice is chosen because it has the lowest degree of freedom (4 possible directions only) which ensures that in each BOA, there will be the lowest number of possible individuals among all other lattices. The number of *success* is also a measure of the capability of exploitation of the basin. Results are shown in Fig. 5. Average number of success for CC is 0.25 and for CCR is 3.06 in Fig. 5(a) whereas 0.11 and 2.4 respectively for CC

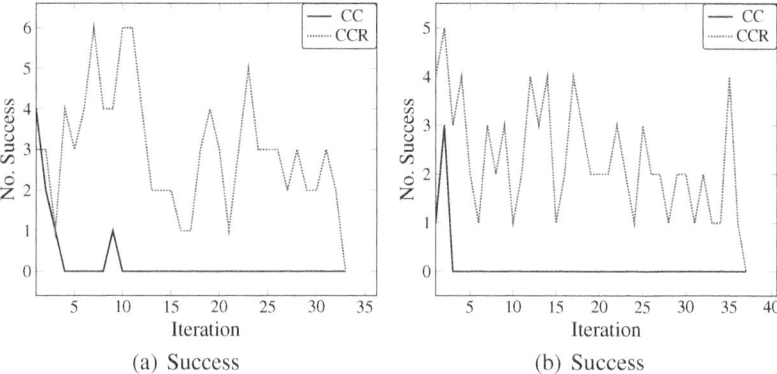

(a) Success   (b) Success

Fig. 5. Crossover features analysis of conventional (CC) and proposed (CCR) techniques in terms of *success*. Success for PDB 1B0F is shown in (a) and for PDB 1BQS in (b).

and CCR in Fig. 5(b), out of the 6 tries. So, the expected success from CCR is almost 45.5% (3 success out of 6).

4.2 Mutation

To closely observe the relationships and differences between mutations using absolute encoding and relative encoding, we have applied 50 mutations at each iteration on same individual in a SMA architecture where population size is 2. We have tested the performance for the long protein sequence *PDB 1B0F* on a 2D square lattice model between traditional mutation and proposed MCR. From Fig. 6(a), it is clear that mutation using relative encoding takes longer time than mutation using absolute encoding which actually supports our theoretical analysis (see Section 3.3). It also suggests that MCR takes relatively less time than conventional mutation. Fig. 6(b) suggests that different mutations do not show any significant impact on fitness value. The time shown here in Fig. 6 is the total time where fitness is the average of the 50 mutations taken at each iteration.

4.3 CMA with Long Proteins

Finally, an experiment has been carried out with our proposed CMA [11, 12] where divide and conquer based niching technique has been introduced using a number of preliminary clusters and a main cluster. Domain knowledge of a BOA, where each preliminary cluster converges due to genetic drift, in the form of substructures (*memes*) are passed to main cluster. Main cluster utilizes this domain knowledge to exploit the BOAs and effectively uses the substructures. In main cluster, the CCR is used in place of CC; and in preliminary clusters, CCR is applied only when CC is failed. MCR is used throughout the simulation in place of conventional mutation. Results are shown for long protein sequences on different lattices in Table 1. Note that we have not concentrated on the extensive simulation here instead focus was on incorporating our proposed methods with CMA. For that, only five long HP sequences found in [20] are tested for all four

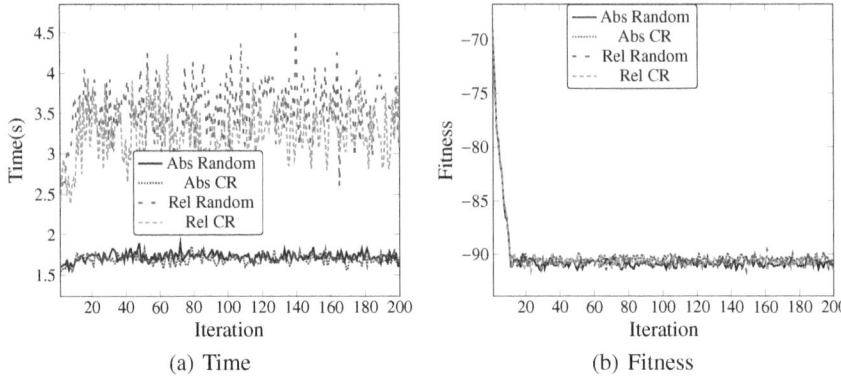

(a) Time   (b) Fitness

Fig. 6. Effect of different mutation techniques on an individual for PDB 1B0F is shown here. (a) and (b) show time and fitness analysis for the individual.

Table 1. Performance analysis for CMA on different lattices. E_{best} is the best result and $E_{avg}(\mu \pm \sigma)$ is the average and standard deviation taken from 5 runs.

		2D Square		2D Triangular		3D Cubic		FCC	
PDB ID	Len	E_{best}	$E_{avg}(\mu \pm \sigma)$	E_{best}	$E_{avg}(\mu \pm \sigma)$	E_{best}	$E_{s-1}(\mu \pm \sigma)$	E_{best}	$E_{avg}(\mu \pm \sigma)$
1B0F	218	-95	-90.6± 3.05	-187	-183.4 ±2.70	-162	-153.4± 6.80	-385	-382.4 ±2.07
1BQS	209	-86	-78.2± 4.38	-165	-158.8± 3.63	-135	-131.6± 2.19	-334	-329.2± 4.87
1CWR	211	-56	-54± 1.41	-106	-103.6± 1.82	-94	-89.2± 3.11	-224	-222.4± 1.67
1NQC	217	-64	-61.2± 1.64	-136	-125.6± 6.15	-112	-107.8 ±3.35	-274	-265.8± 6.30
1RTG	210	-73	-70.2± 1.92	-145	-142.6± 1.82	-128	-122.4 ±4.04	-359	-312.8± 25.96

prominent lattice models [12]. All the result shown in Table 1 based on 5 simulation runs only and have the following CMA parameters — population size = 4, number of preliminary cluster = 5, maximum main cluster iteration = 100. Results with extensive simulation will be shown in our future work.

5   Conclusion

Protein structure prediction on lattice models is both multi-modal and constraint combinatorial problem. By using different niching techniques population based EAs handle multi-modality of a problem efficiently but no effective method is present to handle constraint. The conventional crossover operator which is considered as an effective way to transfer information will not work for the constraint problem like, PSP. This paper points out the reasons of this failure and proposed an effective way to handle this. Theory and experiment suggest that for the first few iterations CC shows good results but as the structures get into their compact state it fails whereas proposed one brings *success*. Though proposed mutation does not have any significant effect still it takes relatively less time.

References

1. Anfinsen, C.B.: Principles that govern the folding of protein chains. Science 181, 223–230 (1973)
2. Cutello, V., et al.: An immune algorithm for protein structure prediction on lattice models. IEEE Trans. Evol. Comput. 11(1), 101–117 (2007)
3. De Jong, K.A.: Analysis of the behavior of a class of genetic adaptive systems. Ph.D. thesis (1975)
4. Dill, K.A.: Theory for the folding and stability of globular proteins. Biochemistry 24, 1501–1509 (1985)
5. Dotu, I., et al.: On lattice protein structure prediction revisited. IEEE/ACM Trans. on Comput. Biology and Bioinformatics 99 (2011)
6. Helles, G.: A comparative study of the reported performance of ab initio protein structure prediction algorithms. J. Roy. Soci. Inter. 5(21), 387–396 (2008)
7. Holland, J.H.: Adaptation in natural and artificial systems. The University of Michigan Press (1975)
8. Hoque, M. T., Chetty, M., Sattar, A.: Genetic Algorithm in*Ab Initio* Protein Structure Prediction Using Low Resolution Model: A Review. In: Sidhu, A.S., Dillon, T.S. (eds.) Biomedical Data and Applications. Studies in Computational Intelligence, vol. 224, pp. 317–342. Springer, Heidelberg (2009)
9. Hoque, M., et al.: Non-Isomorphic Coding in Lattice Model and its Impact for Protein Folding Prediction Using Genetic Algorithm. In: IEEE CIBCB 2006, pp. 1–8 (2006)
10. Islam, M. K., Chetty, M.: Novel Memetic Algorithm for Protein Structure Prediction. In: Nicholson, A., Li, X. (eds.) AI 2009. LNCS, vol. 5866, pp. 412–421. Springer, Heidelberg (2009)
11. Islam, M.K., et al.: Clustered memetic algorithm for protein structure prediction. In: IEEE Con. on Evo. Compu. 2010, pp. 1–8 (2010)
12. Islam, M.K., et al.: Novel local improvement techniques in clustered memetic algorithm for protein structure prediction. In: IEEE Cong. on Evo. Compu. 2011, pp. 1003–1011 (2011)
13. Jones, T.: Crossover, macromutationand, and population-based search. In: Proc. of the 6th Int. Conf. on Genetic Algorithms, pp. 73–80 (1995)
14. Larraañaga, P., et al.: Estimation of Distribution Algorithms: A New Tool for Evolutionary Computation. Kluwer Academic Publishers, USA (2001)
15. Lesh, N., et al.: A complete and effective move set for simplified protein folding. In: Proc. of the 7th Ann. Inter. Conf. on Res. in Comp. Mol. Bio., pp. 188–195. ACM, NY (2003)
16. Mahfoud, S.W.: Niching methods for genetic algorithms. Ph.D. thesis, University of Illinois at Urbana-Champaign, Champaign, IL, USA (1995)
17. Santana, R., et al.: A markov chain analysis on simple genetic algorithms. IEEE Trans. Evol. Comput. 12(4), 418–438 (2008)
18. Shmygelska, A., et al.: An ant colony optimisation algorithm for the 2d and 3d hydrophobic polar protein folding problem. BMC Bioinfo. 6(30) (2005)
19. Spears, W.M., et al.: Crossover or mutation? In: Found. of Gen. Algo., vol. 2, pp. 221–237. Morgan Kaufmann (1992)
20. Thachuk, C., et al.: A replica exchange monte carlo algorithm for protein folding in the hp model. BMC Bioinfo. 8(342) (2007)
21. Zhao, X.: Advances on protein folding simulations based on the lattice hp models with natural computing. App. Soft Comp. 8(2), 1029–1040 (2008)

Personalised Modelling on SNPs Data for Crohn's Disease Prediction

Yingjie Hu and Nikola Kasabov

the Knowledge Engineering and Discovery Research Institute,
Auckland University of Technology, New Zealand
{raphael.hu,nikola.kasabov}@aut.ac.nz

Abstract. This paper presents a study for investigating the feasibility of applying personalised modelling on single nucleotide polymorphisms (SNPs) data for disease analysis. We have applied our newly developed integrated method for personalised modelling (IMPM) on a real-world biomedical classification problem, which makes use of the SNPs data for crohn's disease prediction. IMPM method allows for adaption and monitoring an individual's model and outperforms global modelling methods for the SNPs data classification. Personalised modelling method produces a unique personalised profiling for an individual, which holds the promise of a new generation of analytical tools that can be used for personalised treatment.

Keywords: Integrated method for personalised modelling, IMPM, SNPs, evolutionary computation.

1 Introduction

Being able to accurately predict an individual's disease risk or drug response and using such information to personalised treatment is a major goal of clinical medicine in the 21st century. With the advancement of microarray technologies, collecting personalised genetic data on a genome-wide (or genomic) scale has become quicker and cheaper [1]. Such personalised genomic data may include: DNA sequence data (e.g. Single Nucleotide Polymorphisms (SNPs), gene sequence, protein expression data, etc. Many world-wide projects have already collected and published a vast amount of such personalised data. For example, Genome-wide Association Scan (GWAS) projects have so far published for over 100 human traits and diseases and many have made data available for thousands of people.

However, conventional approaches in medical statistics and computing (or bioinformatics) are not designed to fully utilise the available data banks and incorporate genetic, clinical, environmental, nutritional data to accurately predict the clinical outcome for an individual patient and use this information in clinical practice. We have recently developed an integrated optimisation method for personalised modelling (IMPM) [2] at dealing with such tasks. The dataset available in UK WTCCC data bank (http://www.wtccc.org.uk) is used in this

B.-L. Lu, L. Zhang, and J. Kwok (Eds.): ICONIP 2011, Part I, LNCS 7062, pp. 646–653, 2011.
© Springer-Verlag Berlin Heidelberg 2011

study, which includes multivariate personalised data of DNA SNPs and clinical variables. If this case study is successful, this approach will be used for the development of a prognostic system to accurately predict clinical outcomes and appropriate treatment of Crohn's disease (CD) patients in New Zealand and will be further applied for other diseases.

2   Background and Related Work

2.1   Crohn's Disease

Crohn's disease (CD) is a chronic and debilitating autoimmune disorder of the gastrointestinal tract. It is a major subtype of inflammatory bowel disease (IBD) which is diagnosed endoscopically and characterized by recurring episodes of abdominal pain, diarrhoea and weight loss. As a consequence of ongoing inflammatory "flares", a large number of CD patients will develop strictures and fistulae during the course of disease which can seriously impact the quality of life and often requires surgery [3]. The incidence of CD is increasing dramatically in industrialised countries worldwide, including New Zealand [4,5].

Unfortunately, there is currently no completely effective clinical strategy for treating crohn's disease. Current treatment paradigms used in the clinic are the so-called "step-up" and "top-down" approaches. Whether or not a patient should be given step-up or top-down treatment for IBD is a controversial topic in clinical gastroenterology. The main issue is that it is difficult to accurately predict which of the two approaches will provide the favorable outcome for an individual patient. The inheritance risk probability of Crohn disease is unclear, because a variety of genetic and environmental factors are reported to be involved in literature. Therefore, using accurate predictive tools to identify high-risk patients and give personalised treatment is a major goal for clinicians in CD research.

2.2   SNPs Data Analysis

SNPs genotypes are of great importance for understanding of the human genome, and are the most common genetic variations between human beings. On average, SNPs occur in nucleotides at the rate of $3 \sim 5\%$, which means approximately 10 million SNPs occur in human genome. SNPs are found in the DNA among genes, and most of them have no effect on human health or disease development. However, when SNPs occur within a gene or in a regulatory region near a gene, they may have a direct impact on disease development through affecting genes function. Therefore, some SNPs act as biomarkers that allow scientists to locate the genes associated with disease.

At present, there is no effective way to measure how a patient will respond to a particular drug treatment. In many cases, a treatment can be effective for a group of patients, but is not effective for others at all. The knowledge discovered from SNPs study can help researchers build clinical decision support systems to predict an individual's response to certain drugs and environmental factors (e.g. toxins) and the risk of particular disease development. Also, SNPs offer a new way to track the inheritance of disease genes within societies, especially for studying complex diseases, such as CD, cancer and diabetes.

3   The IMPM Method

We applied our newly developed method (IMPM) [2] for personalised modelling and its implementation for CD risk evaluation using SNPs data in this study. IMPM creates a personalised model for a new input vector and then optimises the aspects of the personalised model (features, neighbouring samples and model parameters) in terms of the training performance achieved from the local neighbourhood of the sample. Next, a personalised model and personalised profile are derived using the selected features and the neighbouring samples with known outcomes. The new sample's profile is compared with average profiles of the other outcome classes in the neighbourhood (e.g. good outcome, or bad outcome of disease or treatment). The difference between the new sample's and average profile's important features may need to be modified through treatment if it is possible. Figure 1 illustrates a functional block diagram of IMPM.

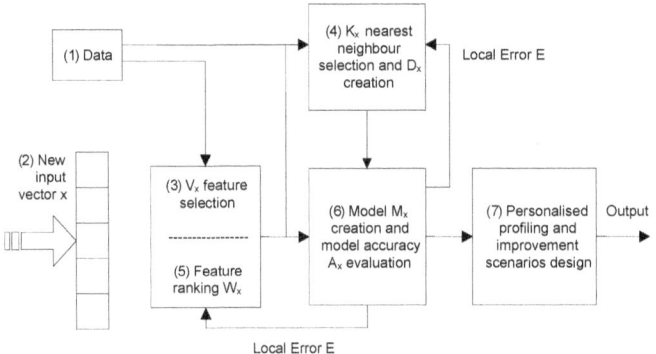

Fig. 1. A functional block diagram of IMPM

Algorithm 1 briefly summarises IMPM method. Procedures 3-8 are repeated a number of iterations or until a desired local accuracy of the model for a local data set D_x is achieved. The optimised parameters of the personalised model V_x, K_x and D_x is global and can be achieved through multiple runs using an evolutionary algorithm (EA).

Initially, the assumption is made that all q variables from a set V have equal absolute and relative importance for a new sample x in relation to predicting its unknown output y. The numbers initially for V_x and K_x may be determined in a variety of different ways without departing from the scope of the method. For example V_x and K_x may be initially determined by an assessment of the global

Algorithm 1. IMPM algorithm

1: Data collection, data filtering, storage and update;
2: Compiling the input vector for a new patient x;
3: Select a subset of relevant variables (features) V_x to the new sample x from a global variable set V;
4: $D_x = f_{sel}(K_x, V_x) \mid K_x \in D, V_x \in V$;
   // Select K_x samples from the global data set D and form a neighbourhood D_x of similar samples to x using the variables V_x;
5: $W_x = f_{rnk}(D_x, V_x)$
   // f_{rnk} is a ranking function for ranking V_x variables within D_x in order of importance to the outcome, and W_x is a weight vector;
6: $M_x = F(P_x, V_x, D_x)$
   P_x is a set of model parameters, F is a function for training and optimising and M_x is a candidate model;
7: Generating a functional profile $\mathfrak{F}(x)$ for the person x using the selected variable set from V_x, along with the average profiles of the samples from D_x belonging to different outcome classes, e.g. $\mathfrak{F}_i$ and $\mathfrak{F}_j$;
8: Performing a comparative analysis between $\mathfrak{F}(x)$, $\mathfrak{F}_i$ and $\mathfrak{F}_j$ to identify the variables $V_x^* \mid V_x^* \in V_x$ are the most important for the person x if a treatment is needed.

dataset in terms of size and distribution of the data. Minimum and maximum values of these parameters may also be established based on the available data and the problem analysis. A classification or prediction module is applied to the neighbourhood D_x of K_x data samples to create a personalised candidate model M_x using the already defined variables V_x, variable weights W_x and a model parameter set P_x. Principally, any types of classification or prediction algorithms can be used, such as: KNN; WKNN; TWNFI [6], etc. A local accuracy (local error E_x), that estimates the local accuracy of the personalised prognosis (classification) for D_x using the obtained model M_x. This error is a local one, calculated in the neighbourhood D_x, rather than a global accuracy, that is commonly calculated for the whole problem space D. The local error can be used for model optimisation and is calculated as follows:

$$E_x = \frac{\sum_{j=1}^{K_x} (1 - d_{xj}) \cdot E_j}{K_x} \qquad (1)$$

where: d_{xj} is the weighted Euclidean distance between the new testing sample x and training sample s_j from D_x, which takes into account the variable weights W_x; E_j is the error between the outcome predicted by M_x for sample s_j and its real output.

The model obtaining best accuracy is stored for the purpose of a future improvement and optimisation. The optimisation procedure iteratively returns to all previous procedures to select another set of parameter values for the parameter vector P_x until the model M_x with the best accuracy is achieved. The method also optimises parameters P_x of the classification/prediction procedure.

Once the optimal model M_x^* is derived, an output value y for the new input vector x is calculated using M_x^*. After the output value y for the new input vector x is calculated, a personalised profile $\mathfrak{F}_x$ of input vector x (a patient) will be derived and assessed against possible desired outcomes for the scenario. The ability offering the unique profiling is a major novelty of this personalised modelling method. The profile $\mathfrak{F}_x$ of new sample x can be formed as a vector:

$$\mathfrak{F}_x = \{V_x, W_x, K_x, D_x, M_x, P_x, t\} \tag{2}$$

where t represents the time of the model M_x creation. At a future time $(t + \Delta t)$ the person's input data will change to x^* (due to changes in variables such as age, weight, protein expression values, etc.), or the data samples in the data set D may be updated and new data samples are added. A new profile $\mathfrak{F}_x'$ derived at time $(t + \Delta t)$ may be different from the current $\mathfrak{F}_x$.

The average profile $\mathfrak{F}_i$ for every class C_i in the data D_x is a vector containing the average values of each variable of all samples in D_x from class C_i. The importance of each variable (feature) is indicated by its weighting in the weight vector W_x. The weighted distance from the person's profile $\mathfrak{F}_x$ to the average class profile $\mathfrak{F}_i$ (for each class i) is defined as:

$$\mathfrak{D}(\mathfrak{F}_x, \mathfrak{F}_i) = \sum_{l=1}^{v} |V_{lx} - V_{li}| \cdot w_l \tag{3}$$

where w_l is the weight of the variable V_l calculated for the data set D_x.

In order to find a smaller number of variables, as global markers that can be applied to the whole population X, IMPM repeats step 2 - step 7 for every individual x. All variables from the derived sets V_x are then ranked based on their likelihood to be selected for all samples. The top m variables (most frequently selected for individual models) are taken as a set of global set of markers V_m. The steps 1-8 will be applied again with the use of V_m as initial variable set (instead of using the whole initial set V of variables). In this case personalised models and profiles are obtained within a set of variable markers V_m that would make treatment and drug design more universal across the whole population X.

4   Experiment Setup

4.1   Data

The raw SNPs data used for Crohn's disease (CD) prediction is accessible from a UK's public data bank - Wellcome Trust Case Control Consortium (WTCCC). The raw SNPs data is originally used in genome-wide association (GWA) studies of 14,000 cases of 7 major diseases and a shared set of 3,000 controls [7]. Our work is a feasibility investigation of personalised modelling on SNPs data for CD prediction, in which a pre-processed subset from raw SNPs data is used. This subset is suggested to be evaluated for computational modelling by the experts and data provider. The SNPs data used in this experiment contains 106 samples (49 controlled samples vs. 57 diseased samples). Each sample is represented by 44 features (42 SNPs plus 2 clinical factors).

5   Experiment Result and Discussion

In this case study, all the experiments are carried out on a PC with Matlab environment. The same SVM algorithm is used for classification in this case study for a fair comparison. The SVM classification model is derived from the widely used LibSVM package [8].

5.1   IMPM vs. Global SVM

The experiment consists of two modelling techniques for SNPs data analysis: (1) global SVM modelling; (2) personalised modelling (IMPM). The classification accuracy obtained by global SVM modelling on the SNPs data is 70% (class 1: 63%, class 2: 75%). The optimised parameters for global SVM model are: $c=200$, $\gamma = 0.01$. IMPM, as the personalised modelling method produces a better classification accuracy than the global SVM model, which achieves 73% classification accuracy (class 1: 76%, class 2: 70%). More importantly, it provides a unique model for each testing sample.

It is clear that we can extract useful information and knowledge from the experiment using IMPM method over this SNPs dataset:

1. The average number of selected features is around 17;
2. The average size of personalised problem space (neighbourhood) is 70;
3. *Age* is an important clinical factor for crohn's disease prediction, which has been selected 98 out of 106 times.
4. Four SNPs are more frequently selected than others (i.e. they are more informative in terms of crohn's disease prediction).
   These 4 SNPs are: X10210302_C, X17045918_C, X2960920_A, X7970701_G.

The discovered information and knowledge are of great importance to create a profile for each patient sample, and can be helpful for tailored treatment design and drug response and unknown types of disease diagnosis.

5.2   Reproducibility Evaluation

The main goal of this experiment is to evaluate the reproducibility of IMPM for SNPs data analysis. Here we are interested in whether the proposed personalised modelling based method is capable of producing highly consistent outcome for one sample? More specifically, this experiment is aiming to answer the questions:

1. What is the performance of proposed personalised modelling based method using global optimisation?
2. What is the variance of the local accuracy calculated from the global optimisation?
3. What is the frequency of each features to be selected during this experiment (20 runs)?
4. How many features should be selected for a successful prediction in general?

Hence, we randomly select one sample (sample 392) from the SNPs data and perform IMPM on it for 20 runs. IMPM creates an applausable prediction outcome: the prediction for sample 392 is always correct through all 20 runs. The average local accuracy for this sample through 20 runs is 82.45%. In addition, the personalised modelling method seems to work effectively on sample 392, as the computed local accuracy through 20 runs is very stable - the highest one is 83% and the lowest is 81%.

Figure 2.a illustrates the selecting frequency of each feature for testing sample 392 during 20 runs. Here *Age* is still selected as an important feature for CD prediction, as it has been always selected in 20 runs. The next top 5 selected features are:

Feature Id	SNP Id	*Selecting frequency(/20times)*
20	X4252400_T	19
24	X2155777_T	18
12	X7683921_A	14
9	X2270308_T	13
23	X10883359_G	13

It seems that SNP X4252400_T and X2155777_T are two decisive factors for predicting CD risk specifically for sample 392.

Figure 2.b summarises the number of selected features in each run. It is easy to elicit that using approximately $15 \sim 16$ SNPs plus the feature of Age could lead to the successful prediction for sample 329. This finding is in agreement with the previous outcome from the experiment reported in last section.

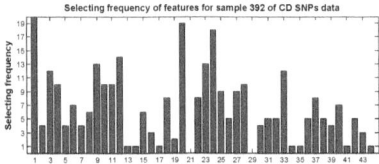

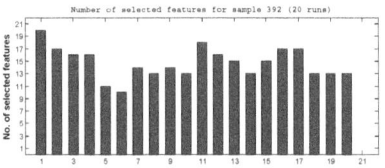

(a) The selecting frequency of each features for sample 392 of SNPs data for CD prediction

(b) The number of selected features for sample 392 (20 runs)

Fig. 2. An example: The experimental results of sample 392 of SNPs datga for CD prediction

Personalised modelling based method works consistently well on a sample (#392) for CD risk prediction. The prediction outcome is reliable and the local accuracy is reproducible. For personalised medical treatment design, this study suggests that our method should run several times over the testing sample, to find the most informative features (SNPs) through different runs, i.e. the most commonly selected features in different testing runs.

6   Conclusion and Future Direction

In this paper, we have demonstrated the strength of personalised modelling over global modelling for Crohn's disease classification over this specific SNPs data. The experiment clearly shows that the proposed integrated method for personalised modelling (IMPM) has a major advantage, when compared to global or local modelling. The proposed IMPM leads to a better prognostic accuracy and a computed personalised profile. With global optimisation, a small set of variables (potential markers) can be identified from the selected variable set across the whole population. This information can be utilised for the development of new efficient personalised treatment or druges.

Our current efforts are on the interdisciplinary collaboration from medicine and computer science for the modelling development of personalised risk probability evaluation. Despite the improved outcome produced by our IMPM method, the method still needs to take into account the probability of risk evaluation. Such risk probability evaluation can be more accurate and realistic for real world biomedical analysis problems.

Acknowledgement. We acknowledge the help with the SNPs data by Dr. Rod Lea and his research team at Environmental Science & Research (ESR) institute in New Zealand, who provided the pre-process SNPs data.

References

1. Hindorff, L.A., Sethupathy, P., Junkins, H.A., Ramos, E.M., Mehta, J.P., Collins, F.S., Manolio, T.A.: Potential etiologic and functional implications of genome-wide association loci for human diseases and traits. Proc. Natl. Acad. Sci. USA 106(23), 9362–9367 (2009)
2. Nikola Kasabov, Y.H.: Integrated optimisation method for personalised modelling and case studies for medical decision support. International Journal of Functional Informatics and Personalised Medicine 3(3), 236–256 (2010)
3. Vermeire, S., Van Assche, G., Rutgeerts, P.: Review article: altering the natural history of crohn's disease. Alimentary Pharmacology & Therapeutics 25(1), 3–12 (2007)
4. Eason, R.J., Lee, S.P., Tasman-Jones, C.: Inflammatory bowel disease in Auckland, New Zealand. Aust. N. Z. J. Med. 12(2), 125–131 (1982)
5. Gearry, R.B., Day, A.S.: Inflammatory bowel disease in New Zealand children - a growing problem. N. Z. Med. J. 121(1283), 5–8 (2008)
6. Song, Q., Kasabov, N.: Twnfi - a transductive neuro-fuzzy inference system with weighted data normalization for personalized modeling. Neural Networks 19(10), 1591–1596 (2006)
7. WTCCC: Genome-wide association study of 14,000 cases of seven common diseases and 3,000 shared controls. Nature 447(7145), 661–678 (2007)
8. Chang, C.C., Lin, C.J.: LIBSVM: a library for support vector machines (2001)

Improved Gene Clustering Based on Particle Swarm Optimization, K-Means, and Cluster Matching

Yau-King Lam, P.W.M. Tsang, and Chi-Sing Leung

Abstract. Past research has demonstrated that gene expression data can be effectively clustered into a group of centroids using an integration of the particle swarm optimization (PSO) and the K-Means algorithm. It is entitled PSO-based K-Means clustering algorithm (PSO-KM). This paper proposes a novel scheme of cluster matching to improve the PSO-KM for gene expression data. With the proposed scheme prior to the PSO operations, sequence of the clusters' centroids represented in a particle is matched that of the corresponding ones in the best particle with the closest distance. On this basis, not only a particle communicates with the best one in the swarm, but also sequence of the centroids is optimized. Experimental results reflect that the performance of the proposed design is superior in term of the reduction of the clustering error and convergence rate.

Keywords: Gene clustering, K-Means, Particle Swarm Optimization (PSO), PK-Means, Vector Quantization (VQ).

1 Introduction

Gene clustering becomes popular nowadays because of matured microarray technology and increasing power of computing. In the DNA microarray experiment, a numerical value of gene expression level in dataset can be attained from the well-prepared the genes of interest through the laser excitation of hybridized targets and software [1]. Microarray technology allows monitoring huge amount of gene expression level simultaneously for whole genome though a single chip only [2]. There is a demand to observe and analyze interactions among thousands of genes in the massive data sets. A cluster analysis plays an important role to extract useful information from the massive raw data. Techniques of class prediction and discovery are discussed [3] in order to intend for applying to mining significant biological process of genome or organ.

Many proposed gene clustering techniques have been reviewed and surveyed recently [4]. The types of these techniques include partitive techniques (such as K-Means and Genetic K-Means Algorithm), neural network techniques (such as Self-Organizing Map) and Graph theoretic methods. State of the art applications invoke certain limitations documented in the literature. For example, new hybrid techniques developed for gene clustering has been slow because of the increased algorithmic complexity. It is not easy to choose suitable method(s) for a given gene expression dataset. However, a classic K-Means [5] is still one of popular methods to cluster

B.-L. Lu, L. Zhang, and J. Kwok (Eds.): ICONIP 2011, Part I, LNCS 7062, pp. 654–661, 2011.

gene expression data effectively in which we are interested. The drawback of K-Means is well known that it is easy to be trapped in local optimized results early. In order to improve performance of K-Means, some advanced evolutionary algorithms based on K-Means are proposed, such as, PK-Means [6] which is integration of Particle Pair Optimizer [6] and K-Means clustering algorithm. The clustering results of the evolutionary algorithm outperform that of K-Means due to a factor of exploration added. It is found that the algorithm may not give an insight into the problem of sequence of the clusters' centroids represented in the particle position. Although a particle can find a better position by sharing information with the best particle position, an arbitrary sequence of the centroids may lead to diverge the clustering results contained in the particle position probably. We find that there should be certain relationship amongst the clusters between the two particles. After we have an insight to this interesting observation, we propose a novel method called cluster matching, which is an extra step to match and rearrange the sequence of clusters in the particle swarm optimization based K-Means clustering algorithm, to enhance the clustering results further. Each cluster's centroid in a particle will match its corresponding centroid in the best particle by the closest distance. As a result, not only a particle can learn from the best particle, but also communication can be done in the level of clusters' centroids contained in the particle position.

This paper is presented as follows. Section 2 describes an overview of the PSO-based K-Means clustering algorithm (PSO-KM). In section 3, a novel scheme of cluster matching, which is used to enhance the PSO-KM, will be described with illustration. This is followed by the experimental evaluation on the proposed scheme to show its significance in the following section. The proposed method has been compared with the original version of PSO-KM, the classic K-Means algorithm and a newer gene clustering algorithm (PK-Means). A summary of key features is drawn in the final section.

2   The PSO-Based K-Means Clustering Algorithm (PSO-KM)

The PSO-based K-Means clustering algorithm, named as PSO-KM hereafter, is the integration of the particle swarm optimization (PSO) [7] and the classic K-means clustering [5]. The former is used to explore the clusters' centroids contained in a particle position and the latter is used to exploit the centroids for a clustering result.

For representation of the clustering problem in the PSO algorithm, each particle position $X_{i,n}$ consists of K clusters' centroids, each of which is a D-dimensional vector, as follows,

$$X_{i,n} = \left(z_{i,n,0}, z_{i,n,1}, \cdots, z_{i,n,K-1} \right) , \tag{1}$$

where n is the iteration number for the i-th particle position. The velocity vector $V_{i,n}$ of this particle towards its next position is denoted by

$$V_{i,n+1} = wV_{i,n} + c_1 r_1 \left(P_{i,n} - X_{i,n} \right) + c_2 r_2 \left(G - X_{i,n} \right) , \tag{2}$$

where W is the inertia weight; $P_{i,n}$ is the best position of the particle; G represents the globally best position for the whole swarm; c_1 and c_2 are called acceleration factors; r_1 and r_2 are two random numbers within $[0,1)$. After the velocity is updated, it is limited to prevent from being in excess of its maximum velocity if any. With the velocity vector available, the particle position is updated by

$$X_{i,n+1} = X_{i,n} + V_{i,n+1} \quad . \tag{3}$$

In order to search the clustering solution locally, one K-Means iteration is performed to the particle position followed and its fitness is evaluated. An overview of the whole algorithm is depicted in Fig 1.

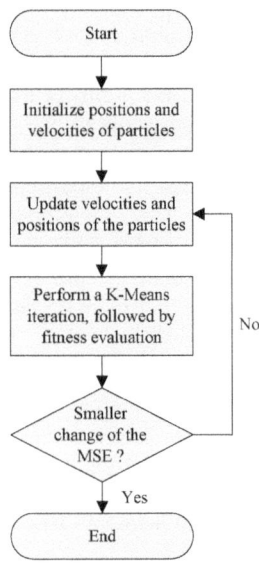

Fig. 1. Summarized steps of the PSO-KM

The objective of the partitioning process is to minimize the error by Mean Squared Error (*MSE*), which is used to measure the compactness inside the clusters, as follows:

$$MSE = \frac{1}{N} \sum_{j=0}^{K-1} \sum_{x_i \in C_j} \left\| y_i - z_j \right\|^2 \tag{4}$$

where y_i is a data point i, z_j is its cluster's centroid and C_j is its cluster.

3   The Proposed Scheme for the PSO-KM Algorithm

The novel method is proposed to increase the quality of exploration conducted by PSO operations. The method is named as Clusters' Matching (CM). This is an extra step, which consist of sequence matching and rearrangement, inserted before conducting the PSO operation (according to Eqs.(2) and (3)) of the PSO-KM. This kind of the clustering scheme is referred to as "PSO-KM(CM)". The First step is to match sequence of the clusters, contained in the particle position, with reference to the clusters in the position of the global best by the nearest distance. The same thing is done again for the particle best position and the global best position which is kept as the reference. Totally, there are two passes of the sequence matching. It aims to raise the quality of exploring region around the two matched clusters which came from the particle position (or its best position) and the global best position. (It is noted that storage order of clusters contained in a particle is not important as a potential solution.)

The key steps of the matching are noted as follows. First, the "Nearest" matrix is constructed as shown in Fig. 2. There are clusters in the current particle position in the column and the clusters in the global best position in the row respectively. There can be any possible matching order between the row and the column. An example of 5 clusters for the matching results is illustrated as shown in Fig. 3. The marker "Nearest" is placed in the cell between the two clusters with the closest distance to indicate the match; other combinations are blocked shown by the marker "X". This kind of the matching will be repeated until all have been matched. Here, the greedy strategy is applied to the matching process. Consequently, the matching results are shown in the Fig. 3. The clusters' pair is one-to-one matching and unique. This is the matching between the particle position and the global best; another matching between the particle best and the global best is done in similar manner.

		The global best position				
		Cluster 0	Cluster 1	Cluster 2	...	Cluster (K-1)
Particle position	*Cluster 0*					
	Cluster 1					
	Cluster 2					
	...					
	Cluster (K-1)					

Fig. 2. The "Nearest" matrix constructed for matching K centroids in both particle position and its reference position (The matching not started yet at that time)

		The global best position				
		Cluster 0	Cluster 1	Cluster 2	Cluster 3	Cluster 4
Particle position	*Cluster 0*	X	X	Nearest	X	X
	Cluster 1	X	Nearest	X	X	X
	Cluster 2	Nearest	X	X	X	X
	Cluster 3	X	X	X	X	Nearest
	Cluster 4	X	X	X	Nearest	X

Fig. 3. An example of matching 5 clusters from the matching results obtained by the "Nearest" matrix

The second step of the proposed method is sequence rearrangement depicted by using the above example as follows. As the matching results are available, the sequence of the clusters contained in the current particle position is rearranged according to the matching results. It is illustrated in Fig. 4. It is reminded that the information inside the particle is unchanged.

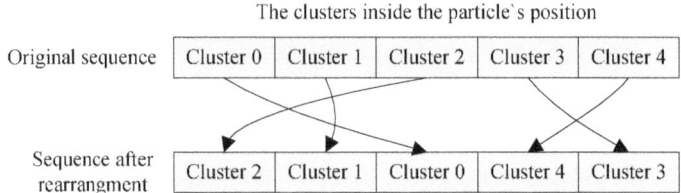

Fig. 4. The difference in the sequence of the clusters contained in the particle position before and after the rearrangement

The overview of the proposed scheme is illustrated in Fig. 5.

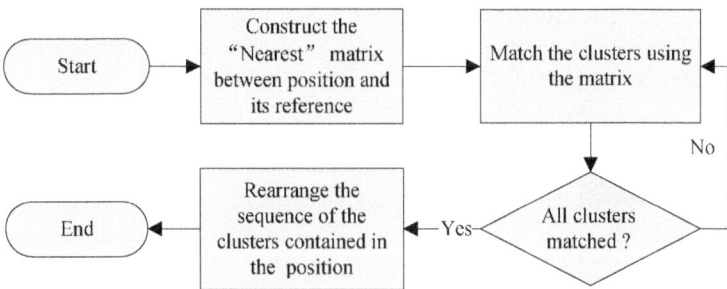

Fig. 5. The proposed scheme for matching clusters's centroids before PSO operation

4   Experimental Evaluation

Two popular gene expression datasets, which are Lymphoma [8] and Yeast cell-cycle [10], are chosen to evaluate the proposed method. The missing values in the test samples are rectified by the KNN algorithm [9].

Performance of the proposed method is compared against its general version of the PSO-KM, the classic K-Means and the PK-Means [6]. The methods are applied to cluster the dataset into 256 clusters. To obtain reliable statistics, a total of 10 repeated trials for the data sets and the schemes are conducted. Experimental settings of the PSO-KM schemes are described in Table 1.

Initial values of the velocity and its maximum velocity are set to be the dynamic range of each dimension respectively. For the K-Means experiment, the termination criterion is set to be the change of the *MSE* less than 0.001. The effects of the parameters in Table 1 are discussed as follows. The w is set to be a typical value between 0 and 1. General values of c_1 and c_2 are about 2 for the general PSO, but there requires smaller values for clustering algorithm. It can be referred to PK-Means [6]. The two values are tuned by experiment. The swarm size of PSO is relatively small generally and it also works for clustering.

Table 1. Key experimental settings for the PSO-KM and the PSO-KM(CM) schemes

Parameters	Values
Swarm size	4
w	0.7
c_1	0.2
c_2	0.2
No. of maximum iterations	40

The results of the *MSE* performance for the methods are listed in Table 2. It can be seen that the results of the classic K-Means clustering is the weakest for the test sets (The lower *MSE* value, the better result). The results of the PSO-KM get improved. The newer gene clustering scheme PK-Means [6] improves the clustering results further. The proposed PSO-KM(CM) amongst the methods is the optimized one.

Table 2. Clustering results of the schemes in term of *MSE*

Scheme	No. of iterations	*MSE* (lymphoma data)	*MSE* (yeast cell-cycle data)
K-Means	14	24.468	5.853
PSO-KM	40	22.067	5.636
PK-Means	52	21.474	5.532
PSO-KM(CM)	40	21.207	5.466

The average variation of the *MSE* performance over iterations of the schemes with the data sets, obtained in a 10 typical runs, are shown in Fig. 6 and 7 respectively. From the results depicted in Fig. 6 and 7, it can be seen that the *MSE* of the K-Means algorithm converges faster mostly amongst all the methods. It is because of simplicity and efficiency of the classic K-Means. However, the *MSE*s for the two kinds of PSO-KM enable to last to improve the *MSE* quality further. Although they can converge at similar time in term of iteration, the proposed method has capability of searching an optimal solution outstandingly at early stage in about the 7-th iteration. It is noted that PK-Means algorithm is converged slowly in time (second) due to 3 K-Means iterations conducted in a PK-Means iteration. Finally, the proposed method of the PSO-KM(CM) can provide the best solution.

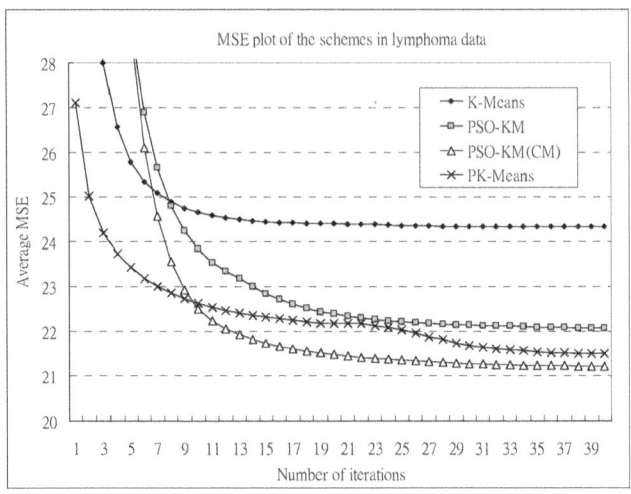

Fig. 6. MSE plots with iterations of the schemes in lymphoma data

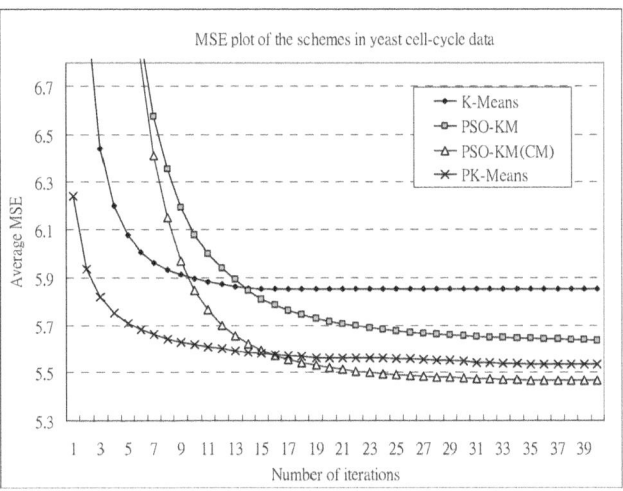

Fig. 7. MSE plots with iterations of the schemes in yeast cell-cycle data

5   Conclusion

Gene cluster analysis is the first step in discovering the function of gene in bioinformatics. Although there are many clustering methods developed, K-Means is one of the popular methods for common gene clustering due to its simplicity. Later, evolutionary algorithm, such as Particle Swarm Optimization (PSO) is incorporated with the K-Means clustering to gain the capability of both exploration and exploitation. This kind of integrated algorithm can give better clustering results in term of compactness within the clusters; it is found that there is still a room for improving the results. It is observed that sequence of clusters' centroids represented in a particle position may cause the normal PSO operations to diverge undesirably. This paper proposes a novel scheme of cluster matching to improve PSO-based K-Means clustering algorithm (PSO-KM). With the proposed scheme, sequence of the clusters' centroids represented in the particle position is matched that of the corresponding ones in the best particle with the closest distance. Thus, not only each particle communicates with the best one in the swarm, but also the sequence of the centroids is aligned optimally. A real two data sets are selected to evaluate the proposed scheme with the existing algorithms. Experimental results demonstrate that the PSO-KM with the proposed scheme can outperform its original version and other methods in term of compactness. Moreover, the proposed scheme makes the PSO-KM to converge faster with good compactness.

References

1. Brown, P., Botstein, D.: Exploring the New World of the Genome with DNA Microarrays. Nature Genetics 21, 33–37 (1999)
2. Brazma, A., Robinson, A., Cameron, G., Ashburner, M.: One-stop Shop for Microarray Data. Nature 403, 699–700 (2000)
3. Asyali, M.H., et al.: Gene Expression Profile Classification: A Review. Current Bioinformatics 1, 55–73 (2006)
4. Kerr, G., Ruskin, H.J., Crane, M., Doolan, P.: Techniques for clustering gene expression data. Computers in Biology and Medicine 38, 283–293 (2008)
5. Hartigan, J.A., Wong, M.A.: A K-Means clustering algorithm. Appl. Stat. 28, 126–130 (1979)
6. Du, Z., et al.: PK-Means: A new algorithm for gene clustering. Comput. Biol. Chem. 32(4), 243–247 (2008)
7. Shi, Y., Eberhart, R.C.: A modified particle swarm optimizer. In: Proceedings of the IEEE International Conference on Evolutionary Computation, pp. 69–73. IEEE Press, Piscataway (1998)
8. Alizadeh, A.A., et al.: Distinct types of diffuse large B-cell lymphoma identified by gene expression profiling. Nature 403, 503–511 (2000)
9. Troyanskaya, O., et al.: Missing value estimation methods for DNA microarrays. Bioinformatics 17, 520–525 (2001)
10. Spellman, P.T., et al.: Comprehensive identification of cell cycle-regulated genes of the yeast. Saccharomyces cerevisiae by microarray hybridization. Mol. Biol. Cell 9, 3273–3297 (1998)

Comparison between the Applications of Fragment-Based and Vertex-Based GPU Approaches in K-Means Clustering of Time Series Gene Expression Data

Yau-King Lam, Wuchao Situ, P.W.M. Tsang, Chi-Sing Leung, and Yi Xiao

Abstract. With the emergence of microarray technology, clustering of gene expression data has become an area of immense interest in recent years. However, due to the high dimensionality and complexity of the gene data landscape, the clustering process generally involves enormous amount of arithmetic operations. The problem has been partially alleviated with the K-Means algorithm, which enables high dimension data to be clustered efficiently. Further enhancement on the computation speed is achieved with the use of fragment shader running in a graphic processing unit (GPU) environment. Despite the success, such approach is not optimal as the process is scattered between the CPU and the GPU, causing bottleneck in the data exchange between the two processors, and the underused of the GPU. In this paper, we propose to realize the K-Means clustering algorithm with an integration of the vertex and the fragment shaders, which enables the majority of the clustering process to be implemented within the GPU. Experimental evaluation reflects that the computation efficiency of our proposed method in clustering short time gene expression is around 1.5 to 2 times faster than that attained with the conventional fragment shaders.

Keywords: Gene clustering, K-Means, Graphics Processing Unit (GPU), General-purpose computation, Vertex shader program.

1   Introduction

With the availability of the microarray technology, which enables large amount of genetic data to be extracted in a short period of time, gene clustering has been identified as an important research in the past two decades. In brief, clustering is a technique to partition a given set of data points into different groups, in a way that each of them will exhibit similar pattern corresponding to certain functionality. Amongst other methods the K-Means clustering [1], on account of its simplicity and effectiveness, has become a popular means for gene clustering as well as other scientific applications. In the study of genetic data, the clustering of time series gene expression data, which could reflect a wide range of biological processes through the change of temporal patterns, is an important topic that has instigated numerous research works. One of the major problems is the size of the gene data set (in the order of ten thousands of genes), and the high dimensionality of the data, which result in generally long computation time. With the advancement of the graphic processing

B.-L. Lu, L. Zhang, and J. Kwok (Eds.): ICONIP 2011, Part I, LNCS 7062, pp. 662–667, 2011.

unit, together with the shader programming technique, the above mentioned problem has been alleviated. It has been demonstrated that the implementation of the K-Means algorithm with the fragment shader is about 20 times faster than a commodity PC. However, we point out that the employing either the fragment or the vertex shader alone may not be the best solution in the context of gene clustering. Notably, the computation efficiency is lowered by the sharing of the task between the CPU and the GPU, leading to underused of the latter, as well as frequent exchange of large amount of data between the two processors. As such, we propose to employ an integration of both shaders [2] so that the majority of computation is conducted in the GPU. Our evaluations reveal that our approach is about 1.5 to 2 times faster than that based on fragment shader alone, and especially favorable for clustering short time-series gene expression datasets with a dimension of under 32. The 32 dimensions is the maximum value due to the hardware limitation of our method. We also note a lot of time-series expression data are short in length (with a dimension of 13 or less) [3], and could be efficiently clustered with our method.

Organization of the paper is given as follows. In section 2, an overview of GPU accelerated K-Means is outlined. Realization of the K-Mean with the integration of the fragment and vertex shaders is presented in sections 3. Application of our porposed method in gene data clustering, with experiments evaluation, is given in section 4. This is followed by a conclusion.

2   Fragment-Based Approach of GPU-Accelerated K-Means

As mentioned in section 1, the K-means [1, 2] clustering is a popular tool for clustering gene expression data because of its simplicity and effectiveness. Grossly speaking, K-Means is a technique to partition a given set of N data points, based on certain similarity metric measurement (usually the Euclidean distance), into a set of K clusters (a.k.a. the centroids) in a D-dimensional Euclidean space. Two key steps, which are iterated repetitively, are involved in the K-Means algorithm. In the first step, each data point is assigned to be a member of the cluster with the nearest distance. In the second step, the centroids are updated from its newly assigned members. The process is repeated until the clustering error, generally measured in terms of the Mean Squared Error (*MSE*), is converged to a steady value.

Although there are various schemes on utilizing the GPU to accelerate the K-means algorithm, the fragment shader-based approach [4] (refer as the fragment-based approach in this paper) is one of the most popular candidates. The method allocates the first (which involves heavy computation) and the second steps of the clustering process to the GPU and the CPU, respectively. On the part carried out in the GPU, each data point is represented by a fragment, and the nearest centroid is determined for each of the fragments. Significant computation advantage is gained by evaluating all the fragments, each conducted with its own fragment shader programming module, in a parallel fashion. The output of each fragment is a membership index pointing to the nearest centroid it belongs, which is then passed to the main memory of the CPU. In the latter, each centroid is recomputed from its newly assigned members.

Evidently, the GPU is not fully utilized in the clustering process, and the relatively slow data transfer between CPU and the GPU further imposes a bottleneck in the fragment-based approach.

3   Vertex-Based Approach of GPU-accelerated K-Means

To overcome the problems in the fragment-based approach, we proposed to employ an integration of the vertex and fragment shaders in realizing the K-Means algorithm. Although such integration has been explored in the past [2], it is the first time being applied in the context of gene expression clustering. The difference between the latter, as compare with other scientific applications such as image compression, is that gene data are rarely perfect. In general, an appreciable proportion of the dataset could be contaminated with noise so that values in some dimensions are missing. In this section, we shall focus on the mechanism of the combined shaders in realizing the K-Means algorithm. Afterwards, in section 4, we shall describe the handling of the defective gene data, and the application of the GPU accelerated clustering process.

In the combined shader approach, one vertex shader is collaborated with two fragment shaders (1 and 2). The centroids are stored in a set of textures, each storing our floating point numbers (commonly denoted by the symbols R, G, B, and A). As such, if each data is represented as a multi-dimensional vector, each texture is capable of storing four of its dimensions. Initially, two texture sets are established.

The source data points (source vectors) are prepared in form of a display list, to be input to the vertex shader. Every time a "drawing" command is launched in the CPU, the source vectors will be input, as the position and texture coordinates in texture set 1, to the vertex shader. Inside the vertex shader, the nearest centroid for each input vector is determined. Each vector is then redirected to an output position in texture set 2, corresponding to the nearest centroid.

Next, the fragment shader 1 is activated, and a blending function is conducted so that vectors in the same location in the texture set 2 will be accumulated. This is equivalent to updating the centroid value, but without averaging the result with the total number of constituting member.   To realize the averaging process, the fragment shader 2, is applied to divide the blended result of each cluster by the number of members, producing a new codebook of centroids. The process repeats and in each iteration the input is derived from texture set 2, which holds the results attained in the previous iteration.

4   Application of the Proposed Method in Clustering Short Time-Series Gene Expression Data

We applied the proposed method to cluster short time-series gene data. Due to noisy contamination, the data sets are generally imperfect and values in some of the dimensions are missing in certain data points. To rectify this defect, we applied the KNN algorithm [5] to recover some of the lost information. However, genes which exhibit over 20% of missing values are discarded. To evaluate our method, we employed two well known short-time series gene expression data; namely the "sporulation" [6], and the "G27 TC1 trial 4 time-series" [7]. For simplicity sake, the latter one is hereafter referred to as "G27". Details of these two sets of gene expression data are listed in Table 1.

Table 1. Read data sets for the experimental evaluation

Real data set	Dimensions	No. of genes
Sporulation	7	6023
G27	5	23663

The abovementioned no. of genes is number of preprocessed gene data. For the K-Means experiment, the termination criterion is set to be the change of the *MSE* less than 0.001.

We then applied the K-Means algorithm, which is realized with the proposed integrated shader implementation, to cluster the two sets of test data. All the evaluations are performed based on a computer equipped with the "Intel Core i7 920" CPU, and the "nVidia GTX260" GPU card. The program are developed with the Visual C++, and the shader programs are written in the nVidia "C for Graphics" (Cg) language. Each data set is clustered into different number of centroids (#Centroids) ranging from 16 to 512, and the computation efficiency in each case (measured in number of iterations per second) based on the CPU, the fragment shader, and our proposed method, are listed in Tables 2 and 3. We observe that the computation time is the longest with the CPU, and linearly proportional to the number of centroids. Our method exhibits the fastest performance, with the fragment-based approach in between. In general, our method is around 1.5 to 2 times faster than the fragment-based approach.

Table 2. Speed of the clustering approaches for sporulation data

#Centroids \ Approaches	CPU-based approach	Fragment-based approach	Vertex-based approach
16	444	514	903
32	239	553	839
64	117	528	702
128	62	423	804
256	31	333	505
512	16	240	360

Note: Speed is measured in number of iterations per second.

Table 3. Speed of the clustering approaches for G27 data

#Centroids \ Approach	CPU-based approach	Fragment-based approach	Vertex-based approach
16	128	491	575
32	64	440	681
64	35	366	570
128	19	270	488
256	10	182	306
512	5	110	198

Next we would like to illustrate a detailed comparison between the performances of our proposed method with the CPU and the fragment-based approach, with Fig. 1 and 2, respectively, showing the speed-up. In Fig. 1, we noted that the fragment-based approach is significantly faster than the computation with the CPU alone. Further observation can obtained from the results. The G27 data with higher data size can gain extra speed-up over that of the sporulation data regardless of their closer number of dimensions. Large data size can benefit processing power of GPU hardware. In addition, larger number of centroids can gain better speed-up for both data sets. As increasing number of centroids, their speed-ups are tending to settle down.

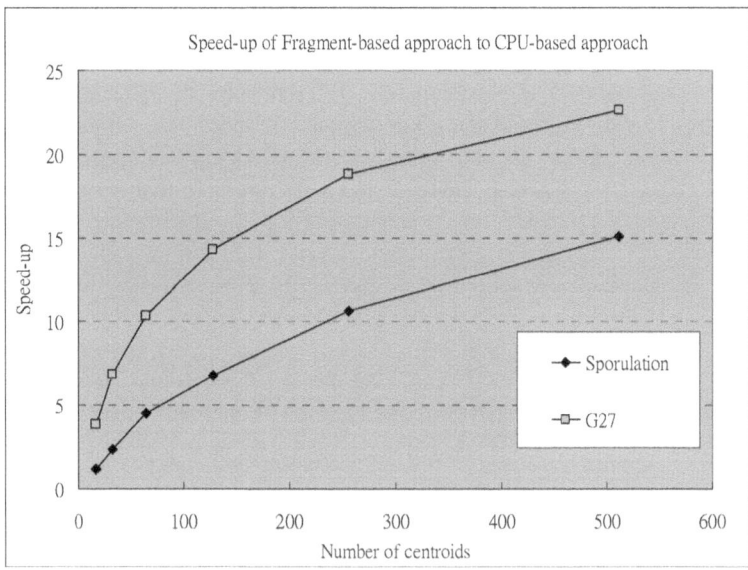

Fig. 1. Speed-up of fragment-based approach to CPU-based approach for the two test sets

In Fig. 2, we observe that our method is always faster than the fragment-based approach. With few exceptions, an improvement between 1.5 to 2 times is achieved. In brief, the vertex-based approach is capable of acceleration over the fragment-based approach in a constant-like ratio regardless of the different data sets.

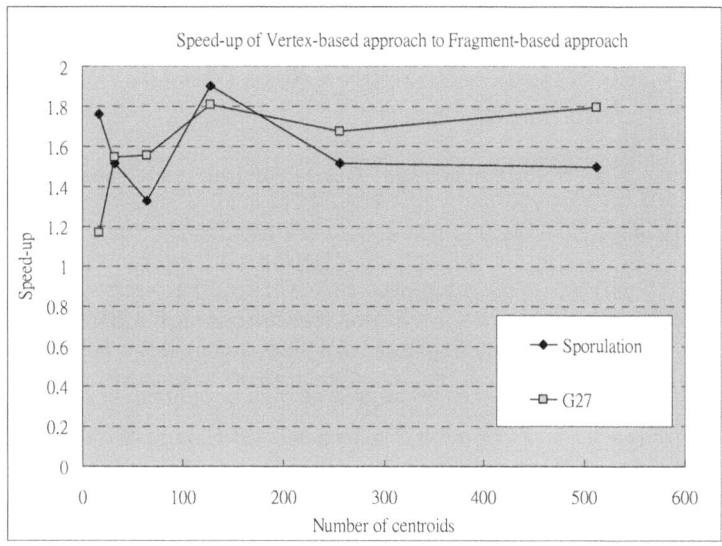

Fig. 2. Speed-up of vertex-based approach to fragment-based approach for the two test sets

5   Conclusion

This paper explores the feasibility of applying GPU based vertex shader in the realization of the K-Means algorithm for clustering short time-series gene expression data. We observe that despite the effectiveness of the approach, the computation efficiency is not optimal as the process, as well as the data, are scattered between the GPU and the CPU. To overcome the problem, we have proposed to apply a method that unifies the vertex and the fragment shaders, so that the majority of operations are conducted in the GPU. Experimental results reveal that our proposed method, when applied to the clustering of gene expression with a dimension of 32 or less, is about 1.5 to 2 times faster than that obtained with the conventional fragment shader.

References

1. Hartigan, J.A., Wong, M.A.: A K-Means clustering algorithm. Appl. Stat. 28, 126–130 (1979)
2. Xiao, Y., Leung, C.S., Ho, T.Y., Lam, P.M.: A GPU implementation for LBG and SOM training. Neural Computing and Applications (2010)
3. Ernst, J., et al.: Clustering short time series gene expression data. Bioinformatics 21, 159–168 (2005)
4. Takizawa, H., Kobayashi, H.: Hierarchical parallel processing of large scale data clustering on a pc cluster with gpu co-processing. J. Supercomput. 36, 219–234 (2006)
5. Troyanskaya, O., et al.: Missing value estimation methods for DNA microarrays. Bioinformatics 17, 520–525 (2001)
6. Chu, S., et al.: The transcriptional program of sporulation in budding yeast. Science 282, 699–705 (1998)
7. Guillemin, K., Salama, N., Tompkins, L., Falkow, S.: Cag pathogenicity island-specific responses of gastric epithelial cells to Helicobacter pylori infection. Proc. Natl. Acad. Sci. USA 99, 15136–15141 (2002)

A Modified Two-Stage SVM-RFE Model
for Cancer Classification Using Microarray Data

Phit Ling Tan[1], Shing Chiang Tan[1], Chee Peng Lim[2], and Swee Eng Khor[3]

[1] Faculty of Information Science and Technology, Multimedia University
Jalan Ayer Keroh Lama, 75450 Melaka, Malaysia
edith_windy@yahoo.com, sctan@mmu.edu.my
[2] School of Computer Sciences, University of Science Malaysia, 11800 Penang, Malaysia
[3] Mimos Berhad, Kulim Hi-Tech Park, 09000 Kulim, Kedah, Malaysia
cplim@cs.usm.my, se.khor@mimos.my

Abstract. Gene selection is one of the research issues for improving classification of microarray gene expression data. In this paper, a gene selection algorithm, which is based on the modified Recursive Feature Elimination (RFE) method, is integrated with a Support Vector Machine (SVM) to build a hybrid SVM-RFE model for cancer classification. The proposed model operates with a two-stage gene elimination scheme for finding a subset of expressed genes that indicate a disease. The effectiveness of the proposed model is evaluated using a multi-class lung cancer problem. The results show that the proposed SVM-RFE model is able to perform well with high classification accuracy rates.

Keywords: Gene selection, microarray, recursive feature elimination, support vector machine, cancer classification.

1 Introduction

Gene expression profiling is a technique that uses the DNA (Deoxyribonucleic acid) microarray technology for measuring the expression (activity) level of various genes in a biological cell under certain conditions. A normal cell if undergoes an uncontrolled growth will result in a cancerous cell. The information obtained from the gene expression profiling in normal and cancerous cells can be compared, and this provides a clue to identify the cancer types accurately. However, such analysis in gene expression profiling, when carried out manually by a human, can be laborious and time-consuming. This is because the microarray often consists of a substantially large number of data samples. In addition, each data sample normally comprises a great amount of genetic information (gene features). As such, intelligent computing techniques can be exploited to facilitate analysis and classification of such data.

From the literature review, many intelligent computing techniques, which include *k*-nearest neighbor, decision trees, artificial neural networks, and Kernel Methods (KMs), have been introduced to classify microarray gene expression data. In this study, a Support Vector Machine (SVM) is utilized as the base classifier to process

B.-L. Lu, L. Zhang, and J. Kwok (Eds.): ICONIP 2011, Part I, LNCS 7062, pp. 668–675, 2011.

microarray gene expression data. SVMs constitute a family of KMs which are able to produce good classification results [1].   The reason is that SVMs, which are based on statistical learning and optimization theories, demonstrate good generalization capabilities in undertaking separable/non-separable classification problems. However, the SVM model may not perform well with a set of gene expression data that are characterized as *imbalanced*, whereby the number of gene features is higher than the available number of samples.   Note that some of the gene features may be irrelevant, redundant, or noisy.   In this case, a gene selection algorithm, which either selects a subset of gene features that are relevant to a disease [2-4] or reserves the relevant gene features from being eliminated at the data level [5-8], is important for assisting the base classifier in learning useful information from the data samples effectively.   Such a gene selection algorithm not only provides a compact set of gene features for keeping the maximum amount of information with regards to a disease, but also produces better classification results.

One of the popular gene selection algorithms is the Recursive Feature Elimination (RFE) method.   RFE has been integrated with SVM, i.e., SVM-RFE [5], to handle binary classification problems.   The SVM-RFE model has been extended to deal with multi-class gene selection problems [9].   In addition, a two-stage gene selection method based on SVM-RFE has been proposed [10].   However, the two-stage SVM-RFE [10] model is applicable to binary classification problems.   In this study, we further extend the two-stage SVM-RFE model to undertake multi-class problems. We also modify the two-stage gene selection algorithm to improve its efficiency.

The organization of this paper is as follows.   In section 2, the training algorithms of SVM, SVM-RFE, two-stage SVM-RFE and its modified version are explained.   In section 3, the classification performances of the modified two-stage SVM-RFE model are evaluated with a multi-class lung cancer problem.   The results are compared with those from other SVM classifiers.   Concluding remarks are presented in section 4.

2   Gene Selection Algorithm and Support Vector Machine

2.1   Support Vector Machine (SVM)

SVM is a kernel-based method that searches for a set of hyperplanes that maximize the margin among themselves and the nearest data samples of arbitrary classes. Assume n data pairs $D = \{x_m, y_m\}, m = 1,...,n$ are available for training, where $x_m \in \Re^p$ is a feature vector denoting the m sample, and $y_m \in \{1,2,...K\}$ is the class label of x_m.   A multi-class linear SVM "one versus one" (SVM-OVO) [11] model, which is based on the construction of $\dfrac{K(K-1)}{2}$ binary linear SVMs, is employed in this study.   In this case, each $\dfrac{K(K-1)}{2}$ binary SVM is trained with $n_{r,s}$ data samples from two classes, where $r, s \in \{1,2,...,K\}, r \neq s$.   SVM-OVO (or simply SVM) can be formulated as an optimization problem, as follows:

Minimize $\wp_{r,s}\left(w_{r,s},\xi_{r,s}\right)=\frac{1}{2}\left\|w_{r,s}\right\|^2+C\sum_{i=1}^{n_{r,s}}\xi_i^{r,s},$ (1)

subject to $z_i^{r,s}[w_{r,s}^T x_i + b_{r,s}] \geq 1 - \xi_i^{r,s},$

$\xi_i^{r,s} \geq 0,$ $i = 1,...,n_{r,s},$

where C is a predefined parameter that controls the trade-off between training accuracy and generalization; $w_{r,s}$ is the weight vectors of SVM trained with data samples from two classes; $b_{r,s}$ is a scalar; $\xi_i^{r,s}$ is the slack variable; $z_i^{r,s} \in \{-1,1\}$ is the class label for the classifier. Given a data sample u, the decision function of each classifier is given as $f_{r,s}(u) = w_{r,s}^T u + b_{r,s}$. All classifiers are combined together to predict the class of a new data sample by using the majority voting strategy, i.e.,

$$\text{class} = \arg\max_{r=1,...,K}\left[\sum_{s=1}^{K} f_{r,s}(u)\right]$$ (2)

2.2   Support Vector Machine-Recursive Feature Elimination (SVM-RFE)

While a microarray data set consists of many gene features, not all gene features in the data set are expressed to indicate a disease. A gene selection algorithm is required to find a subset of gene features that is able to identify a disease accurately. The RFE part of SVM-RFE [5] produces a subset of optimum gene features from a trained SVM by eliminating particular gene features according to a ranking criterion (i.e., the weights). Initially, the data samples with all gene features (i.e., p) are presented to SVM for training. The weights of all existing genes, which are associated with the data samples during the training phase, are determined and sorted. The gene feature with the smallest weight is considered to have the least information, therefore, it is eliminated. The process of training SVM, ranking the weights of all existing genes, and eliminating the least informative gene is repeated for $(p-1)$ iterations. The subset of gene features that leads SVM to give the highest accuracy rate is regarded as the optimal gene subset. Nevertheless, the SVM-RFE model [3] performs gene selection for a binary problem. Its extension to gene selection from a multi-class problem has been reported in [5,12]. The standard procedure of the multi-class SVM-RFE model is summarized as follows:

1. Train a multi-class SVM with the existing gene features.
2. Compute the ranking score of gene features, $v = \sum_{r,s} w_{r,s}^2$.

3. Select the gene feature with the smallest ranking score.
4. Remove the selected gene feature from the training data set.
5. Repeat steps 1 to 4 until only a single gene feature remains in the training data set.

In this study, we adopt the standard multi-class SVM-RFE model as in [12] to further develop a two-stage SVM-RFE model, as explained in section 2.3.

2.3   Two-Stage SVM-RFE and Its Modified Variant

A two-stage SVM-RFE model has been proposed [10] to identify a subset of optimal genes from two consecutive selection processes. The model is applicable to binary data sets. The architecture of the two-stage SVM-RFE model is illustrated in Fig. 1.

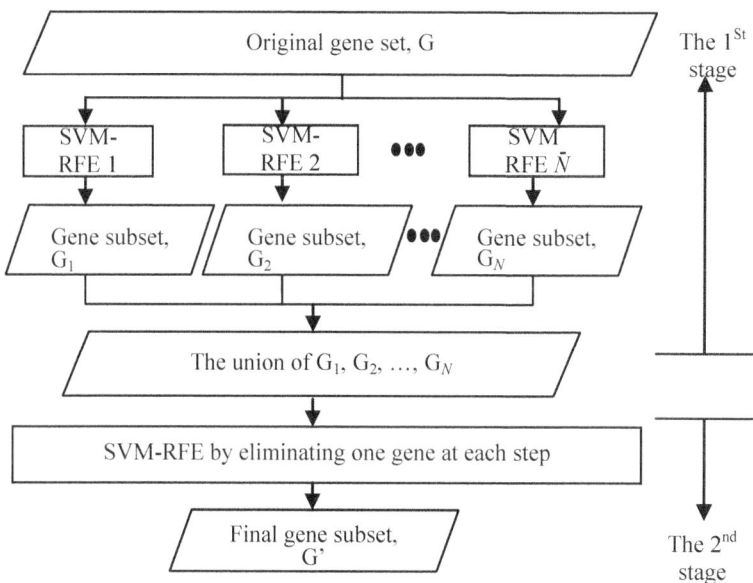

Fig. 1. The two-stage SVM-RFE (figure adapted from [10])

According to [10], a "filter-out" factor, f, is utilized in the first stage to remove an arbitrary number of genes from multiple SVM-RFE models. In this case, a few settings of f are introduced: if $f<0$, $f=\{-1,-2,-3,...\}$, only f unit(s) of bottom-ranked gene will be eliminated at each step; if $0<f<1$, a fraction of f bottom-ranked genes will be eliminated at each step; if $f=0$, the least possible bottom-ranked gene will be eliminated so that the number of remaining genes is the power of 2 at the first step and then half of genes are eliminated in the following steps. All SVM-RFE models in the first stage can take different f values in order to yield multiple gene subsets. It is argued [10] that if a gene is really informative, then it is likely to survive in at least one of the gene subsets; otherwise, it is considered as a redundant, noisy, or irrelevant gene. The gene elimination process in the first stage is safeguarded by a predefined size of the final informative gene subset, S, so as to prevent informative loss. In this case, a careful selection in the setting of S is required in order to ensure the reservation of relevant genes in the first stage, especially when $f=0$ or $0<f<1$. In the second stage, SVM-RFE is set with $f=-1$ so as to find a subset of optimal genes while avoiding a drastic loss in the number of informative genes.

Overall, the two-stage SVM-RFE model is sensitive to the f settings. In addition, one should have a good knowledge on the settings of both f and S for a classification problem. In view of the above limitations, a modified two-stage SVM-RFE model is proposed in this paper. At the first stage of the proposed model, f is replaced by a threshold parameter against a ranking score. Each SVM-RFE model is given a different threshold value. In this case, genes with ranking scores lower than the threshold values are removed. The terminating condition for each SVM-RFE in the first stage is also changed. Each gene elimination process is terminated whenever the ranking of genes remains unchanged in subsequent steps. Another note is that, in the second stage, the gene elimination process stops whenever only E number of genes remain in the data set, where E=the number of target classes (e.g. cancer). This is based on the assumption that each target class is expressed by at least a gene feature. The modified two-stage SVM-RFE model adopts a multi-class gene selection algorithm as explained in section 2.2. Therefore, it is able to overcome the limitations of the standard two-stage SVM-RFE [10] model for which the latter is designed to deal with binary classification problems only.

3   The Experimental Study

A five-class lung cancer data set [13] has been used for experimentation in this study. The data set consists of 203 data samples, which include 139 samples of adenocarcinomas (AD), 21 samples of squamous cell lung carcinomas (SQ), 20 samples of pulmonary carcinoids (CD), 6 samples of small-cell lung carcinomas (SMC), and 17 samples of normal lungs (NL). Each sample is described by 12,600 genes. The performance of the proposed SVM-RFE model is assessed in terms of classification accuracy. For this, a stratified 10-fold cross-validation (CV) has been employed to train and evaluate its performance. In each fold of CV, the proposed SVM-RFE model is trained from scratch with 90% of the data samples and tested with the remaining 10% of the data samples. Both training and test sets contain proportional number of data samples with respect to the distribution of target classes. The training parameters are as follows: for all SVM-RFE models at both stages, C=8, and for each SVM-RFE model in the first stage, its threshold value is changed from 0.0005 to 0.0014. The proposed SVM-RFE model has been developed using the Matlab version of the LIBSVM tool [14].

Table 1 shows the results of the proposed SVM-RFE model trained with different threshold values during its first-stage gene selection. Note that the higher the threshold values, the higher the numbers of eliminated genes; therefore fewer numbers of genes are available for classification. The process of gene elimination continues in the second stage. The classification results of the proposed SVM model are plotted in Fig. 2. As can be seen in Fig. 2, the proposed SVM-RFE model shows improvements in classification results with the use of smaller numbers of genes. The reason is that irrelevant, redundant, and/or noisy genes that would affect the performance of the classifier have been removed. The result starts to deteriorate when the number of relevant genes becomes too small. The best results of the proposed SVM-RFE model and those from other SVM models are shown in Table 2.

Table 1. Accuracy results obtained from the first stage of the proposed SVM-RFE model

Threshold Value	Number of Genes Used (Units)	Accuracy (%)
0.0005	412	93.60
0.0006	354	93.60
0.0007	310	94.58
0.0008	282	94.58
0.0009	254	95.07
0.0010	227	95.07
0.0011	200	95.57
0.0012	188	95.57
0.0013	165	95.07
0.0014	154	95.57

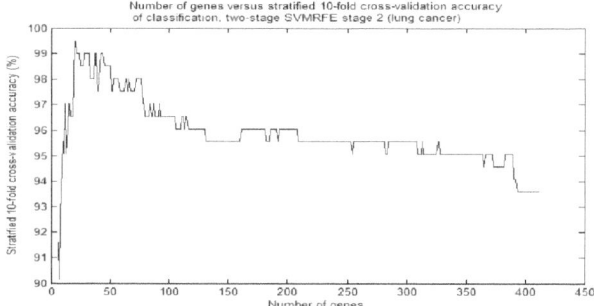

Fig. 2. Classification accuracy of the second stage of the proposed SVM-RFE model

Table 2. Performance comparison of the proposed SVM-RFE model and other SVM models

Model	Time (s)	Number of Support Vector (Units)	Accuracy (%)	Number of Gene Used (Units)
RBF SVM	122.91	203	68.47	12600
Polynomial SVM	64.56	105	93.60	12600
Linear SVM	72.89	95	93.60	12600
Linear SVM-RFE	67724.13	57	99.01	24
Proposed SVM-RFE	594.28	52	99.51	20

The SVM-based models are trained with a full set of gene features utilizing different kernel functions, i.e., linear, polynomial and radial basis function (RBF), and a single SVM-RFE model. The linear, polynomial, and RBF kernel-based SVM models take a shorter training time than the proposed SVM-RFE model; however, the former classifies a data sample using more numbers of gene features (some of them may not be relevant to the disease, hence a lower accuracy rate) and keeps higher numbers of support vectors than the latter. As compared with the SVM models using polynomial (with 93.60% from a stratified 10-fold CV) and RBF kernels (with 68.47%), the proposed SVM-RFE model is able to classify the lung cancer data more accurately, i.e., 99.51%. The average classification accuracy rate of the proposed SVM-RFE is very high. The use of the stratified 10-fold CV method gives less biased results [15]. As such, the issue of over-fitting by the proposed SVM-RFE in its training process can be avoided. Note that the accuracy rates achieved by the linear and polynomial SVM models are similar. Indeed, the linear SVM is a variant of the polynomial SVM. A microarray data set is characterized as imbalanced for which the number of gene features is larger than the number of data samples. The data set, which has high dimensionality, can be deemed as a linear classification task [5]. As such, the data samples can be effectively handled by a linear SVM, without having to map to a higher dimensional feature space in order to improve the results (as per the principle of polynomial SVM).

On the other hand, one can notice that the proposed SVM-RFE model consumes a far shorter training time than that of the linear SVM-RFE model for achieving a slightly better classification accuracy rate. This is because the proposed SVM-RFE model is able to eliminate several genes in each iteration in the first-stage gene selection process whereas the linear SVM-RFE model eliminates only one gene in each iteration before termination. In short, the proposed SVM-RFE model is able to perform with higher classification accuracy (as compared with SVM models with different kernels: linear, polynomial and RBF) with a more compact set of relevant genes (as compared with the aforementioned kernel-based SVM and SVM-RFE models) within a reasonable training duration (as compared with the linear SVM-RFE model which is regarded as the "slowest SVM-RFE" model [10]).

4   Summary

Microarray gene expression data samples are typically characterized as a type of imbalanced data set for which its number of gene features is far higher than its data samples. Not all genes are useful for classifying a disease accurately; this is because the data set may consist of redundant, noisy, or irrelevant genes. In this paper, a modified two-stage linear SVM-RFE model is proposed to identify a subset of optimal relevant genes to a cancer data set that consists of multiple output classes. While the original two-stage SVM-RFE model handles only binary problems, the proposed SVM-RFE model is able to undertake multi-class problems. The effectiveness of the proposed model has been evaluated with a multi-class lung cancer data set. A performance comparison among several SVM models has been conducted. The experimental results have shown that the proposed SVM-RFE model is able to elicit a compact set of gene features within an acceptable training time for making accurate predictions of different types of lung cancers.

For future work, additional experiments will be conducted to evaluate the performance of the proposed SVM-RFE model with other microarray gene expression data sets. This will allow validation and verification of the proposed SVM-RFE model as a useful computing model for disease classification using microarray data.

Acknowledgments. The authors gratefully acknowledge the financial support of the Fundamental Research Grant Scheme (No. 6711195) of Ministry of Higher Education, Malaysia for this research work.

References

1. Lee, J.W., Lee, J.B., Park, M., Song, S.H.: An Extensive Comparison of Recent Classification Tools Applied to Microarray data. Computational Statistics & Data Analysis 48(4), 869–885 (2005)
2. Furey, T., Cristianini, N., Duffy, N., Bednarski, D., Schummer, M., Haussler, D.: Support Vector Machine Classification and Validation of Cancer Tissue Samples Using Microarray Expression Data. Bioinformatics 16(10), 906–914 (2000)
3. Luo, L., Ye, L., Luo, M., Huang, D., Peng, H., Yang, F.: Methods of Forward Feature Selection Based on the Aggregation of Classifiers Generated by Single Attribute. Computers in Biology and Medicine 41, 435–441 (2011)
4. Cai, R., Hao, Z., Yang, X., Wen, W.: An Efficient Gene Selection Algorithm Based on Mutual Information. Neurocomputing 72, 991–999 (2009)
5. Guyon, I., Weston, J., Barhill, S., Vapnik, V.: Gene Selection for Cancer Classification using Support Vector Machines. Machine Learning 46(1-3), 389–422 (2002)
6. Mundra, P.A., Rajapakse, J.C.: SVM-RFE with MRMR Filter for Gene Selection. IEEE Transactions on Nanobioscience 9(1), 31–37 (2010)
7. Yoon, S., Kim, S.: Mutual Information-Based SVM-RFE for Diagnostic Classification of Digitized Mammograms. Pattern Recognition Letters 30, 1489–1495 (2009)
8. Luo, L.-K., Huang, D.-F., Ye, L.-J., Zhou, Q.-F., Shao, G.-F., Peng, H.: Improving the Computational Efficiency of Recursive Cluster Elimination for Gene Selection. IEEE/ACM Transactions on Computational Biology and Bioinformatics 8(1), 122–129 (2011)
9. Zhou, X., Tuck, D.P.: MSVM-RFE: Extensions of SVM-RFE for Multiclass Gene Selection on DNA Microarray Data. Bioinformatics 23(9), 1106–1114 (2007)
10. Tang, Y., Zhang, Y.-Q., Huang, Z.: Development of Two-Stage SVM-RFE Gene Selection Strategy for Microarray Expression Data Analysis. IEEE/ACM Transactions on Computational Biology and Bioinformatics 4(3), 365–381 (2007)
11. Kreßel, U.H.-G.: Pairwise Classification and Support Vector Machines. In: Schölkopf, B., Burges, C.J.C., Smola, A.J. (eds.) Advances in Kernel Methods: Support Vector Learning, pp. 255–268. MIT Press, Cambridge (1999)
12. Knowledge Discovery and Data Mining in Biotechnology,
   http://www.uccor.edu.ar/paginas/seminarios/Software.htm
13. Bhattacharjee, A., Richards, W.G., Staunton, J., Li, C., Monti, S., Vasa, P., Ladd, C., Beheshti, J., Bueno, R., Gillette, M., Loda, M., Weber, G., Mark, E.J., Lander, E.S., Wong, W., Johnson, B.E., Golub, T.R., Sugarbaker, D.J., Meyerson, M.: Classification of Human Lung Carcinomas by Mrna Expression Profiling Reveals Distinct Adenocarcinoma Subclasses. Proc. Natl. Acad. Sci. U.S.A 98(24), 13790–13795 (2001)
14. LIBSVM: A library for Support Vector Machines,
   http://www.csie.ntu.edu.tw/~cjlin/libsvm
15. Kohavi, R.: A Study of Cross-Validation and Bootstrap for Accuracy Estimation and Model Selection. In: Mellish, C. (ed.) Proceedings of the Fourteenth International Joint Conference on Artificial Intelligence, vol. 2, pp. 1137–1143. Morgan Kaufmann, San Mateo (1995)

Pathway-Based Microarray Analysis with Negatively Correlated Feature Sets for Disease Classification

Pitak Sootanan[1], Asawin Meechai[2], Santitham Prom-on[3], and Jonathan H. Chan[4,*]

[1] Individual Based Program (Bioinformatics), [2] Department of Chemical Engineering,
[3] Department of Computer Engineering, [4] School of Information Technology,
King Mongkut's University of Technology Thonburi, Bangkok, Thailand
jonathan@sit.kmutt.ac.th

Abstract. Accuracy of disease classification has always been a challenging goal of bioinformatics research. Microarray-based classification of disease states relies on the use of gene expression profiles of patients to identify those that have profiles differing from the control group. A number of methods have been proposed to identify diagnostic markers that can accurately discriminate between different classes of a disease. Pathway-based microarray analysis for disease classification can help improving the classification accuracy. The experimental results showed that the use of pathway activities inferred by the negatively correlated feature sets (NCFS) based methods achieved higher accuracy in disease classification than other different pathway-based feature selection methods for two breast cancer metastasis datasets.

Keywords: Microarray-based classification, Pathway-based feature selection, Negatively correlated feature sets, Phenotype-correlated genes.

1 Introduction

Contemporary high-throughput technologies used in molecular biology such as microarray allow scientists to monitor changes in the expression levels of genes in response to changes in environmental conditions or in healthy people versus affected patients [1]. It provides a means to classify and diagnose the disease state at the gene expression level [2]. However, a huge number of genes and comparatively small number of samples often leave investigators frustrated when they try to interpret the meanings from the results. To do so, one needs to be able to remove redundant and irrelevant genes and find a subset of discriminative genes from the dataset. Selecting informative genes is crucial to improve the overall effectiveness of the microarray-based diagnostics and classifications [2-5].

Several methods have been proposed to perform the feature selection task [2, 6-7]. However, different methods may produce different gene rankings or gene subsets. It is often difficult to decide which feature selection algorithm best suits a dataset because most of the time the performance of an algorithm varies with different datasets [8]. A number of pathway-based analysis methods have been proposed for disease

---

* Corresponding author.

B.-L. Lu, L. Zhang, and J. Kwok (Eds.): ICONIP 2011, Part I, LNCS 7062, pp. 676–683, 2011.

classification using expression profiles with more precision than using individual genes [9-10]. These methods, however, may lack discriminative power by disregarding member genes that have consistent, but not large, expression changes with different phenotypes. This issue has been alleviated by the use of negatively correlated feature sets (NCFS) for identifying phenotype-correlated genes (PCOG) and inferring their pathway activities at the pathway level. These NCFS-based methods were reported to be more precise and robust than the previous studies [11-12].

In this paper, pathway information from KEGG [13] was used to group the gene inputs according to their pathways and the different feature selection methods were used to identify a subset of member genes. These methods include CORG-based method [10], NCFS-i and NCFS-c method [11], genetic search method with logistic regression classifier subset evaluator, and ranker method with support vector machine (SVM) [12]. We tested these techniques by applying them to classify two breast cancer metastasis datasets and comparing their classification accuracies to identify the method that is the most effective to do this task.

2   Material and Method

2.1   Dataset

We obtained two breast cancer datasets from large-scale gene expression studies by Pawitan *et al.* 2005 [14] and Wang *et al.* 2005 [15]. Pawitan *et al.*'s dataset (GSE1456) contains the gene expression profiles of 159 Swedish patients, where relapse class was detected in 40 patients while the remaining 119 were non-relapse. Wang *et al.*'s dataset (GSE2034) contains the gene expression profiles of 286 breast cancer patients from the USA, where relapse class was detected in 107 of them while the remaining 179 were non-relapse class. In this study, we did not consider the follow-up time or the occurrence of distant metastasis. We retrieved these breast datasets from the public database of Gene Expression Omnibus (GEO) [16]. These datasets can be analyzed readily with their expression levels. For genes with more than one probe in one platform, we chose the probe with the highest mean expression value.

To obtain the set of known biological pathways, we referred to the pathway information from KEGG (Kyoto Encyclopedia of Genes and Genomes) database [13]. We downloaded manually curated pathways containing 204 gene sets. 169 pathways with more than 10 member genes were selected. These gene sets were compiled by domain experts and they provided canonical representations of biological processes.

2.2   Method

Expression data were normalized to z-score form before mapping onto the member genes in each pathway. These data were then used to identify representative set of gene by different feature selection methods. A comparison of the classification performance with the above-mentioned inferring methods was then made.

Data Preprocessing. Expression values g_{ij} were normalized to z-transformed score z_{ij} for each gene i and each sample j. To integrate the expression and pathway datasets, we overlaid the expression values of each gene on its corresponding protein in each pathway.

Identifying Representative Sets of Genes in Pathways and Inferring Pathway Activities. Six pathway-based feature selection methods and three pathway inferring methods were chosen for comparison in this study. Pathway data was used to reduce the dimension of microarray expression matrix and to explain more about how these genes relevance to disease of interest. Within each pathway, a representative set of genes was searched by different pathway-based feature selection methods. A schematic diagram summarizing the overall methodology is shown in Fig. 1.

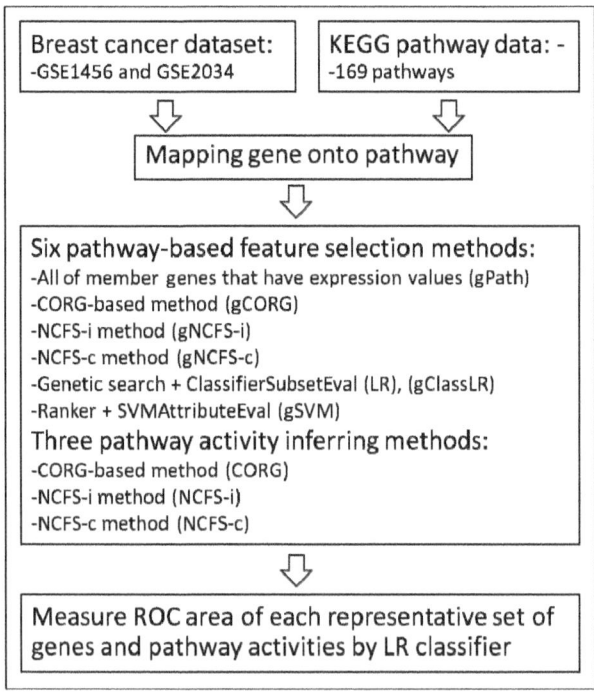

Fig. 1. Schematic diagram of the overall methodology

Expressed member gene (gPath). The expression matrix of all member genes in each pathway was used to measure its classification performance in terms of ROC area.

CORG-based method (gCORG). To identify the representative set of genes (condition-responsive genes, CORG), member genes were first ranked by their t-test scores, in descending order if the average t-score among all member genes was positive, and in ascending order otherwise. For a given pathway, a greedy search was performed to identify a subset of member genes in the pathway for which the discriminative scores (DS, S(G)) was locally maximal [10]. The expression matrix of the representative set of genes and their inferred pathway activities were used to measure its classification performance in terms of ROC area.

NCFS-i method (gNCFS-i). All member genes in each pathway were first ranked by their t-scores in descending and ascending orders, if the average t-score among all member genes was positive and negative, respectively. Then, within each pathway, top ranked genes in these two different gene subsets were used to search for a representative set of genes (phenotype-correlated genes, PCOG) [11]. The expression matrix of the representative set of genes and their inferred pathway activities were used to measure its classification performance in terms of ROC area.

NCFS-c method (gNCFS-c). This method was another modification from CORG-based method [10] by incorporating negatively correlated feature sets (NCFS). The first set of CORG was identified then a second set of CORG which were negatively correlated to the first set was identified. These two set of genes were merged to be a representative set of genes in pathway [11]. The expression matrix of the representative set of genes and their inferred pathway activities were used to measure its classification performance in terms of ROC area.

Genetic search with logistic regression classifier subset evaluator (gClassLR). In this work, genetic search with logistic regression classifier subset evaluator from WEKA [17] was chosen because the classification performance of their pathway markers quite similar to another method [12]. The expression matrix of this representative set of genes was used to measure its classification performance in terms of ROC area.

Ranker with support vector machine attribute evaluator (gSVM). Top pathway markers ranked by SVM showed the highest classification performance in most breast cancer datasets in previous studies [6, 12]. Therefore, the ranker with SVM attribute evaluator was chosen to rank the member genes in each pathway. For a given pathway, a greedy search was performed to identify a subset of member genes in the pathway for which its ROC area was locally maximal. The expression matrix of this representative set of genes was used to measure its classification performance in terms of ROC area.

Classification Performance Evaluation Measure. In this work, we evaluated the performance of a classifier based on the Area Under Curve (AUC) of the Receiver Operating Characteristic (ROC) curve. A final classification performance was reported as the ROC area using ten folds cross validation. WEKA (Waikato Environment for Knowledge Analysis) version 3.6.2 [17] was used to build the classifier by using logistic regression (Logistic). The results in ROC area were used to demonstrate the classification performance of different feature selection methods. Logistic regression (LR) is a standard method for building prediction models with a binary outcome and has been used for disease classification with microarray data [18].

3   Results and Discussion

Breast cancer datasets GSE1456 and GSE2034 obtained from microarray platform GPL96 were prepared and mapped onto 169 pathways. These pathway expression matrices were used in the feature selection step with different methods. Table 1 summarizes the results for the two breast cancer datasets used.

Table 1. The average, standard deviation (SD), maximum and minimum number of representative sets of genes obtained from different pathway-based feature selection methods for two breast cancer datasets GSE1456 and GSE2034

	gPath	gCORG	gNCFS-i	gNCFS-c	gClassLR	gSVM
GSE1456						
Average	62.10	4.03	7.76	7.56	28.14	3.71
SD	79.81	1.62	3.77	2.56	18.08	2.18
Maximum	875	9	20	17	79	10
Minimum	10	1	2	3	3	1
GSE2034						
Average	62.10	4.66	9.76	9.54	33.35	4.43
SD	79.81	2.35	3.98	4.05	28.09	2.69
Maximum	875	14	20	24	128	10
Minimum	10	1	2	2	3	1

The range of member genes in each pathway varied from 10 to 875, with the average and standard deviation being 62 and 80, respectively. The average sets of genes selected by the two NCFS-based methods, NCFS-i method (gNCFS-i) and NCFS-c method (gNCFS-c), were 7.5 and 9.5 for GSE1456 and GSE2034, respectively. These sets were approximately twice as much as the CORG-based method (gCORG) and the ranker method with SVMAttributeEval (gSVM). NCFS-based methods used two subsets of member genes in pathway which had high positive and negative t test scores or negatively correlated feature sets, called PCOG. On the other hand, CORG-based method used only one subset of member genes, called CORG, and disregarded member genes that had negative correlation to the CORG set [10]. gSVM ranked genes by their individual evaluations and evaluated the worth of a gene by using a support vector machine (SVM) classifier. Genes were ranked by the square of the weight assigned by the SVM [6]. The first maximum ROC score of top ranked genes were selected to be a representative set of genes with gSVM. This gene set could have a small size like gCORG. The representative sets of genes selected by genetic search with logistic regression classifier subset evaluator (gClassLR) resulted in the highest number of genes at around 28 and 33 on the average and with 79 and 128 genes as the maximum for GSE1456 and GSE2034, respectively. gClassLR performed a search using the simple genetic algorithm [19] and evaluated gene subsets of training data or a separated hold out testing set. It used logistic regression classifier to estimate the merit of a set of genes. This set of genes could be large. These results suggest that different pathway-based feature selection methods can produce different number of representative sets of genes with the same pathway expression matrix.

In order to use pathway information for classification, representative sets of genes identified by different pathway-based feature selection methods and pathway activities inferred by the three methods were used as feature values in a classifier based on logistic regression (LR). Ten-fold cross-validation experiments were used to test the predictive power of these representative sets of genes. This part of work evaluated all pathway-based feature selection methods using logistic regression and assessed the classification performance using the ROC area. Table 2 shows the average, standard deviation, maximum, and minimum values of the ROC area and the number of maximum ROC scores of the representative sets of genes for six different pathway-based feature selection methods in comparison to the pathway activities inferred by the three methods for two breast cancer datasets GSE1456 and GSE2034. For the GSE1456 dataset, gNCFS-i and gNCFS-c produced average ROC areas that were higher than gPath, gClassLR, and gCORG but lower than gSVM. The highest average ROC area was 0.699 for the gSVM method. For the GSE2034 dataset, the average ROC areas identified by gNCFS-i, gNCFS-c, and gClassLR were similar and higher than gPath and gCORG. The highest average ROC area was 0.643 using the gSVM method. For both breast datasets, use of pathway activities inferred from the three methods including CORG, NCFS-i, and NCFS-c were better than use of expression matrix of the representative sets of genes (Table 2).

Table 2. The average, standard deviation (SD), maximum, and minimum ROC area and the number of maximum ROC scores of the representative sets of genes in six different pathway-based feature selection methods and three pathway inferring methods for two breast cancer datasets GSE1456 and GSE2034

	gPath	gCORG	gNCFS-i	gNCFS-c	gClassLR	gSVM	CORG	NCFS-i	NCFS-c
GSE1456									
Average	0.569	0.686	0.694	0.696	0.636	0.699	0.717	0.747	0.748
SD	0.063	0.052	0.060	0.053	0.060	0.077	0.044	0.046	0.044
Maximum	0.739	0.788	0.814	0.814	0.819	0.872	0.811	0.848	0.847
Minimum	0.379	0.516	0.482	0.516	0.460	0.440	0.565	0.624	0.632
No. Max ROC score	0	1	0	0	0	25	13	71	82
GSE2034									
Average	0572	0590	0.608	0.607	0.608	0.643	0.649	0.681	0.685
SD	0.051	0.044	0.048	0.049	0.050	0.065	0.037	0.047	0.046
Maximum	0707	0676	0.701	0.694	0.731	0.793	0.752	0.806	0.792
Minimum	0443	0425	0.459	0.478	0.463	0.475	0.536	0.538	0.564
No. Max ROC score	0	0	0	0	0	22	12	65	81

The highest classification performances using average ROC area of all 169 pathways were found to be NCFS-based methods. The pathway markers inferred by the NCFS-based methods showed the highest number of maximum ROC scores for both datasets. Note that there were 169 pathways with member genes more than 10, as retrieved from KEGG database [13]. However, the summation of the number of all inferring methods was not equal to 169. The reason being the highest ROC area came from more than one method in some cases. The results from Table 2 show that gSVM was significantly better than the other five pathway-based feature selection methods with more maximum ROC scores for both datasets. The number of maximum ROC scores by the gSVM method was 25 and 22 for GSE1456 and GSE2034, respectively.

These values were higher than the CORG-based method but lower than the NCFS-i and NCFS-c methods. The above results suggest that the use of pathway activities inferred by the NCFS-i and NCFS-c methods were better than the use of subsets of member genes in pathway in classification. Also, gSVM was the best pathway-based feature selection method based on gene expression values.

4   Conclusions

We have demonstrated that pathway activities inferred by NCFS-based methods can be used to effectively incorporate pathway information into expression-based disease classification, especially when comparing it to the use of expression values of representative sets of genes identified by different pathway-based feature selection methods. It can provide better classification performance to discriminate two classes due to the use of both positively and negatively correlated gene sets to represent phenotype of interest. In addition, the representative sets of genes identified by NCFS-based methods provided comparable performance to traditional SVM and better than CORG-based method and the use of genetic search with LR classifier subset evaluator. For future work, further improvements of the NCFS-based methods in terms of selecting reliable representative sets of genes and improvement in classification accuracy by using different classification techniques are planned for more disease types.

Acknowledgments. The main author (PS) gratefully acknowledges financial support from the Royal Thai Government budget, Fiscal year 2010.

References

1. Young, R.A.: Biomedical Discovery with DNA Arrays. Cell 102, 9–15 (2000)
2. Golub, T.R., Slonim, D.K., Tamayo, P., Huard, C., Gassenbeek, M., Mesirov, J.P., Coller, H., Loh, M.L., Downing, J.R., Caligiuri, M.A., Bloomfield, C.D., Lander, E.S.: Molecular Classification of Cancer: Class Discovery and Class Prediction by Gene Expression Monitoring. Science 286, 531–537 (1999)
3. Berns, A.: Cancer: Gene Expression Diagnosis. Nature 403, 491–492 (2000)
4. Su, A.I., Welsh, J.B., Sapinoso, L.M., Kern, S.G., Dimitrov, P., Lapp, H., Schultz, P.G., Powell, S.M., Moskaluk, C.A., Frierson Jr., H.F., Hampton, G.M.: Molecular Classification of Human Carcinomas by Use of Gene Expression Signatures. Cancer Res. 61, 7388–7393 (2001)
5. Lu, Y., Han, J.: Cancer Classification Using Gene Expression Data. Inform. Systems 28, 243–268 (2008)
6. Guyon, I., Weston, J., Barnhill, S., Vapnik, V.: Gene Selection for Cancer Classification using Support Vector Machines. Machine Learning 46, 389–422 (2002)
7. Hamadeh, H.K., Bushel, P.R., Jayadev, S., DiSorbo, O., Bennett, L., Li, L., Tennant, R., Stoll, R., Barrett, J.C., Paules, R.S., Blanchard, K., Afshari, C.A.: Prediction of Compound Signature Using High Density Gene Expression Profiling. Toxicological Sciences 67, 232–240 (2002)
8. Dash, M., Liu, H.: Feature Selection for Classification. Intelligent Data Analysis 1, 131–156 (1997)

9. Guo, Z., Zhang, T., Li, X., Wang, Q., Xu, J., Yu, H., Zhu, J., Wang, H., Wang, C., Topol, E.J., Wang, Q., Rao, S.: Towards Precise Classification of Cancers Based on Robust Gene Functional Expression Profiles. BMC Bioinformatics 6 (2005)
10. Lee, E., Chuang, H.-Y., Kim, J.-W., Ideker, T., Lee, D.: Inferring Pathway Activity Toward Precise Disease Classification. PLoS Comput. Biol. 4, e1000217 (2008)
11. Sootanan, P., Prom-on, S., Meechai, A., Chan, J.H.: Pathway-Based Microarray Analysis for Robust Disease Classification. Neural Comput. & Applic. (2011), doi:10.1007/s00521-011-0662-y
12. Chan, J.H., Sootanan, P., Larpeampaisarl, P.: Feature Selection of Pathway markers for Microarray-Based Disease Classification Using Negatively Correlated Feature Sets. In: 2011 International Joint Conference on Neural Networks, pp. 3293–3299. IEEE Press, New York (2011)
13. Kanehisa, M., Araki, M., Goto, S., Hattori, M., Hirakawa, M., Itoh, M., Katayama, T., Kawashima, S., Okuda, S., Tokimatsu, T., Yamanishi, Y.: KEGG for Linking Genomes to Life and the Environment. Nucleic Acids Res. 36, D480–D484 (2008)
14. Pawitan, Y., Bjohle, J., Amler, L., Borg, A.L., Egyhazi, S., Hall, P., Han, X., Holmberg, L., Huang, F., Klaar, S., Liu, E.T., Miller, L., Nordgren, H., Ploner, A., Sandelin, K., Shaw, P.M., Smeds, J., Skoog, L., Wedren, S., Bergh, J.: Gene Expression Profiling Spares Early Breast Cancer Patients from Adjuvant Therapy: Derived and Validated in Two Population-Based Cohorts. Breast Cancer Res. 7, R953–R964 (2005)
15. Wang, Y., Klijn, J.G., Zhang, Y., Sieuwerts, A.M., Look, M.P., Yang, F., Talantov, D., Timmermans, M., Meijer-van Gelder, M.E., Yu, J., Jatkoe, T., Berns, E.M.J.J., Atkins, D., Foekens, J.A.: Gene-expression profiles to predict distant metastasis of lymph-node-negative primary breast cancer. Lancet 365, 671–679 (2005)
16. Edgar, R., Domrachev, M., Lash, A.E.: Gene Expression Omnibus: NCBI Gene Expression and Hybridization Array Data Repository. Nucl. Acids Res. 30, 207–210 (2002)
17. Hall, M., Frank, E., Holmes, G., Pfahringer, B., Reutemann, P., Witten, I.H.: The WEKA data mining software: An update. SIGKDD Explorations 11 (2009)
18. Liao, J.G., Chin, K.V.: Logistic Regression for Disease Classification using Microarray Data: Model Selection in a Large p and Small n Case. Bioinformatics 23(15), 1945–1951 (2007)
19. Goldberg, D.E.: Genetic Algorithms in Search, Optimization and Machine Learning. Addison-Wesley (1989)

Person Identification Using Electroencephalographic Signals Evoked by Visual Stimuli[*]

Jia-Ping Lin[1], Yong-Sheng Chen[1,2,**], and Li-Fen Chen[3]

[1] Inst. of Biomedical Engineering, National Chiao Tung University, Hsinchu, Taiwan
[2] Dept. of Computer Science, National Chiao Tung University, Hsinchu, Taiwan
[3] Inst. of Brain Science, National Yang-Ming University, Taipei, Taiwan

Abstract. In this work we utilize the inter-subject differences in the electroencephalographic (EEG) signals evoked by visual stimuli for person identification. The identification procedure is divided into classification and verification phases. During the classification phase, we extract the representative information from the EEG signals of each subject and construct a many-to-one classifier. The best-matching candidate is further confirmed in the verification phase by using a binary classifier specialized to the targeted candidate. According to our experiments in which 18 subjects were recruited, the proposed method can achieve 96.4% accuracy of person identification.

Keywords: EEG, person identification, VEP.

1 Introduction

Conventional person identification methods include passwords, smart cards, and a variety of biometric techniques. Passwords and smart cards are widely-used because of the advantage of convenience. However, smart cards might be stolen, simple passwords might be deciphered, and complicated passwords might be forgotten. Current biometric features such as iris, fingerprints, face, voice, palm, and gait do not suffer the above-mentioned disadvantages, but they can be stolen, duplicated, or even provided under violent threats. Brainwave is an emerging biometric feature for person identification because of its uniqueness and consistency. Moreover, brainwave is difficult to steal or duplicate and the characteristics embedded in the brainwave when the subject is under threat are hardly the same as those in normal situation. These advantages promote brainwaves as new keys to safer person identification systems.

---

[*] This work was supported in part by the MOE ATU program, Taiwan National Science Council under Grant Numbers: NSC-99-2628-E-009-088 and NSC-100-2220-E-009-059, and the UST-UCSD International Center of Excellence in Advanced Bioengineering sponsored by the Taiwan National Science Council I-RiCE Program under Grant Number: NSC-99-2911-I-009-101.

[**] Corresponding author.

B.-L. Lu, L. Zhang, and J. Kwok (Eds.): ICONIP 2011, Part I, LNCS 7062, pp. 684–691, 2011.

Among all the non-invasive brainwave acquisition modalities, electroencephalography (EEG) has the advantages of portability, easy operation, high temporal resolution, and low costs. To evaluate the uniqueness and consistency of the
characteristics in EEG signal, the work in [6] confirmed that the inter-subject
variation of EEG spectra where different subjects administered the same task
was larger than the intra-subject variation where the EEG signals of the same
subject were repeatedly acquired for several times. At first resting data was used
for person recognition and the identification rate ranged from 72 to 85% [10].
In 2003, Palaniappan and Ravi investigated the task-related EEG signals. By
extracting features from visual evoked potentials (VEPs), the identification accuracy was improved to be larger than 90% [9]. The features in EEG signals
include autoregressive (AR) coefficients, coherence, and cross-correlation [7]. In
[1] the event-related potentials (ERPs) were utilized for person identification.
This work used the images of self-relevant objects as the visual stimuli and selected prominent channels related to this experiment. Temporal domain features
such as P100, N170, and N250 were used in the signal analysis [4]. For simplicity
and practicability, the work [5] classified subjects simply by thresholding the
EEG power spectrum.

In this paper we present a person identification system using EEG signals.
Because resting state is prone to be more fluctuating, we adopt task-related
EEG signals evoked by visual stimuli in this work. Representative information is
extracted from the EEG signals of subjects and are used to train a many-to-one
classifier for person classification. The best-matching candidate of each classification is further verified by using a binary classifier to exclude the intruder.

2   Materials

2.1   Participants and Paradigm

Eighteen subjects participated in this study (age ranges from 21 to 33 years
with mean 24 years, twelve males). All the subjects have normal or corrected-
to-normal visions. For five participants among all the subjects, EEG data were
acquired two times with an interval of more than one week.

The paradigm of data acquisition in this study is shown in Fig. 1. The subject
was seated comfortably in a silent room and was asked to watch a monitor
screen. The visual stimulus, an image containing either a small disk or a large
one (five times larger than the small one), was presented for one second followed
by another second of fixation image of a cross. The frequency ratio between the
stimulus images is one (large disk) to three (small disk). Around 250 trials were
acquired for each participant.

2.2   EEG Recording and Preprocessing

Thirty-two standard scalp electrodes were placed according to the International
10-20 System of Electrode Placement. We picked the channels related to the

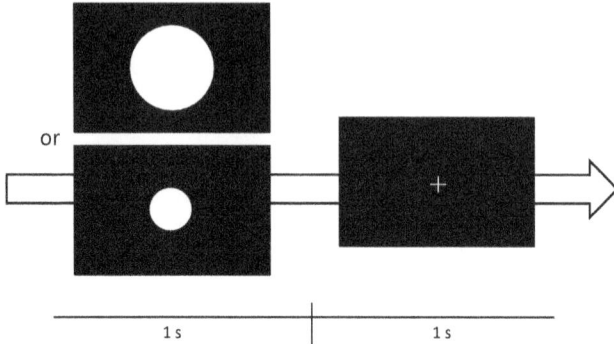

Fig. 1. Paradigm for data acquisition in this study. A trial consists of one-second stimulus, an image containing either a small disk or a large one, and one-second fixation.

visual stimuli and P300 component in the frontal, frontal-central, parietal, and occipital regions [3]. The ten channels we selected were Fz, FCz, Cz, CPz, P3, Pz, P4, O1, Oz, and O2. This process will reduce the quantity of data and eliminate the activities which are not induced by the events. The EEG data were recorded with Scan 4.3 software and the sampling rate for data acquisition was 500Hz. The earlobe electrodes A1 and A2 provided the reference. Signals were digitally filtered within the 5-30 Hz band.

We used EEGLAB 9.0 [2] to perform the following signal preprocessing procedure. The EEG data were first segmented into epochs starting from one second before the stimulus onset to one second after stimulus onset. The baseline correction was applied to remove the DC drift. Epochs with burst activities during the post-stimulus period were rejected (with the threshold values -50μV and 50μV). The trials evoked by the large disk events were used in the following person identification analysis.

3   Methods

3.1   Feature Extraction

For each of the EEG channels, we applied a series techniques to extract features. These techniques, described in the following, include dimension reduction, morphological operation, power spectrum, and stochastic modeling.

Dimension Reduction. Principal component analysis (PCA) is a method for reducing feature dimension. Its main idea is to find a set of basis, usually with a much smaller dimension, to represent the original data set while preserving as much as information measured by the variance of data distribution. If there is an embedded non-linear manifold lying in a high-dimensional space and the dimension of the manifold is relatively low, this manifold can be well represented in a low-dimensional space [8]. Therefore, we also applied the locally linear embedding (LLE) method to transform the data to a low-dimensional space while maintaining the manifold structure manifested in the original high-dimensional space. Firstly, we find a set of nearest neighbors for each data point X_i in D-dimensional Euclidean space. Then we reconstruct, or represent, each data point by a linear combination of its neighbors X_{ij} with weightings W_{ij} as the contribution of the neighbor X_{ij} to this linear combination for X_i. The reconstruction error is:

$$\mathbf{E}(W) = \sum_i |X_i - \sum_j W_{ij} X_{ij}|^2 , \tag{1}$$

where the sum of the weightings for each data point X_i equals one. The data point X_i can be mapped to the corresponding point Y_i in a low-dimensional space as:

$$Y_i = \sum_j W_{ij} Y_{ij} , \tag{2}$$

where the point Y_{ij} is the point in low-dimensional space corresponding to X_{ij} in the original high-dimensional space.

Morphological Features. The latency and amplitude of each EEG epoch were computed as the morphologic features which contain VEPs (with the time interval from 50 ms to 150 ms after stimulus onset) and ERPs (with the time interval from 250 ms to 400 ms after stimulus onset).

Frequency Features. The discrete Fourier transform (DFT) were used to compute the power spectrum for each epoch. In this work we focus on the frequency band from 5 Hz to 30 Hz.

Stochastic Modeling. Considering the EEG signal as an autoregressive (AR) process, we used the Yule-Walker equations to estimate the AR coefficients as the features. To fit a p th-order AR model to the EEG data $X(t)$, we minimize the following prediction error by using the least squares regression:

$$X(t) = \sum_{i=1}^{P} a(i) X(t - i) + e(t) , \tag{3}$$

where $a(i)$ are the auto regression coefficients, $e(t)$ represents the white noise, and the time series can be estimated by a linear differential equation.

Time-Frequency Model. The wavelet transform uses a set of time-scale basis to represent the original signal. Here we applied the Daubechies wavelets to transform the time-domain EEG signals and obtained 250 coefficients as the time-frequency features.

3.2   Classification

For classification, we employed the support vector machine (SVM) and the k-nearest neighbor (kNN) search method (k=9) as the classifier. To fairly evaluate the accuracy of classification, we apply the 8-fold cross validation that separate EEG data into training and testing data to obtain the average classification accuracy for person identification.

3.3   Verification

The purpose of the verification procedure is to reconfirm the best-matching result of classification. For each of the eighteen subjects, we trained a SVM binary classifier by using two groups of training data including EEG data of the targeted subject and those of all others. We evaluate the binary classifier for verification in terms of the true acceptance rate (TAR) and the false acceptance rate (FAR). The best-matching subject from the classification procedure is verified by the corresponding binary classifier. In addition, we modified the false classified data in classification phase through iterative verification. The probability estimate, which is a confidence level of classification, determines an ordered list of candidates having confidence levels larger than 80% of that of the best-matching candidate.

4   Results

4.1   Temporal Characteristics in the Acquired Signals

We first verified whether the resting EEG or ERP is better for distinguishing subjects' identities. By applying the SVM classifier to categorize the pre-stimulus (500 ms before onset) EEG signals among the eighteen subjects, the classification accuracy was 12.2%. When the post-stimulus (500 ms after onset) ERP signals were used for person identification, the classification accuracy achieved 25.3%. Therefore the ERP contains more information for person identification than resting EEG does.

4.2   Accuracy in the Classification Phase

Table 1 shows the classification accuracy comparison among seven features extracted from the 1000ms post-stimulus EEG signals with respect to single trial, average of two trials, SVM, and kNN. The average of two trials can achieve higher classification accuracy compared to single trial data because of higher signal-to-noise ratio. Regarding the classifier, SVM outperforms kNN with respect to various features.

Among the seven kinds of features, power spectrum achieves the best classification accuracy while the latency and amplitude generally lead to poor results. Fig. 2 shows the power spectrum of different subjects with the frequency band ranging from 5 Hz to 30 Hz. We can see that within-subject variation of

Table 1. Results of classification with different features and different classifiers. The data of each subject acquired in the same experiment.

	SVM		kNN	
Feature	Single trial	Avg (2 trials)	Single trial	Avg (2 trials)
Raw data	29.31%	80.86%	23.47%	76.38%
LLE	30.81%	86.69%	28.13%	83.44%
PCA	27.74%	83.48%	25.32%	81.28%
Latency	11.59%	35.23%	10.21%	33.56%
Amplitude	38.53%	50.82%	36.23%	45.19%
Power spectrum	72.03%	91.61%	60.01%	85.92%
AR	53.52%	62.54%	50.96%	60.57%
Wavelet	27.27%	85.41%	22.92%	77.26%

Table 2. TAR in the verification phase, which is the percentage of the best-matching candidates in the classification phase that are accepted in the verification phase

Subject	1	2	3	4	5	6	7	8	9
TAR (%)	97.14	98.70	98.81	100	100	96.43	97.62	100	96.30
Subject	10	11	12	13	14	15	16	17	18
TAR (%)	97.96	100	100	98.57	100	100	100	100	100

the spectra of different trials is smaller than inter-subject variation. In order to accommodate different information of the best two features, we combined the power spectrum and LLE features after normalization and achieve 97.1% of classification accuracy.

4.3   Accuracy in the Verification Phase

The true acceptance rate (TAR) measures the percentage of the best-matching candidates in classification that are accepted by the binary classifier of verification. Table 2 shows the TARs of eighteen subjects and their average is 97.9%. The false acceptance rate (FAR) is zero, that means all the false classified data were successfully rejected in the verification phase. After iterative verification the overall accuracy of our system is 96.4%.

4.4   Classification Accuracy Over Time

For five participants among all the eighteen subjects, EEG data were acquired two times with an interval of more than one week. The goal is to verify whether the EEG data of the same subject is sufficiently stable for person identification over a period of time. The average accuracy of the classification phase is 93.2%. Table 3 shows the TAR, FAR and results of iterative verification. After iterative

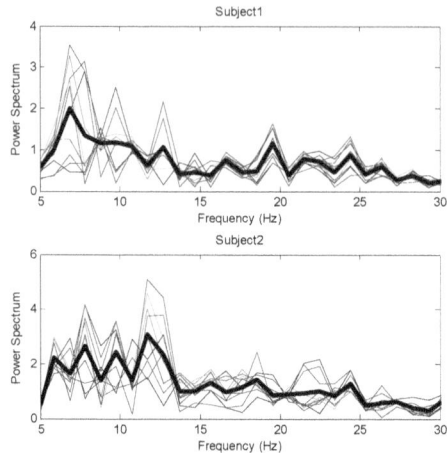

Fig. 2. The power spectrum of ten trials of two subjects (thick black lines represent the averages of ten trials). Each trial shows the average results of ten channels.

Table 3. TAR, FAR, and results of iterative verification of data acquired from different days

Subject	3	5	8	12	13
TAR (%)	78.57	100	97.92	83.33	100
Accepted/False classified	0/1	4/6	0/0	0/0	11/12
FAR (%)	0	66.67	-	-	91.67

verification, the overall identification accuracy of our system is 85.7%, indicating that the the performance of our system slightly degrades over time. One possible remedy is to retrain the classifier by adding the data acquired over time so that the classifier can be adapted to each subject. By using the two sets of EEG data acquired at different times, the average accuracy of the classification phase is improved from 93.2% to 98.4%, TAR is increased from 90.6% to 97.7%, and FAR is decreased from 79.0% to 0%. After iterative verification, the overall identification accuracy of our system is improved from 85.7% to 96.8%.

5 Discussion and Conclusions

The major causes affecting the accuracy of person identification using EEG signals include both external and internal interferences. The external interferences deteriorate the quality of acquired signal whereas the internal interferences result in signal instability over time. From the calculated correlation between EEG trials of different subjects, the EEG data of subjects having high correlation to those of other subjects have more classification errors. Compared with the inter-subject correlation, the intra-subject correlation between EEG trials acquired at different times is higher. Therefore, the brainwave signals are suitable for biometric measures for person identification.

We have proposed a person identification system using visual-evoked EEG signals. According to our experiments, we concluded that the combination of power spectrum and LLE can extract informative features for distinguishing subjects. The identification system contains the classification and verification phases. In the classification phase, we use a multi-class classifier to perform a one-to-many comparison for each acquired data. In the iterative verification phase, the best-matching candidates are furthered verified sequentially by binary classifiers according to their matching levels. The overall person identification accuracy of the proposed system can achieve 96.4%.

References

1. Berlad, I., Pratt, H.: P300 in response to the subject's own name. Electroencephalogr. Clin. Neurophysiol. 96, 472–474 (1995)
2. Delorme, A., Makeig, S.: EEGLAB: an open source toolbox for analysis of single-trial EEG dynamics including independent component analysis. Journal of Neuroscience Methods 134, 9–21 (2004)
3. Donchin, E., Coles, M.G.H.: Is the P300 Component a Manifestation of Context Updating. Behavioral and Brain Sciences 11, 357–374 (1988)
4. Miyakoshi, M., Nomura, M., Ohira, H.: An ERP study on self-relevant object recognition. Brain Cogn. 63, 182–189 (2007)
5. Miyamoto, C., Baba, S., Nakanishi, I.: Biometric Person Authentication Using New Spectral Features of Electroencephalogram (EEG). In: Ispacs 2008, pp. 130–133 (2008)
6. Napflin, M., Wildi, M., Sarnthein, J.: Test-retest reliability of EEG spectra during a working memory task. NeuroImage 43, 687–693 (2008)
7. Riera, A., Soria-Frisch, A., Caparrini, M., Grau, C., Ruffini, G.: Unobtrusive biometric system based on electroencephalogram analysis. Eurasip Journal on Advances in Signal Processing (2008)
8. Roweis, S.T., Saul, L.K.: Nonlinear dimensionality reduction by locally linear embedding. Science 290, 2323–2326 (2000)
9. Singhal, G.K., RamKumar, P.: Person identification using evoked potentials and peak matching. In: Biometrics Symposium, pp. 156–161 (2007)
10. Poulos, M., Rangoussi, M.: Person identification from the EEG using nonlinear signal classification. Methods of Information in Medicine 41, 64–75 (2002)

Generalised Support Vector Machine
for Brain-Computer Interface

Trung Le, Dat Tran, Tuan Hoang, Wanli Ma, and Dharmendra Sharma

Faculty of Information Sciences and Engineering
University of Canberra, ACT 2601, Australia
{trung.le,dat.tran,tuan.hoang,wanli.ma,
dharmendra.sharma}@canberra.edu.au

Abstract. Support vector machine (SVM) and support vector data description (SVDD) are the well-known kernel-based methods for pattern classification. SVM constructs an optimal hyperplane whereas SVDD constructs an optimal hypersphere to separate data between two classes. SVM and SVDD have been compared in pattern classification experiments, however there is no theoretical work on comparison of these methods. This paper presents a new theoretical model to unify SVM and SVDD. The proposed model constructs two optimal points which can be transformed to hyperplane or hypersphere. Therefore SVM and SVDD are regarded as special cases of this proposed model. We applied the proposed model to analyse the dataset III for motor imagery problem in BCI Competition II and achieved promising results.

Keywords: Kernel Methods, Support Vector Machine, Support Vector Data Description, Brain-Computer Interface.

1 Introduction

Brain-Computer Interface (BCI) is an emerging research field attracting a lot of research effort from researchers around the world. Its aim is to build a new communication channel that allows a person to send commands to an electronic device using his/her brain activities [1]. BCI systems have been provided for severely handicapped people using their brain signals, and for patients with brain diseases such as epilepsy, dementia and sleeping disorders [2].

The performance of a BCI system depends on data pre-processing, feature extraction and classification methods used to build the classifier in that BCI system. Currently, numerous pre-processing, feature extraction and classification methods have been proposed and explored for BCI systems. For data pre-processing and feature extraction, the following fetaures have been applied: raw electroencephalograph (EEG) signals [3][4], band powers [5], power spectral density values [6], autoregressive parameters [7] and wavelet features [8]. For classification, perceptron and multi layer perceptron [7], various SVMs [7][9] and linear discriminant analysis [7] methods have been applied.

B.-L. Lu, L. Zhang, and J. Kwok (Eds.): ICONIP 2011, Part I, LNCS 7062, pp. 692–700, 2011.

For feature extraction in EEG classification problems, most of currently used features are uni-variate and extracted from single channels. However EEG signals recorded from multiple channels for a brain activity are correlated, features extracted from the EEG signals should reflect relationships between those channels. To drive this remark, in this paper we propose a feature extraction method [10] that uses windows as seen in the work of Anderson et. al. [11] and Brunner et. al. [12]. For the signal of a pair of channels, instead of using a single window, we use short moving windows having the same size and then combine them together with a pre-defined overlapping window parameter. We calculate bivariate autoregressive model parameters for the current window and then slide to the next window until end of the signal. All BVAR parameters are concatenated from the first window to the last one with a pre-defined moving window step or overlapping window to form a feature vector. Depending on the nature and well known biological knowledge of mental tasks, we can select several pairs of channels, and concatenate their corresponding feature vectors together to form an entire feature vector for a trial.

For classification, Support Vector Machine (SVM) and Support Vector Data Description (SVDD) are widely used in various real-world classification problems. SVM constructs an optimal hyperplane [13] and SVDD constructs an optimal hypersphere [14][15][16] to classify data. The choice of SVM or SVDD depends on the ratio of data in two classes. It is seen that hyperplane is suitable for balanced datasets whereas hypersphere is for imbalanced datasets where a dataset appears as a dominant (major) class. SVM and SVDD have been compared in pattern classification experiments, however there is no theoretical work on comparison of these methods. This paper presents a new theoretical model to unify SVM and SVDD. The proposed model constructs two optimal points which can be transformed to hyperplane or hypersphere. Therefore SVM and SVDD are regarded as special cases of this proposed model. Moreover the proposed model can offer a set of decision boundaries consisting of hyperplane and hypersphere shapes and hence it can provide a better data description to the various datasets than the hyperplane and hypersphere. We apply the proposed model to analyse the dataset III for motor imagery problem in BCI Competition II and achieve promising results.

2   Generalised Support Vector Machine (GSVM)

2.1   Key Lemma

This lemma displays the relevance between the difference of square distances to two points and the margin in the affine space R^d.

Lemma 1. *Let M, A and B be the points in the affine space R^d. Let (H) : $w^T x + b = 0$ be the mid-perpendicular hyperplane of segment AB. The following equality holds:*

$$MA^2 - MB^2 = 2dist(M, \mathrm{H}).AB.sign(B, H).sign(M, H)$$

where $dist(M, H)$ is distance from M to (H) and $sign(P, II) = sign(w^T P + b)$.

2.2   The Idea of GSVM

We consider the original SVM in an alternative view to derive the new model. It can be reformulated in an alternative form as follows:

$$\max_{w,b} \gamma \tag{1}$$

subject to

$$dist(M, H) \geq \gamma \text{ for all } M \in \phi(X) \tag{2}$$

where X is the training set, $\phi(X)$ is image of X through the transformation $\phi(.)$, M is a feature vector in the training set, $(H) : w^T \phi(x) + b = 0$ is a hyperplane in feature space, and $dist(M, H)$ is distance from M to (H).

By referring to Lemma 1, it is able to transform the above optimisation problem to an equivalent form as follows:

$$\max_{A,B} \left(\frac{\gamma}{AB} \right) \tag{3}$$

subject to

$$\left(MA^2 - MB^2 \right) y_M \geq \gamma \text{ for all } M \in \phi(X) \tag{4}$$

where y_M is the label of M, A and B are two points in feature space, and mid-perpendicular hyperplane of segment AB is the optimal hyperplane.

We convert the above optimisation problem to a more concrete form. Note that the function in (3) is margin, i.e. the distance from the closest point in the training set to the mid-perpendicular hyperplane of segment AB and this margin is invariant when stretching or shrinking AB as long as the mid-perpendicular hyperplane is unchanged. Therefore, without losing of generality we assume that:

$$\min_{M} \left(MA^2 - MB^2 \right) y_M = 1 \tag{5}$$

Consequently, we achieve new equivalent optimisation problem:

$$\max_{A,B} \left(\frac{1}{AB} \right) \quad or \quad \min_{A,B} \left(AB^2 \right) \tag{6}$$

subject to

$$\left(MA^2 - MB^2 \right) y_M \geq 1 \text{ for all } M \in \phi(X) \tag{7}$$

Similar to SVM, we have a chance to extend the problem by applying the slack variables as follows

$$\min_{A,B} \left(AB^2 + C \sum_{i=1}^{n} \xi_i \right) \tag{8}$$

subject to

$$\left(M_i A^2 - M_i B^2 \right) y_i \geq 1 - \xi_i \text{ for all } M_i \in \phi(X)$$
$$\xi_i \geq 0, i = 1, \ldots, n \tag{9}$$

where y_i is the label of M_i.

2.3   Formulation of GSVM

To resolve the new optimisation problem (OP), for simplicity we can assume that the equation of hyperplane is of $(H) : w^T x = 0$ and it goes through the origin in feature space. Since the objective function keeps invariant when sliding AB along with its mid-perpendicular hyperplane, we can safely suppose that A and B are symmetric through the origin as in Figure 1. Let us denote the coordinate of points A, B and M in feature space as a, b and $\phi(x)$, respectively. We have $a + b = 0$ and the new OP as follows

$$\min_{a,b,\xi} \left(\|b - a\|^2 + C \sum_{i=1}^{n} \xi_i \right) \tag{10}$$

subject to

$$\left(\|\phi(x_i) - a\|^2 - \|\phi(x_i) - b\|^2 \right) y_i \geq 1 - \xi_i, \ i = 1, \dots, n$$
$$\xi_i \geq 0, i = 1, \dots, n \tag{11}$$

where ϕ is a transformation from the input space to the feature space.

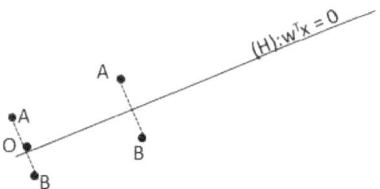

Fig. 1. The hyperplane goes through origin and A, B are symmetric through the origin

We generalize the above OP by introducing a new parameter called "curving degree" k. As explained later, this parameter governs the curving degree of decision boundary. When $k = 1$, it is completely straight. Moreover, when $k \neq 1$, it becomes a spherical form with varied curving degree. We obtain the new OP as follows:

$$\min_{a,b,\xi} \left(\|b - a\|^2 + C \sum_{i=1}^{n} \xi_i \right) \tag{12}$$

subject to

$$\left(\|\phi(x_i) - a\|^2 - k\|\phi(x_i) - b\|^2 \right) y_i \geq 1 - \xi_i, i = 1, \dots, n$$
$$\xi_i \geq 0, i = 1, \dots, n \tag{13}$$

where ϕ is a transformation from input space to feature space.

2.4   Solution

To derive the new OP, we refer to the Karush-Kuhn-Tucker (KKT) theorem. We come up with the following OP:

$$
\min_{\alpha} \left(k \sum_{i=1}^{n} \sum_{j=1}^{n} y_i y_j K(x_i, x_j) \alpha_i \alpha_j \right.
$$
$$
\left. - \sum_{i=1}^{n} \left((k-1) y_i K(x_i, x_i) + 1 \right) \alpha_i \right)
\tag{14}
$$

subject to

$$
\sum_{i=1}^{n} \alpha_i y_i = \frac{k-1}{k} \ and \ 0 \le \alpha_i \le C, i = 1, \ldots, n
\tag{15}
$$

For classifying unknown vector x, the following decision function is used:

$$
\begin{aligned}
f(x) &= sign(\|\phi(x) - a\|^2 - k\|\phi(x) - b\|^2) \\
&= sign((1-k)K(x,x) - 2(1+k)\phi(x)a + (1-k)\|a\|^2) \\
&= sign((1-k)K(x,x) - 2(1+k)\phi(x)\tfrac{-k}{k+1} \sum_{i=1}^{n} \alpha_i y_i \phi(x_i) \\
&+ (1-k)(\tfrac{-k}{k+1} \sum_{i=1}^{n} \alpha_i y_i \phi(x_i))^2) \\
&= sign((1-k)K(x,x) + 2k \sum_{i=1}^{n} \alpha_i y_i K(x_i,x) \\
&+ \tfrac{k^2(1-k)}{(k+1)^2} \sum_{i=1}^{n} \sum_{j=1}^{n} \alpha_i \alpha_j y_i y_j K(x_i,x_j))
\end{aligned}
\tag{16}
$$

2.5   Interpretation of GSVM

The decision boundary is of the following form:

$$
\|\phi(x) - a\|^2 - k\|\phi(x) - b\|^2 = 0
\tag{17}
$$

To realize the shape of decision boundary, we need to interpret the above equation. Two following lemmas are necessary.

Lemma 2. *Let A and B be two distinct points in the affine space R^d. Constructing the fixed point P such that $\vec{PA} = k\vec{PB}(k \neq 1)$. Then for all points M, the following equality holds:*

$$
MA^2 - kMB^2 = (1-k)MP^2 + PA^2 - kPB^2
\tag{18}
$$

Lemma 3. *Let A and B be two distinct points in the affine space R^d. The pilot of equation $MA^2 - kMB^2 = 0$ is one of the following forms:*
i) If $k = 1$ then it is a hyperplane that goes through the midpoint of AB and perpendicular to AB.
ii) If $k \neq 1$ then it is a hypersphere with center P where $\vec{PA} = k\vec{PB}$.

Lemma 3 reveals that the pilot of decision boundary ruled by (17) is either hyperplane or hypersphere according to parameter k. This parameter also governs the curving degree of decision boundary. That is why it is called curving degree parameter. Moreover, the range of expression in GSVM is wider than both SVM and SVDD. Therefore, it appears that new model is easier to fit the real datasets than both SVM and SVDD in terms of hyperplane and hypersphere.

3 Combined Short-Window Bivariate Autoregressive Feature (CSWBVAR)

3.1 Autoregressive model

Univariate autoregressive (UVAR) model is used to model single time-series with assumption that each value of the series can be estimated by taking a weighted sum of its previous values, plus white noise. Whereas, bivariate autoregressive (BVAR) model is used to model two time-series with assumption that each value of the two series can be estimated by taking a weighted sum of not only the previous values of the same series but also values of other series.

Let $X(t) = [X_1(t), X_2(t), ..., X_n(t)]^T$ be n time-series in random process. In BCI system, n is number of channels used for collecting brain signals. Let d be the number of data samples of n channels. The pth-order MultiVariate AutoRegressive (MVAR) model is formulated as follows:

$$X(t) = \sum_{i=1}^{p} A_i X(t-i) + E(t) \tag{19}$$

In (19), A_i, $i = 1 \dots p$ are $n \times n$ coefficient matrices and $E(t)$ is noise vector which is a zero mean uncorrelated with the covariance matrix Σ. We assume that $X(t)$ is a stationary process. If $n = 1$, we have univariate AR model, and $n = 2$, bivariate AR model.

To estimate A_i and Σ, we transfer (19) to Yule-Walker equations by multiplying (19) from the right with $X^T(t-i), i = 1 \dots p$ and then taking expectation:

$$R(-k) + \sum_{i=1}^{p} A_i R(-k+i) = 0 \tag{20}$$

where $R(l)$ is the covariance matrix of lag l of $X(t)$.

To solve these equations with a specific order p, we use Levinson, Wiggins, Robinson (LWR) algorithm. The correct order is identified by minimizing the Akaike Information Criterion defined as in (21).

$$AIC(p) = 2log[det(\Sigma)] + 2\frac{n^2 p}{N_{total}} \tag{21}$$

where N_{total} is the total number of data points from all trials.

In this paper, after visually inspecting the AIC curve, we chose the AR model with order 6 which results to a local minimum.

3.2 CSWBVAR Feature

Given a pair of channels and a time window, we calculate its BVAR coefficients. This sequence will form a feature vector of the window having size of $4 \times p$ coefficients. The feature vector of the trial is defined as the concatenation of these sequences together with a pre-defined moving window step. Let w be the window size and s be the moving window step, the feature vector size of a trial is

$$4 \times \frac{d - w + 1}{s} \times p \qquad (22)$$

Choosing the correct window size and moving window step parameters is the most important task in our model. We can assume EEG signals are stationary by setting the window size to a value between 40ms and 100ms. With the sampling rate of 128Hz, each window has from 5 to 12 data points. They are too small to estimate AR model coefficients. Increasing window size requires us to use adaptive algorithm to estimate AR model parameters. Fortunately, LWR algorithm is an adaptive one. In our experimental design, we considered different window sizes including 12, 32, 64, and 128 data points and the overlapping part between two consecutive windows is 25%, 50% and 75% of the window size.

4 Experimental Results

The aim of our experiment is to demonstrate that the combination of our two methods for feature extraction and classification will significantly improve the performance of BCI systems. The chosen data set which has appropriate number of channels was the well-known data set III provided by Department of Medical Informatics, Institute of Biomedical Engineering, Graz University of Technology for motor imagery classification problem in BCI Competition II [17]. In data collection stage, a female normal subject was asked to sit in a relaxing chair with armrests and tried to control a feedback bar by means of imagery left or right hand movements. The sequences of left or right orders are random. The experiment consisted of 7 runs with 40 trials in each run. There were 280 trials in total and each of them lasted 9 seconds of which the first 3 seconds are used for preparation. The collected data set was equally divided into two sets for training and testing. The data was recorded in three EEG channels which were C3, Cz and C4, sampled at 128Hz, and filtered between 0.5 Hz and 30 Hz. Most of current algorithms only applied to the channels C3 and C4, and ignored the channel Cz. They argued that from brain theory, signals from channel Cz provide very little meaning to motor imagery problem. We truncated the first 3 seconds of each trial and used the rest for further processing. All trials are pre-processed by subtracting the ensemble mean of all trials. For each trial we extracted both CSWUVAR and CSWBVAR parameters with different window sizes and moving window steps. We considered window sizes including 256, 384, and 512 data points, corresponding to 100ms-, 250ms-, 500ms-, and 1s-segments. As with other previous work, we did not try experiments with segment's size greater than 1s due to keeping signal approximately stationary and being comfortable with nature of brain signal. The overlapping part between two consecutive windows was set to 25%, 50% and 75% of the window size.

The popular RBF Kernel $K(x, x') = e^{-\gamma \|x-x'\|^2}$ is applied whereas the parameter γ is varied in grid $\{2^i : i = 2j, j = -4, \ldots, 1\}$. Moreover, for GSVM to ensure that the decision boundary goes through the origin we applied the expanded Gaussian transform $\phi'(x) = [\phi(x), 1]$ which leads to the expanded RBF Kernel where $K'(x, x') = e^{-\gamma \|x-x'\|^2} + 1$. The trade-off parameter C is selected in grid $\{2^i : i = 2j, j = -4, \ldots, 1\}$. The curving degree k is searched in grid $\{0.6 + 0.2i : i = 0, \ldots, 5\}$, and ten folds cross-validation was employed.

We made comparison of three combinations of feature extraction and classification method which are CSUVAR+SVM, CSWBVAR+SVM, and CSWBVAR+GSVM. The chart in Figure 2 shows that the combination of new feature extraction and classification methods significantly improve the performance of classifier.

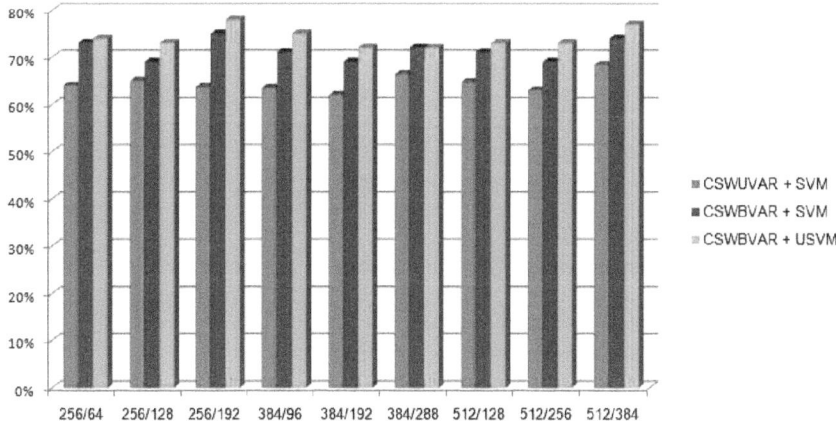

Fig. 2. Experimental results on extracted datasets with various window sizes/moving sizes

5 Conclusion

We have proposed a combination of new feature extraction and new classification methods to improve the performance of BCI systems. The experiment on the dataset III for motor imagery problem in BCI Competition II shows that our proposed methods can provide a significant improvement on classification over BCI datasets.

References

1. Babiloni, F., Cichocki, A., Gao, S.: Brain-Computer Interfaces: Towards Practical Implementations and Potential Applications. Computational Intelligence and Neuroscience, 1–2 (2007)
2. Lotte, F.: PhD thesis: Study of Electroencephalographic Signal Processing and Classification Techniques towards the use of Brain-Computer Interfaces in Virtual Reality Applications (2008)
3. Kaper, M., Meinicke, P., Grossekathoefer, U., Lingner, T., Ritter, H.: BCI competition 2003 data set IIb: support vector machines for the P300 speller paradigm. IEEE Trans. Biomed. Eng. 51, 1073–1076 (2004)
4. Rakotomamonjy, A., Guigue, V.: BCI competition III: dataset II- ensemble of SVMs for BCI P300 speller. IEEE Trans. Biomed. Eng. 55, 1147–1154 (2008)
5. Kalcher, J., Flotzinger, D., Neuper, C., Glly, S., Pfurtscheller, G.: Graz brain-computer interface II: towards communication between humans and computers based on online classification of three different EEG patterns. Med. Bio. Eng. Computing 34(5), 382–388 (1996)
6. Solhjoo, S., Moradi, M.H.: Mental task recognition: A comparison between some of classification methods. In: BIOSIGNAL Int. Conf. EURASIP, pp. 24–26 (2004)
7. Garrett, D., Peterson, D.A., Anderson, C.W., Thaut, M.H.: Comparison of linear, nonlinear, and feature selection methods for EEG signal classification. IEEE Trans. Neural System and Rehabilitation Eng. 11, 141–144 (2003)
8. Bostanov, V.: BCI competition 2003data sets Ib and IIb: Feature extraction from event-related brain potentials with the continuous wavelet transform and the t-value scalogram. IEEE Trans. Biomed. Eng. 51, 1057–1061 (2004)
9. Kuncheva, L.I., Rodriguez, J.J.: Classifier ensembles for fMRI data analysis: an experiment. Magnetic Resonance Imaging 28(4), 583–593 (2010)
10. Hoang, T., Tran, D., Nguyen, P., Huang, X., Sharma, D.: Experiments on using Combined Short Window Bivariate Autoregression for EEG Classification. In: Proc. IEEE Internatinal Conferrence on Neural Engineering, pp. 372–375 (2011)
11. Anderson, C.W., Stolz, E.A., Shamsunder, S.: Multivariate autoregressive models for classification of spontaneous electroencephalographic signals during mental tasks. IEEE Trans. Biomed. Eng. 45, 277–286 (1998)
12. Brunner, C., Billinger, M., Neuper, C.: A Comparison of Univariate, Multivariate, Bilinear Autoregressive, and Bandpower Features for Brain-Computer Interfaces. In: Fourth International BCI Meeting, Poster B-22 (2010)
13. Vapnik, V.: The nature of statistical learning theory. Springer, Heidelberg (1995)
14. Tax, D.M.J., Duin, R.P.W.: Support vector data description. Machine Learning 54, 45–56 (2004)
15. Le, T., Tran, D., Ma, W., Sharma, D.: An Optimal Sphere and Two Large Margins Approach for Novelty Detection. In: Proc. IEEE World Congress on Computational Intelligence (WCCI), pp. 909–914 (2010)
16. Wu, M., Ye, J.: A Small Sphere and Large Margin Approach for Novelty Detection Using Training Data with Outliers. IEEE Trans. Pattern Analysis & Machine Intelligence 31, 2088–2092 (2009)
17. BCI Competition II, http://www.bbci.de/competition/ii/

An EEG-Based Brain-Computer Interface for Dual Task Driving Detection

Chin-Teng Lin, Yu-Kai Wang, and Shi-An Chen

Brain Research Center, Department of Computer Science,
National Chiao-Tung University, Hsinchu, 30010, Taiwan
`ctlin@mail.nctu.edu.tw`, `yukaiwang@cs.nctu.edu.tw`,
`sachen@nctu.edu.tw`

Abstract. A novel detective model for driver distraction was proposed in this study. Driver distraction is a significant cause of traffic accidents during these years. To study human cognition under a specific driving task, one virtual reality (VR)-based simulation was built. Unexpected car deviations and mathematics questions with stimulus onset asynchrony (SOA) were designed. Electroencephalography (EEG) is a good index for the distraction level to monitor the effects of the dual tasks. Power changing in Frontal and Motor cortex were extracted for the detective model by independent component analysis (ICA). All distracting and non-distracting EEG epochs could be revealed the existence by self-organizing map (SOM). The results presented that this system approached about 90% accuracy to recognize the EEG epochs of non-distracting driving, and might be practicable for daily life.

Keywords: driver distraction, SOA, EEG, ICA, SOM.

1 Introduction

The drivers divert their attention away from focused driving to the other event means driver distraction. While driving, drivers must continually allocate their brain resources about attention to both driving and non-driving tasks. There is more electrical equipment in the car to help drivers reduce the overhand and enhance the entertainment, such as navigation system, integration center control system, or in-vehicle entertainment system. They would increase risk since drivers may become easily distracted, thus making it likely that the problem of driver distraction [1]. There is increasing evidence that driver distraction is one major cause of car accidents [2]. The National Highway Traffic Safety Administration (NHTSA) reported that distraction accounts for about 20-30% of traffic accidents [3]. Recognizing driver's attention related brain resources during driving is quite important.

In several brain-computer interface (BCI) studies, most approaches are Electroencephalography (EEG) -based, because the EEG system is small and easy to take [4]. They depict a BCI as a pattern recognition system and emphasize the role of classification [5]. And it is also sensitive to variations in cognitive and behavioral states. In this study, we want to monitor the changing of EEG about distracted driving and identify "patterns" of brain activity through the classification algorithms.

B.-L. Lu, L. Zhang, and J. Kwok (Eds.): ICONIP 2011, Part I, LNCS 7062, pp. 701–708, 2011.
© Springer-Verlag Berlin Heidelberg 2011

Several methods have been proposed to analyzed and classify EEG data. During driving, the human may induce the artifact including eye movement, heartbeat, or breath, and switch attention to different levels when they confront driving and non-driving task. Therefore analyzing the EEG data using independent component analysis (ICA) [6] and self-organization map (SOM) [7] has been proposed in this study. The ICA is extensively applied to separation, identify, and localize the EEG sources. By the results of component maps, ICA could show the brain areas of signal sources even though the noise, so the useful brain sources could be selected to analyze the phenomenon. SOM owns two characteristics which differ from traditional cluster algorithm [14-18]. This artificial neural network offers an easy visualization of topographic to analyze the relationship among data, and the neighboring neurons would be also adjusted to the input stimulus.

2   Methods

2.1   Experiment and Subjects

The most concerned issue in dual task studies is the effect of distraction on driving because it directly related to public safety. With combining the technology of virtual reality (VR), a VR driving environment includes 3D surrounded scenes projected by seven projectors and a real car mounted on a 6-degree-of-freedom platform to provide the kinesthetic stimuli. There were two kinds of tasks were designed: unexpected car deviation and mental calculation to evaluate mathematical addition equations. The car would randomly drift to the right or left side from the third lane of the road during driving, and subject was forced to keep the car in the third lane from left. Calculation task was the two-digit addition equations. If the equation was right (wrong), the right (left) button on the steering wheel would be pressed. The allotment ratio of correct-incorrect equations was 50-50. Every subject was asked to respond these two designed tasks as quick as they can with high accuracy.

To provide different distraction level, the effect of stimulus onset asynchrony (SOA) was considered. Combining these two designed tasks and SOA condition, five particular cases were represented. The five cases are showed as below:

(a) Case 1: Math equation is 400ms earlier than the occurrence of the deviation. (M\D)
(b) Case 2: Car deviation and math equation occur simultaneously. (D&M)
(c) Case 3: Car deviation is 400ms earlier than the appearance of the math equation (D\M)
(d) Case 4: Just math equation occurs. (M)
(e) Case 5: Just car deviation happens. (D)

Fifteen healthy participants (all males), between 20 and 28 years of age, were recruited from the university population. The scalp EEG signals were recorded from 36 channels. The contact impedance between EEG electrodes and the cortex was calibrated to be less than 10 kΩ. Each subject practiced about 15 minute to prevent learning effect. For this four-session experiment, subjects were required to rest for ten minutes between every two sessions to avoid fatigue.

2.2   Signal Processing and Feature Extraction

The procedures of data analysis and feature processing are described here. The EEG data were recorded with 16-bit quantization level at a sampling rate of 500 Hz and down-sampled to sampling rate equaled 250 Hz. The data were cut-off frequency of 50 Hz to remove high frequency noise. One more high-pass filter with a cut-off frequency of 0.5 Hz was utilized to remove DC drift. Because different cases with various combinations of driving and math tasks were designed, EEG signals from five cases were extracted separately. The extracted signal meant one EEG epoch.

EEG source segregation, identification, and localization are very difficult because EEG data are collected from the human scalp induce brain activities within a large brain area. The ICA algorithm had been extensively applied to solve the problem of EEG source separation, identification, and localization since 1990s [8]. Subsequent technical experiments demonstrated that ICA could also be used to remove artifacts from both continuous and event related (single-trial) EEG data [9]. Based on these studies, ICs were selected and clustered semi-automatically based on their scalp maps, dipole source locations, and within subject consistency.

The activation in Frontal areas was induced by mental task and the spectra in Motor component were difference between the single- and dual- task conditions. The brain activities in Frontal and Motor components were extracted to coalesce to form a bigger feature vector. Then the phase part of each EEG epoch was divided into ten intervals with 400ms, and each interval was applied Fast Fourier Transform. In each interval, 0~20 Hz for Fontal and 0~30 Hz for Motor components were reserved. The main difference in power spectra among all five cases were occurred in these low frequency [10]. Thus each case was represented by the power spectrum by 500 dimensions (200 for the Frontal and 300 for Motor). This feature vector implicitly contained the time information of onset of different events.

Although the subjects were asked to try their best to keep the same psychological and physical states during the whole experiment, there might be some variation among the subjects depending on ages, health conditions, and many other factors. The mean feature vector was calculated by every subject, and it was then subtracted from the feature vectors representing different cases for every particular subject to decrease the effect of diversity. The variation among subjects would play a dominant role in forming the maps without applying this procedure [11]. The normalization, Z-score, was applied to the feature vector after decreasing the subject variation. Finally, smoothing the power spectra was by seven EEG epochs, and the window was moved epoch by epoch circularly.

2.3   Classification Algorithm

To evaluate and analyze the phenomenon of driver distraction, one unsupervised method was applied. SOM is implemented through a neural network architecture that is similar with some ways to the biological neural networks [7]. SOM owns input layer and output layer to achieve dimension projection. The input layer receives the incoming data like organism get the stimulus. Not only the stimulated neuron but also the neighboring neurons are adjusted in each training process. When the unsupervised training processes are over, the output layer will adequately represent the input space.

It is similar that the brain under different type stimulus would be organized many special areas to handle variant reaction, such as reading, speaking, or smelling. The neurons in the brain with same functions would be clustered to the same areas during the growth of human. Thus, similar incoming signal will be projected near each other onto the near neurons on the map. SOM could offer an approach to brain activities that provides not only one classification for our distraction data but also a mechanism for visualizing the complex distribution of cognitive states. The high dimension data could also be projected to lower visualization dimension (usually 1-D, 2-D, or 3-D) to represent the distribution of the data.

Three parameters, including neuron number, learning rate, and training steps are needed during SOM algorithm processed. The neuron number directly relates the performance of the trained map. In this study, 625 (25 * 25) neurons were defined to get good performance and less time consuming. Following the usual norm, two phases of training were employer [12]. In first phase, the learning rate was decreased from 1 to 0 in 75000 training steps, while the radius of the neighborhood was decreased from 25 to 1. In the second phase, learning rate decreased from 0.1 to 0 in 50000 steps, while the neighboring radius decreased from 6 to 1. In this study, the maps were initialized, trained, and evaluated by SOM toolbox for MATLAB [13].

2.4   Distracted Detection System

In this study, one detecting driver distraction model is proposed. The classifiers in our model are the majority vote by different classifiers which are nine trained maps. The EEG epochs of one subject were chosen to be the testing data, and the other EEG data were the training data. For each subject, nine maps were generated same as before setting. The unlabeled neurons in these nine maps would be labeled first by the shortest distance from the neighboring neurons to provide complete information about relations among neighboring. Each epoch of testing data located on one neuron of each map and could be estimated case which was this epoch was belonged.

3   Results

The EEG epochs were analyzed by our proposed methods and the main finding could be represented here. The first part is about the phenomenon of driver distraction. By ICA, the interesting components were also found in Frontal and Motor cortex. There are two maps trained by SOM in the second part. The distribution of EEG epochs could be verified the relationship of driver distraction among different cases.

3.1   EEG Results

The Motor component was active when subjects steered the car. Activations simultaneously related to attention in Frontal component appear. Therefore, ICA components, including Frontal and Motor, were selected for IC clustering to analyze cross-subject data based on their EEG characteristics.

At first, IC clustering groups massive components from multiple sessions and subjects into several significant clusters. Cluster analysis, k-means, was applied to the normalized scalp topographies and power spectra of all 450 (30 channels x 15 subjects) components. Frontal and left Motor components were chosen to analyze distraction effects. Fig. 1 shows the scalp maps and equivalent dipole source locations for Fontal and left Motor clusters.

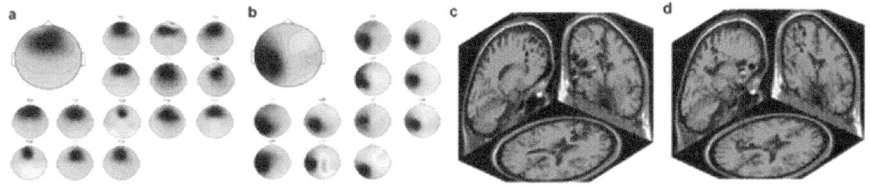

Fig. 1. The scalp maps and equivalent dipole source locations after IC clustering. (a) Frontal Components (b) Motor Components (c) the 14 dipole sources for Frontal components (d) the 11 dipole sources for Motor components.

Fig. 2a shows the cross-subject averaged event related spectral perturbation (ERSP) in Frontal cluster corresponding to the five cases. The theta power increases in three dual task cases are slightly different from each other. Compared to the single math task (Case 4), the power in dual task cases is stronger. Especially, the power increase in case 1 is the strongest. On the beta band, it also shows power increases, which appear only in the math task and time-locked to mathematics onsets. Fig. 2d shows the cross subject average ERSP in the left Motor cluster corresponding to five cases. In Case 4, the alpha and beta power suppressions appear continuously. Compared with Case 4, the alpha and beta power suppressions in Case 5 are stronger and also longer. In other cases, the alpha and beta power suppressions also continue. This phenomenon is suggested to be related to steering the car back to the center of the third lane. The ERSP images mainly show spectral differences after an event since the baseline spectrum prior to event onsets had been removed.

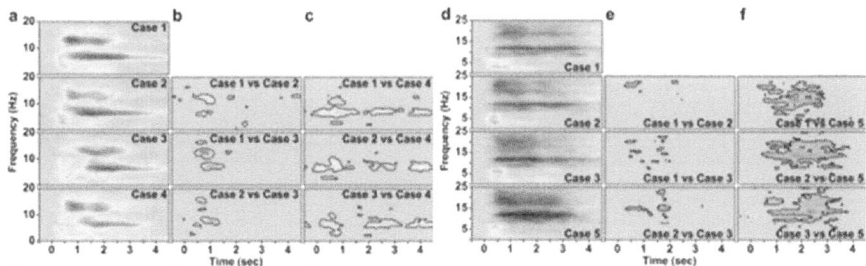

Fig. 2. ERSP without a significance test and the differences between cases

In Fig. 2, columns (b) and (e) show the differences among three single task cases; columns (c) and (f) show the differences between single- and dual- task cases. In columns (b), (c), (e), and (f), a Wilcoxon signed-rank test is used to retain the regions with significant power inside the black circles. Columns (b) and (c) show the comparison of power increases between cases. The remained regions show greater power increases in the single task case than in the dual task case. Columns (e) and (f) show compared power suppressions between cases. The remained regions show greater power suppressions in the dual task cases than in the single task case.

3.2   SOM Results and System Performance

In Fig. 3a, the map was trained by four conditions excluded Case 2 (D&M). The EEG epochs from two single conditions are clustered well to two almost perfect areas on the corner of this map. The other two dual conditions are also clustered to two almost perfect structures. In the second experiment, all EEG epochs are used to train the map shown in Fig. 3b. Two single conditions cover the bottom corner of this map, and the other areas represent the EEG epochs of the three dual tasks with SOA conditions. Here the EEG signals of Case 2 do not form a big compact clusters like the other four cases. The high accuracy of these two maps is exhibited in Table 1 which shows the average accuracy of correct labeling.

Comparing these two maps and the accuracy, the EEG epochs of four or five conditions could be distinguished clearly. This is a good evidence that there are noticeable differences in the EEG spectrum when subjects drive distraction or not even though the distraction levels. There are many sub-areas in Fig. 3b to represent Case 2. When the subjects faced two events simultaneously, they might choose one task to respond first. The brain resources could be occupied by every special case, and the phenomenon of Case 2 could be similar to that special case.

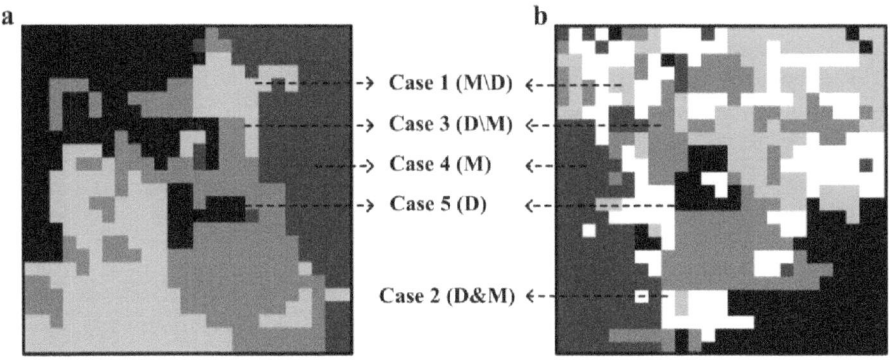

Fig. 3. Trained maps (a) four cases (b) five cases

The SOM could indeed distinguish the difference of brain activities with high accuracy. The accuracy of each case is not so high, but three dual conditions are hard to recognize clearly. By the results of this model, we had two main clusters: single driving (absorbed) and distracted driving (distraction) conditions. The performance of the two conditions is shown in Table 2, especially the accuracy of distracted driving reached 85%. So this system could recognize the EEG epochs of one new coming subject to two clusters: focusing on driving or driver distraction.

Table 1. Average accuracy of each type map

Condition	Case 1	Case 2	Case 3	Case 4	Case 5
4 cases	98.2 (1.2)	X	97.8 (1.6)	99.2 (1.1)	99.7 (0.8)
5 cases	97.1 (2.9)	93.1 (3.1)	95.7 (2.2)	97.9 (1.8)	98.6 (1.6)

Table 2. Average of testing results

Validation	Distracted Driving	Single Driving
Percentage	84.1 (1.8)	91.5 (0.7)

4   Conclusions

This study proposed one detective model for driver distraction under multiple cases and different distraction levels. The major findings include the following: 1) the phasic changes in Frontal and Motor are related driver distraction; 2) one effective feature processing is applied to reduce the subjective variation of brain activities; 3) verifying the EEG epochs of distracting and non-distracting conditions is effective. The EEG epochs of single driving were clearly identified. For such dual tasks, our SOM based exploratory data analysis using EEG suggested existence of distinct signatures among the five cases. The Frontal and Motor components were the main activities area of responding multiple tasks during distracted driving. Furthermore, the recognition of distraction levels will help us to monitor the driver safety and warn them to pay more attention during driving for decreasing and avoiding the traffic accidents in our real-life driving.

Acknowledgments. This work was supported in part by the UST-UCSD International Center of Excellence in Advanced Bio-engineering sponsored by the Taiwan National Science Council I-RiCE Program under Grant Number: NSC-99-2911-I-009-101, in part by the National Science Council, Taiwan, under Contract NSC-99-2220-E-009-026.

References

1. Dukic, T.H., Hanson, L., Falkmer, T.: Effect of drivers' age and push button locations on visual time off road, steering wheel deviation and safety perception. Ergonomics 49, 78–92 (2006)
2. Cross, G.W.: Distracted Driving in Mississippi: A Statewide Survey and Summary of Related Research and Policies. Mississippi (2010)
3. Young, K., Regan, M., Faulks, I.J., Stevenson, M., Brown, J., Irwin, J.D.: Driver distraction: A review of the literature. NSW: Australian College of Road Safety, 379–405 (2007)
4. Lin, C.T., Liao, L.D., Liu, Y.H., Wang, I.J., Lin, B.S., Chang, J.Y.: Novel Dry Polymer Foam Electrodes for Long-Term EEG Measurement. IEEE Transactions on Biomedical Engineering 58(5), 1200–1207 (2011)
5. McFarland, D.J., Anderson, C.W., Muller, K.R., Schlogl, A., Krusienski, D.J.: Bci meeting 2005-workshop on bci signal processing: feature extraction and translation. IEEE Transactions on Neural Systems and Rehabilitation Engineering 14(2), 135–138 (2006)

6. Makeig, S., Bell, A.J., Jung, T.P., Sejnowski, T.J.: Independent component analysis of electroencephalographic data. In: Advances in Neural Information Processing Systems, pp. 145–151 (1996)

7. Kohonen, T.: The self-organizing map. Proceedings of the IEEE 78(9), 1464–1480 (1990)

8. Comon, P.: Independent component analysis, A new concept? Signal Processing 36(3), 287–314 (1994)

9. Jung, T.P., Making, S., Humphries, C., Lee, T.W., Mckeown, M.J., Iragui, V., Sejnowski, T.J.: Removing electroencephalographic artifacts by blind source separation. Psychophysiology 37, 163–178 (2000)

10. Lin, C.T., Chen, S.A., Chiu, T.T., Lin, H.Z., Ko, L.W.: Spatial and temporal EEG dynamics of dual-task driving performance. Journal of NeuroEngineering and Rehabilitation 8(1), 11–23 (2011)

11. Wang, Y.K., Pal, N.R., Lin, C.T., Chen, S.A.: Analyzing effect of distraction caused by dual-tasks on sharing of brain resources using SOM. In: The International Joint Conference on Neural Networks, Barcelona (2010)

12. Joutsiniemi, S.L., Kaski, S., Larsen, T.A.: Self-organizing map in recognition of topographic patterns of EEG spectra. IEEE Transactions on Biomedical Engineering 42(11), 1062–1068 (1995)

13. Laboratory of Computer and Infromation Science, http://www.cis.hut.fi

14. Lv, J.C., Tan, K.K., Yi, Z., Huang, S.: A Family of Fuzzy Learning Algorithms for Robust Principal Component Analysis Neural Networks. IEEE Transactions on Fuzzy Systems, 217–226 (2010)

15. Song, H., Miao, C., Roel, W., Shen, Z., Catthoor, F.: Implementation of Fuzzy Cognitive Maps Based on Fuzzy Neural Network and Application in Prediction of Time Series. IEEE Transations on Fuzzy Systems, 233–250 (2010)

16. Juang, C.F., Hsieh, C.D.: A Locally Recurrent Fuzzy Neural Network With Support Vector Regression for Dynamic-System Modeling. IEEE Transations on Fuzzy Systems, 261–273 (2010)

17. Juang, C.F., Huang, R.B., Cheng, W.Y.: An Interval Type-2 Fuzzy-Neural Network With Support-Vector Regression for Noisy Regression Problems. IEEE Transations on Fuzzy Systems, 686–699 (2010)

18. Han, H., Qiao, J.: A Self-Organizing Fuzzy Neural Network Based on a Growing-and-Pruning Algorithm. IEEE Transations on Fuzzy Systems, 1129–1143 (2010)

Removing Unrelated Features Based on Linear Dynamical System for Motor-Imagery-Based Brain-Computer Interface

Jie Wu[1], Li-Chen Shi[1], and Bao-Liang Lu[1,2]

[1]Center for Brain-Like Computing and Machine Intelligence
Department of Computer Science and Engineering
[2]MOE-Microsoft Key Laboratory for Intelligent Computing and Intelligent Systems
Shanghai Jiao Tong University
800 Dong Chuan Rd., Shanghai 200240, China
bllu@sjtu.edu.cn

Abstract. Common spatial pattern (CSP) is very successful in constructing spatial filters for detecting event-related synchronization and event-related desynchronization. In statistics, a CSP filter can optimally separate the motor-imagery-related components. However, for a single trail, the EEG features extracted after a CSP filter still include features not related to motor imagery. In this study, we introduce a linear dynamical system (LDS) approach to motor-imagery-based brain-computer interface (MI-BCI) to reduce the influence of these unrelated EEG features. This study is conducted on a BCI competition data set, which comprises EEG signals from several subjects performing various movements. Experimental results show that our proposed algorithm with LDS performs better than a traditional algorithm on average. The results reveal a promising direction in the application of LDS-based approach to MI-BCI.

Keywords: motor imagery, brain-computer interface, linear dynamic system, common spatial pattern.

1 Introduction

Brain-computer interfaces (BCIs) are communication systems which enable users to send commands to computers using only their brain activity, which is generally measured by electroencephalography (EEG) [1,2]. BCI technology has been a promising tool for disabled people as well as for healthy people [3,4,5]. Motor imagery is a very popular paradigm in BCI. EEG and event-related synchronization/desynchronization (ERS/ERD) [6] have been employed for research on brain functional activity for many decades and have become the scientific basis of motor imagery. Studies have shown that distinct phenomena such as ERD/ERS are detectable from EEGs for both real and imagined motor movements in healthy subjects [7,8,9]. Common spatial pattern (CSP) [10] is very successful in constructing spatial filters for detecting ERS/ERD. However, the features extracted after CSP still contain unrelated EEG features.

B.-L. Lu, L. Zhang, and J. Kwok (Eds.): ICONIP 2011, Part I, LNCS 7062, pp. 709–716, 2011.

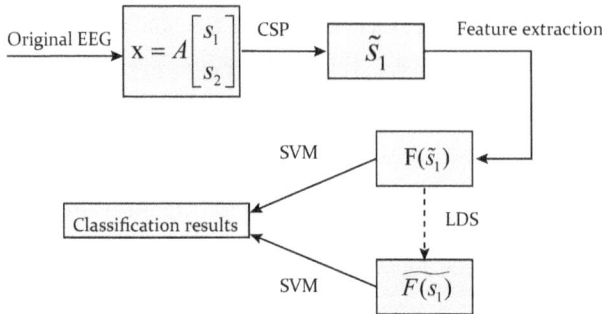

Fig. 1. The flow chart of our proposed algorithm for processing two-class motor imagery EEG recordings

Mental states have the characteristic of continuity. It is a gradual process, and so the EEG features extracted from mental states such as vigilance change continuously [15]. Recently, Shi and Lu [14] have applied linear dynamical system (LDS) approach to vigilance estimation [11,12,13] from EEGs, and their experimental results show that LDS can remove vigilance-unrelated signals effectively. LDS is a kind of state space model, which can effectively remove vigilance-unrelated features using the time dependency of vigilance changes. Motor imagery is also a kind of mental state which has the feature of continuity. In theory, by making use of the time dependency of changes of motor imagery, LDS can filter the motor-imagery-related EEG features more accurately. In this study, we introduce the LDS-based approach to motor-imagery-based brain-computer interface (MI-BCI). By using the LDS-based approach, EEG features are smoothed and the unrelated EEG influences in the EEG features are reduced. Our experimental results show that our proposed algorithm with LDS performs with a higher accuracy than the traditional algorithm on average.

The remainder of this paper is organized as follows. Section 2 describe the methodology and process of our proposed algorithm. Section 3 presents the experimental results. Finally, Section 4 discusses some conclusions.

2   Methodology

2.1   Main Idea

The flow chart of the process for our proposed algorithm with LDS in MI-BCI is shown in Fig. 1. The M recorded EEG signals $x(t) = [x_1(t), x_2(t), ..., x_M(t)]^T$ are assumed to be linear mixtures of the underlying components $s(t) = [s_1(t), s_2(t)]^T$:

$$x = A \begin{bmatrix} s_1 \\ s_2 \end{bmatrix} \tag{1}$$

Suppose s_1 are motor-imagery-related components and s_2 are unrelated components. In traditional algorithms, we use CSP to extract the motor-imagery related components $\tilde{s}_1$ which is an estimate of s_1. Then we extract features $F(\tilde{s}_1)$

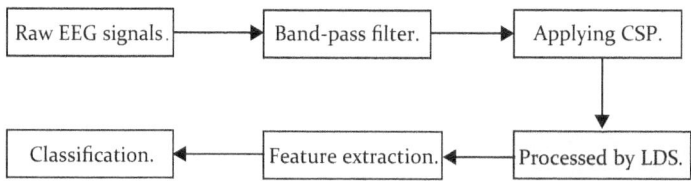

Fig. 2. Architecture of motor imagery-based brain-computer interface with linear dynamical system approach

from components $\widetilde{s}_1$ where $F(\widetilde{s}_1)$ is calculated by feature extraction methods. Suppose $F(s_1)$ are the features extracted from s_1 which are pure motor-imagery related components with no noise, then in the traditional algorithms, we use $F(\widetilde{s}_1)$ to estimate $F(s_1)$ and finally use a classifier such as a support vector machine (SVM) [16] to classify $F(\widetilde{s}_1)$.

In traditional algorithms, a CSP filter is used to optimally separate the motor-imagery related components of left and right motor imagery in statistics [10]. However, for each single trail, the component $\widetilde{s}_1$ filtered by the CSP algorithm may contain unrelated component. So $\widetilde{s}_1$ may not estimate s_1 very precisely. As a result, the features $F(\widetilde{s}_1)$ extracted from $\widetilde{s}_1$ may include features unrelated to motor imagery.

Based on this hypothesis, we try to smooth $F(\widetilde{s}_1)$ into $\widetilde{F(s_1)}$ using LDS to reduce the unrelated features. Compared with $F(\widetilde{s}_1)$, $\widetilde{F(s_1)}$ is expected to be a better estimate of $F(s_1)$. The architecture of the MI-BCI with LDS is shown in Fig. 2.

2.2 Common Spatial Patterns

The CSP algorithm is effective in constructing optimal spatial filters that discriminate two classes of EEG measurements in MI-BCI [10,17,18,19]. The spatial filter maximizes the variance of signals of one class and at the same time minimizes the variance of signals of the other class. Because band power is equal to the variance of band-pass filtered signals, CSP performs very well as a spatial filter for detecting ERS/ERD in EEG measurements and has been well used in in BCI systems [20,21,22].

In this study, we extract one feature every 0.2 second, so a sequence of features $Y(i)$ is obtained in a single trail (7 seconds). The feature $Y(i)$ is obtained by calculating the variance of signals in the time interval of 0.2 second. Let $\bar{Y} = \frac{1}{n}\sum_{i=1}^{n} Y(i)$, so we can choose $\bar{Y}$ as the extracted feature of the trail. Let $\bar{Y}_{classA}$ and $\bar{Y}_{classB}$ be the selected features for two classes which are chosen from four classes (left hand, right hand, tongue, and foot movements). Since CSP maximizes the variance ratio of components of the two classes, we can classify $\bar{Y}_{classA}$ and $\bar{Y}_{classB}$ by SVM.

2.3   Linear Dynamical System

There are some motor-imagery-unrelated influences in the EEG features, so $y(i)$ probably contains noise. We design LDS to reduce these influences as well as to smooth the EEG features.

The motor-imagery-unrelated influences of EEG features result in a difference between the original EEG features Y calculated from traditional methods and the motor-imagery-related EEG features Y_m which can represent the feature of motor imagery more accurately. Because mental state is time dependent, the features Y_m extracted from EEG components are also time dependent. If Y is considered as the observation sequence of the latent state sequence Y_m, we can represent a state space model to filter out the above influences and recover the Y_m from Y in the form of LDS:

$$Y_m(t) = AY_m(t-1) + v(t) \tag{2}$$
$$Y(t) = CY_m(t) + w(t), \tag{3}$$

where A is the state transition matrix, C is the observation matrix, $v(t) \sim \mathcal{N}(0, \Gamma)$ and $w(t) \sim \mathcal{N}(0, \Sigma)$ are the Gaussian variables, and the initial latent state is assumed to be distributed as $Y_m(1) \sim \mathcal{N}(\mu(0), V(0))$, Eqs. (2) and (3) can also be expressed in an equivalent form in terms of Gaussian conditional distributions as follows,

$$p(Y_m(t) \mid Y_m(t-1)) = \mathcal{N}(AY_m(t-1), \Gamma) \tag{4}$$
$$p(Y(i) \mid Y_m(t)) = \mathcal{N}(CY_m(t)\Sigma). \tag{5}$$

The parameters of the LDS model are denoted by $\theta = \{A, C, \Gamma, \Sigma, \mu_0, V_0\}$. According to the LDS model, given the observations, the latent state $Y_m(t)$ can be estimated from the posterior marginal distribution corresponding to $p(Y_m(t) \mid Y)$ and this posterior marginal distribution is Gaussian,

$$p(Y_m(t) \mid Y) = \mathcal{N}(\mu(t), V(t)).$$

The mean $\mu(i)$ is just the maximum a posteriori (MAP) estimation of $Y_m(t)$. For online inference, Y are the observations from $Y(1)$ to $Y(t)$. The parameters of the marginal distribution, $\hat{\mu}(t)$ and $\hat{V}(t)$, can be determined by the following forward recursions:

$$P_{t-1} = A\hat{V}_{t-1}A^T + \Gamma$$
$$K_t = P_{t-1}C^T(CP_{t-1}C^T + \Sigma)^{-1}$$
$$\hat{\mu}_t = A\hat{\mu}_{t-1} + K_t(Y^t - CA\hat{\mu}_{t-1})$$
$$\hat{V}_t = (I - K_tC)P_{t-1},$$

where the initial conditions are:

$$K_1 = V_0C^T(CV_0C^T + \Sigma)^{-1}$$
$$\hat{\mu}_1 = \mu_0 + K_1(Y^1 - C\mu_0)$$
$$\hat{V}_1 = (I - K_1C)V_0.$$

For offline inference, Y is the whole sequence of observations from Y^1 to Y^N. The parameters of the marginal distribution, $\tilde{\mu}(t)$ and $\tilde{V}(t)$, can be determined by the online inference results and the following backward recursions:

$$J_t = \hat{V}_t A^T (P_t)^{-1}$$
$$\tilde{\mu}_t = \hat{\mu}_t + J_t(\tilde{V}_{t+1} - A\hat{\mu}_t)$$
$$\tilde{V}_t = \hat{V}_t + J_t(\tilde{V}_{t+1} - P_t)J_t^T,$$

where the initial conditions are:

$$\tilde{\mu}_N = \hat{\mu}_N$$
$$\tilde{V}_N = \hat{V}_N.$$

Though offline inference is more accurate than online inference, we use online inference in this study, because immediate feedback is required in MI-BCI. Every EEG feature with interval of 0.2 second is smoothed by LDS. The LDS model runs in constant time because the size of input is limited which is the length of a sequence of features obtained in a single trail (7 seconds). So the time complexity of the LDS model is $O(1)$. The parameters of the LDS model can be estimated by the EM algorithm. However, these estimated parameters are locally optimized. Because the form of parameters is relatively simple, we can test by hand and determine the parameters.

3   Experimental Results

This section evaluates the performance of the proposed algorithm on BCI Competition 3 data set 3a [23]. This data set comprises EEG signals from three subjects who performed left hand, right hand, tongue, and foot movements. The four classes of movements that should be discriminated were paired in six groups to yield the 2-class motor imagery data sets. For the first two subjects, there are 90 trails for each class and for third subject, 60. These data sets comprise a training set and a testing set for each subject. Half of each session is the training set and the other half is the testing set. Each trail has a duration of 7 sec. The subjects performed motor imagery from time $t = 3$ sec to $t = 7$ sec of each trail.

We compare the traditional algorithm with the proposed algorithm with LDS to see whether there is an increase in accuracy. The architecture of the CSP algorithm for two-class motor imagery in this study is shown in Fig. 3. We filter the EEG signals in 8-20 Hz and then use CSP to construct the spatial filter. We decompose the signals into 6 bands of 2 Hz, which are 8-10 Hz, 10-12 Hz, 12-14 Hz, 14-16 Hz, 16-18 Hz, and 18-20 Hz, respectively. Signals of each band are filtered by CSP, and features of 6 bands, $Y_{8-10}, Y_{10-12}, Y_{12-14}, Y_{14-16}, Y_{16-18}, Y_{18-20}$, are extracted by calculating the variance of components which are obtained after CSP. Then, $(Y_{8-10}, Y_{10-12}, Y_{12-14}, Y_{14-16}, Y_{16-18}, Y_{18-20})$ as used as the feature of a single trail. And, SVM is used as a classifier with radial basis function (RBF) kernel, and 5-fold cross validation for training. We need to estimate

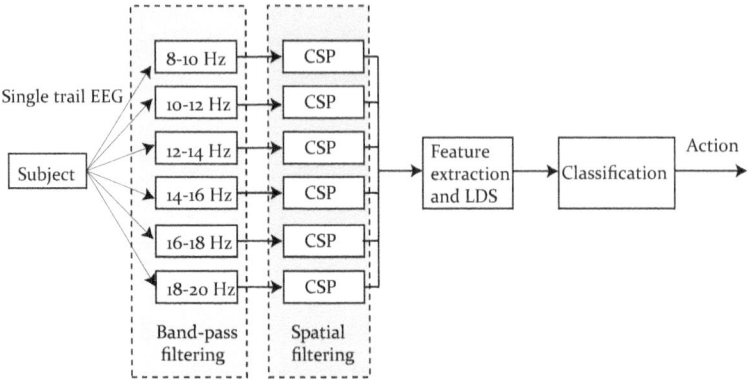

Fig. 3. Architecture of CSP algorithm for two-class motor imagery EEG data

Table 1. Comparison of classification accuracy of the traditional algorithm and that of our proposed algorithm

Data set	Task 1/2	Task 1/3	Task 1/4	Task 2/3	Task 2/4	Task 3/4	Average	Γ/C
k3b(original)	88.9%	85.5%	88.9%	80.0%	78.9%	64.4%	81.1%	
k3b(with LDS)	**91.1%**	**86.7%**	**92.2%**	**81.1%**	**83.3%**	**65.6%**	**83.3%**	0.25
k6b(original)	63.3%	58.3%	86.7%	66.7%	90.0%	90.0%	75.8%	
k6b(with LDS)	**65.0%**	**65.0%**	86.7%	**70.0%**	90.0%	90.0%	**77.8%**	0.03
l1b(original)	65.0%	66.7%	75.0%	68.3%	78.3%	63.3%	69.4%	
l1b(with LDS)	**75.0%**	66.7%	75.0%	**71.7%**	78.3%	**65.0%**	**72.0%**	0.50

the parameters $\theta = \{A, C, \Gamma, \Sigma, \mu_0, V_0\}$ of the LDS model. Considering the form of parameters is relatively simple, they can be determined by hand. We only need to enumerate Γ and C, while the other parameters A, Σ, μ_0 and V_0 have constant values. We set $A = 1, \Sigma = 1, V_0 = 0.1$ and let μ_0 be the the first value of the input sequence to the LDS model.

The experimental results of classification and values of parameter Γ/C are shown in Table 1. The first six columns show the results for the two-class classification of recorded data, while the last column holds the average classification accuracy over all tasks. Comparing the proposed algorithm with the traditional algorithm, the performance for each subject steadily is improved about 2%. However, the overall classification accuracies of l1b and k6b are still relatively low. The reason could be the poor performance of the subject on the practical motor imagery task. In summary, from the experimental results it is clear that LDS is effective for reducing motor-imagery-unrelated EEG features in MI-BCI.

4   Conclusions

In motor imagery, CSP filters can optimally separate the motor-imagery related components of two classes, but for a single trail, the components filtered by

CSP still contain unrelated component. In this paper, we introduced LDS to MI-BCI, which is used to filter the motor-imagery-related EEG features more accurately for a single trail and improve the classification accuracy of MI-BCI. The traditional algorithm and the proposed algorithm with LDS were evaluated on the four-class motor imagery data of BCI competition 3 data set 3a. The experimental results show that after adding the LDS in the traditional algorithm, the average classification accuracy rises about 2%. The experimental results suggest LDS can effectively filter out the motor-imagery-unrelated EEG features and reveal a promising direction in the application of LDS to MI-BCI.

Acknowledgments. This work was partially supported by the National Natural Science Foundation of China (Grant No. 90820018), the National Basic Research Program of China (Grant No. 2009CB320901), the Science and Technology Commission of Shanghai Municipality (Grant No. 09511502400), and the European Union Seventh Framework Programme (Grant No. 247619).

References

1. Wolpaw, J.R., Birbaumer, N., McFarland, D.J., Pfurtscheller, G., Vaughan, T.M.: Brain-computer interfaces for communication and control. Clin. Neurophysiol. 113, 767–791 (2002)
2. Pfurtscheller, G., Neuper, C., Guger, C., Harkam, W., Ramoser, R., Schll, A., Obermaier, B., Pregenzer, M.: Current Trends in Graz Brain-computer Interface (BCI). IEEE Trans. Rehab. Eng. 8(2), 216–219 (2000)
3. Birbaumer, N., Ghanayim, N., Hinterberger, T., Iversen, I., Kotchoubey, B., Kubler, A., Perelmouter, J., Taub, E., Flor, H.: A spelling device for the paralysed. Nature 398, 297–298 (1999)
4. Curran, E.A., Stokes, M.J.: Learning to control brain activity: A review of the production and control of EEG components for driving brain-computer interface (BCI) systems. Brain Cogn. 51, 326–336 (2003)
5. Dornhege, G., del R. Millan, J., Hinterberger, T., McFarland, D., Muller, K.-R. (eds.): Toward Brain-Computer Interfacing. MIT Press, Cambridge (2007)
6. Pfurtscheller, G., Lopes da Silva, F.H.: Event-related EEG/MEG synchronization and desynchronization: basic principles. Clin. Neurophysiol. 110(11), 1842–1857 (1999)
7. Stavrinou, M., Moraru, L., Cimponeriu, L., Della Penna, S., Bezerianos, A.: Evaluation of Cortical Connectivity During Real and Imagined Rhythmic Finger Tapping. Brain Topogr. 19(3), 137–145 (2007)
8. McFarland, D., Miner, L., Vaughan, T., Wolpaw, J.: Mu and Beta Rhythm Topographies During Motor Imagery and Actual Movements. Brain Topogr. 12(3), 177–186 (2000)
9. Pfurtscheller, G., Brunner, C., Schlogl, A., Lopes da Silva, F.H.: Mu rhythm (de)synchronization and EEG single-trial classification of different motor imagery tasks. NeuroImage 31(1), 153–159 (2006)
10. Ramoser, H., Muller-Gerking, J., Pfurtscheller, G.: Optimal spatial filtering of single trial EEG during imagined hand movement. IEEE Trans. Rehab. Eng. 8, 441–446 (2000)

11. Cajochen, C., Zeitzer, J.M., Czeisler, C.A., Dijk, D.J.: Dose-response Relationship for Light Intensity and Ocular and Electroencephalographic Correlates of Human Alertness. Behavioural Brain Research 115, 75–83 (2000)
12. Molloy, R., Parasuraman, R.: Monitoring an automated system for a single failure: vigilance and task complexity effects. Human Factors 38, 311–322 (1996)
13. Weinger, M.B.: Vigilance, Boredom, and Sleepiness. Journal of Clinical Monitoring and Computing 15, 549–552 (1999)
14. Shi, L.C., Lu, B.L.: Off-line and on-line vigilance Estimation based on linear dynamical system and manifold learning. In: Proceedings of 32nd International Conference of the IEEE Engineering in Medicine and Biology Society, Buenos Aires, Argentina, pp. 6587–6590 (2010)
15. Oken, B.S., Salinsky, M.C., Elsas, S.M.: Vigilance, alertness, ro sustained attention: physiological basis and measurement. Clinical Neurophysiology 117, 1885–1901 (2006)
16. Lotte, F., et al.: A Review of Classification Algorithms for EEG-Based Brain-Computer Interface. J. Neural. Eng. 4, R1–R13 (2007)
17. Pfurtscheller, G., Neuper, C.: Motor imagery and direct braincomputer communication. Proc. IEEE 89(7), 1123–1134 (2001)
18. Blankertz, B., Tomioka, R., Lemm, S., Kawanabe, M., Muller, K.-R.: Optimizing Spatial Filters for Robust EEG Single-Trial Analysis. IEEE Signal Proc. Magazine 25(1), 41–56 (2008)
19. Muller-Gerking, J., Pfurtscheller, G., Flyvbjerg, H.: Designing optimal spatial filters for single-trial EEG classification in a movement task. Clin. Neurophysiol. 110, 787–798 (1999)
20. Koles, Z.J.: The quantitative extraction and topographic mapping of the abnormal components in the clinical EEG. Electroencephalogr. Clin. Neurophysiol. 79(6), 440–447 (1991)
21. Guger, C., Ramoser, H., Pfurtscheller, G.: Real-time EEGanalysis with subject-specific spatial patterns for a Brain Computer Interface (BCI). IEEE Trans. Neural Sys. Rehab. Eng. 8(4), 447–456 (2000)
22. Blankertz, B., Dornhege, G., Krauledat, M., Muller, K.-R., Curio, G.: The non-invasive Berlin Brain-Computer Interface: Fast Acquisition of Effective Performance in Untrained Subjects. NeuroImage 37(2), 539–550 (2007)
23. BCI competition III website: http://www.bbci.de/competition/iii/

EEG-Based Motion Sickness Estimation Using Principal Component Regression

Li-Wei Ko[1,2,*], Chun-Shu Wei[1,2], Shi-An Chen[1,2], and Chin-Teng Lin[1,2]

[1] Brain Research Center, National Chiao Tung University (NCTU), Hsinchu, Taiwan
[2] Department of Electrical Engineering, NCTU, Hsinchu, Taiwan
[3] Department of Biological Science and Technology, NCTU, Hsinchu, Taiwan
[4] Institute for Neural Computation, University of California San Diego, San Diego, USA
*lwko@mail.nctu.edu.tw, treeseert@gmail.com, sachen@nctu.edu.tw,
ctlin@mail.nctu.edu.tw

Abstract. Driver's cognitive state monitoring system has been implicated as a causal factor for the safety driving issue, especially when the driver fell asleep or distracted in driving. However, the limitation in developing this system is lack of a major indicator which can be applied to a realistic application. In our past studies, we investigated the physiological changes in the transition of driver's cognitive state by using EEG power spectrum analysis and found that the features in the occipital area were highly correlated with the driver's driving performance. In this study, we construct an EEG-based self-constructed neural fuzzy system to estimate the driver's cognitive state by using the EEG features from the occipital area. Experimental results show that the proposed system had the better performance than other neural networks. Moreover, the proposed system can not only be limited to apply to individual subjects but also sufficiently works in between subjects.

Keywords: EEG, neural networks, fuzzy systems, driving cognition, machine learning.

1 Introduction

Motion-sickness is a common experience to everybody, and it has provoked a great deal of attentiveness in plenty of studies. The sensory conflict theory that came about in the 1970's has become the most widely accepted theorem of motion-sickness among scientists [1]. The theory proposed that the conflict between the incoming sensory inputs could induce motion-sickness. Accordingly, new research studies have appeared to tackle the issue of the vestibular function in central nervous system (CNS). In the previous human subject studies, researchers attempt to confirm the brain areas involved in the conflict in multi-modal sensory systems by means of clinical or anatomical methods. Brandt et al. demonstrated that the posterior insula in human brain was homologous to PIVC in the monkey by evaluating vestibular functions in patients with vestibular cortex lesions [2]. In agreement with previous clinical studies, the cortical activations during caloric [3] and galvanic vestibular stimulation [4] had been studied by functional imaging technologies such as positron

B.-L. Lu, L. Zhang, and J. Kwok (Eds.): ICONIP 2011, Part I, LNCS 7062, pp. 717–724, 2011.

emission tomography (PET) and functional magnetic resonance imaging (fMRI). To overcome the temporal limitation of the two imaging modalities, some studies have investigated the vestibular information transmission in time domain. Monitoring the brain dynamics induced by motion-sickness because of its high temporal resolution and portability De Waele et al., for example, applied current pulse stimulation to patients' vestibular nerve to generate vestibular evoked potentials [5].

The EEG studies related to motion-sickness can be divided into two groups according to the types of stimuli: vestibular and visual. Vestibular stimuli were normally provided to the subjects with rotating chair [6], [7], parallel swing [8], and cross-coupled angular stimulation [9] to induce motion-sickness. Theta power increases in the frontal and central areas were reported to be associated with motion-sickness induced by parallel swing [8] and rotating drum [6], [7]. Chelen et al. [9] employed cross-coupled angular stimulation to induce motion-sickness symptoms and found increased delta- and theta-band power during sickness but no significant change in alpha power. Visually induced motion-sickness is also commonly studied in previous studies. Visually induced sickness can be provoked with an optokinetic drum rotating around the yaw axis. This situation can cause a compelling sense of self-motion (called vection). Vestibular cues indicate that the body is stationary, whereas visual cues report the body is moving. Hu et al. investigated MS triggered by the viewing of an optokinetic rotating drum and found a higher net percentage increase in EEG power in the 0.5-4 Hz band at electrode sites C3 and C4 than in the baseline spectra. [10]. This study employees ahe driving simulator comprised an actual automobile mounted on a Stewart motion platform with six degrees of freedom, providing both visual and vestibular stimulations to induce motion-sickness and accompanied EEG dynamics.

Our past studies [11-14] had investigated the EEG activities correlated with motion sickness in a virtual-reality based driving simulator. We found that the parietal and motor brain regions exhibited significant alpha power suppression in response to vestibular stimuli, while the occipital area exhibited motion sickness related power augmentation in mainly theta and delta bands; the occipital midline region exhibited a broad band power increase. Based on these results, we think that both visual and vestibular stimulations should be used to induce motion sickness in brain dynamic research. Hence, we attempt to implement an EEG-based evaluation system to estimate subject's motion sickness level (MSL) upon the major EEG power spectra from these motion sickness related brain area in this study. The evaluation system can be applied to early detect the subject's MSL and prevent the uncomfortable syndromes occurred in advance in our daily life.

2   Experiment Design and Setup

2.1   Experimental Paradigm

Unlike the previous studies, we provided both visual and vestibular stimuli to participant through a compelling VR environment consisting of 360° projection of VR scene and a motion platform with six degree-of-freedom to induce motion-sickness (shown in Fig. 1). With such a setup, we expected to create motion-sickness in a manner that is close to that in daily life. During the experiment, the subjects were asked to sit inside an actual vehicle mounted on a motion platform, with their hands holding a joystick to report their sickness level continuously. The VR scenes simulating driving in a tunnel were programmed to eliminate any possible visual distracter and shorten the depth of visual field such that motion-sickness could be easily induced. A three-section experimental protocol (shown in Fig. 2) was designed to induce motion-sickness.

(a)

(b)

Fig. 1. VR-based Highway Driving Environment. (a) Snapshot of the virtual reality-based driving scene, (b) six degree-of-freedom motion platform.

First, the baseline section contained a 10-minute straight road to record the subjects' baseline state. Then, a 40-minute motion-sickness section, which consisted of a long winding road, was presented to the subjects to induce motion-sickness. Finally a 15-minute rest section with a straight-road condition was displayed for the subjects to recover from their sickness. The level of sickness was continuously reported by the subjects using a joystick with continuous scale on its side. The experimental setting successfully induced motion sickness to more than 80% of subjects in this study.

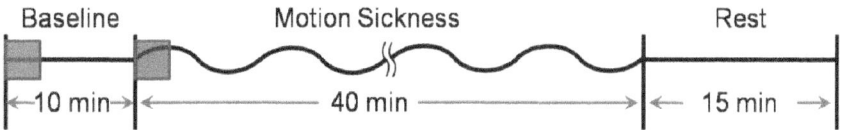

| Baseline | Motion Sickness | Rest |

←10 min→ ←——— 40 min ———→ ← 15 min →

Fig. 2. Experimental design of motion-sickness experiments

2.2 Subjects

Ten healthy, right-handed volunteers with no history of gastrointestinal, cardiovascular or vestibular disorders or of drug or alcohol abuse, taking no medication and with normal or corrected-to-normal vision participated in this experiment. EEG signals were recorded with 500 Hz sampling rate by 32-channel NuAmps (BioLink Ltd., Australia). Simultaneously, during EEG recording, the level of sickness was continuously reported by each subject using a joystick with a continuous scale ranging 0 – 5. The subjects were asked to raise/lower the scale to a higher/lower level if they felt more motion sick comparing to the last condition. This continuous sickness level was reported in real time without interrupting the experiment rather than the traditional motion-sickness questionnaire (MSQ).

2.3   Data Analysis

The acquired EEG signals were first inspected to remove bad EEG channels and then down-sampled to 250 Hz. A high-pass filter with cut-off frequency at 1 Hz and transition band width 0.2 Hz was used to remove baseline-drifting artifacts, and a low-pass filter with cut-off frequency at 60 Hz and transition band width 7 Hz was to remove muscular artifacts and line noise. After the preprocessing procedures, the clean EEG signals will feed into the proposed evaluation system for further analysis.

3   Proposed MS Level Estimation System

The proposed evaluation system to estimate subject's motion sickness level can be divided into five parts: independent component analysis (ICA), component clustering, time-frequency analysis, Feature Extraction by Principal Component Analysis (PCA), and Estimation part by applying linear regression, RBF Neural Network and Support Vector Regression with leave one out (LOO) cross validation. Figure 3 shows the system flowchart of the proposed motion sickness evaluation system.

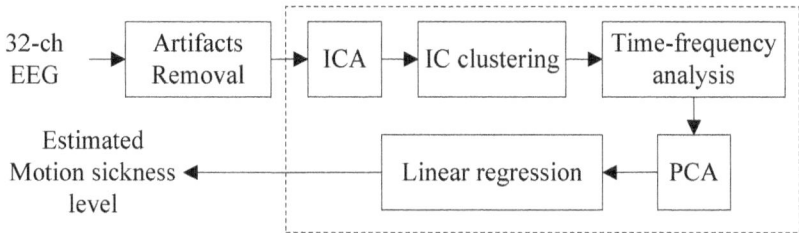

Fig. 3. System flowchart of the proposed motion sickness evaluation system

Independent Component Analysis (ICA) was applied to EEG recordings to remove various kinds of artifacts, including blink artifact and indoor power-line noise, and to extract features of human's cognition. Among components from all subjects, those with similar scalp topographies, dipole locations and power spectra were grouped using k-means clustering.

After doing ICA process, component clustering was analyzed using DIPFIT2 routines, a plug-in in EEGLAB, to find the 3D location of an equivalent dipole or dipoles based on a four-shell spherical head model. Among components from all subjects, those with similar scalp topographies, dipole locations and power spectra were clustered. Ten component clusters recruited more than 10 components from multiple subjects with similar topographic maps were regarded as robust component clusters. In component clustering results, we find that not all subjects have every motion sickness related components because the level of motion sickness induced by vestibular and visual stimuli to each subject had the significant individual difference. According to MSQ results and subject's self-response of motion sickness, we can confirm that each subject indeed felt sickness during the whole experiment session. Consequently, these extracted components are correlated with motion sickness. Then we can feed the ICA signals into the system and do time-frequency analysis.

As previously mentioned, the extracted independent components are different in number and in location through subjects. To maintain the consistency of subject data, we proposed to project these components back on channel domain. The channels of interest (Fp1, Fp2, C3, C4, Pz, and Oz) in this study are those electrodes close to the MS-related component clusters stated in our previous study [11]. C3 and C4 are close to left and right motor region respectively. Pz and Oz mostly represent the activity of parietal, occipital, and occipital midline areas. Frontal electrodes (Fp1 and Fp2) are relatively distant from the MS-related component clusters, but they are included in the channels of interest. The reason is that forehead, which is not covered by hair, is a popular choice to place EEG sensors, and the state-of-art EEG-based BCI devices using dry sensor on forehead have been developed in recent years [15], [16].

Time-frequency analysis was used to investigate the dynamics of the ICA power spectra. In order to provide a temporal resolution of 30 seconds, the spectra of ICA activations were calculated using 30-sec length window with 150-sec overlap, and subdivided into several 250-point sub-windows with 125-point overlaps. Each 125-point sub-window was zero-padded to 256 points for using 256-point fast Fourier transform (FFT) with ~1 Hz resolution in frequency. The linear power spectrum density (PSD) was then converted into a logarithmic scale (dB power).

PCA [17] was then used to summarize the variances and extract first few principal components of the components' PSD's. In this study, the number of selected eigenvector was determined in the PCA training process. Leave-one-subject-out (LOSO) cross validation was performed to evaluate the estimation performance. In LOSO cross validation, each subject's data was prepared as the testing data, and the data from the other 9 subjects were collected as the training data. PCA was performed on the training data to extract the eigenvectors for selection and projection. And then in the training data, another LOSO cross validation was performed repeatedly using from 1 to all 36 eigenvectors. Finally, the number of eigenvectors that support the least average AIC across 9 training subjects in the training process was used in the testing process.

4   Experimental Results and Discussions

In this study, we totally selected 10 subjects that were analyzed and applied to the modeling the estimation of our proposed MS level evaluation system. Figures 4 and 4 were shown the correlation coefficients (CC) results and root mean square errors (RMSE) in comparison with the actual MS level and the estimation performance of our proposed system. To summary the all estimation performance from 10 subjects,

Table 1. Comparison results of the cognitive state estimation systems including correlation and RMSE performances in subject-dependent and cross-subject conditions

Subject	LR		PCR	
	CC	RMSE	CC	RMSE
1	0.6830	0.2454	0.8628	0.1893
2	0.8023	0.2157	0.8477	0.2249
3	0.6842	0.2639	0.7915	0.1601
4	0.4774	0.3989	0.8832	0.2295
5	0.3923	0.5748	0.6605	0.3327
6	0.5086	0.2831	0.7494	0.2302
7	0.8000	0.3355	0.8426	0.2434
8	0.8689	0.1492	0.8101	0.1803
9	0.6922	0.2496	0.7052	0.2313
10	0.5635	0.2952	0.6791	0.2407
Total	0.6472 ±0.1568	0.3011 ±0.1171	0.7832 ±0.0803 *	0.2262 ±0.0468 *
Dimension	36		5.6±1.5	

*: Significant difference between LR's and PCR's performance ($p < 0.05$) tested by Wilcoxon signed rank test

the average correlation coefficients were about 0.6472 and 0.7832 in corresponding to linear regression (LR) and principal component regression (PCR) [18], respectively. As for the RMSE performance, the average estimation results were 0.3011 and 0.2262 in corresponding to LR and PCR, respectively. According to the estimation results in Table 1, we can see that the proposed MS level estimation system using PCR is better than using LR model.

Through the experimental results on the system performance under different conditions, we find that 1 subjects out of 10 subjects except subject 8 had the better CC estimation result via using LR. In conclusion, this study demonstrated that our proposed EEG-based evaluation system could successfully estimate the motion sickness level reported by individual subject, we suggest that SVR model can be utilized to estimate the motion sickness level in the operational environment. Since the potential of real-time application is emerging and desired, nevertheless, we need to consider more about the complexity, instantaneity, and robustness of the system. These results let us open an emerging sight on the potential of real-time application. Nevertheless, the complexity, instantaneity, and robustness of the system still have to be considered for the implementation.

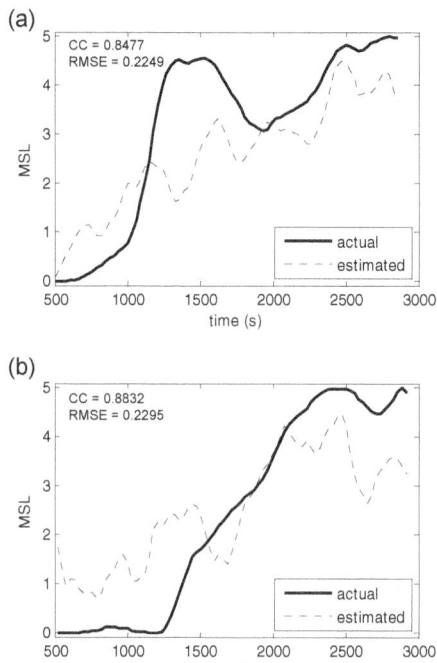

Fig. 4. Subjects 2's (top) and 4's (bottom) estimated MSL performance via using PCR

Acknowledgments. The authors would like to thank Dr. Tzyy-Ping Jung for his advising in this study. This research was sponsored in part by the Army Research Laboratory under Cooperative Agreement Number W911NF-10-2-0022. The views and the conclusions contained in this document are those of the authors and should not be interpreted as representing the official policies, either expressed or implied, of the Army Research Laboratory or the U.S Government. The U.S Government is authorized to reproduce and distribute reprints for Government purposes notwithstanding any copyright notation herein. This work was also supported in part by the National Science Council, Taiwan, on Establishing "International Research-Intensive Centers of Excellence in Taiwan" (I-RiCE Project) under Contract NSC 99-2911-I-009-101, in part by the Aiming for the Top University Plan of National Chiao Tung University, the Ministry of Education, Taiwan, under Contract 99W906, in part by the National Science Council, Taiwan, under Contracts NSC 99-3114-E-009-167 and NSC 99-2628-E-009-091.

References

1. Reason, J., Brand, J.J.: Motion sickness. Academic Press (1975)
2. Brandt, T., Dieterich, M., Danek, A.: Vestibular cortex lesions affect the perception of verticality. Annals of Neurology 35(4) (1994)

3. Fasold, O., von Brevern, M., Kuhberg, M., Ploner, C.J., Villringer, A., Lempert, T., Wenzel, R.: Human vestibular cortex as identified with caloric stimulation in functional magnetic resonance imaging. NeuroImage 17(3) (2002)
4. Lobel, E., Kleine, J.F., Bihan, D.L., Leroy-Willig, A., Berthoz, A.: Functional MRI of Galvanic Vestibular Stimulation. Journal of Neurophysiology 80(5) (1998)
5. De Waele, C., Baudonnière, P.M., Lepecq, J.C., Huy, P.T.B., Vidal, P.P.: Vestibular projections in the human cortex. Experimental Brain Research 141(4) (2001)
6. Wood, C.D., Stewart, J.J., Wood, M.J., Struve, F.A., Straumanis, J.J., Mims, M.E., Patrick, G.Y.: Habituation and motion sickness. Journal Of Clinical Pharmacology 34(6) (1994)
7. Wood, S.J.: Human otolith-ocular reflexes during off-vertical axis rotation: effect of frequency on tilt-translation ambiguity and motion sickness. Neuroscience Letters 323(1) (2002)
8. Wu, J.P.: EEG Changes in Man During Motion-Sickness Induced by Parallel Swing. Space Med. Med. Eng. 5(3) (1992)
9. Chelen, W.E., Kabrisky, M., Rogers, S.K.: Spectral analysis of the electroencephalographic response to motion sickness. Aviation, Space, and Environmental Medicine 64(1) (1993)
10. Hu, S., McChesney, K.A., Player, K.A., Bahl, A.M., Buchanan, J.B., Scozzafava, J.E.: Systematic investigation of physiological correlates of motion sickness induced by viewing an optokinetic rotating drum. Aviation, Space, and Environmental Medicine 70(8) (1999)
11. Chen, Y.-C., Duann, J.-R., Chuang, S.-W., Lin, C.-L., Ko, L.-W., Jung, T.-P., Lin, C.-T.: Spatial and temporal EEG dynamics of motion sickness. NeuroImage 49(3) (2010)
12. Wei, C.-S., Chuang, S.-W., Wang, W.-R., Ko, L.-W., Jung, T.-P., Lin, C.-T.: Development of A Motion Sickness Evaluation System Based on EEG Spectrum Analysis. In: Proceedings of the 2011 IEEE International Symposium on Circuits and Systems (2011)
13. Wei, C.-S., Chuang, S.-W., Ko, L.-W., Jung, T.-P., Lin, C.-T.: EEG-based Evaluation System for Motion Sickness Estimation. In: Proceedings of the 5th International IEEE EMBS Conference on Neural Engineering (2011)
14. Wei, C.-S., Chuang, S.-W., Ko, L.-W., Jung, T.-P., Lin, C.-T.: Genetic Feature Selection in EEG-Based Motion Sickness Estimation. In: Proceedings of the International Joint Conference on Neural Network (2011)
15. Lin, C.-T., Chang, C.-J., Lin, B.-S., Hung, S.-H., Chao, C.-F., Wang, I.-J.: A Real-Time Wireless Brain-Computer Interface System for Drowsiness Detection. IEEE Transactions on Biomedical Circuits and Systems 4(4) (2010)
16. Lin, C.-T., Ko, L.-W., Chiou, J.-C., Duann, J.-R., Huang, R.-S., Liang, S.-F., Chiu, T.-W., Jung, T.-P.: Noninvasive Neural Prostheses Using Mobile and Wireless EEG. Proceedings of the IEEE 96(7) (2008)
17. Jolliffe, I.T.: Principal Component Analysis, 2nd edn. Springer, Heidelberg (2002)
18. Jolliffe, I.T., Ian, T.: A Note on the Use of Principal Components in Regression. Journal of the Royal Statistical Society. Series C (Applied Statistics) 31(3) (1982)

A Sparse Common Spatial Pattern Algorithm for Brain-Computer Interface

Li-Chen Shi[1], Yang Li[1], Rui-Hua Sun[1], and Bao-Liang Lu[1,2,⋆]

[1]Center for Brain-Like Computing and Machine Intelligence
Department of Computer Science and Engineering
[2]MOE-Microsoft Key Lab. for Intelligent Computing and Intelligent Systems
Shanghai Jiao Tong University
800 Dong Chuan Road, Shanghai 200240, China
bllu@sjtu.edu.cn

Abstract. Common spatial pattern (CSP) algorithm and principal component analysis (PCA) are two commonly used key techniques for EEG component selection and EEG feature extraction for EEG-based brain-computer interfaces (BCIs). However, both the ordinary CSP and PCA algorithms face a loading problem, i.e., their weights in linear combinations are non-zero. This problem makes a BCI system easy to be over-fitted during training process, because not all of the information from EEG data are relevant to the given tasks. To deal with the loading problem, this paper proposes a spare CSP algorithm and introduces a sparse PCA algorithm to BCIs. The performance of BCIs using the proposed sparse CSP and sparse PCA techniques is evaluated on a motor imagery classification task and a vigilance estimation task. Experimental results demonstrate that the BCI system with sparse PCA and sparse CSP techniques are superior to that using the ordinary PCA and CSP algorithms.

Keywords: sparse common spatial pattern, sparse principal component analysis, EEG, brain-computer interface.

1 Introduction

Brain-computer interface (BCI) is usually defined as a direct communication pathway between the brain and a computer or a device. And electroencephalogram (EEG) is the most commonly used brain signals for BCIs. Over the last twenty years, with the advances of signal processing, pattern recognition, and machine learning techniques, the field of BCI research has made great progress [1,2]. Through BCIs, people can directly control an external device just by using EEG signals generated from motor imagery, visual evoked potentials, or people's mental states. However EEG signals are very noisy and unstable. Therefore, relevant EEG components selection and feature extraction are very important for BCIs. For traditional BCIs, spatial filters based on common spatial

---

⋆ Corresponding author.

B.-L. Lu, L. Zhang, and J. Kwok (Eds.): ICONIP 2011, Part I, LNCS 7062, pp. 725–733, 2011.

pattern (CSP) are usually used for selecting the relevant EEG components from the linear combination of the original EEG signals of different channels [3], and principal components analysis (PCA) technique is usually used for extracting features from the linear combination of the original EEG features.

However, both the ordinary CSP and PCA algorithms face a loading problem, i.e., their weights in the linear combinations for PCA and CSP are non-zero. That problem makes a BCI system easy to be over-fitted during training process, because not all of the EEG channels or the EEG features are relevant to the given tasks. As a result, to develop efficient algorithms for EEG channel selection and EEG feature selection is highly desirable.

In this paper, we introduce sparse loading representations for both CSP and PCA algorithms. Our proposed sparse technique can accomplish EEG channel selection, relevant EEG component selection, and EEG feature selection. For sparse PCA, Zou's method is adopted [4], where PCA is considered as a regression-type problem and elastic net is used to calculate the sparse loading of PCA. The performance of a BCI system using sparse PCA is evaluated on an EEG-based vigilance estimation task. For sparse CSP, we propose a novel sparse CSP algorithm and consider CSP as a Rayleigh quotient problem. We use sparse PCA and elastic net to calculate the sparse loadings of CSP. The performance of a BCI system with our proposed sparse CSP algorithm is evaluated on a motor imagery task from the BCI Competition III, Data sets IIIa [5]. Experimental results demonstrate that both BCI systems using sparse representation techniques have outperformed the traditional BCI systems.

This paper is organized as follows. In Section 2, the sparse PCA and sparse CSP algorithms are presented. In Section 3, the experimental setups and the EEG data processing of vigilance task and motor imagery task are described. In Section 4, experimental results are presented and discussed. Finally, some conclusions are given in Section 5.

2   Sparse PCA and CSP Algorithms

As both sparse PCA and sparse CSP algorithms are based on elastic net, the elastic net algorithm is briefly introduced first, and then sparse PCA and our proposed sparse CSP algorithms are described.

2.1   Elastic Net

Consider a data set $\{X, Y\}$, here $X = (x_1, ..., x_m)$ is the input set, $x_i = (x_{i,1}, ..., x_{i,n})^T$, $i = 1, ..., m$, is the i-th feature of input set, n is the number of data, m is the feature dimension, and $Y = (y_1, ..., y_n)^T$ is the response set. For linear regression model, a criterion is usually formed as

$$\hat{\beta} = \arg\min_{\beta} |Y - X\beta|^2, \tag{1}$$

where β is the linear coefficients to be estimated. However, the elements of β are typically nonzero, even some features $\{x_i\}$ are almost not correlated with the response set. This makes the linear regression model easy to be overfitted.

To solve this problem, various kinds of methods have been proposed. Lasso is one of the famous methods, which adds a L_1 norm penalty to the ordinary criterion. The Lasso criterion is formed as

$$\hat{\beta} = \arg\min_{\beta} |Y - X\beta|^2 + \lambda|\beta|^1, \tag{2}$$

where λ is the penalty factor and $|\cdot|^1$ stands for L_1 norm.

By tuning λ, Lasso can continuously shrink the linear coefficients toward zero and accomplish feature selection, and then improve the prediction accuracy via the bias-variance tradeoff. Lasso can be efficiently solved by the LARS algorithm [6]. However, LARS has a drawback: the number of selected features is limited by the number of training data or the number of linear unrelated features in the training data. To overcome this promlem, naive elastic net and elastic net have been proposed [7], which add a L_2 norm penalty to the Lasso criterion. The naive elastic net criterion is formed as

$$\hat{\beta} = \arg\min_{\beta} |Y - X\beta|^2 + \lambda_1|\beta|^1 + \lambda_2|\beta|^2, \tag{3}$$

where λ_1 and λ_2 are the penalty factors.

The naive elastic net usually makes too much coefficients shrinkage, and causes more bias to the ELM. But it only reduces a little variances. To correct the bias, elastic net is proposed, whose solution is a rescaled naive elastic net solution with a factor $(1 + \lambda_2)$. The elastic net criterion is formed as

$$\hat{\beta} = (1 + \lambda_2)\arg\min_{\beta} |Y - X\beta|^2 + \lambda_1|\beta|^1 + \lambda_2|\beta|^2. \tag{4}$$

Both naive elastic net and elastic net can be efficiently solved by the LARS-EN algorithm [7]. The elastic net can simultaneously produce an accurate and sparse model without the limitation of LARS.

2.2 Sparse PCA

Sparse PCA used in this paper was proposed by Zou et al. [4]. They reformulate the PCA problem as a regression model and solve it by using the following four theorems.

In theorem 1, let Z_i denote the i-th principal component of X. The corresponding PCA loadings V_i can be calculated from the following regression model,

$$\hat{\beta} = \arg\min_{\beta} |Z_i - X\beta|^2 + \lambda|\beta|^2, \tag{5}$$

where λ can be assigned with any positive value, and $V_i = \frac{\hat{\beta}}{|\hat{\beta}|}$.

In theorem 2, another connection between PCA and a regression model is formed as

$$(\hat{\alpha}, \hat{\beta}) = \arg\min_{\alpha, \beta} \sum_{j=1}^{n} |X_{\cdot,j} - \alpha\beta^T X_{\cdot,j}|^2 + \lambda|\beta|^2 \tag{6}$$

$$\text{subject to } |\alpha|^2 = 1,$$

where $X_{\cdot,j}$ is the row vector of X, α and β are $m \times 1$ vectors, and $V_1 = \frac{\hat{\beta}}{|\hat{\beta}|}$.

In theorem 3, let α and β be $m \times k$ matrices. The connection between PCA and a regression model is formed as

$$(\hat{\alpha}, \hat{\beta}) = \arg\min_{\alpha, \beta} \sum_{j=1}^{n} |X_{\cdot,j} - \alpha\beta^T X_{\cdot,j}|^2 + \lambda \sum_{i=1}^{k} |\beta_i|^2 \tag{7}$$

$$\text{subject to } \alpha^T \alpha = I_k,$$

where $V_i = \frac{\hat{\beta}_i}{|\hat{\beta}_i|}$, for $i = 1, ..., k$.

To achieve sparse loadings, a L_1 penalty is added into (7)

$$(\hat{\alpha}, \hat{\beta}) = \arg\min_{\alpha, \beta} \sum_{j=1}^{n} |X_{\cdot,j} - \alpha\beta^T X_{\cdot,j}|^2 + \lambda \sum_{i=1}^{k} |\beta_i|^2 + \sum_{i=1}^{k} \lambda_{1,i} |\beta_i|^1 \tag{8}$$

$$\text{subject to } \alpha^T \alpha = I_k,$$

where $\lambda_{1,i}$ is the penalty factor. This is a naive elastic net problem, and can be efficiently solved after fixing α.

In theorem 4, suppose the SVD of $X^T X \beta$ is $X^T X \beta = P\Sigma Q^T$. It is proved that the solution of α in (8) should be

$$\hat{\alpha} = PQ^T. \tag{9}$$

Then Eq. (8) can be solved by alternated updating $\hat{\alpha}$ and $\hat{\beta}$ until they converge. When solving Eq. (8), only the covariance matrix of X is need. For more details, please refer [4].

2.3   The Proposed Sparse CSP Algorithm

Let X denote the original EEG signals, where X is a $p(\text{channel}) \times l(\text{time})$ matrix. The CSP-based spatial filter is to determine some linear projections, $y = v^T X$, that can maximize the variance (yy^T or $v^T X X^T v$) of signals of one condition and at the same time minimize the variance of signals of another condition in a specific frequency band. The variance of a specific frequency band is equal to the band-power. Then, CSP can be formulated as a maximum power-ratio problem or a Rayleigh quotient problem as follows:

$$\hat{V} = \{v| \max \frac{v^T R_1 v}{v^T R_2 v} \text{ or } \max \frac{v^T R_2 v}{v^T R_1 v}\} \tag{10}$$

where R_i is the covariance matrix of original EEG signals on condition i, and $\hat{V}$ are the projection vectors or loadings of CSP.

Equation (10) can be solved as follows. Let

$$v^T R_2 v = u^T u, \tag{11}$$

and then,

$$v = P \Sigma^{-1/2} u, \tag{12}$$

where P and Σ are the PCA decomposition of R_2, $R_2 = P \Sigma P^T$.

By applying Eqs. (11) and (12), $\frac{v^T R_1 v}{v^T R_2 v}$ can be reformed as

$$\frac{u^T D u}{u^T u}, \tag{13}$$

where $D = \Sigma^{-1/2} P^T R_1 P \Sigma^{-1/2}$.

It is easy to show that the i-th largest value of Eq. (13) is the i-th largest eigenvalue of D, and u is the corresponding eigenvector. The i-th smallest value of Eq. (13) corresponds to the i-th largest value of $\frac{v^T R_2 v}{v^T R_1 v}$. Usually, not two projections but several projections corresponding to the large values of $\frac{v^T R_1 v}{v^T R_2 v}$ and $\frac{v^T R_2 v}{v^T R_1 v}$ are used for EEG spatial filtering. The loadings, v, of CSP can be calculated by using Eq. (12) together with the eigenvectors corresponding to some large eigenvalues or small eigenvalues of D.

To achieve sparse loadings of CSP, we can reformulate Eq. (12) as an elastic net problem as follows:

$$\hat{v} = \arg\min_v |u - \Sigma^{1/2} P^T v|^2 + \lambda_1 |v|^1 + \lambda_2 |v|^2, \tag{14}$$

and solve it by the LARS-EN algorithm.

2.4 Complexity Analysis of the Proposed Sparse CSP Algorithm

In EEG data analysis, the number of features, m, is usually less than the number of data, n. Therefore, the complexities of the proposed spare CSP algorithm can be analyzed only on $m < n$ condition.

For elastic net, the time cost is $O(m^3 + nm^2)$ [7], which is equivalent to the cost of least square problem. For sparse PCA, the time cost is $nm^2 + pO(m^3)$ [4], where p is the number of iterations when solving the sparse PCA. As a result, the cost of sparse PCA is comparable with the cost of the ordinary PCA, $O(m^3)$.

For our proposed sparse CSP algorithm, the extra time cost is $kO(nm^2 + m^3)$ in comparison with the ordinary CSP algorithm, where k is the number of components extracted by CSP. The cost of ordinary CSP algorithm is $O(m^3 + nm^2)$. Therefore, the total cost of the proposed sparse CSP algorithm is $(k+1)O(m^3 + nm^2)$, which is comparable with the cost of the ordinary CSP algorithm.

3   Experiment

3.1   Experimental Setup

Vigilance Task. This is a monotonous visual task [8,9,10]. The subjects are asked to sit in a comfortable chair, two feet away from the LCD. There are four colors of traffic signs being presented in the LCD randomly by the NeuroScan $Stim^2$ software. Each trial is 5.5~7.5 seconds long, including 5~7 seconds black screen and 500 millisecond traffic signs presented. The subjects are asked to recognize the sign color, and press the correct button on the response pad. A total of 11 healthy subjects have participated in this experiment. After training, each subject has finished at least two sessions (one for train, and others for test). For each session, a total of 62 EEG channels are recorded by the NeuroScan system sampled at 500Hz. Each session continues for more than one hour, during 13:00~15:00 after lunch. The local error rate of the subject's performance is used as the reference vigilance level, which is derived by computing the target false recognition rate within a 2-minute time window at 2-second step.

Motor Imagery Task. This data set comes from BCI Competition III, data sets IIIa, provided by the Laboratory of Brain-Computer Interfaces (BCI-Lab), Graz University of Technology [5]. It is a 4 classes (left hand, right hand, foot, and tongue) cued motor imagery experiment from 3 subjects. After trial begin, the first 2s were quite, at t=2s an acoustic stimulus indicated the beginning of the trial, and a cross + is displayed; then from t=3s an arrow to the left, right, up or down was displayed for 1 s; at the same time the subject was asked to imagine a left hand, right hand, tongue or foot movement, respectively, until the cross disappeared at t=7s. There are 60 trials per class for each subject. A total of 60 EEG channels are recorded by the NeuroScan system sampled at 250Hz.

3.2   Data Processing

Vigilance Task. Six EEG channels (P1, Pz, P2, Po3, Poz, Po4) are used for the vigilance estimation task, which are measured from the posterior regions of the scalp. The vigilance estimation process consists of the following five main components: a) a bandpass filter (1Hz-50Hz) is used to remove the low-frequency noise and the high frequency noise; b) the power spectral density (PSD) of each channel is calculated by every 2 seconds with a 2 Hz frequency resolution as the original features; c) the features are smoothed with a 2 min moving-average filter; d) the top 10 principal components of the PSD are calculated by the sparse PCA algorithm as features; and e) a least square regression model is adopted for vigilance estimation by every 2 seconds.

Each subject has an individual vigilance estimation model. For each vigilance estimation model, one session of a subject is used for training, while other sessions of this subject are used for test.

Motor Imagery Task. All 60 EEG channels are used for motor imagery classification. The 4-class motor imagery data sets are paired into 6 groups of 2-class motor imagery data sets for classification. The classification process consists of the following four main components: 1) a bandpass filter (8Hz-32Hz) is used to remove the noises and EEG signals which are unrelated to motor imagery; 2) the top 10 motor imagery related EEG components are extracted by the proposed sparse CSP algorithm; 3) the variance of each component in each single motor imagery trial is calculated as the feature; and 4) SVMs with RBF kernel is adopted as the motor imagery classifiers.

The classification model is trained for each subject and each pair of 2-class motor imagery separately. For each classification model, half of each 2-class motor imagery data set is used for training, while the other half is used for test. The parameters used in SVMs are fine tuned by 5-fold cross validation.

4   Experimental Results

The performance of BCI system using sparse PCA is evaluated on the vigilance estimation task. The parameter λ_1 in sparse PCA is used to control the sparseness of loadings. Instead of tuning λ_1, we directly set the number of nonzero coefficients in the loadings of sparse PCA. An early stopping strategy is used for the LARS-EN algorithm. When the number of nonzero coefficients of β_i meets the predefined number, the LARS-EN algorithm used for solving the naive elastic net in sparse PCA is stopped. In this study, without fine-tuning, λ is assigned to 10^{-5}, and the number of nonzero coefficients in each principal component loading is set to 20.

For comparison, another BCI system with using the ordinary PCA is used for vigilance estimation. There are totally 30 pairs of training and test data set from the 11 subjects. The linear correlation coefficient and mean square error between the estimated vigilance level and the reference vigilance level are used for performance evaluation. The experimental results of vigilance estimation is

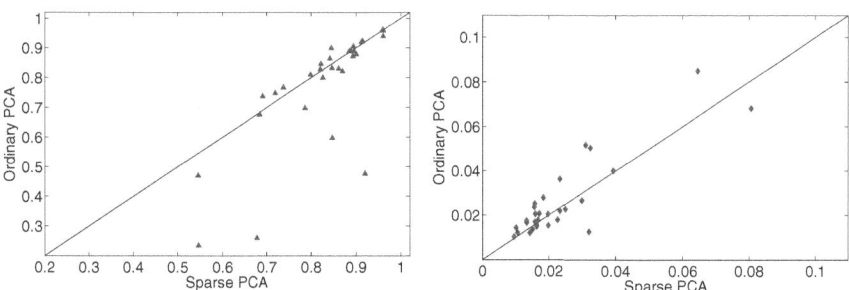

Fig. 1. The result of linear correlation coefficient between the estimated vigilance level and the reference vigilance level (left), and the result of mean square error between the estimated vigilance level and the reference vigilance level (right)

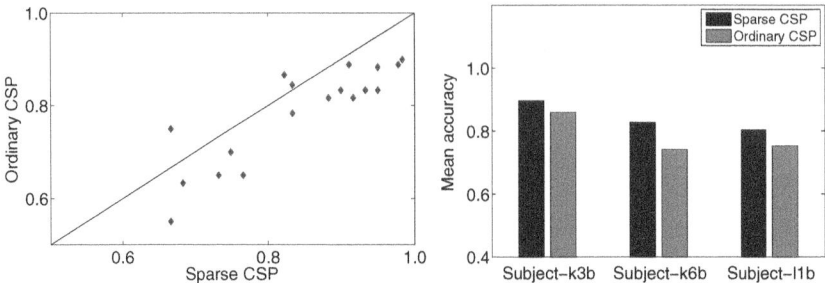

Fig. 2. Comparison of classification accuracies of all 2-class motor imagery data set from 3 subjects (left), and the means of two-class classification accuracies for each subject (right)

shown in Fig. 1. From this figure it can be seen that the average performance of the BCI with sparse PCA is better than that of the BCI system with the ordinary PCA. For those data set the BCI with the ordinary PCA performed well, and the BCI with sparse PCA also performed well. But for those data set the BCI with the ordinary PCA didn't perform well, and the BCI with sparse PCA still performed well, or at least performed much better than the BCI with ordinary PCA.

The performance of the BCI system with the proposed sparse CSP algorithm is evaluated on the motor imagery task. There are totally 6 pairs of training and test data set for each subject. The LARS-EN algorithm used in the sparse CSP algorithm also adopts an early stopping strategy. The number of nonzero coefficients in each CSP loading is set to 30, and λ_2 is assigned to 0.01.

For comparison, a BCI system with the ordinary CSP algorithm is also applied to the motor imagery classification. The experimental results are shown in Fig. 2. From this figure it can be seen that, for most two-class data sets, the BCI system with the proposed sparse CSP algorithm performed better than the BCI system with the ordinary CSP algorithm; and for each subject, the average performance of the BCI system with the proposed sparse CSP algorithm is better than that of the BCI system with the ordinary CSP algorithm.

5   Conclusions

In this paper, sparse PCA and sparse CSP techniques are introduced to EEG-based BCIs. And a novel sparse CSP algorithm has been proposed. The performance of BCI systems with sparse PCA and sparse CSP algorithms have been evaluated on a vigilance estimation task and a motor imagery classification task. Experimental results demonstrate that the BCI systems with sparse PCA and CSP techniques have outperformed the ordinary BCI systems. This result indicates that sparse subspace learning technique is very useful for EEG data processing, which can improve the robustness of EEG-based BCI systems.

In addition, as the solution of LARS-EN is global optimal in comparison with other spare subspace learning techniques such as non-negative matrix factorization [11], the solutions of sparse PCA and sparse CSP can be much more stable.

Acknowledgments. This work was partially supported by the National Natural Science Foundation of China (Grant No. 90820018), the National Basic Research Program of China (Grant No. 2009CB320901), and the European Union Seventh Framework Programme (Grant No. 247619).

References

1. Lotte, F., Congedo, M., Lecuyer, A., Lamarche, F., Arnaldi, F.: A review of classification algorithms for EEG-based brain-computer interfaces. Journal of Neural Engineering 4, R1–R13 (2007)
2. Brunner, P., Bianchi, L., Guger, C., Schalk, G.: Current trends in hardware and software for brain-computer interfaces. Journal of Neural Engineering 8(2) (in press, 2011)
3. Koles, Z.: The quantitative extraction and topographic mapping of the abnormal components in the clinical EEG. Electroencephalogr. Clin. Neurophysiol. 79(6), 440–447 (1997)
4. Zou, H., Hastie, T., Tibshirani, R.: Sparse principal component analysis. Journal of Computational and Graphical Statistics 15(2), 265–286 (2006)
5. BCI Competition III, http://www.bbci.de/competition/iii/
6. Efron, B., Hastie, T., Johnstone, I., Tibshirani, R.: Least angel regression. Annals of Statistics 32(2), 407–499 (2004)
7. Zou, H., Hastie, T.: Regularization and variable selection via the elastic net. Journal of the Royal Statistical Society: Series B 67(2), 301–320 (2005)
8. Shi, L.C., Lu, B.L.: Dynamic clustering for vigilance analysis based on EEG. In: Proce. of the 30th International Conference of the IEEE Engineering in Medicine and Biology Society, Vancouver, Canada, pp. 54–57 (2008)
9. Shi, L.C., Lu, B.L.: Off-Line and On-Line Vigilance Estimation Based on Linear Dynamical System and Manifold Learning. In: Proce. of the 32nd International Conference of the IEEE Engineering in Medicine and Biology Society, Buenos Aires, Argentina, pp. 6587–6590 (2010)
10. Ma, J.X., Shi, L.C., Lu, B.L.: Vigilance estimation by using electrooculographic features. In: Proce. of the 32nd International Conference of the IEEE Engineering in Medicine and Biology Society, Buenos Aires, Argentina, pp. 6591–6594 (2010)
11. Lee, D.D., Seung, H.S.: Learning the parts of objects by non-negative matrix factorization. Nature 401, 788–791 (1999)

EEG-Based Emotion Recognition Using Frequency Domain Features and Support Vector Machines

Xiao-Wei Wang[1], Dan Nie[1], and Bao-Liang Lu[1,2,*]

[1] Center for Brain-Like Computing and Machine Intelligence
Department of Computer Science and Engineering
[2] MOE-Microsoft Key Lab. for Intelligent Computing and Intelligent Systems
Shanghai Jiao Tong University
800 Dong Chuan Road, Shanghai 200240, China
`bllu@sjtu.edu.cn`

Abstract. Information about the emotional state of users has become more and more important in human-machine interaction and brain-computer interface. This paper introduces an emotion recognition system based on electroencephalogram (EEG) signals. Experiments using movie elicitation are designed for acquiring subject's EEG signals to classify four emotion states, joy, relax, sad, and fear. After pre-processing the EEG signals, we investigate various kinds of EEG features to build an emotion recognition system. To evaluate classification performance, k-nearest neighbor (kNN) algorithm, multilayer perceptron and support vector machines are used as classifiers. Further, a minimum redundancy-maximum relevance method is used for extracting common critical features across subjects. Experimental results indicate that an average test accuracy of 66.51% for classifying four emotion states can be obtained by using frequency domain features and support vector machines.

Keywords: human-machine interaction, brain-computer interface, emotion recognition, electroencephalogram.

1 Introduction

Emotion plays an important role in human-human interaction. Considering the proliferation of machines in our commonness, emotion interactions between humans and machines has been one of the most important issues in advanced human-machine interaction (HMI) and brain-computer interface (BCI) today [1]. To make this collaboration more efficient in both HMI and BCI, we need to equip machines with the means to interpret and understand human emotions without the input of a user's translated intention.

Numerous studies on engineering approaches to automatic emotion recognition have been performed. They can be categorized into two kinds of approaches.

---

* Corresponding author.

B.-L. Lu, L. Zhang, and J. Kwok (Eds.): ICONIP 2011, Part I, LNCS 7062, pp. 734–743, 2011.

The first kind of approaches focuses on the analysis of facial expressions or speech [2][3]. These audio-visual based techniques allow noncontact detection of emotion, so they do not give the subject any discomfort. However, these techniques might be more prone to deception, and the parameters easily vary in different situations. The second kind of approaches focuses on physiological signals, which change according to exciting emotions and can be observed on changes of autonomic nervous system in the periphery, such as electrocardiogram (ECG), skin conductance (SC), respiration, pulse and so on [4,5]. As comparison with audio-visual based methods, the responses of physiological signals tend to provide more detailed and complex information as an indicator for estimating emotional states.

In addition to periphery physiological signals, electroencephalograph (EEG) captured from the brain in central nervous system has also been proved providing informative characteristics in responses to the emotional states [6]. Since Davidson et al. [7] suggested that frontal brain electrical activity was associated with the experience of positive and negative emotions, the studies of associations between EEG signals and emotions have been received much attention.

So far, researchers often use two different methods to model emotions. One approach is to organize emotion as a set of diverse and discrete emotions. In this model, there is a set of emotions which are more basic than others, and these basic emotions can be seen as prototypes from which other emotions are derived. Another way is to use multiple dimensions or scales to categorize emotions. A two dimensional model of emotion is introduced by Davidson et al. [8]. According to this model, emotions are specified by their positions in the two-dimensional space as shown in Figure 1, which is spanned by two axes, valence axis and arousal axis. The valence axis represents the quality of an emotion ranging from unpleasant to pleasant. The arousal axis refers to the quantitative activation level ranging from calm to excited. The different emotional labels can be plotted at various positions on a 2D plane spanned by these two axes.

Since emotional state corresponds to a separate subsystem in the brain, EEG signals can reveal important information on their functioning. The studies of associations between EEG activity and emotions have been received much attention. Bos used the international affective picture system (IAPS) and international affective digitized sound system (IADS) for eliciting emotional states [9]. They achieved an average classification accuracy of 65% for arousal and 68% for valance by using alpha power and beta power as features and fisher's discriminant analysis (FDA) as classifiers. Takahashiet et al. used EEG signal to recognize emotion in response to movie scenes [10]. They achieved a recognition rate of 41.7% for five emotion states. In our previous work, we proposed an emotion recognition system using the power spectrum of EEG as features [11]. Our experimental results indicated that the recognition rate using a support vector machine reached an accuracy of 87.5% for two emotion states .

Despite much efforts have been devoted to emotion recognition based on EEG in the literature, further research is needed in order to find more effective feature extraction and classification methods to improve recognition performance. In this paper, we deal with all of the essential stages of EEG-based emotion recognition

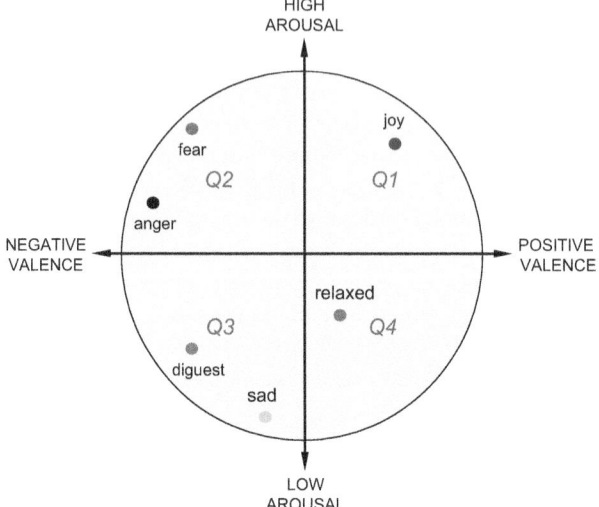

Fig. 1. Two-dimensional emotion model

systems, from data collection to feature extraction and emotion classification. Our study has two main purposes. The first goal is to search emotion-specific features of EEG signals, and the second goal is to evaluate the efficiency of different classifiers for EEG-based emotion recognition. To this end, a user-independent emotion recognition system for classification of four typical emotions is introduced.

2 Experiment Procedure

2.1 Stimulus Material and Presentation

To stimulate subject's emotions, we used several movie clips that were extracted from Oscars films as elicitors. Each set of clips includes three clips for each of the four target emotions: joy (intense-pleasant), relax (calm-pleasant), sad (calm-unpleasant), and fear (intense-unpleasant). The selection criteria for movie clips are as follows: a) the length of the scene should be relatively short; b) the scene is to be understood without explanation; and c) the scene should elicit single desired target emotion in subjects and not multiple emotion. To evaluate whether the movie clips excite each emotion or not, we carried out investigation using questionnaires by human subjects who don't take part in the experiment to verify the efficacy of these elicitors before the experiment.

2.2 Participants

Five right-handed health volunteers (two males, three females), 18-25 years of age (mean = 22.3 and SD = 1.34), participated in the study. All subjects had no per-

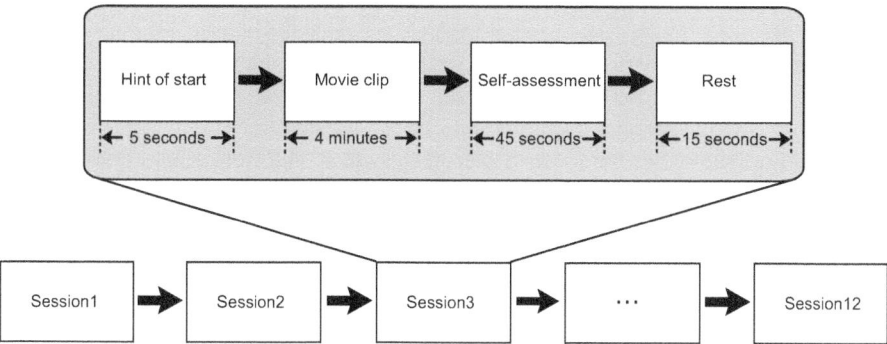

Fig. 2. The process of experiment

sonal history of neurological of psychiatric illness and had normal or corrected-normal vision. All subjects were informed the scope and design of the study.

2.3 Task

In order to get quality data, subjects were instructed to keep their eyes open and view each movie clip for its entire duration in the experiment. Movie clips inducing different emotion conditions were presented in a random order. Each movie clip was presented for 4 to 5 minutes, preceded by 5 s of blank screen as the hint of start. At the end of each clip, subjects were asked to assign valence and arousal ratings and to rate the specific emotions they had experienced during movie viewing. The rating procedure lasted about 45 seconds. An inter trial interval (15 s) of blank screen lapsed between movie presentations for emotion recovery. Valence and arousal ratings were obtained using the Self-Assessment Manikin (SAM) [12]. Four basic emotional states, joy, relax, sad, and fear, describing the reaction to the movie clips were also evaluated at the same time. The given self-reported emotional states were used to verify EEG-based emotion classification.

2.4 EEG Recording

A 128-channel electrical signal imaging system (ESI-128, NeuroScan Labs), SCAN 4.2 software, and a modified 64-channel QuickCap with embedded Ag/AgCl electrodes were used to record EEG signals from 62 active scalp sites referenced to vertex (Cz) for the cap layout. The ground electrode was attached to the center of the forehead. The impedance was kept below 5 k Ω. The EEG data are recorded with 16-bit quantization level at the sampling rate of 1000 Hz. Electrooculogram (EOG) was also recorded, and later used to identify blink artifacts from the recorded EEG data.

3   Feature Extraction

The main task of feature extraction is to derive the salient features which can map the EEG data into consequent emotion states. For a comparison study, we investigated two different methods, one based on statistical features in the time domain, and the other based on power spectrum in the frequency domain. First, the EEG signals were down-sampled to a sampling rate of 200 Hz to reduce the burden of computation. Then, the time waves of the EEG data were visually checked. The recordings seriously contaminated by electromyogram (EMG) and Electrooculogram (EOG) were removed manually. Next, each channel of the EEG data was divided into 1000-point epochs with 400-point overlap. Finally, all features discussed below were computed on each epoch of all channels of the EEG data.

3.1   Time-Domain Features

In this paper, we use the following six different kinds of time-domain features [12].

a) The mean of the raw signal

$$\mu_X = \frac{1}{N} \sum_{n=1}^{N} X(n) \tag{1}$$

where $X(n)$ represents the value of the nth sample of the raw EEG signal, $n = 1, \ldots N$.

b) The standard deviation of the raw signal

$$\sigma_X = \left(\frac{1}{N-1} \sum_{n=1}^{N} (X(n) - \mu_X)^2 \right)^{1/2} \tag{2}$$

c) The mean of the absolute values of the first differences of the raw signal

$$\delta_X = \frac{1}{N-1} \sum_{n=1}^{N-1} |X(n+1) - X(n)| \tag{3}$$

d) The mean of the absolute values of the second differences of the raw signal

$$\gamma_X = \frac{1}{N-2} \sum_{n=1}^{N-2} |X(n+2) - X(n)| \tag{4}$$

e) the means of the absolute values of the first differences of the normalized signals

$$\tilde{\delta}_X = \frac{1}{N} \sum_{n=1}^{N-1} |\tilde{X}(n+1) - \tilde{X}(n)| = \frac{\delta_X}{\sigma_X} \tag{5}$$

where $\tilde{X}(n) = \frac{X(n)-\mu_X}{\sigma_X}$, μ_X and σ_X are the means and standard deviations of X.

f) the means of the absolute values of the second difference of the normalized signals

$$\tilde{\gamma}_X = \sum_{n=1}^{N-2} |\tilde{X}(n+2) - \tilde{X}(n)| = \frac{\gamma_X}{\sigma_X} \tag{6}$$

3.2 Frequency-Domain Features

The frequency-domain features used in this paper are based on the power spectrum of each 1000-point EEG epochs. Analysis of changes in spectral power and phase can characterize the perturbations in the oscillatory dynamics of ongoing EEG. First, each epoch of the EEG data is processed with Hanning window. Then, windowed 1000-point epochs are further subdivided into several 200-point sub-windows using the Hanning window again with 100 point steps, and each is extended to 256 points by zero padding for a 256-point fast Fourier transform (FFT). Next, the power spectrum of all the sub-epochs within each epoch is averaged to minimize the artifacts of the EEG in all sub-windows. Finally, EEG log power spectrum are extracted in different bands such as delta rhythm, theta rhythm, alpha rhythm, beta rhythm, and gamma rhythm.

After these operations, we obtaine six kinds of time domain features and five kinds of frequency features. The dimension of each feature is 62, and the number of each feature from each subject is about 1100.

4   Emotion Classification

For an extensive evaluation of emotion recognition performance, classification of four emotional states is achieved by using three kinds of classifiers, kNN algorithm, MLPs, and SVMs. These classifiers have been separately applied to all of the aforementioned features. In this study, the Euclidean distance method is used as the distance metric for kNN algorithm. In MLPs, a three-layer neural network is adopted, and the activation function is the sigmoidal function. In SVMs, a radial basis function kernel is used.

In order to perform a more reliable classification process, we constructed a training set and a test set for each subject. The number of training set, which formed by the data of the former two sessions of each emotion, is about 700 for each subject. The number of test set, which formed by the data of the last session of each emotion, is about 400 for each subject.

Given the fact that a rather limited number of independent trials were available for each class, we apply cross-validation to select common parameters for each classifier, and pick the parameters that led to the highest average result in the training sets. For cross-validation, we chose a trial-based leave-one-out method (LOOM). In kNN training, we searched the number of neighbors

Table 1. Classification accuracy using time domain features

Subject	Classifier	μ_X	σ_X	δ_X	γ_X	$\bar{\delta}_X$	$\tilde{\gamma}_X$	All
1	kNN	21.36	30.17	32.20	35.47	28.93	26.30	39.17
	MLP	23.15	29.32	34.38	36.23	29.71	29.47	40.21
	SVM	26.54	32.15	38.41	37.31	30.45	32.37	**42.44**
2	kNN	32.41	21.77	23.73	22.21	23.21	24.31	35.63
	MLP	33.58	23.51	25.39	24.17	26.42	28.90	37.40
	SVM	36.52	26.53	29.32	26.56	31.95	32.64	**45.35**
3	kNN	25.34	21.04	25.12	23.92	24.55	31.54	32.57
	MLP	26.73	22.25	27.35	26.84	26.78	33.35	35.79
	SVM	27.98	23.78	30.34	27.74	27.07	36.57	**41.42**
4	kNN	29.89	26.84	27.96	32.29	22.97	34.90	35.58
	MLP	28.25	26.86	29.08	33.32	24.62	35.52	38.07
	SVM	29.84	27.67	31.49	36.91	27.79	37.43	**42.13**
5	kNN	23.02	29.93	33.76	27.18	34.98	37.92	40.10
	MLP	26.91	28.47	35.82	30.85	36.26	39.76	41.98
	SVM	30.97	34.46	39.49	33.24	40.19	43.91	**45.62**
Average	kNN	26.40	25.95	28.57	28.21	26.93	30.99	36.61
	MLP	27.72	26.08	30.40	30.28	28.75	33.40	38.69
	SVM	30.37	28.92	32.95	32.35	31.49	36.58	**43.39**

Table 2. Classification accuracy using frequency domain features

Subject	Classifier	Delta	Theta	Alpha	Beta	Gamma	All frequency
1	kNN	25.09	35.62	42.62	45.35	42.09	48.89
	MLP	26.91	36.36	45.13	47.64	44.45	51.18
	SVM	27.74	43.18	52.07	50.61	49.96	**55.09**
2	kNN	31.43	47.10	60.84	64.03	67.94	72.13
	MLP	40.51	54.62	61.83	63.90	70.74	80.67
	SVM	41.21	55.47	65.78	70.82	80.91	**82.45**
3	kNN	29.22	32.28	43.38	46.52	42.49	59.92
	MLP	35.35	45.84	46.42	49.69	47.28	61.34
	SVM	34.27	42.16	47.26	57.49	55.35	**65.43**
4	kNN	23.08	34.31	43.51	42.72	40.45	55.79
	MLP	27.38	36.52	45.02	45.82	42.37	57.91
	SVM	33.71	43.94	47.75	49.74	47.17	**58.83**
5	kNN	26.46	28.45	52.39	45.78	46.93	62.45
	MLP	28.17	30.48	53.20	46.57	45.09	64.26
	SVM	32.59	34.97	63.63	52.75	48.80	**70.74**
Average	kNN	27.05	35.55	48.54	48.88	47.98	59.84
	MLP	31.66	40.76	50.32	50.72	49.60	63.07
	SVM	33.90	43.94	55.29	56.28	56.43	**66.51**

k. In MLP training, we searched the number of hidden neurons assigned to the MLPs. In SVM training, we searched the cost C and γ of the Gaussian kernel.

The experimental results of classification with different classifiers for statistical features in the time domain are given in Table 1. From this table, we can see that the classification performance of using all statistical features is evidently better than those based on individual features under the same conditions.

Table 2 shows the averaged classification performance of different classifiers using six frequency-domain features at different EEG frequency bands. From this table, we can see that the classification performance of using all frequency bands is evidently better than those based on individual frequency bands under the same condition. In addition, an interesting finding is implied that the frequency bands of alpha, beta, and gamma are more important than the frequency bands of delta and theta to the emotion classification.

From the results shown in Tables 1 and 2, we can see that the classification performance based on the frequency domain features were better than those based on time domain features. We can also found that the performance of SVM classifiers is better than those of kNN and MLPs, which proved true for all of the different features.

We tried to identify the significant features for each classification problem and thereby to investigate the class relevant feature domain and interrelation between the features for emotions. Feature extraction methods select or omit dimensions of the data that correspond to one EEG channel depending on a performance measure. Thus they seem particularly important not only to find the emotion-specific features but also expand the applicability of using fewer electrodes for practical applications. This study adopted minimum redundancy-maximum relevance (MRMR), a method based on information theory for sorting each feature in descending order accounting for discrimination between different EEG patterns. Since the best performance was obtained using power spectrum across all frequencies, MRMR was further applied to this feature type to sort the feature across frequency bands.

Table 3 lists top-30 feature rankings of individual subject, which are obtained by applying MRMR to the training data set of each subject. The common features across different subjects are marked with a grey background. As we can see, the top-30 ranking features are variable for different subjects. Person-dependent differences of the EEG signals may account for the differences among different subjects [14]. Moreover, every subject experiences emotions in a different way, which is probably another reason for this inter-subject variability. Nevertheless, there are still many similarities across different people. We can find that the features derived from the frontal and parietal lobes are used more frequently than other regions. This indicates that these electrodes provided more discriminative information than other sites, which is consistent with the neurophysiologic basis of the emotion [15].

Table 3. Top-30 feature selection results using MRMR

Rank	Subject				
	1	2	3	4	5
1	AF4, Alpha	T7, Gamma	C6, Beta	F6, Beta	F8, Alpha
2	C5, Beta	PZ, Delta	CP3, Theta	F8, Theta	O1, Gamma
3	P3, Gamma	OZ, Alpha	F3, Gamma	TP8, Beta	FC3, Alpha
4	FC3, Delta	T8, Gamma	F7, Alpha	FT7, Beta	C5, Theta
5	T8, Alpha	C6, Delta	CP4, Delta	AF4, Alpha	F4, Alpha
6	F1, Theta	TP8, Gamma	FT7, Beta	FC4, Delta	TP8, Theta
7	C3, Theta	T8, Theta	P3, Alpha	C1, Gamma	T7, Gamma
8	C4, Gamma	AF4, Alpha	CP5, Beta	FC3, Alpha	CP6, Beta
9	F1, Gamma	F1, Gamma	POZ, Gamma	AF3, Delta	FC6, Gamma
10	F3, Gamma	CP2, Theta	PO5, Gamma	CB1, Alpha	F8, Gamma
12	T8, Beta	CP3, Theta	C4, Beta	F3, Gamma	CP2, Alpha
13	AF3, Delta	CP6, Beta	FCZ, Theta	P6, Theta	C6, Delta
14	F2, Beta	F3, Gamma	T7, Gamma	TP7, Theta	TP8, Beta
15	FC5, Theta	FP2, Alpha	FO7, Alpha	C3, Theta	CP4, Gamma
16	FC3, Theta	AF3, Delta	FCZ, beta	FT7, Theta	FP2, Alpha
17	CP4, Beta	P2, Alpha	AF4, Alpha	F7, Gamma	F1, Delta
18	CP6, Beta	C4, Theta	FC5, Alpha	F3, Delta	FCZ, Alpha
19	T7, Gamma	FCZ, Alpha	C6, Delta	P3, Theta	OZ, Gamma
20	FCZ, Alpha	C5, Gamma	F1, Delta	CP4, Gamma	AF3, Delta
21	C4, Beta	OZ, Gamma	TP8, Beta	FP2, Alpha	CP5, Beta
22	P1, Beta	PO8, Alpha	FC3, Theta	FZ, beta	AF4, Alpha
23	TP8, Gamma	F8, Gamma	AF3, Delta	T7, Beta	P2, Alpha
24	FP2, Alpha	PZ, Beta	FC1, Theta	FC3, Beta	C3, Theta
25	CP5, Gamma	FP2, Gamma	C3, Theta	P6, Beta	C4, Beta
26	CP4, Delta	CB2, Theta	FC2, Beta	O1, Alpha	FC3, Theta
27	FT7, Beta	CPZ, Theta	FP2, Alpha	T7, Gamma	P3, Alpha
28	PZ, Beta	C3, Theta	P2, Gamma	CP6, Beta	FC2, Beta
29	TP8, Theta	CP3, Gamma	F4, Alpha	FC6, Alpha	F3, Gamma
30	F7, Delta	CP4, Gamma	FC6, Gamma	FT8, Alpha	FC5, Theta

5 Conclusion

In this paper, we presented a study on EEG-based emotion recognition. Our experimental results indicate that it is feasible to identify four emotional states, joy , relax, fear and sad, during watching movie , and an average test accuracy of 66.51% is obtained by combining EEG frequency domain features and support vector machine classifiers. In addition, the experimental results show that the frontal and parietal EEG signals were more informative about the emotional states.

Acknowledgments. This research was supported in part by the National Natural Science Foundation of China (Grant No. 90820018), the National Basic

Research Program of China (Grant No. 2009CB320901), and the European Union Seventh Framework Program (Grant No. 247619).

References

1. Picard, R.: Affective computing. The MIT press (2000)
2. Petrushin, V.: Emotion in speech: Recognition and application to call centers. Artificial Neu. Net. In Engr., 7–10 (1999)
3. Black, M., Yacoob, Y.: Recognizing facial expressions in image sequences using local parameterized models of image motion. International Journal of Computer Vision 25(1), 23–48 (1997)
4. Kim, K., Bang, S., Kim, S.: Emotion recognition system using short-term monitoring of physiological signals. Medical and Biological Engineering and Computing 42(3), 419–427 (2004)
5. Brosschot, J., Thayer, J.: Heart rate response is longer after negative emotions than after positive emotions. International Journal of Psychophysiology 50(3), 181–187 (2003)
6. Chanel, G., Kronegg, J., Grandjean, D., Pun, T.: Emotion assessment: Arousal evaluation using eegs and peripheral physiological signals. Multimedia Content Representation, Classification and Security, 530–537 (2006)
7. Davidson, R., Fox, N.: Asymmetrical brain activity discriminates between positive and negative affective stimuli in human infants. Science 218(4578), 1235 (1982)
8. Davidson, R., Schwartz, G., Saron, C., Bennett, J., Goleman, D.: Frontal versus parietal ecg asymmetry during positive and negative affect. Psychophysiology 16(2), 202–203 (1979)
9. Bos, D.: Eeg-based emotion recognition. The Influence of Visual and Auditory Stimuli
10. Takahashi, K.: Remarks on emotion recognition from bio-potential signals. In: The Second International Conference on Autonomous Robots and Agents, pp. 667–670. Citeseer (2004)
11. Nie, D., Wang, X.W., Shi, L.C., Lu, B.L.: EEG-based emotion recognition during watching movies. In: The Fifth International IEEE/EMBS Conference on Neural Engineering, pp. 186–191. IEEE Press, Mexico (2011)
12. Bradley, M., Lang, P.: Measuring emotion: the self-assessment manikin and the semantic differential. Journal of Behavior Therapy and Experimental Psychiatry 25(1), 49–59 (1994)
13. Picard, R.W., Vyzas, E., Healey, J.: Toward machine emotional intelligence: Analysis of affective physiological state. IEEE Transactions on Pattern Analysis and Machine Intelligence 23(10), 1175–1191 (2001)
14. Heller, W.: Neuropsychological mechanisms of individual differences in emotion, personality, and arousal. Neuropsychology 7(4), 476 (1993)
15. Schmidt, L., Trainor, L.: Frontal brain electrical activity (eeg) distinguishes valence and intensity of musical emotions. Cognition Emotion 15(4), 487–500 (2001)

Author Index

GPSR Compliance

The European Union's (EU) General Product Safety Regulation (GPSR)
is a set of rules that requires consumer products to be safe and our
obligations to ensure this.

If you have any concerns about our products, you can contact us on
ProductSafety@springernature.com

In case Publisher is established outside the EU, the EU authorized
representative is:

Springer Nature Customer Service Center GmbH
Europaplatz 3
69115 Heidelberg, Germany

Batch number: 09478952

Printed by Printforce, the Netherlands